Fishes

An Introduction to Ichthyology

FIFTH EDITION

Peter B. Moyle
Joseph J. Cech, Jr.

Department of Wildlife, Fish, and Conservation Biology
University of California, Davis

PEARSON

Benjamin
Cummings

San Francisco   Boston   New York
Cape Town   Hong Kong   London   Madrid   Mexico City
Montreal   Munich   Paris   Singapore   Sydney   Tokyo   Toronto

Library of Congress Cataloging-in-Publication Data

Moyle, Peter B.
    Fishes : an introduction to ichthyology / Peter B. Moyle, Joseph J. Cech, Jr.—5th ed.
      p. cm.
    Includes bibliographical references and index.
    ISBN 0-13-100847-1
     1. Fishes. 2. Ichthyology. I. Cech, Joseph J. II. Title.

QL615.M64 2004
597—dc21   2003054863

Assistant Editor: Colleen Lee
Executive Editor: Teresa Chung
Editorial Assistant: Mary Kuhl
Production Editor: Kevin Bradley
Manufacturing Manager: Trudy Pisciotti
Manufacturing Buyer: Lynda Castillo
Art Director: Jayne Conte
Cover Designer: Bruce Kenselaar
Cover Photograph & Diagram: Hammerhead sharks. Photo by Paul H. Humann.
   Diagram by Richard Ellis.
Art Editor: Jessica Einsig
Market Manager: Shari Meffert

Printed in the United States of America

10 9 8

ISBN 0-13-100847-1

Pearson Education Ltd., *London*
Pearson Education Australia Pty., Limited, *Sydney*
Pearson Education Singapore, Pte. Ltd.
Pearson Education North Asia Ltd., *Hong Kong*
Pearson Education Canada, Ltd., *Toronto*
Pearson Educación de Mexico, S.A. de C.V.
Pearson Education—Japan, *Tokyo*
Pearson Education Malaysia, Pte. Ltd.

CONTENTS

11 BEHAVIOR AND COMMUNICATION   187

PART III: THE FISHES   209

12 SYSTEMATICS, GENETICS, AND SPECIATION   209

13 EVOLUTION   221

18 MINNOWS, CHARACINS, AND CATFISHES   299

19 SMELT, SALMON, AND PIKE   319

20 ANGLERFISH, BARRACUDINAS, CODS, AND DRAGONFISHES   331

21 MULLETS, SILVERSIDES, FLYING FISH, AND KILLIFISH   349

22 OPAHS, SQUIRRELFISH, DORIES, PIPEFISH, AND SCULPINS   361

23 PERCIFORMES: SNOOKS TO SNAKEHEADS   377

24 FLOUNDERS, PUFFERS, AND MOLAS   405

PART IV: ZOOGEOGRAPHY   413

25 ZOOGEOGRAPHY OF FRESHWATER FISHES   413

26 ZOOGEOGRAPHY OF MARINE FISHES   437

PART V: ECOLOGY   455

27 INTRODUCTION TO ECOLOGY   455

28 TEMPERATE STREAMS   469

PREFACE

Ichthyology has traditionally emphasized the systematics, anatomy, and distribution of fishes. In the past, most prominent names associated with the field made their major contributions in these areas. Today, however, people who study fish have more far-reaching interests. They study fish to find ways to improve fisheries or aquaculture, to determine the effects of human activities on aquatic environments, and to test ideas in rapidly developing fields, such as ecology, physiology, behavior, and evolution. Growing numbers of sophisticated amateur ichthyologists desire to increase their understanding of fish they keep in aquaria or of those they pursue with hook and line. Regardless of why fish are studied, those studying them still need the basic vocabulary and understanding of fish biology that traditional areas of emphasis provide and that are found in this book, integrated with recent developments in other areas. Our goal is to provide some feeling for the excitement engendered by recent research on fishes. We also want to promote a sense of urgency for the need to protect fishes and aquatic ecosystems. It is critical that a high diversity of fishes continue to be around to fascinate future generations.

In large part, this book is designed to serve as a text in classes on fish biology. The large number of chapters and the cross-references within chapters provide instructors of such courses with flexibility when assigning readings in the text. The students we had in mind while writing were junior- and senior-level university students. Our goal, however, is also to provide a useful and palatable summary of recent developments in ichthyology for individuals who have been away from the college classroom for some time and for anyone else who wants an introduction to the most numerous, diverse, and fascinating of all vertebrate groups.

This book would not have been possible without the encouragement and help of many people. Initial stimulation and support in fish biology was provided by John B. Moyle, Evelyn W. Moyle, and James C. Underhill (to P.B.M.) and by Donald E. Wohlschlag (to J.J.C.). Gary D. Grossman, Donald M. Baltz, and Robert A. Daniels were especially helpful in developing the first versions of many chapters. Numerous graduate and undergraduate students contributed valuable comments on various chapters and/or helped to keep our research programs going while we devoted time to writing. We benefited from discussions with Jeff Graham, Fred White, Tony Farrell, Alan Heath, Carl Schreck, Monica Choi, Stephanie Chun, Hiram Li, Ken Gobalet, Dave Randall, Mikko Nikinmaa, George V. Lauder, Laurie Sanderson, Chris Myrick, Cincin Young, Tina Swanson, Carlos Crocker, Shana Katzman, Ann Houck, Ryan Mayfield, Marianne Brick, Keith Marine, Michael Karogosian, Peter Wainwright, and Serge Doroshov. The expert editorial assistance of Chris Myrick and Julie Roessig was especially appreciated during preparation of the fourth and fifth editions, respectively. Ms. Roessig also contributed several original figures to the fifth edition. Trilia Chen also contributed an original figure and editorial assistance to the fifth edition. The reviews of selected chapters in previous editions by Eugene Balon, Michael Bell, David Ehrenfeld, Dale Lott, John Radovich, Arnold Sillman, Randolph Smith, and Paul Webb are appreciated, as are comments by Brooks Burr, Barbara Block, Alfred Ebeling, Kurt Fausch, Malcolm Gordon, Bruce Herbold, Mark Hixon, Paul James, Douglas Markle, John McEachran, Lawrence

Page, Theodore Pietsch, Howard Reisman, Frank J. Schwartz, Jerry J. Smith, Timothy Tricas, Linda A. Ward, and Ronald M. Yoshiyama. Theodore Pietsch, Joseph Eastman, Leonard J. Compagno, and Tim M. Berra kindly shared with us their photographs and illustrations of curious fishes. Marjorie Kirkman-Iverson and the staff of our department assisted us in many ways—but especially by keeping the departmental office running efficiently, making it much easier to accomplish our regular duties while the book was in progress. Finally, we are exceedingly grateful to our wives, Marilyn Moyle and Mary Cech, for permitting our marriages to survive and even grow stronger during the many hours over the years we have worked on fish, and to our now-grown children, Petrea and Noah Moyle and Scott and Gregor Cech, for continuing to accept us despite our sometimes obsessive interest in fish.

The following reviewers were generous in providing comments and criticism of various editions of the book: Gary J. Atchison, Iowa State University; Dan Beckman, Southwest Missouri State University; Giacomo Bernardi, University of California, Santa Cruz; William Falls, Hillsborough Community College; Ronald A. Fritzsche, Humboldt State University; Kurt D. Fausch, Colorado State University; Malcolm S. Gordon, University of California, Los Angeles; Paul Grecay, Salisbury University; David W. Greenfield, University of Hawaii, Honolulu; Ralph J. Larson, San Francisco State University; Douglas E. Markle, Oregon State University; Andrew Martin, University of Nevada, Las Vegas; John D. McEachran, Texas A & M University; Karina Mrakovcich, U.S. Coast Guard Academy; Jay Nelson, Towson University; Douglas B. Noltie, University of Missouri, Columbia; Steven M. Norris, Miami University of Ohio; J. Michael Parrish, Northern Illinois University; Anne Phelps, Morehead State University; Theodore W. Pietsch, University of Washington; Howard M. Reisman, Southampton College; Charles G. Scalet, South Dakota State University; Andrew L. Sheldon, University of Montana; Jerry J. Smith, San Jose State University; Ronald L. Smith, University of Alaska, Fairbanks; and Timothy C. Tricas, Florida Institute of Technology.

PETER B. MOYLE
Department of Wildlife, Fish, and Conservation Biology
University of California, Davis

JOSEPH J. CECH, JR.
Department of Wildlife, Fish, and Conservation Biology
University of California, Davis

C H A P T E R 1
Introduction

1.1　MODERN FISHES

The fishes are the most numerous and diverse of the major vertebrate groups. They dominate the waters of the world through a marvelous variety of morphological, physiological, and behavioral adaptations. Their diversity is reflected in the large number of living fish species. A recent compilation of fish species (Eschmeyer 1998) lists 23,250 species with valid descriptions but estimates the number is actually around 25,000. Because about 200 new fish species are being described each year, the total may reach 30,000 or more!

　　Fish occupy an extraordinary array of habitats. They can be found thriving in seasonal ponds, intermittent streams, tiny desert springs, the vast reaches of the open oceans, deep oceanic trenches, cold mountain streams, saline coastal embayments, and so on through a nearly endless list of aquatic environments. Physiologically, this means that species of fish can be found living at temperatures ranging from $-1.8°C$ to nearly $40°C$, in water with pH values ranging from 4 to 10+ and dissolved oxygen levels ranging from close to zero to saturation, in salinities ranging from 0 to 90 ppt, and at depths (with associated pressures) from 0 to 7,000 m (Davenport and Sayer 1993).

　　The diversity of fish and their habitats reflects their long evolutionary history and the willingness of ichthyologists to include the jawless hagfishes and lampreys within their definition of *fish.* Modern fishes (and vertebrates in general) consist of three major groups that have been going their own evolutionary ways for at least 500 million years (Fig. 1.1). The line with the most ancestral[1] characteristics (Myxini) is that of the hagfishes (Myxiniformes, 40+ species), which arguably are not even vertebrates. They may represent, however, an offshoot of the group that did give rise to the vertebrates. The first unquestioned vertebrates were a loose group of jawless fishes (ostracoderms) that are represented in the modern fauna by lampreys (Petromyzontiformes, 40+ species). Their heyday was 350 to 500 million years ago, and they had surprising diversity despite their small size and lack of jaws and paired fins. These forms, in turn, presumably gave rise to the jawed vertebrates (Gnathostomata), which are the vertebrates that dominate our planet today. The jawed fishes divided into three distinct groups early in their evolution: the now-extinct placoderms (Placodermi), the cartilaginous fishes (Chondrichthyes),

---

[1]The terms *ancestral* and *derived* are used throughout this book instead of the more traditional *primitive* and *advanced* to avoid connotations of "inferior" and "superior," respectively.

1

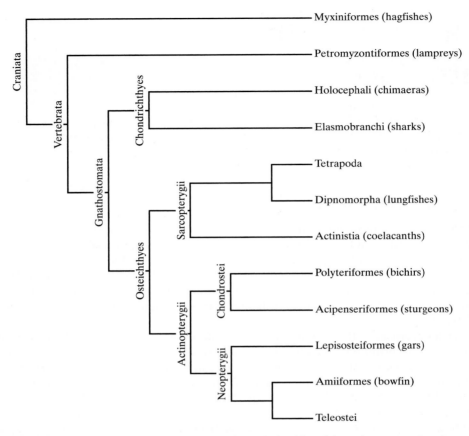

FIGURE 1.1 Branching diagram showing the interrelationships of the major groups of modern fishes, lampreys, and hagfishes (craniates). In this scheme, each branching represents a major evolutionary divergence, with each split consisting of two *sister groups*. Thus, the hagfishes (Myxiniformes) are regarded as the sister group to the lampreys (Petromyzontiformes) plus all jawed vertebrates, whereas the lampreys are the sister group to all other fishes.

and the bony fishes (Osteichthyes). The aberrant bony fishes that decided to leave the water and invade the land became modern tetrapods.

The cartilaginous fishes split into two distinct lineages early in their history: the sharks and rays (Elasmobranchi), and the ratfishes and chimaeras (Holocephali). There are more than 800 living species of sharks and rays and approximately 30 species of chimaeras and ratfishes. The bony fishes also split into a number of evolutionary lineages early in their history, although the exact nature of these lineages is debated. The major groups of bony fishes represented in the modern fauna are the lungfishes (Dipnoi, six species), the coelacanths (Coelacanthiformes, two species), and the ray-finned fishes (Actinopterygii). Living ray-finned fishes represent two distinct lines. The Chondrostei (bichirs, sturgeons, and paddlefishes; 36 species) have many ancestral characteristics and are the specialized survivors of the earliest bony fish groups. The Neopterygii are the rest of the 22,000+ known species of modern bony fishes. The high diversity of fishes reflects not only their ancient history but also their ability to speciate rapidly in response to isolation, environmental change, interactions with other species, and other factors.

It might be expected that most of this diversity of fishes would be contained in the oceans, because salt water covers more than 70% of the surface of the earth and fresh water covers only approximately 1%. By volume, 97% of all water is in the oceans, and 0.0093% is in freshwater lakes and streams (the remainder is in ice, atmospheric water, etc.) (Horn 1972). Surprisingly, only 58% of modern fish species are marine, whereas 41% are freshwater inhabitants and 1% move on a regular basis between the two environments (Cohen 1970). Fresh water consists largely of thousands of distinct "islands" of water in a sea of land, which helps to promote speciation. In contrast, most of the saltwater habitat consists of open ocean, which is rather unproductive and lighted only in the surface layer. Only 13% of all fish species are associated with the open ocean: 1% in the surface layer (epipelagic fishes), 5% in the unlighted sections of the water column (deepwater pelagic fishes), and 7% on the bottom (deepwater benthic fishes). A majority (78%) of marine fish species (making up 44% of all fishes) live in the narrow band of water less than 200 m deep along the margins of land masses (Cohen 1970). Thus, a majority of fishes live in close association with land—and with humans. An additional factor affecting the number of fish species is the annual temperature regime. In both fresh and salt water, a majority of the species are found in warmer environments where annual temperature fluctuations are comparatively small.

Despite the diversity of habitats and of adaptations enabling fish to live in these habitats, most fish are readily recognizable as fish. The reason is that the physical and chemical characteristics of water impose a number of constraints on the functional design of fish. Most of the characteristics we recognize as fish-like are adaptations to allow the most efficient use of the aquatic medium by mobile vertebrate predators. The characteristics of water exerting the greatest influence on fish design are its density, low compressibility, properties as a solvent, and transparency.

Water is approximately 800 times denser than air. This greatly reduces the effects of gravity on fish and enables them to remain suspended in the water column with little effort compared to that required by most birds to stay airborne. In fact, most fish are neutrally buoyant, so virtually all muscular effort can be devoted to movement and little is wasted on counteracting the pull of gravity. In addition, more thrust can be obtained by pushing against water than against air—as the futile flapping of a fish on land demonstrates. On the other hand, the high density of water means that it resists movement through it. Fish typically solve this problem by being streamlined to lower the resistance of the water to their swimming, by having a high proportion of their bodies devoted to the muscles needed for forward motion, and by having an efficient means of shoving the body and tail fin against the water column to thrust themselves forward.

The resistance of water to motion results not only from its density but also from its virtual incompressibility. Movement through air is made considerably easier because air compresses slightly along bodies moving through it, flows by smoothly, and so does not have to be completely displaced. Water, in contrast, literally has to be pushed out of the way by organisms moving through it. This creates turbulence along the sides and in the wake of the organism, which increases further the drag of the water. The incompressibility of water creates problems for fish, but they are also able to take advantage of it. Each fish has an extremely sensitive sensory system, the lateral line system, that can detect small amounts of turbulence and water displacement created either by its own motion or by that of other organisms or objects. Using this system, fish can detect nearby stationary objects, other fish, and food organisms. The lateral line system is particularly useful when visual cues are lacking. Fish (mainly teleosts) also use the incompressibility of water to help them in eating and breathing: Quick expansion of the mouth and gill chambers allows water to rush in, carrying food and oxygen into the chambers along with the water. This "pipette effect" is particularly well developed in fishes that feed on small, mobile organisms, and it results in jaw structures very different from those of terrestrial vertebrates.

Water behaves as if it is incompressible for the purposes of movement, feeding, and breathing of fish, but it can be compressed slightly. This is fortunate, because it enables sound to be carried. In fact, sound is carried farther and faster in water than it is in air (1,433 m/s versus 335 m/s) thanks to the greater density of water. As a consequence, most fish have an excellent sense of hearing, even though they lack the external ears associated with hearing in terrestrial vertebrates. External ears are not needed, because fish tissue is roughly the same density as water and so is nearly transparent to sound waves. However, the sound waves can be intercepted internally by structures that are either much denser than water (otoliths) or much less dense than water (swim bladders). Because most fish can hear well, it is not surprising that many also make sounds for communication. A single toadfish, for example, when engaging in its courtship rites can produce sound that may reach 100 dB in loudness. The "silent world" of the oceans is largely a myth, at least as far as fish are concerned!

Perhaps the most important characteristic of water that enables it to support life, including fish, is its property as the nearly universal solvent. The waters of the world contain complex mixtures of dissolved gases, salts, and organic compounds, many of which are needed by fish to sustain life and are taken up either directly through the gills or indirectly through food organisms. The most important of the gases is oxygen, which is present in extremely small amounts compared with that found in air (1 to 8 ml of oxygen per liter of water versus 210 ml of oxygen per liter of air). Although the gills of fish are incredibly efficient at extracting oxygen from water, the low availability of oxygen in water can limit their metabolic activity. To maximize their ability to extract oxygen from the water, fish expose a large gill surface area to the water, bringing their blood in close contact with the environment to facilitate gas exchange. Not surprisingly, other substances carried in the blood also pass between the environment and the fish. Fish eliminate waste products (especially carbon dioxide, ammonia, and heat) through the gills, but they also take in harmful substances, including pollutants such as heavy metals and pesticides. The large area of exposed gill also makes fish very sensitive to changes in the concentration of salts in the water, accounting, in part, for the difficulty most fish have in moving between fresh and salt water.

A final important characteristic of water to fish is its low penetrability to light compared with that of air. Even in the clearest water, light seldom penetrates deeper than 1,000 m, and in most water, the depth of penetration is considerably less. Because the lighted (photic) zone is where algae and aquatic plants grow and where grazing invertebrates are concentrated, it is not surprising that most fish are found there. Indeed, a majority of fish rely primarily on sight for prey capture. However, many fishes have sensory structures, such as barbels and electric organs, that allow them to find their way about at night, in muddy water, or in other low-light situations. Below the lighted zone of the oceans are complex communities of fish that produce their own light, both to signal each other and to attract prey.

As the following chapters will reveal, there are many exceptions to the preceding generalities about fish. Each fish species has its own unique combination of adaptive traits that enables it to survive in its own particular environment. These traits reflect the rich evolutionary history of fishes. They provide the notes for the complex sonata of systematics, whose playing has traditionally been the showpiece of ichthyology. Today, the concert has been joined by other equally complex pieces featuring the physiology, ecology, and behavior of fish. Increasingly, these pieces are merging into a grand symphony of modern ichthyology. This book is only a program guide to that dynamic, but always unfinished, symphony.

1.2 HISTORY OF ICHTHYOLOGY

Ichthyology has its origins in the writings of Aristotle (384–322 B.C.). Aristotle made observations that allowed him to distinguish fish from whales and to recognize about 117 species of fish. He was the first to set down a myriad of basic facts about fish, such as that the sex of sharks can be determined from the structure of the pelvic fins and that certain sea basses change sex as they grow older. Unfortunately, for nearly 2,000 years after Aristotle, few original observations about fish were recorded, because Aristotle was considered to have completely covered all areas of natural history. Aristotle's grip on ichthyology (and science) was finally broken in the sixteenth century by natural historians such as Pierre Belon (1517–1564), H. Salviani (1514–1572), and G. Rondelet (1507–1566). Belon published a natural history of fishes in 1551, in which he classified approximately 110 species according to anatomical characteristics, especially body size and shape. From 1554 through 1557, Salviani published sections of a treatise on Italian fishes in which 92 species were described and illustrated. In 1554 and 1555, Rondelet published books that summarized much of what was known about fishes at the time. Essentially, Belon published the first modern systematic treatise on fish, Salviani the first regional faunal work, and Rondelet the first ichthyology text.

Knowledge of fishes subsequently expanded rapidly, stimulated in good part by the discoveries and reports of naturalist explorers. Preeminent among them was Georg Marcgrave of Saxony (1610–1644), whose *Natural History of the Fishes of Brazil* was published posthumously in 1648. The knowledge of European fishes also continued to grow, so when the English naturalists John Ray (1627–1705) and Francis Willughby (1635–1672) published their *Historia Piscium* in 1686, they could describe 420 species, including 178 new species, and arrange them in a reasonable classification system. Works such as these are perhaps most important because they were the foundation on which Peter Artedi (1705–1734) built the classification system of fishes that earned him the title

"Father of Ichthyology." Artedi also critically reviewed the previous literature on fishes and recommended standard measurements and counts that remain the basis of fish taxonomy. Artedi drowned before any of his studies were published, but his notes and manuscripts were purchased by his good friend Carolus Linneaus (1707–1778), who edited this material and issued it in 1738. Linneaus adopted (and eventually changed, not always for the better) Artedi's classification of fishes for the 12 editions of his own extraordinary *Systema Naturae*, which became the basis for all future classification systems.

The first attempt after Artedi's to organize the expanding knowledge of fishes was that of Marc Elieser Bloch (1723–1799) of Berlin. Remarkably, Bloch did not begin publishing on fishes until he was 56 years old, and he then published a series of volumes on the fishes of Germany and of the world. His *Systemae Ichthyologiae*, with a revised classification system, was edited by J. G. T. Schneider (1750–1822) and issued in 1801. The massive volumes of Bloch were the standard reference works on fishes until publication of the comprehensive volumes written by Georges Cuvier (1769–1832) and his pupil A. Valenciennes between 1828 and 1849. Cuvier and Valenciennes not only compiled and classified the known fishes but also conducted many detailed studies of fish anatomy, which greatly improved the understanding of fish interrelationships (Fig. 1.2). In the first volume of *Histoire Naturelle des Poissons* (1828), Cuvier provided a fascinating and accurate early history of ichthyology. This history was recently translated into English and edited,

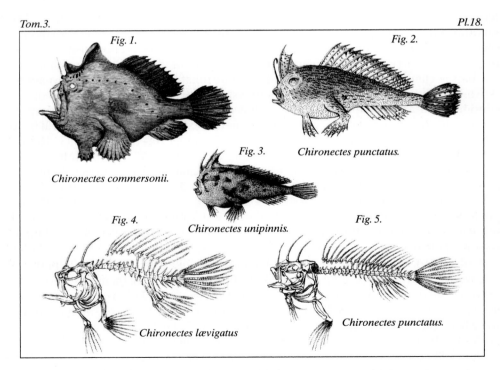

Tom.3. *Pl.18.*

Fig. 1.

Fig. 2.

Chironectes punctatus.

Chironectes commersonii.

Fig. 3.

Chironectes unipinnis.

Fig. 4.

Chironectes lævigatus

Fig. 5.

Chironectes punctatus.

FIGURE 1.2 Frogfishes and their skeletons (Antennariidae) as depicted by Georges Cuvier in an 1817 publication. Photograph by S. A. Shockey, University of Washington, Seattle, courtesy T. W. Pietsch.

with updated footnotes, by T. W. Pietsch (1995). The influence of Bloch and Cuvier on ichthyology would probably have been less if the French Revolution had not cut short the studies of B. G. E. Lacepède (1756–1826), who was consequently obliged to compose his five-volume work on fishes largely from memory and rough notes. Much of his immediate successors' time seems to have been spent correcting the errors he made. While such general works on fishes were being composed, the detailed studies of fish anatomy and physiology were also proceeding, resulting in Alexander Monro's *The Structure and Physiology of Fishes Explained and Compared to that of Man and Other Animals* (1785).

In the early nineteenth century, the most important ichthyological works were descriptions of regional fish faunas. The first major work for North America was *The Fishes of New York, Described and Arranged* (1815), by Samuel L. Mitchill (1764–1831); however, the most detailed of the early accounts was *Ichthyologia Ohiensis* (1820) by the eccentric Sicilian naturalist Constantine Rafinesque (1783–1840). Many of the fishes of the Pacific coast of North America were first described by a British navy surgeon, Sir John Richardson (1787–1865), in his *Fauna Boreali-Americana* (1836). In 1829, another treatise on the fishes of Brazil appeared, which was the first major work of Louis Agassiz (1807–1873). Only 21 at the time, Agassiz classified the fishes collected by Johann Baptiste von Spix (1781–1826), who had died after spending three years pursuing them in the rivers and streams of the Brazilian jungle. Agassiz's next works (1833–1844) were a series of volumes on fossil fishes, which laid the foundation for evolutionary studies of fishes. This is ironic, because Agassiz himself was a fervent opponent of evolution. In 1846, Agassiz moved from Switzerland to Harvard University, where, besides convincing the American public of the importance of science in general, he continued to make many contributions to ichthyology.

Meanwhile, in Germany, Johannes Müller (1801–1858) revised Agassiz's ideas on fish classification to produce a system that recognized most of the major groups that scientists still recognize today. His system was improved by such leading scientists of that era as Ernst Haeckel (1834–1919), Thomas H. Huxley (1825–1895), Edward D. Cope (1840–1897), and Theodore Gill (1837–1914), who were all greatly influenced by the ideas of Charles Darwin. Curiously, one of the most prominent ichthyologists of the late nineteenth century did not accept evolutionary ideas. This was Albert Günther (1830–1914), who labored in the British Museum with its magnificent collections of fishes from all over the British Empire to produce the eight-volume *Catalogue of the Fishes in the British Museum*. This was the last time anyone tried single-handedly to describe the fish fauna of the world species by species. Günther's successor at the museum, George A. Boulenger (1858–1937), recognized the impossibility even of revising Günther's volumes after reworking the volumes on one major group. Instead, he devoted much of his time to the fishes of Africa, producing the four-volume *Catalogue of the Freshwater Fishes of Africa* (1909). Working with Boulenger in the British Museum was A. S. Woodward, who reviewed what was known about fossil fishes and set the knowledge in an evolutionary framework.

Perhaps the one person who could have revised the British Museum catalogue was David Starr Jordan (1851–1931). He, however, turned down the curatorship at the museum to continue his career in the United States, where he eventually became the founding president of Stanford University. For much of his career, Jordan occupied an Agassiz-like role as a great popularizer of science; at the same time, he was extraordinarily productive as an ichthyologist. Although Jordan worked with fishes from all over the world, he is best known for *Fishes of North and Middle America* (four volumes, 1896–1900), coauthored with B. W. Evermann (1853–1932). Jordan was also the author of the standard ichthyology text of the early twentieth century, *Guide to the Study of Fishes* (1905).

As the twentieth century progressed, ichthyology became more and more diverse. Ecology, physiology, and behavior shared the field with the more traditional areas of anatomy and systematics. Yet if there remains a common thread that holds these diverse areas together, it is evolution and systematics, which perhaps accounts for the fact that the best-known ichthyologists of our century—such as Charles T. Regan (1878–1943), Leo S. Berg (1876–1950), and Carl L. Hubbs (1894–1979)—have made their major contributions in this area. Regan's work on teleost anatomy and classification is the foundation on which much of our more-contemporary systems rest. Berg described and organized the information on much of the fish fauna of Russia and nearby areas. In addition, his *Classification of Fishes* (1940) was the standard reference work on the subject for many years. Hubbs, like Jordan, worked mainly on the fishes of North and Central America, smoothing off the rough edges of Jordan's work. His *Fishes of the Great Lakes Region* (1964; authored with Karl F. Lagler), besides being a classic regional work, remains the standard reference on taxonomic techniques for fishes. Since then, regional works on fishes have proliferated, and most states, provinces, and countries have guides to their fishes. The growing interest in fish has led, finally, to a new compilation of fish species of the world, by William N. Eschmeyer (1998) and colleagues—a work made possible, in part, by modern computers and communications.

BOX 1.1
Pursuing Ichthyological Knowledge

The history of ichthyology contains many harrowing tales of expeditions that sometimes led to severe injury, sickness, or even death of the investigator, for reasons which seem hard for us to imagine today. This is well illustrated by John Samuel Budgett (1872–1904), who made four expeditions into the swamps of tropical Africa, under the worst possible conditions, to seek embryos of bichirs (*Polypterus*, see Chapter 16). At that time, bichirs were known from only few specimens. They have a mixture of characteristics—such as lungs, lobed fins, spiracles, and ganoid scales—that baffled nineteenth-century biologists. It was speculated that bichirs might be "primitive" ray-finned fish, or lungfish, or sharks, or even amphibians (Hall 2001).

Budgett was among the biologists of his era who thought that much could be learned about the early evolution of vertebrates, including humans, by studying the embryology of such a primitive fish, under the assumption that embryological development mirrored evolutionary change. He mounted four expeditions in search of bichir embryos, going to the tropics during the rainy season and acquiring a number of tropical diseases in the process. He returned to England with a series of carefully preserved embryos from the fourth expedition but died of blackwater fever after his return. Another scientist published descriptions of the embryos, but few of Budgett's burning questions on evolution were answered by the study (Hall 2001).

1.3   FISH CLASSIFICATION

Although describing new taxa of fishes—and organizing the taxa into systems that demonstrate interrelationships—is no longer the primary occupation of most biologists who work with fish, it is nevertheless of fundamental importance. To understand the significance of the ecological, physiological, behavioral, and morphological adaptations of fish, scientists must also understand the evolutionary relationships among fishes. Modern classification schemes are generally presumed to reflect evolutionary relationships, because common structural features (on which most schemes are based) generally reflect common ancestry. Because our knowledge of most fishes is far from complete, however, refinements and changes to accepted classification systems are continually being proposed. The most controversial but far-reaching changes are often those based on advances in biochemical techniques. Eventually, these changes lead to new systems that may bear little resemblance to the old systems on which they are based. Indeed, the most prominent figures in much of the history of ichthyology are individuals who organized the recent advances in the knowledge of fishes into "new" classification schemes. Most classification schemes today are ultimately rooted in those of Regan (1929) and Berg (1940).

More recently, the provisional classification of the teleosts by Greenwood et al. (1966) has had a major impact on the thinking of systematic ichthyologists and has stimulated many further attempts at revising taxa within the bony fishes. Considerable attention is being paid by modern systematists to other groups of fishes as well, so controversies over arcane but important details of classification are still slowly raging in the literature. Nelson (1994) put much of this recent work together in a classification system that will be largely followed in this book, except where we prefer other schemes (especially for higher levels of classification). Some of the more significant differences include:

1. We follow the analysis of Forey and Janvier (1993) for classification of the earliest fishes, because they separate the hagfishes from the rest of the vertebrates, thus eliminating the artificial category Agnatha that places hagfishes and lampreys together.

2. We retain Osteichthyes rather than the word Teleostomi, as the all-encompassing word for bony fishes. This makes the tetrapods a subdivision of the bony fishes, but as Nelson (1994) states, tetrapods are "a divergent sideline within the fishes that ascend onto land and into the air and secondarily return to the water" (p. 68).

3. For the Chondrichthyes, the classification schemes in Stiassny et al. (1996) are followed, which have a debt to Compagno (1973).

4. In a number of instances, we follow various authorities in Stiassny et al. (1996) in updating the classification of bony fishes. These changes are indicated in the text.

For the sake of simplicity, we usually will only present one scheme of classification for any group of fishes, even where controversy and alternative schemes exist. Systematics is a dynamic field, so continuous changes are likely as new studies provide a better understanding of the interrelationships among fishes. Thus, the classification system used in this book should be viewed as a guide to fish diversity rather than an immutable entity.

1.4 INTRODUCTORY LESSONS

Each chapter of this book will end with a short section presenting what we think are some of the more important concepts or generalities to be gained from the chapter: the knowledge remaining in the brain after the details have faded away. In this case, some of the basic lessons are: (1) the extraordinary diversity of fishes is related to their ability to adapt to almost every aquatic environment on Earth through their long evolutionary history, (2) fish systematics has a long and noble history and is still a dynamic and exciting field, and (3) humans are not the pinnacle of evolutionary progress but only an aberrant side branch of fish evolution.

SUPPLEMENTAL READINGS

Berg 1940; Boulenger 1910; Cohen 1970; Compagno 1973; Davenport and Sayer 1993; Horn 1972; Hubbs 1964; Jordan 1922; Myers 1964; Nelson 1994; Patterson 1977; Pietsch 1995; Stiassny et al. 1996; Wooton 1990.

WEB CONNECTIONS

GENERAL
www.fishbase.org
www.flmnh.ufl.edu/fish
www.tolweb.org/tree
www2.biology.ualberta.ca/jacksonhp/iwr
www.aquatic.uoguelph.ca/fish
www.calacademy.org/research/ichthyology

CHAPTER 2
Form and Movement

The great ecological diversity of fishes is reflected in the astonishing variety of body shapes and means of locomotion they possess. Indeed, much can be learned about the ecology of a fish simply by examining its anatomical features or by watching it move through the water. Equally important to students of ichthyology (if not to the fish) is that these features also form the basis for most schemes of classification and identification. The purpose of this chapter, therefore, is to provide an overview of (1) external anatomy, (2) internal support systems (the skeleton and muscles), and (3) means of locomotion.

2.1 EXTERNAL ANATOMY

Although life in water puts many severe constraints on the "design" of fish, the thousands of species living in a wide variety of habitats means that these constraints are pushed to their limits. This results in many unlikely forms, such as seahorses and lumpfishes. Understanding the significance of the peculiar external anatomy of such forms requires study on a case-by-case basis. On the other hand, species that are more recognizably fish-like (Fig. 2.1) can usually be placed in some sort of functional category through an examination of body shape, scales, fins, mouth, gill openings, sense organs, and miscellaneous structures (Fig. 2.2).

2.2 BODY SHAPE

Most fishes fall into one of six broad categories based on body configuration: (1) rover-predator, (2) lie-in-wait predator, (3) surface-oriented fish, (4) bottom fish, (5) deep-bodied fish, and (6) eel-like fish (Fig. 2.1).

Rover-predators have the body shape that comes to mind when most people think of fish: streamlined (*fusiform*), with a pointed head ending in a terminal mouth and a narrow caudal peduncle tipped with a forked tail. The fins are more or less evenly distributed about the body, providing stability and maneuverability. Such fish typically are constantly moving and searching out prey, which they capture through pursuit. Examples include many species of minnows (Cyprinidae), bass, tuna, mackerel, and swordfish. The rover-predator body shape is also characteristic of stream fish, such as trout, which spend much of their time foraging in fast water.

Lie-in-wait predators are mainly *piscivores* (fish eaters) that have a morphology well suited for the ambushing of fast-swimming prey. The body is fusiform, but it is also

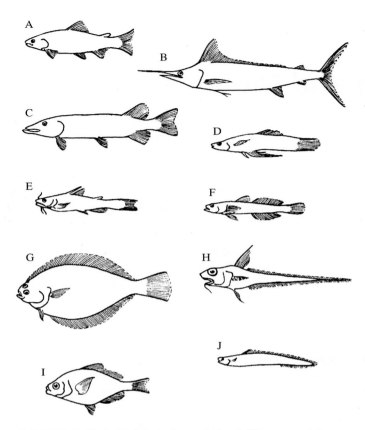

FIGURE 2.1 Typical fish body shapes: (**A**) and (**B**) rover-predator; (**C**) lie-in-wait predator; (**D**) surface-oriented fish; (**E**) bottom rover; (**F**) bottom clinger; (**G**) flatfish; (**H**) rattail; (**I**) deep-bodied fish; (**J**) eel-like fish.

elongate, often torpedo-like. The head is flattened and equipped with a large mouth filled with pointed teeth. In many species, the mouth is largely contained in a long, pointed snout. The caudal fin tends to be large, and the dorsal and anal fins are placed far back on the body, often in line with each other. This arrangement of the fins gives a fish the large amount of thrust it needs to launch itself at high speed toward passing fish. The narrow frontal profile that these fish present, coupled with their cryptic coloration and secretive behavior, also makes them less visible to their prey. Members of this group include the freshwater pikes (Esocidae), barracuda (Sphyraenidae), gars (Lepisosteidae), needlefish (Belonidae), and snook (Centropomidae).

Surface-oriented fish are typically small in size, with an upward-pointing mouth, a dorsoventrally flattened head with large eyes, a fusiform to deep body, and a dorsal fin placed toward the rear of the body. The morphology is well suited for capturing plankton and small fishes that live near the water's surface or insects that land on the surface. In stagnant water, the surface-oriented morphology is particularly suitable for taking

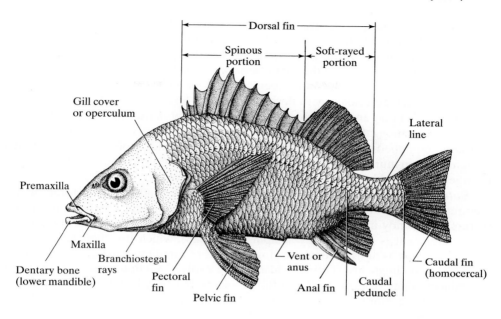

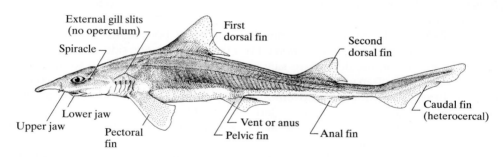

FIGURE 2.2   External features of a bony fish (snapper, top) and cartilaginous fish (smoothhound shark, bottom). From Moyle (1993); © 1993 by Chris Mari Van Dyck.

advantage of the thin layer of oxygen-rich water that exists at the air–water interface. The mouth of the fish can be placed in the layer and the water then pumped across the gills. Most surface-oriented fish are stocky-bodied fresh or brackish water forms, such as mosquitofish (*Gambusia*), many killifish (Fundulidae), and the four-eyed fish (*Anableps anableps*), but a number of elongate marine forms, such as the halfbeaks and flying fishes (Exocoetidae), have similar adaptations.

   Bottom fish possess a wide variety of body shapes, all of them adapted for a life in nearly continuous contact with the bottom. In most such fish, the swimbladder is reduced or absent, and most are flattened in one direction or another. Bottom fish can be divided into five overlapping types: (1) bottom rovers, (2) bottom clingers, (3) bottom hiders, (4) flatfish, and (5) rattails. ***Bottom rovers*** have a rover-predator-like body, except that the head tends to be flattened, the back humped, and the pectoral fins enlarged. Examples include forms as varied as (1) North American catfishes (Ictaluridae) with large mouths at the end of the snout (***terminal mouths***); (2) small, armored catfishes (Loricariidae) with small mouths beneath the snout (***subterminal mouths***); and (3) sturgeons (Acipenseridae) with fleshy, protrusible lips located well below the snout (***inferior mouths***) that are used to suck plant and animal matter off the bottom. Many bottom rovers among the bony fishes have small eyes and well-developed ***barbels*** ("whiskers" equipped with tastebuds) around the mouth, indicating their ability to find prey at night or in murky water. Many sharks, with their inferior mouths, flattened heads, and large pectoral fins, can also be classified as bottom rovers.

Bottom clingers are mainly small fish with flattened heads, large pectoral fins, and structures (usually modified pelvic fins) that allow them to adhere to the bottom. Such structures are handy in swift streams or intertidal areas with strong currents. The simplest arrangement is possessed by sculpins (Cottidae), which use their small, closely spaced pelvic fins as antiskid devices. However, other families of fishes, such as gobies (Gobiidae) and clingfishes (Gobiesocidae), have evolved suction cups. ***Bottom hiders*** are similar in many respects to the bottom clingers, but they lack the clinging devices and tend to have more elongate bodies and smaller heads. These forms usually live under rocks or in crevices or lie quietly on the bottom in still water. The darters (Percidae) of North American streams are in this category, as are many blennies (Blenniidae). However, the latter family contains species that range in form from "good" bottom hiders to more eel-like forms.

Flatfish have the most extreme morphologies of the bottom fish. Flounders (Pleuronectiformes) are essentially deep-bodied fish that live with one side on the bottom. In these fish, the eye on the downward side migrates during development to the upward side, and the mouth often assumes a peculiar twist to enable bottom feeding. In contrast, skates and rays are flattened dorsoventrally (***depressiform***) and mostly move about by flapping or undulating their extremely large pectoral fins. Not only is the mouth completely ventral on these fish, the main water intakes for respiration (the spiracles) are located on the top of the head.

The ***rattail shape*** is another type of body shape that has independently evolved in both the Osteichthyes and Chondrichthyes. Groups such as the grenadiers (Macrouridae), brotulas (Ophidiidae), and chimaeras (Holocephali) have bodies that begin with large, pointy-snouted heads and large pectoral fins and end in long, pointed, rat-like tails. These fish are almost all bottom-dwelling (***benthic***) inhabitants of the deep sea, but exactly why this peculiar morphology is so popular among them is poorly understood. The fishes live by scavenging and by preying on benthic invertebrates.

Deep-bodied fish are laterally flattened (***compressiform***) fish, with a body depth usually at least one-third that of the standard length (distance from snout to structural base of caudal fin). The dorsal and anal fins are typically long, and the pectoral fins are located high on the body, with the pelvic fins immediately below. The mouth is usually small and protrusible, the eyes large, and the snout short. Deep-bodied fish are well adapted for maneuvering in tight quarters, such as the catacombs of a coral reef, dense beds of aquatic plants, or tight schools of their own species. They are also well adapted for picking small invertebrates off the bottom or out of the water column. A majority of deep-bodied fish possess stout spines in the fins, presumably because during the course of their evolution they sacrificed speed for maneuverability and developed spines for protection from predators. Although most deep-bodied fish are closely associated with the bottom, many open-water plankton feeders (*planktivores*; e.g., herring) are also moderately deep-bodied. This is largely the result of a sharp ventral keel, which functions to camouflage these silvery fish by eliminating ventral shadows, thus making them less visible to predators approaching from below.

Eel-like fish have elongate bodies, blunt or wedge-shaped heads, and tapering or rounded tails. If paired fins are present, they are small, whereas the dorsal and anal fins are typically quite long. Scales are small and embedded or absent. In cross-section, their bodies can range from compressed to round. Eel-like fishes are particularly well adapted for entering small crevices and holes in reefs and rocky areas, for making their way through beds of aquatic plants, and for burrowing into soft bottoms. However, a surprising number are also found swimming about in the open ocean, so this body shape is useful for other purposes as well. Examples of this group include the many eels (Anguilliformes), loaches (Cobitidae), and gunnels (Pholididae).

2.3 SCALES

The type, size, and number of scales can reveal much about how a fish makes its living. The scales of bony fish range from a heavy coating of mail-like armor to a few large bony plates on the back to a dense covering of thin, flexible scales to no scales at all (Fig. 2.3). Bony plates are large, modified scales that serve as armor on a number of bottom-oriented fishes, such as sturgeons (Acipenseridae), many South American catfishes, poachers (Agonidae), and pipefishes and seahorses (Syngnathidae). Most such fish are rather slow in their movements. In contrast, typical scales usually cover the bodies of most free-swimming fish, apparently providing some degree of protection from predators while not excessively weighing down the fish. Fish that are fast swimmers or regularly move through the fast water of streams typically have many fine scales (e.g., trout), whereas those that live in quiet water and do not swim continuously at high speeds tend to have rather coarse scales (e.g., perch and sunfish).

Scales evolved independently in cartilaginous and bony fish (see Chapter 13), as indicated by their fundamentally different structure. The *placoid scales* of sharks are tiny, tooth-like structures, whereas the scales of bony fish are layered plates, with bone as one of the layers. The ancestral condition for bony fish is represented by the heavy *ganoid scales* of gars (Lepisosteidae; Fig. 2.3) and the more derived condition by the *bony ridge (elasmoid) scales* of teleosts. The latter are of two basic types: cycloid, and ctenoid. *Cycloid scales* are the round, flat, thin scales found on such fishes as trout, minnows, and herrings. *Ctenoid scales* are found mainly on spiny-finned teleosts (Acanthopterygii) and are similar to cycloid scales except for the tiny, comb-like projections (*ctenii*) on the exposed (posterior) edge of the scales. The ctenii apparently improve the hydrodynamic efficiency of swimming. Some kinds of fish possess both types of scales. Curiously, the tiny placoid scales of sharks may be an independently evolved solution to the same "problem," because these scales, like ctenoid scales, make the exterior of the fish rough to the touch.

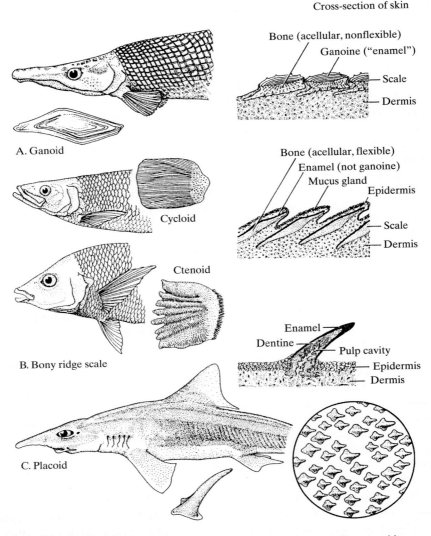

FIGURE 2.3 Examples of ganoid scales on gar, cycloid scales on sardine, ctenoid scales on snapper, and placoid scales on a shark. Ganoid scales after Dareff (1971), copyright © Fratelli, Fabbri Editori, Milan; cross-sections after Hildebrand (1974), copyright © John Wiley. From Moyle (1993); © 1993 by Chris Mari Van Dyck.

Although scales are usually considered to be an integral part of any fish, a surprising number of species lack them altogether or have just a few that are modified for other purposes. Such fish are by and large bottom dwellers in moving water (e.g., sculpins); fish that frequently hide in caves, crevices, and other tight places (many catfish and eels); or fast-swimming pelagic fish (swordfish and some mackerels). However, many fish that appear to be scaleless in fact have a complete coating of deeply embedded scales (most tunas and anguillid eels). Many of the bottom-dwelling skates and rays also do not have placoid scales, except as patches of bony armor or as spines (in stingrays).

2.4 FINS

Like scales, fins reflect the independent evolutionary history of bony and cartilaginous fishes (Fig. 2.2). In both groups, the fins are supported internally by sturdy fin rays. In sharks and rays, the fin rays are called **ceratotrichia** and are fairly stiff, unbranched, and unsegmented. In contrast, the **lepidotrichia** of bony fish are flexible, segmented, and branched. Lepidotrichia seem to start as embryonic spines (**actinotrichia**) that become covered with embryonic scales, which then replace the actinotrichia completely before emerging as fin rays. When this developmental process does not occur, **true spines** emerge, which are stiff, round in cross-section, and unsegmented. In all fish, the various combinations of location, size, and shape of the fins are closely associated with the different body shapes. Although the paired fins (pectorals and pelvics) and the unpaired fins (dorsal, anal, caudal, and adipose) evolved together as a system that simultaneously propels, stabilizes, and maneuvers each fish, they are discussed here separately for convenience.

   Pelvic fins are the most variable of the fins in terms of position. In more ancestral bony fishes, such as salmon, shad, and carp, as well as in sharks, the fins are located ventrally, toward the rear of the fish (**abdominal position**). Most of these fish have rover-predator body shapes, and the fins assist in steering and braking. In more derived teleosts, many of which are deep-bodied, the pelvics are more anterior, below the pectoral fins (**thoracic position**; Fig. 2.1); occasionally, they are even in front of the pectorals (**jugular position**). In eels and eel-like fish, the pelvic fins are absent or greatly reduced in size, in part for ease of squeezing through tight places. In bottom-dwelling fish, the pelvics are frequently modified into organs for holding on to the substrate.

   Pectoral fins are generally located high up on the sides of deep-bodied fish, which depend on precise movements for picking prey from the bottom or the water column. In rover-predators, these fins tend to be more toward or below the midline of the fish. In very fast-swimming fish, such as tuna, and in very deep-bodied fish that pick prey from the substrate (bluegill and many cichlids), the fins tend to be long and pointed. In slower-moving rover-predators or other fish that need more surface area for stability while swimming, the fins tend to be more rounded. The pectoral fins of bony fish that rest on the bottom, such as suckers (Catostomidae) and sculpins (Cottidae), are usually broad, rounded, ventral in position, and spread out laterally. Other fish use enlarged pectoral fins for gliding (flying fish, Exocoetidae) or, in the case of many rays (eagle rays, Myliobatidae), "flying" in the water. In some fish, such as the "flying" gurnards (Dactylopteridae) and the tropical lion and turkey fishes (Scorpaenidae), enlarged pectoral fins are apparently used mainly for display, either to startle predators when suddenly opened (gurnards) or to signal predators (and conspecifics) to stay away from poisonous spines. In contrast, the pectoral fins of sharks are rather like rigid wings that can be moved but not collapsed. These fins operate not only as stabilizers but also as "diving planes." Because of the latter function, they are normally set at an elevated angle of attack to generate lift for the anterior part of the body; the heterocercal tail provides lift for the posterior part of the body (Lauder 2000).

Dorsal fins and *anal fins* are generally long on rover-predators and deep-bodied fish to provide stability while swimming. In fast-swimming pelagic fish, such as tuna and mackerel, the rearmost portions of both fins are frequently broken up into numerous finlets. When such fish are swimming at high speed, the forward portion of the dorsal fin may fold into a dorsal slot to reduce the resistance the additional surface area creates; likewise, the pectoral fins of such fish lie down into shallow pockets. However, even bony fish that lack these specializations will collapse the dorsal and anal fins and fold back the pectoral and pelvic fins when putting on a burst of speed. Interestingly, this method of reducing drag is not available to sharks, although a number of the mackerel sharks are capable of swimming at quite high speeds.

Another group of fish with long dorsal and anal fins are the eel-like fish. Their fins frequently run most of the length of the body and may unite with the caudal fin; such a configuration is necessary for the anguilliform locomotion discussed later in this chapter. In eel-like electric fish, only one of the two long fins is developed. In such cases, the fin is the principal means of propulsion, operated by sending waves of movement down the fin, because electric fish must keep their bodies rigid while swimming to maintain a uniform electrical field about them.

The *caudal fin* (*tail fin*) has a shape that is strongly related to the normal swimming speed of a fish. The tails of most bony fish are *homocercal*, with upper and lower lobes being about the same size. The fastest-swimming fish, such as tuna and marlin, have a stiff, quarter-moon-shaped (*lunate*) fin attached to a narrow caudal peduncle. Fish whose survival depends on frequent, sustained swimming have forked tails, with the deepest forks occurring on the most active fish. Deep-bodied fish as well as most surface and bottom fish have tails that are square, rounded, or only slightly forked. When a homocercal tail lacks well-defined lobes, it is referred to as *isocercal*. In the Chondrichthyes, the tail is usually *heterocercal*, with the upper lobe being longer than the lower lobe. These two basic tail types (homocercal and heterocercal) reflect the evolutionary history of the fishes that have them (see Chapter 13). In the homocercal tail, the vertebral column ends in modified vertebrae that support the fan-like tail structure; in the heterocercal tail, the vertebral column actually extends into the upper lobe of the tail. Functionally, whereas (stiffer) heterocercal caudal fins in sharks act to lift the posterior body, the heterocercal caudal fin of sturgeon is extremely flexible and the upper tail lobe trails the lower lobe during the tailbeat cycle (Lauder 2000). Working with white sturgeon (*Acipenser transmontanus*), Wilga and Lauder (1999) showed that their caudal undulations function best when the sturgeon's body is oriented at a positive (anterior end upward) angle of 8° to 25° to the flow, propelling the entire sturgeon forward without lifting the posterior end. Thus, digitized kinematic analyses show that the similarity of heterocercal shape between the tails of a shark and a sturgeon is not mirrored by a similarity of function (Lauder 2000).

The *adipose fin* is a fleshy, dorsal appendage that is found in the trouts (Salmonidae), smelts (Osmeridae), lanternfishes (Myctophidae), and various catfish and characins. Although located between the dorsal and caudal fins toward the caudal peduncle, the small size and lack of stiffening rays make the function of this fin a mystery. It may have an important function in the swimming of fish during the postlarval stage of development, when other fins are poorly developed.

One of the most important attributes of the fins of fish is the presence or absence of **spines** on the dorsal, anal, and pectoral fins. The importance of spines is indicated by the fact that they have developed independently in several different groups of fish. In the dominant group of spiny-finned teleosts (Acanthopterygii), spines are solid bony structures without any segmentation and are round in cross-section. In contrast, the spines of catfish, carp, goldfish, and similar fishes are just stiffened, thickened rays, which are segmented, dumbbell-shaped in cross-section, and often branched. The spines that precede the dorsal fin in some sharks are modified placoid scales, as are the stings of stingrays. Regardless of their structure and origin, spines are an effective and lightweight means of protection against predators. Dorsal, pectoral, and opercular spines are often located at the fish's center of mass, the usual target point of piscivorous fish (Webb 1978a). Besides being uncomfortable for a predator to bite down on, spines greatly increase the effective size of a small fish, because once the dorsal, anal, and pectoral spines are locked into place, the fish can be grabbed only by a predator that can get its mouth around the spines. By increasing its effective size through the use of spines, a small fish reduces the number of predators that can prey on it, because large predators are almost always fewer in number than small predators. As a consequence, well-developed spines are found mainly in small to medium-sized fishes that actively forage for their food. As an additional disincentive to predators, many spines have poison glands associated with them, such as those found on scorpionfish (Scorpaenidae), some catfish, and stingrays.

2.5   OTHER STRUCTURES

The **mouth** reveals much about the habits of a fish by its position, shape, and size. Not surprisingly, bottom-feeding fish have downward-pointing (**inferior**) mouths, whereas surface-oriented fish have upward-pointing (**superior**) mouths. For most fish, however, the mouth is at the end of the snout (**terminal**). The size and shape of the mouth is usually directly related to the size of the preferred food organisms. Thus, fish that feed on small invertebrates by suction have a small mouth surrounded by protractile "lips" that, when protruded, form an O-shaped opening (the shape with the maximum ratio of area to perimeter length). However, fish that feed on large prey typically have an inflexible rim to the mouth, which is oval in cross-section and frequently lined with sharp teeth. Beyond these general types, the mouths of fish can show extraordinary shapes that reflect specialized modes of feeding, as discussed in the chapters on the fish of tropical reefs (see Chapter 33) and tropical lakes (see Chapter 30).

The **gill openings** of most bony fish are covered by a thin and flexible bony **operculum** that is an important component of the "two-pump" respiratory system possessed by most fish (see Chapter 3). As a consequence, the gill openings and cover do not show a large amount of variability, except to be smaller on fish with low activity levels than on fish that are more active. In eels, the openings are typically reduced to a small hole, presumably because of the problems associated with raising and lowering the operculum under confined conditions. In sharks, skates, and rays, the **spiracles**, which are small openings used for the intake of water for respiration (Fig. 2.2), are dorsally located in the bottom-dwelling rays and skates and laterally located in most sharks. In the most active species of sharks, however, the spiracular opening is greatly reduced in size, because most of the respiratory water is forced through the mouth and across the gills (ram gill ventilation; see Chapter 3) while swimming.

Of the externally visible ***sense organs*** of fish, the ***eyes*** perhaps reveal the most about habits. The size of the eyes, relative to the size of the fish, vary according to the methods used to capture food and the light levels under which food is taken. Well-developed eyes are found in most fish that are diurnal predators, although the largest eyes are present on fish that feed mainly at dusk and dawn, such as walleye (*Stizostedion vitreum*), or fish that live near the limits of light penetration in the oceans (see Chapter 35). Moderately small eyes are characteristic of fish that do not rely on vision for feeding, but especially of fish that feed on the bottom or at night. Such fish often have well-developed barbels around the mouth for taste and touch (catfish) or sensitive, fleshy lips (suckers, Catostomidae). Many cave-dwelling and deep-sea fish have either tiny eyes or none at all but typically do have extremely well-developed ***lateral-line systems***. The lateral-line system is visible on most fish as a faint line of pores along the midline and as a series of lines on the head. This system detects movement of the water along the fish (see Chapter 10).

2.6   SKELETAL SYSTEM

For convenience, the skeletal system is considered here to have three main components: (1) the vertebral column, (2) the skull, and (3) the appendicular skeleton (Fig. 2.4).

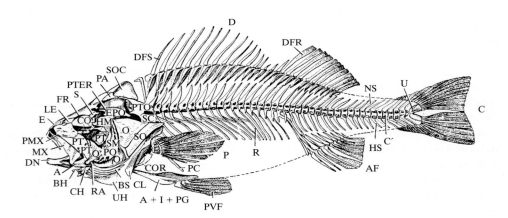

FIGURE 2.4   The teleost (*Perca*) skeleton: A, Anguloarticular; A + I + PG, actinosts + interneural + pelvic girdle; AF, anal fin; BH, basihyal; BS, branchiostegal rays; C, caudal fin; C', caudal vertebral centrum; CL, cleithrum; CH, ceratohyal; CO, circumorbitals; COR, coracoid; D, dorsal fin; DN, dentary; DFR, dorsal ray fin; DFS, dorsal fin spine; E, ethmoid; EPO, epiotic; FR, frontal; HM, hyomandibular; HS, hemal spine; IO, interoperculum; LE, lateral ethmoid; MX, maxillary; NS, neural spine; O, operculum; P, pectoral fin; PA, parietal; PC, postcleithrum; PMX, premaxillary; PO, preoperculum; PT, PT', PT", ectometapterygoid; PTER, pterotic; PTO, posttemporal; PVF, pelvic fin; Q, quadrate; R, rib; RA, retroarticular; S, sphenotic; SC, supracleithrum; SM, symplectic; SO, sub-operculum; SOC, supraoccipital crest; U, urostyle; UH, urohyal. Modified from Dean (1895).

The ***vertebral column*** in fish ranges in structure from a sheath of cartilage around a ***notochord*** (the ancestral supporting rod of vertebrates; found in hagfish) to a cartilaginous vertebral column to one of partially ossified cartilage (ratfish) to one of solid bone (teleosts). Although the ***centra*** (round centers of the vertebrae) are aligned with one another in fish, their vertebrae lack the elaborate interlocking processes found in terrestrial vertebrates to counteract gravitational forces. The gars (Lepisosteidae) are an exception to this rule: They have interlocking vertebrae that resemble those of reptiles. The vertebral column of fish, of course, provides the structural basis for swimming movements, and there is generally one vertebra per body segment. This system is admirably suited for withstanding the heavy compression load put on it by contractions of the large muscle masses during swimming.

Despite their apparent simplicity in structure compared with the vertebrae of terrestrial vertebrates, the vertebrae of fish are structurally specialized. The anteriormost vertebrae (the ***atlas*** and ***axis***) are structured for articulation with the skull. At the other end of the spinal column, the posteriormost vertebrae are modified into a series of flattened elements (***penultimate vertebrae, hypurals, epurals***, and finally, ***urostyle***), which articulate with the rays of the caudal fin. Between these two sets of highly modified vertebrae are the ***trunk vertebrae***, most of which bear ***ventral (pleural) ribs***. These ribs extend ventrally from centra and extend between adjacent muscle masses. In addition to the ventral ribs, most bony fish also have ***dorsal ribs*** that extend between dorsal muscle masses but usually are just loosely associated with the vertebral column. When such ribs or ***intermuscular bones*** are well developed, as in minnows, suckers, and pikes, they can be a real nuisance—and an occasional hazard—to individuals who eat fish.

Trunk vertebrae of most fish also feature ***dorsal (neural) processes*** that form an arch to accommodate the spinal cord. The neural arches of all vertebrae together compose the ***neural canal***. In bony fishes, a ***neural spine*** prominently extends from each neural arch to provide points of attachment for ***dorsal (epaxial) musculature***. In like manner, fusion of the transverse processes of the vertebrae creates a ***hemal canal*** on the ventral side of the vertebrae located posterior to the body cavity. The hemal canal carries the primary blood vessels supplying and draining all the caudal musculature (see Chapter 4). In many cases, a ***hemal spine*** extends ventrally from each ***hemal arch*** and provides a site of ***ventral (hypaxial) body*** musculature attachment. In laterally flattened teleosts, both the neural and hemal spines are very long to give adequate support to the body musculature. They resemble a delicate "double comb."

The ***skull*** of fish (Fig. 2.4) is an extremely complex structure that represents a design compromise among frequently conflicting uses of the head region. It is (1) the entry point for food and for water for respiration; (2) the site of major sensory organs; (3) a protective container for brain, gills, and other organs; (4) the attachment site for many major muscle masses; and (5) a streamlined entry point necessary for efficient swimming. In lampreys and hagfishes, the skull is little more than a cartilaginous trough for the brain from which other cartilages are suspended that support the mouth parts and the gills. Superficially, the skull of the sharks, skates, and rays is also relatively simple; it consists of (1) a solid-appearing ***chondrocranium*** molded around the brain and sense organs of the head; (2) the jaws, which largely consist of the ***palatoquadrate cartilage*** (upper jaw) and ***Meckel's cartilage*** (lower jaw) supported by elements of the hyoid arch; and (3) **branchial cartilages** supporting the gills. Despite its simple appearance, the chondrichthyan skull is quite complex in many subtle ways (e.g., the position and size of foraminal and fenestral openings and the development of rostrum) and is important in the study of taxonomy and evolution of these fishes.

The skull of bony fish is an elaborate puzzle of articulating bones that differs from the elasmobranch skull in many fundamental ways: (1) In general, the optic rather than the olfactory areas are well developed; (2) the relative positions of many of the elements are quite different; and (3) the important parasphenoid and opercular bones have no equivalents in the elasmobranches. The skull of bony fish is highly variable. Bones important in the skull of one group may be totally absent from another, and equivalent bones are often difficult to define among major taxonomic groups. However, for convenience, the skull can be divided into five elements: (1) the neurocranium, (2) the suspensorium, (3) the jaws, (4) the opercular bones, and (5) the branchiohyoid apparatus.

The **neurocranium** is the braincase and the most solid portion of the skull, yet it typically consists of 40 to 50 bones (counting the small bones around the eyes and optic region). In ancestral bony fishes, including the modern gars, the neurocranium is a solidly fused unit with strong connections (also often fused) to the other portions of the skull. In more derived bony fish, a solid "core" of bone remains around the brain (the various frontal, parietal, occipital, optic, and sphenoid bones), but the rest of the neurocranium and the skull are rather loosely articulated to provide the expansion capabilities needed for suction feeding and the two-pump respiratory system. Another development in derived bony fishes is the presence of crests on the parietal and pterotic bones and the fossae ("trenches") between them on which trunk muscles insert. This reflects changes in methods of swimming.

Connecting the neurocranium and the jaws is a series of bones (hyomandibular, symplectic, quadrate, pterygoids, etc.) collectively called the **suspensorium**. This series of bones has undergone rather dramatic changes in shape, size, and position during the course of evolution from ancestral fishes to modern teleosts and tetrapods (see Chapter 13). The **jaws** have also undergone rather dramatic changes in bony fishes that relate to the change from a firm, biting mouth to a flexible, sucking mouth. During the course of this change, the principal bones of the upper jaw have changed from the maxillas to the premaxillas (see Chapter 13); the principal bones bearing teeth have also changed. In contrast, the principal bones of the lower jaw (**dentaries**) have remained fairly constant. The **opercular bones** have remained fairly constant in bony fish, although the teleosts have added an interopercular bone on each side to increase the efficiency of the operculum as a respiratory pump. In more ancestral bony fish, the interopercular bone is just another branchiostegal bone (ray), part of the **branchiohyoid apparatus** that makes up the floor of the mouth and the support for the gills. The **branchiostegal rays** are a fan-like series of bones that make up the floor of the branchial chamber and that provide much of its expansion capabilities, which are so important for suction feeding and respiration.

The **appendicular skeleton**, compared with the skull, is relatively simple, consisting of the internal supports for the various fins. In the Chondrichthyes, the support for the pectoral fins consists of a series of **coracoid** and **scapular cartilages** that form a U-shaped bar attached to the fins on either end. The pelvic girdle is even simpler, consisting solely of a connecting bar, the **ischiopubic cartilage**. In bony fish, the girdles are more complex. In the pectoral fins, the rays articulate with a series of (usually five) radial bones, which in turn articulate with the scapula and coracoid. These bones are attached to those of the cleithral series, but particularly to the ***cleithrum*** itself, a large bone that is firmly united with the body musculature and that is joined, via the supracleithrum, with the skull. The girdle of the pelvic fins of bony fish is relatively simple, usually consisting of just one basipterygial bone on each side. These bones are usually united with each other or, in more derived teleosts, with the cleithrum of the pectoral girdle (when the fins are in the thoracic or jugular position). For the dorsal and anal fins, the internal supports consist of basal cartilages in the Chondrichthyes and a series of small bones (***pterygiophores***) in the bony fishes (one pterygiophore for each ray or spine in most teleosts). Because they interdigitate with the neural spines of the vertebrae, the dorsal pterygiophores are frequently called ***interspinous bones***.

2.7   MUSCULAR SYSTEM

In almost all fish, the large muscles of the body and tail comprise the majority of the body mass, although many other muscles are associated with the head and fins. The body muscles are divided vertically along the body length into sections called the ***myomeres*** (or ***myotomes***), which are separated by sheets of connective tissue. The myomeres are shaped like a W on its side, so that they fit into one another like a series of cones. The myomeres on the right and left halves of the body are separated by a vertical ***septum***. A horizontal septum separates the muscle masses on the upper and lower halves of the body. The upper muscles are called the epaxial muscles and the lower muscles the hypaxial muscles (Fig. 2.5). In addition, a lateral band of muscles usually lies along or slightly below the midline of the fish. On inspection, fish muscles can often be divided into red (slow), white (fast), and pink (intermediate) muscle (Fig. 2.5).

    Red muscle is infused with capillaries and appears red in color because of the high concentrations of red, oxygen-binding pigments in the blood (***hemoglobin***) as well as in the muscle tissue itself (***myoglobin***). The high capillary density and presence of the pigments ensure that the red muscle receives adequate oxygen for its abundant mitochondria to metabolize fat (lipids) to sustain high levels of continuous (aerobic) swimming. Therefore, in continuously active fish (e.g., bonito and marlin), a large proportion of their muscle mass is red muscle. Fish of "intermediate" activity levels often have the lateral band of muscles, which is always red in color and well developed. These fishes may also have red muscle fibers scattered in the white muscle that makes up most of the body mass. For example, in bluefish (*Pomatomus saltatrix*) and striped bass (*Morone saxatilis*), both of which are open-water predators, the greatest proportion of red muscle occurs in the lateral bands in the caudal peduncle. The red bands average 19% and 11%, respectively, of the total musculature in this region (Freadman 1979). In some cases, the position of the red bands within the body also reflects swimming ability. Thus, the fastest-swimming tunas (e.g., *Thunnus*) carry their red muscle bands deep in the body core. This arrangement aids the stiff-body (thunniform; discussed later) swimming mode and permits the conservation of metabolic heat, which in turn allows faster muscular contractions and higher swimming velocities (see Chapter 5). At the opposite end of the spectrum of temperatures at which red muscle operates (i.e., at low temperatures), red muscle shows increased (1) capillary densities (enhancing blood flow through muscle tissues),

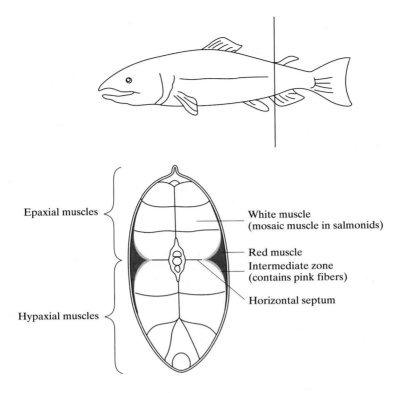

FIGURE 2.5   Cross-section of salmon posterior to vent, indicated by line in top drawing, showing position of locomotory muscles (bottom). Redrawn from Stoiber et al. (1999).

(2) cellular mitochondrial densities (increasing aerobic potential), and (3) lipid droplet densities (which may accelerate O_2 flux via increased O_2 solubility) (Johnston 1982, Egginton and Sidell 1989). These adaptations presumably maintain swimming performance in the face of slowed metabolic and chemical processes at cold temperatures (Pörtner 2002).

White muscle fibers are thicker than those of red muscle, have a poorer blood supply, and lack red, oxygen-carrying pigments such as myoglobin. Not surprisingly, white muscle contraction is not as dependent on oxygen supply. White muscle usually converts glycogen to lactate via anaerobic pathways. Thus, white muscle is most useful for short bursts of swimming and dominates the muscle mass of moderately active to "sluggish" swimmers. For example, dogfish sharks have almost all white muscle and can use up 50% of available muscle glycogen in approximately 2 minutes (Bone 1966). The fast (white) fibers exert tensions that are 2.7 times greater than the slow (red) ones, although there is a four-fold increase in energy cost to the dogfish (Altringham and Johnston 1986). After this burst of activity, the fish "repays its oxygen debt" by aerobic renewal of its energy sources through the conversion of lactate to glycogen and glucose by red muscle, heart

> ### BOX 2.1
> ## Hormones "Remodel" Salmon Swimming Muscle
>
> Salmonid fishes (trout and salmon) have red and pink muscle fibers mixed with the white fibers as **mosaic** locomotory muscles. Juveniles of Anadromous species migrate downstream to the ocean as they transform from the parr to smolt life stages (see Chapters 6, 11, and 19), but the mechanisms have remained elusive. Katzman and Cech (2001) recently demonstrated that the mosaic locomotory muscle of coho salmon (*Oncorhynchus kisutch*) parr, carrying implanted thyroid hormone pellets, show faster contraction velocities and shorter contraction times (i.e., more like pure white muscle) compared with sham (no hormone in pellets) or control (no implantation surgery) cohos. Thus, the same hormones that potentiate the needed saltwater tolerance (see Chapter 6) and odor-imprinting (for return migrations as adults; see Chapter 10) also remodel locomotory muscles for anaerobic bursts rather than aerobic endurance. Such changes presumably favor salmon survival, as the juvenile does not "hold station" in streams well (migrates) and needs to catch elusive estuarine prey (rather than feeding on drifting insects and plankton in streams).

muscle, and liver (Hochachka and Somero 1984) or, sometimes, by the white muscle itself (Batty and Wardle 1979). Recent evidence (Richards et al. 2002) demonstrates that oxidation of lipids also fuels recovery from exhaustive exercise in rainbow trout (*Oncorhynchus mykiss*).

Pink muscle, containing fibers intermediate in character between those of white and red muscle, is typically used at swimming velocities too high for red muscle to sustain but too low for the effective use of white muscle (Johnston et al. 1977). However, in some fish, pink muscle is used similarly to red muscle at slow velocities; whereas in other fish, pink muscle is used for swimming at fairly high velocities (Davison and Goldspink 1984).

The pink (or red) color of these muscles should not be confused by the "pink" or "red" color imparted to white muscles from the carotenoid pigments in the fish's diets. For example, salmon (especially sockeye salmon) that forage on crustaceans will have white muscle that is heavily tinted in pink or red shades because of the carotenoid pigments in the prey. This carotenoid-based pink color disappears when salmon are fed diets without carotenoid pigments.

Not surprisingly, the powering of fish swimming rarely is strictly a matter of red and white—or even pink. The different kinds of muscle fibers are typically used in concert as fish change swimming speed (Roberts and Graham 1979). Thus, white muscle fibers in chub mackerel (*Scomber japonicus*), mosaic red and white mixed fibers in salmon, and pink and white fibers in common carp (*Cyprinus carpio*) are recruited for only moderate increases in swimming speed. In addition, shifts in muscle pH, oxygen partial pressure (see Chapter 3), temperature, and biochemical substrates can temporarily alter enzyme activities in white muscle, thereby changing its ability to function at different swimming speeds (Guppy and Hochachka 1978). Thus, although some species maintain a fairly strict segregation in red and white muscle function (e.g., bluefish and striped bass), others (e.g., chub mackerel, rainbow trout, and common carp) typically mix the activity of the different muscle types (Freadman 1979).

2.8 LOCOMOTION

Fish move by a variety of means. The simplest is the passive drift of many larval forms, but such drifters quickly metamorphose into forms capable of active, directed movement. Although various fishes have evolved the abilities to burrow, walk, crawl, glide, and even fly, swimming is by far the most important means of locomotion. To swim forward (or backward!), most fish utilize rhythmic undulations of part or all of their bodies or fins. The sides of the body and the fins exert force on the relatively incompressible surrounding water through the sequential contraction of myomeres. The relatively stiff vertebral column provides compression resistance, so the body bends from side to side rather than shortening. In sharks, a helical network of stiff collagen in the skin, to which the myomeres attach distally, acts as an external tendon for more efficient transmission of muscular force to the tail (Wainwright et al. 1978). Changes in the collagen fiber angle during swimming prevent loss of tension and keep the skin from wrinkling on the concave side of sharks.

Figures 2.6A and 2.7 show how lateral flexures of the body muscles at the appropriate angle of attack propel fish forward. These flexures typically move backward along the body with increasing amplitude and at a speed somewhat greater than the forward progress of the fish. As this propulsive wave moves posteriorly, the water adjacent to the fish is accelerated backward until it is shed at the posterior margin of the caudal fin, producing thrust (Lighthill 1969). The more undulatory waves a fish can exert against the surrounding water, and the faster and more exaggerated the waves are, the more power the fish can generate. If other factors such as **drag** (resistance to movement) from body features and shape are held constant, fish that generate more power can accelerate more quickly and swim faster (Blake 1983a, Webb and Weihs 1983). The close correlation between the frequency of locomotory muscle contraction and the velocity of swimming in lake trout (*Salvelinus namaycush*) and brown trout (*Salmo trutta*) allows swimming velocities to be estimated from radiotransmitted electromyogram signals (Thorstad et al. 2000). Using electromyographic electrodes positioned at 35% and 65% along the length of the pollock's (*Pollachius virens*) body, Altringham et al. (1993) demonstrated the alternating (left and right) sequence of the shortening, lengthening, and relaxation of the lateral muscles as they were stimulated to contract from the anterior toward the caudal end. The fish's progress through the water (approximately two-thirds of a body length) is a function of the thrust it generates and the slippage between its body surfaces and the water (Fig. 2.7).

Webb (1971) measured the mean wavelength of rainbow trout to be 0.76 times the fish's length, which is a constant at all swimming velocities greater than 0.3 body lengths per second. To increase swimming velocity, trout increase both **tail-beat frequency** (lateral movements per minute) and **amplitude** (lateral deflection per movement). The **thrust** (forward-directed force) generated by this propulsive wave is a function of forces

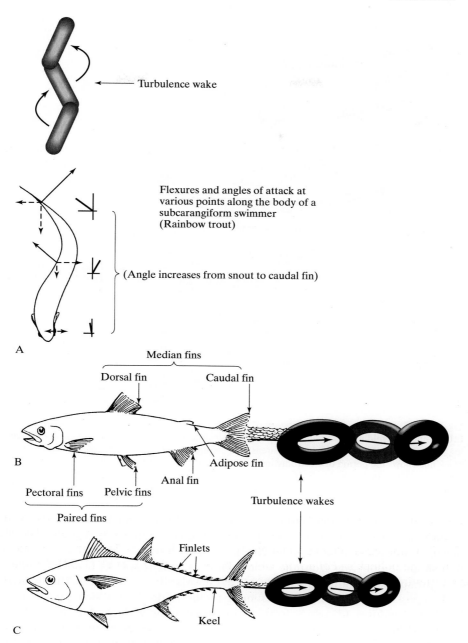

Turbulence wake

Flexures and angles of attack at
various points along the body of a
subcarangiform swimmer
(Rainbow trout)

(Angle increases from snout to caudal fin)

A

Median fins

Dorsal fin

Caudal fin

Adipose fin

Anal fin

Pectoral fins   Pelvic fins

Paired fins

B

Turbulence wakes

Finlets

Keel

C

FIGURE 2.6 Generalized swimming features of fishes.

produced by the body flexures and the rearward velocity of the propulsive wave, and it is
limited by decreased hydrodynamic efficiency at high tail-beat frequencies and amplitudes.
An inevitable consequence of this forward movement is a trail of *vortices* (turbulence) in
the water behind the fish (Fig. 2.6B and 2.6C). Recent evidence from experiments using

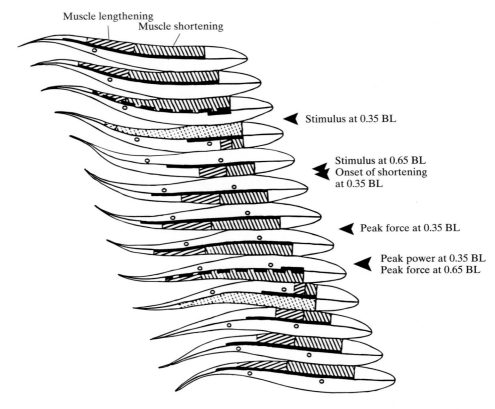

Muscle lengthening
Muscle shortening

Stimulus at 0.35 BL

Stimulus at 0.65 BL
Onset of shortening
at 0.35 BL

Peak force at 0.35 BL

Peak power at 0.35 BL
Peak force at 0.65 BL

FIGURE 2.7 A schematic representation of the events of a single tail-beat, showing progressive fish outlines and midlines from above. The more anterior hatched areas indicate muscle activity during shortening; the more caudal hatched areas indicate muscle activity during lengthening. The thick black line shows the period of electromyographic (EMG) activity (0.2–0.8 body length [BL]); the dashed black line indicates that EMG activity stops at this time. Circles represent muscle bundle locations at 35% and 65% of BL. Modified from Altringham et al. (1993).

sliver-coated glass microbeads and light reflected from a laser has shown that the wake of a swimming chub mackerel consists of a series of linked elliptical vortex rings, each with a central jet flow (Nauen and Lauder 2002).

It should be obvious from the wide variety of body shapes of fish that, despite the basic approach to swimming just discussed, considerable variation exists in how fish swim. It is possible to divide these swimming methods into four basic types: (1) anguilliform, (2) carangiform, (3) ostraciform, and (4) swimming with the fins alone (Fig. 2.8).

Anguilliform swimming is characteristic of flexible, elongate fish, such as eels (Anguilliformes). The whole body of such fish is flexed into lateral waves for propulsion. Just as an oar surpasses a pole for rowing a boat, the flattened posterior surface of most eels improves their swimming efficiency. Typical eel median "fins" consist of a continuous dorsal-caudal-anal fin extending around the posterior half of the fish. A similar fin configuration and swimming style is also found in such diverse groups as marine gunnels

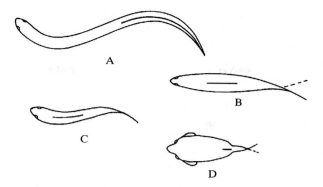

FIGURE 2.8   Swimming modes of fishes: (**A**) anguilliform;
(**B**) carangiform; (**C**) subcarangiform; (**D**) ostraciform.

(Pholididae), anadromous lampreys (Petromyzontidae), and lungfish (Dipnomorpha). Anguilliform swimmers without a strict eel-like shape include certain elasmobranches (e.g., nurse sharks, Ginglymostomatidae) as well as some teleosts, such as the cods (Gadidae), when swimming slowly (Wardle and Reid 1977).

   Carangiform swimming is intermediate between the anguilliform and ostraciform extremes and is named after the jacks (Carangidae). Carangiform swimming involves throwing the body into a shallow wave (up to one-half wavelength within the body length), with the amplitude increasing from very small at the head and anterior two-thirds of the body to large at the posterior edge of the caudal fin (Webb 1975a). The body shape of carangiform swimmers is typically fusiform, tapering to a narrow caudal peduncle and then broadening to a large, forked caudal fin. Species normally thought of as carangiform types (e.g., jacks, drums, snappers, tunas, and mackerels) are swift swimmers. Slower fish that undulate their bodies into less than one full wave length yet more than one-half wavelength (at speeds greater than one body length per second) are often placed in a separate movement category: *subcarangiform swimming*. At the other extreme, the fastest-cruising fish (tunas, billfish, and lamnid sharks) are often placed in a separate *thunniform swimming* category. These fish have a low-drag fusiform shape and undulate a narrow caudal peduncle stiffened by a keel and a large, slim, lunate (moon-shaped) caudal fin for propulsion (Fig. 2.6C). With its "swept-back," tapered tips and "scooped-out" center, the lunate tail provides efficient power for fast underwater movement. Frictional drag is minimized by the high-aspect-ratio design of the tail. By having a small surface area per span length or distance between lobe tips (high-aspect ratio), propulsive force is maximized, and energy wasted on lateral displacement of water (and consequent creation of excessive vortex wakes) is minimized. The vertically large tail allows a greater mass of water to be accelerated to the rear, thus increasing forward thrust.

   Ostraciform swimming, named after the boxfish family (Ostraciidae), also involves flexing the caudal peduncle, but not to generate high velocities. By contracting the entire muscle mass on one side of their bodies and then the other, these swimmers oscillate the caudal fin to produce a sculling type of locomotion. Fish with this type of locomotion, mainly the trunkfishes, cowfishes, and boxfishes (Tetraodontiformes), rely on armor (plus spines and toxins) rather than on speed to protect them from predators. The small isocercal caudal fin also is in keeping with their low swimming velocities.

Swimming with fins alone is characteristic of a surprising number of teleosts, including forms that use their body musculature for swimming when high-speed or sustained swimming is necessary. The ray-and-membrane fin design allows these fishes to undulate individual fins or fin pairs rather than their bodies to achieve precise movements. Examples of this type of swimming are found among, but are not limited to, fish inhabiting areas of dense vegetation or coral or rock reefs. Many species can utilize a rowing/flapping action of the pectoral fins for forward propulsion. This is sometimes termed **labriform swimming**, from the wrasses (Labridae) of tropical coral reefs. Figure 2.9 shows the pattern of pectoral fin adduction and abduction (propulsive stage) and the subsequent pause (refractory stage) that striped surfperch (*Embiotoca lateralis*) use for their usual locomotion. Pectoral fin movements directed toward the bottom provide lift to negatively buoyant fish (see Chapter 5). If such movements are made while very close to the bottom, the virtual incompressibility of water and the consequent "ground effect" significantly improve the efficiency of such hovering. For example, mandarin fish (*Synchiropus picturatus*) save 30% to 60% of the power needed for hovering if they are within approximately one pectoral-fin-width's distance from the bottom (Blake 1979). This power savings exceeds that of a conventional helicopter (Blake 1983a). **Rajiform** locomotion is the unique style of batoid fishes (skates and rays) in which thrust is generated by undulatory waves passing down the enlarged, muscular pectoral fins (Rosenberger and Westneat 1999). These fishes increase pectoral fin undulatory wave frequency to increase swimming velocity (Rosenberger and Westneat

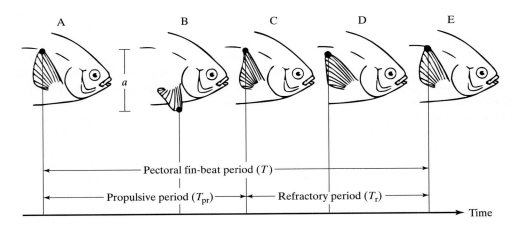

FIGURE 2.9 Sequential lateral images of striped surfperch during one complete pectoral-fin stroke cycle at a constant labriform swimming speed. The propulsive period of pectoral fin abduction and adduction (T_{pr}; **A–C**) is followed by a pause or "refractory" period (T_r; **C–E**) during which the fins rotate posteriorly and then slightly anteriorly while fully adducted against the body. **a**, Pectoral fin-beat amplitude. From Drucker and Jensen (1996).

1999). In contrast to the pectoral fin movements, triggerfish (Balistidae) use alternate dorsal and anal fin undulations with some pectoral adjustments, whereas gymnotiform fishes use anal fin undulations for propulsion. The beautiful lionfish (*Pterois volitans*) of shallow tropical waters moves by deliberate fanning of the caudal, anal, and second dorsal fins when the feathery pectorals and dorsal spines, capable of injecting painful venom, are held erect.

2.9 ENERGETICS OF SWIMMING

The energy costs of swimming have long fascinated investigators, both because fish seem to move through water so effortlessly and because so much of their body mass is devoted to muscle. The pioneering efforts were those of Houssay (1912), who harnessed fish to a device that measured swimming power by determining how much weight a fish could lift. In a more sophisticated effort, Webb (1971) attached wire grids or plastic plates of various resistances to the dorsal surfaces of fish swimming in a tunnel respirometer (see Chapter 3). Bainbridge (1958) showed that for goldfish, trout, and dace, the swimming speed at any tail-beat frequency increased with length. As a general rule, a doubling of body length corresponds to a 150% increase in velocity per body flexure. The relationship worked out for small fish is $V = kL^b$, where V is the mean swimming velocity (in m/s), L is fish length (in m), k is a constant, and the exponent b has been determined experimentally, usually between 0.5 and 0.6 (Wu and Yates 1978). When sudden bursts of swimming are required, the value of the exponent b typically is approximately 0.8, a value determined for fish as varied as dace, goldfish, and barracuda. One factor that complicates this relationship is that smaller fish are typically capable of faster tail-beat frequencies, which also can increase swimming velocities (Wardle 1975). However, when speed is normalized as body lengths per second, the relationship between speed and tail-beat frequency becomes predictable for a range of fish lengths (Fig. 2.10A). When this information is coupled with the relationship between energetic costs and swimming velocities (Fig. 2.10B) (Brett 1964, Wohlschlag et al. 1968), estimates of fish metabolic demands can be made on free-swimming animals by measuring tail-beat frequencies (Feldmeth and Jenkins 1973). Briggs and Post (1997) estimated swimming costs by implanting electromyographic (EMG) electrodes into the axial muscle of rainbow trout (0.9–1.2 kg body wt). The frequencies of radiotransmitted muscle contraction were compared with those determined for fish, carrying the same transmitters, for which energy costs were determined in the laboratory (see Chapter 3). Beddow and McKinley (1999) showed that the optimal position for these EMG electrodes in another salmonid, Atlantic salmon (*Salmo salar*), is in the red muscle toward the fish's tail, at 70% of the distance from the snout to the end of the caudal fin (total length). As future technological advances allow the development of smaller transmitters, these estimates will become available for smaller fishes.

   Temperature is an important factor in the swimming dynamics and energetics of fish. A 10°C rise in temperature (from 13°C to 23°C) is accompanied by a 22% decrease in water viscosity, which presumably reduces hydrodynamic drag. Moreover, warm muscles tend to operate more efficiently by exerting more force per contraction. In cold water, therefore, water viscosity increases, and potential reductions in (especially white)

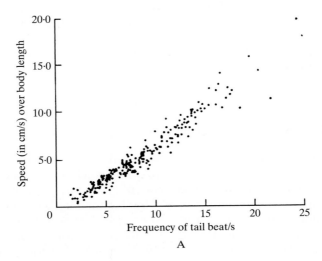

A

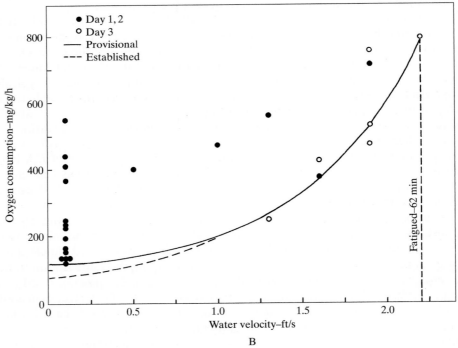

B

FIGURE 2.10 (**A**) Relationship between swimming velocity and frequency of caudal undulations (tail-beats) for dace (*Leuciscus leuciscus*). From Bainbridge (1958). (**B**) Relationship between swimming (water) velocity and oxidative metabolic (oxygen consumption) rate for juvenile sockeye salmon. From Brett (1964).

muscle efficiency would be expected to slow acceleration. Johnson et al. (1998) determined that muscle-related effects of temperature were more significant than changes in water viscosity in accelerating guppies (*Poecilia reticulata*) and goldfish. Stevens (1979) found that temperature affected the swimming of rainbow trout and largemouth bass (*Micropterus salmoides*). At a constant swimming velocity, tail-beat frequencies were lower (i.e., longer stride lengths) for both species acclimated to higher temperatures. Similarly, Webb (1978b) found that trout benefit from warmer environments by increased acceleration rates, which enhance their attack success on prey. As indicated earlier, fish also have various ways of improving the efficiency of the operation of red muscle tissue at both high and low temperatures.

Regardless of temperature, the ***energetic cost*** of swimming increases with swimming speed. Thus, the power required to propel a trout is exponentially proportional to the swimming speed and is approximately 2.9 times greater than that required to propel a rigid trout model of equal dimensions (Webb 1971). Approximately equal contributions of carbohydrate and lipid oxidation fuel swimming at sustainable velocities in rainbow trout, whereas their nonsustainable (burst) swimming is supported primarily by carbohydrate oxidation (Richards et al. 2002). In steady swimming, the average thrust equals the average total drag on the fish. For the fish to keep moving, the inertial forces of motion must be larger than the viscous forces of the surrounding water. Thus, the ***Reynold's number*** (Re, Re $= VL/v$, where V is the mean swimming velocity, L is the length of the fish, and v is the coefficient of kinematic viscosity of the fluids) is always greater than one. Except for larval fishes, values of the Reynold's number typically range from 10^4 to 10^8 for fish swimming normally (Wu and Yates 1978). Viscosity effects at such high Reynold's numbers are mostly confined to a thin boundary layer adjacent to the body surface, especially in more streamlined fish. In torpedo-shaped species (tunas, marlin, etc.) with small turbulence wakes (Fig. 2.6C), the maximum thickness of the boundary layer (at the tail end) is generally not more than a few percent of the body thickness (Wu and Yates 1978). However, fish that are less efficiently designed for swimming find various means to decrease the energetic cost. For example, evidence suggests that the mucous coating on a fish reduces frictional drag (Rosen and Cornford 1971). There is also evidence that the swim-and-glide swimming pattern seen in some fishes may be more efficient than continuous swimming in some situations (Weihs 1974, Blake 1983b). In 3- to 7-cm-long delta smelt (*Hypomesus transpacificus*), the swim-and-glide pattern allows individuals to efficiently maintain position in relatively low-velocity (10 cm/s) currents (Swanson et al. 1998). Kramer and McLaughlin (2001) review the evidence for several potential advantages of intermittent locomotion, including energetics, recovery from fatigue, stabilization of sensory fields, and detectability by potential predators.

Sometimes efficiency of movement is superceded by maximal performance capabilities, especially when survival is threatened. For example, Webb (1975b) found that rainbow trout expend 18% of the total energy required for rapid acceleration in overcoming frictional drag. However, both trout and green sunfish (*Lepomis cyanellus*) were found to display uneven rates of acceleration during events such as pursuing prey or escaping predators. These species initially tended to accelerate maximally. Even though both velocity and distance covered increased with time, the acceleration rate actually decreased in the milliseconds after a fast start. To capture prey or avoid predation from larger predators, this behavior would seem to have high survival value.

BOX 2.2

Exercise Is Good for Fish Too!

The swimming performance of fish responds to conditioning, much like that of human athletes. Exercise conditioning increases aerobic capacity and improves swimming performance (e.g., U_{crit}) in several fishes. Thus, captive largemouth bass increase U_{crit} and aerobic capacity with regular exercise (Farlinger and Beamish 1978). Likewise, the aerobic capacity of rainbow trout red and white muscle increases with exercise conditioning (Davie et al. 1986). Leopard sharks (*Triakis semifasciata*) respond to endurance training with increases in white muscle fiber diameter and metabolic enzyme activities (Gruber and Dickson 1997). Striped bass significantly increase their U_{crit} when conditioned to swim at higher velocities for 50 to 60 days (Fig. 2.11). Improvement in swimming performance can persist for 2 months after conditioning has stopped. Thus, exercise conditioning may equip hatchery-reared fish with increased survival abilities, such as greater ease of escaping predators. For example, exercise-conditioned Atlantic salmon (*Salmo salar*) return to spawn in greater numbers than nonconditioned fish (Wendt and Saunders 1973).

Greater velocities would be attained and increased distances covered in a shorter time than with a constant acceleration rate, at an increased energy cost of only 2% to 3% of the total expended.

One measure of fish swimming performance is ***critical swimming velocity*** (U_{crit}), which is the velocity at which a fish becomes fatigued after its activity is increased at regular intervals in a variable-speed swimming tunnel (Brett 1964). The U_{crit} of fishes can change with the seasons. The seasonally related effects of temperature, photoperiod, and reproductive state influenced the U_{crit} of wild smallmouth buffalo (*Ictiobus bubalus*), with U_{crit} values in the fall and winter generally being less than those in the spring and summer (Adams and Parsons 1998). Whereas U_{crit} could be expected to correlate with (aerobic) red muscle mass, burst (fast acceleration) swimming performance should correlate with (anaerobic) white muscle mass. Reidy et al. (2000) found that in individually tested Atlantic cod (*Gadus morhua*), U_{crit} negatively correlated with acceleration performance, presumably reflecting red and white muscle trade-offs.

2.10   LESSONS FROM FORM AND MOVEMENT

It is said that the great nineteenth-century scientist Louis Agassiz would check out a student's potential for science by placing the student alone in a room with a single fish for a day and then pop in periodically to quiz the student about what he had learned from studying that single fish. This chapter indicates that, indeed, you can learn a great deal about how a fish lives from a simple examination of its exterior traits. Likewise, close examination of the skeleton and muscles of a fish can reveal much about how that fish moves, including how much exercise it has been getting. Because most fish devote a majority of their musculature to swimming, they have diverse adaptations to improve swimming efficiency and performance.

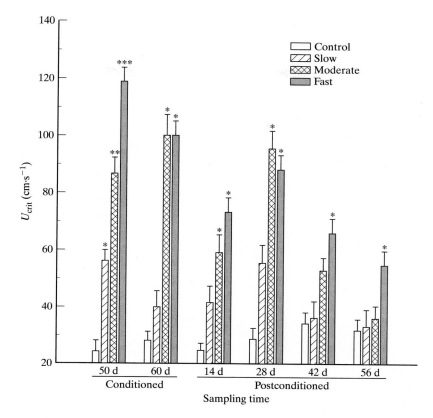

FIGURE 2.11   Mean ($\pm$SE, $n = 12$) critical swimming velocities (U_{crit}) of juvenile striped bass that were exercise conditioned at different flow velocities for 50 and 60 days and at 14, 28, 42, and 56 days postconditioning: *Significantly greater than control; **significantly greater than * and control; ***significantly greater than **, *, and control; $P < 0.05$. From Young and Cech (1994).

SUPPLEMENTAL READINGS

Blake 1983b; Maddock et al. 1994; Pörtner 2002; Videler 1993; Webb 1975a; Webb and Weihs 1983.

WEB CONNECTIONS

FISH SCALE PICTURES, AUSTRALIAN MUSEUM
www.amonline.net.au/fishes/students/scales

FISH LOCOMOTION
www.csuchico.edu/~pmaslin/ichthy/loco.html
www.fisheriesmanagement.co.uk/Fish%20Studies/locomotion_study.htm

C H A P T E R 3
Respiration

Respiration in an aquatic environment presents different problems compared to respiration in air. Most terrestrial vertebrates have internal lungs that must be ventilated through bidirectional (tidal) movement of air to replenish the oxygen supply at the gas-exchange surfaces. In contrast, most fish have external *gills* that are *ventilated* by a unidirectional flow of water, which is created either by branchial pumping or by simply opening the mouth and operculi while swimming forward. Not having to move the water back and forth obviously saves the fish valuable energy. The fine sieve structure of the gills (Fig. 3.1) enables them to extract oxygen very efficiently from the water. This efficient oxygen uptake is vital to fish because of (even air-equilibrated) water's low oxygen *solubility* characteristics. Dissolved oxygen solubility is decreased by increases of both *temperature* and *salinity*. For example, fresh water (fully equilibrated with air) contains only 1/23 as much oxygen content per unit volume as does the atmosphere above it at 5°C—and only 1/43 as much oxygen as air at 35°C. Figure 3.2 shows these freshwater solubility differences, with the shaded area describing the oxygen content (mg/L) range for the entire temperature range of 0°C to 35°C—the thermal limit that encompass those of most freshwater fishes. Note that whereas the oxygen content value of 35°C water (6.94 mg/L) is less than that of 5°C water (12.76 mg/L), both are fully equilibrated with air and are considered to be *normoxic* (normal oxygen). Normoxic conditions are those near air saturation, that is, near 156 mm Hg *partial pressure of oxygen* (*PO$_2$*), regardless of the temperature and salinity conditions. Increasingly hypoxic (low-oxygen) conditions are depicted by the deeper shading on Figure 3.2 at decreasing PO$_2$ values, or toward *anoxia* (lack of oxygen) at the origin. Air-equilibrated *sea water* at 35°C contains only 1/52 as much oxygen as air, with the added dissolved solutes in sea water decreasing the oxygen solubility a further 17% compared with fresh water at the same temperature. Water's low oxygen solubility undoubtedly has contributed to the evolutionary development of gills, which are characterized by large surface areas and extremely efficient gas exchange, and to the many—and often bizarre—mechanisms that some fishes use to extract oxygen directly from the air. The low availability of oxygen also has placed limits on the rates of oxygen uptake and, consequently, on fish metabolism.

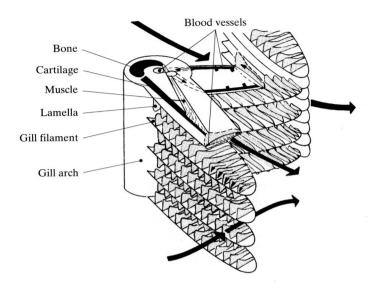

FIGURE 3.1 Diagram of teleost gill structure. Large arrows indicate direction of water flow; small arrows show direction of blood flow. Modified from Hughes and Grimstone (1965).

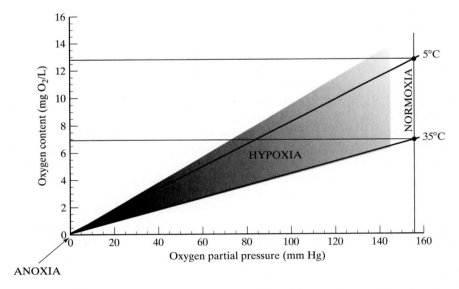

FIGURE 3.2 Relationship of oxygen partial pressure (PO_2) and oxygen content in fresh water at 5°C and 35°C. Whereas the white zone at high PO_2 values approximates air-saturation levels (normoxia), the shaded area shows sub–air-saturation levels (hypoxia), with the degree of shading representing the extent (severity) of hypoxia at lower PO_2 values for freshwater fishes for 0°C to 35°C down to the absence of oxygen (anoxia). Note that at the same PO_2, the oxygen content is lower at the higher temperature.

3.1 GILLS

Gills are the main site of gas exchange in almost all fishes. The gills consist of bony (Fig. 3.1) or cartilaginously stiffened arches that anchor pairs of *gill filaments*. In sharks, the pairs of gill filaments are separated by a fleshy septum. The numerous, minute *lamellae* that protrude from both sides of each filament are the primary sites of gas exchange; however, not all the blood flow in the gills is directed to the lamellae. "Nonrespiratory" basal blood channels or venolymphatic sinuses (Fig. 3.3) may carry a significant fraction of the gill blood. Booth (1979) found that resting rainbow trout perfused approximately 58% of their lamellae with blood. Trout in hypoxic (low dissolved oxygen) water or injected with *epinephrine* (e.g., simulating stress or excitement) perfused more than 70% of their lamellae. Conversely, injections of *acetylcholine* decreased perfused lamellae to approximately 43%. These results indicate that rainbow trout can increase the number of lamellae that are used in respiration as the dissolved oxygen level decreases and that they regulate blood flow distribution in the gills by *cholinergic* (acetylcholine-sensitive) and *adrenergic* (epinephrine-sensitive) receptors (Booth 1979). Sundin and Nilsson (1998) as well as Stenslokken et al. (1999) demonstrated that this redistribution of blood flow in the gills of rainbow trout and Atlantic cod (*Gadus morhua*) may result from the contraction of lamellar *pillar cells* facilitated by endothelin, a potent vasoconstrictor (Evans and Gunderson 1999).

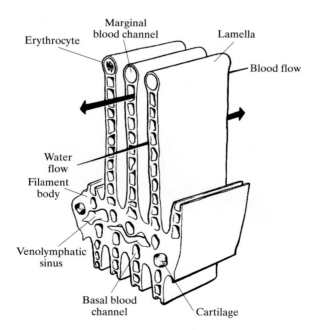

FIGURE 3.3   A section through a gill filament and several lamellae of a teleost, drawn approximately to scale. From Randall (1982).

Olson (2002) reviews the roles of regulatory substances (including endothelin) in fish gill circulation, and Sundin and Nilsson (2002) review sensory and motor innervation of fish gills.

The lamellae are made up of thin *epithelial cells* on the outside and thin *basement membranes* and supportive pillar cells on the inside (Fig. 3.4), allowing blood cells to flow through the interior. Rainbow trout and roach (*Rutilus rutilus*) erythrocytes are substantially deformed into C- or S-shapes—and sometimes even Y-shapes—as they progress through the narrow lamellar channels (Nilsson et al. 1995). Oxygen is taken up by *diffusion* across the thin lamellar membranes. The erythrocytic deformations may play a role in oxygen uptake at the gills by diminishing the diffusion boundary layer of fluid (a component of the diffusion distance) along the inside of the lamellar vessel wall (Nilsson et al. 1995). Gas-exchange efficiency is maximized by the opposing flows (*countercurrent exchange*) of water and blood along the lamellae (Figs. 3.1 and 3.3). The countercurrent flow ensures a steady oxygen tension gradient along the entire diffusion surface. Thus, the blood can be oxygen saturated to a greater degree than with a parallel-flow arrangement. Van Dam (1938) reported oxygen utilizations (extraction

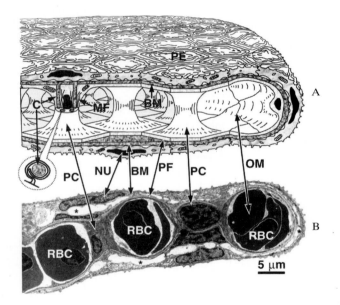

FIGURE 3.4   A. Schematic cross-section through lamella near outer marginal channel (OM). **B.** Transmission electron micrograph of rainbow trout (*Oncorhynchus mykiss*) lamella. Dotted inset shows pillar cell (PC) membrane enveloping a collagen strand. Subepithelial sinuses are evident (*). The microridges on the lamellar surfaces are thought to hold mucus, which may protect the surface of these thin-walled epithelial cells. *BM*, Basement membrane; *C*, collagen strands; *MF*, microfilaments; *NU*, cell nucleus; *PE*, pavement epithelial cell; *PF*, pillar cell flange; *RBC*, red blood cell. From Olson (2002).

efficiencies) as high as 80% by a rainbow trout. When a parallel flow was experimentally induced by reversing the ventilatory water in the tench (*Tinca tinca*), utilization dropped to less than 10% (Hughes 1963).

All oxygen is taken up via diffusion, and the uptake rate depends on the ***surface area*** of the lamellae, the thickness (***diffusion distance***) of the gill epithelia across which the oxygen diffuses, and the ***PO_2 gradient*** (note that this is *not* the oxygen content gradient) across the membranes. Consequently, highly active, fast-growing fishes typically have larger gill surface areas and extra-thin gill epithelial layers (Table 3.1) in meeting their increased demand for oxygen from the water. To increase gill surface area, fishes have two main evolutionary "options": increasing the number of lamellae by spacing them more closely together, or increasing the length of the lamellae. The latter option is seldom found because of the fragility of the delicate lamellae. Some tunas have evolved lamellae that are fused at the tips between adjacent filaments to add rigidity and, hence,

TABLE 3.1   COMPARISON OF GILL DIMENSIONS IN SEVERAL TELEOST FISHES

Species	Thickness of lamellae (μm)	Lamellae (per mm)	Distance between lamellae (μm)	Distance between blood and water (μm)	
Icefish *Chaenocephalus aceratus*	35	8	75	6	
Brown bullhead *Amerius nebulosus*	25	14	45	10	Sluggish Species
European eel *Anguilla vulgaris*	26	17	30	6	
Antartic renotheniid *Notothenia tessellata*	20	17.5	35	2	
Sea scorpion *Ancanthocottus scorpius*	15	14	55	3	
Brown trout (5 kg) *Salmo trutta*	15	20	40	3	
Flounder *Platichthys flesus*	10	14	70	2	
Icefish *Champsocephalus esox*	10	18	40	1	Active Species
Brown trout (400 g) *Salmo trutta*	12	23	35	3	
Roach *Rutilus rutilus*	12	27	25	2	
Pollock *Pollachius virens*	7	21	40	<1	
Perch *Perca fluviatilis*	10	31	25	<1	
Atlantic herring *Clupea harengus*	7	32	20	<1	Very Active species
Atlantic mackerel *Scomber scombrus*	5	32	25	<1	

to protect the lamellae against damage from the high ventilatory flows (Muir and Kendall 1968). Both the close lamellar spacing and the thin epithelia usually associated with the lamellae of active fish decrease the distance across which oxygen in the water must diffuse to enter the blood and, thereby, increase the rate of diffusion. The PO_2 gradients are maintained across the lamellar membranes by *gill ventilation*. As oxygen diffuses from the water through the lamellar surfaces and into the relatively oxygen-depleted blood, diffusion slows or stops if fresh, oxygen-rich water is not provided continuously to these surfaces via gill ventilation.

Gill Ventilation

In most bony fishes, ventilation is accomplished by synchronous expansion and contraction of the buccal and opercular cavities to provide a nearly continuous, unidirectional flow of water over the gill surfaces. During the first phase of the pumping cycle, water enters the mouth by expansion of the buccal cavity (see Fig. 7.4). The water is then accelerated over the gills by a simultaneous contraction of the buccal cavity and abduction ("expansion") of the operculi. After the branchial cavity contracts, thereby expelling the water out the opercular openings, the cycle begins again. Interruptions of this cycle produce brief reversals of flow, or "coughs," that fish use to clear foreign matter or excess mucus from the gills. The frequency of these coughs in brook trout (*Salvelinus fontinalis*) has been used as a sublethal indicator of excessive copper concentrations in fresh water (Drummond et al. 1973). The greater the volume of water passed over the gills, the faster the "boundary" water is replaced at the lamellar surfaces, thereby maximizing the PO_2 gradient and, therefore, the oxygen diffusion rate. Instead of using bony operculi, sharks, skates, and rays use their gill ventilatory muscles with fleshy flaps of skin to create a ventilatory current through the branchial cavity. The ability to ventilate their gills via either mouth or spiracles provides a presumed flexibility for these fishes when hiding in soft sediments, engulfing prey, or reversing ventilatory flows to clear detritus from gill surfaces (Ferry-Graham 1999, Summer and Ferry-Graham 2001).

Varying the *ventilation volume* is one adjustment that fish can make to influence the rate of gas exchange at the gills. Roberts (1975) has shown that at least one species from each of eight teleostean families ceases branchial movements and passively ventilates (*ram-gill ventilation*) at a high-threshold swimming velocity. Presumably, these threshold velocities are evolutionarily determined to maximize energetic efficiency by using the swimming musculature both to propel the fish and to ventilate the gills adequately. Freadman (1981) measured 8.1% and 8.4% gains in efficiency as striped bass (*Morone saxatilis*) and bluefish (*Pomatomus saltatrix*) initiated ram-gill ventilation at 1.6 and 1.1 body lengths/s swimming velocity, respectively.

Fish also increase their (pumped) gill ventilation volume to maintain gas-exchange (oxygen-uptake) rate in hypoxic water. In this case, the PO_2 of the inspired water is reduced, and diffusion of oxygen across the gills slows as a result of the decreased water-to-blood-oxygen gradient. To increase the volume of water pumped over the gills, either the number of buccal and opercular strokes per minute (*ventilatory frequency*) or the volume pumped with each stroke (*ventilatory stroke volume*) must increase. Commonly, a fish will increase both in response to hypoxic conditions. By rapid displacement of water next to the lamellar surfaces, increases in ventilation volume maximize the environmental hypoxia-depleted, water–blood PO_2 gradient. Even though the oxygen extraction efficiency per volume of water (***percentage utilization of oxygen***) typically is reduced at high ventilatory flows, a sufficient amount of oxygen normally can be taken up to maintain respiratory homeostasis (Table 3.2). On the other hand, when environmental PO_2 decreases below the ***critical PO₂***, these ventilatory (and circulatory, see Chapter 4) adjustments are insufficient to maintain respiratory homeostasis, thereby resulting in oxygen consumption rate decreases (discussed later; see *Fish Oxygen Requirements*).

Aquatic Cutaneous Respiration

Whereas most juvenile and adult fishes use the skin for some aquatic gas exchange, early life-history stages, such as larvae, may use it almost exclusively for respiration. For example, the larvae of the swamp eel (*Monopterus albus*) of southeast Asia display an extensive respiratory capillary network just below the epithelial surfaces of the median fins, the pectoral fins, and the yolk sac before the gills have developed. Interestingly, this species also increases the movement of water posteriorly over the body surfaces, countercurrent to the anteriorly flowing blood, to optimize oxygen uptake when exposed to hypoxic water (Liem 1981). Cutaneous surfaces account for 96% of the respiratory surface area in larval (0.045 g wet wt, 3.7 days posthatch) chinook salmon (*Oncorhynchus tshawytscha*). The branchial area of developing gills exceeds the cutaneous area only when the young salmon reach 2.5 to 4.0 g (Rombough and Moroz 1990). In fact, surface area probably is not the limiting factor in larval gas exchange, at least until the larvae reach 100 mg. The gills probably are more important in their ionic regulatory and acid–base balancing functions (see Chapter 6) than for respiration, at least in developing walleye (*Stizostedion vitreum*) (Rombough and Moroz 1997). Significant cutaneous respiration also has been measured in several adult fishes. Measurements of cutaneous respiration in six freshwater teleosts mostly matched the oxygen demand of the skin itself. Thus, in crucian carp (*Carassius carassius*), yellow perch (*Perca fluviatilis*), brook trout, and brown trout, the skin is not an oxygen exchanger for the benefit of other tissues. Cutaneous oxygen diffusion also matches cutaneous oxygen consumption in the marine plaice (*Pleuronectes platessa*) (Steffensen and Lomholt 1985), and cutaneous oxygen consumption of carp (*Cyprinus carpio*) skin increased with increased PO_2 gradients from increased water flows over its surface (Takeda 1996). Only in the scaleless black bullhead (*Ameiurus melas*) does the skin function as a minor respiratory organ, supplying approximately 5% of the total oxygen demand (Nonnotte 1981).

TABLE 3.2 MEANS OF STRIPED MULLET (BODY WEIGHT, 100 g) VENTILATORY RESPONSES TO HYPOXIA

Variable	Units	Ambient conditions	Hypoxic conditions
Inspired dissolved concentration	mg O_2/L	8.81	3.48
Oxygen consumption rate	mg O_2/h	12.05	11.94
Ventilation volume	ml water/min	36	171
Ventilation frequency	strokes/min	60	95
Ventilary stroke volume	ml water/stroke	0.60	1.80
Percentage utilization of oxygen	%	66	39

Modified from Cech and Wohlschlag (1973), with permission, from *Journal of Fish Biology*. © by Academic Press, Inc. (London) Ltd.

3.2 AIR-BREATHING FISHES

Behavioral means also have evolved with corresponding respiratory structures in fish to cope with hypoxic water. Whereas some species swim to the surface and inspire the oxygen-rich water next to the atmosphere (***aquatic surface respiration***) (Kramer and McClure 1982), others have evolved the ability to breathe air—some by actually leaving the water! At least 374 species of air-breathing fishes (in 49 families) have been described (Graham 1997). Almost all air-breathing fishes retain the ability to breathe water and are categorized as ***bimodal breathers***. Some of the bimodal breathers leave the water to disperse or to seek better water quality (e.g., *Clarias batrachus*; see Chapter 18), and some are left in a drying habitat as the water recedes because of seasonal drought (e.g., *Protopterus* spp., see Chapter 25) or receding tide in the intertidal zone (Horn et al. 1998). Air-breathing fish in aquatic habitats often ***breathe synchronously*** (Kramer and Graham 1976), presumably to minimize individual notice by terrestrial predators (Gee 1980). Kramer et al. (1983) determined that overall survival of water-breathing fishes was significantly higher than that of bimodal breathers when exposed to a predatory green heron (*Butorides striatus*) when water oxygen levels approximated 20% of air saturation (PO_2 = 30 mm Hg) because of the water breathers' superior ability to breathe hypoxic water, thereby avoiding the water surface. The survival of bimodal breathers, however, exceeded that of the water breathers when the water held insufficient oxygen ($\approx$6% of air saturation, or PO_2 = 9 mm Hg) for sustaining their metabolic demands (critical PO_2).

Other fish that migrate short distances over land or are exposed to long-term dewatering or severe drought also have evolved aerial respiratory adaptations. These adaptations range from modifications of the gills to use of the skin to special respiratory structures in the mouth and gut to true lungs.

Modified Gills

The walking catfish (*Clarias batrachus*) of Southeast Asia and, more recently, of Florida (see Chapter 18), represents an example of a fish with modified gills. These catfish feature thickened, widely spaced lamellae on the dorsal side of the filaments and branched, bulbous, dendritic structures dorsally that emanate from the second and fourth gill arches. These dendritic structures resemble "***respiratory trees***" and reside in ***suprabranchial cavities*** (dorsal extensions of the branchial cavity that trap air bubbles) (Jordan 1976). The thickened and bulbous shapes of these modified gills ensure adequate support in air. Ordinary gills, having numerous closely spaced filaments and lamellae, tend to stick together and functionally lose much of their surface area when removed from water. Another requirement for respiratory surfaces is that they remain properly moist. The walking catfish moves on land mainly when it is raining (Jordan 1976). Juvenile monkeyface prickleback (*Cebidichthys violaceus*) of the California intertidal zone can survive air exposure for periods of several hours (e.g., during low tides). Using "standard" gills and, probably, cutaneous surfaces, they lower their overall oxygen demand by becoming very quiescent, as they hide beneath rocks (Edwards and Cech 1990). Several other intertidal fishes, such as sculpins (Cottidae), apparently have similar abilities to survive short-term air exposure under damp conditions (Yoshiyama and Cech 1994, Horn et al. 1999).

Skin

The extent to which fish use cutaneous respiration for aerial respiration has been studied in approximately 15 species (Graham 1997). The percentage of respiratory gas that is exchanged across the skin ranges from approximately 10% to 41% (European eels, *Anguilla anguilla*) to 63% and 94% for the mudskipper (*Periophthalmus sobrinus*) and the synbranchid eel (*Synbranchus marmoratus*), respectively (Graham 1997). Mudskippers actually carry air to subsurface burrow chambers for their aerial respiratory needs while avoiding predators (Ishimatsu et al. 1998). European eels can migrate short distances across land. By diffusion through a well-vascularized skin and, to some extent, across the gills, they are able to use the atmosphere for respiration (Berg and Steen 1965). Desiccation of the body surface is avoided by limiting their terrestrial sojourns to nocturnal movements through moist grass.

Mouth

Electric eels (*Electrophorus electricus*), in contrast to "true" eels, are among the obligate air-breathing fishes. This species has a well-vascularized area in the buccal cavity, where most of its required oxygen is taken up. Whereas this region has a large surface area, resulting from surface convolutions and papillae, the gills have degenerated over evolutionary time. The electric eel surfaces at intervals of approximately 1 minute to replenish the oxygen supply in its mouth, and it will drown if it is forcibly kept immersed (Johansen et al. 1968).

Two other genera that use modified areas of the mouth for aerial gas exchange are the Asian climbing perch (*Anabas testudineus*) and the North American mudsucker (*Gillichthys mirabilis*). Both have evolved bimodal-breathing capabilities to feed or to escape predators by moving out of the water.

Gut

Hoplosternum, Ancistrus, and *Plecostomus* are three genera of tropical catfishes that have parts of their gut specialized for oxygen uptake by actually swallowing air. In these and most other air-breathing fishes, elimination of respiratory carbon dioxide occurs primarily at a site different from that of oxygen uptake. Because the gut is not closely associated with the external environment, the highly water-soluble carbon dioxide is excreted primarily at the gills.

Lungs and Swimbladders

Another example of a gas-specific exchange location comes from the famous lungfishes (Dipnomorpha; see Chapter 16). The South American (*Lepidosiren paradoxa*) and African (*Protopterus* spp.) lungfishes are obligate air breathers. The latter forms have adapted to extensive drought conditions, which may completely dry up their environments. By breathing air through a small vent to the atmosphere, these fish survive extensive dry periods in the mud of dried lakes and rivers in an estivated state. When their habitats refill with water, they surface to inspire air into well-sacculated and heavily vascularized lungs. On the other hand, most of the carbon dioxide is eliminated directly into the water through vestigial gills. The Australian lungfish (*Neoceratodus forsteri*) is not subjected to such lengthy drought conditions in its natural environment, and it will perish if experimentally denied access to water for extensive periods. Other facultative air breathers that use a modified swimbladder for some gas-exchange function include bichirs (*Polypterus* spp.), the bowfin (*Amia calva*), and gars (*Lepisosteus* spp.).

3.3   FISH OXYGEN REQUIREMENTS

Fish need energy to move, to find and digest food, to grow, to reproduce, and to maintain the body and internal environment. Energy stored in food must be metabolically converted to power these various bodily functions. Oxygen, along with an organic substrate, is needed for all *aerobic* (oxidative) metabolic processes. Aerobic metabolic pathways are dominant in organisms that have a fairly reliable oxygen source, because they are biochemically more efficient (18- to 19-fold) than anaerobic pathways. When sufficient oxygen is unavailable, however, such as during extreme hypoxia (i.e., environmental $PO_2 <$ critical PO_2) or burst swimming, some *anaerobic metabolism* is likely. Many fish use *glycolysis* and accumulate *hydrogen ions* (become *acidotic*) and *lactate* (Box 3.1).

The amount of oxygen that a fish requires for these processes over a given length of time is called its *oxygen consumption rate* and is a commonly used measure of aerobic metabolism. The oxygen consumption rate can be affected by a variety of factors. Six of the most significant are (1) *life stage*, (2) *body weight*, (3) *level of activity*, (4) *environmental temperature*, (5) *dissolved oxygen decreases below the critical PO_2*, and (6) *feeding*. In general, fish eggs use very little oxygen until hatching. Both oviparous fish eggs in the environment (e.g., medaka, *Oryzias latipes*) (Marty et al. 1990)

BOX 3.1
Living Without Oxygen

Goldfish, and their congener Crucian carp (*Carassius carassius*), can survive anoxia for periods varying from hours to days. Walker and Johansen (1977) and Hyvärinen et al. (1985) showed that survival is limited by liver glycogen stores. Crucian carp liver glycogen decreased from 35% of total liver weight at the beginning of the long Finnish winter to 2% after enduring prolonged hypoxic and anoxic periods in ice-covered ponds (Hyvärinen et al. 1985). Goldfish hepatocytes also decrease their metabolic requirements (*metabolic depression*) during anoxia (Krumschnabel et al. 2000), and their red and white muscle convert the lactate to *ethanol*

presumably to minimize the adverse effects of *metabolic acidosis* (Shoubridge and Hochachka 1980). The ethanol also triggers the behavior of seeking cooler water, thereby lowering total energy requirements via the thermoregulatory center in the brain (*preopticus periventricularis*) (Crawshaw et al. 1989). Finally, goldfish activate appropriate enzyme systems to counteract possible damage from reactive oxygen species (e.g., free radicals and hydrogen peroxide) during tissue reoxygenation, when oxygen again becomes available after the anoxic period (Lushchak et al. 2001).

and viviparous fish eggs in the mother (e.g., yellowtail rockfish, *Sebastes flavidus*) (Hopkins et al. 1995) show significantly increased oxygen consumption rates after hatching. In contrast, Darken et al. (1998) showed that grunion (*Leuresthes tenuis*) eggs increase their oxygen consumption rates during the first 8 days after being deposited in wet beach sand by the mother (see Chapter 21). If waves do not reach the eggs' location during the next spring (extreme) tide event, then the eggs' viability remains intact until the subsequent extreme tide by shifting to a more constant oxygen consumption rate, thereby maintaining the embryos in an arrested developmental state. Among juvenile and adult fishes, larger fish use more total oxygen per hour than smaller fish (Fig. 3.5) because of the greater summed metabolic demands of more tissue mass. Swimming fish use more oxygen than resting animals (Fig. 3.6) because of the extra metabolic demands of exercising red swimming muscles (see Chapter 2).

With adequately oxygenated environments, fish in warmer water generally have higher oxygen consumption rates than those in cooler water (right side of Fig. 3.7), because warmer temperatures typically increase the cellular nutrient and oxygen (metabolic) demands of ectothermic organisms. The rate of change in oxygen consumption (or other biochemical reaction) rate over a 10°C temperature interval is termed the Q_{10}. Schmidt-Nielsen (1975) presents a good discussion and formulae for calculating these values, which typically are two (a doubling over a 10°C increase in temperature) to three (a tripling) and can change depending on the temperature range. For example, the largemouth bass Q_{10} changes from 2.41 over the 20°C to 25°C range to 2.05 over the 25°C to 30°C range under near-air-saturation conditions (Fig. 3.7). Hopkins and Cech (1994) measured Q_{10} values ranging from 1.85 (20–26°C) to 6.81 (14–20°C) in the bat ray

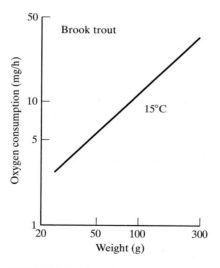

FIGURE 3.5 The influence of body weight (mass) on total oxygen consumption rate for brook trout at 15°C. Note that both axes are logarithmic. The slope of this relationship is 1.01. Modified from Beamish (1964).

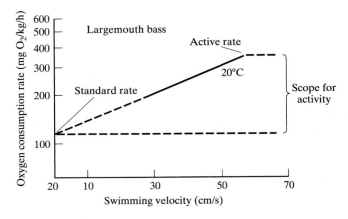

FIGURE 3.6 The effect of swimming velocity on the oxygen consumption rate of largemouth bass. Extrapolation back from the data to 0 cm/s estimates the standard metabolic rate. Subtraction of the standard metabolic rate from the active rate gives the scope for activity. Modified from Beamish (1970).

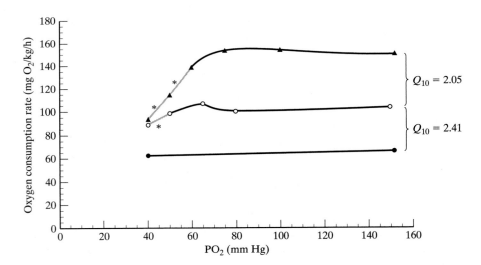

FIGURE 3.7 The influence of temperature and hypoxia on the routine oxygen consumption rate of largemouth bass at 20°C (solid circles), 25°C (open circles), and 30°C (solid triangles). Gray lines and asterisks indicate significantly decreased rates (below critical PO_2 values). The Q_{10} values indicate slightly higher temperature sensitivity over the 20°C to 25°C range compared with the 25°C to 30°C range. Overall, $Q_{10}(20-30°C) = 2.23$. Data from Cech et al. (1979).

(*Myliobatis californica*). The Q_{10} analysis quantifies the bat ray's high metabolic temperature sensitivity over the 14°C to 20°C range.

When PO_2 decreases below the critical PO_2 (e.g., <60 mm Hg at 30°C and <50 mm Hg at 25°C for largemouth bass) (Fig. 3.7), oxygen consumption rates will decrease because of insufficient oxygen for aerobic metabolic pathways. Note that for largemouth bass, the critical PO_2 is apparently less than 40 mm Hg (no significant decrease in oxygen consumption rate) at 20°C. Thus, the critical PO_2 increases with increasing temperature, at which more oxygen is required (Fig. 3.7).

Finally, oxygen consumption rates increased by 41% to 48% and by 130% over resting rates after feeding in Pacific cod (*Gadus macrocephalus*) (Paul et al. 1988) and juvenile northern pike (*Esox lucius*) (Armstrong et al. 1992), respectively. This extra energy (usually termed *specific dynamic action*) is needed for digestion and growth (see Chapters 7 and 8).

The apparatus generally used to measure fish oxygen consumption rates is called a *respirometer*. *Sealed* and *open* (flow-through) respirometers may be used to measure *routine, active*, and—with special precautions—*standard* or *resting* metabolic rates of fishes (Cech 1990). Measuring active rates of fish swimming at various velocities commonly is used to estimate standard metabolic rates by extrapolation back to a swimming speed of 0 body lengths/s (Fig. 3.6). Subtraction of the standard metabolic rate from the active metabolic rate yields the *scope for activity*, which is a useful index for determining the relative amount of nonmaintenance energy reserves. Fish with more reserves are better able to move, to grow, to reproduce, and to resist diseases and parasitism.

3.4 LESSONS FROM RESPIRATION

Water severely limits oxygen solubility, yet fishes have evolved respiratory structures and mechanisms that allow them entry into virtually all aquatic—and even some terrestrial—environments. Most fish breathe water using efficient gills, and oxygen diffuses across respiratory membranes following the PO_2 gradient. Fishes' requirements for oxygen vary with species, size, activity level, and important water conditions, including temperature. Several fishes have evolved the ability to breathe oxygen-rich air, thereby freeing them from hypoxic water constraints but also forcing them to deal with terrestrial predators when they breathe.

SUPPLEMENTAL READINGS

Cech 1990; Graham 1997; Hoar and Randall 1984; Horn et al. 1998; Hughes 1963; Olson 2002; Perry and Tufts 1998; Sundin and Nilsson 2002.

WEB CONNECTIONS

RESPIRATION IN FISHES
www.csuchico.edu/~pmaslin/ichthy/fshrsp.html

FISH GILLS
www.rhul.ac.uk/electronmicroscopy-unit/fishgills.html

C H A P T E R 4
Blood and Its Circulation

All fish tissues require a delivery of nutrients (including oxygen) to function properly. The circulatory system, with blood as its medium, delivers these essential nutrients throughout the body and removes both metabolic wastes and pathogens before harmful concentrations are reached. Under various conditions of temperature or activity, nutrient demands may change, so delivery rates must keep pace (within physiological limits). Delivery rates for oxygen are changed via temperature, pH-influenced hemoglobin–oxygen dynamics, and the output of the heart, which is regulated by an elegant array of aneural as well as neural controls.

4.1 BLOOD

Fish blood, like that of other vertebrates and many invertebrates, is composed of blood cells suspended in plasma that are circulated throughout the body tissues. ***Blood volumes*** constituting the primary circulation usually range from 3% to 7% of the total body weight of bony fishes, 4% to 8% of elasmobranchs, and 8% to 20% of agnatha (Thorson 1958, Satchell 1971, Tort et al. 1991, Olson 1992, McCarthy and Conte 1996). The cells are of two basic types; ***erythrocytes*** (red blood cells [RBCs]), and ***leukocytes*** (white blood cells [WBCs]). In this chapter, we first describe the nature and function of erythrocytes and leukocytes and discuss in detail the structure and function of ***hemoglobin***, the oxygen-carrying pigment of the blood.

Both erythrocytes and leukocytes are formed from ***hemocytoblast*** precursor cells that originate from a variety of organs and usually mature after they enter the bloodstream. In hagfish, the primary blood-forming site is the ***mesodermal envelope*** surrounding the gut (Jordan and Speidel 1930). Percy and Potter (1976) have shown that adult lampreys (*Lampetra fluviatilis*) synthesize blood cells from the fatty tissue dorsal to the nerve cord. Elasmobranch fishes may produce blood cells from the ***Leydig organ*** (situated in the esophagus), the ***epigonal organ*** (around the gonads), and the ***spleen***. The Leydig organ produces leukocytes in the deepwater shark *Etmopterus spinax* (Mattisson and Fänge 1982), whereas the Atlantic nurse shark (*Ginglyostoma cirratum*) lacks this organ and produces these cells from the epigonal organ. The ***white pulp*** portion of the spleen produces more lymphocytes, and the ***red pulp*** portion produces erythrocytes and some ***granulocytes*** (a class of leukocytes) in the Atlantic nurse shark (Fänge and Mattisson 1981). The spleen of elasmobranch and teleost fishes is supplied by ***splanchnic (autonomic) innervation*** that stimulates this organ to contract during stressful states (adding blood cells to the circulation), such as during severe exercise (Yamamoto 1988) or during hypoxia (Fänge and Nilsson 1985). Besides direct nervous stimulation, hormonal (***adrenergic*** or ***cholinergic***) stimulation also causes contractions of the Atlantic cod (*Gadus morhua*) spleen (Nilsson and Grove 1974). Teleostean ***hemopoietic*** (blood-forming) sites are primarily the ***kidney*** and the spleen (Satchell 1971), whereas ***lymphomyeloid*** (leukocyte-producing) tissue also is found in the cranium of holocephalans (*Chimaera monstrosa*) and sturgeons (*Acipenser ruthenus*) as well as surrounding the heart of sturgeons (Fänge 1984). The ***thymus*** constitutes another WBC-producing tissue in the juvenile stages of many jawed fishes, but it often regresses in sexually mature individuals (Fänge 1984). Fish bone has no marrow for blood formation.

4.2 ERYTHROCYTES

Erythrocytes, or RBCs, usually are the most abundant cells in fish blood (up to 4 million cells/mm^3). They contain hemoglobin and carry oxygen from the gills to the tissues. Like RBCs of other nonmammalian vertebrates, fish RBCs generally are nucleated and show a wide range of sizes among different species. Elasmobranchs typically have RBCs that are larger, although fewer in number, than those of teleosts (Table 4.1). Even within the teleosts, species having more erythrocytes per milliliter of blood generally have smaller RBCs. Within this group, the more active species tend to have more RBCs than do sedentary species (Table 4.1). Perhaps the smaller cell size in these active species presents a shorter mean diffusion distance and a greater overall surface area for essential respiratory gases, such as oxygen. Shorter diffusion distances and more RBCs having a greater overall surface area would increase the efficiency of oxygen uptake at the gills and its delivery to the oxygen-requiring swimming muscles. Open-ocean pelagic species, such as albacore (*Thunnus alalunga*) and Atlantic shortfin mako sharks (*Isurus oxyrinchus*), which use physiological thermoregulation to warm red muscle masses (see Chapter 5), characteristically have the highest ***hematocrit*** (packed cell volume) and hemoglobin levels (Table 4.1).

Because the oxygen demands of fish vary with the life-history stage and environmental conditions, the number of RBCs per milliliter varies as a way of balancing the energy costs of producing RBCs with those of pumping blood to the tissues. Blood with a low erythrocytic concentration (***anemia***) obviously has to be pumped through the body at a greater rate than blood with a higher erythrocytic concentration if the oxygen demand is high. Because the teleost ***heart*** requires as much as 4.4% of the total energy of the fish (Cameron 1975), the RBC concentration can have a significant effect on its overall energy balance, including growth (see Chapter 8). Indeed, rainbow trout experimentally made anemic show significant increases in the volume of blood pumped by the heart (***cardiac output***) (Cameron and Davis 1970). When oxygen demands of tissues are relatively low, however, such as when water temperatures are low and the fish are not very active, large numbers of RBCs are not required, and the concentration decreases. Thus, ***seasonal*** changes often are found in RBC production. For example, winter flounder (*Pleuronectes americanus*) have an erythrocytic production peak during late spring

TABLE 4.1 HEMATAOLOGICAL CHARACTERISTICS OF VARIOUS FISHES

Species	Erythrocyte number ($\times 10^6$ cells/mm³)	Hematocrit (%)	Hemoglobin concentration (g/dl)	Mean erythrocytic volume (μm³)	Blood oxygen capacity (ml/dl)	Source
Spiny dogfish (*Squalus acanthias*)	0.09	18.2	—	650–1,010	—	Thorson (1958) Wintrobe (1934)
Blue shark (*Prionace glauca*)	—	22.3	5.70	—	—	Johanson-Sjöbeck and Stevens (1976)
Oyster toadfish (*Opsanus tau*)	0.69	—	6.84	—	—	Eisler (1965)
Winter flounder (*Pleuronectes americanus*)	2.21	23	5.36	107	8.30	Bridges et al. (1976)
Common carp (*Cyprinus carpio*)	1.43	27.1	6.40	186	12.50	Houston and DeWilde (1968)
Striped mullet (*Mugil cephalus*)	3.08	26.9	7.14	88	8.36	Cameron (1970a)
Pinfish (*Lagodon rhomboides*)	2.66	32.9	7.59	124	7.78	Cameron (1970a)
Spotted sea trout (*Cynoscion nebulosus*)	3.25	32.3	6.99	99	8.56	Cameron (1970a)
Bluefish (*Pomatomus saltatrix*)	3.85	—	13.41	—	—	Eisler (1965)
Atlantic shortfin mako shark[a] (*Isurus oxyrinchus*)	—	40.8	14.30	—	—	Emery (1986)
Albacore[a] (*Thunnus alalunga*)	—	53	17.20	—	21.80	Barret and Williams (1965) Cech et al. (1984)

[a]Species that use physiological thermoregulation (see Chapter 5).

and early summer in Maine coastal waters (Fig. 4.1). Cameron (1970b) has shown that changes in erythrocytic concentration (and in total hemoglobin concentration) in pinfish (*Lagodon rhomboides*) are of some significance in meeting seasonal changes in respiratory demands. Other adjustments, however, such as in RBC size and in the rate of blood circulation, also are required to meet the nearly 10-fold change in respiratory metabolism that is associated with seasonal temperature extremes. In striped mullet (*Mugil cephalus*), changes in erythrocytic and hemoglobin concentrations are associated not only with seasonal temperature changes (Cameron 1970b) but also with spawning activity and its high energy demands (Fig. 4.2). Increases in erythrocytic concentration during the spawning season have been recorded for fish as diverse as common carp and the redbelly tilapia (*Tilapia zilli*), so the phenomenon may be widespread among teleosts. It also is worth noting that erythrocytic concentrations may be affected by other environmental factors, particularly pollutants. Destruction of RBCs through the inhibition of vital metabolic pathways in the cell appears to be one of the reasons why chlorine in water is so harmful to fish (Grothe and Eaton 1975, Buckley 1977).

4.3 LEUKOCYTES

Leukocytes, or WBCs, are less abundant (20,000–150,000 cells/mm^3) than RBCs in fish blood. The principal types of WBCs are lymphocytes, thrombocytes, monocytes, granulocytes, and nonspecific cytotoxic cells. Often, several types of WBCs are found in fish blood, and different roles have been attached to their presence. Leukocytes provide a mechanism for blood clotting and help to rid the body of foreign material, including invading pathogens via immune system and other responses (Iwama and Nakanishi 1996).

Measurements of change in total leukocytic concentration or in the percentages of the various WBC types often can lead to a better understanding of the physiological or

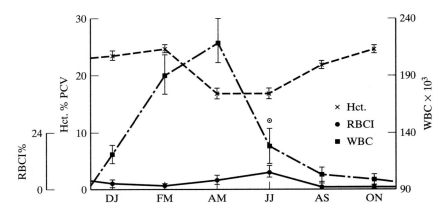

FIGURE 4.1 Bimonthly mean values (±standard error) of microhematocrit (Hct.), immature red blood cells (RBCI), and leukocytes (WBCs) for winter flounder (*Pleuronectes americanus*) from Casco Bay, Maine. The circled dot signifies the highest mean RBCI, which occurred during the third week in June. From Bridges et al. (1976).

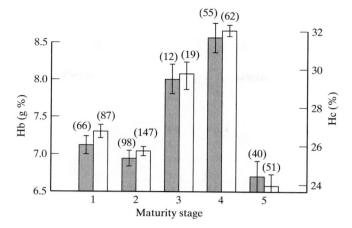

FIGURE 4.2   Mean hemoglobin (Hb) and hematocrit (Hc) levels at different reproductive maturity stages. Histogram bar heights indicate mean values; vertical lines indicate ± one standard error. Numbers in parentheses indicate sample size. Stippled bars indicate total hemoglobin concentration; open bars indicate the percentage of packed cell volume (hematocrit). Maturity stages: 1, immature (gonads very small or absent); 2, mature (gonads mature but small); 3, prespawning (gonads enlarged but not secreting reproductive products); 4, ripe (reproductive products flow easily from fish); 5, postspawning (gonads somewhat enlarged but empty). From Cech and Wohlschlag (1982).

pathological state of the animal. Circulating leukocytic concentration also can vary through the year in some fish species. Bridges et al. (1976) describe a pattern of total and differential (separate) WBC counts in winter flounder that generally shows an inverse relationship with fish condition or health. Sick individuals presumably make more WBCs to synthesize antibodies or phagocytize bacteria. Sample photomicrographs of winter flounder blood cells are shown in Figure 4.3.

   Lymphocytes can vary in size (diameter, 4.5–12 μm) among species. Their morphology, on the other hand, is more consistent. They are dominated by the nucleus, with only a narrow rim of basophilic cytoplasm, in which a few mitochondria and ribosomes are found. The number of lymphocytes varies among species—and with counting technique! Ellis (1977) found that 12×10^3 lymphocytes/mm^3 was the representative value for plaice. The number of lymphocytes can vary with the season, however, following general WBC seasonal trends (Fig. 4.1). Teleostean lymphocytes appear to be produced by the thymus, spleen, and kidney (Fänge 1984), although specific antigen responses may differ among cells produced by each organ (Ellis 1977).

   The primary function of fish lymphocytes seems to be to act as the executive cells of immune mechanisms via ***antibody production***. Klontz (1972) reported a large increase in small lymphocytes in the kidney of rainbow trout that correlated with high antibody production, 2 to 3 days after antigen injection. Some evidence also indicates that fish lymphocytes may demonstrate ***phagocytic activity*** (engulfing foreign cells) or

Mature erythrocytes

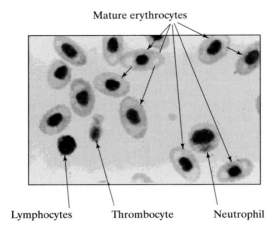

Lymphocytes   Thrombocyte   Neutrophil

"Early" immature erythrocytes

"Later" immature erythrocytes

Neutrophils

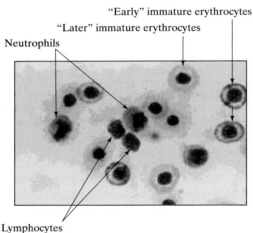

Lymphocytes

FIGURE 4.3 Winter flounder blood cells.

give rise to cells (e.g., macrophages) that have this ability (Klontz 1972, Weinreb and Weinreb 1969).

Thrombocytes of fishes function in the *clotting* of circulating fluids. Clots are produced by spreading of the thrombocytic cytoplasm into long threads that cross-link the denuded nuclei, thereby forming a fibrous network that traps circulating cells. They appear as spiked, spindle, oval, and lone nucleus forms in stained blood smears on microscopic slides. The various thrombocyte shapes may well be different forms of the same cell, because changes in shape can be observed in live preparations. From his work on plaice, Ellis (1977) believes that thrombocytes originate in splenic tissue.

Monocytes/macrophages comprise a small proportion of the WBC population, unless foreign substances are present in the tissues or bloodstream. Thought to originate in the kidney, monocytes concentrate at and phagocytize foreign particles. Morphologically, therefore, the cell outline may be quite irregular (from pseudopod formation). They are capable of killing a variety of pathogens, including bacteria and helminth larvae (Secombes 1996).

Granulocytes are leukocytes with conspicuous cytoplasmic granules. Three basic types are found: (1) eosinophils, (2) basophils, and (3) neutrophils. The names of the three types reflect the comparative visibility of the granules when the cells are treated with acidic (e.g., eosin), basic, and neutral stains, respectively. Neutrophils are the most commonly encountered fish granulocyte, comprising as much as 25% of all WBCs in brown trout (Blaxhall and Daisley 1973). In addition to the neutral-staining (e.g., gray) cytoplasm, fish neutrophils often have an eccentric nucleus, which is divided into several lobes. Granulocytes are formed in the kidney and spleen in teleosts (Ellis 1977) and in the Leydig organ in elasmobranchs (Fänge 1968). Neutrophils apparently migrate to sites of bacterial infection, where they may be phagocytic (Finn and Nielson 1971) or otherwise bactericidal (Fänge 1984, Secombes 1996). In rainbow trout, neutrophilia (i.e., having increased numbers of circulating neutrophils) often corresponds to stress (Weinreb 1958).

Basophils are present in goldfish, Australian lungfish (*Neoceratodus forsteri*), common carp, and Pacific salmon, but they are absent in the blood of anguillid eels, plaice, yellow perch (*Perca flavescens*), brown and rainbow trouts, and lampreys (Ellis 1977). Much confusion exists in the fish hematological literature regarding the presence or absence of eosinophils in fish. Although still unclear, basophil and eosinophil functions seem to be related to antigen sensitivity, stress phenomena, and phagocytosis (eosinophils) (Ellis 1977).

Nonspecific cytotoxic cells are considered to be the equivalent of mammalian natural killer cells (Secombes 1996). They are capable of lysing (breaking) a wide variety of tumor cells and fish protozoan parasites (Evans and Jaso-Friedmann 1982). While in elasmobranchs the cells responsible for this type of cytotoxicity are considered to be macrophages, in teleosts a much smaller, lymphocyte-like cell is involved. The teleostean nonspecific cytotoxic cells differ from mammalian analogues. They do not contain cytoplasmic granules, and their nuclei can take different shapes (Secombes 1996).

4.4   HEMOGLOBIN

Hemoglobin is a respiratory pigment that vastly increases the ***blood oxygen-carrying capacity***. For example, in the Port Jackson shark (*Heterodontus portjacksoni*) at 20°C, 93% of the oxygen carried by the blood is reversibly bound to the hemoglobin, whereas 7% is physically dissolved in the plasma at saturation (Grigg 1974). In colder environments, the plasma percentage may increase (12% in the Antarctic nototheniid *Trematomus bernacchii* at −1.5°C). Indeed, the Antarctic crocodile icefishes (family Channichthyidae) carry no hemoglobin in their blood at all. The icefishes survive because (1) their metabolic oxygen requirements are low and the environmental dissolved oxygen is high in the consistently cold Antarctic waters, (2) their sluggish activity levels are adequate to catch sufficient quantities of plentiful krill and small fish, and (3) special cardiovascular adaptations (e.g., comparatively large heart and blood volume with relatively low-resistance capillaries) promote efficient movement of their blood (Holeton 1970).

Despite this evidence that some fish can get along without hemoglobin, its importance to most fish is difficult to overstate. Hemoglobin, however, is not just a single type of molecule. It is a class of structurally similar molecules that vary in their structure and in their affinity for oxygen under different conditions. This variability will become evident in the sections that follow on (1) hemoglobin structure, (2) blood oxygen affinity, and (3) factors affecting blood oxygen affinity. The important role of hemoglobin as an intraerythrocytic buffer is discussed in Chapter 6.

Hemoglobin Structure

Fish hemoglobin is of two basic types; *monomeric*, and *tetrameric*. Monomeric hemoglobins consist of single-heme polypeptide molecules, each with a molecular weight of approximately 17,000 daltons. They are characteristic of lampreys and hagfish. Tetrameric hemoglobins are characteristic of all other fishes and are composed of four chains of amino acids (two α and two β chains), much like mammalian hemoglobins, and have molecular weights of approximately 65,000 daltons. There are many different kinds of tetrameric hemoglobins, and several (*polymorphic*) kinds may be found in one fish! For example, four kinds of hemoglobin are found in rainbow trout blood (Binotti et al. 1971), two in American eel (*Anguilla rostrata*) blood (Poluhowich 1972), and three in goldfish blood (Houston and Cyr 1974). The significance of synthesizing more than one hemoglobin type appears to relate to the different functional properties of each, so different combinations of hemoglobin types reflect adaptations to different environments or ways of life.

Multiple hemoglobins may be especially adaptive in migratory species, which experience considerable environmental variation. For example, the catadromous American eel has one hemoglobin with a high oxygen affinity in saltwater conditions and one with a high oxygen affinity in freshwater conditions. Poluhowich (1972) suggests that the polymorphic hemoglobins assist in the acclimation of these eels to environments of different salinity by maintaining an approximately constant blood oxygen affinity. In contrast, anadromous chinook salmon develop "adult" hemoglobins after conclusion of their parr–smolt transformation (see Chapter 6), and this change in hemoglobin types seems to be unrelated to success in seawater environments (Fyhn et al. 1991).

Goldfish hemoglobins differ in their responses to temperature (Houston and Cyr 1974). Goldfish acclimated to 2°C had two different hemoglobins; others held at 20°C and 35°C featured three. Because the observed concentration of the third hemoglobin did not exceed 12.5% of the total concentration in any individual, its physiological importance may be minor, and a warm-temperature function for this component has yet to be demonstrated. Houston and Rupert (1976), however, have shown that the third hemoglobin can be made to appear and to disappear with temperature changes from 3°C to 23°C and vice versa, respectively, within *3 hours*. Thus, this rapid synthesis of the third hemoglobin in goldfish probably stems from rearrangement of the α and β subunits in other hemoglobins rather than from synthesis of a new hemoglobin or production of a new type of erythrocyte (Houston and Gingras-Bedard 1994).

Hemoglobin polymorphism for activity levels also has been hypothesized for species of suckers (Catostomidae). The desert sucker (*Catostomus clarki*) possesses a pH-insensitive hemoglobin that maintains a high oxygen affinity even when those of other hemoglobins are drastically reduced because of increases in circulating lactic acid from violent muscular activity (Powers 1972). This species typically lives in fast water. The sympatric Sonora sucker (*Catostomus insignis*), which does not possess this hemoglobin, is found mainly in the slower-water portion (e.g., quiet pools) of their streams. These hemoglobin types also may expand the realm of available habitats for these species by increasing intracellular buffer capacity (see Chapter 6) in RBCs (Weber 1990).

Changes in hemoglobin types with age have been demonstrated in fishes. Coho salmon (*Oncorhynchus kisutch*), for example, show changes with the progression from alevin to fry or presmolt stages. Giles and Vanstone (1976) believe that these changes are controlled genetically and may relate to known changes in the organ of origin of the erythrocytes during development. Certainly, the pattern of hemoglobins seems to be more fixed in developing cohos than in goldfish, because exposure of the salmon fry and presmolts to extremes of temperature, salinity, and dissolved oxygen produces no detectable variation in hemoglobin types. Finally, the presence of multiple hemoglobins is not always correlated with our perception of enhanced performance. Turbot (*Scophthalmus maximus*) with two hemoglobin types (and a corresponding intermediate oxygen affinity) show slower growth rates than those Turbot with either single hemoglobin type having a higher or lower oxygen affinity, respectively (Imsland et al. 1997). The mechanisms behind the differential growth rates await further research.

Blood Oxygen Affinity

Blood oxygen affinity varies widely among fishes, reflecting the environments in which they live. Figure 4.4A shows the ***blood oxygen equilibrium curves*** of a Sacramento blackfish (*Orthodon microlepidotus*), and Figure 4.4B shows those of a rainbow trout. ***Hyperbolic*** curves like those from blackfish blood result from the paucity of interaction among the four subunit ***hemes*** (O_2–binding sites) that are characteristic of tetrameric molecules. This subunit independence results in curves that are similar to those of the monomeric hemoglobin of agnathans. The steep, hyperbolic curve of blackfish hemoglobin displays the ability of the hemoglobin to be 50% saturated with oxygen at very low partial pressures of oxygen (Po_2) in the water (2 mm Hg at 20°C). This is termed the ***half-saturation value*** (or P_{50}) and reflects the ***affinity*** of the hemoglobin for oxygen. The low P_{50} of the Sacramento blackfish (2 mm Hg) indicates a ***high*** blood oxygen affinity. By comparison, the rainbow trout blood's more ***sigmoid***-shaped curve features a higher P_{50} (17 mm Hg), or a ***lower*** blood oxygen affinity.

The importance of this difference is especially apparent in ***hypoxic*** environments. For example, if the Po_2 of the water is only 32 mm Hg ($\approx$20% of air saturation), fish can raise their arterial Po_2 only to approximately 25 mm Hg in the gills, despite the efficiency of this countercurrent gas exchanger. At 25 mm Hg, Sacramento blackfish can saturate their arterial blood to approximately 90% (Fig. 4.4A), whereas rainbow trout can saturate their blood only to approximately 65% (Fig. 4.4B).[1] This ability to saturate blood to the 90% level in poorly oxygenated water obviously is an advantage to a fish

---

[1]Note that 2 mm Hg Po_2 = partial pressure (p) of oxygen equal to 2 mm mercury (Hg) pressure.

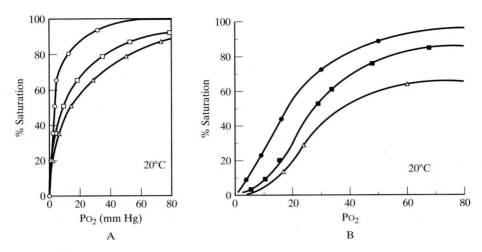

FIGURE 4.4   **A**. Blood oxygen equilibrium curves for Sacramento blackfish equilibrated at 20°C and three P_{CO_2} levels: <1 mm Hg (○), 5 mm Hg (□), and 10 mm Hg (△). **B**. Blood oxygen equilibrium curves for rainbow trout blood equilibrated at 20°C and three P_{CO_2} levels: 0 mm Hg, (●), 3 mm Hg (■), and 7–8 mm Hg (△). **A** from Cech et al. (unpublished data); **B** from Cameron (1971a).

living in warm sloughs or stagnant pools, deep water of lakes, or hypoxic zones of coastal oceans. The higher-*percentage saturation* achieved by the blackfish means that a greater *content* of oxygen is reversibly bound by the hemoglobin for transport to the tissues to meet the metabolic oxygen (content) requirements of the fish.

The evolutionary advantages that are inherent in the sigmoid curve of trout include the unloading of oxygen at the tissues at a fairly high P_{O_2}. Typifying quite active fish in well-oxygenated waters, the high P_{O_2} of the inspired water ensures full oxygenation of the blood while traversing the gills. Fish with sigmoid curves having a fairly *steep* middle segment can unload and reload large quantities (contents) of oxygen over a quite narrow range of P_{O_2} values. This steep portion of the curve represents the *physiologically most efficient P_{O_2} range* for oxygen uptake and delivery for the species. Because the steep portion of sigmoid curves usually is shifted to the right (compared with hyperbolic curves), the active fish with sigmoid curves operate most efficiently in well-oxygenated environments (e.g., streams, well-mixed lakes, and shallow ocean areas).

Oxidized (rather than oxygenated) hemoglobin, which cannot function as a respiratory pigment, is termed *methemoglobin* and may occur in significant quantities in fish blood. Cameron (1971b) found that 11% of the total hemoglobin in pink salmon (*Oncorhynchus gorbuscha*) was methemoglobin, as was 22% of the hemoglobin in hatchery steelhead (sea-run rainbow) trout (Meade and Perrone 1980). Although why high percentages of methemoglobin occur in the few species examined is unclear, energy is required to reduce the methemoglobin back to the less stable, functional form (Cameron 1971b).

Factors Affecting Blood Oxygen Affinity

Numerous factors can influence the blood oxygen affinity. Among the most important are (1) pH, (2) partial pressures of carbon dioxide (P_{CO_2}), (3) temperature, and (4) organic phosphate concentrations.

Carbon dioxide and pH effects are often interrelated and, physiologically, are the most important factors affecting blood oxygen affinity. Figure 4.5 shows the effect of pH and P_{CO_2} on winter flounder blood oxygen affinity. The decrease in ***affinity*** with decreasing pH or increasing P_{CO_2} ***(Bohr effect)*** normally works to "drive off" oxygen from the hemoglobin, thereby raising the plasma P_{O_2} and facilitating its diffusion to surrounding tissues. The Bohr factor (quantification of the Bohr effect) is calculated by dividing the shift or change in log P_{50} by the change in pH associated with the shift. Table 4.2 shows that in general, more active fish species tend to have larger (absolute value) Bohr factors. This tendency is adaptive, because exercise provokes greater oxygen demands by the red swimming muscles. The larger Bohr effect increases the O_2 diffusion rate across the capillary walls to meet this demand. Moreover, violent or high levels of sustained exercise activate the primarily anaerobic white muscles, thereby incurring an oxygen debt. The hydrogen ions resulting from this (glycolytic) anaerobic metabolism decrease blood pH even further, possibly magnifying the Bohr shift. Nikinmaa et al. (1984), however, showed that striped bass *(Morone saxatilis)* maintain their ***intracellular*** pH in somewhat swollen RBCs despite marked ***extracellular*** (plasma) acidification from exercise-induced lactic acid increases. The intracellular pH and arterial oxygen content apparently were protected via endogenous epinephrine (adrenaline) secretions during the exercise, because both of these variables were lower in exercised striped bass injected with the β-adrenergic antagonist ***propranolol***.

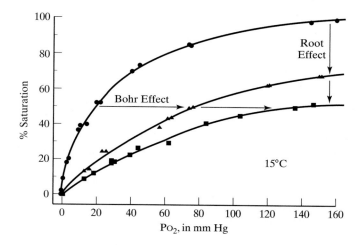

FIGURE 4.5   Blood oxygen equilibrium curves for winter flounder *(Pseudopleuronectes americanus)* blood equilibrated to three levels of P_{CO_2} at 15°C: <1 mm Hg (●; mean pH, 8.02), 8 mm Hg (▲; mean pH, 7.48), and 24 mm Hg (■; mean pH, 7.17). Modified from Hayden et al. (1975).

TABLE 4.2   BOHR EFFECTS OF FISHES CHARACTERIZED VARIOUS ACTIVITY LEVELS

Activity level	Species	Temperature (°C)	Bohr factor[a]	Source
Lowest	Brown bullhead (*Ameiurus nebulosus*)	9, 24	−0.31	Grigg (1969)
Lower	Flounder (*Platichthys flesus*)	15	−0.55	Weber and DeWilde (1975)
Higher	Rainbow trout (*Oncorhynchus mykiss*)	15	−0.57	Eddy (1971)
Highest	Atlantic mackerel (*Scomber scombrus*)	25	−1.2	Hall and McCutcheon (1938)

[a] $\Delta \log P_{50}/\Delta pH$.

Figure 4.6 shows a model for this adrenergic response in rainbow trout RBCs (Nikinmaa 1986). This ***adrenergically mediated Na^+/H^+ exchange*** across the rainbow trout RBC is stimulated under stress conditions when ***cortisol*** secretions (see Chapter 6) increase the number of ***membrane adrenoreceptors*** (labeled as 1 in Fig. 4.6), thereby enhancing epinephrine binding (Reid and Perry 1991). The exchange apparently also shows variability between species in the same river systems (Val et al. 1998) and between seasons in the same species (Cossins and Kilby 1989, Lecklin and Nikinmaa 1999).

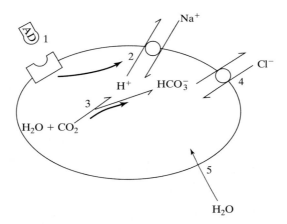

FIGURE 4.6   A model for the adrenergic response of rainbow trout red cells. Binding of adrenaline (epinephrine) to membrane receptor (*1*) activates the sodium/proton exchange (*2*),which extrudes protons and shifts (*3*) the reaction $H_2O + CO_2 = H^+ + HCO_3^-$ to the right by removing one of the end products. Accumulation of bicarbonate (*4*) leads to increased efflux of HCO_3^- and influx of chloride. The accumulation of sodium and chloride draws (*5*) osmotically obliged water into the cell, and cell volume increases. From Nikinmaa (1986).

The **Root effect** (Root 1931) is another pH- or P_{CO_2}-induced shift (displacement of the oxygen equilibrium curve), except that blood oxygen **capacity** rather than **affinity** is affected (Fig. 4.5). The Root shift is now thought of primarily as an extreme Bohr shift, and the molecular basis of this phenomenon may be associated with a special hemoglobin within a complement of multiple hemoglobins. Increases in P_{CO_2} or decreases in pH lower the oxygen capacity (oxygen content at 100% saturation). This effect typifies species with **swimbladders** and **retia mirabilia** (vascular countercurrent exchangers). Sharks have no swimbladders, so no Root effect is associated with their hemoglobins. As described in detail in Chapter 5, the unique features of these retial countercurrent exchangers and the Root effect makes swimbladder inflation possible at great depths. Baines (1975) found a large Root effect characterizing the hemoglobin of relatively deepwater rockfishes *(Sebastes,* Scorpaenidae) off the California coast. Rockfish from shallower water, which have less extensive vertical migrational patterns, had smaller Root effects. In the related but strictly shallow-water California scorpionfish *(Scorpaena guttata)*, which has no swimbladder, no Root shift could be detected. Curiously, winter flounder also have no swimbladder, yet they display a significant Root effect (Fig. 4.5). Like many other teleosts that depend on vision to feed, winter flounder possess a **choroid rete**, or countercurrent vascular organ, behind the retina of the eye (Wittenberg and Haedrich 1974). The Root effect may play a significant role in delivering sufficient oxygen to the retinal tissue, which has a high oxygen demand (Hayden et al. 1975).

Aspects of carbon dioxide transport in the blood are covered in more detail in Chapter 6, in the section concerning acid–base balance. Regardless, it is germane here to explain the carbon dioxide equilibrium curve, which relates carbon dioxide loading (content) in the blood with the partial pressure of carbon dioxide (P_{CO_2}) Typically, significant carbon dioxide loading occurs with little change in P_{CO_2} within the normal physiological range of P_{CO_2} values (<10 mm Hg), with some flattening out at higher tensions (Fig. 4.7). Carbon dioxide–combining power is greater in fishes (e.g., carp) that are adapted to living in stagnant, high P_{CO_2} environments and is lower for fishes typifying low carbon dioxide habitats (e.g., trout and mackerel). Figure 4.7 also shows the increased carbon dioxide–combining power of deoxygenated hemoglobin (**Haldane effect**, shown as the vertical distance between paired curves) because of the pH rise accompanying blood deoxygenation (Grigg 1974). The magnitude of the Haldane effect also varies among species.

Temperature effects on blood oxygen affinity and capacity of fish hemoglobins are most noticeable in fish that have narrow temperature tolerances **(stenothermal)**. Figure 4.8 shows that increases in temperature depress both oxygen affinity and oxygen capacity of tench blood. The extra oxygen delivery to the respiring tissues (e.g., from temperature-induced loss of blood oxygen affinity) when oxygen demand is elevated by increased temperatures would seem to be an adaptive advantage. Instead of exerting a selective effect, however, like the Bohr shift (i.e., working primarily at the tissue sites where P_{CO_2} is high and/or pH is low), the temperature effect in these ectotherms works equally well at the gills! Thus, large temperature effects would not appear to be advantageous to species inhabiting environments that exhibit large temperature fluctuations or to species that move quickly from one temperature to another. The effect of temperature on the oxygen affinity and capacity of hemoglobin

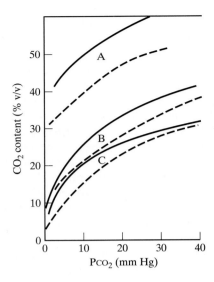

FIGURE 4.7   Representative carbon dioxide equilibrium curves for (**A**) common carp, (**B**) Atlantic mackerel, and (**C**) rainbow trout. Solid lines and dashed lines represent deoxygenated and oxygenated blood, respectively. Modified from Grigg (1974).

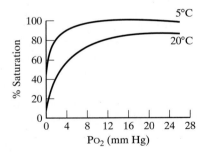

FIGURE 4.8   Oxygen equilibrium of tench *(Tinca tinca)* blood at 5°C and 20°C. Modified from Eddy (1973).

can be quantified by the ***apparent heat of oxygenation (ΔH)*** from a form of the van't Hoff equation (Riggs 1970). The "living fossil" coelacanth *(Latimeria chalumna)*, which resides at the quite thermally stable depth of 200 to 400 m in the ocean, has hemoglobin displaying a ΔH of −10.42 kcal/mol (Wood et al. 1972). The more eurythermal winter flounder of coastal marine environments has a hemoglobin ΔH of −7.7 (Hayden et al. 1975). Species confronting wide temperature variations, such as the globally migratory bluefin tuna *(Thunnus thynnus)* (Rossi-Fanelli and Antonini 1960), display a hemoglobin ΔH of -1.8. Cech et al. (1984) showed a reverse temperature effect ($\Delta H = +1.72$) for the warm-bodied albacore, in which oxygen is bound to hemoglobin (rather than released) despite rapid warming of the blood in countercurrent retia (see Chapter 5).

   Organic phosphates also are important to the reversible binding of fish hemoglobins with oxygen. Gillen and Riggs (1971) have shown that concentrations of naturally occurring organic phosphates can profoundly influence hemoglobin–oxygen affinity, primarily via changes in the hemoglobin molecular structure associated with organic phosphate binding (Weber 1979, Nikinmaa 1990). *Adenosine triphosphate* (ATP) is the primary phosphorylated compound in Rio Grande perch *(Cichlasoma cyanoguttatum)*, and ATP additions depress oxygen affinity, increase the Bohr effect, and modify heme–heme interactions. In carp RBCs, *guanosine triphosphate* (GTP) plays a greater role than ATP in blood oxygen affinity regulation (Weber and Lykkeboe 1978). Fish such as carp that encounter hypoxia in their natural habitats have relatively high concentrations of GTP, and they appear to use this phosphate selectively to modulate oxygen affinity (Weber and Jensen 1988). Fishes that normally inhabit well-oxygenated habitats (e.g., salmonids) mainly use ATP (Jensen et al. 1993). Decrease of organic phosphate concentration inside the RBC enhance the efficiency of oxygen uptake in fishes exposed to environments that are warm (Grigg 1969) or hypoxic (Greaney and Powers 1978, Weber and Lykkeboe 1978). In the spiny dogfish *(Squalus acanthias)*, GTP also exerts a greater effect than ATP, but urea *increases* oxygen affinity (Weber et al. 1983).

4.5   CIRCULATION

The *primary circulatory system* of most fish is a "closed" system that typically consists of a single heart as the pump in line with *branchial* (gill) and *systemic* (body) capillary beds connected by *arteries* and *veins* (Fig. 4.9). Recent evidence shows numerous fluid connections with the *secondary circulation system*. Hagfish circulatory systems have *accessory* inline hearts. Lungfish also differ by the presence of a *pulmonary circulation* and partial mixing of oxygenated and deoxygenated blood in the heart (Randall 1970). A variety of circulatory adaptations can be found in fishes having accessory respiratory surfaces in cavities, at the skin, or in the gut (Satchell 1976).

   Given the diversity of circulatory adaptations in fishes, it is instructive to examine aspects of the cardiovascular anatomy to better understand circulatory function. Of paramount importance is the heart.

4.6   HEART STRUCTURE

The pulsed *(systole*: emptying; *diastole*: filling) propulsion of blood through the circulatory system of the most fishes is accomplished by a four-chambered heart (Fig. 4.10). All four chambers are in line and pump only venous blood. Except for a few air-breathing fishes, all flow of blood is to the gills. The heart and gills are closely associated, because fish hearts are located the farthest anterior relative to those all the vertebrates. The heart is enclosed in the fluid-filled *pericardial cavity*, which is separated from the

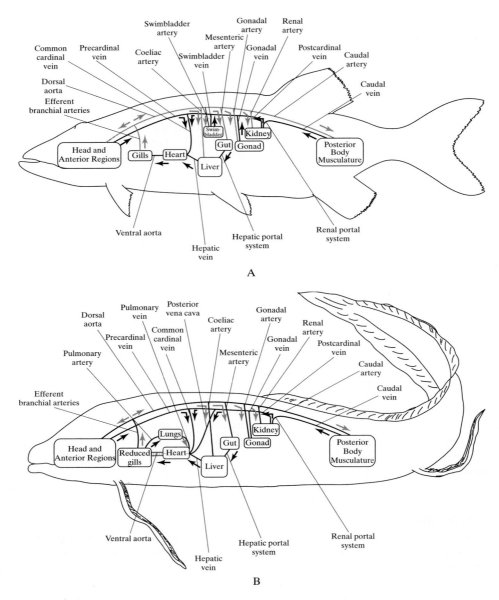

FIGURE 4.9 Diagramatic representation of the major vascular pathways (arterial, gray; venous, black) in (**A**) a teleostean fish and (**B**) an African lungfish. Arrows represent the direction of blood flow, but distances and lengths of vessels and location of organs are not to scale.

peritoneal cavity by the ***pericardial wall***, which is more rigid in elasmobranchs than in teleosts (Satchell 1971). The mean pericardial fluid volume in the horn shark *(Heterodontus francisci)* was found to be 2 ml/kg body weight, although pericardial fluid was ejected via the pericardial wall's ***pericardioperitoneal canal*** into the peritoneal

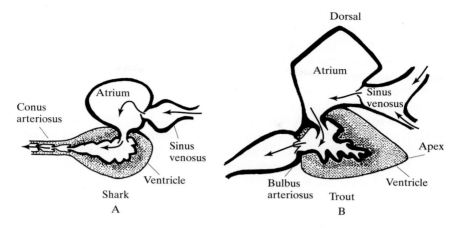

FIGURE 4.10   Diagrams of the heart in (**A**) a shark (elasmobranch) and (**B**) a trout (teleost). Modified from Randall (1968).

cavity when the shark was handled or when it began vigorous swimming (Abel et al. 1986). This fluid ejection lowers intrapericardial pressures to negative values, presumably facilitating diastolic filling of the heart.

The first chamber of the fish heart is the ***sinus venosus***, which functions as a manifold; that is, venous blood from the liver and the ducts of Cuvier are collected in this relatively thin-walled chamber. From there, the blood is directed to the ***atrium***, in many species through a ***sinoatrial valve*** (Fig. 4.10). Whereas the sinus venosus provides the initial transition from smooth to pulsed flow, the atrium provides the first significant circulatory acceleration of the blood. Compared with the sinus venosus, the atrium is a relatively large chamber that empties ventrally to the ***ventricle***, via the ***atrioventricular valve***, in a biphasic manner. ***Biphasic filling*** implies that blood starts to fill the ventricle during ***ventricular diastole*** (early phase, which constitutes 42–73% of the total flow among three example teleosts) and continues to fill immediately after ***atrial systole*** (atrial phase) (Lai et al. 1998).

The ***ventricle*** also is a relatively large chamber featuring heavy walls of cardiac muscle (Fig. 4.10). It is pyramid-shaped in elasmobranchs and conical in teleosts, with the apex pointing posteriorly. The heavy muscle and efficient geometry of the ventricle provide the main propulsive force for circulatory flow. Ventricular walls may be composed of two layers of muscle: the cortex, and the spongy myocardium. The ***cortex*** is relatively dense cardiac muscle ***(compact myocardium)*** that generally receives oxygen and nutrients from the ***coronary artery***. The cortex is well developed in active species, such as skipjack tuna *(Euthynnus pelamis)* and rainbow trout (Cameron 1975). Santer and Greer Walker (1980), however, found that 73 of 93 teleostean species they examined had no ventricular cortex. These were generally more sluggish species compared with the 20 that had ventricular cortexes. All 14 sharks and rays examined had a cortex. As rainbow trout increased swimming velocity from 0.15 to 1.0 body length/s in normoxic water, ***coronary blood flow*** increases (≈110%) paralleled those of cardiac power. Under hypoxia, resting coronary blood flow was 35% higher than that under normoxia, and it increased further (to ≈150% of resting levels) with a velocity at 1.0 body length/s swimming (Gamperl et al. 1995). The *spongy myocardium* (the more universal inner layer of the ventricular myocardium), consists of a spongy mesh that is supplied with oxygen and nutrients only by the venous blood that it pumps. Indeed, cardiac performance in fishes without a compact myocardium may be limited by the low venous oxygen supply under hypoxic conditions (Farrell 1991).

In contrast to the atrium or ventricle, the fourth chamber (*conus arteriosus* in elasmobranchs, lampreys, hagfish, and holosteans, and *bulbus arteriosus* in teleosts) does not increase the acceleration of the blood. It functions as an elastic chamber to dampen the pulses of pressure and intermittent flow from the ventricle into a more continuous flow to the ventral aorta and the gills. The bulbus wall consists only of elastic tissue and layers of smooth muscle; it features no valves (Priede 1976). Conversely, the conus can have many valves (up to 72 in gars [Lepisosteus]!) as well as cardiac musculature in the walls. The more rigid membrane (pericardium) that houses the heart in elasmobranchs is thought to produce a more active "rebound" between contractions of the heart, which assists in its filling from the veins. The conal valves ensure that significant reverse flow of blood back into the heart does not occur during this rebound. The conus is more poorly developed in lampreys and hagfish, which have a single pair of valves. Despite the presence of inline hearts (Fig. 4.11), hagfish have the lowest "vertebrate" blood pressures (e.g., 8 mm Hg in the ventral aorta) (Forster et al. 1991).

Johansen and Hanson (1968) have summarized the differences in heart structure in lungfish and amphibians associated with the evolutionary transitions from aquatic to atmospheric breathing. The most dramatic difference is the partial division in the lungfish heart that is associated with the separate return of blood (via the *pulmonary vein*) on the left side of the heart from the lungs. This separate return emanates from a special pulmonary vascular circuit, which is coupled in parallel with the normal systemic and branchial circulation. The *partially divided* ventricle moves the arterial (from the lungs) and the venous (from the body) blood through a *bulbus cordis*, which largely maintains separation of the two flows by *spiral ridges* that twist throughout its length. The arterial blood is conveyed to the body, whereas the venous blood passes through the functional gills, some of it subsequently entering the lung. The extent of the separation of flows in the heart correlates with dependence on air breathing. The Australian lungfish, which cannot withstand lengthy air exposure, displays the least separation. African and South American lungfish, however, which must withstand periodic droughts in their natural habitats, show more complete separation of flows.

4.7   MYOCARDIAL ELECTRICAL ACTIVITY

As is typical for a vertebrate, most fish hearts are *myogenic* (i.e., no nervous signal from the brain is necessary for each heartbeat) and show a complex *electromyogenic waveform*. Although the actual sites of the primary *pacemaker* nodes remain obscure, evidence gathered for some teleosts suggests islets of pacemaker cells in the sinus venosus and atrium (Satchell 1971). Kisch (1948) demonstrated that many areas of the myocardium can show pacemaker activity.

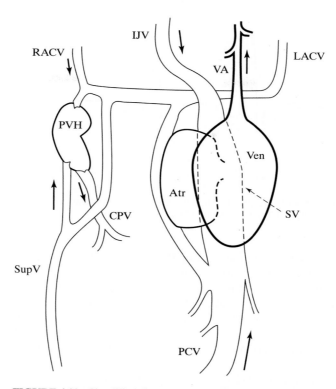

FIGURE 4.11 Simplified diagram showing the arrangement of the heart, portal vein heart, and venous system in the hagfish *(Myxine glutinosa)*. Atr, atrium; CPV, common portal vein; IJV, inferior jugular vein; LACV, left anterior cardinal vein; PCV, posterior cardinal vein; PVH portal vein heart; RACV, right anterior cardinal vein; SupV, supraintestinal vein; SV, sinus venosus; VA ventral aorta; Ven, ventricle. From Forster et al. (1991).

Fish *electrocardiograms* showing the progression of electrical phenomena through a cardiac cycle are obtained by implantation of bipolar electrodes under the skin, commonly spanning the pericardial area. The electrodes are wired to an appropriate amplification and display system, consisting of an oscilloscope, physiological recorder, or computer. Figure 4.12 shows the electrocardiogram of an elasmobranch, the Port Jackson shark. The chronological sequence of chamber depolarizations (contractions) shows the synchronous movement of blood through the heart. The T wave usually is the only electrocardiographic wave visibly indicating chamber muscle repolarization (ventricle).

4.8   CARDIAC FLOW

The ventricular pumping rhythm induces the *systolic* (contraction)/*diastolic* (relaxation) rhythmic flow in the *ventral aorta*. The bulbar pressure pulse-dampening effect,

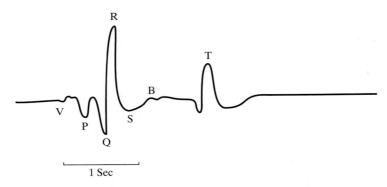

FIGURE 4.12   Electrocardiogram of the Port Jackson shark *(Heterodontus portusjacksoni).* From Satchell (1971).

as measured in lingcod *(Ophiodon elongatus)*, is seen in Figure 4.13. The systolic/diastolic waveform of pressure also is detectable in the dorsal aorta, even though the branchial vascular network of the gills drops the blood pressure by one-quarter to one-third. For example, in winter flounder, bulbus arteriosus mean pressure is 29 mm Hg, and dorsal aorta mean pressure is 22 mm Hg. This difference (7 mm Hg) represents a 24% **pressure drop** through the gills (D'Amico Martell and Cech 1978). The various hearts in hagfish work independently of each other to perfuse various parts of the body with blood (Randall 1970).

Teleost hearts are generally two- to five-fold more powerful as pumps than those in elasmobranchs—and 10- to 25-fold more powerful than the hagfish's branchial heart (Sidell and Driedzic 1985). The hagfish heart is fueled by **carbohydrates**, the elasmobranch heart by **carbohydrates** and **ketone bodies**, and the teleost heart by **carbohydrates**

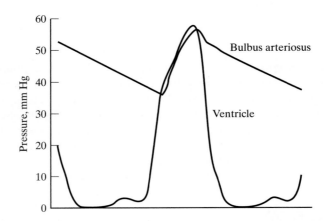

FIGURE 4.13   Rectilinear records of blood pressure during one heartbeat of a lingcod (2.0 kg, 15°C). Modified from Stevens et al. (1972).

and *fatty acids* (Sidell and Driedzic 1985). The hagfish caudal heart consists of a *cartilaginous plate* and attached skeletal musculature along with a pair of *lateral sacs with valves* to regulate the direction of blood flow. Movement of the muscles forces blood anteriorly.

Body undulations during swimming also have been implicated in the blood flow of higher fishes. Satchell (1965) recorded increased blood flows in the caudal vein of a Port Jackson shark trunk preparation when electrical stimulation produced swimming-type movements. He also found valves in the caudal vein that allowed only anterior movement of blood. Sutterlin (1969) described similar valves in the ventral segmental veins of brown trout. Isolated heart studies by Bennion (1968) revealed that increased filling volume (*venous return*) produces an increased force of contraction (*inotropism*) by the heart. Mediated by stretch receptors in the myocardial wall, this response (*Starling's law of the heart*) acts to increase the volume of blood pumped per contraction (*cardiac stroke volume*, or Q_{sv}). The Q_{sv} and frequency (*heart rate*, or *HR*) determine the circulatory flow rate (*cardiac output*, or *Q*).

Cardiac output and blood pressure are influenced by several factors. Increases in temperature typically increase *Q* by increasing HR (*tachycardia*; under aneural cardiovascular control). Hypoxia (at a threshold P_{O_2} of 40–70 mm Hg) typically decreases HR (*bradycardia*), and some species show a compensatory increase in Q_{sv}. The decreased HR and the increased Q_{sv} in some species (e.g., rainbow trout and spotted dogfish [*Scyliorhinus canicula*]) result in no significant change in *Q*. Other species, however, show a smaller Q_{sv} increase (e.g. lingcod), decreasing *Q*, and tuna are unusually hypoxia-sensitive, decreasing *Q* via bradycardia at a P_{O_2} of 85 to 104 mm Hg with no change in Q_{sv} (Bushnell 1988, Brill and Bushnell 1991, Farrell 1993). Although increases in *Q* would increase blood pressure if all other factors stayed constant, many other factors control blood pressure in fishes. Total flow through a vessel (or pipe) is proportional to the *fourth* power of its diameter (Hagen-Poiseuille relationship) (Vogel 1983). Consequently, small decreases in vessel diameter (e.g., from *arterial*

BOX 4.1

What Are the Cardiac Performance Limits of Fishes?

Maximum aerobic swimming effort demands maximum performance from the cardiovascular (oxygen delivery) system, especially from the cardiac pump. Fish that are reasonably "athletic" can virtually double their cardiac output (*Q*) when swimming maximally. Rainbow trout increase their *Q* to a maximum (48.7 ml blood/min/kg body wt) at 97.3% of the maximum sustainable velocity when forced to swim in a variable-speed swim tunnel (see Chapter 2). This rate is greater than

their resting *Q*, stemming from a 38% increase in heart rate (HR) and a 25% increase in cardiac stroke volume (Q_{sv}) (Thorarensen et al. 1996). In contrast, smallmouth bass (*Micropterus dolomieu*) increased their *Q* by 100%, via a 100% increase in HR alone (no change in Q_{sv}), when stimulating to swim maximally in a similar tunnel, thereby simulating a 2- to 3-min angling event (Schreer et al. 2001).

vasoconstriction—the contraction of smooth, constrictor muscles in the arterial wall) can dramatically increase the resistance to flow and decrease the flow through them if the pressure drop remains constant (Ohm's law). Factors such as ***angiotensin II*** (of the ***renin-angiotensin system***) are potent vasoconstrictors, whereas ***atrial natriuretic peptides*** (ANPs) are potent ***vasodilators*** and can inhibit release of angiotensin II (Evans 1990). As the name implies, ANPs typically originate in the atrium of the heart. Cousins and Farrell (1996) showed that increased cardiac filling pressure (from increased venous return) produced increases in Q (Starling's law of the heart) and in ANP secretion from the atria of the isolated rainbow trout hearts. Evans et al. (1993) demonstrated that one ANP type is a potent vasodilator of vascular smooth muscle of spiny dogfish (*Squalus acanthias*). Recent studies of rainbow trout (Hoagland et al. 2000) and the spiny dogfish (Evans 2001) show that many endogenous substances (e.g., acetylcholine, endothelin, nitric oxide, and prostaglandins), in addition to natriuretic peptides, have vasoactive effects in fishes.

The degree of cardiovascular control that is possible in fish is demonstrated by the spiny dogfish. Surgical closure of various gill slits prevents water ventilation of isolated gill arches. By selective vasodilation and constriction, blood is largely directed away from the nonventilated arches and toward those receiving ventilatory irrigation (Cameron et al. 1971). Thus, the dogfish successfully maintains an efficient ratio of water flow (ventilation) to blood flow (cardiac output). ***Ventilation:perfusion ratios*** of approximately 10:1 to 20:1 have presumably evolved in gill breathers, because fish blood oxygen capacities approximate 10- to 20-fold the oxygen capacity of water. The matched flows ensure efficient oxygen diffusion. Absolute matching (stroke to stroke) of the ventilatory water pump and the cardiac pump is relatively rare in fish, with few exceptions (e.g., rainbow trout in hypoxic water).

4.9   CARDIOVASCULAR CONTROL

Fish control their hearts and cardiovascular systems with several aneural and neural mechanisms. Venous return is one of the controls that can be produced by a variety of factors.

Aneural Cardiovascular Control

Aneural cardiovascular control is effected by changes in blood volume, by direct responses of heart muscle to temperature changes, and by secretions from various organs. Along with swimming movements in certain species, changes in blood volume also affect venous return. Mobilization of blood into the general circulation has been associated with the spleen, liver, or a blood sinus in various species. For example, Yamamoto (1988) described the contraction of the spleen (decrease in weight and hemoglobin concentration by 47–85% and 56–93%, respectively) in five species of freshwater teleosts after 3 to 5 min of severe exercise. As explained earlier (see Section 4.1), an autonomic nervous signal triggers this contraction. The extra erythrocytes would increase blood oxygen and buffering (see Chapter 6) capacities, thereby increasing the fish's ability to be active. Temperature acts as another aneural regulator of circulation by ***direct*** action on the pacemakers in the myocardium (Randall 1970).

Table 4.3 shows the positive **chronotropic effect** (increased HR) of temperature on winter flounder as the water temperature increases from 10°C to 15°C. These HR increases cause an elevated Q at these temperatures, even though the Q_{sv} does not change. This increase in blood flow provides an increased delivery of oxygen throughout the body, which is operating at a higher metabolic rate at the warmer temperatures (Cech et al. 1976). A temperature effect on the heart is apparent from the very beginning. For example, temperature affected the pacemakers of the minnow's (*Phoxinus phoxinus*) embryonic heart at 2- to 4-days postspawning, the initiation of cardiac activity (Schönweger et al. 2000).

Secretions of hormones that affect either the heart or the relative constriction or dilation of the blood vessels represent an important category of aneural cardiovascular regulators in fishes. Catecholamines, such as epinephrine, affect both the heart and the resistance to flow in various vascular beds. Nakano and Tomlinson (1967) have shown that levels of circulating catecholamines (**epinephrine** and **norepinephrine**) rise with exercise in rainbow trout. Epinephrine stimulates increases in HR at relatively low temperatures (e.g., 6°C), and Q_{sv} increases in warmer water (e.g., 15°C) (Bennion 1968). Studies using blocking agents (e.g., **inderal**) have shown that the heart has receptor sites that primarily are epinephrine sensitive. The interplay of temperature, myocardial stretching, and epinephrine as controlling mechanisms determines the frequency/stroke volume characteristics under the various conditions. Other studies have shown that primarily norepinephrine-sensitive receptors predominate in constricting systemic vascular beds, whereas epinephrine-sensitive receptors dilate gill vasculature in rainbow trout (Wood and Shelton 1975). Removal and metabolism of circulating catecholamines apparently occurs in the gills (Colletti and Olson 1988).

Neural Cardiovascular Control

The hearts of all fish except hagfish are innervated by a branch of the 10th cranial nerve (**vagus**) (Randall 1970). Stimulation of the lamprey vagus produces an increased HR,

TABLE 4.3 **VALUES OF CARDIOVASCULAR VARIABLES OF WINTER FLOUNDER AT SPRING AND AUTUMN TEMPERATURES**

| Variable[a] | Units | Mean ± standard error (with number of fish) | | Level of significant difference[b] |
		10°C	15°C	
HR	bpm	35 ± 1 (18)	62 ± 2 (18)	<0.01
Vo₂	ml/min	0.25 ± 0.01 (20)	0.45 ± 0.02 (20)	<0.01
Q	ml/min	14.7 ± 1.0 (16)	24.6 ± 3.1 (16)	<0.01
Q_{SV}	ml/beat	0.43 ± 0.05 (15)	0.41 ± 0.05 (15)	NS[c]
Body weight	g	635 ± 40 (20)	681 ± 42 (20)	NS

Modified from Cech et al. (1976).

[a]HR, heart rate; V_{O_2}, oxygen consumption rate; Q, cardiac output; Q_{SV}, cardiac stroke volume

[b]As determined by the Student's *t* test (Snedecor and Cochran 1967)

[c]NS, not significant

whereas elasmobranch or teleostean vagal stimulation slows the HR. Because these stimuli mimic the effects of acetylcholine, these fibers are termed *cholinergic*. Several factors may alter the level of *vagal tone* (level of fiber excitement), although stimuli threshold levels vary considerably between species. Light flashes, sudden movements of objects or shadows, touch, or mechanical vibrations usually promote a bradycardia in teleosts and elasmobranchs by increasing the level of vagal tone. *Atropine* injections sufficient to block cholinergic innervation (the vagal nerve supply to the heart) attest to the neural origin of these responses. Stevens and Randall (1967) found no vagal tone in resting rainbow trout, but they did find that atropine injections into a large-scale sucker (*Catostomus macrocheilus*) increased the resting HR from 38 to 55 bpm. Swimming in this sucker also reduced the inhibitory vagal tone, thereby elevating the HR. Several teleosts also possess *(adrenergic) stimulatory* fibers from the vagus as an additional neural control mechanism (Gannon and Burnstock 1969, Holmgren 1977, Donald and Campbell 1982). Changes in coronary blood flow with exercise also appear to be under adrenergic (not cholinergic) control in rainbow trout (Gamperl et al. 1995).

Armed with such a variety of aneural and neural control mechanisms, fish have much built-in flexibility for circulatory adjustments to environmental or other changes (Farrell 1984, 1991, 1993). For example, environmental hypoxia invokes a bradycardia and an increased Q_{sv} (see above), elevated peripheral resistance of both branchial and systemic blood vessels, and enhanced gas-exchange efficiency in many teleosts and elasmobranchs. Changes in gas-exchange efficiency may be linked to changes in blood distribution in the gills (e.g., lamellar recruitment; see Chapter 3). The reflex bradycardia has been described for the California grunion (*Leuresthes tenuis*) and the California flyingfish (*Cypselurus californicus*) during exposure to air (Garey 1962). Both of these species have "standard" gills and are exposed to the atmosphere for short time periods (e.g., only 42 s for the flying fish and up to a few minutes for the spawning grunion; see Chapter 9). In contrast, the Indian climbing perch (*Anabas testudineus*), which has labyrinthine organs specialized for air breathing and spends much of its time out of the water, shows an initial tachycardia (increased HR) just after taking an air breath. Hypoxic exposure increases catecholamine (including epinephrine) concentrations in spotted dogfish (Butler et al. 1978), and epinephrine protects myocardial contractility in rainbow trout during anoxia (Gesser et al. 1982).

These circulatory patterns may be linked to maximization of oxygen uptake and delivery and, possibly, conservation of cardiac energy in the case of the bradycardic responses. Of course, to best understand the physiological response pattern of a fish to hypoxia (or any other change), the ventilatory, hematological, circulatory, and other affected systems must be considered as well. Apparently, the receptors that detect low oxygen levels and that provoke the bradycardia and increases in gill ventilation (see Chapter 3) are located on the dorsal part of the first gill arches in rainbow trout (Daxboeck and Holeton 1978, Smith and Jones 1978). Singh (1976) has pointed out the importance of carbon dioxide and pH (along with oxygen) receptors in the air-breathing climbing perches and catfishes.

4.10   SECONDARY CIRCULATION

Neither the anatomy nor the physiological functions of the *secondary ("venolymphatic") circulation* of fishes is clearly defined (Olson 1998). Secondary circulation vessels and anastomoses carry few blood cells and may serve osmoregulatory (see Chapter 6) and nutritive functions (Olson 1996). These low-pressure vessels have been reported in teleostean gills (see Chapter 3), skin, fins, linings of the peritoneal and oral cavities, and the heat-exchanging rete of tunas (see Chapter 5). They have not been identified in most skeletal muscle, splanchnic (visceral) circulation, brain, or kidney tissues (Olson 1998). By using plasma protein (total circulation fluid) and erythrocytic (primary circulation) markers, the rainbow trout's secondary circulatory volume (4.9% of body wt) has been estimated to be approximately 1.5-fold that of the primary circulation (3.4% of body wt) (Steffensen and Lomholt 1992). Finally, the rate of fluid movement between the primary and the secondary circulatory systems is apparently slow, with equilibration of the plasma protein marker between the two "compartments" taking 10 to 12 hours (Steffensen and Lomholt 1992).

4.11   LESSONS FROM BLOOD AND CIRCULATION

Fish circulatory systems transport blood to every part of the body to deliver nutrients, gases, and hormones and to remove metabolic wastes and pathogens. Interactions of hemoglobins with carbon dioxide, H^+, temperature, and organic phosphates influence both oxygen uptake at the gills and delivery to metabolizing tissues. The body's changing demands for oxygen and food-derived nutrients stimulate changes in cardiac output.

SUPPLEMENTAL READINGS

Ellis 1977; Fänge 1984; Farrell 1984, 1991, 1993; Hoar et al. 1992a, 1992b; Holeton 1970; Iwama and Nakanishi 1996; Jensen 1991; Jensen et al. 1993; Laurent et al. 1983; Olson 1996, 1998; Randall 1970; Riggs 1970; Satchell 1971, 1991; Weber 1990; Weber and Lykkeboe 1978.

WEB CONNECTIONS

CIRCULATORY SYSTEM AND BLOOD
www.bishops.ntc.nf.ca/wells/vfish/projects/circ.index.html

BLOOD CONSTITUENTS, COLLECTION, AND HEMATOLOGICAL TECHNIQUES
www.aqualex.org/html/onedin/fish_haematology/

C H A P T E R 5
Buoyancy and Thermal Regulation

The gas-secreting structure of the **swimbladders** found in derived teleosts and the structure of the **heat-exchange** organs found in some large, active, oceanic fish are morphologically very similar. Both involve exchanges, of **gas** or **heat**, across blood-vessel walls in a **countercurrent** (opposite direction) exchange network. This common factor brings these functionally disparate areas together in this chapter.

5.1 BUOYANCY

Neutral buoyancy (weightlessness) allows fish to minimize the energy cost of staying at a particular depth to feed, hide, reproduce, or migrate. Because an active fish can exert a **propulsive** force greater than 25% to 50% of its body weight only for brief periods, continuous effort to support its body by muscular power alone would be energetically costly (Marshall 1966). It is not surprising, therefore, that various ways of achieving neutral buoyancy have evolved among fishes. Essentially, four strategies[1] are recognizable: (1) incorporation of large quantities of low-density compounds in the body; (2) generation of lift by appropriately shaped and angled fins and body surfaces during forward movement; (3) reduction of heavy tissues, such as bone and muscle; and (4) incorporation of a swimbladder as a low-density, gas-filled space.

The use of low-density compounds to reduce the overall density of the body is characteristic of most sharks and of a few teleosts. In many sharks, large quantities of **lipids** (specific gravity, 0.90–0.92) and of the hydrocarbon **squalene** (specific gravity, 0.86), found especially in their large livers, bring the total body mass toward neutral buoyancy in sea water (specific gravity, 1.026). Furthermore, the characteristic **heterocercal** tail of sharks combined with the **positive angles of attack** of the leading edges of the pectoral fins and the surface of the head provide additional lift while swimming. Hydrodynamic drag is minimized in the more pelagic sharks, which have relatively smaller fins and larger, fatty livers.

Among teleosts, only a few marine species, such as sablefish (Anoplopomatidae), pelagic medusa fishes (Stromateidae), and shallow-water rockfish (Scorpaenidae), use low-density (triglyceride) oils to lessen negative buoyancy (Lee et al. 1975). These oils

---

[1]The term **strategy**, in this sense, implies a direction dictated by the principles of natural selection. It does *not* imply a conscious mode of action by the fish either at birth or during its life.

are found mainly in the bones. At least one deepsea fish, *Acanthonus armatus* (Ophidiidae), has an enlarged cranial cavity ($\approx$10% of the head volume), which is mostly filled with watery fluid (Horn et al. 1978). This fluid has a total osmotic concentration that is almost half of those found in the plasma or perivisceral fluid and approximately one-quarter of that of the seawater environment. The cranial position of the "light-fluid" reservoir nicely balances most of the heavy body components (otoliths, gill rakers, pharyngeal teeth, and cranial spines) that are also located there.

Fish of deep ($>$1,000 m) oceanic midwaters characteristically have **reduced skeletal and muscular tissues**. Food is scarce in these environments, and tissues that are energetically expensive to maintain—such as swimbladders—and compounds that are expensive to synthesize—such as body lipids—usually are reduced or absent. It also is likely that the cartilaginous skeleton (specific gravity, 1.1) of elasmobranchs and some bony fish is partly an adaptation to decrease body density. Bone has a specific gravity of 2.0.

The major problem with the preceding methods of regulating density is that they either greatly restrict the activity of fish (reduced tissues) or make it difficult for a fish to regulate its density in response to changes in pressure, temperature, and salinity. The swimbladder is an "invention" of bony fish that overcomes these problems and no doubt has been largely responsible for their success.

Swimbladders allow precise control of buoyancy, because the volume of gas that they contain can be regulated with comparative ease. Because of the increased density of sea water, fish are more buoyant in it than in fresh water. Hence, with typical teleostean skeletal and body composition, swimbladders occupy approximately 5% of the marine teleost's body volume and approximately 7% of the freshwater forms.

Swimbladders are of two basic types; **physostomous,** and **physoclistous**. Physostomous swimbladders have a connection (**pneumatic duct**) between the swimbladder and the gut (Fig. 5.1). Physoclistous swimbladders lack this connection. Fish with physostomous swimbladders include many of the more ancestral, soft-rayed teleosts, including herrings, salmonids, osteoglossids, mormyrids, pikes, cyprinids, characins, catfishes, and eels. Physostomous fish inflate their swimbladders by gulping air at the water's surface (Box 5.1) and then forcing it through the pneumatic duct and into the swimbladder by a buccal force mechanism (Fänge 1976). It is, therefore, not surprising that physostomes largely are shallow-water forms. Additional swimbladder inflation is needed for neutral buoyancy at greater depths. Swimbladder volume at the surface, 1 atmosphere gas pressure, is proportionately decreased with each 10 m ($=$ 1 atmosphere hydrostatic pressure) that are descended. Thus, a fixed swimbladder volume at the surface would be reduced to half at 10 m (2 atmospheres total pressure), to one-third at 20 m, to one-quarter at 30 m, and so on. Therefore, the amounts of gas needed to achieve neutral buoyancy at depth would be so great that it would be impossible for the fish to submerge!

Noninflation of the swimbladder during the available time period (e.g., 5–8 days after hatching in striped bass; Bailey and Doroshov 1995) leads to poor development (including bent vertebral columns) and early mortality (Bennett et al. 1987)—at significant cost to fish culturists. Goolish and Okutake (1999) showed that zebrafish (*Danio rerio*) can initially inflate their physostomous swimbladder by gulping air at the surface (air–water interface) or by ingesting tiny bubbles within their environment. Similarly, Doroshev and Cornacchia

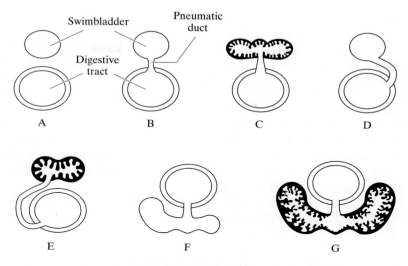

FIGURE 5.1   Cross-sections of swimbladders or lungs (as viewed from the anterior end) showing relationships to the digestive tract in various fishes. Many of these fishes (**C–G**) can use the swimbladder or lung as an air-breathing organ (see Chapter 3). **A.** Typical (water-breathing) phystoclistous species. **B.** Typical (water-breathing) phystostomous species. **C.** Gars (*Lepisosteus* spp.) and bowfin (*Amia* sp.). **D.** Climbing perches (*Erythrinus* sp.). **E.** Australian lungfish (*Neoceratodus* sp.). **F.** Reedfish (*Erpetoichthys* sp.) and bichirs (*Polypterus* spp.). **G.** African and South American lungfish (*Protopterus* spp. and *Lepidosiren* sp., respectively). Adapted from Dean (1895) and Graham (1997).

(1979) recommended strong aeration and turbulent water in the rearing container to maximize striped bass' success at initial swimbladder inflation. Deflation of the physostomous

---

BOX 5.1

How Do Fish Inflate Their Swimbladder for the First Time?

Most larvae of bony fishes develop a functional swimbladder either before or during initiation of external feeding. Hydrostatic regulation is essential for efficient swimming, prey capture, and maintaining the position of pelagic larvae in the water column (Doroshev and Cornacchia 1979). Different fish use two distinct methods to initially inflate the swimbladder: secretion of gas from gas gland–type tissue, as in Mozambique tilapia (*Tilapia mossambica*); and gulping of atmospheric gas, as in

striped bass (*Morone saxatilis*) (Doroshev and Cornacchia 1979). Although both of these fish function as physoclistous species after metamorphosis, the striped bass (and many other physoclists) maintains a rudimentary pneumatic duct for initial swimbladder inflation as larvae. Furthermore, successful swimbladder inflation in striped bass larvae is correlated with high levels of thyroid hormone, with the hormone presumably being supplied from the mother (Brown et al. 1988).

swimbladder is accomplished by a reflex action, the ***gass-puckreflex*** (gas-spitting reflex), which is initiated when reduced external pressure makes the fish too light, thereby releasing gas via the pneumatic duct into the esophagus. ***Pneumatic sphincter*** muscles (both smooth and striated muscle tissue), which are under nervous control, guard the entrance to the pneumatic duct. Fänge (1976) attributes the release of gas to a relaxation of the sphincter muscles, a contraction of smooth muscles in the swimbladder wall (under similar nervous control), the elasticity of the swimbladder wall, and the contractions of body wall muscles. The importance of gas diffusion mechanisms (i.e., with the blood) for physostome swimbladder inflation or deflation probably is minimal. Although the number of species that have been investigated is comparatively low, only a few species (e.g., eels [*Anguilla sp., Conger sp.*] and whitefish [*Coregonus sp.*]) show the richly vascularized wall structures specially adapted for gas exchange (Fänge 1976).

In contrast to physostomous fish, fish with "closed" (physoclistous) swimbladders have special structures associated with the circulatory system for inflating or deflating the swimbladder. Presumably, because these structures "free" fish from dependency on the surface, more than two-thirds of all teleosts (especially the more derived, spiny-rayed species) are physoclistous. A ***rete mirabile*** ("wonderful net") and associated ***gas gland*** are the source of inflation gas in these swimbladders. Afferent blood flows through the rete capillaries (Fig. 5.2) into the gas gland (Fig. 5.3). Cell-culture studies of

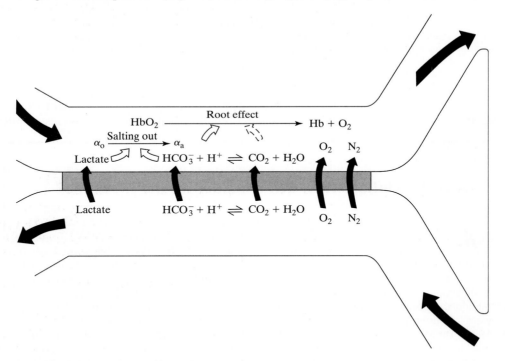

FIGURE 5.2 Movements of gases and solutes in the rete mirabile (smaller solid arrows) and the resulting countercurrent concentration of gases and solutes. Open arrows indicate influences on inert gas solubility and hemoglobin(Hb)–O_2 binding characteristics. Larger solid arrows indicate circulation of blood. From Pelster and Scheid (1992).

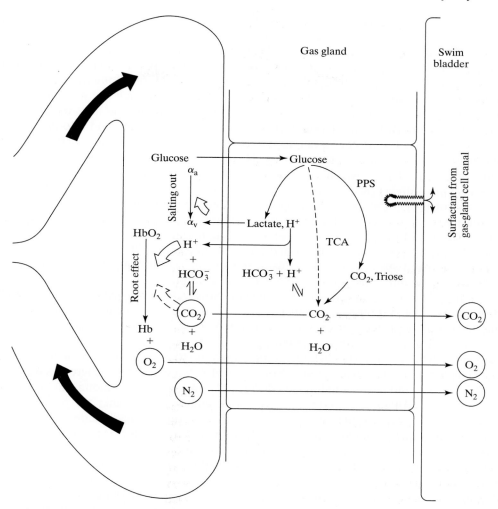

FIGURE 5.3 Gas-gland cell metabolism and its influence on the physical solubility of inert gases or the release of gas from chemical binding sites in the blood (open arrows). Solid arrows indicate movement of substances. *Hb*, Hemoglobin; *PPS*, pentose phosphate; *TCA*, tricarboxylic acid cycle shunt. From Pelster and Scheid (1992) and Prem et al. (2000).

swimbladder gas glands show that lactic acid is produced and secreted at these cells' basolateral membranes, where it is closely associated with the blood circulating through the gland (Prem and Pelster 2001). Interestingly, a surfactant is produced at these cells' apical (luminal) membranes and is secreted, via canals, into the swimbladder lumen (Prem et al. 2000) (Fig. 5.3). By biochemical additions of *acid* (hydrogen ions, via Na^+-H^+ and other exchangers; Sötz et al. 2002) and *solutes* (e.g., lactate ions) to the blood entering the gas gland, dissolved gas partial pressures increase and, thereby, inflate the swimbladder via *gas diffusion*. The gas-gland tissue produces acid via the glycolytic transformation of *glucose* (Pelster and Niederstätter 1997), and it produces CO_2

via dehydration of HCO_3^- and the transformation of glucose via the tricarboxylic acid cycle and the pentose phosphate shunt (Fig. 5.3). The H^+ apparently moves across the cell membranes into the afferent blood, acidifying it. Some of the CO_2 also moves into the afferent blood, where it hydrates, forming more H^+ and HCO_3^-. Some of the H^+ and CO_2 bind to the hemoglobin in the red blood cells, driving O_2 off of the hemoglobin via *rapid* (50 ms) ***Bohr*** and ***Root ("Root off")*** effects and increasing the plasma oxygen partial pressure (PO_2). The resulting PO_2 gradient between the plasma and the ***swimbladder lumen*** moves O_2, via diffusion, through the gas gland and into the lumen (Fig. 5.3). Some CO_2 also diffuses into the lumen down the PCO_2 gradient (Pelster and Scheid 1992). In addition, inert gases (e.g., N_2) diffuse into the lumen via a reduced gas solubility in the blood, which is induced by the addition of solutes, such as lactate (***"salting out"*** effect) (Kuhn et al. 1963). The rete consists of a tight ***bundle*** (rather than net) of thousands of afferent (running toward) and efferent (running from) capillaries surrounding each other, with only 1 μm separating afferent and efferent blood (Steen 1970). This structure provides for an efficient countercurrent exchange of blood gases and HCO_3^-, H^+, and lactate ions from efferent to afferent capillaries, as shown for one capillary pair in Figure 5.2. These diffusive movements result from elevated gas partial pressures and solute concentrations in the efferent blood.

High efferent PO_2 is facilitated by the comparatively *prolonged* (10 s) ***"Root-on"*** shift (return of full hemoglobin capacity for O_2), thereby minimizing hemoglobin–O_2 binding in the rete (Steen 1970). The resulting buildup of gas and solutes in afferent capillaries "multiplies" (***countercurrent multiplier effect***) the gas-secretory functions in the gas gland such that exceedingly high gas pressures (up to 300 atm in some deepsea fish) can be generated in the swimbladder. Kobayashi (1990) measured PO_2 increases from 40 to 281 mm Hg and corresponding pH decreases from 7.82 to 7.33 along the length of European eel (*Anguilla anguilla*) afferent rete capillaries. The higher the partial pressures and the lower the pH, the greater the effect of further acidification as afferent blood reaches the gas gland. Kobayashi (1990) found that PO_2 decreases from 293 to 44 mm Hg and that pH increases from 7.10 to 7.64 along the eel's efferent capillaries. These eel data demonstrate both the PO_2 and H^+ (efferent to afferent) gradients and the high efficiency of the countercurrent arrangement. The *longer* the retial capillaries, the *more complete* the gas exchange and the *more efficient* the filling of the swimbladder at depth. Investigations of deepsea fish show a good correlation between retial capillary length and habitat depth. Fish found between 1,500 and 3,500 m may have retial capillaries 25-fold longer than those of fish found between 150 and 500 m (Marshall 1971). Multiple layers of ***guanine*** crystals dramatically reduce the O_2 conductance of the swimbladder wall (Lapennas and Schmidt-Nielsen 1977).

Deflation of the physoclistous swimbladder is accomplished by diffusion of gas back into the bloodstream via a richly vascularized area adjacent to the enclosed gases. The more soluble gases, including CO_2 and O_2 in blood, are preferentially reabsorbed, thereby leaving a higher percentage of less-soluble gases (e.g., N_2) in the lumen. This area may be an ***"oval"*** patch of densely packed capillaries on the dorsal wall of the swimbladder, or it may consist of the posterior lining of the swimbladder wall, the functional area of which is regulated by anterior/posterior movements of a mucosal ***"diaphragm."***

5.2   THERMAL REGULATION

Most fish, as ectotherms, have body temperatures that are close to that of their environments because of metabolic heat losses via the skin or gills. The low rate of metabolic heat production (compared to mammals) and the high heat capacity of water result in continuous heat losses and cool bodies for most fish. The sea raven (*Hemitripterus americanus*) remains at a temperature very close to its environment and probably loses most of its heat conductively through its skin (Stevens and Sutterlin 1976). The albacore tuna (*Thunnus alalunga*), in contrast, is one of several fish that can physiologically maintain warmer parts of its body by the use of heat-exchanging retia and primarily loses heat convectively through its gills (Graham 1983). Field and laboratory studies have shown that many fishes prefer water of particular temperatures (see the review by Reynolds and Casterlin 1979). Thus, examples of both behavioral and physiological temperature-regulating mechanisms are found in fishes.

Behavioral thermoregulation concerns the movements of fish from one water mass or area to another that is characterized by a warmer or a cooler temperature. Because temperature affects the rates of metabolism and digestion so profoundly (see Chapters 3 and 7), some fish may "select" a particular temperature to conserve energy or to run their metabolic machinery (i.e., enzymes) at its most efficient temperature. For example, Neverman and Wurtsbaugh (1994) found that juvenile Bear Lake sculpin (*Cottus extensus*) migrate after feeding on the bottom of Bear Lake (5°C) to the warmer surface layers (13–16°C) to accelerate their digestion (from 3%/h to 22%/h) at night. This pattern allows the juvenile fish to empty their guts for feeding during the following day, thereby increasing their growth by 300% compared with those reared at 5°C. In contrast, Brett (1971) found that sockeye salmon (*Oncorhynchus nerka*) select a warm temperature (15°C) at a depth of approximately 11 m to digest the food eaten at their dusk feeding during the short Canadian summer nights. Interestingly, the fish go deeper (37 m) to 5°C water between the dawn and dusk feeding periods. Thus, during the longer daytime period, the fish conserve energy by lowering their body-maintenance energy requirements in the colder water (see Chapter 3).

Human-induced thermal changes in aquatic environments have stimulated further investigations of fish thermoregulatory behavior. For example, Neill and Magnuson (1974) consistently captured bluegill, largemouth bass, longnose gar (*Lepisosteus osseus*), small rockbass (*Ambloplites rupestris*), pumpkinseed (*Lepomis gibbosus*), large yellow bass (*Morone mississippiensis*), and common carp in the heated outfall plume waters of an electrical power–generating plant located on Lake Monona, Wisconsin. Although localized food sources (e.g., zooplankton for the bluegill and small fish for the longnose gar) influenced the preference of some of these species for the 2°C to 4°C warmer thermal plume area in summer, concurrent laboratory studies using a shuttle-box apparatus showed that the consistent plume residents that were tested had higher preferred temperatures than the one species (yellow perch) that consistently avoided the thermal plume.

Physiological thermoregulation in fish, to a significant degree, is exhibited only by several continuous-swimming species. Each of the "warm-bodied" species leads a pelagic marine existence, and most have *heat-exchanging retia mirabilia* to conserve the heat that is produced by the fish's metabolism (Carey et al. 1971). These fish also have large **cutaneous arteries** and **veins** for blood transport between the heart and gills and the heat exchanger (Fig. 5.4). This enables them to transport the cool (i.e., near water temperature) blood to and from the heat exchanger without absorbing much of the heat that is produced by the swimming muscles. The structure of these retia is similar to that of the physoclistous swimbladder (i.e., the gas-secreting retia described in the buoyancy section of this chapter). Essentially, by means of a bundled arrangement of afferent and efferent blood vessels (**arterioles** and **venules** rather than capillaries), **heat** (instead of gas) is exchanged by **convection** across the walls of these many vessels running parallel to each other. The absence of localized blood acidification, such as in the physoclistous swimbladder gas gland, minimizes gas exchange in the heat-exchange retia. Although hemoglobin usually loses its affinity for O_2 with temperature increases such as those experienced by arterial blood in the rete (see Chapter 4), O_2 delivery to the red swimming muscles of albacore is maintained despite retial warming because of a reverse temperature effect of the albacore hemoglobin. This special hemoglobin increases its O_2 affinity with warming between 10°C and 30°C and its CO_2 affinity with cooling, thereby minimizing O_2 loss to

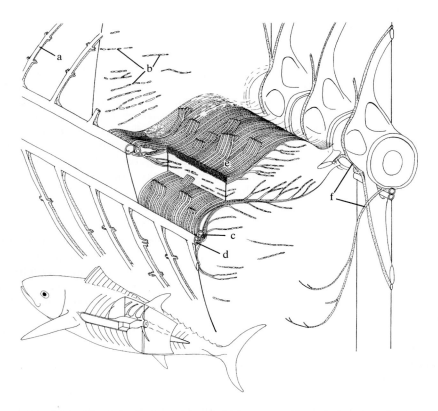

FIGURE 5.4   Circulation in the muscles of a bigeye tuna: *a*, Segmental artery and vein; *b*, vascular bands; *c*, cutaneous artery; *d*, cutaneous vein; *e*, rete mirabile; and *f*, arterial branches from dorsal aorta. From Carey and Teal (1966).

the venous blood in the rete (Cech et al. 1984). Also, the larger vessel diameter (10-fold as large: 0.1 vs. 0.01 mm) and thicker vessel walls (compared with the swimbladder rete capillaries) further slow the diffusion of O_2 molecules, which diffuse at a rate 10-fold slower than heat. Because of the countercurrent flow, the metabolic heat is efficiently *conserved* in the rete, which surrounds the red swimming muscles (Fig. 5.4). The efficiency of this heat exchanger is 95% as a thermal barrier between the gills and red muscle in skipjack tuna (*Katsuwonus pelamis*) (Neill et al. 1976).

Although the number and position of heat-exchanging retia vary among the different tunas and mackerels (Scombridae) and lamnid sharks, all of them are capable of fast, continuous swimming. Warm muscles confer swimming performance advantages to these species. Stevens and Carey (1981) showed that the facilitated diffusion of O_2 by myoglobin and, consequently, red muscle metabolism are greatly enhanced at warmer temperatures. Because warm muscles can contract faster than cool ones, the heat exchanger presumably allows these predatory fishes to exert more swimming thrust and, thus, to outswim the squid and smaller fishes that compose their diet. For example, the grouper (*Epinephelus*), with special circulatory adaptations for metabolic heat conservation, has an internal temperature of 0.3°C above that of the water it lives in, whereas the swimming muscles of albacore tuna show a 12°C elevation (Carey et al. 1971). The red swimming muscles (located very close to the vertebral column) of the endothermic yellowfin tuna (*Thunnus albacares*) and those (located more peripherally) of the ectothermic Pacific bonito (*Sarda chiliensis*) both show increased contraction power with warming temperatures (≈twofold per 10°C increase) (Altringham and Block 1997). The tuna muscle was able to operate very efficiently at a higher temperature, but its performance was more sensitive to temperature decreases, thereby indicating evolutionary selection for operation at more constant temperatures compared with the bonito's muscles. "Warm-bodied" fish, however, do not have constant body temperatures (as in mammals or birds), but they do have temperatures that fluctuate with that of the environment. The core temperature of the largest of these "warm-bodied" species, the bluefin tuna, seems to be least affected by environmental temperature, apparently because of the great thermal inertia that is inherent in large bodies. In some situations, this may work to their disadvantage by causing overheating of the muscle mass when exercising. Thus, larger tuna adopt a cyclical pattern of depth distribution to "cool off" below the thermocline if surface waters are too warm for continuous occupancy. Although the largest of the warm-bodied tunas may have the most stable internal temperatures, black skipjack tuna (*Euthynnus lineatus*) as small as 207 mm (fork length) can elevate their muscle temperatures significantly above that of the ambient water (Dickson 1994). The ability of "warm-bodied" fish to detect changes in their thermal environment may operate either by neural comparisons of the water temperature (external surface of the fish) with the temperature deep inside the body or by the blood temperature gradient across the heat exchanger (Neill et al. 1976). Regardless, bigeye tuna (*Thunnus obesus*) can sense thermal differences between their body and the water—and can respond with 100- to 1,000-fold changes in whole-body thermal conductivity (Holland and Sibert 1994). Carey (1982a), Graham (1983), and Brill et al. (1994) provide useful reviews of heat-transfer processes in fishes.

Carey (1982b) described a ***"brain heater"*** in swordfish (*Xiphias gladius*) from highly ***modified eye muscles*** that locally warm the brain and eye tissues and, presumably, contribute to foraging success in cold, deep (to 600 m) oceanic depths (Block 1991a). Subsequent studies (e.g., through sequencing the cytochrome *b* gene) (Block et al. 1993) indicate that this type of regional endothermy has evolved twice, once in billfish (Istiophoridae) and once in scombrids (Block 1991b, Block et al. 1993). The relative relatedness of the partially endothermic fishes is reviewed by Block et al. (1993) and by Block and Finnerty (1994). The brain-heater tissue has an exceedingly high *mitochondrial volume* (a measure of aerobic metabolic capacity and activity) and high activity of the aerobic-enzyme *citrate synthase* (Block 1991a, Tullis et al. 1991). The high percentages of mitochondria (55–70% of cell volume, depending on the species), *sarcoplasmic reticulum* (SR; the highly organized intracellular membrane system that regulates calcium uptake, release, and storage, 25–30%), and *transverse (T) tubules* as well as the lack of organized contractile elements equip the brain-heater cells for both rapid and sustained *calcium cycling* and *heat production* (Block 1994). In skeletal muscle, the signal for calcium release (and consequent muscle contraction) is initiated at the neuromuscular junction, and the action potential spreads down the T tubules, thereby initiating a molecularly mediated release of Ca^{2+} from the SR along the *excitation-contraction (EC) coupling pathway*. It is thought that the similar *depolarization* of the T tubules in brain-heater cells conformationally changes its voltage-sensing molecule (*dihydropyridine receptor*) and, via a molecular bridge, the SR Ca^{2+} release channel (*ryanodine receptor*), regulating SR Ca^{2+} release (Fig. 5.5). The Ca^{2+} is rapidly "pumped" back into the SR via an SR Ca^{2+} adenosine triphosphatase-mediated system that requires energy, increasing relative

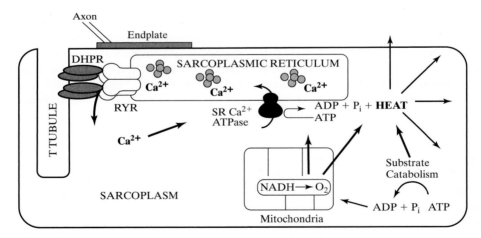

FIGURE 5.5 Excitation–thermogenic coupling in the billfish heater cell. A nervous impulse stimulates thermogenesis via the same molecular components found in the excitation-contraction coupling pathway in "standard" skeletal muscle. Heat would be produced as a consequence of the Ca^{2+} release and reuptake at the sarcoplasmic reticulum (SR) and the resulting stimulation of substrate oxidation at the mitochondria. The high mitochondria and myoglobin content provides ample adenosine triphosphate (ATP) and O_2 for the Ca^{2+}-mediated thermogenic cycle. The Ca^{2+} may stimulate the heater cell's mitochondrial mechanism, contributing to heat generation. *ADP*, adenosine diphospate; *ATPase*, adenosine triphosphatase; *DHPR*, dihydropyridine receptor; *P_i*, inorganic phosphate; *RYR*, ryanodine receptor. Reproduced, with permission, from Block, B.A., *Annual Review of Physiology*, Volume 56, © 1994, by Annual Reviews, Inc.

adenosine diphosphate (ADP) and inorganic phosphate concentrations in the cytosol. Increased mitochondrial uptake of ADP stimulates oxidative processes and substrate catabolism, thereby producing more heat (Block 1994). Block et al. (1993) explain that this type of partial endothermy probably allowed the successful expansion of these fishes' ranges into cooler waters.

5.3   LESSONS FROM BUOYANCY AND THERMAL REGULATION

As aquatic ectotherms, buoyancy and temperature are two primary concerns of fishes. Without a means of regulating their buoyancy, negatively buoyant fishes would either sink or be forced to continuously expend energy (via swimming) to maintain their position in the water column. The evolution of swimbladders, however, including those with gas-exchanging retes, and other low-density structures has given fish the ability to achieve near-neutral buoyancy at a relatively low metabolic cost. Most fish are capable of behavioral rather than physiological thermoregulation. A few species, however, have evolved the structures and mechanisms to warm critical parts of their bodies. A heat-exchanging vascular rete design conserves metabolically produced heat in muscles in some pelagic tunas and sharks, whereas a modified-muscle "brain heater" represents another evolutionary solution to cold, oceanic water.

SUPPLEMENTAL READINGS

Alexander 1993; Block 1991a; Block 1994; Block and Finnerty 1994; Brill et al. 1994; Carey 1982a; Carey et al. 1971; Fänge 1976, 1983; Gee 1983; Graham 1983; Hazel 1993; Kuhn et al. 1963; Neill et al. 1972, 1976; Neill and Magnuson 1974; Pelster 1998; Sharp and Dizon 1978; Steen 1971.

WEB CONNECTIONS

COUNTER CURRENT HEAT EXCHANGER ANIMATED MODEL
www.biology.ualberta.ca/courses/zool241/index.php?Page=1815

BUOYANCY
www.csuchico.edu/~pmaslin/ichthy/byncy.html

C H A P T E R 6
Hydromineral Balance

Living cells require an environment that is characterized by particular concentrations of certain substances (including ions) dissolved in water. Thus, in fish, the internal environment must have the necessary "balance" of ionized salts, alkalinity, and dissolved organic compounds despite an external environment that may have a very different combination of these factors. Special problems, such as movement between freshwater and saltwater environments, stress induced by capture or transport, or survival in habitats subject to freezing, add to the complexity of maintaining the appropriate internal environment among fishes. The following sections concerning osmoregulation, ionic regulation, stress responses and effects, freezing resistance, and acid-base balance address the "strategies," dynamics, and diversity of hydromineral balance in fish.

6.1 OSMOREGULATION

Fish can be divided into four groups, or "strategies," of regulation of internal water and total solute concentrations. The first osmoregulatory strategy is used by the hagfish (Myxiniformes) and is characterized by no regulation at all (*osmoconforming*). Hagfish are all strictly marine fish and are *stenohaline* (able to tolerate only a narrow range of salinities). Thus, the total salt concentration in their body fluids is very similar to that of sea water; hagfish are the only "vertebrates" (see Chapter 1) with this characteristic (Schmidt-Nielsen 1975). They can be described as osmoconformers rather than *osmoregulators*. Hagfish do not have to withstand large changes in internal *osmolality* (total dissolved solute particles), however, because they live only in marine environments of quite constant salinity. As Table 6.1 shows, hagfish still show some individual (Na^+) *ion-regulatory* ability (see Section 6.2).

The second strategy encompasses all marine elasmobranchs. Like most vertebrates, the elasmobranchs maintain an internal *inorganic* salt concentration equal to approximately one-third that of seawater (Table 6.1). Large quantities of *organic* salts (primarily *urea*, and secondarily *trimethylamine oxide* [*TMAO*]) in their blood, however, bring the total osmotic concentration up to that of seawater (Table 6.1). Yancey and Somero (1980) found that the typical 2:1 concentration ratio of urea:methylamines (e.g., TMAO) protects the elasmobranchs' enzymes from denaturation and consequent

TABLE 6.1   PLASMA SOLUTE CONCENTRATIONS (MMOL/L) IN MARINE (M) AND FRESHWATER (F) SPECIES

Habitat	Species	[Na$^+$]	[Ca^{2+}]	[K$^+$]	Urea	Total salts (mOsm/L)
M	Hagfish (*Myxine glutinosa*)[a]	549	5	11	—	1152
F	Lamprey (*Lampetra fluviatilis*)[b]	120	2	3	—	270
M	Dogfish (*Squalus acanthias*)[c]	263	7	4	357	1,007
M	Anglerfish (*Lophius americanus*)[d]	198	2	3	—	—
M	Moray eel (*Murena helena*)[b]	212	4	2	—	—
F	Bass (*Micropterus dolomieu*)[e]	120	3	3	—	—
F	Whitefish (*Coregonus clupoides*)[b]	141	3	4	1	—
	Seawater[fg]	~450	~20	10	—	1,000
	Fresh water[hi]	<1	<1	<1	—	1–10

From [a]Bellamy and Chester-Jones (1961), [b]Robertson (1954), [c]Murdaugh and Robin (1967), [d]Forster and Berglund (1956), [e]Shell (1959), [f]Schmidt-Nielson (1975), [g]von Arx (1962), [h]Hutchinson (1957), and [i]Royce (1972).

inactivation that usually is associated with high urea levels (Kay 1998). A unique combination of active urea transport and modification of the gill membrane is responsible for decreasing the gill's permeability to urea and facilitating urea retention by spiny dogfish (*Squalus acanthias*) (Fines et al. 2001). Despite a total salt concentration that approximates that of the sea, elasmobranchs possess considerable abilities to regulate the concentrations of individual ions. The coelacanth (*Latimeria chalumnae*), a bony fish, uses this osmoregulatory strategy as well. Fish using either of these two strategies of osmoregulation have solved a major problem regarding *water balance*. Water diffuses quite easily across thin membranes, such as the skin and especially those in the gills. Elasmobranch gills are quite *water-permeable* (Evans 1984). Because the internal *total* (inorganic + organic) salt concentration of elasmobranchs mimics that of their environment, however, passive water *influx* (inflow) or *efflux* (outflow) is minimized. Passive Na$^+$ and Cl$^-$ effluxes are minimized because of the *low permeability* of elasmobranchs to these ions (Evans 1979).

The third osmoregulatory strategy is that of the marine teleosts. The salt concentration of their internal environment is approximately one-third that of their external environment (Table 6.1). Thus, they operate *hyposmotically* and tend to diffusively *lose* water to the more saline environment and to *gain monovalent* ions from it, especially across the thin membranes of the *gills*. These teleosts and chondrosteans, such as seawater-acclimated sturgeon (McEnroe and Cech 1985, Altinok et al. 1998), continually replace lost water by drinking (ingesting) seawater. Naturally, this also results in a large intake of salts, which must be excreted at a concentration higher than that ingested. Special (*chloride*) cells in the *gill filament* and *opercular skin* epithelia eliminate much of the excess salt via *active transport*. Teleost *kidneys* excrete primarily *divalent* ions (see Section 6.2) but cannot produce a urine that is more salty than the blood (Schmidt-Nielsen 1975).

The fourth strategy has evolved in freshwater teleosts and elasmobranchs that operate *hyperosmotically*. Because their internal environment (one-fourth to one-third the salt concentration of sea water) is more concentrated than their external environment

(Table 6.1), they continually *gain* water by diffusion. This uptake between cells (paracellular pathways) is regulated via **calcium** concentrations (i.e., hard water), whereas transmembrane water permeation (through cells) is regulated by membrane cholesterol (Robertson and Hazel 1999). The excess water is continually excreted by well-developed kidneys as a large volume of dilute urine (up to one-third of the body weight per day). Control of **diuretic** (urine-producing) processes is influenced by many factors, including blood pressure, glomerular filtration, and kidney tubular reabsorption changes that are mediated by hormones such as the **renin-angiotensin system** (Perrot et al. 1992, Fuentes et al. 1996, Fuentes and Eddy 1996), **atrial natriuretic peptide** (Evans 1990, Evans and Takei 1992, Olson and Duff 1992, Loretz 1995), and **arginine vasotocin** (Sawyer et al. 1976). Some salts are unavoidably lost through the urine and by diffusion through the gill tissues. Diffusional losses of monovalent ions (e.g., across the gill membranes) are reduced in external environments having higher calcium concentrations (Grizzle et al. 1985) and from secretions of the pituitary hormone, **prolactin** (Potts and Fleming 1971, Wendelaar Bonga et al. 1983, Flik et al. 1989). Lost internal solutes are replaced by those taken in with food or taken up at the gills using active transport mechanisms. Thus, an energy-requiring salt pump operates in chloride cells of the gills in these fishes as well, except that ions are pumped *inward* rather than outward, as exhibited by the marine teleosts. In especially ion-deficient (soft) mountain water, chloride cells are well developed on gill lamellae as well as filaments in rainbow trout (Laurent et al. 1985). Few freshwater elasmobranchs are found, possibly because of their high permeability to water. The remarkable Amazon stingray (*Potamotrygon*), a stenohaline freshwater species, displays an interior milieu strikingly similar to that of the freshwater teleosts with essentially *no* urea. The more **euryhaline** (meaning "broad salt tolerance") sharks, such as the bull shark (*Carcharhinus leucas*), which can ascend rivers, have a urea concentration of approximately one-third that found in the marine elasmobranchs (Thorson et al. 1973).

Whereas most bony fishes are stenohaline and have evolved the osmotic structures (e.g., the appropriate sorts of chloride cells and intercellular junctions between cells) and mechanisms (e.g., hormonal controls) that are needed to cope with the relatively constant salt concentration of either marine or freshwater habitats, **euryhaline** fishes have broader ranges of salt tolerance. Examples of euryhaline fishes include the sheepshead minnow (*Cyprinodon variegatus*), Mozambique tilapia, and striped bass (*Morone saxatilis*). In contrast, **diadromous** (meaning "to run across") fishes are euryhaline species that move between freshwater and marine environments at discrete life stages. Examples of diadromous species include the Pacific lamprey (*Lampetra tridentata*) Pacific salmon (*Oncorhynchus* spp.), American eel (*Anguilla rostrata*), and American shad (*Alosa sapidissima*). American shad larvae and juveniles increase their salinity tolerance (i.e., ability to hyposmotically regulate) during development (seawater survival increases from 0% at 36 days post-hatch to 89% at 45 days). Their enhanced survival at the older age is associated with increased gill *Na$^+$-K$^+$-dependent adenosine triphosphate [Na$^+$-K$^+$–ATPase]* activity (a critical, energy-requiring enzyme system) during their freshwater-rearing period near their autumnal downriver migration (Zydlewski and McCormick 1997).

Regardless of a fish's range of salinity tolerance and how it may change during its life cycle, osmoregulation requires energy. Indeed, growth rates and swimming performance of fishes may be affected by how much energy they must allocate to osmoregulation or other energy costs, such as activity. For example, euryhaline sciaenid fishes (drums) are able to maximize metabolic efficiency (e.g., for optimal growth [Brocksen and Cole 1972] or swimming performance capacity [Wohlschlag and Wakeman 1978]) when held near the midpoints of their salinity tolerance ranges. In contrast, the cultured milkfish (*Chanos chanos*) of southeast Asia grows best at very high salinities (55 ppt) compared to lower salinities (15 and 35 ppt), because it reduces its voluntary activity rates at these high salinities (Swanson 1998). The cost of osmoregulation generally is considered to be a *small* fraction of the overall energy costs of fishes (Morgan and Iwama 1991), although it probably increases with increases in activity (Febry and Lutz 1987). Gibbs and Somero (1990) calculated that the energetic cost of osmoregulation is proportional to total metabolic rate across a variety of marine teleosts: higher for shallow-water forms with higher metabolic rates, and lower for deep-living fish with lower metabolic rates. The exception was the presumed high metabolic rates in fish living near deep, hydrothermal vents, where the sea water is geothermally warmed and food is abundant.

6.2 IONIC REGULATION

Even if a fish has blood with nearly the same osmotic concentration as sea water, energy still must be expended in regulation of solutes, because the concentrations of individual ions will differ between the internal and external environments. To maintain an optimal ionic composition, active, energy-consuming processes are needed. These processes—and the organs involved—vary considerably among fish. In terms of general "strategies" of ionic regulation, however, fish fall into five main groups: (1) hagfish, (2) elasmobranchs, (3) marine bony fish and lampreys, (4) euryhaline and diadromous teleosts, and (5) freshwater teleosts.

Hagfish

Hagfish are not only *isosmotic* with seawater but have a rather similar ionic composition as well, although some differences are detectable. For example, *Myxine glutinosa* demonstrates a sodium concentration ($[Na^+]$) somewhat greater than that of sea water (Table 6.1). The low Na^+ concentration in the secreted *slime* that coats its body probably helps maintain plasma $[Na^+]$, because McFarland and Munz (1965) could find no evidence of Na^+ active transport across the gut, gills, or skin. The divalent ions Ca^{2+}, SO_4^{2-}, and Mg^{2+} are all present at lower concentrations in hagfish than in sea water. However, Mg^{2+}, K^+, SO_4^{2-}, and PO_4^{2-} are secreted into the glomerular filtrate by the mesonephric duct cells and appear in the *urine* at higher concentrations than in the plasma (Munz and McFarland 1964, McFarland and Munz 1965). Moreover, the slime has high concentrations of Ca^{2+}, Mg^{2+}, and K^+. Thus, hagfish ionic regulation largely is controlled via ionic secretions into the urine and the slime.

Elasmobranchs

The retention of urea and other compounds has provided an efficient solution to the problem of water balance in sharks and their relatives, but they still must eliminate the excess Na^+ and Cl^- they ingest. Excretion of these ions is, in fact, the principal function of the ***rectal gland***, which is found only in elasmobranchs (Burger and Hess 1960) and the coelacanth. The rectal gland secretes a fluid that has Na^+ and Cl^- concentrations approximating those of sea water (twice those in the plasma) (Silva et al. 1977a). In spiny dogfish, rectal glandular secretions can be produced by hypertonic injections of NaCl into the bloodstream (Burger 1962), and removal of the rectal gland provokes a steady increase in plasma $[Na^+]$ (Forrest et al. 1973a). Silva et al. have shown that the Cl^- secretion rate into the lumen of dogfish rectal glands is dependent on $[Na^+]$ (1977a), and is regulated via cardiac natriuretic peptide (1999). Those authors hypothesize that the movement of Na^+ and Cl^- from the blood (or perfusate) across the glandular wall into the lumen results both from active transport "sodium pumps" catalyzed by the special enzyme, Na^+-K^+-ATPase, and from electrical forces inducing movement of these charged ions toward electrical homeostasis. This mechanism is essentially the same as that found in the chloride cells, which are so numerous in the gills of marine teleosts. Miller et al. (1998) showed that spiny dogfish (*Squalus acanthias*) rectal gland tubules have another function: removal of foreign substances (e.g., some pollutants) from the shark's body. Rectal glands show evolutionary *regression* in elasmobranchs adapted to *fresh* water (Oguri 1964). Marine elasmobranchs also use their kidneys and gill chloride cells to excrete excess salts, but the density of gill chloride cells probably is only 10% to 20% of that found in marine teleosts (Jampol and Epstein 1970, Shuttleworth 1988). For euryhaline Atlantic stingrays, acclimation to freshwater conditions promoted the activation of more Na^+-K^+-ATPase–rich (chloride) cells in the gills, but fewer such cells in the rectal gland, compared with seawater-acclimated stingrays (Piermarini and Evans 2000). Those authors concluded that the gills were especially important for active ion uptake in fresh water, whereas the rectal gland was important for active NaCl secretion in sea water. Wilson et al. (1997) localized a H^+-ATPase system in spiny dogfish gills for ionic (and acid–base) regulation.

Marine Bony Fish and Lampreys

Marine teleosts, chondrosteans, and lampreys maintain a total ionic concentration in the plasma approximately one-third that of seawater. Because most of the ions needed by the fish are present in excess in the environment, the principal method used by marine teleosts for maintenance of ionic balance is ***selective excretion***, particularly of Na^+ and Cl^- (Fig. 6.1). The gills have a relatively high permeability to *monovalent* ions, allowing Na^+ and Cl^- to move passively from sea water into the plasma. In addition, when sea water is ingested to replace water that has diffused into the environment, monovalent ions, as well as water, are absorbed in the ***intestine***. Teleostean kidneys are of little help in the excretion of these ions, because they do not have the ability to form a urine that is more concentrated than the blood. Indeed, many marine teleosts (e.g., oyster toadfish [*Opsanus tau*] and plainfin midshipman [*Porichthys notatus*]) have an "evolutionary regressive" *aglomerular* kidney to minimize water losses! Instead, special large cells in the gills actively transport the excess monovalent anions against the concentration gradient back into the environment. Catlett and Millich (1976) describe

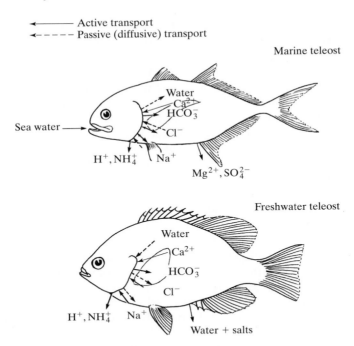

←———— Active transport
←- - - - - Passive (diffusive) transport

FIGURE 6.1   Passive and active routes of salt and water exchange.

these "chloride cells" as being larger than the flat cells of the gills that are specialized for respiratory gas exchange. The chloride cells are termed **alpha chloride cells** (Pisam et al. 1987) to distinguish them from the **beta chloride cells** found in freshwater teleosts. They are located primarily on the gill filaments at the base of the lamellae and, in some species, on the skin lining the inside of the operculum. They feature an abundance of energy-providing mitochondria, display extraordinary development of cytoplasmic microtubules, and contain the Na^+-K^+-ATPase system (Fig. 6.2) also found in elasmobranch rectal gland secretory tissue. The Na^+-K^+-ATPase is located along the basolateral areas and in the extensive microtubular system of the chloride cell, and it actively transports Na^+ out of the cell in exchange for K^+ (Karnaky 1986). This enzyme system functions to maintain a high Na^+ gradient, with high $[Na^+]$ in the tubules and adjacent plasma and low $[Na^+]$ in the chloride cell cytoplasm (Silva et al. 1977b). This high Na^+ gradient drives a linked Na^+-K^+-$2Cl^-$ carrier system, increasing the cytoplasmic Cl^- concentration ($[Cl^-]$). The buildup of Cl^- inside the chloride cell also increases the cell's **electronegativity**, and Cl^- follows its electrochemical gradient (including a negative-to-positive **transmembrane potential**) by passively moving out of the apical pit area and into sea water (Fig. 6.2). The Na^+ probably exits passively to sea water via **shallow tight junctions** between chloride cells or between chloride cells and **accessory cells**, driven by the positive-to-negative **transepithelial potential** (Karnaky 1986). Utida and Hirano (1973) have shown that both Na^+-K^+-ATPase concentrations and numbers

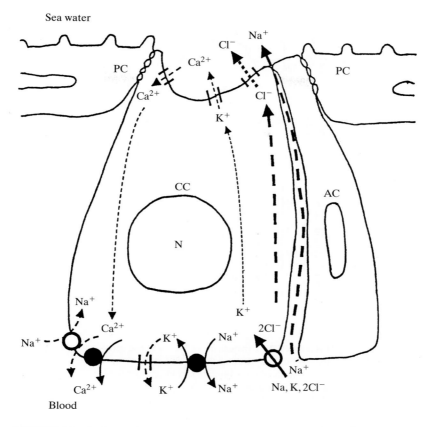

FIGURE 6.2   A schematic model for the movement of Na⁺, Cl⁻, and Ca²⁺ by chloride cells of seawater-adapted teleosts. An α-type chloride cell (CC) is shown with a nucleus (N) and is connected by a shallow tight junction to an accessory cell (AC). Pavement cells (PC) are connected by deep tight junctions. Solid lines indicate active transport; dashed lines show diffusion or exchange through membrane channels or across the cell. The ATPase pumps are represented by solid circles; cotransporter and exchangers by open circles; and ion channels by parallel lines. Ionic exchanges shown at the basolateral membrane (bottom of drawing) are actually widely spread along an extensive microtubule system, which dramatically increases the basolateral membrane surface area. Cellular mechanisms are explained in the text. Modified from Marshall (2002).

of chloride cells increase with increasing environmental salinity in gill preparations of Japanese eel (*Anguilla japonica*) (Fig. 6.3).

While the gills are the principal site of monovalent ion excretion, the kidneys typically eliminate excess *divalent* ions, such as Mg^{2+} and SO_4^{2-}. Such ions are present only in small amounts, so they do not pose the problems of the abundant monovalent ions. Calcium ions are the exception, with a Ca^{2+} ion channel (at the apical membrane) and a Ca^{2+}-ATPase-facilitated pump (with possible Na^+ exchange) at the basolateral membrane of the chloride cells in the gills (Fig. 6.2).

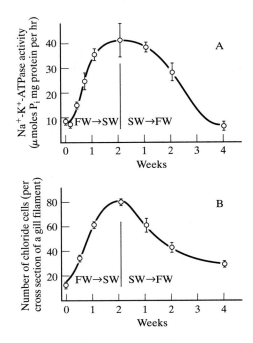

FIGURE 6.3   Changes in Na^+-K^+-ATPase activity and number of chloride cells following transfer of Japanese eel (*Anguilla japonica*) from fresh water to sea water (FW → SW) and vice versa (SW → FW). From Utida and Hirano (1973).

Euryhaline and Diadromous Teleosts

The study of euryhaline and diadromous teleosts can provide considerable insight regarding the mechanisms and energy costs of ionic regulation, because such fish often endure dramatic changes in their external environments. Euryhaline forms typically are estuarine and intertidal inhabitants that experience continual shifts in external salinity, which are dictated by tidal rhythms as well as by wind, storm, or river-flow variations. Most diadromous species spend part of their life cycle in fresh water and part in salt water, as rather stenohaline inhabitants, except during the hormone-mediated transitions. Evans (1967a, 1967b) studied an example of the former group, the intertidal black prickleback (*Xiphister atropurpureus*), whose plasma $[Na^+]$ and $[Cl^-]$ are stable when individuals are immersed in dilutions of sea water as low as 31%. The $[Na^+]$ and $[Cl^-]$ fall approximately 15% after transfer to environments with salinities between 10% and 31% that at seawater. Salinities less than 10% that of sea water are not in the tolerable range, apparently because of the prickleback's inability to retard Cl^- losses from passive diffusion.

Whereas ***prolactin*** can be thought of as the "freshwater" hormone, ***cortisol*** (a steroid hormone from the interrenal tissue associated with the kidney) can be thought of as the "seawater" hormone. Anguillid eels are ***catadromous*** (meaning "to run down") fish that migrate down rivers as adults to spawn in the ocean. Forrest et al. (1973b) noted that plasma cortisol concentration showed a 5- to 7-day increase in freshwater-adapted American eels (*Anguilla rostrata*) when they were transferred to sea water. Experiments with cortisol injection support the notion that this transient peak morphologically prepares diadromous or euryhaline fishes for survival in hypertonic environments (Evans 1984). For example, injection of cortisol into freshwater-acclimated Mozambique tilapia, a euryhaline teleost, stimulated an increase in chloride cell density (Foskett et al. 1981). Conclusively, direct exposure of freshwater-acclimated tilapia opercular membranes to cortisol increased chloride cell density, size, and Na^+-K^+-ATPase activity (McCormick 1990). Adaptive mechanisms such as increased Na^+ excretion and Na^+-K^+-ATPase activity at the gills, increased water permeability of the ***urinary bladder*** (to retain water), and increased uptake of ions and water in the gut (related to drinking sea water) have been associated with such cortisol-stimulated changes (Matty 1985).

Changes in diadromous fishes' ion-regulatory abilities typically are associated with ontogenic changes mediated by hormones. Adult salmon and some trout are ***anadromous*** fish that migrate up rivers from the ocean (or a lake) to spawn. Young salmon and trout from the spawn migrate down the rivers to salt water after changing from a freshwater (***parr***) form to a saltwater-capable (***smolt***) form. Saltwater readiness in young trout and salmon ranges from a modest springtime rise in resistance, shown by underyearling steelhead trout and other trouts and charrs, to complete tolerance and survival, shown by chum (*Oncorhynchus keta*) and pink (*O. gorbuscha*) salmon alevins! For most salmonids, tolerance of marine conditions develops during the spring, before the seaward migration of the silvery smolts (see Chapter 19). Hoar (1976) pointed out the importance of lengthening springtime ***photoperiods*** for timing these changes. On a finer scale, Grau et al. (1981) showed that plasma ***thyroxine*** surges, which occur in coho salmon (*O. kisutch*) during their smoltification period, correspond to ***new-moon phases*** during spring. New-moon–related, swift-ebbing currents and dark nights may decrease predation vulnerability of the young coho during their movements down rivers to the estuary. A few weeks after the thyroxine peaks, increased plasma ***cortisol*** concentrations (Specker and Schreck 1982) and the cortisol-related changes in chloride cells, urinary bladder, and intestine are observed (Loretz et al. 1982), and the coho smolts migrate to the sea. Gill Na^+-K^+-ATPase activity and Na^+-K^+-$2Cl^-$ cotransporter abundance both increased in α-type Cl^- cells (see Fig. 6.2) on the gill filaments of Atlantic salmon during both seawater acclimation and smolting (Pelis et al. 2001). Besides cortisol, the pituitary hormone, ***growth hormone (GH)***, also plays an important role in the parr–smolt transformation among anadromous salmonid species (Barrett and McKeown 1988c). For example, freshwater-acclimated, sea-run brown trout (*Salmo trutta*) yearlings regulate plasma ions better and survive in greater numbers after seawater exposure with injections of both GH and cortisol compared with noninjected controls (Madsen 1990). Furthermore, the effectiveness of GH regarding a young salmonid's ionic regulation in hypertonic environments may be linked with ***insulin-like growth factor-I*** (McCormick et al. 1991, Björnsson 1997). Finally, environmental factors (e.g., temperature) can significantly impact the parr–smolt transformation (Box 6.1).

Studies of several species have demonstrated the role of ***prolactin*** in preventing Na^+ diffusive loss in freshwater-adapted fish and in minimizing increases in passive Na^+ loss as euryhaline forms pass from sea water to fresh water. For example, removal of the

BOX 6.1

Fish in "Hot Water": Warm Temperature-Related Losses of Smolt Characteristics

Increased water temperatures (e.g., as would be experienced toward the end of a springtime migration in warmer, more southern rivers of New England but not in more northern rivers of the Atlantic Maritime Provinces of Canada) were associated with decreased gill Na^+-K^+-ATPase activity in Atlantic salmon (*Salmo salar*) smolts (McCormick et al. 1999). When those authors held hatchery- and stream-reared smolts in a laboratory, both exhibited a more rapid loss of physiological smolt characteristics when held at a higher temperature and a higher mortality rate when subjected to a seawater challenge test. Because the loss of most characteristics was a function of holding time at the warmer temperature, McCormick et al. (1999) concluded that significant delays (e.g., from hydrological or climatic conditions or from barriers to migration, e.g., dams) in the smolts' migration in the warmer rivers will have negative impacts on the smolts' capacity to survive in sea water and to return as adults.

anterior lobe of the pituitary (**hypophysectomy**)—where prolactin is synthesized—of freshwater-acclimated mummichogs (*Fundulus heteroclitus*) promotes a marked drop in plasma electrolytes compared with sham-operated controls (Maetz et al. 1967). The many ion-regulatory roles of fish prolactin are essentially *opposite* those of cortisol (discussed earlier) regarding various physiological and morphological changes in the gills, gut, kidney, and urinary bladder of various species (see the review by Hirano 1986).

It is apparent that gill ionic permeabilities in many species also are affected by the concentration of **calcium** ($[Ca^{2+}]$) in the water, possibly through Ca^{2+}-binding at paracellular (between cells) pathways. Addition of 10 mM Ca^{2+} to fresh water will reduce prolactin synthesis rates by 50% in Mozambique tilapia (Wendelaar Bonga et al. 1985). The sodium efflux from the plains killifish (*F. kansae*) in fresh water is reduced by 50% when 1 mM Ca^{2+} is added to the water (Potts and Fleming 1971). In sea water, calcium also can reduce Na^+ permeability across the branchial epithelia of fish, such as European eels (Cuthbert and Maetz 1972). Carrier and Evans (1976) demonstrated that the euryhaline marine pinfish (*Lagodon rhomboides*) tolerates essentially fresh water (5 mM Na^+) if 10 mM Ca^{2+} also is present. Transfer of the pinfish to calcium-free fresh water stimulates substantial Na^+ efflux. Pinfish left in calcium-free fresh water for 2.5 hours died with less than 50% of the body Na^+ concentrations found in those pinfish acclimated to calcium-supplemented sea water. These results help to explain why Breder (1934) observed several marine fish species living in a freshwater lake on Andros Island of the Bahamas. Analysis of this lake water showed it to have an unusually high (1.0–1.5 mM) $[Ca^{2+}]$.

Freshwater Teleosts

The hyperosmotic state of the freshwater teleosts dictates that small ions, such as Na^+ and Cl^-, are continually lost to the environment by diffusion across the thin epithelia of the gills (Fig. 6.1). Solutes also are continually lost in the large volumes of dilute urine that are produced to expel the excess water that is passively taken up by diffusion across the gills. Some of the salts are regained via food sources, and much of the Na^+ and Cl^- needed to regain internal ionic homeostasis is taken up by active transport mechanisms in the gills.

From the pioneering work of Krogh (1939) and of Maetz and Garcia Romeu (1964), a model describing ion-exchange mechanisms across the gills in freshwater teleosts has been formulated. These ion-exchange mechanisms apparently reside in epithelial cells, including chloride cells. These ***beta chloride cells*** (Pisam et al. 1987) occur on gill filaments between lamellae and, in very soft (low dissolved ion concentrations) water, even on gill lamellae. They resemble chloride cells in marine fish, in that they apparently are activated by cortisol (McCormick 2001, Sloman et al. 2001) and contain mitochondria, the tubular system, and Na^+-K^+-ATPase, but they are less numerous, usually occur singly, and generally assist movement of Na^+ and Cl^- into rather than out of the fish (Eddy 1982). Figure 6.4 diagramatically shows current ideas concerning these Na^+ and Cl^- uptake mechanisms. The Na^+-K^+-ATPase enzyme that is so abundant in both alpha and beta chloride cells is modulated by cyclic adenosine monophosphate in euryhaline brown trout (Tipsmark and Madsen 2001). Some of the Cl^- uptake probably occurs in pavement epithelial cells, as well as in beta chloride cells (Fig. 6.4). These ion-exchange mechanisms serve several functions besides maintenance of $[Na^+]$ and $[Cl^-]$ in the fish. The Na^+ exchange for NH_4^+ conveniently rids the fish of part of its ***ammonia*** production (see Fig. 6.6), the principal waste product of protein digestive breakdown (see Chapter 7). Injections of NH_4^+ into freshwater goldfish thereby stimulated Na^+ influx (Maetz and Garcia Romeu 1964). Both the Na^+ exchange for H^+ and Cl^- exchange for HCO_3^- tend to maintain internal ***acid–base homeostasis*** (see Section 6.5). Thus, just two ion-exchange mechanisms provide for maintenance of appropriate internal $[Na^+]$ and $[Cl^-]$, elimination of some potentially toxic NH_3 (as NH_4^+), elimination of some metabolic CO_2 (as HCO_3^-), adjustment of internal $[H^+]$ and $[OH^-]$, and maintenance of ionic electrical balance. The divalent Ca^{2+} probably also is taken up from the water via these beta chloride cells (Patrick et al. 1997a Verbost et al. 1997), (Fig. 6.4).

Evans (1977) reported results indicating the same exchange in four species of *marine* fish. Apparently, *active* (besides passive) uptake of Na^+ and Cl^- via these active transport ion-exchange pumps may be *necessary* for adequate excretion of NH_4^+, H^+, and HCO_3^-. If this mechanism to take up Na^+ and Cl^- is functional in marine teleosts, one may ask why all marine species are not more euryhaline. After drinking rates, urine flows, and permeability differences have been considered, the relative inefficiency of NaCl *uptake* by the marine fish compared with diffusional NaCl losses in fresh water probably represents the limiting factor (Evans 1975). Interestingly, the widely euryhaline mummichog shows a greater disparity of Na^+ and Cl^- dynamics than is illustrated in Fig. 6.4 when challenged with systemic acidosis, indicating that alternate models of "freshwater" ionic regulation exist and opening avenues for future research (Patrick et al. 1997b).

6.3 STRESS RESPONSES AND EFFECTS

Stressors, such as extremely vigorous exercise, netting and handling, and pronounced hypoxia (including air exposure), stimulate physiological changes (including ***increased gill***

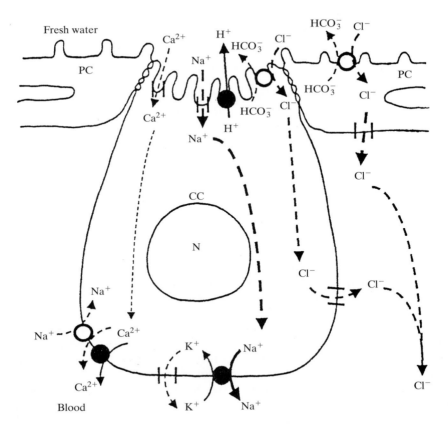

FIGURE 6.4   Model of the mechanism(s) for Na^+, Cl^-, and Ca^{2+} uptake through a β-type chloride cell (CC), with its nucleus (N), and pavement cells (PC) of the gills of a freshwater fish. Solid lines indicate active transport, and dashed lines show diffusion or exchange through membrane channels or across the cell. The ATPase pumps are represented by solid circles, co-transporter and exchangers by open circles, and ion channels by parallel lines. Passive diffusion of HCO_3^-, OH^-, H^+, and NH_4^+ to the fresh water, via paracellular pathways, are not shown. Movements of NH_3 and CO_2 across gill epithelia are shown in Figure 6.6. Modified from Marshall (1994, 2002).

permeability and possible *hydromineral imbalances*) in fish. These changes are either *adaptive*, allowing fish to respond to an emergency, or *detrimental*, leading to adverse effects and possible mortality. Less-severe and short-term stressors usually lead to responses such as mobilization of glucose for metabolic needs (Barton and Iwama 1991, Reid et al. 1992) and induction of heat shock proteins (Box 6.2). They also promote *lamellar recruitment* for enhanced gas exchange (Booth 1979) (see Chapter 3), and mobilization of more *epinephrine receptors on red blood cell membranes* for maintenance of blood O_2 transport (Reid and Perry 1991) (Fig. 6.5, see Fig. 4.6). Long-term or especially severe stressors lead to decreased growth, increased metabolic exhaustion and disease incidence, decreased reproduction (Pankhurst and Van Der Kraak 1997), and

BOX 6.2

Responses to Heat Shock (and Other) Stresses: Expression of Molecular Chaperones

Heat shock proteins (HSPs) have been found in the blood, brain, liver, red and white muscle, gills, and heart of various fishes (Iwama et al. 1998, Currie et al. 2000). They function widely among fishes as molecular chaperones, interacting with other proteins (e.g., as in the assembly, correct folding, and translocation of proteins in cells) (Iwama et al 1998). The induction of HSPs in various tissues has been reported in the mummichog (*Fundulus heteroclitus*) (Koban et al. 1991), fathead minnow (*Pimephales promelas*) (Dyer et al. 1991), and in several marine species (Dietz and Somero 1992, 1993) exposed to temperature increases (Fig. 6.5). Furthermore, exposure to water pollutants also induces HSPs in fishes (Vijayan et al. 1998), thereby enhancing their value as a "biomarker" of toxic exposure. Finally, the possible relationship between HSP induction and the neuroendocrine stress–related responses of fishes (Fig. 6.5) represents an interesting area of current research (Basu et al. 2001).

possible mortality (Barton and Iwama 1991). Mediated by ***corticosteroids***, such as ***cortisol*** (in teleosts), and by ***catecholamines***, such as ***epinephrine***, stress responses and effects typically are accompanied by increased permeability to water and small ions (Pic et al. 1974). This permeability is especially acute in the ***gills***, where most of the catecholamine uptake and removal from the circulation occurs (Nekvasil and Olson 1986a, 1986b). Thus, stress will invoke a *loss* of water (and some gain in ions) in *marine* fish and a *gain* of water (and some loss of ions) in *freshwater* fish. These increased diffusive movements probably are caused by the epinephrine-stimulated increases in "exposure" of blood (at higher pressures because of increased cardiac output [Farrell and Jones 1992, Randall and Perry 1992]) to the environment across thin membranes (because of lamellar recruitment; see Chapter 3). Exercised Nile tilapia (*Tilapia nilotica*) adapted to sea water tended to increase plasma osmotic pressure (from water loss). The same fish tended to decrease osmotic pressure (from water uptake) with exercise after adaptation to fresh water (Farmer and Beamish 1969). This trade-off between enhanced O_2 uptake and gill permeability probably is more costly for less-active species (e.g., sunfish) from lakes than for more active ones (e.g., trout and shiner minnows) from streams (Gonzalez and McDonald 1994). Those authors showed that the former group has a much higher Na^+ ion loss–to–O_2 gas gained ratio after vigorous exercise (chasing in a tank) than the latter group. Whiteley and Egginton (1999) showed that Antarctic fishes living under the permanently cold conditions of the Southern Ocean (see Chapter 36) demonstrated a limited capacity to mount catecholamine and cortisol-mediated stress responses.

Minimization of the external-to-internal osmotic gradient and incorporation of exercise conditioning can minimize capture, handling, and transport stress–induced osmotic problems under fish culture conditions. When a stressed striped bass is in fresh water, water rapidly diffuses into the fish (as revealed by a rapid gain in water weight), overwhelming osmotic and ionic regulatory controls. The consequent mortality is high compared with a group kept in brackish (near-isosmotic) salt concentrations (10 ppt salinity) (Mazik et al. 1991, Cech et al. 1996). When intensively exercised and stressed from netting and handling, subadult striped bass incur an acidotic state. During exercise/handling stress recovery in a hypertonic environment, both plasma $[Cl^-]$ and mortality rate increase compared with a group recovered in 10 ppt salinity (near-isosmotic) conditions (Cech et al. 1996). Addition of some salt (NaCl, i.e., "rock salt," at 3–8 ppt) to the tanks or transport vessels holding stressed, freshwater fish minimizes the osmotic gradient (and, consequently, water and ionic diffusion rates) between the environment and the fish's plasma, thereby increasing survival (Box 6.3).

Hypercapnic (high dissolved CO_2) stress can characterize fish in some poorly circulated natural systems, and it is becoming more common in high-density culture systems. In high-density fish culture tanks that incorporate O_2-injection systems to hyperoxygenate the water, dissolved O_2 levels remain high at high fish densities. The larger respiring biomass of more densely cultured fishes, however, produces excess excreted CO_2, thereby creating the hypercapnic conditions. The CO_2 is hydrated in the water, making *carbonic acid*, which partially dissociates to H^+ (and HCO_3^-) ions, thereby *lowering water pH*. The H^+ ions diffuse into the fish, thereby lowering their internal pH. Fish respond by conserving their metabolically produced HCO_3^- ions (to *buffer* the H^+) and increasing their *excretion of H^+* to *restore (increase) pH* toward that found under *normocapnic* (low, "normal" CO_2 levels) conditions. The H^+ efflux occurs mostly at the gills, but some of the acid excretion (e.g., 16% in rainbow trout [Perry et al. 1987a]) can be accounted for in the urine. Interestingly, conserving HCO_3^- ions decreases HCO_3^- efflux, thereby *decreasing Cl^- influx*, according to the ionic exchanges outlined earlier (Fig. 6.4). Thus, plasma Cl^- decreases in freshwater fish, such as Arctic grayling (*Thymallus arcticus* [Cameron 1976]), rainbow trout (Perry et al. 1987b), and common

---

BOX 6.3
Stressful Rides: Fish Responses to Transport and Handling

Transport stress of both largemouth bass (Carmichael et al. 1984) and the delicate delta smelt (*Hypomesus transpacificus*) was minimized after addition of salt and other water conditioners (Swanson et al. 1996). Coral trout (*Plectropomus leopardus*) responded to capture and transport stress with increases in plasma cortisol, lactate (from stuggling-associated glycolysis in white muscles) (see Chapters 2 and 3), and hematocrit (from erythrocyte release from the spleen and from adrenergic swelling of the erythrocytes) (see Chapter 4) (Frisch and Anderson 2000).

These fish also decreased their circulating lymphocytes and granulocytes, which are important immune system components. These decreases may account for the increased disease susceptibility (Fig. 6.5) observed in these fish during live transport (Frisch and Anderson 2000). Finally, both cultured and wild young-of-the-year striped bass that had been exercise conditioned (trained in flowing water) for 60 days at 1.2 to 2.4 body lengths/s regulated plasma ions significantly better than unexercised striped bass after capture and net-confinement stress (Young and Cech 1993b).

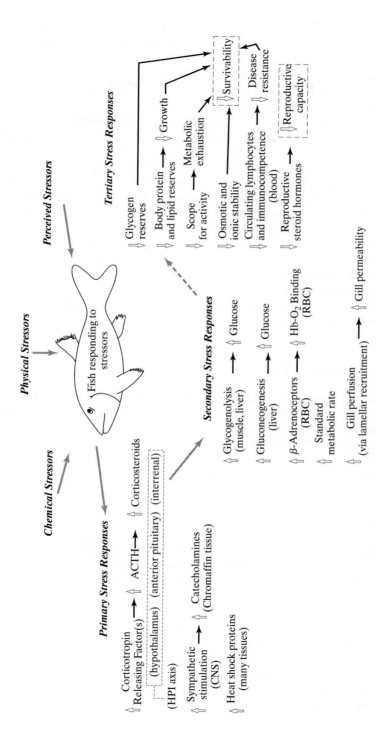

FIGURE 6.5 Adaptive (mostly consisting of the primary and secondary responses) and detrimental (mostly consisting of the tertiary responses) aspects of the major stress response pathways in fish. Dashed arrow indicate possible extention of stress response (to the tertiary level) with exposure to long-term or especially severe stress. Modified from Barton (1988, 2002).

carp (*Cyprinus carpio* [Claiborne and Heisler 1986]). The magnitude of this response varies among species. Channel catfish (*Ictalurus punctatus*) show progressive plasma Cl^- decreases as environmental hypercapnia intensifies (Cameron and Iwama 1987), whereas freshwater American eels show very slight Cl^- decreases with hypercapnic exposure (Hyde and Perry 1989). Simultaneous with the plasma Cl^- decreases, increases in H^+ efflux also *increase Na^+ influx*, according to the previously discussed ionic exchanges (Fig. 6.4), thereby increasing plasma Na^+. These increases vary with species, but they return to normal levels faster in a seawater environment, at least in rainbow trout (Iwama and Heisler 1991). The plasma Na^+ increases also are associated with decreases in plasma ammonia (from protein metabolism) effluxes in seawater-acclimated toadfish (*Opsanus beta* [Evans 1982]), although the hypercapnia-induced pH decrease lowers the proportion of dissolved ammonia that is in the un-ionized (more toxic) form. The result of these plasma Cl^- decreases and Na^+ increases is a widening difference between these two important plasma ions' concentrations under hypercapnia. The longer-term physiological implications of these compromises between the regulation of ionic concentrations versus acid–base status (see Section 6.5) associated with hypercapnia in fishes have yet to be resolved (Cameron and Iwama 1989). Long-term (weeks to months) exposure to hypercapnia, however, is known to decrease the growth rate of juvenile white sturgeon (*Acipenser transmontanus* [Crocker and Cech 1996]) and to produce nephrocalcinosis in rainbow trout (Smart et al. 1979).

6.4 FREEZING RESISTANCE

Because their body fluids are either hyperosmotic or isosmotic with their external environments, hagfish, marine elasmobranchs, and freshwater teleosts are not subject to freezing as long as their external environment remains unfrozen. Because the *marine* environment has a higher salt concentration than their body fluids, however, and, consequently, a lower freezing point, marine teleosts can freeze to death even though the water around them is still liquid. To prevent this, many cold-water marine teleosts *increase* the **osmolality** of their blood and tissues. For example, rainbow smelt (*Osmerus mordax*) increase their glycerol, urea, and TMAO concentrations, which depress their freezing point, during winter (Raymond 1994, Driedzic et al. 1998). These smelt and many other species also synthesize special **macromolecular antifreeze compounds** in their blood. These compounds can be categorized into **antifreeze glycopeptides (AFGPs)** or one of three types of **antifreeze peptides (AFPs)** having molecular masses of 2.6 to 34 kDa (Table 6.2). These compounds prevent freezing of fish tissues by bonding to the surface of nascent ice crystals in the tissues, thereby interfering with the *ice crystal growth* and *depressing* the freezing point of the blood (DeVries 1984). Apparently, the linearly arranged (a helical formation) and the globularly arranged AFPs bond in somewhat different ways (Sicheri and Yang 1995, Jia et al. 1996). Consequently, demersal Antarctic fish, such as the naked dragon fish (*Gymnodraco acuticeps*), *Trematomus bernachii*, and *T. hansoni*, can be found resting on anchor ice (DeVries and Wohlschlag 1969). This mechanism does *not* change the melting point, thereby leading to a *difference* between the melting and freezing points (typically of 1–1.6°C), which is termed **thermal hysteresis.** The smaller AFPs are found in the skin, thereby increasing its effectiveness in

TABLE 6.2   OCCURRENCE AND CLASSIFICATION OF FISH ANTIFREEZE PROTEINS (AFP) AND ANTIFREEZE GLYCOPROTEINS (AFGP)

Property	AFGP	AFP type I	AFP type II	AFP type III
General feature	Carbohydrate-rich	Alanine-rich	Cystine-rich	Slightly proline- and methionine-rich
Molecular mass	2.6–34 kDa	3.3–4.5 kDa	11.3–24 kDa	6 kDa
Primary structure	(Ala-Ala-Thr)$_n$ dissacharide	(11 Amino acid repeat)	Disulfide-linked	Unremarkable
Secondary structure	Expanded	Amphipathic α-Helical	Contains β-Sheet	Predominantly β-sheet
Fish species	Nototheniids	Right-eye flounders	Sea raven	Eel pouts
	Northern cods Polar cods	Sculpins	Smelt Atlantic herring	Wolffish

Modified from Davies et al. (1993).

preventing ice crystal propagation (Valerio et al. 1992). An aglomerular kidney is considered to be part of the freezing-resistance system in Antarctic fish, because the AFGPs are conserved rather than filtered out of the blood. Thus, energy is not needed for AFGP reabsorption, thereby lowering the energetic cost of low-temperature osmoregulation in these forms (Dobbs et al. 1974).

Duman and DeVries (1974a) showed that winter flounder (*Pleuronectes americanus*) from Nova Scotian waters display seasonal changes in serum freezing point. This flounder's serum freezing point is lowered from −0.69°C during the summer (water temperature, 17°C) to −1.47°C in the winter (water temperature, −1.2°C). Most of its circulating AFPs are synthesized in the ***liver*** via ***AFP mRNAs***. The accumulation of liver AFP mRNAs is *negatively* regulated by GH, the production of which (pituitary) is controlled by the photoperiod's effect on the central nervous system (Fletcher et al. 1989). During the short-photoperiod winter months, GH production is low, and substantial AFP production adequately depresses the winter flounder's freezing point. Removal of the pituitary (hypophysectomy) during the summer months causes, via an increase in AFP gene transcription (Vaisius et al. 1989), an increase in liver AFP mRNAs to the level of the winter months (i.e., 700-fold increase!) (Fourney et al. 1984). Other AFPs, which are synthesized (as revealed by the presence of AFP mRNAs) in skin, scales, and gills (Gong et al. 1992), appear *not* to be controlled via the pituitary (Gong et al. 1995). A combination of both warm temperatures *and* long photoperiods was necessary as a "fail-safe" system for winter flounder to lose the antifreeze completely over a period of 3 weeks (Duman and DeVries 1974b). From studies of the high cockscomb (*Anoplarchus purpurescens*) taken from cold Alaskan waters and from milder Californian intertidal waters, genetically based population differences in antifreeze production are apparent. With acclimation to cold water, the Alaskan individuals were capable of producing the antifreeze compounds, whereas the Californian population could not (Duman and DeVries 1974b). In contrast, Goddard et al. (1999) showed that a more northerly located (51°30′N) Newfoundland population of juvenile Atlantic cod developed approximately 50% higher antifreeze levels compared with those from three bays located further to the south $(47°30'-49°30'N)$, which exhibited very similar antifreeze levels. Those authors concluded that the physiological distinctness of the northern population evolved by genetic amplification of antifreeze production capacity rather than by selective mortality in these particularly harsh conditions (Goddard et al. 1999).

6.5 ACID–BASE BALANCE

The control of internal acid–base conditions within a certain range is essential for life in fish. For example, the catalytic properties of enzymes and the "regulatory responsiveness" of intracellular proteins to various enzyme regulatory functions are highly sensitive to solution conditions, including pH and ionic strength (Hochachka and Somero 1984, Somero 1986). Because the *ionization constant of water* changes with temperature, the *pH* of water and weakly buffered aqueous solutions (including blood) also changes with temperature. For example, Cech et al. (1984) measured a 0.016 pH per unit decrease for each 1°C increase in blood temperature for albacore *(Thunnus alalunga)* between 5°C and 35°C. Thus, rather than trying to keep a constant blood pH, fishes subjected to diurnal or seasonal temperature changes maintain a constant $OH^-:H^+$ ratio *(relative alkalinity)*. Hydrogen ions are continually produced by the fish's metabolic processes (e.g., from the hydration of metabolically produced CO_2). Dissolved carbon dioxide is *hydrated* and *dehydrated* in aqueous solutions according to the following reaction:

$$CO_2 + H_2O \rightleftharpoons H_2CO_3 \rightleftharpoons H^+ + HCO_3^-$$

These hydrogen ions must be buffered in the circulation until they can be excreted. Figure 6.6 shows the pathways for H^+ and CO_2 excretion in fish. The lower part of Figure 6.6 shows what happens at the tissue (e.g., red muscle) capillaries; the upper part shows what happens at the gill lamellae. As CO_2 is produced at the tissues, it diffuses across the capillary walls into the plasma. A small portion of this plasma CO_2 forces the preceding equilibria to the right, thereby slowly forming HCO_3^- and lowering blood pH (increased $[H^+]$). A much larger portion of the CO_2 diffuses across the membranes of the red blood cells (RBCs), and a fraction of this binds with the *hemoglobin* to form *carbaminohemoglobin*. Because much of the hemoglobin may be in the reduced state (deoxygenated), the hemoglobin's CO_2 capacity is relatively high (see the discussion of the Haldane effect in Chapter 4). Most of the CO_2 in the RBCs is very *quickly* converted to HCO_3^- and H^+ by the enzyme *carbonic anhydrase* (Fig. 6.6). The hemoglobin binds and, therefore, buffers the H^+. Because the HCO_3^- level in the plasma is much lower in the absence of the enzyme catalysis, "excess" HCO_3^- produced in the RBCs diffuses into the plasma, toward an HCO_3^- equilibrium across the RBC membrane. Because of the HCO_3^- movement across the membrane, the RBCs become more positively charged than the plasma, and Cl^- diffuses into the RBCs to alleviate the electrical disequilibrium. Cameron (1978) demonstrated this "chloride shift" in RBCs of a marine snapper *(Lutjanus campechanus)* and the freshwater rainbow trout. The chloride shift brings more osmotically active particles (Cl^-) from the plasma in the RBCs. Thus, water enters

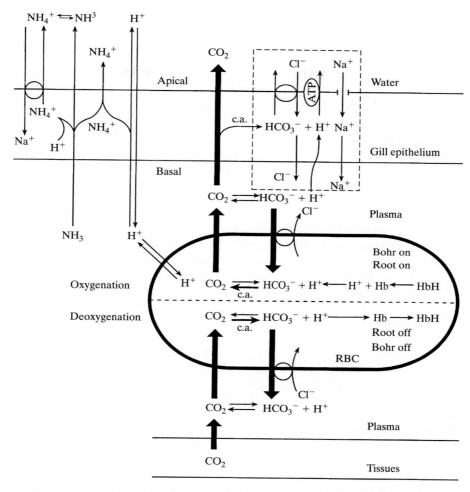

FIGURE 6.6 A schematic diagram of CO_2, H^+, and ion movements across fish gills. The lower part of the figure (i.e., deoxygenation-related pathways in erythrocyte) takes place at tissue sites; the upper part of the figure (oxygenation) takes place at the gill. Carbonic anhydrase is denoted by c.a. The area enclosed by the dotted line is shown in greater detail in Figure 6.4. Modified from Randall (1982) and Lin and Randall (1991).

these RBCs along with the Cl^-, thereby ensuring an osmotic equilibrium. This induces a slight swelling of the RBCs and causes the RBC volume (hematocrit) in venous blood to exceed that of the arterial blood by 2% to 3% in the same fish (Stevens 1968).

As the venous blood enters the gills, 95% of the carbon dioxide is in the form of plasma bicarbonate (Cameron 1978). Dissolved CO_2, which moves easily across gill epithelia, *diffuses* across the lamellar epithelium into the water, which usually has a high absorbing capacity for CO_2. This decrease in plasma CO_2 activates the dehydration of

CO_2, relatively slowly in the plasma and very rapidly in the RBCs. Consequently, most of the plasma HCO_3^- diffuses back into the RBCs (in exchange for Cl^-), where the catalyzed dehydration produces dissolved CO_2, which rapidly moves across the RBC membranes and the gill epithelia. A small fraction of the *epithelial* CO_2 is catalytically hydrated to HCO_3^- and H^+, which ultimately are "exchanged" for Cl^- and Na^+, respectively (Fig. 6.6). Lin and Randall (1991) found evidence for an ATP-powered, H^+-excreting pump on the outside membrane of rainbow trout gills. The H^+ (possibly as NH_4^+, from combination with NH_3) produced in the plasma or RBCs also is excreted in exchange for either Na^+ or H^+ (Fig. 6.6).

Although **hyperventilation** acts to "wash out" blood CO_2 and to raise arterial pH—for example, after exercise in striped bass (Nikinmaa et al. 1984)—acid–base balance primarily is controlled by adjustments of the **bicarbonate equilibrium system** (Randall and Cameron 1973). In another example, exposure of rainbow trout to **hypercapnic** (high P_{CO_2}) water induced an initial *"respiratory acidosis"* with a high blood P_{CO_2} (matching that of the environment) and, consequently, lowered blood pH (see Section 6.3). Over the next few days, blood pH almost completely recovered (i.e., acidosis almost completely **compensated**) by plasma $[HCO_3^-]$ increases at the elevated P_{CO_2} (Janssen and Randall 1975). Subsequent studies of white sturgeon exposed to "high-density fish farm" (high P_{CO_2} low pH) conditions showed similar (though only partial [35% pH recovery]) compensation responses after 96 hours of hypercapnic exposure (Crocker and Cech 1998).

The "exchanges" of H^+ for Na^+ and of HCO_3^- for Cl^- generally balance the acid–base and ionic requirements in elegant fashion (Fig 6.4). This link, however, can force unwanted adjustments in one system while compensating for a disturbance in another. The osmotic problems that fish confront in maintaining their acid–base balance are among the main reasons why they have a hard time surviving in highly *acidic waters*, such as in streams draining many mines or in lakes contaminated by acid rain or snow. Rainbow trout show significant increases in $[H^+]$ and decreases in total CO_2 after 2 days of exposure to water with pH 4. Compensatory rises in hemoglobin concentration apparently offset Root-shifted losses (see Chapter 4) in blood O_2 capacity (Neville 1979). Another major problem in low-pH environments concerns excessive *Na$^+$ losses* from the body. Because high environmental $[H^+]$ can inhibit H^+ excretion associated with Na^+ uptake across gill epithelia (Fig. 6.6), acid-exposed teleosts often die from insufficient plasma NaCl. For example, Leivestad and Muniz (1976) attributed the death of brown trout exposed to low-pH conditions in the Tovdal River in southern Norway to extreme reductions in plasma NaCl. Lampreys have lower overall blood-buffering capacities compared with teleosts, thereby leading to more dramatic blood pH decreases, and their mortality increases when they are immersed in water at pH 4.0 (Mattsoff and Nikinmaa 1988). In contrast, the cardinal tetra (*Paracheirodon axelrodi*), from the Rio Negro in the Amazon Basin, suffered *no* body Na^+ loss when exposed to water at pH 6.0, 4.0, or 3.5. Exceptional acid tolerance seems to be a general characteristic of fishes inhabiting the very "soft" and acidic (pH 3.5–5.5) Rio Negro (Gonzalez et al. 1998, 2002). Most fish from neutral or slightly alkaline environments will avoid low-pH environments if possible (Peterson et al. 1989).

6.6 LESSONS FROM HYDROMINERAL BALANCE

Most fishes have internal salt concentrations that differ dramatically from those in their external environments. Because of the permeability of gills to water and small ions, fishes experience constant water and solute fluxes. These fish continuously regulate water and solute levels via active and passive processes in organs such as gills (including chloride cells), kidney, gut, urinary bladder, and rectal gland. Marine elasmobranchs also use high concentrations of urea and TMAO to minimize the external–internal osmotic gradient, thereby slowing water movements. Stress and environmental changes in salinity or pH invoke a cascade of hormonal, metabolic, and ionic adjustments in fishes. Many cold-water marine teleosts possess antifreeze peptide compounds that limit ice crystal growth, thereby permitting survival in subzero temperatures.

SUPPLEMENTAL READINGS

Barton and Iwama 1991; Claiborne 1998; Evans 1979, 1984, 1993, 1998, 2002; Hazel 1993; Heisler 1988, 1993; Hoar 1976; Iwata 1995; Karnaky 1986, 1998; McCormick 2001; Sherwood and Lew 1994; Shuttleworth 1988; Wedemeyer 1996; Wood and Marshall 1994; Wood and Shuttleworth 1995.

WEB CONNECTIONS

GILL ION REGULATION ANIMATION
www.biology.ualberta.ca/courses/zoo1241/index.php?Page=1815

ENDOCRINE CONTROL OF OSMOREGULATION
www.st-andrews.ac.uk/~seeb/ferg/

C H A P T E R 7
Feeding, Nutrition, Digestion, and Excretion

ish must have an ***energy source*** to run the body machinery (i.e., metabolism). They also require an adequate amount of ***essential amino and fatty acids*** plus ***vitamins*** and ***minerals*** to sustain life and to promote growth. This chapter examines feeding, food requirements, and the resulting dynamics in fish.

7.1 FEEDING

Fish can be classified broadly on the basis of their feeding habits as ***detritivores***, ***herbivores***, ***carnivores***, and ***omnivores***. Within these categories, fish can be characterized further as (1) ***euryphagous***, or having a mixed diet; (2) ***stenophagous***, or eating a limited assortment of food types; and (3) ***monophagous***, or consuming only one sort of food. A majority of fish, however, are ***euryphagous carnivores***. Often, the feeding mode and food types are associated with the body form and digestive apparatus. For example, longer guts with greater surface areas typify species that feed on detritus and algae and that take in a high percentage of indigestible material, such as sand, mud, or cellulose. In contrast, carnivorous species tend to have shorter ***gut lengths***. Among carnivorous fish, however, gut lengths are often greater in those that prey on small (relative to their own size) organisms than in those that prey on large organisms. Thus, the herbivorous, euryphagous Sacramento blackfish (*Orthodon microlepidotus*) has a vastly longer gut than the carnivorous Sacramento pikeminnow (*Ptychocheilus grandis*), which feeds largely on other fish. The Sacramento hitch (*Lavinia exilicauda*) has an intermediate gut length (Fig. 7.1), corresponding to its diet of small zooplankters (Kline 1978). Another example comes from tropical species. Kramer and Bryant (1995) examined the diets and measured the gut lengths of 21 fish species from the forest streams of Panama. They found that gut lengths, relative to standard lengths of the fish, had values of 0.7 to 0.9 for carnivores, 1.1 to 2.2 for omnivores, and 5.4 to 28.7 for herbivores. ***Digestive and absorptive area*** can also be increased through the use of ***spiral valve intestines***, which are found in the Chondrichthyes and in ancestral bony fishes, such as sturgeons and lungfishes. A spiral valve is a longitudinal fold that spirals down the length of the intestine, much like a spiral staircase down the inside of a lighthouse (Fig. 7.2).

Overall, ***food demand*** is a direct function of a species' ***metabolic rate*** and is regulated, in part, by brain peptides (Box 7.1). Chapter 3 discussed the effects of body size,

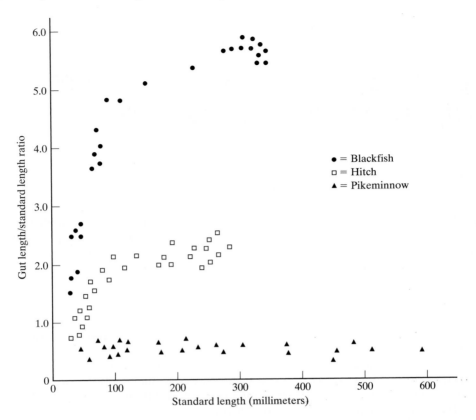

FIGURE 7.1 Relationship between the relative length of gut and fish standard length for Sacramento blackfish, hitch, and Sacramento pikeminnow. From Kline (1978).

temperature, and activity on aerobic metabolic rate. Feeding rates can also be measured directly in the laboratory. Measurements of *gastric evacuation rates* quantify feeding rates, incorporating the *relative digestibility* of prey (Bromley 1994). For example, Hopkins and Larson (1990) showed that marine rockfish (*Sebastes chrysomelas*, a euryphagous predator) digest fish faster than shrimp and shrimp faster than crab. They determined that *friability*, the ease with which a food item is fragmented in the stomach, may be an important factor in determining evacuation patterns. Chitinous exoskeletons are found on both shrimp (lighter exoskeleton) and crab (heavier exoskeletons); this may delay *enzymatic attack* or interfere with the emptying of *chyme* (partly digested food) from the stomach. A combination field and laboratory study on Bear Lake, Utah, sculpins (*Cottus extensus*) showed that increases in temperature decreased gastric evacuation time, allowing more frequent feeding and faster growth in the juveniles (Neverman and Wurtsbaugh 1994).

Structures in the buccal-pharyngeal cavity often correlate with food type and feeding habits. For example, the *pharyngeal pad* (or *palatal organ*), which is situated dorsally at the entrance to the esophagus, has been implicated in removing excess water from

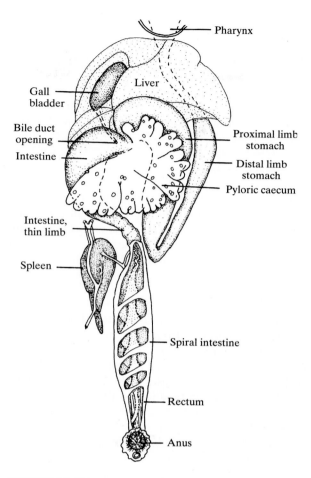

Pharynx

Liver

Gall
bladder

Bile duct
opening

Intestine

Proximal limb
stomach

Distal limb
stomach

Pyloric caecum

Intestine,
thin limb

Spleen

Spiral intestine

Rectum

Anus

FIGURE 7.2 Viscera of the paddlefish showing the spiral
valve intestine. The proximal limb of the stomach normally
lies in the midline of the coelom. The diffuse pancreas is omit-
ted from this illustration. From Wiesel (1973).

the ingested food of common carp (Jara 1957). In contrast, the ***pharyngeal valve***, which
hangs from the roof of the pharynx of parrot fish (Scaridae), probably assists in the
placement of pieces of coral for grinding by the ***pharyngeal teeth*** and lubrication from
epithelial mucous cells (Kapoor et al. 1975b). Likewise, the bony or cartilaginous ***gill
rakers*** protruding from each gill arch can be specialized for modes of feeding (Fig. 7.3).

Many of these gill-raker specializations can be found in the sunfish family (Centrarchidae). In piscivorous members of this family, such as the largemouth bass
(*Micropterus salmoides*), the gill rakers are short, stout, widely spaced, and pointed.
They function mainly in preventing the prey from escaping through (and damaging) the
gills, but they also may partially descale the prey as it passes by the rakers. The shortest
and stubbiest gill rakers in the family are possessed by members such as the redear

---

BOX 7.1

Brain Peptides Help Regulate Appetite in Fishes

Kah et al. (1989) showed that a type of neuropeptide Y (NPY), a potent appetite-stimulating factor in mammals, is present in goldfish. Injections of goldfish NPY into the goldfish brain caused an increase in food intake (Narnaware et al. 2000). In addition, NPY stimulates growth hormone secretions in the goldfish pituitary (Peter and Chang 1999).

---

sunfish (*Lepomis microlophus*), which are specialized for feeding on snails. Gill rakers of intermediate length, thickness, and spacing are found in euryphagous species, such as bluegill (*L. macrochirus*). The longest and finest gill rakers in the family are those of the crappies (*Pomoxis* spp.), reflecting the importance of zooplankton in their diets, which they "pick" individually from the water column (Box 7.2).

Prey-capture methods

Liem (1980) identified three major prey-capture methods among fishes: (1) **ram feeding**, (2) **suction feeding**, and (3) **manipulation**. A ram-feeding fish overtakes its prey by rapid swimming, thereby ramming water through its open mouth and operculae. A suction-feeding fish creates, while stationary, a strong, inward-directed water current by rapid expansion of the buccal cavity. If these methods can be thought of as two ends of a continuum, then most fish use a combination of these two methods to capture whole prey. Fish using manipulation (e.g., biting, scraping, clipping, gripping, grasping) to feed use their true or dermal teeth on their upper and lower jaws. Also, brook trout (*Salvelinus fontinalis*) use their tongue-bite apparatus to manipulate different types of prey (e.g., crickets, fish, or worms) via raking and open-mouth chewing behaviors (Sanford 2001).

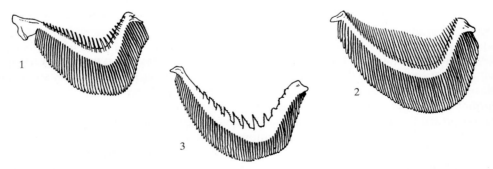

FIGURE 7.3   Gill rakers of planktophagic and predatory fishes: (**1**) powan (*Coregonus lavaretus*), (**2**) *Coregonus muksun* (Pall.), and (**3**) pike-perch (*Lucioperca lucioperca* L). The first two species feed on planktonic crustacea, and the pike-perch ingests larger prey. From Nikolsky (1963), with permission, from *The Ecology of Fishes*. Copyright by Academic Press Inc. (London) Ltd.

BOX 7.2

Fish Use Polarized Light to Locate Prey

The photons of polarized light, in contrast to diffuse light, vibrate in the same plane (Shurcliff 1962). Sunlight becomes partially polarized by molecular scattering, reflection at interfaces, and passage through optically active materials such as water (Hecht and Zajac 1974). Some fish species appear to be sensitive to polarized light. Juvenile rainbow trout were able to locate *Daphnia magna* prey at greater distances in polarized, compared with diffuse, light (Novales Flamarique and Browman 2001).

Preferences regarding prey size dictate further classification of prey-capture methods. For example, fish that remove quantities of small prey (e.g., plankton) suspended in the water are termed ***suspension feeders***. These fish typically have the finest, most closely spaced gill rakers for "filtering" (e.g., by sieving or using sticky surfaces) prey. The ***ram-suspension-feeding*** fishes directly ***sieve*** plankton out of the water while swimming and ramming water over their gill rakers. This group includes the basking shark (*Cetorhinus maximus*), paddlefish (*Polyodon spathula*), and Atlantic menhaden (*Brevoortia tyrannus*). Large schools of Atlantic menhaden decrease the phytoplankton (via suspension feeding) as well as the dissolved oxygen concentrations (via gas exchange across gill lamellae; see Chapter 3) of the waters they swim through (Oviatt et al. 1972). Dense concentrations of the food organisms in the water stimulate faster swimming by the menhaden (to 2.5 body lengths/s). Durbin and Durbin (1975) showed that the feeding response of menhaden is linked to the presence of comparatively large zooplankton or phytoplankton rather than to much greater densities of small phytoplankters, which are filtered less efficiently.

Sanderson et al. (1996) showed that juveniles of the normally ***piscivorous*** (fish eating) greater amberjack (*Seriola dumerili*) can also ram-suspension-feed, possibly capturing planktonic prey on the fine denticles of their gill rakers while swimming with mouth open and operculae flared open. ***Pump-suspension-feeding*** fishes use pulsatile gill ventilatory currents (Chapter 3) to capture food, even while stationary in the water. For example, subadult and adult Sacramento blackfish use pumped ventilatory flows and gill rakers that resemble brushy tufts to ***guide*** rather than sieve food particles to a sticky palatal organ on the roof of the mouth (Sanderson et al. 1991). Food and mucus are apparently rubbed or sloughed off the palatal organ into the esophagus. Feeding efficiency ($> 95\%$) via particle guiding in blackfish is not adversely affected by hypoxic water conditions when spreading gill rakers from hyperventilation (increased ventilatory stroke volume; see Chapter 3) would compromise direct particle sieving (Cech and Massingill 1995). Juvenile Sacramento blackfish switch to suspension feeding from suction feeding on individual particles at a small metabolic cost (ca. 1% of total metabolic energy costs), presumably to maximize energy intake (Sanderson and Cech 1992).

Mouth structure is closely related to the feeding modes and habits of fish, especially for suction feeding. Mouth structure is highly variable, and this variability explains, in part, the evolutionary success of both teleosts and elasmobranchs (see Chapter 12). The ancestral mouth consists of firm jaws lined with sharp teeth for grasping active prey. Such jaws are still possessed by many piscivorous fishes (e.g., barracuda and pike).

More common among modern fish are jaws modified for suction feeding. In suction-feeding fish, the *jaw* is *shortened* to limit the *gape*, whereas the *expansibility* of the buccal (mouth) cavity is maintained, resulting in *increased water velocity* through the smaller mouth when the cavity is expanded or contracted. In elasmobranchs, such as skates (Rajidae), eagle rays (Myliobatidae), and nurse sharks (Orectolobidae), the strong oral suction that is created allows them to feed effectively on benthic invertebrates. In teleosts, three primary motions are involved in the expansive phase of suction feeding: the *elevation of the neurocranium*, the *depression of the mandible*, and the *depression of the "floor" (hyoid) of the mouth* (Lauder 1985). The major musculoskeletal couplings involved in these movements are shown in (Fig. 7.4).

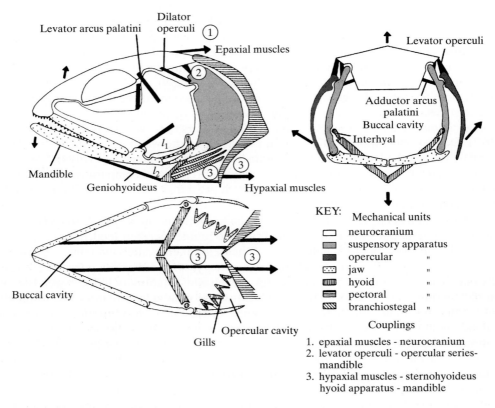

FIGURE 7.4   The major muscles and mechanical units in the head of a teleost fish. Three major musculoskeletal couplings are involved in the expansive phase of prey capture: (**1**) the epaxial muscles coupling, which causes cranial elevation; (**2**) the levator operculi coupling, which mediates mandibular depression via the opercular apparatus and interoperculomandibular ligament (ligament l_1); and (**3**) the hyoid coupling, which also governs mandibular depression via the mandibulohyoid ligament (l_2). From Lauder (1985); reprinted by permission.

Lauder (1983) showed that bluegill draw their gill bars together to firmly close the connection between the buccal and branchial (gill) cavities during suction feeding. This protects the delicate gill lamellae from hydraulic damage during suction feeding. Nemeth (1997a, 1997b) measured the feeding behavior and buccal negative pressures in kelp greenling (*Hexagrammos decagrammus*), a versatile predator. She found that this fish used a greater ram-feeding component (i.e., approached at higher speed [1,500 vs. 1,150 mm/s] and from a greater distance [13.5 mm vs. 10 mm]) and developed a more negative mean buccal pressure (-36 kPa vs. -10 kPa) to take elusive prey (e.g., live shrimp) compared with passive prey (e.g., pieces of shrimp), respectively.

The feeding of many sharks represents an unusual type of manipulation prey capture, that of taking **bites** from prey that is *larger* than the predator, such as whales or surfboards. The sawtooth edges on the awl-shaped teeth of the lower jaw and the blade-like teeth of the upper jaw teeth, coupled with a head-shaking action, provide for efficient cutting through flesh of larger, slow-moving, or disabled animals. The well-developed jaw musculature and the unique position of the hyomandibular cartilages provide support and positioning for the jaws when the mouth is open to effect a deep, gouge-like bite (Moss 1977). Studies of other species using manipulation prey capture (e.g., algae-scraping catfishes [Schaefer and Lauder 1986] or biting reef fishes [Turingan et al. 1995]) are just beginning to reveal structural and functional relationships.

7.2 NUTRITION

Most of our knowledge concerning fish **dietary requirements** comes from experimental nutrition studies conducted on cultured species, primarily salmonids. These studies have demonstrated the relative importance of dietary **proteins**, **lipids**, and **carbohydrates** for **growth (anabolism)** and for energy to run the bodily machinery (**catabolism**). Proteins, which consist of chains of amino acids, seem to be essential mainly for growth, although they also may be used for catabolic functions. The importance of proteins for growth has been shown in numerous nutritional studies that omit proteins containing **amino acids** the fish are not capable of synthesizing themselves. For example, Halver (1957) fed experimental groups of chinook salmon (*Oncorhynchus tshawytscha*) diets that were devoid of single amino acids and compared their growth rates with those of control animals that were fed diets containing all the amino acids. He found that the experimental growth rates were greatly reduced, because new structural proteins (for muscle, bone, etc.) could not be synthesized when one or more amino acids composing the specific protein chain were missing. Missing amino acids can also provoke developmental vertebral abnormalities, such as **scoliosis** and **lordosis** (Halver and Shanks 1960). These nonsynthesizable amino acids therefore become "essential" in the diet of the fish. Ten amino acids have been shown to be essential for fish: (1) **arginine**, (2) **histidine**, (3) **isoleucine**, (4) **leucine**, (5) **lysine**, (6) **methionine**, (7) **phenylalanine**, (8) **threonine**, (9) **tryptophan**, and (10) **valine**. The *quantities* of the various required amino acids that are needed, however, vary among species, and excessive amounts of any one acid also may be detrimental to growth and survival.

In wild fish, proteins are often an important source of energy for meeting metabolic demands. For example, rainbow trout in the wild feed largely on aquatic and terrestrial invertebrates, making protein a high percentage of their natural diet—far beyond what is needed for growth. In cultured fishes, however, the protein fraction of the diet usually comes from fish meal and is a comparatively expensive part of the feed. To minimize their monetary costs of operation, fish culturists include protein in quantities sufficient only for anabolic processes, and they substitute lipids or, especially, relatively inexpensive carbohydrates for a source of energy. Kieffer et al. (1998) recently determined that cultured, juvenile rainbow trout require only 30% of their diet to be protein at 15°C, even while swimming at 75% of their critical swimming velocity (see Chapter 2).

A significant *energy cost* is incurred in *breaking down (hydrolyzing)* the large, complex protein molecules to amino acids and in *synthesizing* new proteins (i.e., growth) following a meal. This cost is termed the *specific dynamic action (SDA)*, and it *increases* with the *amount* of protein in the diet. Thus, Schalles and Wissing (1976) calculated that 12.6% to 16.1% of the ingested energy was used by bluegills to digest, assimilate, and resynthesize protein from diets containing 23.9% to 45.3% protein, respectively. Boyce and Clarke (1997) showed that the ingested energy used for SDA decreased with increasing ration size in Antarctic plunderfish (*Harpagifer antarcticus*), from 56% of the ingested energy at "low" (2.5% of wet body mass) to only 10% at satiation rations. Those authors concluded that the absolute energetic cost of processing a meal was largely independent of the size of the meal across these two ration levels. Brown and Cameron (1991) showed that growth was an important component of the SDA process by infusing a normal ration of essential amino acids directly into the bloodstream of channel catfish (*Ictalurus punctatus*) and measuring a peak 56% increase in metabolism (measured as O_2 consumption rate) 4 hours after infusion. Significantly, the excreted ammonia accounted for only 21% of the total nitrogen from the infused amino acids, indicating tissue incorporation of most of the nitrogen. Conclusively, infusion of either (nondigestible) D-stereoisomer essential amino acids or a protein synthesis inhibitor (*cycloheximide*) into the catfish failed to produce the metabolic elevation.

Carbohydrates and lipids constitute the other available energy sources in foods. In natural aquatic environments, lipids are found in both animal and plant sources, whereas carbohydrates are found almost exclusively in plants. The *low digestibility* of carbohydrates by carnivores such as trout and salmon contributes to the low energy value gained from their ingestion of food such as pasta. A salmonid extracts only 1.6 kcal of energy from 1 g of carbohydrate, but they gain 3.9 kcal/g for ingested protein and 8.0 kcal/g for lipids. While being cultured, juvenile rainbow trout required only 15% of their diet to be carbohydrate at 15°C under resting conditions, this percentage increased to 25% while swimming at 15°C and to 40% while swimming at 5°C (Kieffer et al. 1998). Thus, the salmonid culturist must balance the low feed costs of carbohydrate sources (e.g., grains and cereals) with their relatively low nutritional value (Box 7.3). Of the carbohydrates, *monosaccharides* are the most digestible, followed (in order) by *disaccharides, simple polysaccharides, dextrins, cooked starches*, and *raw starches* (Halver 1976). Some herbivorous and omnivorous fishes, such as anchovies (Engraulidae), sea catfish (*Arius felis*), channel catfish, common carp, and gizzard shad (*Dorosoma cepedianum*), may utilize *gut microbes* to break down cellulose, the plant structural carbohydrate (Stickney and Shumway 1974, Smith et al. 1996). The bacteria having this cellulase activity are either maintained in the gut or regularly brought in with ingested detritus (Prejs and Blaszczyk 1977).

BOX 7.3

Adrenaline Regulates Fish Blood Sugar

As in humans, an increase in adrenaline levels in the blood of fish will increase blood sugar levels, thereby increasing the availability of energy for quick action. Thus, Weber and Shanghavi (2000) showed that epinephrine (adrenaline) infusion into intact rainbow trout caused hyperglycemia (high circulating glucose concentrations) from increased glucose production in the liver. In contrast, infusion of propranolol (a blocker of adrenergic stimulation) decreased hepatic glucose production. Basal circulating levels of epinephrine help to maintain resting glucose production from the liver. Because no epinephrine is present, isolated hepatocytes produce only half the glucose that is produced in intact trout livers (Weber and Shanghavi 2000). The mobilization of glucose (hyperglycemia) measured in stressed fish also is mediated by epinephrine increases that are associated with fish stress responses (see Chapter 6).

The seaweed-eating sea chubs (Kyphosidae) have apparently taken this ability to full fermentative capacity. ***Volatile fatty acids (VFAs)*** are the nutritive end products of ***microbial fermentation*** of plant materials such as celluloses, fiber, starches, and sugars (Bergman 1990). Two species of *Kyphosus* from Australian marine waters produce high VFA concentrations in ***caecal pouches***, containing a diverse microflora, located near the posterior end of their intestines (Rimmer and Wiebe 1987). Two Hawaiian *Kyphosus* species, along with the warm-temperate halfmoon (*Medialuna californiensis*), produce a full complement of six VFAs (Kandel et al. 1994), which are rapidly absorbed at the site of production and contribute substantially to these herbivores' energy requirements (Bergman 1990). Smith et al. (1996) found that wild, omnivorous common carp and gizzard shad as well as the largely piscivorous largemouth bass also produce intestinal VFAs, the concentrations of which decrease during the cool season, presumably because of ***temperature limitations*** on fermentation. These authors propose that low temperatures pose a limitation on fermentation as a digestive strategy among fish.

Lipids represent a rich source of energy for fish in general and are an important component in the diet of predaceous fish. Besides their ***high specific energy value*** (8.0 kcal/g), they also are almost completely digestible (Halver 1976). Kieffer et al. (1998) found that rainbow trout ***lipid use*** exceeds 50% of the total fuel used while at rest at 15°C. This percentage decreases to approximately 40% in swimming fish and to 35% in resting rainbows at 5°C. Rapid growth rates typically are achieved by predaceous fishes, such as the mackerels, billfishes, salmon, pikes, and sharks. The high lipid content of a diet consisting of small fish maximizes growth by ***sparing*** the ingested protein for tissue synthesis. Besides being an energy source, lipids provide ***essential fatty acids***, such as ***linolenic acid*** (Millikin 1982). Fatty acids are used in the construction of fats or oils (***triglycerides***) to be stored by a fish for later use as an energy source. A classic example are Pacific salmon, which accumulate lipids at sea and expend them while ***fasting*** during migrations upstream to spawn. Experiments with catfish (*Ictalurus punctatus*) have shown that body lipids synthesized by fish for energy storage parallel those ingested in terms of saturation (completeness of hydrogen bonding of constituent fatty acid carbon chains) (Andrews and Stickney 1972).

The relative importance of lipids and proteins as energy sources is also shown by their mobilization by fishes during periods of *starvation*, which can be a regular occurrence in the life cycle of many fish. For example, winter flounder (*Pleuronectes americanus*) inhabiting coastal Maine waters fast while in deeper water from January to May (Bridges et al. 1976). Pacific salmon as well as Atlantic salmon and steelhead trout fast during their spawning migration. Because no carbohydrates, protein, or fat are taken into the body during a fast, the fish must use compounds that are *stored* in body tissues. Savitz (1971) showed that bluegill utilize body protein and, especially, fat to meet the body's energy demands when fasting. The quantities of body fat and protein are significantly reduced in the starved fish, whereas inorganic content (ash) remains about constant. Protein depletion is presumably accomplished by the high concentration of *proteolytic enzymes* in fish muscle (Siebert et al. 1964). Losses in body fat or protein, as seen in bluegill, rarely are reflected by significant changes of body weight in the fish. Instead, the metabolized fat or protein is replaced by *water* to make up the body weight difference. For example, the whole-body water content of sockeye salmon was shown to increase from approximately 60% to 77% during the spawning migration, whereas the sum percentage of lipid plus water was constant at approximately 80% (Idler and Bitners 1959). The familiar phenomenon of ascending salmon possessing an atrophied gut but still striking an angler's bait represents an interesting contradiction. Even if food were swallowed, the degeneration of the digestive tract and the significant decrease in gastric enzyme secretion indicate that very little of the food could be digested.

At the other end of the feeding spectrum, what happens when fish have unlimited food available to them? In their investigation of unrestricted feeding in juvenile rainbow trout, Grayton and Beamish (1977) found that trout held at 10°C would consume only just under 4% of their wet body weight per day of dry, prepared trout pellets. The trout would consume this quantity whether they were offered pelletized food in unlimited quantities twice daily or up to six times daily. As one would expect, growth rate also did not vary with feeding frequencies from two to six feedings daily. In contrast, Balon (1977) described deep-bodied, *obese body shapes* associated with an extreme abundance of food. Deep-bodied salmonids (including rainbow trout), pikes (Esocidae), carps, and others have been described in cultured or natural environments where food is very abundant. As described in Chapter 8, fish tend to grow throughout their lives, but they increase more in *girth* than in *length* toward the end of their lives.

As in other animals, the metabolic conversion of biochemical compounds, either to provide energy or to synthesize other compounds (e.g., enzymes, structural proteins, and stored triglycerides), requires particular cofactors to proceed. These cofactors, which largely are unavailable in the body, constitute the vitamins. The vitamin requirements of fish probably vary somewhat with the species, although only a few species (mostly of commercial value) have been investigated (Table 7.1). Dietary deficiencies of the vitamins that are essential for life and growth provoke a variety of physiological disturbances (Table 7.2). As with the research into essential amino acids, most of the vitamin deficiencies were determined by single-vitamin deletions in an otherwise complete diet.

TABLE 7.1 VITAMIN REQUIREMENTS FOR GROWTH[a]

Vitamin	Rainbow trout	Brook trout	Brown trout	Chinook salmon	Coho salmon	Carp	Eel	Gold fish	Yellow tail	Channel catfish
					Requirement (mg/kg dry diet)					
Thiamine	10–12	10–12	10–12	10–15	10–15	R[b]	R	R	R	R
Riboflavin	20–30	20–30	20–30	20–25	20–25	7–10				R
Pyridoxine	10–15	10–15	10–15	15–20	15–20	5–10				R
Pantothenate	40–50	40–50	40–50	40–50	40–50	30–40		R	R	R
Niacin	120–150	120–150	120–150	150–200	150–200	30–50				R
Folacin	6–10	6–10	6–10	6–10	6–10	?				R
Cyanocobalamin	R	R	R	0.015–0.02	0.015–0.02	?				R
myo-Inositol	200–300	R	R	300–400	300–400	200–300				R
Choline	R	R	R	600–800	600–800	1,500–2,000				R
Biotin	1–1.2	1–1.2	1.5–2	1–1.5	1–1.5	1–1.2	R			R
Ascorbate	100–150	R	R	100–150	50–80	R		R	R	R
Vitamin A	2,000–2,500	R	R	R	R	1,000–2,000		R	R	
Vitamn E[c]	R	R	R	40–50	R	R		R	R	R
Vitamin K	R	R	R	R	R	80–100			R	R

From Halver (1972).

[a]Fish fed at reference temperature with diets at about protein requirement.

[b]R, Required.

[c]Requirement directly affected by amount and type of unsaturated fat in the diet.

TABLE 7.2 VITAMIN DEFICIENCY SYNDROMES

Vitamin	Symptoms in Salmon, Trout, Carp, and Catfish
Thiamine	Poor appetite, muscle atrophy, convulsions, instability and loss of equilibrium, edema, poor growth
Riboflavin	Corneal vascularization, cloudy lens, hemorrhagic eyes–photophobia, dim vision, inco-ordination, abnormal pigmentation of iris, striated constrictions of abdominal wall, dark coloration, poor appetite, anemia, poor growth
Pyridoxine	Nervous disorders, epileptiform fits, hyperirritability, ataxia, anemia, loss of appetite, edema of peritoneal cavity, colorless serous fluid, rapid postmortem rigor mortis, rapid and gasping breathing, flexing of opercles
Pantothenic acid	Clubbed gills, prostration, loss of appetite, necrosis and scarring, cellular atrophy, gill exudate, sluggishness, poor growth
Inositol	Poor growth, distended stomach, increased gastric emptying time, skin lesions
Biotin	Loss of appetite, lesions in colon, coloration, muscle atrophy, spastic convulsions, fragmentation of erythrocytes, skin lesions, poor growth
Folic acid	Poor growth, lethargy, fragility of caudal fin, dark coloration, macrocytic anemia
Choline	Poor growth, poor food conversion, hemorrhagic kidney and intestine
Nicotinic acid	Loss of appetite, lesions in colon, jerky or difficult motion, weakness, edema of stomach and colon, muscle spasms while resting, poor growth
Vitamin B_{12}	Poor appetite, low hemoglobin, fragmentation of erythrocytes, macrocytic anemia
Ascorbic acid	Scoliosis; lordosis; impaired collagen formation; altered cartilage; eye lesions; hemorrhagic skin, liver, kidney, intestine, and muscle
p-Aminobenzoic acid	No abnormal indication in growth, appetite, mortality

From Halver (1972).

7.3   DIGESTION AND ABSORPTION

Digestion in fish concerns the **breakdown** of foods by enzymatic and, in many cases, **acidic secretions** in the gut. The diversity of foods found in the guts of fish attests to the variety of morphological and chemical adaptations that have evolved for digestion. The esophagus of fish often contains many **mucous cells** and functions as a lubricated transit tube between the buccal-pharyngeal cavity and the lower gut. The lower gut of many fish (especially carnivorous ones) contains a true stomach that is characterized by a layer of smooth muscle (**muscularis mucosa**) tissue. On the other hand, development of a **gizzard** (for masticatory as well as secretory digestive processes), as found in mullets and shad, is a stomach specialization for microphagous food habits.

The gastric mucosa of the stomachs of carnivorous fish produces a protease (protein breakdown) enzyme (e.g., **pepsin**) with an optimal activity at a pH of 2 to 4. **Hydrochloric acid** is also secreted by the gastric mucosa glands in these species, creating the low-pH environment. Gastric acid secretions are stimulated by **stomach distension**, which apparently activates cholinergic (mimicking acetylcholine response) neural fibers. The "secretory" signals of these fibers to the acid glands can be blocked with injections of the neural-blocking agent **atropine**. The rates of both gastric acid and pepsin secretion are influenced by **temperature**. As temperatures increase (up to a point), the rates of secretion also increase. These increased secretions largely account for the three- to fourfold increases in digestion rate that follow 10°C increases in temperature (Kapoor et al. 1975b).

Proteins are also broken down in the **alkaline** medium of the intestine by the enzyme **trypsin**. Trypsin is secreted by **pancreatic tissue**, which may be concentrated in a compact organ, as in mackerel, or diffusely located in the mesenteric membranes surrounding the intestine and liver. Some fish possess one or more ***pyloric caeca*** (see Fig. 9.1), which are blind pouches of secretory tissue located near the pyloric valve at the stomach–intestine junction. The caeca apparently are filled by relatively rapid, strong, retrograde (reverse-direction) contractions, instead of the normal, orthograde, peristaltic contractions, in the pyloric region of the gut (Rønnestad et al. 2000). Trypsin may be secreted from the caecal or the pancreatic tissue, which commonly envelopes the caeca.

Fish also have enzymes that break down carbohydrates (***carbohydrases***) and fats (***lipases***). The pancreas appears to be the primary site of carbohydrase (e.g., ***amylase***, which breaks down starch) production, although the intestinal mucosa and pyloric caeca represent additional production sites in various species. Presumably the pancreas is the primary site of lipase production as well. Lipase activity has been found, however, in extracts of the pyloric caeca and upper intestine as well as in the pancreas in mackerel, menhaden, scup (*Stenotomus chrysops*), and searobins (Triglidae) (Chesley 1934).

Both the presence and the quantity of digestive enzymes seem to correlate with the diet. The diet of marine pinfish (*Lagodon rhomboids*, Sparidae) switches from a primarily carnivorous one in juveniles to a primarily herbivorous one in adults. Gallagher et al. (2001) found that liver enzyme activities accompanied this dietary shift. Alanine amino transferase is involved in the primary protein deamination pathway in fish (Cowey and Walton 1989), and its activity decreases significantly in adult compared with juvenile pinfish (Gallagher et al. 2001). Conversely, pyruvate kinase is an important enzyme in energy-producing pathways from glucose substrates, and its activity is significantly increased in adult compared with juvenile pinfish (Gallagher et al. 2001). Many herbivorous and some omnivorous fish lack a true stomach. These species also lack pepsin as a low-pH, proteolytic enzyme (Kapoor et al. 1975b). Omnivorous species, however, have amylase activities in the gut many times the level of that found in carnivorous species (Volya 1966). Initiation of gastric secretions of pepsin and acid in white sturgeon (*Acipenser transmontanus*) larvae is concomitant with the switch from ***endogenous*** (yolk) to ***exogenous*** (feeding) nutrition (Buddington and Doroshov 1986).

Compounds that are broken down by the actions of pharyngeal teeth, gizzards, and/or secretions of acid and enzymes are subsequently absorbed through the intestinal wall. Absorption (or assimilation) can be estimated by the difference between the quantity and quality (energy value in kilocalories or joules[1]) of the food ingested and of the feces excreted. These estimates are even more precise when fecal nitrogen and calories from nonfood sources (e.g., sloughed gut wall) are considered.

Absorption of nutrients across fish intestinal walls closely parallels the process described for mammals, although we know more about fish mucosal (brush-border) intestinal surfaces compared with the basolateral membranes of these absorptive epithelia. Nutrients may enter cells via ***diffusion*** (i.e., nonmediated) or via ***mediated*** processes employing a ***membrane transporter protein***. The active entry of monosaccharides and amino acids into absorptive cells is driven by a Na^+ gradient powered by Na^+ pumps (incorporating Na^+-K^+-ATPase; see Chapter 6). The transporters can be very specific, and some substrates competitively inhibit the uptake of others. For example, the mediated entry of

---

[1]Note that 4,183 Joules = 1 kilocalorie (kcal).

D-glucose across the brush border can be inhibited by D-galactose, and leucine transport may be inhibited by L-form neutral amino acids (Collie and Ferraris 1995). Nutrient uptake can be influenced by dietary composition and hormones (e.g., growth hormone and androgens), but the associated mechanisms remain the subjects of intense research. Collie and Ferraris (1995), Jürss and Bastrop (1995), and Watanabe et al. (1997) provide interesting reviews on nutrient absorption.

7.4 EXCRETION

Digestive breakdown of either **lipids** or **carbohydrates** yields **water** and **carbon dioxide** as the end (i.e., waste) products. Water is either conserved, excreted, or diffused away, depending on the salinity of the environment (see Chapter 6). Carbon dioxide enters into the bicarbonate equilibrium system, and most is excreted at the gills (see Chapter 6). **Protein** digestion yields **nitrogenous compounds** in addition to carbon dioxide and water. In teleosts, these nitrogenous wastes typically take the form of ammonia. Thus, teleosts are primarily **ammoniotelic**. Ammonia must be readily excreted, however, because it is acutely toxic. Acute toxicity levels (lethal concentration to 50% of test organisms after a 96-hour exposure, LC_{50}) range from 0.068 to 2.0 mg/L for freshwater fishes and from 0.09 to 3.35 mg/L for marine fishes (Handy and Poxton 1993). Despite its toxicity, ammonia has many advantages compared with urea or uric acid as the chief excretory product of nitrogen metabolism—as long as the animal resides in an environment with abundant water. First, the small molecular size and high lipid solubility permits un-ionized ammonia (NH_3) to diffuse easily across the gills (see Fig. 6.6). Second, ionized ammonia (NH_4^+) is exchanged for Na^+ at the gills for the maintenance of relative alkalinity and internal ion balance (see Chapter 6). Third, conversion of ammonia to either urea or uric acid requires energy. Thus, in contrast to terrestrial forms, less energy is required to complete nitrogenous compound catabolism, and in teleosts, the end products resulting from this catabolism largely are released at the **gills** rather than the kidney. For example, common carp and goldfish excrete 6- to 10-fold as much nitrogen at the gills as at the kidney. Of the total nitrogenous excretion, 90% is typically in the form of ammonia, and only 10% consists of urea (Smith 1929, Wood 1993). Intertidal fish, such as the tidepool sculpin (*Oligocottus maculosus*), also excrete their nitrogenous wastes in similar proportion, with 8% to 17% being excreted as urea (Wright et al. 1995). The amphibious mudskipper (*Periophthalmodon schlosseri*) lives in burrows during high tides in brackish, mud flat habitats in Malaysia and Singapore. This fish does not produce urea (Morii 1979), but it is unusual in that it can tolerate much higher environmental ammonia levels compared with other teleosts (Peng et al. 1998) and can actively excrete NH_4^+ against a concentration gradient (Randall et al. 1999).

Elasmobranchs as well as coelacanths excrete **urea** as the primary nitrogenous end product (i.e., are **ureotelic**). As discussed in Chapter 6, much of the urea is retained in these marine fish, thereby giving their body fluids a near-isosomotic relationship with their environment. The elasmobranch **kidney** filters urea from the blood plasma at the glomerulus. Much of the urea is subsequently recovered from the filtrate by active tubular resorption, which prevents major losses of urea in the urine (Schmidt-Nielsen 1975).

Lungfish and, to varying degrees, teleosts possess the biochemical machinery to be *either* ammoniotelic or ureotelic. For example, the African lungfish sometimes must endure extensive droughts. When the aquatic environment is drying up, the lungfish constructs a cocoon of mucus in the bottom mud and estivates there until the water returns. The African lungfish is mostly ammoniotelic while aquatic, but it shifts to complete ureotelism while *estivating* and survives by metabolizing the proteins in its muscles. This shift is made possible by high concentrations of the necessary *enzymes* for urea production in its liver tissue (Janssens and Cohen 1966). Nontoxic urea may accumulate in the blood of lungfish to concentrations of 500 mmol/L after a 3-year estivation (Smith 1961). In contrast, the liver of Australian lungfish possesses only 1% of the concentration of urea-synthesizing enzymes found in African lungfish. This finding is in accord with the obligatory aquatic habits of the Australian lungfish, which does not estivate.

In contrast, some water-breathing teleosts, such as toadfish (*Opsanus beta*) and the Magadi tilapia (*Oreochromis alcalicus*), also have the necessary liver enzymes for substantial ureogenesis (Randall et al. 1989, Mommsen and Walsh 1991). Toadfish are *intermittently ureotelic*, such as during *confinement* in a small volume of water (Walsh et al. 1994). When confined, toadfish excrete most of the urea in irregular pulses (one or two per day) from the gill region (Wood et al. 1998). Experiments using *phloretin* (a urea transporter inhibitor) and observations of *bidirectional* movement of urea when external urea concentrations were artificially increased led these authors to suggest that these urea excretion events may be linked to the periodic activation of a facilitated urea transporter in the gills (similar to the vasopressin-regulated urea transporter in the mammalian kidney). Pilley and Wright (2000) showed that bidirectional urea transporters also exist in rainbow trout embryos, which excrete as much as 26% of their nitrogenous wastes as urea (Wright and Land 1998). The Magadi tilapia are *exclusively ureotelic*, apparently to survive in their extremely *alkaline* (pH 10) lake ecosystem (Randall et al. 1989). The relative dearth of hydrogen ions in such a highly alkaline environment makes excretion of NH_3 virtually impossible. Three reasons for this phenomenon are the potentially low blood-to-water NH_3 gradient, the high buffer capacity of Lake Magadi water (inhibiting NH_3-to-NH_4^+ conversion in water passing over the gills), and the inhibition of Na^+/NH_4^+ exchange (see Chapter 6) across the gill membranes (Wood et al. 1989). Nitrogen excretion and metabolism in fishes is reviewed by Walsh (1998).

7.5 LESSONS FROM FEEDING, NUTRITION, DIGESTION, AND EXCRETION

Like all organisms, fishes require an energy source to fuel their body machinery and processes, including growth, metabolism, and reproduction. Fishes have evolved feeding structures and mechanisms that allow them to exploit a vast array of plant and animal food sources, ranging from the indiscriminate filtering of a large, ram-feeding planktivore to the precision biting of a manipulating carnivore. Similarly, the guts of fishes incorporate numerous adaptations for the efficient breakdown and absorption of essential nutrients, including appropriate enzymes and absorptive surface areas. Finally, nitrogenous wastes must be excreted, principally as ammonia (or urea in strongly alkaline environments). Typically, the various nutritional, digestive, and excretory components are tightly linked with the relevant dietary characteristics. For example, predatory fishes with a high-protein diet usually have a relatively short gut (featuring a true stomach with acidic secretions), protease (and other enzymatically) assisted food breakdown, and substantial ammonia secretions.

SUPPLEMENTAL READINGS

Budker 1971; Collie and Ferraris 1995; Donald 1998; Goldstein et al. 1967; Halver 1972, 1976; Holmgren et al. 1983; Jürss and Bastrop 1995; Kapoor et al. 1975b; Lauder 1983, 1985; Love 1970; Narnaware et al. 2000; Nikolsky 1963; Nilsson and Holmgren 1993; Savitz 1971; Stickney and Shumway 1974; Walsh 1998; Watanabe et al. 1997; Wood 1993.

WEB CONNECTIONS

FISH DIGESTIVE ANATOMY, FLUIDS, AND ENZYMES
www.aqualex.org/html/onedin/fish_feeding/digestion/index.html

COD AND HADDOCK FEEDING, DIGESTION, DIGESTIVE ENZYMES, AND EXCRETION
www.umaine.edu/aquaculture/GeneralInfo/Biology/Anatomy/digestion.htm

AQUACULTURE-ORIENTED NUTRITION, FEEDING RATE, AND FEEDS
www.ext.vt.edu/pubs/fisheries/420-256/420-256.html

CHAPTER 8
Growth

Most fish continue to grow throughout their lives. Consequently, ***growth*** has been one of the most intensively studied aspects of fish biology, because it is a good indicator for the health of both individuals and populations. Rapid growth indicates abundant food and other favorable conditions, whereas slow growth is likely to indicate just the opposite. Growth can be defined as the ***change in size (length and/or weight)*** over time or, energetically, as the ***change in calories*** that are stored as somatic and reproductive tissue. The energetic definition is particularly useful for understanding factors that affect growth in fishes, because ***ingested food energy (I)***, measured in calories, must emerge either as energy that is expended for ***metabolism (M)*** or ***growth (G)*** or as ***energy excreted (E)*** (Brett and Groves 1979). This can be simply expressed as

$$I = M + G + E \tag{1}$$

As explained in Chapter 3, metabolic energy expenditures can include calories that are expended for body maintenance and repair, for digesting food, and for movement. Excreted energy can take the form of feces, ammonia, urea, and the small quantities of mucus and sloughed epidermal cells of the skin. The remaining factor in the energy equation is growth.

8.1 FACTORS AFFECTING GROWTH

Because growth is usually *positive* (e.g., increase in weight over time), a positive energy balance in metabolism is indicated. Metabolism is the sum of ***anabolism*** (the tissue synthesis or "building up" aspect of metabolism) plus ***catabolism*** (the energy-producing breaking of chemical bonds or "tearing down" aspect). Thus, the rate of anabolism exceeds that of catabolism in a growing fish. The principal factors controlling anabolic processes are ***growth hormones*** secreted by the ***pituitary*** and ***steroid hormones*** from the ***gonads*** (see Section 8.2). The growth rate of fish is highly variable, however, because it is greatly dependent on a variety of interacting environmental factors, such as water temperature, levels of dissolved oxygen and ammonia, salinity, and photoperiod. Such factors interact with each other and with other factors, such as the degree of competition, the amount and quality of food ingested, and the age and state of maturity of the fish, to influence growth rates.

Temperature is among the most important environmental variables. For example, growth in desert pupfish (*Cyprinodon macularius*) increases with temperature up to 30°C before falling off some what at 35°C. Brett et al. (1969) measured the weight gain in fingerling sockeye salmon (*Oncorhynchus nerka*) at several ration (meal size) levels. As for pupfish, maximal growth rates in the young salmon are achieved at intermediate temperatures (15°C). It is also noteworthy in the study by Brett et al. that the maintenance ration (ration at zero growth) increased with temperature, reflecting the increased standard or maintenance metabolism (*M* in Eqn. [1]) at warmer temperatures (see Chapter 3). At any particular temperature, there often is an *optimal ration* for maximum growth. For example, juvenile mosquitofish (*Gambusia affinis*) show increased food consumption and growth with increasing temperature to a 30°C to 35°C peak, with *ad libitum* (to satiation) rations (Fig. 8.1). Growth rate decreases dramatically with reduced (20% of ad libitum) rations, and the growth peak *shifts* to 25°C, presumably because of increased metabolic (*M*) demands. *Gross efficiency* (*G/I*) is maximal at 25°C to 30°C (Fig. 8.1). Fish that vertically migrate on a diel basis between strata of different temperatures can optimize their feeding and digestive efficiencies to maximize growth. Juvenile Bear Lake sculpin (*Cottus extensus*) shorten the gut evacuation time in warmer (13–16°C) surface waters of Bear Lake at night, thereby allowing them to feed at the bottom (30–40 m deeper, 5°C) the following day. This alternation of temperatures increases their growth by 300% compared with fish held at a constant 5°C (Neverman and Wurtsbaugh 1994). Overall, fish tend to prefer temperatures at which their growth is maximal (Jobling 1981).

Dissolved oxygen levels, although temperature-dependent, are often by themselves an important factor affecting the growth rates of fishes. Stewart et al. (1967) measured a significant reduction in growth rate and food conversion efficiency in juvenile largemouth bass when dissolved oxygen concentrations fell below approximately 5 mg/L at 26°C. Presumably, the reduced oxygen below this threshold precludes "extra," aerobic, energy-requiring activities, such as growth and reproduction above maintenance energy costs. These fishes (called *oxygen regulators* and including largemouth bass, channel catfish, striped mullet, Sacramento blackfish, and others) maintain a homeostatic level of metabolism as oxygen levels are reduced (see Chapter 3). In some cases, attempts to swim to more favorable environments will be made by these species.

Ammonia is the primary excretory product of fish, but if it is present at high concentrations, it will slow growth rates. For example, juvenile channel catfish display a linear drop in weight gain with increasing ammonia in their water (Fig. 8.2). Although the mechanism of growth inhibition by ammonia is still unknown, there may be a relationship with food consumption rates. Juvenile lake trout (*Salvelinus namaycush*) have been found to decrease their food consumption rate along with their growth rate at un-ionized ammonia concentrations greater than 198 μg/L (Beamish and Tandler 1990). Obviously, this information has important applications in fish culture systems. Culture systems designed to maximize growth rate must have either high flows of fresh water to carry away excreted ammonia or have ammonia-removal systems, such as green plants or "biological filters" utilizing appropriate bacteria. It is generally acknowledged that *un-ionized*

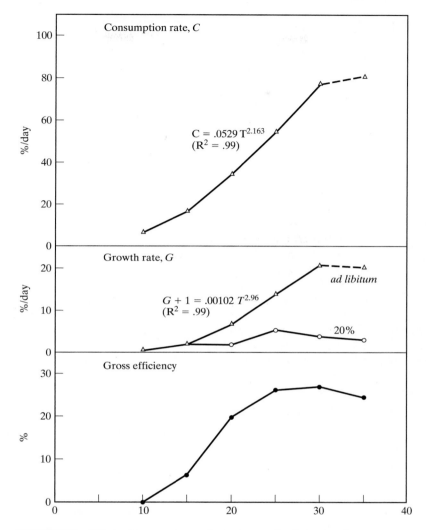

FIGURE 8.1   Effects of temperature on instantaneous food consumption rate, instantaneous growth rate, and gross efficiency of juvenile mosquitofish fed ad libitum. Growth rates of fish fed at 20% of ad libitum also are shown. All data are based on dry weights of fish and food. Regression lines for consumption and growth rates were fitted for temperatures from 10°C to 30°C. From Wurtsbaugh and Cech (1983).

ammonia (NH_3) in the water produces more toxic effects on fish (see Chapter 7) than an equal concentration of the ionized form (NH_4^+). Because the relative proportion of the two forms depends on water pH, regular monitoring of pH is an essential part of the operation of intensive freshwater-fish culture systems. Although ammonia is a "natural" compound, its effects on fishes are typical of many pollutants, which also reduce growth rates when present at sublethal levels.

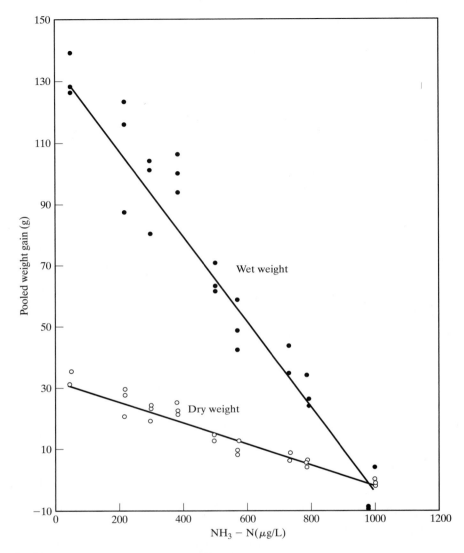

FIGURE 8.2 Effects of un-ionized ammonia on growth as measured in terms of pooled weights of the fish in each aquarium. From Colt and Tchobanoglous (1978).

Salinity also affects growth rate. The euryhaline (broad salt tolerance) desert pupfish shows maximum growth rate at 35 ppt salinity compared with both higher and lower salinities (Kinne 1960). Growth is altered as other energy-demanding components (e.g., ion and osmoregulatory active transport systems) respond to environmental characteristics. These responses increase maintenance energy requirements (M), which will decrease growth rate (G) if ingested food energy (J) and energy excreted (E) remain constant (Brett 1979). This topic is discussed more completely in Chapter 6.

Competition, either within or among species, for limited food supplies may slow growth. Swingle and Smith (1940) showed that bluegill, a species in which both adults and young eat virtually the same aquatic invertebrates and are not cannibalistic, become stunted when the population size reaches a particular level. Fertilization of the pond will increase the invertebrate food base and, consequently, the total biomass of bluegill. The average size of the bluegills, however, remains small as growth slows and some reproduction continues.

Food availability and quality also interact with other factors, particularly temperature, to affect the growth of fishes on a seasonal basis. For example, Gerking (1966) found marked seasonal differences in growth (length increases) in northern Indiana bluegill populations (Fig. 8.3). Bluegill growth was accelerated during the warmer months of plentiful food. Figure 8.3 also shows the reduced rate of growth (especially gains in length) that is consistent with advancing age in bluegill and also is typical of most other fishes. Striped mullet from southern Texas coastal waters show cycles of seasonal growth similar to those of bluegill, except that growth virtually ceases during the warmest months of midsummer through midautumn. This leveling off of growth when food is abundant probably can be attributed to excessive water temperatures reducing *assimilation efficiency* (Cech and Wohlschlag 1975, 1982). *Compensatory growth* occurs after feeding is temporarily

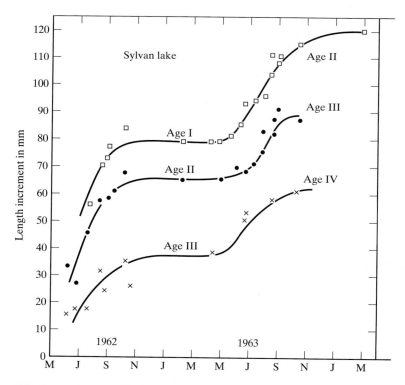

FIGURE 8.3   Seasonal gains in length of a population of bluegill (*Lepomis macrochirus*) by age groups. The growth increment varies with season and decreases with age and size. From Gerking (1966).

stopped, and growth accelerates when feeding is resumed. Hayward et al. (1997) showed that green sunfish (*Lepomis cyanellus*) × bluegill (*L. macrochirus*) hybrids denied food for 14- and 2-day cycles displayed compensatory growth that exceeded ad libitum ration growth by 1.4- to 2-fold, respectively! Compensatory growth may be associated with higher post–food-deprivation feeding rates. Gibel carp (*Carassius auratus gibelio*) increased their food consumption rate and gross efficiency after only 1 week of food deprivation (Qian et al. 2000).

Regarding food quality, a **complete diet** (with essential **amino acids**, **fatty acids**, and **vitamins**) is required for high growth rates in fishes (see Chapter 7). High dietary protein content often stimulates increased growth. Dwarf gourami (*Colisa lalia*) and largemouth bass (*Micropterus salmoides*) are two species in which increased **dietary protein** (up to 45% and 37%, respectively, of their respective diets) increased growth to maximal levels (Shim et al. 1989, Brecka et al. 1996).

Photoperiod (day length) also may affect seasonal growth phenomena. For example, Hogman (1968) found a close association between the growth of lake whitefish (*Coregonus clupeaformis*) and seasonal photoperiod but no relationship between the spring water temperatures and growth.

Age and maturity usually are the best predictors of **relative growth rates** in fishes, although the absolute growth rates are influenced strongly by environmental factors. Thus, fish typically grow very rapidly in length during the first few months or years of life until maturation. Then, increasing amounts of energy are diverted from growth of **somatic tissues** to growth of **gonadal tissues**. As a consequence, growth rates of mature fish are much slower than those of immature fish. Partly because of the amount of gonadal tissue, however, mature fish are typically heavier than immature fish per unit of length. This is reflected in their higher **condition factor (K)**, an index of "plumpness":

$$K = \frac{W(100)}{L^3} \qquad (2)$$

where W is the weight of the fish in grams and L is the length of the fish in centimeters. The condition factor frequently is used by fishery biologists as an indicator for the health of a fish population. If the fish in a population have high K values, then plenty of food probably is available to support both somatic and gonadal growth.

Exercise conditioning (training) has been shown to increase growth in several fishes. For example, Young and Cech (1993a) found that young-of-the-year striped bass, continuously swimming for 60 days at 1.2 to 2.4 body lengths/s, grew 7% to 10% faster than unexercised (<0.2 body lengths/s) control fish. Houlihan and Laurent (1987) report that rainbow trout continuously swimming at 1 body lengths/s for 42 days grew 118% faster than unexercised controls! Increased growth rates associated with exercise conditioning have been attributed to various factors in different species. These include (1) better food conversion efficiency (Davison and Goldspink 1984), (2) increased growth hormone levels (Barrett and McKeown 1988a, 1988b), (3) increased protein synthesis rates (Houlihan and Laurent 1987, Christiansen et al. 1989), (4) decreased antagonistic behavior (East and Mangan 1987), and (5) decreased stress levels (Woodward and Smith 1985). Young and Cech (1994) found that significant conditioning-related growth increases were still detectable in striped bass 56 days after conditioning (60 days of swimming at 1.5–2.4 body lengths/s) had stopped.

8.2   GROWTH REGULATION

Photoperiodic and other factors affecting growth rates may well act through variations in hormone secretions. Fish *growth hormone (GH)* is synthesized in the *alpha cells* of the *pars distalis* (anterior lobe) of the *pituitary gland* (Donaldson et al. 1979). Removal of this tissue (hypophysectomy) results in a cessation of growth in the species investigated so far (including poeciliids, salmonids, and sharks). On the other hand, mammalian GH injections increase growth rates of juvenile coho salmon (Markert et al. 1977). Two major metabolic effects of GH in salmonid fishes are *protein accretion* and *lipid mobilization* (Björnsson 1997). The GH produced increased protein synthesis rates in rainbow trout, thereby leading to documented increases in growth rate. Also, GH stimulated increases in *triacylglycerol lipase* activity in the liver, breaking down lipids into usable *fatty acids* and *glycerol*. This latter effect probably accounts for the observed decreases in condition factor that accompany GH-stimulated increases in growth (Björnsson 1997, Markert et al. 1977). The growth-promoting actions of GH are now known to be mediated by *insulin-like growth factor I (IGF-I)* from the *liver*, including negative feedback on GH release from the pituitary (Fig. 8.4) (Duan 1997). Growth rates in transgenic (with modified GH genes) salmonid fishes exceed those of nontransgenic individuals (Devlin et al. 1994), in part because of a significantly increased intestinal surface area (Stevens and Devlin 2000).

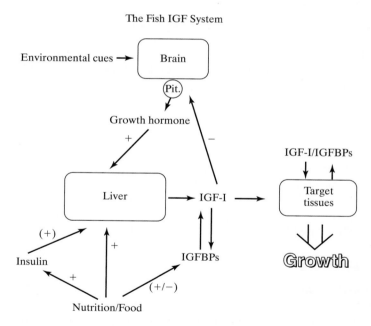

FIGURE 8.4   Insulin-like growth factor (IGF) and growth regulation. The endocrine axis controlling growth in teleost fish is shown. Multiple hormonal and nutritional factors may stimulate (+) the production and/or modify (+/−; e.g., via IGF-binding proteins [IGFBPs]) the activity of IGF-I. Negative feedback (−) by IGF-I inhibits growth hormone secretion by the pituitary (pit.). From Duan (1997).

Thyroid and gonadal steroid hormones also have been used to increase growth in various species. The thyroid hormone *triiodothyronine* (*T₃*) increased growth of under-yearling coho salmon when added to their diet, whereas *thyroxine* additions increased food consumption but not growth (Higgs et al. 1979). Additions of T_3 increased both growth and food consumption of rainbow trout (Matty 1986). Two synthetic androgens, *dimethazine* and *norethandrolone*, increased the growth rate of juvenile rainbow trout when either compound was added to a pelleted diet. These weight gain increases result-ed from the increased protein synthesis rate and the improved food conversion efficiency (Matty and Cheema 1978). Two natural androgens, *testosterone* and *11-ketotestosterone*, improved the food-conversion efficiency of common carp when added to their diet (Lone and Matty 1982).

8.3   GROWTH RATE MEASUREMENTS AND MODELS

Growth rate in fish can be determined by measuring *changes in size over time*. In prac-tice, it usually is measured by changes in body weight (mass) or length per unit of elapsed time. Growth rates of fish generally are measured or predicted by one of six methods.

Raise in a Controlled Environment

A fish (or egg or larva) of known age is placed in a tank, small pond, or cage in a larger body of water. Its length or body weight is measured at time intervals for growth rate calculations. This method is especially valuable for assessing the growth of cultured fish-es, in which feeding rates, temperatures, and other factors may be controllable. For ex-ample, *instantaneous* (or specific) *growth rates (GR)* in terms of percentage of body weight/day can be calculated as:

$$GR = 100 \times (\log_e W_f - \log_e W_i)/(t_f - t_i) \tag{3}$$

where W and t are body weights and times (usually in days), respectively, and i and f are initial and final experimental times, respectively (Ricker 1979, Wurtsbaugh and Cech 1983). This relationship is especially descriptive of growth over relatively short time in-tervals (e.g., during a single growing season). For estimating growth over multiple grow-ing seasons, sequential calculations can be made to describe changes in growth with age, or von Bertalanffy–type growth estimates can be made.

Mark and Recapture

A fish is *marked* (or *tagged*) and then *released* after an initial measurement of size is made. The fish is *recaptured* at a future date and measured again. The growth rate is cal-culated from the change in size over the time period that the marked animal spent in its habitat. It should be established that the marking method does not significantly alter the behavior, feeding rate, etc. *Marks* may consist of clipped fin rays, liquid nitrogen "cold brands," pigmented epidermis from high-pressure spray painting, or fluorescent rings on bones or scales (visible under ultraviolet light) from incorporation of tetracycline or 2,4-*bis* (*N*, *N*′-dicarboxymethylaminomethyl) fluorescein in the diet (Weber and Ridgway

1962, Hankin 1978). ***Tags*** also may vary considerably, from externally attached disks, plates, and streamers to small, implantable metal rods that are detectable in a magnetic field, to passive integrated transponders (PIT tags) that emit a unique code to a nearby receiver. Although these marked fish provide more realistic data concerning growth rates in their natural setting, they are more difficult to recapture compared with those in a controlled environment.

Length–Frequency Distribution

Length–frequency distributions are produced by measuring the ***lengths*** of individuals sampled from a population and then plotting the number of fish (***frequency***) of each length caught. This is especially useful as a technique with *young* fish, and the individual peaks often separate by ***age classes*** (Fig. 8.5). By comparing the ***mean lengths*** between age classes, one can determine approximate growth rates at various ages. For example, in Figure 8.5, the difference between age class 0 fish (young of the year) and age class I fish (between first and second "birthdays") is 49 mm − 20 mm = 29 mm. Thus, these grow at a mean rate of 29 mm (length) per year during their first year of life (when sampled in late July). In this example, the growth increment declines with increasing age from 29 to 24 mm per year (between age classes I and II).

Back-Calculation from Rings on Hard Structures

Juvenile fish have the same number of ***bones*** and ***scales*** as very old, much larger individuals of the same genetic stock when rearing takes place under similar conditions. For

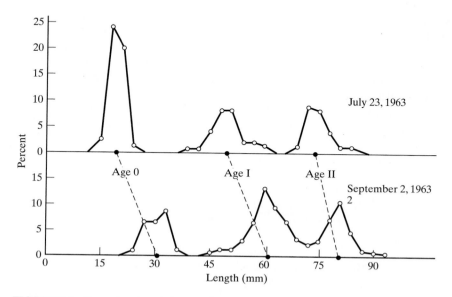

FIGURE 8.5 Percentage length frequencies of pond smelt (*Hypomesus olidus*) from Black Lake, Alaska, in middle and late summer. Note that growth slows (increased negative slope of dashed lines) as age increases. From Royce (1972).

many species, the rate of growth in the diameter of bones, spines, and scales also is proportional to the growth rate (length) of the fish. For many reasons, scales are the hard structure that is most commonly examined. The relative transparency of scales, the ease of sampling them, and the minimal damage to the fish make them desirable for use. Scales grow by *fibroblast cells* in the *fibrillar plate region* supplying *collagen* (a protein) and by *calcification* ($CaCO_3^-$) of the outer surface. Figure 8.6 shows the proliferation of cells at the scale margin, which enlarges the diameter of the scale. These additions to the scale diameter are formed repeatedly at a relatively constant rate over time and are distinguishable as *growth rings (circuli)* when the scale is magnified (Fig. 8.7). Periods of *slow* growth are discernible in magnified scales as areas of *closely spaced* circuli. Closely spaced circuli may occur on an annual basis from:

1. Decreased metabolism and appetite in cold seasons, especially in temperate water.
2. Fasting periods associated with spawning and unavailability of food.
3. Partial scale decalcification (resorption) in females with developing eggs and young.

These annual variations in the circuli pattern are termed *annuli*. Thus, age can be determined by counting the annuli, and fish lengths at each year can be back-calculated by measuring the linear distance *(radius)* from the *"focus"* of the scale to each annulus (Fig. 8.7). For example, the length at n years can be calculated by using the following formula (Lee 1920):

$$L_n = a + \frac{(L - a)(V_n)}{V_r} \qquad (4)$$

where L_n is the fish length at n years, a is a constant that often approximates fish length at time of scale formation, L is the fish length at time of capture, V_n is the scale radius distance from focus to n-th annulus, and V_r is the scale radius from focus to scale edge. From these back-calculations of length at different years, growth rates can be estimated. Table 8.1 shows back-calculated standard lengths for striped mullet from the southern Texas coastal waters. These techniques also can be used on opercular bones, otoliths, spines, or other hard structures that exhibit annuli. Borkholder and Edwards (2001) showed that dorsal fin spines were easier to interpret than scales for

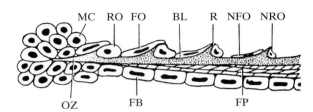

FIGURE 8.6   Schematic diagram showing the structure of the anterior margin of a teleostean fish scale and the scale-forming cells. *BL*, bony layer; *FB*, fibroblast; *FO*, flattened osteoblast; *FP*, fibrillary plate; *MC*, marginal cell; *NFO*, necrotic flattened osteoblast; *NRO*, necrotic round osteoblast; *OZ*, osteoid zone; *R*, ridge; *RO*, round osteoblast. From Kobayashi et al. (1972).

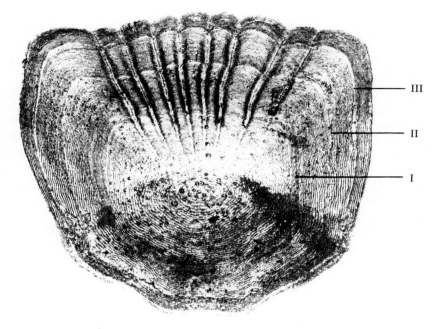

FIGURE 8.7   A scale of a kelp perch (*Brachyistius frenatus*) from Bodega Bay, California. Shown are three annuli (magnification, ×14). Photo by Samuel Woo, Illustration Services, University of California, Davis.

back-calculated age and growth estimates in walleyes (*Stizostedion vitreum*), especially for older specimens.

   Daily increments of growth have been detected on fish ***otoliths*** ("ear stones") (Pannella 1971). The widths of these increments average 1 to 2 μm in larval anchovies (*Engraulis mordax*) and 3 to 4 μm in larger hake (*Merluccius* spp.), but they are difficult to count where annual rings are formed. Thus, using daily growth increments for age

TABLE 8.1   MEAN BACK-CALCULATED STANDARD LENGTHS IN MILLIMETERS OF STRIPED MULLET

Age class	$(n)^{a}$	Standard lengths (mm)				
		I	II	III	IV	V
1+	(39)	132				
2+	(71)	127	194			
3+	(54)	122	188	234		
4+	(12)	125	178	227	266	
5+	(3)	127	183	213	249	286
Means		127	186	225	258	286
Growth increments		59	39	33	28	

From Cech and Wohlschlag (1975).

[a] Sample size in each age class.

estimates is particularly effective in *young* fish (less than 1 year) and in *tropical species* having poorly differentiated annuli (Brothers et al. 1976). Recent advances in sectioning otoliths, sampling discrete layers, and determining the layers' chemical compositions give clues about the chemical nature of the fish's environments when the layers were formed (Box 8.1).

Either daily or annular rings on hard structures need to be ***validated***. Using methods such as ***oxytetracycline*** baths (for larvae) or injections to make a fluorescent marker ring or by sampling a population over an extended period, the time period between annular, daily, or other rings can be determined (Summerfelt and Hall 1987). Data from this method (or from length–frequency plots) can be used to estimate past or future growth patterns or to calculate theoretical limits to fish size for a particular sampled population using appropriate growth models. For example, the classic von Bertalanffy models (von Bertalanffy 1938) were used to compare gulf toadfish (*Opsanus beta*) growth and theoretical size limits with its congener, the oyster toadfish (*O. tau*) (Serafy et al. 1997), and they were used to model California killifish (*Fundulus parvipinnis*) growth in a Mexican lagoon (Pérez-España et al. 1998). Everhart and Youngs (1981) as well as Busacker et al. (1990) have reviewed these models (including corresponding Walford plots).

Radiocarbon Uptake Method

Ottaway and Simkiss (1977) describe a method in which living fish scales, plucked from the epidermis, are incubated in a medium containing the simple amino acid ***glycine***, which has been radiotagged with ***^{14}C***. The rate at which the [^{14}C] glycine is incorporated into the collagenous structure of the scale (protein synthesis rate) after an incubation of less than 4 hours is measured by the level of ***beta radiation*** emitted by the scale. Faster growth rates of fish should accompany increased ^{14}C incorporation by the scale, as measured in a scintillation counter.

BOX 8.1
Ear Stone Analyses Reveal Life History

Advances in otolith microchemistry (e.g., microspectrometric analysis) have shown the potential to record previous events in the life of the fish. For example, strontium (Sr) is present in higher concentrations in sea water than in fresh water. Increased Sr content in otolithic layers indicates marine, rather than fresh water, habitation of Japanese eels (*Anguilla japonica*) during the period the layers were deposited (Tzeng and Tsai 1994). Otolithic Sr, from scanning proton microscopic analyses, differentiated populations of anadromous (having oscillating, high, and low Sr otolithic layers) and fresh water (having uniformly low Sr throughout the otolithic layers) inconnu (*Stenodus leucichthys*) of the Mackenzie River system (Northwest Territories, Canada) (Howland et al. 2001).

RNA:DNA Ratio Method

The RNA:DNA ratio method is another measure of the protein synthesis (tissue elaboration) rate as an index of ***instantaneous growth rate***. Whereas ***DNA*** content is constant per cell and ***RNA*** content is a function of the cell's protein synthesis rate, the RNA:DNA ratio reflects this rate per number of cells in a tissue sample (Haines 1980). This ratio correlates well with measured weight gains in golden shiners *(Notemigonus crysoleuca)* (Bulow 1970). Growth rate comparisons based on RNA:DNA ratios, however, should be limited to the same species and to restricted size and life-history stages (Bulow 1987).

8.4   LESSONS FROM GROWTH

Growth constitutes the conversion of excess energy into biomass gains, and healthy fish usually demonstrate significant growth rates. Because fish continue to grow throughout their lives, this process has become a well-studied integrator of fish well-being in aquatic habitats and an important measure of production success to the fish culturist. Important influences of diet and environmental and social factors work via hormone-mediated mechanisms to regulate fish growth. Growth is determined through laboratory and field-applicable techniques using a variety of direct and indirect methods.

SUPPLEMENTAL READINGS

Borkholder and Edwards 2001; Brett 1979; Brett and Groves 1979; Busacker et al. 1990; Donaldson et al. 1979; Everhart and Youngs 1981; Kinne 1960; Matty 1986; Ottaway and Simkiss 1977; Ricker 1979; Royce 1972; Summerfelt and Hall 1987; Weatherley 1972; Webb and Brett 1972.

WEB CONNECTIONS

FISHBASE
www.fishbase.org

MATHEMATICAL MODEL OF FISH GROWTH
www.ife.ac.uk/fishbiology/mathmod.html

CHAPTER 9
Reproduction

The success of any fish species ultimately is determined by the ability of its members to reproduce successfully in a fluctuating environment. One of the defining features of a species therefore is its reproductive strategy, as reflected in anatomical, behavioral, physiological, and energetic adaptations.

9.1 REPRODUCTIVE ANATOMY

The reproductive strategies of fishes often are reflected in anatomical differences between the sexes. Internally, of course, the sexes of most fishes can be distinguished easily by examining the gonads, at least during the spawning season. Both the testes of males and the ovaries of females typically are paired structures that are suspended by mesenteries across the roof of the body cavity and in close association with the kidneys (Fig. 9.1). During the spawning season, the *testes* are smooth, white structures that rarely account for more than 12% of the weight of a fish, whereas the *ovaries* are large, yellowish structures, granular in appearance, that may be 30% to 70% of a fish's weight. The testes are rather similar among the various groups of fishes, although the path by which the sperm exit may show considerable variation. Thus, in lampreys, hagfish, and salmon, sperm are shed directly into the body cavity, and they exit through an abdominal pore or pores. In contrast, in the Chondrichthyes, sperm pass through a duct that is shared with the kidney, and they may be stored in a *seminal vesicle* for a short period of time before being expelled. A similar situation exists in most nonteleost bony fishes, although a seminal vesicle usually is lacking. Teleosts usually have special, separate sperm ducts. The means of passing eggs from the ovary to the outside are similar to those for sperm, although special modifications of the oviducts (and/or ovaries) are common in fishes that retain the fertilized eggs and/or young. This is particularly true if nutrients are supplied to the developing young by the mother (see the next section).

The internal differences between the sexes generally are obvious in mature fish, but it frequently is difficult to distinguish sexes externally. Indeed, many fishes show virtually no sexual *dimorphism* (differences in body shape or size) or *dichromatism* (color differences), even when spawning. The opposite end of the spectrum includes fishes that are permanently and obviously dimorphic or dichromatic. Many of these species have internal fertilization, so the males have penis-like *intromittent organs*. In most such cases, such as sharks or poeciliid fishes (e.g., guppies), the intromittent organ is a modified fin. Because intromittent organs evolved independently in many groups, they show

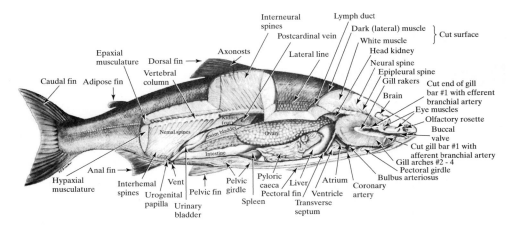

FIGURE 9.1   An adult female salmon, with portions cut away, showing the location and identity of various external and internal features. From Smith and Bell (1975).

considerable anatomical variation, ranging from the barely noticeable thickening of the anal spine region of the surfperches (Embiotocidae) to the elongate, flexible anal rays of the Poeciliidae to the complex claspers of the pelvic fins of the sharks and rays.

The most widespread type of sexual dimorphism is size. With egg-laying fish in which the males are territorial during the breeding season, the males often are larger than the females, as in salmon. In some such species, however, nonterritorial males typically are smaller than the females; these males use alternative reproductive strategies (discussed later).

In most fishes with sexual differences in size, the female is larger—or at least achieves a larger size. Thus, the record-sized individuals of many game fishes, such as striped bass and sturgeon (Acipenseridae), are females. Among live-bearing fishes, females, as a rule, are larger than males. What males may lack in size they often make up for in bright coloration. Sexual dichromatism is mostly a seasonal phenomenon among fish, because bright colors that are likely to increase reproductive success by attracting mates also are likely to attract predators. Exceptions to this "rule" are found primarily among tropical fish in which the breeding season lasts through much of the year, so the males are more or less continuously on display. Males of dichromatic fishes often have temporary structural modifications as well. Best known are the bizarre, hooked mouths (kypes) and humps that are characteristic of many spawning salmon and trout. More widespread, however, are breeding tubercles and contact organs, which Wiley and Collette (1970) noted were found in at least 25 families of fishes. *Breeding tubercles* are tiny, hard bumps of keratin that grow on the fins, head, and body scales during the breeding season. *Contact organs* are similar but have an internal core of bone as well. Both types of structure are found primarily in males, on body parts that are likely to come in contact with females or other males during the spawning season. These tubercles assist the males in maintaining contact with females during spawning, stimulate the females during spawning, and add in the defense of territories and nests. For ichthyologists, the breeding tubercles and contact organs have proved to be useful taxonomic tools, because the number and pattern differ among species.

9.2   BREEDING BEHAVIOR

Fish have a fascinating array of reproductive behavioral patterns. Not surprisingly, such behavior reflects both the evolutionary heritage of each species and the environment in which it lives. Thus, it is possible to classify fishes on the basis of their methods of reproduction and the means by which fish protect their developing embryos and young (Balon 1975a, 1984), as shown in Table 9.1. In this system, fish can be (1) nonguarders, (2) guarders, or (3) bearers.

Nonguarders are fish that do not protect their eggs and young once spawning has been completed. They fall into two basic groups: those that simply scatter their eggs in the environment (**open-substrate spawners**), and those that hide the eggs as part of their spawning behavior (**brood hiders**). By and large, the open-substrate spawners spawn in groups, without elaborate courtship behavior or specialized reproductive structures. Often, the spawning groups are quite large, with males outnumbering females. The spawning behavior is very difficult to describe precisely, because all that can be observed typically is a swirling mass of fish. These fish fall into two groups; pelagic spawners, and benthic (bottom) spawners.

Pelagic spawners are very common among marine fishes. They spawn in open water, often near the surface. Many such spawners are schooling fishes, such as tuna (Scombridae) and sardines (Clupeidae). Some river-dwelling, anadromous fishes, such as shad (*Alosa*), also are pelagic spawners, often seeking out areas where the flows are most favorable for successful spawning. Although pelagic spawning most often is associated with pelagic fishes, many benthic fishes temporarily rise off the bottom to spawn. This is particularly true of fishes that are associated with coral reefs, such as wrasses (Labridae) and parrotfishes (Scaridae). One of the main functions of pelagic spawning seems to be to assure that the young become widely dispersed by water currents.

To become widely dispersed, the eggs, embryos, and larvae of pelagic spawners have to be buoyant, and buoyancy is achieved either through the presence of oil globules or high water content. One of the problems with this, however, is that the mortality of eggs and young is extremely high, either because they are carried to unfavorable areas or because they are eaten by pelagic predators. To compensate, females produce extremely large numbers of eggs, and spawning periods often are protracted. The shortest spawning periods are found in temperate-zone pelagic fishes, because optimal conditions for embryo and larval survival are likely to be highly seasonal. In pelagic spawners associated with tropical reefs, spawning may be nearly a daily activity for months, with each female producing a small number of eggs each day. In these fishes, the breeding behavior may be extremely complex and variable. Indeed, some of the most complex mating systems are found in such fishes, involving territoriality, harems, leks, and sex changes.

Benthic spawners typically are species without elaborate courtship rituals. Very often, when spawning is observed, one female is closely followed by several males, who fertilize the eggs as the female releases them. The eggs and embryos are adhesive and stick to whatever surface they are laid on, are laid in long strings that are wrapped around objects, or drop into interstices and cracks where they take on water, thereby swelling enough to wedge themselves in place. The free embryos and larvae of these fishes may be either pelagic or benthic. Those that are pelagic (e.g., those of

---

TABLE 9.1 A CLASSIFICATION OF REPRODUCTIVE STRATEGIES OF FISHES BASED ON SPAWNING HABITS

I. Nonguarders
  A. Open-substrate spawners
    1. Pelagic spawner
    2. Benthic spawners
      a. Spawners on coarse bottoms (rocks, gravel, etc.)
        (1) Pelagic free embryo and larvae
        (2) Benthic free embryo and larvae
      b. Spawners on plants
        (1) Nonobligatory
        (2) Obligatory
      c. Spawners on fine substrates
    3. Terrestrial spawners
  B. Brood hiders
    1. Benthic spawners
    2. Crevice spawners
    3. Spawners on invertebrates
    4. Beach spawners
II. Guarders
  A. Substratum choosers
    1. Rock tenders
    2. Plant tenders
    3. Terrestrial tenders
    4. Pelagic tenders
  B. Nest spawners
    1. Rock and gravel nesters
    2. Sand nesters
    3. Plant-material nesters
      a. Gluemakers
      b. Nongluemakers
    4. Froth nesters
    5. Hole nesters
    6. Miscellaneous-materials nesters
    7. Anemone nesters
III. Bearers
  A. External bearers
    1. Transfer brooders
    2. Auxiliary brooders
    3. Mouth brooders
    4. Gill-chamber brooders
    5. Pouch brooders
  B. Internal bearers
    1. Facultative internal bearers
    2. Obligate internal bearers
    3. Livebearers

After Balon (1975a, 1984).

smelt [Osmeridae]) become active or buoyant immediately after "hatching," whereas those that are benthic stay close to the spawning area until they can swim freely. In lakes or shallow ocean waters, plants often are important substrates for the adhesive embryos of benthic spawners. Some fishes always spawn on such plants. For example, common carp and pikes (Esocidae) generally spawn only on vegetation that is flooded by high water in the spring. If such material is not available, spawning either does not take place or is much less successful. Once the embryos hatch, the young remain among the plants; some of them have adhesive organs on their heads, with which they can stick to the stems of the plants.

Brood hiders hide their eggs (in one way or another), but they do not show any parental care beyond spawning (Box 9.1). Most of these species are benthic spawners that bury the fertilized eggs. The females of salmon and trout, for example, excavate nests (*redds*) by digging with their tails. The redds are defended by both sexes from other members of the same species. Once the eggs are laid, fertilized, and buried (by the female), the nest site is abandoned. Many North American minnows (Cyprinidae) build nests as well, although a number of species build piles of stones rather than dig depressions. The males of these species also defend nest sites and have well-developed tubercles on the head and body to assist with defending the nest site.

Guarders go a step further than the brood hiders in their reproductive behavior, because they guard the embryos until they hatch and, frequently, tend the larval stages as well. Because tending the embryos usually means being tied to a specific location, territoriality (often implying competition for egg-laying sites) and elaborate courtship behavior are the usual state of affairs among the fishes of this group. Except among the cichlids, the embryos almost always are guarded by males. These males protect the embryos against predators (including other members of the same species), maintain high oxygen levels around the embryos by fanning currents of water across them, and keep the embryo mass free of dead embryos and debris. The amount of time that is spent guarding can range from a few days to more than four months (Antarctic plunderfish

---

BOX 9.1

High and "Dry": Beach Spawning in the Grunion

California grunion (*Leuresthes tenuis*) leave the ocean to spawn at night on sandy beaches during extreme high tides in the summer (Darkin et al. 1998). The fish are washed high onto the beaches by waves, and then each female wriggles tail-first into the sand, depositing her eggs. Each male wraps himself around a female to release sperm and fertilize the eggs. Grunion embryos develop out of the water and are able to hatch within 9 to 13 days. Embryos hatch on immersion and agitation by waves at the next extreme high-tide series. If the waves do not reach them, however, and hatching is not induced, the embryos remain viable in the sand for several weeks. During this period, grunion embryos will delay hatching, but they will hatch at any time if immersed and agitated in seawater. During the delayed period (days 13–26 postspawning), however, hatching success is significantly decreased. Grunions use their oil droplet for an energy source during this delay, as is evident from the linear decrease in oil-droplet diameter in larvae following hatching (Darkin et al. 1998). These embryos develop quickly and remain continuously ready to hatch over an extended period in response to a somewhat unpredictable environmental timetable (Darkin et al. 1998).

[*Harpagifer bispinis*]). Guarding species can be broken into two groups: those that do not build nests (**substratum choosers**), and those that do. The differences, however, between substratum choosers, which merely clean off a suitable area of bottom, and fish that build slightly more elaborate structures often are slight.

Nest spawners construct some sort of structure, cavity, or pit in which the eggs are laid and fertilized and the embryos are defended. The young often are defended in the nest as well, and in some species, the parents even herd schools of young around for a period of time. Pit nests can be made of a wide variety of material, but the most common are nests of gravel and rock. Such nests typically are shallow depressions that are carefully constructed and tended by territorial males. Most sunfishes and black basses (Centrarchidae) construct nests of this sort, often in colonies. The males defend the embryos and the young until the young become too active to be kept in the nest. In some cichlids, eggs are laid and incubated in one nest depression, but the young are tended (by both parents) in one or more additional depressions that are constructed nearby. A few species construct nests in sandy bottoms. Either the eggs and embryos in such nests have special adaptations (e.g., semibuoyancy) to reduce the possibility of being smothered, or the parent fish must spend considerable time handling them. In areas where bottoms are muddy, nest-building fish commonly build nests of plant material. Usually, plant-nesting fishes construct loose aggregations of material in which the eggs are laid. One special type of plant-nest builder uses a kidney secretion to glue the pieces of plant together to form a tight nest. The main examples of this type of nesting are the sticklebacks (Gasterosteidae), in which the male builds a tube, which usually is attached to rooted plants or is placed in a small pit. By means of a complicated courtship ritual, a female is enticed into a nest, where she deposits some eggs after being nudged by the male. After the eggs are laid, he chases the female out and fertilizes them. This process is repeated several times, until enough embryos are in the nest and the male begins to incubate them by fanning currents of water across them. After the embryos hatch, the male guards the young for several days, first in the nest and then in a shoal outside the nest.

Perhaps the most common type of nests among fish are those in caves, cavities, or burrows. These nests are characteristic of fish that live in active habitats, such as rocky intertidal zone or streams. For stream-dwelling fishes, the typical cave nest is the underside of a flat rock in midstream, which has plenty of current to keep the water well oxygenated. Sculpins (Cottidae) and many darters (Percidae) deposit their eggs in clusters on the roofs of such shelters and then guard the embryos until they hatch. In quieter water, minnows, such as the fathead minnow (*Pimephales promelas*), lay their eggs on the underside of a log or a board in an area that has been cleaned off by the male. Spawning males develop a thick, horny pad on their heads, which is used it for rubbing algae and other material off the spawning surface. Many large catfishes (Siluriformes) lay their eggs in cavities ranging from hollow logs to old muskrat burrows to oil drums, where the embryos are guarded by the males. These catfish also guard schools of the young for one or more weeks after they emerge from the nest.

Bearers are fish that carry their embryos (and, sometimes, their young) around with them, either externally or internally. Fishes that carry their embryos externally have developed a wide variety of ways for doing so, ranging from the short-term attachment of embryos to the adult until a suitable place is found to put them (**transfer brooders**) to the carrying of embryos and young in special pouches. Good examples of various kinds of bearers can be found among pipefishes and seahorses (Syngnathidae). In this family, only the males do the brooding. After the eggs are fertilized, the female places the embryos on the male. Herald (1959) notes a continuum of forms in the Syngnathidae, starting with skin-brooding pipefishes and progressing through pipefishes with open brood pouches on their bellies and ending with seahorses, which have closed pouches with only one opening. In seahorses, after an extended courtship period, the female deposits eggs in the pouch (marsupium) of the male by means of a penis-like oviduct. The eggs are fertilized as they enter the marsupium, and 25 to 150 embryos are provided with nutrients, oxygen, and protection (Jones and Avise 1997). After hatching, the free embryos are carried in the pouch until they are capable of fairly active swimming, at which time they are expelled by the brooding male.

Another way of carrying internally young that are spawned externally is found in **mouth brooders**. This method is found in families as diverse as the sea catfishes (Ariidae), cichlids (Cichlidae), cardinal fishes (Apogonidae), and bonytongues (Osteoglossidae). Mouth brooders carry the large, yolky embryos until they hatch, and these brooders typically carry the free embryos about as well. Even after the young become active, they usually are closely associated with the parent fish for a period of time and may flee back into the mouth cavity when threatened. In cichlids, both sexes may participate in some species, but the female usually carries the embryos, which she picks up quickly after spawning.

Internal bearers have many similarities to pouch brooders, although fertilization is internal and females always carry the embryos and/or young. Internal bearers typically produce only a small number of large, active young, which is a strategy characteristic of all sharks and rays (see Chapter 15) and a few bony fish families. There seems to be no strong correlation between internal bearing and the elaborateness of courtship behavior. In guppies, mosquitofish, and other live-bearing poeciliids, the most important trait of the courtship behavior of males that successfully copulate with females seems to be persistence (although subtle interactions among competing males and other courtship-related behaviors do occur). In contrast, some surfperches (Embiotocidae) have fairly elaborate courtship behavior, with males establishing breeding territories to attract passing females.

The first step in the evolution of live-bearing fishes can be seen in **facultative internal bearers**. These consist of a few species of **oviparous** (egg-laying) killifishes, in which eggs retained by the female are accidentally fertilized during the normal process of spawning on the substrate. These eggs usually have only a short period of embryonic development before being laid. The next step in the evolutionary process is found in **obligate internal bearers**, in which all the embryos are retained by the female. The only source of nutrition for these embryos, however, is the egg yolk, as in externally spawned eggs. This situation, also referred to as **ovoviviparity**, is characteristic of marine rockfishes (Scorpaenidae) and the Lake Baikal sculpins (Comephoridae). This strategy allows these fish to have fecundities approaching those of pelagic fish with external fertilization, but it also enables them to protect the young during their most vulnerable stage of development. In contrast, sharks and rays using this strategy produce a relatively small number of embryos and retain them for a few weeks to 16 months or longer. The shorter time spans are characteristic of species that eventually deposit their embryos in the environment, surrounded by a horny capsule; whereas the longer periods are characteristic of sharks that retain the embryos until they are ready to emerge as actively swimming young (Wourms 1981).

Given the advantages of retaining the embryos internally, it is not surprising that a number of fishes have developed means to provide additional nutrition for their young while the female is carrying them and then give birth to large, active young (*viviparity*). Viviparous fishes provide nutrition for their young in varying ways. The greatest variety is found in the sharks (see Chapter 15), but the teleosts also show a considerable range of methods. In the surfperches, the young develop in the mother's ovary and obtain nutrients through the close contact of the extralarge fins with the ovarian wall. Many poeciliid embryos have highly vascular pericardial tissue that loops around the neck and is in close contact with the ovarian wall of the mother, through which nutrients are exchanged. Nutrients, however, can be passed to the embryos even without special structures. Thus, largespring gambusia (*Gambusia geiseri*) were thought to be ovoviviparous until Marsh-Mathews et al. (2001) demonstrated experimentally the uptake by embryos of nutrients from the mother. Clearly, many fish do not fit nicely into human-made categories!

9.3   DEVELOPMENT

The eggs, embryos, larvae, and juveniles of a fish species reflect the reproductive behavior and ecology of the parent (Blaxter 1974, Russell 1976, Braum 1978, Balon 1981b). When describing the differences among species, it is useful to divide the life history of a fish into five major developmental periods: (1) embryonic, (2) larval, (3) juvenile, (4) adult, and (5) senescent (Balon 1975b). For convenience, these periods can be further divided into more or less arbitrary developmental phases. This classification scheme was developed to provide some consistent terminology for the enormously confusing literature on fish development and to support the *saltational theory* of development (Balon 1979, 1981b). According to this theory, development proceeds gradually through each period until a point is reached when a rather abrupt change in habits is possible, such as the change from an embryo that cannot capture food to a larva that can. The change is accompanied by further, rather rapid changes in morphology and physiology. Once a new developmental threshold is reached through this process, development proceeds gradually once again. The contrasting theory of development is that the process is continuous and gradual, so that any designated periods or phases are arbitrary.

Embryonic period

The embryonic period is the period during which the developing individual is entirely dependent on nutrition provided by the mother, either by means of yolk from the egg, a direct placenta-like connection (in viviparous fishes), or some compromise between the two methods. The period begins at fertilization, and it can be usefully divided into three phases: (1) cleavage egg, (2) embryo, and (3) free embryo. The *cleavage-egg phase* is the interval between the first cell division and the appearance of recognizable predecessors of the organ systems (but, especially, the neural plate). The *embryo phase* is the interval during which the embryo becomes recognizable as a vertebrate, because the major organ systems begin to appear. It ends at hatching. Although hatching is a major, recognizable event in the life of a fish (at least from our perspective), the exact state of development of an embryo at hatching not only varies among but also within species, depending on environmental conditions. In any case, once the embryo is free of the egg membranes, the ***free-embryo phase*** begins. During this phase, the embryo ceases to be curled up and becomes increasingly fish-like, but it continues to rely on its yolk or mother for nutrition and usually remains in its original environment. This phase may range from very lengthy (e.g., in salmon embryos buried in gravel) to nonexistent (e.g., in cyprinodont fishes with rapid life cycles).

Larval period

The beginning of the larval period is signified by the appearance of the ability to capture food organisms. During this period, special larval structures may develop, often related to respiration. The period ends when the axial skeleton is formed and the embryonic median finfold is gone (Fig. 9.2). In marine fishes, the period often is lengthy, lasting from 1 to 2 weeks (e.g., sardines and rockfishes) to many months (e.g., anguillid eels). For fishes with pelagic larvae (and free embryos), the larval period often is the major period of dispersal. It also is the period of highest mortality because of the vulnerability of larvae to predation and starvation (see Chapter 34). In freshwater fishes, pelagic larvae are present mainly in lake- or river-dwelling forms, such as sunfishes (Centrarchidae), pikeperches (Percidae), and whitefishes (Salmonidae), but they are absent from most stream-dwelling forms. Such forms typically have benthic larvae (Cottidae, many Cyprinidae) that exist only for short periods and often live in heavy cover, such as flooded beds of vegetation. In live-bearing fishes, the larval period may be absent or very short (and internal). In salmon and trout, the ***alevins*** that emerge from the gravel and begin feeding seem to be vestigial larvae, because, although they possess many larval characteristics, they are more like true juvenile fishes in their overall appearance and behavior (Balon 1980).

Juvenile period

Although the change from larva to juvenile may involve a dramatic metamorphosis, as in the transformation of the leptocephalus larva of an eel to an elver, usually the change is more subtle. The period begins when the organ systems are fully formed (or nearly so). Juveniles are recognizable by the presence of fully formed fins as well. In appearance, they are miniature adults, although they often possess distinctive color patterns that reflect the distinct habitats they typically occupy. The juvenile period lasts until the gonads become mature and usually is the period of most rapid growth in the life of a fish.

Adult period

Once the gonads are mature, a fish is an adult. The onset of this period is reflected in spawning behavior and, often, in the development of reproductive structures and color patterns.

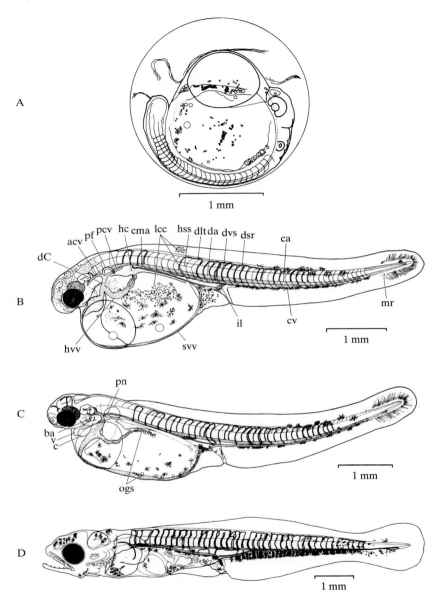

FIGURE 9.2 Some stages in the development of the walleye (*Stizostedion vitreum*). A. Embryo at 3 days, 50 minutes. B. Embryo at 6 days, 18 hours. The embryo has been removed from the egg membrane to show structures (*acv*—anterior cardinal vein, *ca*—caudal artery, *cv*—caudal vein, *da*—dorsal aorta, *dC*—duct of Cuvier, *dlt*—dorsal longitudinal trunk of the segmental muscles, *dsv*—dorsal segmental vein, *dsa*—dorsal segmental artery, *hc*—hepatic capillaries, *hss*—horizontal skeletogenous septum, *hvv*—hepatic vitelline vein, *il*—intestinal loop of caudal vein, *lcc*—large clear cells, *mr*—mesenchyme rays, *pcv*—posterior cardinal vein, *pf*—pectoral fin bud, *svv*—subintestinal vitelline vein). C. Embryo at 8 days, 1 hour (immediately after hatching). D. Larva at 17.5 days. From McElman and Balon (1979).

Senescent period

Few fish reach this period of "old age," when growth has virtually stopped and the gonads are degenerate and, usually, not producing gametes. This period may last for years in sturgeon, but it only lasts for a few days in Pacific salmon.

9.4   PHYSIOLOGICAL ADAPTATIONS

The complex behavioral adaptations that fishes have evolved to ensure reproductive success obviously are of little use if reproduction takes place when environmental conditions are unfavorable for the survival of the young. Thus, the reproductive cycles of fishes are closely tied to environmental changes, particularly seasonal changes in light and temperature. These two factors often are the most important, because they can act, directly or through sense organs, on glands that produce hormones, which in turn produce the appropriate physiological or behavioral responses. This can be demonstrated by examining the reproductive cycle of the longjaw mudsucker (*Gillichthys mirabilis*) in San Francisco Bay, California (DeVlaming 1972a, 1972b, 1972c).

In the mudsucker, the testes and ovaries develop between late September and mid-November, although ovarian development typically continues through early December (Fig. 9.3). Spawning occurs from December to June, with each individual spawning more than once. Gonads of both sexes regress abruptly in July, and they remain

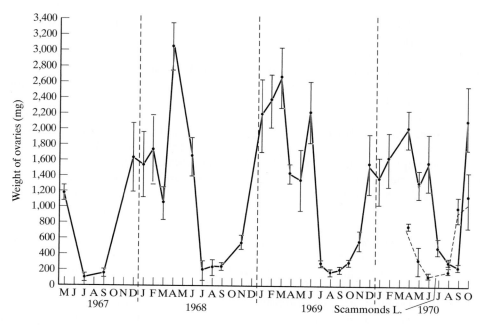

FIGURE 9.3 Seasonal variation in ovarian weight (mean ± standard error) of the Alviso population (solid line) and Scammons Lagoon population (broken line) of longjaw mudsucker. From De-Vlaming (1972a).

so through August and September. Laboratory studies show temperature to be the primary environmental factor regulating the reproductive cycle in the mudsucker, although photoperiod also plays a role. For example, temperatures of less than 20°C are required for gonadal development under any experimental light regime. At low temperatures, however, short photoperiods (equivalent to those of late autumn) accelerate gonadal development. Temperatures higher than 22°C to 24°C inhibit spawning in mudsuckers regardless of the photoperiod, because they inhibit *spermatogenesis* in males, which is the development of *spermatocytes* (the predecessors of sperm). High temperatures also inhibit the formation of the *vitelline membrane* of eggs in females, which is the final stage of *vitellogenesis*, the process of providing high-energy stores of yolk for the developing *oocytes*. Following vitellogenesis, the oocytes are stored in the ovary for maturation and eventual release from the follicle (*ovulation*). After ovulation, the oocytes must be spawned within a short period of time if they are to be viable. Thus, both ovulation and the final stages of sperm maturation must be precisely timed for successful spawning. Both processes are stimulated not only by temperature and photoperiod but also by the presence of suitable substrates for spawning (muddy bottoms for burrows in the case of mudsuckers) and the presence of mates that are in a state of reproductive readiness.

Indeed, one of the functions of courtship behavior is to stimulate ovulation and spermatogenesis (Stacey 1984). Males of some species actually produce pheromones to stimulate ovulation in females (Chen and Martinich 1975). Evidence suggests that the terminal nerve, which is situated alongside the olfactory nerve, mediates these responses to pheromones (Demski and Northcutt 1983).

The relative importance of the various external factors in stimulating reproduction varies considerably among species. In threespine sticklebacks (*Gasterosteus aculeatus*), increasing daylight is the single most important factor in gonad maturation, although appropriate temperatures and substrates are necessary for spawning. In contrast, for common carp, photoperiod is of minor importance compared to temperature regime (Davies et al. 1986). Water chemistry also can affect maturation; low pH created by acid rain, for example, can create severe problems in the gonadal maturation of rainbow trout (Weiner et al. 1986), as can toxic chemicals (Box 9.2).

Internally, the gonadal maturation processes are regulated by the production of *gonadotropic-releasing hormone* by the hypothalamus and of *gonadotropic hormones* by the anterior portion of the pituitary gland (*neurohypophysis*) (Fig. 9.4). Thus, removal of the neurohypophysis in mudsuckers inhibits the production of sperm cells in males and blocks vitellogenesis in females. These same processes can be stimulated, regardless of the temperature, by the injection of gonadotropic hormones into the fish. This latter fact is now taken advantage of by fish culturists, who routinely inject various carps and catfishes with pituitary extracts to induce spawning on a prescribed schedule. In fact, some species can be induced to spawn with pituitary extracts from other animals—even mammals.

When the gonadotropic hormones stimulate the gonads to develop, they also stimulate the gonads to produce *gonadal steroid hormones*. These steroid hormones include the *androgens* (testosterone and 11-ketotestosterone), which stimulate spermatogenesis in the testes. Females produce primarily estrogenic steroid hormones, although testosterone often can be detected in their plasma as an intermediate steroid. The dominant *estrogen* (17β-estradiol; E_2) acts to stimulate *vitellogenin* (VTG; the phosphoglycolipopeptide yolk precursor) production in the liver. Vitellogenin circulates back to the

---

BOX 9.2

Feminization of Male Fish by Endocrine-Disrupting Chemicals

Endocrine-disrupting chemicals (EDCs) are hormone-mimicking compounds and their metabolites, which can alter reproductive cycles in fish. EDCs include: (1) persistent chemicals once used heavily in industry and agriculture, such as polychlorinated biphenyls and organochlorine pesticides; (2) a number of currently used chemicals, such as plasticizers and surfactants; and (3) synthetic estrogens (Bell 2001, Fossi et al. 2002, Petrovic et al. 2002). Because of the lipophilic (high lipid affinity) and persistent (slow chemical breakdown) nature of most EDCs, they accumulate in animals by becoming concentrated through food chains. They are best known for their feminization effects on male fish, changing reproductive development, behavior, and performance (Fossi et al. 2002). For example, vitellogenin (the yolk precursor normally synthesized in the liver of female fish on stimulation by estrogen) accumulated in male carp that were collected downstream of a sewage treatment plant outfall carrying EDCs in Spain (Petrovic et al. 2002). Spermatogenesis was significantly inhibited in male rainbow trout after exposure to an agricultural fungicide (prochloraz; Le Gac et al. 2001). Few of the hermaphroditic rivulus (*Rivulus marmoratus*, Rivulidae) developed testicular tissue after exposure to nonylphenol, an estrogenic degradation product of surfactants that are used in industrial processes (Tanaka and Grizzle 2002). Male threespine sticklebacks exposed to synthetic estrogen (ethinyl estradiol, the active ingredient in human contraceptive and postmenopausal hormone-replacement treatments) decreased their aggressive responses to other males, thereby putting them at a reproductive disadvantage (Bell 2001). These effects of EDCs likely are not limited to fish!

ovary (Fig. 9.4), where yolk platelets accumulate in the developing oocytes, during *vitellogenesis*. During vitellogenesis, the oocytes increase in size as they advance through their developmental stages (Fig. 9.5). As vitellogenesis concludes, E_2 concentrations decrease in the plasma. At this point, concentrations of another important estrogenic steroid (17α-hydroxy, 20β-dihydroprogesterone, a *progestin*) increases, thereby stimulating final oocyte maturation just before spawning (and birth processes in live-bearing fishes) (Fig. 9.5). Gonadal steroid hormones also stimulate reproductive behavior and development of secondary (nongonadal) sexual characteristics. In males, the production of nuptial colors, breeding tubercles, and other secondary sexual characteristics is stimulated by androgens that are produced in the testes (or in cells adjacent to them). The gonadal steroid hormones regulate their own production through inhibition (negative feedback) of the hypothalamus and pituitary (Fig. 9.4).

Estrogen has other functions as well. Thompson et al. (2001) showed that E_2 injections into female squirrelfish triggered transient increases in plasma zinc and VTG, which followed similar time courses. The results suggest that E_2 is responsible for the large liver-to-ovary transfer of zinc in female squirrelfish, with VTG as the zinc vehicle. Squirrelfish have the highest known zinc levels in their livers because of high expression of the zinc-binding protein, metallothionein, so it presumably has some importance in their development.

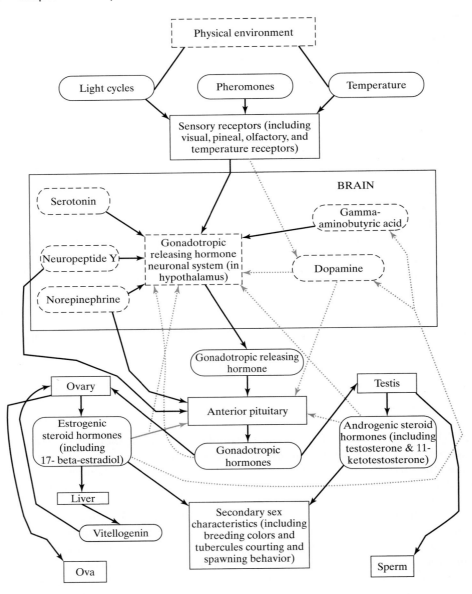

FIGURE 9.4 Pathways for reproductive stimulation (solid lines) and inhibition (dashed lines), consisting of substances or factors (circled) that regulate structures or behaviors (boxed) among various adult fishes. Inhibitory pathways shown here are often stimulatory in juveniles. Portions modified from Peter and Yu (1997).

9.5 BIOENERGETICS

It should be obvious from the previous sections that fish invest a great deal of energy in reproduction, both in behavior and in the production of gonadal tissue. Indeed, life-history strategies of fish can be viewed as mechanisms to balance the amount of energy

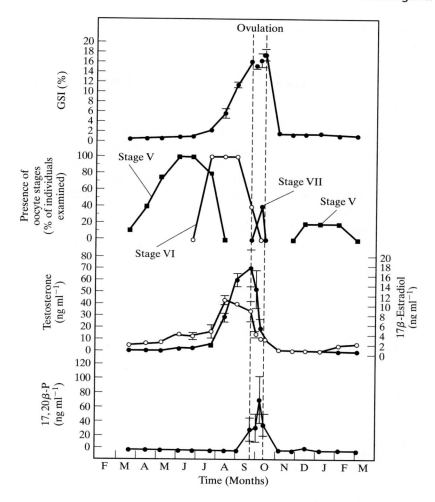

FIGURE 9.5   Changes in the percentage of body weight made up of gonadal tissue (gonadosomatic index [GSI]), development of oocyte stages, and changes in plasma levels of three steroid hormones: 17β-estradiol (○), testosterone (●), and 17,20β-dihydroxy-4-pregnen-3-one (17,20β-P), during the reproductive cycle of female Arctic charr. The values of the oocyte stages represent the percentage of females (in proportion to the number of females examined) with ovaries containing oocytes in the last four stages of development. From Frantzen et al. (1997).

put into reproduction with that put into growth and metabolism (**bioenergetics**; Williams 1966). Among and even within fish species is enormous variation in the amount of energy invested in various aspects of reproduction, such as spawning, migration, courtship behavior, parental care, and egg and sperm production. Many factors contribute to this variation, and in this section, we examine some of the more general ones: (1) reproductive effort, (2) age of onset of reproduction, (3) fecundity, (4) survivorship rates, and (5) frequency of reproduction. A key concept to understanding the importance of these factors is *fitness*, which is a measure of how likely an individual is to have offspring that also will reproduce successfully.

Reproductive effort is a measure of the amount of energy (or time) that is invested in the production of offspring. The most easily measured index of this factor is the size of the gonads compared with the body size. Because ovaries are much larger than testes, it generally is assumed that females invest more energy than males in reproduction. Even in pelagic fishes, where males have to produce enormous amounts of sperm to assure the fertilization of eggs, the ovaries are several times larger than the testes. A consequence of this disproportionate investment is, in theory, that females are much more careful than males in their choice of mates (to avoid wasting reproductive effort on less-fit males). This idea goes back at least to Charles Darwin. The differences in energetic investment may be less than usually are thought, however. In the lemon tetra (*Hyphessobrycon pulchripinnis*), for example, the amount of sperm extruded by males decreases with each successive spawning act, and females therefore discriminate against recently spawned males in their selection of mates. This greatly limits the number of matings that are possible for males, and it indicates that they, too, have to be fairly discriminating in their choice of mates (Nakatsuru and Kramer 1982).

In fact, evidence suggests that sperm production in males is a limiting factor in their reproductive success (Warner 1997). In pelagic fishes, sperm released into the environment can yield low fertilization rates if it is not in precise proximity to eggs. Even then, competition may occur among sperm from different males that are spawning at the same time (Taborsky 1998). In addition, spawning may take place multiple times during a day. All three situations require high sperm output from each male. Thus, considerable evolutionary pressure is placed on males to increase the success of the sperm that they produce. One way is to allocate more energy to sperm production, thereby producing more sperm. In wrasses (Labridae) on tropical reefs, dominant males that control females have much smaller testes than smaller, subordinate males, which must sneak in to spawn while the dominant male is spawning (see Section 9.7). In fact, the sneaker males release 15-to 20-fold more sperm per spawning than do the dominant males (Warner 1977). Another way to improve fertilization success under competitive situations is to have better sperm. In salmonids, the sperm of small, sneaker males lives longer and is more active than that of the large males with which the small males compete against for matings.

Differences in gonadal investment between sexes also are partially evened out by the greater investment by males in reproductive structures (e.g., breeding tubercles, etc.), and in reproductive behavior (e.g., courtship and territoriality). The investment of energy in reproductive behavior, coloration, and structures can be interpreted as a way for both males and females to ensure that their reproductive effort is not wasted by matings with less-fit individuals of the same species or with members of closely related species. In addition, in many species, many males are excluded from mating. Mortality of males also often is higher than that of females because of their conspicuousness (as the result of color patterns or behavior) during the spawning season. In wild guppies, the females of which prefer gaudy males, higher mortality rates of the brightly colored males occur even *before* they develop the bright coloration, thereby indicating they represent a less-fit genotype (Brooks and Endler 2000). In turn, this suggests that conventional explanations of females always selecting the "most-fit" males do not always apply. The territorial behavior of many fishes, such as the sunfishes (Centrarchidae) and the stick-lebacks (Gasterosteidae), combined with external fertilization of the eggs, predisposes the males to provide parental care as a way of further protecting their investment in the zygotes. The fact that parental care in fishes generally is performed by males (including those of nonterritorial species, e.g., pipefishes and seahorses) indicates that this is another way of equalizing the energetic investments by the sexes in reproduction. Parental care in such fishes apparently is more "cost effective" than merely increasing the output of the gonads (Blumer 1979, Gross and Sargeant 1985).

The ***age of onset of reproduction*** is another factor that varies with sex, because males typically mature at smaller sizes and younger ages than do females. For both sexes, however, the age at first reproduction depends, in good part, on the condition of the environment as well as on the nature of the population itself. Where the environment favors growth and high adult survival, fishes tend to delay reproduction. If the conditions are unfavorable, so that growth and adult survival are low, reproduction tends to take place at a younger age. For example, in Europe, two distinct forms of the brown trout *(Salmo trutta)* are found whose life-history tactics reflect the environments in which they live (Alm 1949). The lake-dwelling form inhabits a fairly predictable and productive environment, so it grows to a large size and spawns first when 5 to 7 years old. The stream-dwelling form, in contrast, lives in a much less predictable and productive environment, so it grows more slowly and matures in 3 to 5 years. On an energetic basis, it appears that in a more predictable environment, natural selection favors females that delay reproduction to invest their energy in large numbers of large eggs, because egg size and number tend to increase with size of the female. In a less predictable environment, natural selection favors females that reproduce as quickly as possible, because the probability of survival from one year to the next is low. A similar pattern has been found for the number of young that are produced in different environments by live-bearing fish (Baltz and Moyle 1982).

The nature of the population to which a female belongs also influences the age of reproduction. Females in an expanding population tend to reproduce at an earlier age than those in more stable populations. Part of the reason for this is that expanding populations tend to occur in favorable environments, so a larger size can be achieved at a younger age. Natural selection also may favor, in this situation, females that can fill "empty" space most quickly with the most young, especially if juvenile mortality is exceptionally low.

Fecundity is the most common measure of reproductive potential in fishes, because it is relatively easy to measure. It is the number of eggs in the ovaries of a female fish. In general, fecundity increases with the size of the female, with the relationship

$$F = aL^b$$

where F = fecundity, L = fish length, and a and b are constants derived from the data (Bagenal 1978). With this relationship, larger fishes produce considerably more eggs than do smaller fishes, both absolutely and relative to body size (i.e., number of eggs per gram of body weight). This means that the energetic investment in reproduction tends to be higher in the larger members of a species (i.e., they have larger ovaries in relation to body weight). Smaller females tend to invest more in growth (particularly during the first year or so of life). In contrast, in males, the size of the testes tends to increase in a linear fashion throughout life. Sargent and Gross (1986) argue that this is a major reason that males usually are the sex that provides parental care in fishes; males have more energy to invest in parental care in comparison to females, who are putting most of their energy into production of eggs.

The exponential relationship between fecundity and size in females seems to hold true for many species (but, especially, for species that produce large numbers of eggs and that spawn only once each year), but many factors complicate the interpretation of fecundity data, especially in relation to the investment of energy. Some of these complications are the relationship between fecundity and (1) fertility, (2) the frequency of spawning, (3) parental care, (4) egg size, (5) environmental conditions, and (6) population density (Bagenal 1978).

Fertility contrasts with fecundity in that fertility is the actual number of young that are produced rather than the number of eggs. In measuring reproductive success, the number of young that are produced is what really counts, so fertility should be a better measure than fecundity. Unfortunately, it is extremely hard to measure, because young fish typically disperse immediately after hatching. For some species, fecundity probably is a fairly close approximation of fertility. This is especially true in viviparous fish. In many other species, however, the relationship between fecundity and fertility is not clear, either because many eggs may be laid but not fertilized or because eggs may develop in the ovary but then be resorbed. In fishes that spawn repeatedly over long periods of time (called *fractional spawners*), eggs may be in the ovary in several stages of development, and it is very difficult to determine the contribution of the immature eggs to the season's spawning. In fishes with *parental care*, the relationship between fecundity and fertility is obscured by limitations that are placed on parents by the number of young for which they can effectively care. Thus, in mouth-brooding cichlids, fecundity often is considerably higher than the number of embryos that are brooded (Welcomme 1967).

Parental care is just one of a number of factors related to *egg size*. Most important of these factors is size of the female, because egg size often (but not always) increases with size of the parent. The advantage of larger eggs lies in the greater ability of larger young to survive. As Svardson (1949) pointed out, each population of fish seems to achieve a balance between the size and number of eggs that produces the most young. Rounsefell (1957) found that within the Salmonidae, river spawners produce larger eggs than lake spawners, and in anadromous forms, species with the longest freshwater stages produce the largest eggs, thereby demonstrating fine evolutionary "tuning" to environmental factors.

Although the evolutionary adjustment of fecundity to *environmental conditions* (e.g., food supply) is important, equally important is the ability of fishes to adjust egg production on an annual basis to environmental variation. Usually, this adjustment is related to the food supply. When food is plentiful, plenty of energy is available for investment in reproduction beyond what is required for maintenance and growth. As a result, well-fed fishes produce more and, frequently, larger eggs. Wooten (1973) found that threespine sticklebacks are exceptionally fecund in food-rich environments. A female in such situations produces two to three times her weight in eggs during a season, despite the limitations placed on egg production by the abilities of males to care for the embryos and the young. High availability of food leading to high fecundities often is related to a lack of intense intraspecific competition (population density). Thus, when population levels are low, reproductive success is likely to increase; whereas when population levels are high, reproductive success is likely to decrease, thereby providing an effective feedback mechanism for density-dependent population regulation (see Chapter 34). This mechanism also allows fishes to adjust their populations to conditions that reduce the availability of food, such as high water temperatures and low oxygen levels.

Survivorship rates are inversely related to fecundity. Fishes with high fecundities have very high death rates, especially through the free embryo and larval stages. Indeed, much research on commercially important fishes with pelagic larvae has focused on factors affecting larval survival, because a very small increase or decrease in the survival rate of larvae can have enormous impact on the adult fish population. In these fishes, any individual larva has a very low probability of survival. At the opposite extreme, viviparous fishes, such as the poeciliids, surfperches, and many sharks, produce a small number of young that have a fairly high probability of survival. Curiously, if all appropriate measurements could be made, the amount of parental investment per successful offspring is likely to be about equal in the two groups—or at least to show equal variability.

What are the advantages of these two divergent reproductive strategies if the energetic investment is approximately the same? Williams (1975) argues that highly fecund fishes are capable of making rapid genetic adjustments to environmental change, because even with such high mortality rates, larval deaths are not entirely random. Therefore, survivors are likely to have high fitness for the particular set of local or short-term environmental conditions that exist. In addition, highly fecund fishes are capable of rapidly adjusting their populations to environmental changes. For example, under conditions that favor high larval survival rates, even a small number of adults can produce a large number of successful offspring. Alternatively, under unfavorable conditions, only a small number of fish will survive the larval stages, regardless of the number of spawning adults. Essentially, in such fishes, the number of young that survive to adulthood is largely independent of the number of parents. This strategy is likely to be characteristic of fishes that live in environments where food availability fluctuates in an unpredictable fashion from year to year and in which competition usually is not a major problem.

In contrast, populations of large, low-fecundity fishes are not as capable of responding genetically to short-term environmental fluctuations, but their young have either much greater capabilities as individuals for adjusting to the fluctuations or for avoiding unfavorable conditions. Because of their large size, the young of these fishes also are much more capable of avoiding predators and potential competitors. The high frequency of small pelagic eggs among teleost fishes, however, argues that the high-fecundity strategy has been a major factor contributing to their success.

Frequency of reproduction is another characteristic of fish life-history strategies that appears to reflect the predictability of the environment in which the fishes live. The two basic strategies here are semelparity and iteroparity. *Semelparity* is "big bang" reproduction, in which the adults spawn and die (as among Pacific salmon), whereas *iteroparity* is repeated reproduction (characteristic of most fishes). The reason for the commonness of iteroparity is that most environments are unpredictable enough that the young in any particular spawn have no guarantee of surviving to adulthood (Box 9.3). If conditions are favorable, a semelparous fish may have very high reproductive success,

---

BOX 9.3

American Shad: Why Some Die, and Why Some Do Not

A study that demonstrates nicely the advantages of iteroparity and semelparity is that of Leggett and Carscadden (1978) on American shad (*Alosa sapidissima*). They examined the reproductive strategies of this anadromous species in rivers of the Atlantic coast of North America, from Florida to New Brunswick, and they found that the degree of iteroparity increased with the latitude of the river. All fish in Florida populations die after spawning, but 55% to 77% of shad in New Brunswick streams spawn more than once. Intermediate populations having intermediate values for the frequency of repeat spawners. The fecundity of the northern populations is lower than that of the southern populations, indicating that they devote less energy to egg production at one time compared with the southern shad and, therefore, have more in reserve for survival after spawning. The differences in strategy reflect differences in environment; Northern rivers are a harsher and more variable environment for eggs and larvae, whereas southern rivers are more benign. Thus, it "pays" the northern fish not to put all their eggs into one spawn but to save some energy so that individuals can reproduce again, in case the first spawning has no survivors.

---

but if conditions are unfavorable, it may lose out completely. Thus, most fishes increase their fitness by adopting a "bet-hedging" strategy of not putting all their energy into one spawn (Murphy 1968). An unusual example of this occurs in capelin (*Malotus villosus*), an arctic smelt (Osmeridae) in which males are semelparous but females are iteroparous (Huse 1998). For males, it pays to spend all their energy on spawning with multiple females in a single season, whereas for females, which can produce only a limited number of eggs each season, it pays to spawn over several years.

9.6   MATING SYSTEMS

As the previous section indicates, fish have many ways of achieving reproductive success (i.e. producing offspring that also survive to reproduce). This diversity also is reflected in the mating systems of fishes. Here, mating systems are defined rather narrowly to mean the various ways in which conflicting interests of males and females are resolved to result in reproduction. Berglund (1997) divides mating systems into four types: (1) monogamy, (2) polygyny, (3) polyandry, and (4) promiscuity (polygynandry).

Monogamy is where one male and one female mate exclusively. It is uncommon in fishes, and it often alternates with other mating systems. It is most likely to occur when both sexes care for the young, territories for feeding and breeding are small, or low encounter rates between the sexes are found (Berglund 1997). Monogamy is characteristic of a number of tropical cichlids in which both sexes rear their young together in areas that must be defended vigorously against competitors and predators—often of the same species as the monogamous pair (Barlow 2000). In some pipefishes and seahorses (Syngnathidae), development of eggs takes a long time before the female can place them in the brood pouch of a male, where they are fertilized. While the male is pregnant, the female starts a new batch of eggs, which are ready at about the same time that the male gives birth to the young from the previous mating. This close timing of development promotes monogamy, especially if the likelihood of encountering another potential mate is low (Berglund 1997).

Polygyny is the name for one male mating with several females. This usually involves a large, conspicuous male either defending some turf (to which females are attracted) or defending females directly from other males (Berglund 1997). In sculpins (Cottidae), males defend prime sites for incubation of embryos, which are "caves" underneath rocks. Females seem to choose males as the result of a combination of the quality of the breeding site and the size of the male. Males attempt to obtain exclusive mating rights with multiple females in many other ways as well. One method is the use of *leks*, or places where many males gather together and display to one another and where females come to choose the highest-ranking males. The most spectacular example of this occurs in the cichlid *Cyrtocara eucinostomus* in Lake Malawi, Africa. As many as 50,000 large, colorful males display together on a lek that is 4 km long! Females, which are mouth brooders, presumably choose the "best" males for fertilizing their relatively small batch of eggs (McKaye 1983).

Polyandry is the opposite of polygyny; One female seeks to mate with several males. This is relatively uncommon, but it occurs when females are wont to change sex (as in anemone fishes; see Section 9.8) or males do the brooding but can take care of fewer eggs than the female can produce (some pipefishes). In deepsea angler fishes (Lophiiformes), males of some species are parasitic on females, attaching to them, and females may have more than one male attached (see Chapter 35).

Promiscuity (polygynandry) is presumably the original fish mating system, the result of external fertilization, in which many males and females mate simultaneously. The most obvious examples are species with huge mating aggregations, such as herrings (Clupeidae), where shallow water may become white with sperm and the bottom covered by millions of fertilized eggs.

9.7 ALTERNATIVE REPRODUCTIVE STRATEGIES

In this chapter, so far we have largely discussed fish reproduction as if all species had two well-defined sexes and 1:1 sex ratios. In fact, this is not the case. Some alternatives include: (1) alternate male strategies, (2) hermaphroditism, and (3) unisexuality.

Alternative male strategies are likely to develop among species in which large, aggressive males can dominate spawning, thereby creating opportunities for small males to engage in cuckoldry. In salmon and trout, for example, *jack* males are common. These are small, silvery males that migrate upstream along with the standard, large, hooknosed males and that spawn by sneaking into redds to release sperm simultaneously with a mated pair. This behavior is an *evolutionarily stable strategy* for reproduction, because it is favored by natural selection just like the "standard" strategy of large males (Gross 1984). Such alternative methods also have been recognized in many other species, including parrotfishes (Scaridae) and wrasses (Labridae) on tropical reefs and the bluegill sunfish (*Lepomis macrochirus*) in fresh water. Bluegill have two alternatives to being a large male who defends a nest in a colony of similar males and lures passing females into the nest (Gross 1982). One method is *sneaking*, as described for salmon, in which a small male hides near an active nest and dashes in to release sperm when the resident male is spawning with a female. Another method is for a small male (called a *satellite male*) to hover above a nest containing a pair of courting sunfish and slowly descend into it, reaching the pair in time for spawning. The satellite males manage to do this because they mimic females in coloration and behavior. Both sneaker and satellite males usually are badly chewed up by irate parental males by the end of the spawning season, but their strategy is nevertheless a successful one (giving them high fitness). Because these smaller males do not have to spend any energy on parental care, they can spawn at a much younger age than the parental males. Parental males may be 6 or 7 years old before they are large enough to compete successfully for the best nest sites, but the sneaker and satellite males may be only 2 or 3 years old.

Hermaphroditism, in which one individual can be either male or female, is known from at least 14 families of teleost fishes (Shapiro 1984). Hermaphrodites can be either *synchronous*, in which individuals possess both ovarian and testicular tissue, or *sequential*, in which individuals change sex. Synchronous hermaphrodites are uncommon. Among the best studied is the black hamlet (*Hypoplectrus nigricans*, Serranidae), a species in which individuals take turns releasing sperm and eggs during spawning. Because such egg trading is advantageous to both individuals, hamlets typically are monogamous for short periods of time—an unusual situation in fishes (Fischer and Peterson 1987). In sequential hermaphrodites, the most common pattern is for a female to change into a male (*protogyny*). This often happens when a large, dominant male controlling a harem of females is removed by a predator (or experimenter). Within a few days, the largest female in the harem becomes a dominant male and takes over the missing male's function (Shapiro 1984). This pattern is common in coral reef fishes, such as the parrotfishes (Scaridae), wrasses (Labridae), and groupers (Serranidae). Less common than protogyny is *protandry*, in which a male converts to a female. The best-known example of this occurs in anemone fishes (*Amphiprion*, Pomacentridae), in which the male of pair guarding an anemone will change sex if the larger female disappears (see Section 9.8).

Unisexuality is characteristic of a number of fishes, and it takes a number of forms (Box 9.4). True *parthenogenesis*, in which females produce only female offspring, with no involvement from males, is rare in fishes, but the Texas silverside (*Menidia clarkhubbsi*) appears to be one such all-female species (Echelle et al. 1983). Other all-female species are best known from a complex of Mexican mollies (Poeciliidae). The first of these all-female species to be discovered was the Amazon molly (*Poecilia "formosa"*; Hubbs and Hubbs 1932). Studies revealed that the Amazon molly and similar unisexual species were sexual parasites on males of bisexual species of the same genera from which they originally were derived as hybrids (Schultz 1989). The sperm from males of the host species is required to activate development of the eggs of the Amazon molly, but union of the male and female chromosomes does not occur. Therefore, only genetically uniform females are produced (*gynogenesis*). In *hybridogenesis*, another variation on this theme (Schultz 1989), the mating between a female molly (*Poeciliopsis*) and the host male results in fertilization of the egg, and a true hybrid is formed. During oogenesis in the hybrid females, however, the chromosomes contributed by the host male are

BOX 9.4

The Mystery of An All-Male "Species"

In several river basins of Portugal and Spain exists a distinctive minnow (Cyprinidae) that is known only from males. Because the all-male form is diploid (i.e., has the "normal" set of paired chromosomes in the nucleus of cells) and also is not hermaphroditic, it appears to be an all-male species, which would seem to be impossible. So, how can this be? The all-male form is part of a complex of forms that together are labeled *Squalius alburnoides*. This Iberian minnow originated as a hybrid between two species and is found in both polyploid forms (triploid and tetraploid) and diploid forms, including all-female forms that reproduce mainly through hybridogenesis. One of the ancestral forms apparently is extinct as a "normal" species but may still maintain a genetic presence through the all-male line. These nonhybrid males maintain themselves through matings with triploid hybrid females, which produce haploid eggs with one set of chromosomes that is identical to that of the males! These matings apparently produce only males, so the appropriate female genome has to be maintained through crosses among other forms. In fact, "nuclear non-hybrid males of the *S. alburnoides* complex seem to be unique, as they essentially represent a stable all-male lineage nested within an almost all-female lineage" (Alves et al. 2002, p. 659). The origin of this all-male lineage is still not fully worked out, but its presence demonstrates the potential complexity of mating systems—and throws further cold water on the idea that species must be diploid, sexually reproducing organisms.

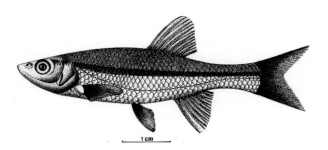

Male *Squalius alburnoides*. From Collares-Pereira, (1989).

lost in meiosis, so only the female genes are passed on to the next generation. The result is a self-perpetuating strain of all-female fish. To make matters even more complicated, some unisexual populations of *Poeciliopsis* exist that are actually trihybrids. These trihybrids apparently result when hybridogenetic females mate with males from a third species, resulting initially in fish that are hybrids between the original maternal species and the new species. In this new hybrid, however, some chromosomal reassortment takes place, resulting in eggs containing at least some genetic material from the third species. The new hybrid females then mate with the males that produced the original hybridogenetic fish, resulting in female progeny with characteristics of all three original species.

One of the interesting questions posed by the unisexual "species" (i.e., clones) is: Why are they so successful? There is little question that they are, indeed, successful. They are widespread, and they often are more abundant than the parent species. This is attributed to a combination of factors: (1) *heterosis* (hybrid vigor) exhibited by the hybrids (e.g., larger size and higher survival rates), (2) increased reproductive potential of an all-female population, and (3) specialization of clones for particular aspects of the environment, so they can thrive under highly localized conditions. These factors, however, offer only a partial explanation, because unisexual species have two other problems to overcome: low genetic variability that occurs in the absence of genetic recombination, and continued dependence of unisexual forms on bisexual males for reproduction. The first problem is minimized by the fact that unisexual populations apparently have arisen repeatedly, so each "species," in fact, represents many different clones, each with a different origin and different ecological requirements. When individuals from two clones coexist, they may segregate ecologically from one another (as reflected in dentition and feeding behavior) as well as from the parental bisexual species (Balsano et al. 1989). The continued dependence of the unisexual fish on the bisexual males is a problem, however, because unisexual fish cannot afford to be so successful in their competitive interactions with bisexual fish that they eliminate them. In addition, there should be strong selection pressures on the host species to develop mechanisms for not wasting reproductive effort on the parasitic unisexual fishes. Males of bisexual species do have a preference for the appropriate females, but they also mate freely with unisexual females. Sperm does not seem to be a limited resource for which females compete—at least in these fishes (Balsano et al. 1989).

9.8 SEX CHANGE IN FISH

As the discussion of hermaphroditic fish indicates, the sex of many species of fish is not necessarily fixed; rather, it is determined by the social and physical environment in which the fish lives (Chan and Yeung 1983). Sex change occurs mainly in fishes in which one sex has higher survival and reproductive rates, usually as the result of larger size. In such situations, it may pay for an individual of the smaller sex to change to the larger sex when the opportunity arises (Kuwamura and Nakashima 1998). For example, anemone fishes (Pomacentridae) live as monogamous pairs in single anemones, protected from predators by the stings of the anemone. Female anemone fish typically are larger than males, because a male does not have to vigorously compete for mates once he has paired with a female. If the female dies, however, a juvenile anemone fish quickly moves in; such fish are always male. The resident male then turns into a female and reproductive advantages of the large female–small male combination continue (Fricke and Fricke 1977). The willingness of a male to convert to a female is determined partly by the densities of the anemones and their fishes. With high densities, a widowed male can find a new mate more easily by moving than by changing sex (Kuwamura and Nakashima 1998). Curiously, in such areas, small, immature males live between the territories of pairs and will opportunistically become either male or female if the opportunity arises. In some fishes, sex change is reversible, depending on the social situation. Thus, some gobies (Gobiidae) will switch sex when kept in what initially were single sex groups of either sex (Kuwamura and Nakashima 1998).

An example of environmentally determined sex occurs in the Atlantic silverside (*Menidia menidia*). Silverside larvae that develop at low temperatures are more likely to become females than are larvae that develop at high temperatures (Conover and Kynard 1981). In the southern brook lamprey (*Ichthyomyzon gagei*), males become more frequent when larval densities are high and temperatures are low (Beamish 1993). More disturbingly, evidence is growing that pollutants may cause sex changes in fish by disrupting endocrine cycles (see Box 9.3).

9.9   LESSONS FROM FISH REPRODUCTION

One of the most fascinating ideas in recent decades is that fish (and other organisms) have a ***life-history strategy***, in which traits that are related to reproduction and development play a central role. The life-history strategy of a species is a complex of evolved traits that are related to the allocation of energy; along with morphology, physiology, and behavior, it defines the species. Although the life cycle of a species is fine-tuned to the environment, as illustrated by the longjaw mudsucker, different species use different life-history strategies to reach the same "place." For example, among the top predators of the open ocean, sharks use a low-fecundity strategy of viviparity, tunas a high-fecundity strategy of oviparity, and salmon a strategy involving moderate fecundity and some parental care. The evolutionary diversity of fishes has meant the development of (at least to us) unusual life-history strategies involving such things as multiple sexes, unisexuality, sex change, and hermaphroditism. Although the life-history strategies of fish often are surprisingly robust, human activities, ranging from overfishing to the release of chemicals that mimic reproductive hormones, can disrupt life cycles and cause populations to collapse.

SUPPLEMENTAL READINGS

Balon 1984, Berglund 1997, Breder and Rosen 1966, Gross and Sargent 1985, Pitcher 1993, Potts and Wooton 1984, Wooton 1990.

WEB CONNECTIONS

GENERAL
www.fishbase.org/fishonline/FOL_reproduction.htm

ENDOCRINE DISRUPTORS
www.water.usgs.gov/pubs/FS/FS-081-98/pdf

C H A P T E R 10
Sensory Perception

Fish sense the world around them in diverse ways. Most fish have the "terrestrial" senses of *sight, hearing, smell*, and *taste* which we also comprehend from our own experience, but they also possess sensory means for detecting stimuli such as *water particle displacement* and *electrical currents*, for which we have little empathy. Such modes take advantage of the physical and chemical properties of water and work in conjunction with the more "conventional" sensory modes. In this chapter, therefore, we cover the aspects of *chemoreception* (smell and taste), *mechanoreception* (hearing, orientation, and lateral-line detection of water disturbances), *electroreception*, and *vision*.

10.1 CHEMORECEPTION

Odors and tastes are quite distinguishable to the great majority of terrestrial animals. *Olfactory organs*, stimulated by airborne molecules, are more sensitive and chemical-specific than *gustatory organs*, which are generally stimulated by contact with dilute solutions. In fish, both types of organs have similar sensitivities to "contact" stimuli of chemicals dissolved in water, but they are distinguished by the location of the sensory receptors and the processing centers in the brain.

10.2 OLFACTION

The *olfactory receptors* are usually located in *olfactory pits*, which have incurrent and excurrent channels (*nares*) divided by a flap of skin. Water is induced to flow through olfactory pits by the movements of *cilia* within the pit, the muscular movement of the branchial pump, swimming, or a combination of these. Odors are perceived when the *dissolved chemical* makes contact with the *olfactory rosette*, a multifolded epithelium that is rich in receptor cells and located in the olfactory pit (Fig. 10.1). Fish that rely heavily on olfactory cues have elongated rosettes with many receptor cells located in elongate olfactory pits. In anguillid eels, such elongate olfactory rosettes can detect certain chemicals, such as β-phenylethanol in concentrations of 1×10^{-13} M (Teichmann 1962)! Seasonal variations in sensitivity suggest that *cyclical hormone action* may influence the threshold level of detection.

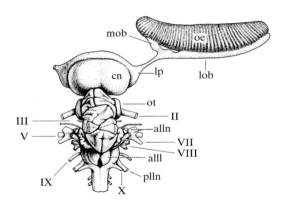

FIGURE 10.1 Dorsal view of the brain of the bonnethead shark (*Sphyrna tiburo*). alll, *Anterior lateral-line lobe;* alln, *anterior lateral-line nerve;* cn, *central nucleus;* lob, *lateral division of the olfactory bulb;* lp, *lateral pallium;* mob, *medial division of the olfactory bulb;* oe, *olfactory epithelium (organ);* ot, *optic tectum;* plln, *posterior lateral-line nerve;* II, *optic nerve;* III, *oculomotor nerve;* V, *trigeminal nerve;* VII, *facial nerve;* VIII, *statoacoustic nerve;* IX, *glossopharyngeal nerve;* X, *vagal nerve.* From Northcutt (1978).

At least two types of olfactory receptor cells detect and encode chemical signals (Hara 1986). Olfactory stimuli are communicated via *lateral or medial divisions* of the *olfactory bulb* (Fig. 10.1) to the *olfactory lobe* of the brain via the *first cranial nerve*. Species that rely heavily on olfactory information, such as anguillid eels, moray eels, and sharks, display oversized olfactory bulbs and lobes. A few fish, such as some of the puffers (Tetraodontidae), have greatly reduced bulbs and lobes that are consistent with the evolutionary loss of the olfactory organs and pits. These puffers probably rely entirely on sight for feeding.

Olfactory cues are important to some adult fish in locating spawning streams. For example, once they are in the vicinity of the river mouth, salmon apparently use olfaction in locating their natal stream (Hasler and Scholtz 1983). Thus, salmon are thought to be imprinted with odors while presmolts and smolts as they migrate down rivers and streams to a larger body of water, where they spend the majority of their adult life (Box 10.1) (see Chapter 19). The chemical nature of the attracting odors is not well known. Bodznik (1978) recorded sensory responses from the olfactory lobes of sockeye salmon (*Oncorhynchus nerka*) and found that the inorganic ion calcium could be used as an upstream odorant cue substance. Hara et al. (1984) demonstrated that salmon skin mucus (e.g., from juvenile fish of a previous spawn) contains *species-specific amino acid combinations* that stimulate olfactory neurons. Li et al. (1995) showed that adult sea lampreys (*Petromyzon marinus*) are olfactorily stimulated by two *bile acids, allocholic acid* and *petromyzonol sulfate*, that are released from conspecific larvae in the streambed.

BOX 10.1

Finding the Way Back Using Familiar Odors

The parr–smolt transformation (transition from stream-dwelling, freshwater "parr" juvenile salmon to a migrating, salt water–capable "smolt") is a crucial period for the olfactory imprinting that salmon require if they are to find their way back to their natal areas as adults. Imprinting is a delicate process. At least part of the process occurs at the olfactory (rosette) epithelium (Nevitt et al. 1994), and receptivity to olfactory imprinting appears to be related to thyroid hormone (thyroxine) surges in the salmon after exposure to novel water (e.g., at confluence points of tributary streams) during downstream migration (Dittman et al. 1994). So, can the salmon be artificially imprinted to lure them back to new areas or to a hatchery? The answer is yes. Juvenile coho salmon exposed to β-phenylethyl alcohol, an artificial odorant, at the smolt stage demonstrate an increased attraction to this chemical as adults (Dittman et al. 1996), and other compounds have been used, via smolt exposure, to bring fish into hatcheries.

Studies on nurse sharks (Ginglymostomatidae) show a distinct gradient-searching activity (***klinotaxis***) for food detection. By detecting a stronger chemical concentration in the olfactory rosette on one side of its head, the shark turns in that direction. Thus, tacking motions through the odorant field lead the shark to the odor source. Experimentally plugging the water channels on one side of the shark's head will produce a continuous circling in the opposite direction. The extremely wide head of hammerhead sharks (Sphyrnidae) should be particularly useful in locating odor sources when only dilute concentrations are detectable (because of the increased separation of the nares). Particular ***amines*** and ***amino acids*** attract teleosts as well as elasmobranchs. Sutterlin (1975) found ***glycine*** and ***alanine*** to be especially attractive to winter flounder (*Pleuronectes americanus*), whereas ***alanine*** and ***methionine*** attracted Atlantic silversides (*Menidia menidia*) most effectively.

10.3 TASTE

Taste, or ***gustatory chemoreception***, is especially useful in the identification of both ***food*** and ***noxious substances***. Whereas olfactory receptors are localized in the nares, ***taste buds*** are commonly located on several ***exterior surfaces*** of the fish in addition to the ***mouth***. Bottom fishes, such as catfish, have considerable numbers of taste receptors on their ***skin, fins***, and ***barbels***. The high gustatory sensitivity of the barbels extends the usefulness of these chemoreceptors for a reasonable distance from the fish, presumably to aid in finding food in murky water. A similar "distant touch" sense might be inferred from the presence of taste buds on the free ***pelvic fin rays*** of some codfish (Gadidae) and threespot gourami (*Trichogaster trichopterus*). Other structures bearing recognizable taste buds include the well-developed ***lips*** of some minnows (Cyprinidae) and the suckers (Catostomidae). Suckers may have 41 to 57 taste buds in a 1.3-mm^2 field (Miller and Evans 1965). Taste buds have also been described on the ***palatal organs*** in the buccal cavities of various cyprinids, catostomids, cobitids, and salmonids as well as on the gill rakers and arches, primarily in freshwater fish (Kapoor et al. 1975a).

Whereas the **cutaneous taste buds** are innervated by branches of the **facial (cranial nerve VII)** nerve (Fig. 10.1), the sensory signals from the **internal taste buds** (e.g., those in the pharyngeal cavity and palate) are transmitted to the **glossopharyngeal (IX)** and **vagus (X) nerves**. In species with a palatal organ, the total number of taste buds is considerably augmented, and the terminal centers in the visceral sensory column of the **medulla** are correspondingly enlarged, as **vagal lobes** (Kapoor et al. 1975a). Just as the olfactory lobes of the brain are grossly enlarged in fishes that rely heavily on olfactory sensory information (e.g., Atlantic eels and sharks), the facial and/or vagal lobes of the medulla display a marked enlargement in species that find food by taste. For example, the vagal lobes are larger than the remainder of the brain in suckers (Catostomidae), reflecting the large number of taste buds on the palatal organs and their "mouth-tasting" behavior (Miller and Evans 1965). These correlations between brain morphology and feeding habits have prompted researchers to categorize teleostean taxa into one of (usually) three overlapping groups: (1) fish that feed by *sight* and *taste* and show prominent *optic lobes, facial lobes*, and relatively large *vagal lobes*; (2) those that detect food using *barbels* and that have enlarged *facial lobes* with less prominent *vagal lobes*; and (3) *sight-feeding species* possessing well-developed *optic lobes* and poorly developed *facial* and *vagal lobes* (Khanna and Singh 1966).

The functional importance of taste receptors to catfish is demonstrated by the fact that surgical removal or sectioning of optic and olfactory senses does not prevent catfish from swimming to a food source in still water. When taste reception is surgically blocked on one side of the catfish, it continually circles toward the side of intact reception until, eventually, the food is found. Thus, the bilateral taste receptors in catfish function much as the sensitive olfactory system does in some other fish (i.e., to indicate the directionality of a food source). Furthermore, surgical removal of the facial lobes of the medulla prevents catfish from locating and ingesting food. In contrast, removal of the vagal lobes does not inhibit catfish from finding food or taking it into its mouth, but it does prevent swallowing, which is controlled via the glossopharyngeal and vagus nerves (Atema 1971). Interestingly, sharks and rays show no sensory elaboration (including enlarged medullary lobes) for taste functions (Kapoor et al. 1975a). Elasmobranchs rely on olfaction, hearing, vision, and electroreception to locate their food. Hara (1993) and Sorensen and Caprio (1998) have summarized advances in chemoreception research on fishes.

10.4   ACOUSTICOLATERALIS SYSTEM

The **acousticolateralis system** of fish senses **sounds**, **vibrations**, and other **displacements** of water in the environment. It has two main components; the **inner ear**, and the **neuromast/lateral-line system**. Besides sound detection, the inner ear functions to **orient**, or "balance," the fish in three-dimensional space, giving it a feeling of the direction in which **gravity** is acting even when suspended, neutrally buoyant, in lightless, pelagic habitats. In this chapter, hearing is considered first, followed by aspects of **spatial equilibrium** and balance and a discussion of the lateral-line system.

10.5   HEARING

The nature of sound transmission in water has had an important influence on the evolution of hearing in fish. Because of its greater **density**, water is a much more efficient conductor than air of **sound pressure waves**. Therefore, sound carries much farther and travels 4.8-times faster underwater than in air. A propagated sound wave results from the compression of particles, which rebound after being compressed and impart this energy to neighboring particles. Each particle first moves in the direction of the wave propagation and then in the opposite direction during these pressure fluctuations. Thus, the sound pressure wave generated by the sound source causes individual particles to oscillate along the axis of wave propagation. It is important to note that the particles themselves have not moved from their original position once the sound wave has passed (Popper and Coombs 1980).

Structure of the Inner Ear.   Figure 10.2A and 10.2B shows diagrammatic representations of teleostean inner ears. The dorsal part (pars superior) includes three **semicircular canals** (through the horizontal, lateral, and longitudinal planes) with their respective **ampullae** (fluid inertia-sensing chambers). The **utriculus**, with its **utricular otolith** (ear stone), completes the **pars superior**, which mainly functions as an **equilibrium system** and **gravity detector**. The ventral part (**pars inferior**) consists of the **sacculus** and **lagena**, which also contain otoliths, and functions primarily in **sound detection**. As sound vibrations (e.g., in the near field) impinge on a fish, the whole fish moves to and fro with the particle displacement of the water. The **inertia** of the comparatively dense otoliths ($\approx$3-fold the total fish density) in their chambers causes the otoliths to lag somewhat behind the movements of the fish. The otoliths are suspended in fluid and are surrounded by **ciliary bundles** emanating from **sensory hair** cells on the chamber walls.

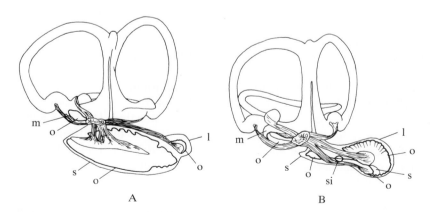

FIGURE 10.2   Drawing of a medial view of the inner ear of (**A**) zander (*Stizostedion lucioperca*) and (**B**) ide (*Leucisus idus*) modified from drawings by Retzius. Only the auditory portions of the ear are labeled. The nerves shown in both figures are the auditory portions of the eighth nerve. *lagena,* m *utriculus,* o *otolith* of each otolithic organ, s *sacculus,* si *transverse canal.* From Popper and Fay (1973).

This layer of hair cells constitutes a sensory epithelium (***macula***) in each chamber. Thus, the ***differential amplitude*** and ***phase motions*** of the otolith with respect to its chamber cause the otolith to bend some of these cilia mechanically. Bending the cilia causes the sensitive hair cellular membranes to deform, thereby stimulating ***neural transmissions*** to the auditory center of the brain, where they are processed and comprise the sense of "hearing."

At higher frequencies, the amplitude of fish displacement decreases, and more sound energy is needed for otolithic stimulation. The sensitivity of hearing in the far field is increased by anatomical devices that can transform sound pressure into displacement movements to provide a movement differential between the sacculus/lagena and their respective otoliths. The ***gas bubble*** in the ***swimbladder*** provides this acoustical transformation in many bony fish. Because the gas is far more compressible than water, it *pulsates* when exposed to sound. The pulsating surface of the swimbladder acts to vibrate the tissues of the fish surrounding it, which provides the necessary particle movement for otolithic/auditory nervous stimulation, especially in species having a *close association* of the swimbladder and the auditory apparatus (pars inferior). Among three species of squirrelfish (Holocentridae), the species having the shortest distance between the swimbladder and inner ear also has the most sensitive (lowest-intensity threshold) hearing (Coombs and Popper 1979). Certain squirrelfishes, along with the tarpon (Elopidae), featherbacks (Notopteridae), deepsea cods (Moridae), and sea breams (Sparidae), all have a forked, forward extension of the swimbladder that end close to the ear. Atlantic cod (*Gadus morhua*, Gadidae), which also have anterior extensions of the swimbladder, can discriminate between high and low repetition rates of ultrasonic pulses, potentially to their benefit in detecting sonar pulses from feeding toothed whales (Astrup and Møhl 1998). Herrings (Clupeidae) and mormyrids feature similar swimbladder extensions that actually enter the cranial auditory capsule and lie close to the inner ear. Inflation of the sphere-shaped tips (***auditory bullae***) of these paired anterior swimbladder extensions during larval development significantly improved the responsiveness of herring larvae to an attacking predator (Blaxter and Fuiman 1990).

Minnows, catfish, and other otophysan teleosts connect the auditory system to the swimbladder with a chain of small bones called ***Weberian ossicles***. The ossicles connect the pulsating swimbladder wall (***tunica interna***) with a Y-shaped lymph sinus (***sinus impar***) that abuts the lymph-filled ***transverse canal*** joining the sacculi of the right and left ears (Fig. 10.3). In studies of goldfish, deflation of the swimbladder significantly decreased the hearing sensitivity at frequencies greater than 150 Hz. Swimbladder deflation of catfish (*Ictalurus*) revealed an even greater loss of high-frequency sensitivity, whereas the same procedure on the perciform *Tilapia*, which lacks Weberian ossicles and a close swimbladder–ear connection, had no significant effect on its already poor sense of hearing.

Sharks, skates, and rays also show evidence of hearing sounds or vibrations, especially at lower frequencies. Sharks seem particularly attracted to *irregularly* pulsed sounds, perhaps signaling crippled prey. However, they seem to have little sensitivity to high-frequency sounds. This low-frequency range correlates well with their auditory sensitivity and the fact that elasmobranchs have no swimbladders, which may act as a sound pressure transducer. For example, the auditory sensitivity of the horn shark (*Heterodontus*

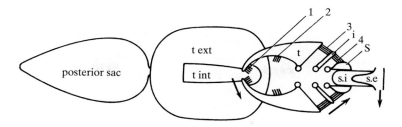

FIGURE 10.3   Diagrammatic dorsal view of the swimbladder and Weberian apparatus of a typical member of the Cyprinidae. The anterior parts (right) are drawn to a larger scale than the posterior ones. Arrows indicate movements resulting from enlargement of the swimbladder. The axes about which the ossicles pivot are indicated by circles. t ext, *Tunica externa;* t int, *tunica interna* as seen through the slit in the tunica externa; 1,2,3,4, *ligaments;* t, *tripus;* i, *intercalarium;* s, *scaphium;* s.i, *sinus impar;* s.e, *sinus endolymphaticus.* From Alexander (1966b).

francisci) is maximal at 40 Hz. Cartilaginous fishes generally have ***endolymphatic ducts*** that connect the inner ear with the environment. Some of these species, which cannot produce calcareous otoliths, apparently use ***exogenous material*** for otoliths. Fänge (1982) found sand grains held together by a gelatinous material in the endolymph of two electric rays (Torpedinidae).

Many shark species also show an ability to directionally locate pulsed, low-frequency sounds from as far as 100 m, which is well into the far field (Popper and Fay 1977). The taut skin over the ***parietal fossae*** of the shark's ears may serve in "drum-like" fashion to induce fluid movements on the canal-based sensory organ, the ***macula neglecta*** (Fig. 10.4). The functional similarities between this auditory system and the tympanic membrane and associated structures of higher vertebrates are striking (Myrberg 1978). Presumably, the ***differences in amplitude or time*** of the sounds received at the right and left ears give the shark a general direction to the source of the sound. Because of the vertically oriented axis of the macula neglecta in sharks, sound that has bounced off the surface of water bodies may also be useful in determining direction. In this case, the air–water interface would mimic the sound transduction of the swimbladder wall in some teleosts. Thus, sharks (and other fish) might increase their auditory sensitivity and directionality by swimming just below the water surface (Popper and Fay 1977).

10.6   EQUILIBRIUM AND BALANCE

The dorsal part (pars superior) of the inner ear is concerned with spatial equilibrium and balance in fish. The utriculus connects to three semicircular canals in teleosts and elasmobranchs (Fig. 10.2). Lampreys have only two such canals, and hagfish make do with one. The three canals (also termed the ***labyrinth***) in jawed fish are filled with lymph (***endolymph***) and function to inform the fish as to ***angular accelerations*** through space. Because the three canals are more or less aligned in the horizontal, vertical, and lateral planes, changes in ***pitch*** (i.e., head up or down), ***yaw*** (i.e., head from side to side), ***roll***

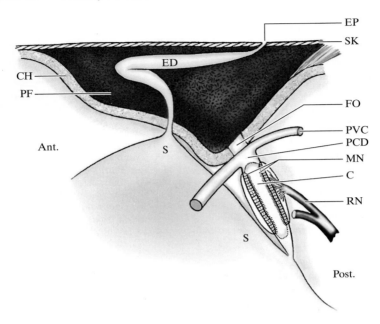

FIGURE 10.4 Schematic section of portions of the ear of the shark. C, Cupula; CH, chondrocranium; ED, endolymphatic duct; EP, endolymphatic pore; FO, fenestra ovalis; MN, macula neglecta; PCD, posterior canal duct; PVC, posterior vertical canal; PF, parietal fossa; RN, ramus neglectus nerve; S, sacculus; SK, skin covering fossa. From Fay et al. (1974).

(i.e., rotation about the head-to-tail axis), ***straight-line acceleration or deceleration***, or any combination of these are detectable. Each canal has an ***ampulla*** (bulbous area) containing a sensory hair cell area and a ***gelatinous cupula*** attached to the hair cell cilia. The cupula extends into the canal path and partially blocks the flow of endolymph. Angular accelerations, either from swimming/turning movements of the fish or from water currents moving the fish, are detected as the endolymph lags behind the movements of the labyrinth (Marshall 1966). The appropriate cupula is bent by the pressure of the endolymph, thereby stimulating the hair cells. Thus, the sensory hair cell system, as a mechanoreceptor, parallels that of the hearing and lateral-line systems. Changes in the pattern of the continuous trains of nervous impulses from the hair cells to the balance/equilibrium center in the medulla provoke the appropriate motor responses by the fish, such as eye movements to maintain a stable visual field and fin movements to restore body equilibrium. Even at rest in quiet water, some fishes, such as northern pike (*Esox lucius*), constantly make pectoral fin movements to maintain balance. This species, among others, has denser tissues in the dorsal part of its body (bone and muscle) than in the ventral part (swimbladder and viscera). Thus, its center of gravity (balance point) is dorsal to its roll axis (midline through body profile), and the fish always tends to roll to one side or the other to reach stability—ventral side up!

10.7 LATERAL LINE

The lateral line of fish provides a ***ferntastsinn,*** or "distant touch," sense (Dijkgraaf 1962). By means of mechanoreceptors similar to those in the auditory and equilibrium systems, water movements around the fish can be detected. The receptors are called ***neuromasts,*** and each consists of individual hair cells with an attached cupula (Fig. 10.5). Water movements bend the protruding cupula, which stimulates the hair cell by bending the attached cilia (sense hairs). All fish, including hagfish and lampreys, have at least some free (individual) neuromasts on the body surface or at the bottom of shallow pits or grooves (Dijkgraaf 1962). Most teleosts and elasmobranchs have also developed ***lateral-line canals***, in which the neuromasts lie between ***canal pores*** that open to the environment (Fig. 10.6). The cupulae of the canal neuromasts are sensitive to movements of the watery endolymph fluid through the canal. Like the auditory and equilibrium hair cells, the neuromasts continually send neural impulses to the brain—even when they are undisturbed. As in the sensory cells of the auditory and equilibrium systems, the neural impulse frequency is increased as the cupulae are flexed in one direction and decreased when the cupulae are flexed in the other direction (Roberts 1978). Thus, the pattern of impulses from the free or canal neuromasts imparts a direction to the

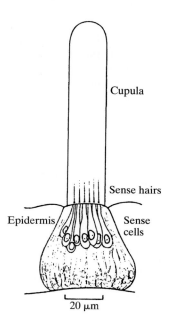

FIGURE 10.5 Superficial neuromast of a bony fish (*Phoxinus* sp.). Characteristic features are the bottle-shaped sense cells, each bearing a hair, and the jelly-like cupula ensheathing the hairs. From Dijkgraaf (1962).

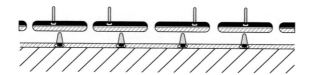

FIGURE 10.6   Longitudinal lateral-line canal section. Black denotes epidermis, spacious striation denotes subepidermal tissues, and gray-shaded structures denote cupulae. From Dijkgraaf (1962).

disturbance. Most of the lateral-line organs of the head region are innervated by sensory fibers of the *lateralis anterior root* of *cranial nerve VII (facialis)*. The remaining organs of the system are innervated by the *lateralis posterior root* of the *vagus nerve (X)*. The fibers of both roots unite with those of the *labyrinth nerve (VIII)* in the *acoustic tubercle* of the *medulla oblongata*.

The lateral-line system is developed and used in various ways by fish exhibiting different modes of life. For example, both the roach (*Rutilus rutilus*) and the stone loach (*Noemacheilus barbartulus*), which inhabit streams, have extensive canal neuromast systems. In contrast, another loach that lives in still water has no canals (Alexander 1967). In general, more active fish have a greater percentage of canal neuromasts compared to free neuromasts. Presumably, the canals offer some "protection" from the continuous stimulation of water rushing past the laterally located neuromasts. Thus, the canal-based receptors can still function to detect weak, local water displacements during rapid swimming (Dijkgraaf 1962). Figure 10.7 shows a dorsal displacement of the lateral line in the region of the pectoral fins in two species where this fin may drive water against the canal during locomotion. In these cases, a straighter lateral line would generate more "noise" from locomotion alone and provide less sensitivity to external

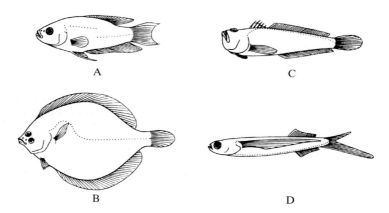

A

B

C

D

FIGURE 10.7   Secondary displacement of lateral line on trunk in connection with dorsal shift of pectoral fin in (**A**) paradisefish (*Macropodus viridiauratus*), and (**B**) flounder (*Pseudorhombus* sp.) or as an adaptation to special modes of living in (**C**) stargazer (*Uranoscopus scaber*) and (**D**) flying fish (*Exocoetus volitans*). From Dijkgraaf (1962).

water disturbances. Similarly, lateral-line placements on the dorsal side of the demersal stargazer (Fig. 10.7C) and the ventral side of the flying fish (Fig. 10.7D) show selection toward the direction in which water movements would be critical for survival.

Parts of the lateral-line system are often *specialized* for detecting prey. The lateral line of the Antarctic fish *Pagothenia borchgrevinki* seems to be tuned to the frequencies produced by the swimming motions of its planktonic prey (Montgomery and Macdonald 1987). Surface-oriented fish, such as killifish (*Fundulus*), sense surface ripples from struggling insects by using the short canal neuromast system on the flattened dorsal surface of their head. Blind cavefish, such as *Typhlichthysi,* have rows of free neuromasts on the head that stand out in ridges, presumably to aid the cavefish in locating food so precisely that it can be ingested without sight (see Chapter 20). The deepsea gulper eel (*Eurypharynx*) features groups of neuromasts on the ends of stalks projecting from its body, which aid in lightless detection of prey (Marshall 1966).

The function of neuromasts can also be useful to fish regarding stimuli other than localized water disturbances from distant objects. As fish swim, they push some water in front of their heads, like a boat creates a bow wave. The amount of water pushed or displaced depends on the size of the fish and the shape of the head. The water displaced forward rebounds from objects in front of the swimming fish, and the rebounds or increases in resistance in front of the fish are apparently sensed by the neuromasts on the head, allowing the obstacle to be avoided. Fish with very sensitive cephalic neuromast systems seem to be especially skilled in object avoidance. Hahn (1960) found that blind cavefish could navigate better through a barrier of thick, fixed rods than thin rods when placed equal distances apart. Presumably, this system also accounts for the avoidance of transparent aquarium walls when naive fish are first introduced into an aquarium, even in total darkness.

The lateral line can also mediate behavioral orientation to water currents *(rheotaxis)*, as shown in three teleostean species (Montgomery et al. 1997). Experiments with adult pollock (*Pollachius virens*) fitted with opaque eyecovers showed that schooling behavior continued when these fish were placed among nonblinded individuals as long as their lateral line remained intact. Five blinded pollock with their lateral lines cut at the operculum failed to school at all (Pitcher et al. 1976). In addition, vibrations are detected by lateral-line neuromasts at frequencies up to 200 Hz because of the accompanying water displacements.

On the other hand, sensitivity to *temperature* change or *physical touch* stems from *general cutaneous endings of spinal nerves*, not from lateral-line receptors. Trained minnows (*Phoxinus* sp.) could not distinguish between warm and cold jets of water from a pipette located posterior to the point where the spinal cord was sectioned, even though the water movements (of either temperature) were detected by the lateral-line neuromasts. Popper and Platt (1993) as well as Schellart and Wubbels (1998) have summarized advances regarding fishes' auditory and mechanosensory lateral-line systems.

10.8   ELECTRORECEPTION/MAGNETORECEPTION

The primary function of the *external pit organs* of teleosts is the reception of minute electrical currents in the water. These pit organs open to the surrounding water via canals that are filled with an ***electrically conductive gel***. Specializations for electrore-ception are widespread among fishes—*except* among teleosts, which are represented by relatively few groups (Bullock et al. 1983). The freshwater examples, including the gym-notids (Gymnotidae), electric catfishes (Malapteruridae), and African electricfishes (Mormyriformes), have very short canals (300 μm). In contrast, marine catfish (*Plotosus* sp.) have longer canals, resembling similar structures found in marine elasmobranchs called ***ampullae of Lorenzini***, which range from 5 to 160 mm in length. The longer ducts in the marine species compared with those in freshwater species relate to the ***electrical conductivity differences*** between the environments compared with those of the skin and body tissues. Specifically, an increased electrical conductivity gradient (related to salti-ness) exists between freshwater fish and their low-conductivity freshwater environment compared with the reduced conductivity gradient between marine species and their high-conductivity seawater environment. Thus, freshwater-fish electroreceptors need be only skin deep to sample a maximum change in potential. In both cases, the gel-filled canals terminate at sensory cells in the more bulbous pit portion of the organ.

An array of electroreceptors (Fig. 10.8) capable of detecting weak electrical cur-rents can be very helpful to a fish in perceiving aspects of its environment. For example, sensitivity to the electrical phenomena generated by movement through the Earth's ***magnetic force field***, that resulting from the depolarization and repolarization of a con-tracting muscle, or even the electrical transmissions from a conspecific individual may have adaptive value. Lissman (1963) described both electrical reception and transmis-sion used by the mormyrid and gymnotid fishes to probe their turbid African and South American stream environments (see Chapter 11). The rigid body posture of these elec-tric fishes, with locomotion generated by undulation of the long anal (gymnotid) or dor-sal (mormyrid) fins, may make the electrolocation process easier (Alexander 1967). Body undulations would continually change the distances and orientations among the electric organs and receptors, thereby complicating the processing of electroreception information. The neural pathways in the large mormyrid brain (Box 10.2) have been re-viewed by von der Emde (1998).

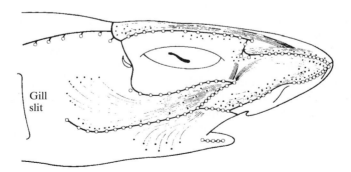

FIGURE 10.8 Location of sense organs on the head of the spotted dogfish *(Scyliorhinus* sp.*)*. Openings of the ampullae of Lorenzini are shown by black dots. The open circles show the pores of the lateral-line system, and the heavy black lines show its location. From Dijkgraaf and Kalmijn (1963).

BOX 10.2

Do Large Brains Make Smarter Fish?

The relative brain weight of the mormyrid *Gnathnemus petersii* (3.1% of body weight) exceeds that of other fishes (most are <1% of body weight) and even humans (2.3%). This large brain features a huge cerebellum and consumes 60% of the total oxygen requirement (Nilsson 1996, von der Emde 1998). Presumably, this large brain coordinates the relatively complex electrical transmission and reception functions in these freshwater fish. This species' "intelligence" (beyond electrical communication capabilities) has yet to be determined, although mormyrids are known for their "playful" behavior in aquaria.

Kalmijn (1971) demonstrated that predaceous marine elasmobranchs can locate prey by electroreception alone. For example, the spotted dogfish shark (*Scyliorhinus canicula*) could locate a flounder buried in sediment, even when that flounder was concealed by an electrically conductive agar plate. The dogfish was not attracted to buried chopped fish or to live flounder covered by an electrically insulating sheet of polyethylene. Conclusively, the shark tenaciously dug out two electrodes, emitting a biological-strength electrical current, that were buried in sediment. Nighttime field studies showed that smooth dogfish (*Mustelus canis*) also responded more strongly to short-term electrical rather than olfactory stimulation. As an applied use of this information, the U.S. Navy's antishark screen (a large polyvinyl bag suspended from an inflatable collar) provides good electrical as well as visual and olfactory insulation for a mariner with skin cuts in shark-infested waters (Kalmijn 1978a).

In marine environments, the resting discharge activity of electrosensory neurons (e.g., of marine elasmobranchs) uniquely functions in *both* the information transmission (ampullae to brain) of electrical stimuli as well as the signaling of others nearby because of sea water's high electrical conductivity and consequent transmission of these impulses. This *single* electrical sensory/transmission system differs from those with which fishes receive and transmit other sensory media, which use *separate* structures and mechanisms. For example, light is received by the eyes but is transmitted via bioluminescent cells, sounds are received via the acousticolateralis system but are transmitted via swimbladder muscle contractions, and dissolved chemicals are received via gustatory and olfactory sensory epithelia but are transmitted via secretions from broken skin (schreckstoff in cyprinids, see Chapter 11). The adult clearnose skate (*Raja eglanteria*) provides an example in which the sensory receiver system appears to be tuned to the appropriate stimulus (Sisneros et al. 1998). This skate's mean electrical pulse frequency (2.5 Hz) is aligned with their peak electrical frequency sensitivity (2–3 Hz). This match may represent an adaptation of the adult electrosensory system to facilitate communication during social interaction and mating (Sisneros et al. 1998).

Any electrical conductor moving through a magnetic field induces an electrical field through that conductor. Thus, a fish with dorsally and ventrally located electroreceptors that is swimming across the Earth's north–south magnetic field should be able to detect the ***induced electrical currents*** (Fig. 10.9). These induced currents are of sufficient strength for detection by sharks swimming as slowly as 2 cm/s, thereby proving the feasibility of an electromagnetic compass sense (Kalmijn 1978b). When the eastbound shark depicted in Figure 10.9 turns toward the north or the south, the potentials vanish; turns toward the west induce potentials of the opposite polarity. Recent studies on the barndoor skate (*Raja laevis*) and the white shark (*Carcharodon carcharias*) present neuroecological evidence for different functions (i.e., prey detection and navigation) for specific clusters of ampullary organs located on these species' heads (Tricas 2001).

The discovery of ***magnetite crystals*** in the ***ethmoid region*** of skulls suggests another avenue of research concerning geomagnetic orienteering in fishes. Walker et al. (1984) found magnetite crystals in a ***dermethmoid bone sinus*** of the yellowfin tuna (*Thunnus albacares*), a species known to behaviorally respond to Earth-strength magnetic fields. Possible links of a suspected magnetoreceptive system with the acousticolateralis system have been drawn by the discovery of magnetite in the lateral line (and associated nerve tissues) of Atlantic salmon, another species that undertakes open-ocean migrations (Moore et al. 1990). Heiligenberg (1993) and von der Emde (1998) have summarized advances concerning electrosensation in fishes.

10.9   VISION/PHOTORECEPTION

The ***eye*** of a fish is the primary receptor site of light from its surroundings. It is worth noting as well that the ***pineal organ*** is also sensitive to light in some fish and appears to have importance in the control of ***circadian rhythmicity*** (Kavaliers 1979). The eyes of all vertebrates have many similarities. A notable feature of the typical teleost eye is a ***cornea*** of constant thickness (Fig. 10.10). This cornea imposes no optical alterations (convergence or divergence) on incoming light. Thus, all the ***focusing*** of light occurs at the ***spherical lens***, which has the highest (1.65) effective ***refractive index*** (light-bending ability) among the vertebrates (Marshall 1966). As in other vertebrate eyes, the lens consists of water and structural proteins. High concentrations of soluble protein in the lens confer the high refractive index.

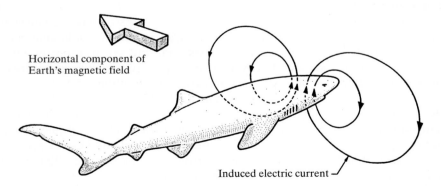

Horizontal component of
Earth's magnetic field

Induced electric current

FIGURE 10.9   A shark swimming through the Earth's magnetic field induces electric fields that provide the animal with the physical basis of an electromagnetic compass sense. From Kalmijn (1974).

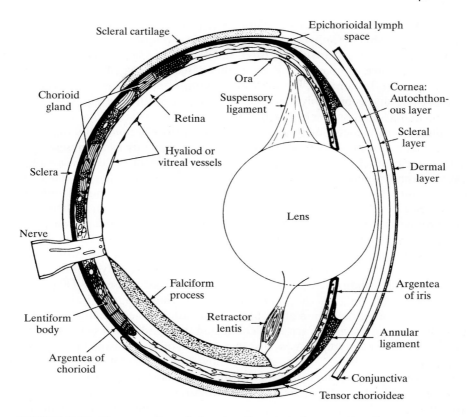

FIGURE 10.10 Diagrammatic vertical section of a typical teleost eye. Not all structures shown are present in every teleost eye; for example, hyaloid vessels are not present in conjunction with a falciform process. From Walls (1942).

The teleost eye lens protrudes through the pupilar opening in the *iris*, and the eye bulges from the body surface. Therefore, the field of view includes a considerable arc forward, which continues laterally to almost directly behind the fish. The alternating head movements of anguilliform or subcarangiform swimmers with co-ordinated eye movements (Harris 1965) tend to eliminate the blind spot to the rear. Light focused by the lens is projected in the appropriate pattern of light, shade, and (often) color on the *retina*, the light-sensitive cell layer in the eye. Because the spherical lens is eccentrically located in the elliptical eye of teleosts (Fig. 10.10), the lens–retinal distance varies, determining relatively *near-field* (close-up) vision in front of the fish and comparatively *far-field* vision to the side. Presumably, this "nearsightedness" to the front, coupled with the *binocular depth-perception* capabilities, allows fish to seize their prey accurately and avoid obstacles.

The elasmobranch lens, on the other hand, is slightly flattened on the optical axis (light path). Most elasmobranchs have a unique ability to *dilate* and *constrict* the pupil of the eye, thereby comprising another mechanism to regulate incoming light levels, although some deepsea sharks have fixed pupils (Kuchnow 1971). The constricted, light-adapted pupil takes the shape of a pinhole in the blacktip shark (*Carcharhinus limatus*), a vertical slit in the lemon shark (*Negaprion brevirostris*), a horizontal slit in the bonnethead shark (*Sphyrna tiburo*), and an oblique slit in the nurse shark (*Ginglystoma cirratum*). Sawfish (*Pristis*) and skates (*Raja*), among others, have an **operculum pupillare**, or opaque projection, that descends ventrally to shield part of the pupillary opening. In contrast, the **nictitating membrane** (semiopaque eyelid) that is present in some sharks probably functions more to protect the eye when feeding than to reduce incoming light levels (Gruber 1977).

Adjustments in focus for near or far vision *(optical accommodation)* are accomplished through moving the lens without changing its shape by using muscles within the eye. The specific muscles within the eye that are responsible for these movements differ in lampreys, elasmobranchs, and teleosts. For example, in teleosts, the lens is pulled *inward* by a **retractor muscle** (Fig. 10.10), whereas in elasmobranchs, it is pulled *outward* by a **protractor muscle**. The four-eyed fish (*Anableps anableps*) displays an aspherical lens and two-part retina as structural adaptations associated with its life at the air–water interface. The lens-thickness differences allow for the density differences of the two media.

The retina is made of a dense packing of (often) both **rod** and **cone** cells for discriminating light images. Incoming light must penetrate a clear layer of nerve cells and fibers to reach the photochemically active tips of the rods and cones (Fig. 10.10). The closer spacing of the rods and the connection of many rods to one neural fiber allows for reasonably fine image definition of "lightness and darkness," especially in dim light. The more widely spaced cone cells are each connected to an individual nerve fiber and provide vision of higher resolution, as well as in color, but only in reasonably brightly lit surroundings (Marshall 1966). Coral reef fishes show developmental changes in retinal rod and cone densities during their transition from planktonic, late-larval stages to settled, reef-living juvenile stages. Whereas cone densities decrease and rod densities increase in nocturnal species during this transition, high cone densities are retained in diurnal species (Shand 1997). Some of the cones in the retina of rainbow trout are sensitive to ultraviolet (UV) and polarized light, whereas the rods were not (Parkyn and Hawryshyn 1993). Gruber et al. (1975a) measured a rod:cone cell ratio of approximately 10:1 in the retinas of four species of requiem sharks (Carcharhinidae) and of approximately 6:1 in three species of the more active mackerel sharks (Lamnidae).

The ability to use UV wavelengths (UV-A:320–400 nm) has been demonstrated in many fishes (Novales Flamarique and Hawryshyn 1998). The UV sensitivity can persist throughout the fish's lifetime (e.g., goldfish) (Neumeyer 1985, Fratzer et al. 1994), or it may be restricted to certain life stages (e.g., presmolting and reproductive stages in salmon, *Oncorhynchus* sp.) (Beaudet et al. 1993, Novales Flamarique and Hawryshyn 1996, Beaudet et al. 1997). Because the UV-sensitive cones are located primarily in areas of low-photoreceptor density (Beaudet et al. 1997), UV sensitivity probably does not function, significantly, to enhance the fish's spectrum. However, UV sensitivity probably enhances contrast of planktonic targets (Loew et al. 1993, Browman et al. 1994), increases reflective communication among schooling fishes (Denton and Rowe 1994), and provides sensitivity to polarized light (Hawryshyn and McFarland 1987), except in species, such as white suckers (*Catostomus commersoni*), with a random (rather than ordered) cone mosaic (Fig. 10.11) in the centrotemporal retina (Novales Flamarique and Hawryshyn 1998). Detection of polarized light (e.g., by rainbow trout) is especially important during crepuscular (dawn and dusk) periods, when atmospheric conditions

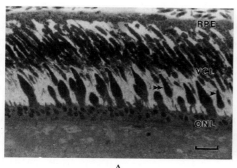 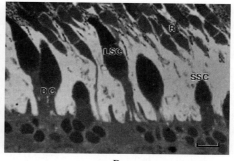

A   B

FIGURE 10.11   Radial section through the retina of the common white sucker (**A**, bar = 20.7 μm, arrowhead points to SSC and double arrowhead to LSC; **B**, bar = 8.32 μm). RPE, Retinal pigment epithelium; VCL, visual cell layer; ONL, outer nuclear layer; DC, double cone; LSC, long single cone; SSC, short single cone; R, rod. From Novales Flamarique and Hawryshyn (1998).

for polarizing shorter wavelengths (UV-A and blue parts of the spectrum) are optimal (Novales Flamarique and Hawryshyn 1997). Finally, the cornea and lens of the zooplanktivorous tropical reef damselfish (*Dascyllus albisella*) accumulate UV-blocking structural proteins as the fish ages, presumably sacrificing some short-wavelength UV sensitivity while preserving the more visually effective long-wavelength UV as its eye increases in size (Losey et al. 2000).

As light strikes the rods or cones, it is absorbed by the ***light-sensitive pigment*** (e.g., ***rhodopsin*** and ***porphyropsin***) in the cell. Visual pigments are organic dyes consisting of an ***opsin protein*** complexed onto a short-chain ***prosthetic*** or ***chromophore molecule*** (e.g., retinal) related to vitamin A (Gruber and Cohen 1978). These pigments, which are now distinguishable using RNA probes (Helvik et al. 2000), are located in the photoreception area of the retina and provide its characteristic purple or pink coloration. For example, rhodopsin constitutes up to 35% of the dry weight of the rod outer segments. The pigments reversibly bleach out when exposed to light, and each has a characteristic absorption spectrum. The maximum absorption of the visual pigments of "typical" shallow-water dogfish sharks (*Squalus*), skates, and rays (Myliobatidae) occurs at ***wavelengths*** of approximately 500 nm, which is typical for rhodopsin (Gruber and Cohen 1978). Denton and Warren (1956) reported high densities of a golden-colored pigment (***"chrysopsin"***) in the retinas of three teleosts living below 500 m. As one would predict from the spectral shift in light penetrating to deeper strata, the maximum absorbance of chrysopsin is shifted toward the blue (shorter-wavelength) end of the spectrum by approximately 20 nm. Parallel evolution of "deepwater" visual pigments is found among three elasmobranch species (caught at 1,150 m) containing pigments that absorb maximally from 472 to 484 nm (Denton and Shaw 1963). Most temperate-freshwater teleosts have three cone pigments that absorb light maximally at approximately 455, 530, and 625 nm (Loew and Lythgoe 1978). Rainbow trout show similar spectral sensitivity peaks and, most likely, can distinguish different colors (Douglas 1983). In all cases, pigmental absorption of light stimulates the retinal cells to send impulses via the ***optic nerve*** (Fig. 10.10) to the ***optic lobe*** on the same side of the brain.

The reception and integration of the visual image in the brain then provokes the appropriate motor response (e.g., to the muscles and/or fins).

The **choroid coat** underlies the retinal layer and primarily functions to supply **nutrients** and **oxygen** to the retina for its high metabolic demand. Most teleosts possess a pigmented choroid projection into the posterior portion of the interior of the eye called the **falciform process** (Fig. 10.10). The falciform process is highly vascularized and, probably, serves a nutritive function (Munz 1971). In addition, a **choroid gland** (or rete) is present in most teleosts and the bowfin (*Amia calva*) behind the retina. The countercurrent arrangement of blood vessels in the choroid rete makes it eminently suited for oxygen delivery to the retina in a fashion similar to that and the rete mirabile of physoclistous swimbladders (see Chapters 4 and 5). Wittenberg and Haedrich (1974) described the presence and relative size of choroid retia in several species of North Atlantic fishes. Invariably, the fishes that rely heavily on sight, such as bluefish (*Pomatomus saltatrix*), have the best-developed choroid retia. Presumably, localized modification (e.g., acidification) of the blood next to the rete drives oxygen off the hemoglobin via the Root shift, much as in the swimbladder rete (see Chapter 5). Hayden et al. (1975) attributed the significant Root shift in the swimbladderless winter flounder to choroid rete function in this demersal sight feeder. Wittenberg and Wittenberg (1962) measured the partial pressure of oxygen (PO_2) in the **vitreous humor** of living marine fish. Those species with the best-developed choroid rete also had the highest vitreous PO_2 (250–820 mm Hg). Teleosts with smaller retia had lower PO_2 values (20–210 mm Hg), whereas elasmobranchs and those teleosts that lacked the choroid rete had the lowest PO_2 values (10–20 mm Hg).

The choroid layer of elasmobranchs also contains the **tapetum lucidum**, or reflecting layer, which produces eyeshine when a direct beam of light is trained on the eye, especially under darkened conditions. Many teleosts also have tapeta lucida, but most are located in the retinal pigment epithelium (Nicol and Zyznar 1973). The reflecting material of the tapetum varies among fish. **Guanine** crystals comprise the primary reflecting substance in the bigeye (*Priacanthus arenatus*), the bream (*Abramis brama*), the bay anchovy (*Anchoa mitchilli*), and all of the elasmobranchs so far investigated. A lipid-reflecting substance is found in at least six teleost families; Yellow **"melanoid"** substances occur in gars (Lepisosteidae) and catfish, and a **pteridine** is found in the gizzard shad (*Dorosoma cepedianum*) (Zyznar and Nicol 1973).

In all cases, the tapetal layer increases the visual sensitivity of retinal pigments by reflecting most of the transmitted light back through the retina. The arrangement of choroidal tapetal plates reflects light back in the same direction from which it first stimulates the retina, thereby preserving image clarity. Whereas most benthic and deepsea elasmobranchs have tapeta lucida that function continuously, pelagic species from well-lit waters can cover the reflecting layers by moving dark pigment between the tapetal plates. Occlusion of the reflecting layer prevents "overloading" of the retinal pigments under bright conditions. With the onset of darkness, the occluded tapetum becomes completely shiny after 1 hour. Gruber and Cohen (1978) point out that the light reflection from elasmobranch tapeta approaches 90% at certain wavelengths. This layer gives elasmobranchs a light sensitivity approximately equal to that of sympatric teleosts (without tapeta), which typically have twice the concentration of visual pigment per unit area of retinal surface.

Fish tissues other than the eyes also may show receptivity to light. The pineal gland, which is dorsally located on the brain, is one of these tissues. Gruber et al. (1975b) showed that the ***chondrocranium*** of three shark species is modified for light transmission, in that 7-fold more light impinges on the pineal receptors than on the surrounding areas of the brain. Using immunocytological techniques, Forsell et al. (2001) showed the presence of green wavelength-sensitive pigments (opsins) in the pineal organs, as well as in the retinas, of several marine and freshwater fish embryos and larvae. Hamasaki and Streck (1971) found evidence that the pineal gland of *Scyliorhinus* sp. responds to changes in illumination approximating 4×10^{-6} lumens/m^2, which is far below the intensity of moonlight at the water surface. Thus, the pineal gland may function as an ultrasensitive light sensor closely connected to the brain, and it may be used to cue the fish's behavior to changes in light intensity on a daily and a seasonal basis. Larval sea lampreys show ***photosensitivity*** (escape response) when the lateral skin of the tail is illuminated. Apparently, the photoreceptor signals are carried by a ***trunk lateral-line nerve***, because bilateral sectioning of this nerve significantly diminishes this sensitivity (Ronan and Bodznick 1991).

10.10   LESSONS FROM SENSORY PERCEPTION

The characteristics of water have shaped the evolution of a variety of sensitive receptors by which fish perceive their surroundings. Dissolved chemicals are detected using either the paired olfactory rosettes or the widely scattered taste buds. Sounds, vibrations, and other water displacements or pressures are sensed by the consequent bending of sensory hair cells (e.g., by otoliths) or neuromasts of the allied hearing and lateral-line systems. Electrical pulses are detected by certain families of fishes via jelly-filled, pit-based receptors, and light is detected through a variety of structures, with the most discriminating being the eyes.

SUPPLEMENTAL READINGS

Bullock et al. 1983; Dijkgraaf 1962; Fernald 1993; Hara 1986, 1993; Hasler and Scholtz 1983; Hawryshyn 1998; Heiligenberg 1993; Kalmijn 1978a; Losey et al. 2000; Novales Flamarique and Hawryshyn 1998; Popper and Platt 1993; Schellart and Wubbels 1998; Sisneros et al. 1998; Sorensen and Caprio 1998; Sorensen and Scott 1994; von der Emde 1998.

WEB CONNECTIONS

SHARK SENSORY SYSTEMS
www.mote.org/~rhueter/sharks/shark.phtml
www.flmnh.ufl.edu/fish/Sharks/sharks.htm
www.floridamarine.org

C H A P T E R 11
Behavior and Communication

T he anatomical and physiological characteristics of a fish are best understood when explained in relation to their effects on its behavior. Thus, much of the chapters on the major fish groups (see Chapters 14 through 24) is spent describing behavior patterns. Similarly, descriptions of fish behavior are important to the chapters on ecology (see Chapters 27 through 36), because ecology is the study of organism–environment interactions—and the most conspicuous manifestation of these interactions is behavior, particularly behavioral change in response to environmental change. For this chapter, it is convenient to divide the subject of behavior into seven categories (which are not necessarily mutually exclusive): (1) migratory behavior, (2) shoaling behavior, (3) feeding behavior, (4) aggressive behavior, (5) resting behavior, (6) reproductive behavior, and (7) interspecific interactions. Reproductive behavior was covered in Chapter 9; interspecific interactions, which include predator–prey, mimicry, and symbiotic relationships, are covered in the ecology chapters. Thus, only the first five areas are discussed in this chapter. Because so much behavior involves interactions with other fish of the same or different species, it is important to understand how fishes communicate. Therefore, the latter half of this chapter is devoted to communication using visual, auditory, chemical, tactile, and electrical signals.

11.1 MIGRATORY BEHAVIOR

Mass movements of fish from one place to another, on a regular basis, are common. Such migratory behavior can range in occurrence from seasonal to daily. Daily migrations typically are for feeding and/or predator avoidance (see Chapters 33 and 35), so only long-term movements that result in major shifts in the location of a population are discussed here. The general patterns usually are broken up into three basic patterns: (1) oceanodromy, (2) potamodromy, and (3) diadromy. *Oceanodromy* refers to fish that migrate entirely within salt water. *Potamodromy* refers entirely to species that migrate within fresh water. *Diadromy*, the phenomenon of migrating between the two environments as a regular part of the life cycle, is broken into three basic types (McDowall 1997) (Fig. 11.1):

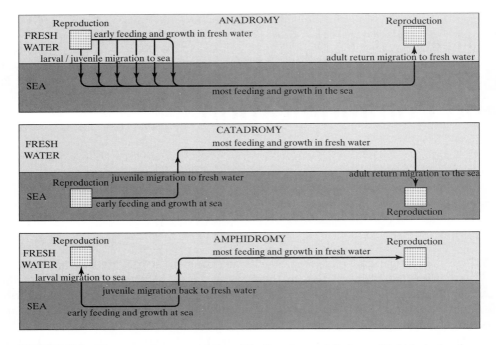

FIGURE 11.1 Diagrammatic representation of the three forms of diadromy; life histories involving migrations between fresh and salt water.

1. **Catadromy.** Catadromous fishes spend most their life in fresh water, where most growth occurs, and migrate into salt water to spawn. Usually, adults do not feed once they enter salt water. The best known examples are anguillid eels, which can migrate enormous distances to their spawning areas in the oceans (see Chapter 17).

2. **Anadromy.** Anadromous fishes divide their lives between fresh and salt water, with most of their growth taking place in salt water. Maturing fish move from salt to fresh water to spawn, and no significant feeding usually takes place once the migration has begun. The best-known examples are various salmon and trout, which show enormous variability in this life-history pattern (see Chapter 19).

3. **Amphidromy.** Amphidromous fishes move between salt and fresh water for purposes other than spawning (McDowall 1988). As larvae, they move out to sea soon after hatching, where they feed and grow for a few months before returning to fresh water as small juveniles. They then grow to adulthood in fresh water, where they spawn. Most amphidromous fishes occur on islands (including Australia) and are members of families such as the Gobiidae, Eleotridae, and Galaxiidae.

Whatever the label, fish migrations are often quite spectacular. Tuna, salmon, and eels, for example, may migrate thousands of kilometers in relatively short periods of time, often arriving within a relatively narrow time span at a fairly specific locality. Two questions will be addressed here:

- "Why do they do it?"
- "How do they do it?"

Why fish migrate is a fascinating question. In many cases, the energetic investment in migration is quite high, and often, migrating fishes do not feed. The catadromous eels that migrate from Europe to the Sargasso Sea near Bermuda cease feeding once they enter salt water, as do anadromous salmon once they enter fresh water, some to swim hundreds of kilometers upstream. Most such migrations are for spawning, allowing the eggs to be laid in places favorable for the development and survival of the newly hatched larval stages. As Harden-Jones (1968) points out, many of these migrations have a regular pattern. The adults swim against a current (ocean or stream) to the spawning grounds: The currents then carry the helpless young to favorable feeding areas. Once the young reach a certain size and are active swimmers, they migrate to feeding grounds and, eventually, join the adult population. This pattern has several advantages:

1. It greatly increases the probability of the larval stages finding their way to the proper habitats.
2. It reduces the likelihood of intraspecific competition for food among different age classes, a problem most likely to affect plankton-feeding fishes such as herring.
3. It puts each life stage in habitats where feeding and growth can be maximized.
4. It reduces the probability of cannibalism.

The use of migration to separate the different life-history stages seems to be one of its main characteristics, regardless of the environment in which the migration occurs. River- and lake-dwelling fishes often migrate into small tributary streams to spawn in riffle areas (Lucas and Baras 2001). The young may then use the small streams as nursery areas before moving down to the adult habitat. Many lake-dwelling species migrate only as far as suitable spawning grounds in shallow waters of the lake itself. In fishes such as the sunfishes (*Lepomis*) and crappies (*Pomoxis*), the young become pelagic after hatching and drift about in the surface waters for several weeks before settling to the bottom and moving inshore, where they shoal in large numbers in shallow, weedy areas or other protected habitats. A similar pattern is typical of many marine benthic species, such as flounders (Pleuronectidae) and rockfishes (Scorpaenidae). Some of the most prolonged larval stages, however, are found in amphidromous gobies that live in the streams of Hawaii and other islands. The larvae wash downstream as soon as they hatch and then spend a month or more at sea before returning to the streams as 20-mm post-larvae (McDowall 1988). Not only do these tiny fish manage to find their way to a stream, but once there, they proceed to move upstream long distances—even ascending major waterfalls. The gobies climb over the falls through the use of their specialized pelvic fins, which are modified into a sucking disk.

A majority of fish migrations relate to reproduction and separation of the life-history stages, but many are also in response to changing environmental conditions, particularly temperature, and both the movements and abundance of food organisms. The seasonal movements of albacore (*Thunnus alalunga*), for example, appear to follow the development of the 14°C isotherm in the North Pacific. In herring (*Clupea harengus*), the northward movement during the spring in the North Sea is related to the warmth of the inflowing Atlantic waters; the warmer the water, the greater the blooms of plankton on which the herring feed. Migrations across the North Pacific by bluefin tuna (*T. thynnus*) are tied in part, to the abundance of food; many apparently choose not to migrate in some years if a major prey, such as sardines, is abundant off Japan (Polovina 1996;

BOX 11.1

The Complex Movements of Atlantic Bluefin Tuna

Atlantic bluefin tuna (*Thunnus thynnus*) are, kilo for kilo, the most valuable fish in the ocean—and in desperate need of conservation. It has been thought that distinct populations exist in different parts of the ocean, with each needing separate management. To see if this was true, Barbara A. Block, from Stanford University, and her colleagues recorded the movements of tuna using special pop-up tags that archive the behavioral patterns of the tuna for a period of time and then are released, floating to the surface and downloading their information to a satellite (Block et al. 2001). By combining this information with that from other kinds of tags, they found that tuna

from the same general area, off North Carolina, exhibited four distinct migratory patterns. One group simply cruised about the western Atlantic for a year. Another group also stayed mainly in the western Atlantic but migrated to the Gulf of Mexico for spawning. A third group moved back and forth across the ocean. Yet another group moved to the eastern Atlantic and into the Mediterranean Sea, another spawning area. This study strongly suggests that essentially one intermixing population exists that uses the entire north Atlantic Ocean, the Gulf of Mexico, and the Mediterranean Sea, although there may be some segregation by spawning area.

Box 11.1). Increasingly, knowledge about the responses of migratory fish to temperature, salinity, and other oceanographic conditions is being used to predict how good fishing is likely to be in a given area, sometimes several months in advance.

One factor that greatly increases the predictability of the migrations of many fishes is their ability to find their way back to a "home" area, particularly a spawning area. Homing has been well documented in nonmigratory as well as migratory fishes. Many tide pool fishes, for example, can find their way back to their home pool, using olfactory and visual clues, after being displaced by several hundred meters (Gibson 1993). Weakfish (*Cynoscion regalis*) return to the same Atlantic coast estuaries in which they were hatched, although in this case, the homing mechanism is not known (Thorrold et al. 2001). The most studied of homing fishes have been Pacific salmon (*Oncorhynchus*). They are able to return, with an amazing degree of accuracy, to the stream and area in which they were spawned after wandering for several years and thousands of kilometers through the Pacific Ocean. A. D. Hasler and his students showed, rather conclusively, that the precision of homing results from the ability of the salmon to recognize the distinctive odor of their home stream, and that the salmon become imprinted on this odor as they transform into smolts, just before their outward migration (Hasler and Scholtz 1978). Not only are migrating salmon capable of recognizing the odor of their home stream, they also recognize, and use as navigational cues, the odors of other streams they pass on their way up the main rivers. Furthermore, evidence suggests that salmon also may respond to chemicals (pheromones) given off by conspecifics, such as juveniles inhabiting a stream, and are able to discriminate between water that contains fish from the same population and that which contains fish from other populations (Groot et al. 1986). Such abilities can permit a great deal of precision in homing. Mechanisms that other fishes use for the precise location of home areas have not been as well studied, but in addition to odor, topographic cues, currents, and both salinity and temperature gradients likely are important. Yellowfin tuna (*T. albacares*) will use topographic features, such as anchored buoys, as "way points" in their migrations, passing by them on a regular basis in a fairly precise fashion. Two tuna tracked by Klimley and Holloway (1999) passed by a marker buoy together twice in 8 months—both times at approximately 9 A.M.!

The recognition that migratory fishes can home precisely has led to widespread acceptance of the idea that they can also navigate precisely over long distances. This has resulted in an extensive search for mechanisms of orientation. The most likely mechanisms so far proposed have been: (1) Orientation to gradients of temperature, salinity, and chemicals; (2) Sun orientation; and (3) Orientation to geomagnetic and geo-electric fields (Leggett 1977).

As the previous discussion has indicated, fish can detect gradients in water and do orient to them, often timing migrations on the basis of seasonal changes in such factors as temperature and salinity. Such phenomena, however, fluctuate from year to year and frequently cover wide areas; thus, they seem to be poor candidates as cues for the precise nature of the homing that has been observed. It is possible that migrating salmon may be able to detect sequential changes in the chemical content of the oceanic water masses through which they move, much as they detect changes in the streams through which they pass, but no direct evidence supports this. The evidence that fish can navigate by orienting to the position of the sun is somewhat better. For example, Hasler et al. (1958) tied small floats onto white bass (*Morone chrysops*). On sunny days, bass displaced from their spawning grounds had a strong directional bias when released; on cloudy days, such bias was lacking. Evidence that fish widely use the sun for navigation, however, is limited. Indeed, for both salmon and American shad (*Alosa sapidissima*), correctly oriented migrants have been observed both at night and on cloudy days. In addition, salmon followed with electronic tags were observed to have strong directional movement both at night and in water much too deep for celestial cues to be used (Ogura and Ishida 1995).

Evidence is stronger that migratory fishes can orient to the Earth's magnetic fields. These fields are weak but can be detected by fishes, particularly when additional electric fields are generated by the movement of oceanic currents through them. The strongest evidence supporting the use of geomagnetic fields for navigation is for anguillid eels, which migrate long distances in deep water, have an extreme sensitivity to the fields, and have different directional responses to them depending on their developmental stage (Leggett 1977).

Despite precise mechanisms of fish orientation, migrations are not always very accurate. Tagged salmon frequently show up in the wrong streams, many miles from their home streams. Tagging studies of various migratory fishes frequently note that dispersal from the tagging site is essentially random in direction, even though most marked fish end up in the home area. Furthermore, in orientation experiments, outright "mistakes" by fish are common, and the overall orientation of fish to the factor being tested (e.g., sun azimuth) tends to be rather general and not exact. In fact, Saila (1961) demonstrated that homing of winter flounder to inshore spawning areas could be explained largely on the basis of random search movements of fish that will not go deeper than 40 m.

A majority of these fish find the coastline during the spawning period, then find their way to the spawning area, partially through random searching and partially through the use of cues such as topography and water chemistry. Similarly, salmon may not need highly precise navigational abilities to reach a general area of coastline that contains their spawning stream. All they may need is some cue, or multiple cues, to get them swimming in the right general direction. Cues can include food availability, currents, temperature and chemical gradients, as well as sun direction and geomagnetic fields (Patten 1964). By using a combination of cues, most of the fish can reach their home region and, once there, use more precise mechanisms—particularly olfaction—to find the home stream. The jury on this subject is still out, however, because it does not seem to make a great deal of sense for fish to waste time and energy on generalized movement if more precise navigational methods are possible.

Migrations of fish generally seem to have evolved as a mechanism to place adult fish in a favorable place for feeding and to place larval and juvenile fish in a favorable place for survival. Most salmon, for example, feed on fish in the ocean and grow much larger there than would be possible with the limited resources of the streams and rivers in which they spawn. This results in a great increase in fecundity, giving anadromous forms a tremendous advantage over resident forms. Gross et al. (1988) provide evidence for this pattern by showing that anadromous fish are most common in the northern latitudes, where ocean productivity exceeds freshwater productivity, whereas catadromous fish are most common in the tropical latitudes, where ocean productivity is typically low. It is not always advantageous to migrate, however, as indicated by the commonness of resident populations of fish in the headwaters of coastal streams that are derived from anadromous forms. For example, Snyder (1991) found that the inland populations of threespine stickleback (*Gasterosteus aculeatus*) in coastal streams had lost an inherited ability to migrate that characterized populations only a few kilometers downstream. Apparently, the reason for this loss was that energy requirements of migration were just too costly for upstream populations of this tiny fish. The upstream populations were less robust and fecund than anadromous sticklebacks, but they were persisting in places where the anadromous forms clearly could not.

11.2   SHOALING BEHAVIOR

Shoaling is perhaps the most fascinating social behavior possessed by fish, although most attention has been focused on the most spectacular aspect of shoaling: schooling. A school of fish often seems to have a mind of its own, moving in co-ordinated fashion through complicated maneuvers with the members precisely spaced within it. A *shoal* is any group of fishes that remains together for social reasons, whereas a *school* is a polarized, synchronized shoal (Pitcher and Parrish 1993). This distinction is necessary, because experimental studies indicate that schooling is just one extreme in the entire phenomenon of shoaling and that fish shift in and out of schools on a regular basis. A shoal can go from the classic, polarized organization to an amorphous mass in a matter of seconds, depending on whether its members are traveling, feeding, resting, or avoiding predators (Fig. 11.2). The shape of a shoal can also vary widely. Traveling shoals, for example, range in shape from long, thin lines to ovals and squares to more amoeboid

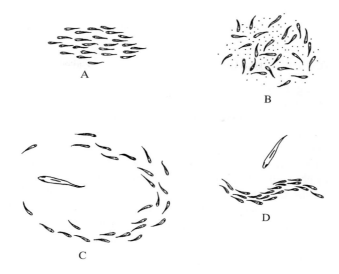

FIGURE 11.2   Common patterns of shoaling. **A.** Traveling school.
B. Planktivores feeding. **C.** Encirclement of a predator. **D.** Streaming
school avoiding predator. (After Radakov 1972.)

shapes. Fast-moving schools, however, typically assume a wedge shape, whereas feeding
shoals usually are more or less circular.

Shoaling is of considerable interest because of its prevalence. Shaw (1978) estimated that 25% of all fishes shoal throughout their lives and that approximately half
of all fishes spend at least part of their lives shoaling. Most of the important commercial fish species also shoal, and their behavior greatly increases their vulnerability to
capture in large numbers. In many commercial species, the largest shoals form during
migrations, when smaller shoals join together. Radakov (1972) mentions "chains" of
shoals of migrating mullet (*Mugil*) in the Caspian Sea that were 100 km long! In the
North Atlantic, herring shoals are sometimes encountered in winter that occupy 279 million to 4,580 million m^3, with densities of 0.5 to 1.0 fish/m^3 (Radakov 1972).

How do fish school? Obviously, the precise structure and movements of polarized
schools require precise sensory contacts among the individuals within the schools. The
fact that most schools (but not necessarily shoals) break up at night indicates that vision
is the main sensory link. Many laboratory experiments with temporarily blinded fish
also demonstrate the importance of vision, although other senses can also be important.
Because the lateral-line system is so sensitive to the movement of water, fish quite likely use the regular turbulence created by neighboring fish as another cue to help maintain their spacing. It is also possible that pheromones or sounds may play a role in
schooling, although no direct evidence has been found for either one.

Why do fish shoal? Various hypotheses have been put forward: (1) increased hydrodynamic efficiency, (2) increased efficiency of food finding, (3) increased reproductive success, and (4) reduced risk of predation. Landa (1998) argues that the

combined benefits of these factors provide strong selective incentives for fish to join shoals and become part of a highly co-operative group.

Increased Hydrodynamic Efficiency

The idea that shoaling increases the efficiency of swimming applies mainly to schooling and is very appealing, both because of the regular spacing that seems to characterize fish in schools and because fish in shoals tend to be uniform in size (Hoare et al. 2000). To gain hydrodynamic advantages, however, each fish must maintain rather precise positioning within a school to use the hydrodynamic lift created by its neighbors. By and large, measurements of fish positions within schools in the laboratory have not found this to be true (Pitcher and Parrish 1993). Nevertheless, the regular spacing of fish in schools observed in the wild indicates that being in a school probably does have a hydrodynamic advantage—at least for fish that are behind other fish. Landa (1998) argues that the leadership of a school is constantly changing, because while being immersed in a school incurs hydrodynamic advantages to individuals, leaders of schools are first to find food, which is also advantageous.

Increased Efficiency of Food Finding

Schooling by predators increases the probability of their detecting a school of prey, just as shoaling in plankton-feeding fish increases the probability of detecting suitable patches of plankton. The reason for this is the presence of many searching eyes over a large area. Fish in shoals "share" information by monitoring each other's behavior closely. Feeding behavior in one fish quickly stimulates food-searching behavior in others (Pitcher and Parrish 1993). For planktivores, however, this advantage may be decreased by having to share the patch with many other fish: Fish in the rear of a school are likely to encounter much lower densities of plankton than those in the front as well as higher concentrations of waste products from their schoolmates.

Increased Reproductive Success

For fish that shoal throughout their lifetimes, little energy has to be expended in finding a mate when spawning time comes. Presumably, shoaling also allows fish to closely synchronize their reproductive cycles (through behavioral and hormonal cues), which is particularly important for fish that are mass spawners. In addition, for fish that migrate long distances to breed, schooling may increase the accuracy of homing, because the average direction sought by fish in a school is more likely to be a better choice than that chosen by an individual migrating alone. For tunas, with habitual migration routes, this means that groups of individuals may stay together for long periods (Klimley and Holloway 1999).

Reduced Risk of Predation

Shoaling reduces predation risk in two main, interacting ways: dilution, and confusion (Pitcher and Parrish 1993). The ***dilution effect*** relies on simple safety in numbers (Parrish

1993). When predators attack, large shoals are particularly advantageous because of the reduced probability that any one individual will be eaten; in any given attack, a smaller percentage of a large shoal will be eaten compared to a small shoal (Major 1978). The ***confusion effect*** is based on the common observation that shoaling fish tend to get eaten mostly when they have been separated from the shoal; many attack strategies of predators can thus be best interpreted as efforts to break up shoals so that individuals can be picked off. Shoaling fish are the same size and silvery, so it is difficult for a visually oriented predator to pick an individual out of a mass of twisting, flashing fish and then have enough time to grab its prey before it disappears into the shoal. Shoaling fishes also are fairly uniform in size and appearance, a further factor confusing predators. Individuals that stand out (e.g., because of heavy parasite infestations) are quickly picked off (Krause and Godin 1996). Even when shoals are made of several species, individual fish are similar in size, although the species may remain segregated within the school. Thus, Krause et al. (1996) noted that banded killifish (*Fundulus diaphanus*), golden shiner (*Notemigonus chrysoleucas*), and white sucker (*Catostomus commersoni*) shoal together, but that the killifish occupy the top of the shoal, the shiners the middle, and the suckers the bottom. These shoal positions reflect the preferred feeding areas of the three species (surface, midwater, and bottom, respectively).

Shoaling fish that are captured are individuals that have become separated from the main shoal (Major 1978). Thus, when a predator approaches, a shoal will typically bunch up into a tight, weaving school. One reason that attacks on shoaling fish are often more successful at dawn or dusk than during the day is the reduced effectiveness of flashing scales in confusing the predators at those times. Shoals of fish also confound predators by complicated maneuvering. In one maneuver, the "flash expansion," a shoal of fish explodes in all directions and then reassembles quickly, never giving the predator a chance to focus on any given individual (Pitcher and Parrish 1993). In a more common maneuver, a school divides in two, turning away from the predator in two directions and finally joining again behind it. This leaves the predator isolated in a large, empty pocket in the school, from which it has great difficulty in attacking effectively, because every movement is closely monitored by individuals in the school. Indeed, a general advantage of being an individual swimming in a shoal is the large number of eyes around you, which make it virtually impossible for a predator to approach during the day without being detected. Thus, whereas shoals appear to be co-operative ventures, they most likely result from individuals maximizing their own chances of survival by hiding in a group of similar individuals (Parris and Edelstein-Keshet 1999).

Despite the many antipredator advantages, shoaling fishes remain the favorite prey of most pelagic predators, and many effective hunting strategies have been evolved to overcome the defenses of the shoal. One strategy for the predator is simply to attack at low light levels, when shoals are more dispersed and the predator is more difficult to spot. Another strategy is to swim along with a school and grab fish that either make defensive errors and become separated from it or are sick or injured. This strategy, which is used by stalking predators such as cornetfish (Fistularidae), has a relatively low success rate (Parrish 1993). A more effective strategy is to attack in schools. When a school of jacks (Carangidae) attacks a school of herring (Clupeidae), the school of herring cannot respond as effectively as they could to a single predator, especially when the jacks hunt co-operatively and separate clusters of fish from the main shoal or literally ram into the shoal. Shoals of small fish can be devastated when a school of predators drives them to the surface and flocks of seabirds then dive down on them from above.

11.3 FEEDING BEHAVIOR

The feeding behavior of fish is as diverse as their morphology, which more often than not reflects their general method of feeding (see Chapter 2). Most fish are specialized to a greater or lesser degree, but their actual diet at any given time depends on the availability of both typical and atypical food items as well as on the presence of potentially competing fish species. When a major hatch of mayflies occurs in a lake, for example, almost every fish that is large enough to capture them will do so, although normally, adult mayflies are taken mainly by fish with adaptations for surface feeding. Similarly, fish that normally filter-feed on plankton will, on occasion, selectively feed by "picking" large individual zooplankters, including larval fish. Rainbow trout usually feed on drifting organisms in streams, but they may become benthic feeders when drift is hard to see, even though this mode of feeding is generally less efficient (Tippets and Moyle 1978). Other fishes may regularly switch between two or more modes of feeding, depending on the relative abundance of different kinds of prey. Bay gobies (*Lepidogobius lepidus*), for example, will hop up from the bottom to seize swimming prey or bite the substrate to obtain benthic prey, especially polychaete worms (Grossman et al. 1980). Obviously, the feeding behavior of fishes is influenced by many factors, including the characteristics of the environment, the predators, and the prey.

To understand just how and why predators select their prey, a general theory of *optimal foraging* has been developed. The idea is that natural selection favors predators that maximize the efficiency of prey capture. Thus, the feeding "strategy" adopted by a fish at any time should be the one that allows it to take in the most or highest-quality food with the least amount of effort (i.e., energy expenditure). Because the ways that a predator expends energy in feeding involve searching for and pursuing prey, handling the prey, and then digesting it, an optimal foraging strategy should balance these factors in such a way that the total energy expenditure is minimized—even if the cost of one of the factors is higher than it otherwise could be. For example, it does not "pay" a carnivorous largemouth bass to browse on algae, even though the algae may be very easy to find and handle, because the fish will expend more energy passing algae through the gut than it can extract from it. On the other hand, bass select slender, soft-rayed minnows over deep-bodied, spiny sunfish, because the handling time is likely to be considerably less. Although many characteristics of prey can influence prey selection, the most important factors usually are size and availability. For example, bluegill sunfish in aquaria will select all size classes of zooplankton prey equally if all are present at low densities; they simply grab the prey as they encounter them. If the densities of all sizes of the prey are greatly increased, however, the fish select only the largest sizes (Werner and Hall 1974). An interpretation of such an experiment is that when food is scarce, a fish captures any prey item that likely will provide a net gain of energy, no matter how small, whereas when food is abundant, the fish selects the prey that produce the most energy for the least amount of effort (resulting in more energy available for growth and reproduction). Likewise, Fausch (1984) showed that juvenile trout and salmon select places to forage in a stream where the energy gain can be maximized. Dominant fish therefore choose the very best spots to hang out; the slow-moving places near fast water that delivers lots of food. Ringler (1979), however, noted that brown trout in aquaria would abandon, for short periods of time, areas in which prey was most abundant to check out other areas. They also continued to take small amounts of "suboptimal" prey long after preferred, larger prey became abundant. Such behavior kept trout from completely optimizing their diets in the aquaria. In the wild, however, it would be highly adaptive, because prey size, availability, and location are likely to fluctuate considerably.

As these experiments indicate, several factors influence the choice of prey. Obviously, an optimal foraging strategy for a fish must include other factors besides maximizing the energy obtained per prey item; perhaps most important is reducing the risk of predation. Thus, an optimal strategy for a small, crevice-dwelling bottom fish might be to minimize the time spent foraging to minimize exposure to predation (***risk-sensitive foraging***). Such a fish could presumably obtain more food by foraging more continuously but would increase its own probability of being eaten as it spent more time away from cover (Noakes 1992). An optimal foraging strategy here would balance energy gains against survival probabilities. In fact, balancing these two factors is a continuous problem for fish, because diets and predators change with the life-history stage. In streams, the earliest life stages of many species are spent on the stream edges, where food is abundant and the water is too shallow for larger fishes that might prey on them. As they grow larger, however, the fish become increasingly desirable as prey for herons and other wading birds, which forces the fish into deeper water. Their increased size has made them less vulnerable to medium-sized piscivorous fish but more desirable to large ones that hang out in deeper water. As a result, intermediate-sized prey fish tend to be most abundant at intermediate depths, near suitable cover (Power 1987). In another variation on this theme, Harvey (1990) found that, in an Oklahoma stream, larval cyprinids were most abundant in pools that contained large, predatory bass. The bass kept the pools free of fish small enough to prey on the larvae, but the bass were too large to prey on the larvae themselves.

11.4 AGGRESSIVE BEHAVIOR

Aggressive interactions among fish are frequently observed, particularly in aquaria. Such interactions range from direct charges of one individual at another, which may result in bites or fights, to elaborate ritualistic displays involving modified swimming, flaring of gill covers and fins, and changes in color to barely detectable movements. The best-known aggressive interactions among fishes are those related to reproduction and the defense of breeding territories, because they are part of elaborate sets of conspicuous, ritualized behavior (see Chapter 9). Just as important to many fishes—but much less conspicuous—is aggressive behavior associated with the defense of food and space. Trout and juvenile salmon, for example, frequently defend feeding territories in streams. Such territories generally center around a rock or some other object that shelters the fish from currents but provides a good view of food organisms as they drift downstream. Proximity of the territory to cover (for protection from predators) also may be important. The most frequently observed form of territory defense is the "nip," in which a defending fish charges an intruder and attempts to nip its caudal peduncle or side. When

---

BOX 11.2

Do Fish Sleep?

Most fishes spend a good part of each 24-hour period in an inactive state (sleep), even though they cannot close their eyes. On coral reefs, marked differences occur between the day and night fish faunas, in large part because day-active fishes hide in crevices and caves in the reef at night, where they rest quietly, whereas night-active fishes are inactive in the same crevices during the day. Many species even change color patterns when at rest. In lakes, many fish become pale at night and rest quietly on the bottom, under logs or on beds of aquatic plants. Often, such fish are so torpid that a diver can swim up to one and touch it, something that is impossible during the day (Emery 1973). Most schooling fishes, such as herring, tend to remain in a quiescent state in the water column, milling about slowly. The polarized school of the day becomes an unpolarized shoal at night, but the members remain close enough to each other so that schooling resumes once enough light is present. Some pelagic fishes, such as tuna, that must swim to keep water moving across the gills are active all night long. Likewise, blind cave fishes are continuously active (Kavanau 1998). The inactivity of many fish at night is presumably an energy-conserving measure, although Kavanau (1998) suggests that the main function of sleep is to "permit the refreshment of memory circuits" in the brain, essentially "down time" from the continuous input of visual stimuli during the day. According to Kavanau, fishes that do not sleep tend to live in visually impoverished environments (e.g., the open ocean and caves).

---

the intruder is the same size or smaller than the defender, it usually flees quickly. Occasionally, however, an intruder will not flee at once, and more elaborate displays result, most conspicuously lateral displays in which the two fish swim side by side in a rigid fashion, with opercula and fins flared. Three factors seem to be of greatest importance in deciding which fish wins a bout: (1) previous residency, (2) size, and (3) results of previous encounters. A fish that already lives in a territory is likely to defeat an intruder, unless that intruder is considerably larger than the defender. The displacement of small fish by larger fish can result in frequent shifts in the patterns of territories among trout and salmon streams, because larger fish also require larger territories. In trout, several fish of different sizes may coexist in the same area, with the larger fish being dominant over the smaller fish but not driving them away (Jenkins 1969). Presumably, such dominance hierarchies exist, in part, because the smaller trout feed on smaller prey and so are not competing too much with the larger trout for food and because the smaller trout may actually help the larger trout to defend the area against intruders.

11.5 COMMUNICATION

Communication is, according to E. O. Wilson (1975), "the action on the part of one organism that alters the probability pattern of behavior in another organism in a fashion that is adaptive to one or both of the participants." This definition implies that the signals given off by a fish, as well as the mechanisms to receive those signals, are highly adaptive. Therefore, fish employ a rich variety of signals to communicate with one another: (1) visual, (2) auditory, (3) chemical, and (4) electrical.

Visual Signals

For most fish, vision is the most important sense in finding food and communicating with other fish. Thus, the enormous variety of visual signals among fish, from subtle movements of the body and fins to bright colors arranged in elaborate patterns, is not surprising. Because colors and color patterns are so important for visual communication, this section discusses (1) how they are produced by fish, (2) their general significance, and (3) the specific significance of some of the most common types of color patterns.

Colors in fish are of two basic types; pigments (biochromes), and structural colors. ***Pigments*** are colored compounds that are located primarily in ***chromatophores***, which are cells located mainly in the dermis of the skin (but also in the epidermis, peritoneum, eyes, and various organs). Chromatophores are highly irregular in shape but usually appear as a central core that gives out many branched processes. This shape permits the rapid color changes that are frequently observed in fish; when the pigment granules are concentrated in the center portion, the overall appearance of the fish will be paler than when the granules are dispersed through the branches. The hues change through the combined action of chromatophores containing pigments of different colors overlying one another or by the internal changes of chromatophores containing more than one pigment.

The internal control of the chromatophores—and, consequently, of the color patterns—is complex, involving both hormones and nerves. As might be expected, the initiation of a color change usually comes from visual cues. A flounder placed with its head on one background and its body on another will have a body color matching that of the background around the head. Likewise, body coloration can be experimentally altered by fitting the fish with colored, transparent eye lenses or by changing the intensity of illumination. Visual stimuli are carried to the chromatophores by sympathetic nerves, special hormones secreted by the pituitary, or both. The method used depends on the type of chromatophore and on the species of fish. Nonvisual cues also may cause color changes, as indicated by the sensitivity of the nerves associated with chromatophores to such substances as adrenalin (epinephrine) and acetylcholine. Adrenalin and its relatives cause pigment granules in the chromatophore to aggregate, which rapidly makes the fish lighter in color, whereas acetylcholine makes them disperse slowly. Thus, fish that are frightened may undergo rapid color changes and then slowly revert to their original colors when the danger is past.

Many different kinds of pigments are found in fish (Fox 1978), but the most frequently encountered are carotenoids, melanins, and purines. ***Carotenoid pigments*** are responsible for the bright reds and yellows frequently seen in fish—and for greens as well, when yellow carotenoids overlie a blue structural color. ***Melanins*** are mainly dark red, brown, and black, and they form the background coloration of most fishes. ***Purines*** (mostly guanine in fish) are crystalline substances that, perhaps, should not be classified as pigments, because they are colorless and often nonmotile in the chromatophores. They have the peculiar reflective qualities that produce the silvery sheen of pelagic fish, however, and the iridescent colors of many tropical reef and freshwater fishes. In fact, most of the **structural colors** of fish (i.e., colors produced by light reflecting from structures rather than from pigments) are caused by light reflecting from purine crystals in special chromatophores. The silvery color of pelagic fish, for example, is created by the **stratum argenteum**, which is a sheet made of layers of guanine-containing cells (**iridophores**) that are capable of reflecting most of the light that hits them, much like a mirror. Because the light is scattered as it is reflected, it appears silvery. If the crystals or iridophores are aligned in such a way that the light waves are reflected back in parallel, then iridescent colors are created.

What is the adaptive significance of the complex color patterns of fish? Color patterns in general seem to have three main purposes: thermoregulation, intraspecific communication, and evasion of predators (Endler 1978). The role of color patterns in thermoregulation is not likely to be great, given the closeness of fish body temperatures to environmental temperatures. Most color patterns can be viewed as compromises between the "need" to communicate with other members of the species and the "need" to avoid being eaten. The former need favors bright, distinctive, and conspicuous coloration; the latter need favors subdued coloration. Whether a particular color or color pattern is conspicuous or cryptic depends on such factors as background colors, water clarity, and visual capabilities of the predators. Thus, the significance of color patterns must be interpreted carefully. Interpretation of the significance of color patterns is also complicated by the ability of many fishes to change colors, often quite rapidly, in response to changing conditions but particularly in relation to reproduction (Fig. 11.3). The multipurpose nature of color patterns in fishes can be best demonstrated by examining some of the more common patterns in fishes: (1) red coloration, (2) poster colors, (3) disruptive coloration, (4) countershading, (5) eye ornamentation, (6) eye spots, (7) lateral stripes, and (8) polychromatism. It is always good to be careful about interpreting color patterns in fish, because what we humans see through our eyes (or through flash photographs) may be quite different from what fish see through their eyes.

Red coloration.   The wavelengths in the red region of the spectrum are the first to be filtered out as light passes through water, yet bright red fishes are common. Marine fishes that are solid red in color are generally either nocturnal (e.g., the cardinal fishes, Apogonidae; or the squirrel fishes, Holocentridae) or live at moderate depths (e.g., many rockfishes, Scorpaenidae). In both situations, red light is virtually absent from the water, so the fish are, in fact, dark colored, blending in with their dark surroundings. Red can also be a cryptic color in tidepool fishes, because many of them are found in close association with red algae. Many shallow-water fishes, however, do have red bands and other markings that seem to make them quite conspicuous. Such coloration is particularly common in the spawning males of freshwater fishes, such as minnows, salmon, trout, and sticklebacks, and it appears to result from strong sexual selection by females (Bakker and Milinski 1993). It can be argued that red is a good compromise color for spawning fish, because it is highly visible at short distances—important for males trying to attract females for spawning—yet likely is difficult to see laterally for any distance, especially in turbid or shaded water. The importance of hiding even small patches of red color that do not function in communication is demonstrated by sturgeons, which have a small dorsal portion of their gill exposed. While the covered part of the gill is red, the exposed portion is darkly pigmented, like the skin of the sturgeon (Burggren 1978).

FIGURE 11.3 Color patterns of tilapia *(Oreochromis mossambica)*: (1) neutral pattern, (2) male territorial pattern, (3) aggression pattern, (4) arousal pattern, (5) female spawning pattern, (6) female brooding pattern, (7) frightened juvenile, and (8) and (9) frightened adult. The arrows show probable direction of change of color pattern. (From Lanzing and Bower 1974.)

Poster colors.   Konrad Lorenz gave the name "poster colors" to the bright, complex patterns so characteristic of coral reef fishes, because he thought their primary function was advertisement of territorial ownership. The most conspicuous—and most studied—of the poster-colored fishes are the butterfly fishes (Chaetodontidae), which typically are yellow or white with dark, contrasting stripes and other markings. For many reef fishes, the bright colors may, indeed, be important for advertising territories, but it is now apparent that this is only one of several possible functions (Ehrlich et al. 1977). Many highly colored reef fishes are not territorial, whereas many plain fishes, such as some damselfishes (Pomacentridae), are strongly territorial. For more gregarious species, the bright colors may serve to keep foraging individuals in contact with each other. Ehrlich et al. (1977) noted that when a pair of butterfly fish becomes separated, one of them may rise off the bottom in a brief display that helps to bring the two together again. Coloration also may be closely tied to sex and courtship; the complex polymorphic patterns of wrasses (Labridae) and parrotfishes (Scaridae), for example, reflect sex, status, and maturity.

Still other possible functions of poster colors lie in the realm of predator avoidance. Often, fishes with bright colors can hide very effectively in the reef, partly because the reef itself is quite colorful and partly because the patterns may be disruptive in nature (see the next section). In fact, the patterns—and not the colors—may be most important for concealment, because being invisible to reef predators is most important when light levels are low, the predators are most active, and colors hard to see. Thus, the patterns could serve different purposes at different light levels. Brightly colored fishes also may be able to escape predators through the "flash effect," whereby a predator is confused when a fish approached from the side suddenly turns and "disappears," because only the narrowest profile is visible. On the other hand, bright colors may function as a warning in announcing a fish's presence to a predator, because the fish is too poisonous or spiny to be worth pursuing (*aposematic coloring*). Indeed, the spiny nature of butterfly fishes is thought to account, at least in part, for their rarity in the stomachs of predatory reef fishes.

Disruptive coloration.   Colors and patterns that disrupt the outline of fish make that fish less visible. Thus, fishes that associate with beds of aquatic plants, such as sunfish (*Lepomis* spp.) and many cichlids, often have vertical bars on their sides that help them to blend in with the vertical pattern of the plants. Part of the effect of such patterns may be "flicker fusion" in the predator (Endler 1978), whereby the rapid movement of the vertical bars on the side of the fish across a field of vertical bars (i.e., the plant stems) may cause them all to blend together in the eyes of the predator. The intriguing aspect of flicker fusion as camouflage is that it may permit a fish to have a color pattern that is conspicuous at rest but confusing when the fish is swimming rapidly. Many of the bright but irregular patterns of coral-reef or freshwater fishes of the tropics may function in this way, because they are often associated with irregular, dappled backgrounds.

The real masters at matching their backgrounds, however, are slow-moving bottom fishes, such as sculpins (Cottidae), darters (Percidae), blennies (Blenneidae), and flounders. Flounders are famous for their ability to match their background—even to the point of coloring themselves in a fair imitation of a checkerboard when placed on one in the laboratory. Other fishes may break up their outlines by having cirri and other irregular growths that resemble seaweed, especially on the head.

Countershading.   Countershading is the most common way that fish disguise themselves. Being dark on top helps to hide them from predators attacking from above; being light on the bottom helps them to blend in better with the light streaming down from above. The importance of countershading is demonstrated nicely by the upside-down catfish (*Brachysynodontis batensoda*) of the Nile River, which swims upside down and so has a dark belly and a light back.

Eye ornamentation.   The eye in fishes, being naturally conspicuous, seems to provide a focus for both the attacks of predators and for intraspecific communication. Thus, two trends in eye ornamentation have evolved: to disguise the eye, and to emphasize the eye. Disguising the eye is accomplished in various ways, such as minimizing the contrast between the iris and pupil and then surrounding the eye with a matching background, having a line run through the eye that matches the pupil, and having a group of spots surrounding the eye of a size similar to that of the pupil. Eye lines are the most common form of eye disguise, and Barlow (1972) notes that such lines tend to be vertical on deep-bodied but horizontal on slender-bodied species, so that they are consistent with body patterns.

Although disguising the eye is a common practice, particularly in reef fishes, much more common is emphasizing the eye with a striking pattern around it or with bright colors in the eye itself. In Caribbean reef fishes, most species have eyes with conspicuous coloration, predominately black, blue, and yellow (Thresher 1977). Usually, the dark pupil is emphasized by a light-colored iris and often has some supplementary markings, such as eye rings, as well.

Eye spots. Among the most common markings on the bodies of fish are single spots about the size of the pupil of the eye and often surrounded by light-colored pigmentation for emphasis. Such eye spots most commonly are located at the base of the caudal peduncle and most often are found in the juveniles of tropical characins, cyprinids, and cichlids. Although such spots may function as orientation marks for fishes in a school, their principal function seems to be to confuse predators into aiming attacks at the caudal area rather than at the head, thereby giving the fish greater opportunities to escape (McPhail 1977a). In some species, eye spots serve as recognition signals that inhibit cannibalism of juveniles by adults. In the poeciliid fish *Neoheterandria formosa*, the midlateral spot that apparently serves such a function is retained during adulthood and is used as an appeasement signal during courtship (McPhail 1977b).

Lateral stripes. Lateral stripes typically are a single midlateral band and are best developed in schooling fishes. They seem to serve the dual function of keeping school members oriented to each other and of confusing predators. Presumably, the latter aim is achieved by visual fusion of the stripes of members of the school, thereby increasing the difficulty of picking out individuals.

Polychromatism. Nowhere is the conflict between the need to be cryptically colored to avoid predators and the need to be bright to attract mates more apparent than in polychromatic fishes, which have different patterns predominating in different populations depending on the degree of predation. In the midas cichlid (*Cichlasoma citrinellum*), gold morphs are dominant over plainer morphs, particularly in contests regarding food (Barlow 1973), yet failure of the gold morphs to "take over" their habitats in the wild reflects their greater vulnerability to predation. They typically dominate only in deep water, where they are less visible. In the annual killifish (*Nothobranchius guntheri*), brightly colored morphs are dominant over dull morphs and so have greater reproductive opportunities early in the season. They are gradually picked off by predators, however, thereby giving the dull morphs greater reproductive opportunities later in the season (Haas 1976a, 1976b).

Special patterns. In addition to general patterns, such as those just discussed, many species have special color patterns that fit their own particular lifestyles. The males of some cichlids have circles on their anal fins that resemble eggs. These are displayed just after a female has laid some eggs and taken them into her mouth; when she tries to pick up the dummy eggs as well, the male releases sperm, and the eggs in the mouth are fertilized (Barlow 2000). On coral reefs, many fishes have special color patterns, such as the bright colors of cleaner fishes that advertise their presence to fish wanting to be "cleaned" and the patterns of fishes that mimic other fishes or invertebrates (see Chapter 33). In the deep sea, photophore patterns play much the same roles as color patterns in lighted environments.

Auditory Signals

The use of sound for communication among fish is common but far from universal; however, the hearing of fish is usually keen (Hawkins 1993). In some species, the sounds produced can be very loud and continuous and are fundamental to their way of life, especially during reproduction, when courtship "singing" may occur. Fish produce sounds in three main ways: (1) by rubbing of hard structures together (*stridulation*), (2) by vibration of the swimbladder, and (3) as incidental to other activities. Stridulation produces mainly low-frequency sounds, but this method is the most common means of deliberate sound production, presumably because it often requires little modification of existing structures. Thus, some fishes can communicate just by grinding the pharyngeal or jaw teeth together in a regular fashion. Some filefishes (Balistidae) even have special ridges on the backs of their front teeth that are used for sound production. Another common stridulatory mechanism is to rub the specially roughened base of a fin spine (usually pectoral spines) against its socket; sea catfishes (Ariidae) are particularly well known for their use of this mechanism. Many, and perhaps most, stridulatory sounds are amplified by the fish's swimbladder, and muscular connections often exist between the sound-producing structure and the swimbladder.

In fishes with the most complex auditory signals, the swimbladder itself usually serves as the sound-producing organ. This is done in many ways, often involving major modification of the swimbladder, but the basic means is to vibrate the swimbladder, either by using special muscles attached to it or by rubbing it with adjacent structures. The special musculature may be muscles that insert onto the swimbladder and originate on the skull or vertebral column or that are intrinsic to the swimbladder itself. In the former case, the sound produced by the muscles depends on their tension and rate of vibration. This mechanism is found in the noisier members of many families of spiny-rayed fishes and in some catfishes. The loudest and most elaborate sounds, however, seem to be made by fishes with intrinsic swimbladder musculature for sound production, such as the toadfishes (Batrachoididae), searobins (Triglidae), and gunnards (Dactylopteridae). The drums (Sciaenidae), however, get their name and sound-producing reputation by vibrating muscles in the body walls next to the swimbladder.

Compared to the sounds of many terrestrial animals, the sounds produced by fish are not particularly elaborate. In general, fish have only limited abilities to make and detect sounds of different frequencies—perhaps the most important means of varying sound patterns in mammals and birds. Nevertheless, sounds produced by fish can be varied by changes in loudness (amplitude), duration, rate of repetition, and number of pulses within a signal (Fine et al. 1977). The loudest calls produced by fishes seem to be those associated with agonistic behavior, especially territorial defense. During the courtship season of plainfin midshipman (*Porichthys notatus*), people living on houseboats in Sausalito, California, find their peace and quiet disrupted when the loud and persistent humming of many fish is transmitted through the hulls of their boats! Anemone fishes (*Amphiprion*) have an extremely loud call that is used in defense of an anemone against conspecifics, but a much quieter call is associated with close encounters and fighting between two fish. The calls of anemone fish can also be distinguished on the basis of their duration: The more intense a threat, the longer a call will last. The repetition rate of a signal and the number of pulses within it (i.e., its complexity) also may increase with the intensity of the interaction. In drums (Sciaenidae), the drumming sounds are produced when the fish are in schools, and they increase in intensity during certain times of the day. As the intensity of the drumming increases, the length of each call increases as well, as does its frequency of repetition. Such variation in calls may serve to keep closely related species reproductively isolated from each other. For example, Gerald (1971) found that the courtship grunts of six species of sunfishes could be distinguished from each other on the basis of duration and number of pulses per second.

Chemical Signals

As the discussion about the use of odors as migrational cues indicated, most fish have a well-developed sense of smell. Thus, it is not surprising to find that chemicals produced by fish, even in very small amounts, are often important during intraspecific communication, especially in fish without keen eyesight (e.g., catfish). The chemical nature of most substances used in communication by fishes is not known, but many of them likely are produced specifically for that purpose. This is particularly true of compounds that, when released into the water, produce an immediate and fairly specific reaction in other fish of the same species (***pheromones***). The general areas in which pheromones have their most important uses are in reproductive behavior, individual recognition, and predator avoidance.

For reproductive behavior, pheromones are often important in the recognition of both sex and sexual condition, and they may be necessary for the release of certain behavior patterns. In frillfin gobies *(Bathygobius soporator)*, for example, an ovarian pheromone will elicit courtship behavior in males, even in the absence of any females (Tavolga 1956). Similarly, blind gobies *(Typhlogobius californiensis)* react very differently to the invasion of their burrows by conspecifics, depending on their sex; Gobies of the same sex produce aggressive behavior, whereas those of the opposite sex produce more passive behavior. This sexual recognition is accomplished largely through the detection of pheromones (MacGintie 1939). In some species of fish, odors may be used not only to determine sex but also to identify individuals. In yellow bullheads *(Ictalurus natalis)*, individuals recognize each other by odors, and they also associate the odors with rank in social hierarchies. Thus, a subordinate fish will avoid an area of an aquarium into which the water from the tank of a dominant fish has been introduced, even though the dominant fish is not present (Bardach and Todd 1970). Some cichlids can identify their own young on the basis of odor as well, at least during the 3 weeks or so when parental care is necessary (McKaye and Barlow 1976).

Predator avoidance by means of chemical signals is accomplished through the use of fear scents, or ***Schreckstoff*** (Smith 1982). These compounds are present in epidermal cells and are released into the water when the skin is broken. When the compound is detected, other fish of the same (or closely related) species immediately adopt some sort of predator avoidance behavior, such as diving into vegetation, bunching up as a school, or becoming motionless. This ability is known mainly from fishes of the superorder Otophysi, although a few other fishes possess it as well. Presumably, fear scents not only enhance the ability of the fishes to avoid predators (especially in turbid environments) but also inhibit cannibalism. The observation that predatory pikes (Esocidae) defecate in special places is thought to help them avoid being detected by their cyprinid prey through release of the fear scent in feces containing the remains of previous cyprinid meals (Brown et al. 1995).

Electrical Signals

Every time a muscle contracts, it gives off a small electrical discharge. Water containing dissolved minerals is a good conductor of electricity. Together, these two facts mean that electricity is a means of communication readily available to aquatic organisms. Many fishes have specialized organs for detecting electrical impulses, and the capacity to generate electricity for communication has evolved independently in five diverse orders of fishes: (1) Rajiformes (Rajidae, Torpedinidae), (2) Mormyriformes (Mormyridae, Gymnarchidae), (3) Gymnotiformes (six families), (4) Siluriformes (Malapteruridae), and (5) Perciformes (Uranoscopidae) (Möller 1995). Of these fishes, the mormyrids and the gymnotids, both of which inhabit turbid, tropical fresh waters, are most dependent on electricity for intraspecific signaling. The signals produced by the electric organs of these fishes can be modified in complex ways by varying such factors as discharge frequency, waveform, and the times between discharges. The signals are received in specialized lateral-line organs known as *electroreceptors*. As a consequence, electrical discharges can have all the functions that visual and auditory signals have in other fishes, including courtship, agonistic behavior, and individual recognition. The electric eel *(Electrophorus electricus)* preys on other electrical fish by detecting their electrical fields while shutting off its own; in turn, the prey may try to hide by ceasing to emit electrical signals when they detect an eel nearby (Westby 1988).

11.6   LESSONS FROM FISH BEHAVIOR AND COMMUNICATION

Like structural and physiological traits, behavioral traits are evolved characteristics, with each species having its own distinctive suite of behaviors. From our land-locked, mammalian perspective, it is often hard to appreciate some of the complex behavior patterns of fish. The ability of salmon, eels, and tuna to "home" to precise areas after migrating thousands of miles through the oceans is difficult to comprehend, because doing so requires a sensitivity to environmental cues, such as the Earth's geomagnetic fields, that we as humans do not have. Likewise, the ability of these fish to memorize subtle cues ("signposts"), such as the odor of streams or local anomalies in the Earth's electrical field, that help them to find their way home is difficult for us to comprehend. In addition, it is difficult for us to appreciate how fish shoal, because we have little concept of how it must feel to have a sensory system that is sensitive to turbulence and pressure or of how it is to be so sensitive to the movements of companions that thousands of individuals can wheel and turn like a single organism. The alien nature of some fish behavior is what makes them so fascinating to study. Great care must be taken, of course, when interpreting fish behavior from our perspective. The bright colors of reef or deepwater fishes, for example, which might be interpreted as advertising their owners, turn out to be (for the most part) colors that help to hide the fish instead.

SUPPLEMENTAL READINGS

Dingle 1996; Endler 1978; Godin 1997; Harden-Jones 1968; Huntingford and Torricelli 1993; Leggett 1977; Lucas and Baras 2001; McDowall 1997; McKeown 1984; Pitcher 1993.

WEB CONNECTIONS

GENERAL

www.calacademy.org/research/ichthyology
www.fishbase.org
www2.biology.ualberta/jackson.hp/IWR
www.flmnh.ufl.edu/fish/default.htm
www.people.clemson.edu/~jwfoltz/wfb418

C H A P T E R 12
Systematics, Genetics, and Speciation

The study of the evolutionary relationships among organisms is termed ***systematics***. In modern usage, it has become nearly synonymous with ***taxonomy***, the science of describing and classifying organisms, because modern classification systems reflect evolutionary relationships. Modern taxonomy had its beginnings, however, in the work of Carolus Linnaeus, who believed each species to be an unchanging entity created by God. The binomial system of nomenclature he developed for species and the hierarchical classification system he used for higher categories proved to be exceedingly useful in organizing the rapidly expanding knowledge of animals and plants. Although Linnaeus' distinctions among species and, especially, higher categories were often rather arbitrary by our standards, later workers (including many prominent ichthyologists; see Chapter 1) modified his system to reflect presumed evolutionary relationships among the various taxa. By the late nineteenth century, evolutionary trees were being drawn. Ancestral, presumably primitive, forms were placed near the base of such trees, and successively higher branches led to more advanced, or at least more modern, forms. Such trees had two main problems. First, they implied that evolution is more or less linear, with each lineage somehow striving toward some perfect organism (e.g., humans). Second, the location of the branches depended, in good part, on how a particular investigator felt about the relationships of the organisms or groups making up the tree.

To make the decision-making process for branching diagrams less arbitrary, the methods of ***phylogenetic systematics*** were developed by Willi Hennig and adopted by systematic ichthyologists (Mayden 1992b). This method uses branching diagrams, called ***cladograms***, in which each branch represents a monophyletic group of organisms (e.g., populations, species, families, orders). A ***monophyletic*** group is one that is assumed to have a common ancestor; therefore, all members share one or more distinctive ***derived characters***. Branches that split from a common "trunk" (burdened with the name ***synaptomorphies***) share common ***ancestral characters***. Divergent groups at each branching are called, appropriately, ***sister groups***. Derived and ancestral characters can be defined precisely, and their presence or absence in groups allows the systematist to construct a cladogram of the groups according to specific rules rather than whims. Each cladogram essentially represents a series of evolutionary hypotheses. An example of a cladogram showing the phylogeny of the major groups of fishes is presented in Figure 12.1 along with some of the characters used to define each group. Detailed explanations of the method can be found in Lundberg and McDade (1990) and in Mayden and Wiley (1992).

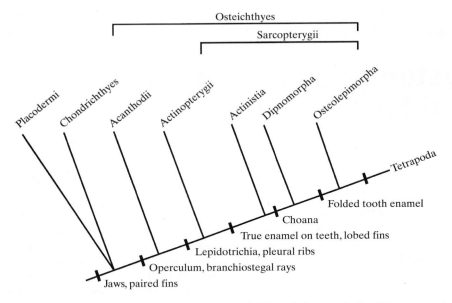

FIGURE 12.1   Simplified cladogram showing probable evolutionary relationships (phylogeny) among major groups of fishes, plus the tetrapods. The dark crossbars indicate ancestral characters shared by most members of the group ahead of each bar but not by the group behind the bar. The characters listed for each bar are only a few of the more obvious ones. The tetrapods are the uppermost group in the diagram because they were the last to evolve. The various groups are discussed in detail in Chapter 13. Based on cladograms in Lauder and Liem (1983) and Bemis et al. (1986), which also show some alternative arrangements.

12.1   TAXONOMIC METHODS

The information used to develop an understanding of evolutionary relationships among fishes comes largely from detailed taxonomic studies, particularly the careful descriptions of species. Such studies are the foundation of ichthyology (see Chapter 1) and today are more necessary than ever, not only for evolutionary, ecological, and physiological studies but for conservation as well. Major decisions regarding land and water use are being made on the basis of arcane aspects of taxonomy that indicate whether an endangered population of fish is an entire species (and, therefore, unique) or is part of a widely distributed species (and, therefore, less worthy of protection). The basic methods used in taxonomic studies fall into six categories: (1) morphometric measurements, (2) meristic traits, (3) anatomical characteristics, (4) color patterns, (5) karyotypes, and (6) biochemical (genetic) characteristics.

Morphometric measurements include any standard measurement that can be made on a fish, such as standard length, snout length, length of longest ray on the dorsal fin, or depth of caudal peduncle. General instructions for making these measurements are found in Hubbs and Lagler (1964) and in Strauss and Bond (1990). Because these measurements change as a fish grows, they are usually expressed as ratios to standard length (or some other measurement that is easily made). Even such ratios, however, are most useful if comparisons are made between samples of fish of approximately the same size and sex, because the growth of a fish is not always proportional in all directions and sexual dimorphism is common (but often not obvious). To standardize morphometric measurements and make them more comparable among species, a ***truss protocol*** has been developed in which a systematic set of measurements is made between anatomical "landmarks" on a fish, such as the distance between the base of the dorsal fin and the base of the anal fin (Strauss and Bond 1990).

Meristic traits are often considered to be the most reliable taxonomic characteristics, because most are easy to determine. Meristic traits include anything on a fish that can be counted, such as vertebrae, fin rays and spines, scale rows, pyloric ceca, or lateral-line pores. Because these characteristics often vary considerably within a species, it is important to make the counts on enough individuals so that means, ranges, and variances can be determined. One of the biggest sources of variation in meristic counts is human error, especially with small fish. Such error is reduced if standard methods are used. Another source of variation is the conditions under which larval fish develop. Any factor that affects larval growth, such as temperature, dissolved oxygen concentration, salinity, or food availability, is likely to affect meristic characters (Barlow 1961; Johnson and Barnett 1975).

Anatomical characteristics are hard to quantify but are nevertheless important for species descriptions. They include shape, completeness, and position of the lateral line; position and size of internal organs; special anatomical features (e.g., airbreathing and electrical organs); secondary sexual characteristics (e.g., breeding tubercles on males); and the shapes, sizes, positions, and interrelationships of bones, nerves, and muscles. Most of these are yes-no characteristics: Either a fish has them, or a fish does not. Consequently, they can be definitive characteristics, useful for separating not only species but also higher taxa. Even these characteristics are not always absolute, however, because individuals of one species may possess characteristics that are supposedly definitive for another, closely related species. Thus, basibranchial teeth, usually a reliable feature for separating cutthroat trout from rainbow trout, are occasionally found in "good" rainbow trout and are occasionally absent from "good" cutthroat trout.

Color patterns are perhaps the most variable characteristics of species, because they may change with age, time of day or year, and the environment in which a fish is found. Nevertheless, they should be an important part of every species description, because color patterns are species-specific, reflecting habitat, reproductive condition, sex, and methods of communicating with conspecifics (see Chapter 11). The main problem with using color as a taxonomic tool is that it fades in preservatives, and descriptions of living fish tend to be highly subjective. Some pigmentation patterns, such as dark spots and stripes, tend to be fairly permanent and can be used judiciously.

Karyotypes are descriptions of the number and morphology of chromosomes. The number of chromosomes per cell seems to be a rather conservative characteristic, so it may be used as an indicator for the closeness of species interrelationships within families. The number and position of the arms of chromosomes is even more conservative than chromosome number and is often equally useful in taxonomic studies. Thus, during the course of an evolutionary change, two one-armed (acrocentric) chromosomes may fuse into a single two-armed (metacentric) chromosome, thereby decreasing the chromosome number but not the number of arms. Another way that chromosome number may increase is through polyploidy, which results from the failure of an entire set of chromosome pairs in a diploid cell to separate. The production of diploid gametes by this method, although infrequent, is common enough so that triploid and tetraploid individuals may result (Ohno 1974). Triploidy has been shown to be of significance in the origin of unisexual "species" of poeciliid fishes in Mexico. Tetraploidy appears to have been important in the evolution of some species (Ohno 1974) and, hence, an important taxonomic indicator. For example, carp, goldfish, and barbel all have approximately double the number of chromosome arms of other cyprinids, as do all members of the derivative family Catostomidae (Uyeno and Smith 1972), thus indicating that tetraploidy may have played an important role during the early evolution of these successful species. Similarly, salmon, trout, whitefish, and graylings (Salmonidae) all have double the number of chromosome arms of more ancestral members of the order Salmoniformes.

Biochemical characteristics are patterns in proteins and genetic material that are typically used to quantify genetic differences among populations. These techniques have come into vogue because of the rapid development of molecular biology during past decades. The principal biochemical techniques used are electrophoresis and various methods of analyzing DNA composition and structure.

Electrophoresis evaluates the protein (allozyme), DNA, peptide, and other biochemical similarities of species. Protein electrophoresis was the first major application of this technique. Proteins are encoded by loci on DNA, so variants in proteins encoded at different loci (*alleles*) represent alternate gene complexes for specific traits. Analysis of the number and type of variants of each allele becomes a measure of genetic structure and differentiation. In this technique, a tissue sample is treated mechanically to disrupt the membrane structure of the cells, thereby releasing water-soluble proteins. The resulting solution is placed in a gel, usually made of starch or agar, and is then subjected to an electric current. Each protein migrates in response to the current at a rate that depends on its molecular size and electric charge. The proteins (alleles) can then be identified, and the genetic similarity of individuals and species can be compared by noting the presence and absence of proteins as well as differences between their positions in the gels. A slight difference in the molecular structure of a protein can result in its having a different net charge and, hence, in its migration to a different position in the gel. The number of such differences encountered between two samples is considered to be an index of genetic similarity and, if the samples involve different species, of the closeness of their evolutionary relationship. Usually, species have unique alleles or combinations of alleles, which can be useful for species identification, especially when doubts exist about the identity of morphologically similar species (Leary and Booke 1990).

Although protein electrophoresis remains a powerful tool in systematics, more direct measures of genetic structure, techniques that make some measurement of DNA sequencing, are displacing it for many studies. These techniques have the advantages of using small tissue samples (so the fish does not have to be killed to be characterized) as well as tissue from preserved or even fossil material and of having the potential to measure variation of more than 3 billion DNA sequences (Wirgin and Waldman 1994).

Two basic kinds of DNA are present in cells; mitochondrial DNA, and nuclear DNA. Mitochondrial DNA is particularly useful for analyzing populations within a species, because it lacks recombination, has a simple inheritance pattern (through the mother), and evolves faster than nuclear DNA. On the other hand, nuclear DNA is much more abundant and contains approximately 3 billion nucleotide subunits, as opposed to 20,000 for mitochondrial DNA (but virtually all of them encode proteins). One technique for analyzing DNA is to compare the exact sequence of base pairs on a standard section of DNA, which can now be done (after some preparation!) by automated DNA sequencers. Another technique is to break up the DNA strands using standardized enzymes (restriction endonucleases) and then compare the number and size of the fragments using electrophoresis; this technique is known as **DNA *fingerprinting***. Wirgin and Waldman (1994) point out that many uses are being found for DNA techniques in taxonomy and fisheries management, such as identification of distinct stocks of striped bass, salmon, and other fishes even when they are mixed in the ocean.

12.2 GENETIC VARIABILITY

Fish in general show much greater ***phenotypic*** variability than other vertebrates; large variations in body size, growth rates, color patterns, and other features are common both between and within populations. Much of this variation is related to the environment, because most fish are very plastic in their ability to respond to favorable or unfavorable environmental conditions. Bluegill that may reach large sizes under favorable conditions will "stunt" and reproduce at small sizes under unfavorable conditions. Recognition is growing, however, that this phenotypic variability has a genetic basis and that ***genotypic*** diversity is important to maintain it (Carvalho 1993). Recently, a genetic basis has been found in fishes for variability in traits such as color patterns, life-history patterns, ability to migrate, tolerance to low pH and toxicants, shoaling behavior, and resistance to diseases. In general, fish exhibiting high ***heterozygosity*** (having high allelic diversity) are more likely to produce offspring that also reproduce (have high ***fitness***), especially in a variable environment (Carvalho 1993).

The patterns of genetic diversity within species vary tremendously from species to species (Carvalho 1993). The catadromous American eel *(Anguilla rostrata)* appears genetically to be one population up and down the Atlantic coast (Avise et al. 1986). In contrast, anadromous salmon tend to have hundreds of genetically distinct populations, one in each stream to which they call home. Among freshwater fishes, species that are specialized for living in narrow habitats, especially headwater streams or isolated lakes, and that have poor dispersal abilities are likely to have many genetically distinct populations, whereas species that are less specialized and more vagile are likely to be more genetically uniform (Tibbets and Dowling 1996). In contrast, sedentary fishes inhabiting isolated tropical reefs tend to show little genetic divergence over wide areas, regardless of life-history characteristics, including length of the planktonic larval period (Shulman and Bermingham 1995). This indicates that past and present ocean currents have carried fish to and from the isolated reefs on a regular basis, thereby promoting gene flow. The amount of genetic diversity both within and between populations therefore can be a good reflection of both the life-history traits of a species and the history of the environment in which it lives. In general, however, freshwater and anadromous fishes show a greater degree of genetic fragmentation (i.e., development of genetically distinct local populations) than do marine fishes, because the freshwater environment itself is more fragmented than the marine environment (Carvahalo 1993). A correlation of this trend is that individual freshwater fish tend to have lower hetereozygosity than marine fish, with anadromous fish being in between regarding variation (DeWoody and Avise 2000).

Contributing to the fragmentation of gene pools in freshwater fishes is the phenomenon of ***local adaptation***, which is defined as "a process that increases the frequency of traits which enhance the survival and reproductive success of individuals in a particular environment" (Carvalho 1993, p. 58). Every local environment has its own unique combination of characteristics on which natural selection can operate, resulting in distinct gene complexes that reflect adaptation to that environment. For example, guppy *(Poecilia reticulata)* populations in Trinidad streams show marked differences in coloration, morphology, and behavior, which have a genetic basis. The major driving force behind the differences among populations is predation (Endler 1978, Carvalho 1993). Male guppies will show bright-red color patterns where the predominant predators are large shrimp, which do not perceive red well, but these guppies tend to favor blue color patterns where the main predators are fish. In Pacific and Atlantic salmon, most populations seem to show evidence of genetically based local adaptation, reflecting the importance of being adapted to variable local conditions that influence such things as the timing of spawning, the length of time the young stay in the stream, and the timing of ocean entry by the young. When juvenile salmon are transplanted to streams varying distances from their home stream, survival rates decline with the distance of separation (Reisenbichler 1988). Likewise, introduction of large numbers of salmon of hatchery origin into a stream to "enhance" the local population may have the opposite effect. The hatchery fish, by their sheer numbers, may disrupt local breeding and rearing of locally adapted salmon, but these hatchery fish are genetically less able to reproduce successfully (Waples 1991, Carvalho 1993).

12.3 SPECIATION

Local adaptation is the short-term reflection of natural selection acting on the genetic variability of a species. It can therefore be the first step in the process of developing new species (speciation). Obviously, given the fluctuating nature of the environment, most populations exhibiting local adaptation do not develop into distinct species, partly because many other factors also contribute to speciation and because the high dispersal rates of most fishes assure gene flow among populations.

In most cases, it is assumed that the most important factor for species formation is geographic isolation of one population from related populations (***allopatric speciation***). With isolation, different selection pressures, combined with intrinsic factors such as size and makeup of the original gene pool of the isolated population, result in the development of genetically distinct populations. An isolated population that originates from a small number of individuals may show differences that result from ***genetic drift***, in which the limited genetic makeup of the "founding" organisms and the loss of genes after isolation are essentially stochastic events that result in a subset of the species' genes being present in the isolated population. Given sufficient time, this divergence will continue until the population can retain its unique genetic identity even after it comes into contact again with other, similarly derived populations, even if some hybridization occurs. For example, Nelson (1968) found that two species of suckers (Catostomidae) evolved from a common ancestral population during the Pleistocene, after advancing glaciers isolated one portion of the population in the Columbia River drainage and the other in the eastern United States. Recently, one species has, by natural means, entered the range of the other. The two species hybridize, but there appears to be little gene flow between them even though they remain ecologically and reproductively quite similar. In situations such as these, the two species, given time and continued lack of backcrossing, can be expected to continue to diverge ecologically in the region of overlap, exaggerating whatever adaptive differences in morphology and behavior now exist (***character displacement***).

In some situations, character displacement can occur very rapidly. In Lake Michigan, the native bloater *(Coregonus hoyi)* switched from plankton feeding to benthic feeding in the face of competition from the introduced alewife *(Alosa pseudoharengus)*. In less than 20 years, the selection pressures for more efficient foraging on the bottom resulted in bloaters with both fewer and shorter gill rakers (Crowder 1984). This rapid shift seems to have been a key to the bloater's continued success in Lake Michigan when other plankton-feeding fishes have suffered drastic declines because of competition from alewives. Morphological shifts that result from competition are common in fishes. Robinson and Wilson (1994) note that, especially in lakes, many examples of competition resulting in the development of pelagic and benthic forms are found. Groups of closely related, but not formally described, species are found in threespine sticklebacks, various whitefish, salmon, charr, and trout (Salmonidae) and in rainbow smelt *(Osmerus mordax)*, mainly within lakes in recently glaciated areas (Schluter 1996) (Box 12.1).

How did these "species" evolve? The conventional explanation is that the divergence resulted from allopatric speciation, whereby each form became differentiated in isolation (e.g., in another lake) through processes such as genetic drift, founder effects, or different selection pressures, followed by character displacement once the forms came into contact. The evidence, however, increasingly supports ***sympatric speciation***, which occurs when two forms diverge without geographic isolation (Via 2001). Schluter (1996) suggests that most sympatric speciation is ***ecological speciation***, through which "reproductive isolation evolves from divergent selection, to emphasize the vital contribution of the resource environment" (p. 807). In other words, in an environment with diverse resources (e.g., zooplankton vs. benthic insects in lakes), natural selection may favor different phenotypes specialized for using different resources. Schluter (1996, 2001) suggests the following lines of evidence for ecological speciation:

- Distinct forms exist in environments that are of a young geological age, requiring rapid evolution. Thus, divergent fishes are especially common within lakes of recently glaciated areas.
- Strong morphological differences exist between the forms, reflecting major differences in feeding behavior, habitat use, and in some species, spawning habitat.
- The ecologically distinct forms are maintained despite some evidence of gene flow between them. Thus, characters must be associated with differences in resource use that also influence choice of mates (e.g., size and color in sticklebacks).
- In the laboratory, the species readily hybridize, and the hybrids are fertile. The rarity of interbreeding in the wild indicates strong selection pressure against the hybrids because of their intermediate characteristics (Hatfield 1997).
- Under similar conditions, the phenomenon is repeated.

Ecological speciation may be more common in fishes than is generally realized. Agreement is growing, for example, that the flocks of cichlid species in the Great Lakes of eastern Africa evolved sympatrically, through sexual selection (see Chapter 30). In northern lakes of Eurasia and North America, distinct morphs of sticklebacks (*Gasterosteus*), charrs *(Salvelinus)*, and whitefish *(Coregonus)* have evolved repeatedly. The frequency of rapid character displacement within or among similar species of fishes from many taxonomic groups (Robinson and Wilson 1994) also suggests that ecological speciation is common.

Whether speciation is allopatric or sympatric, it often can occur very rapidly in fishes. Usual estimates for the amount of time required for adaptive differentiation of fish to reach the species level are between 100,000 and several million years (e.g., Avise et al. 1975). Yet, some species and subspecies of fish in both temperate and tropical areas apparently evolved in response to Pleistocene or post-Pleistocene conditions in less than 20,000 years. For example, Moodie and Reimchen (1976) indicate that populations of threespine sticklebacks on islands off the British Columbia coast diverged to the species level in 8,000 to 10,000 years. The species flocks in the Great Lake of Africa may have developed in less than 12,000 years—with some species evolving in just a few hundred years (see Chapter 30).

BOX 12.1

The Nature of Species: Arctic Charr

Despite increasingly sophisticated molecular tools, defining a species often remains difficult. It seems that the more we know, the fuzzier species concepts become. Hey (2001), for example, lists more than 20 species concepts. Yet, managers often want precise definitions of species so they can set regulations for fishing or can develop conservation strategies. This problem is well illustrated by Arctic charr (*Salvelinus alpinus*), which is widely distributed in lakes and streams across northern Eurasia and North America. In many of the lakes, two to four varieties (morphs) of the charr show strong ecological and reproductive isolation. Lake Thingvallavatn, Iceland, for example, has four morphs: large and small benthic feeding charr, a planktivore, and a piscivore. Genetic analysis shows that all these forms are very closely related to one another, yet they behave as "good" species, with little or no natural interbreeding (Danzmann et al. 1991).

Such morphs develop as the result of intense intraspecific competition, usually in the absence of other species of salmonids. Jonsson and Jonsson (2001) point out that the remarkably similar morphs in different lakes (e.g., the benthic feeding morph) cannot belong to one biological species because of their separate origins in each lake. In any case, they state that the lakes are transitory features on the landscape and that genetic differences among the morphs are low, so persistence of the distinct lineages through a long period of time is unlikely. In human time frames, however, each morph in each lake behaves as a classic biological species. One option, therefore, would be to describe dozens of new species of charr. Jonsson and Jonsson (2001), however, recommend a more practical option: Recognize the evolutionary phenomenon, and manage each lake ecosystem to maintain the morphs as if they were, indeed, separate species.

12.4 HYBRIDIZATION

Hybridization is common among fish species, especially in fresh water (Hubbs 1955, Schwartz 1972), and its significance in the evolution of fishes is much debated. In some situations, it can lead to the formation of new species, such as unisexual mollies (see Chapter 9). Alternatively, it can lead to the elimination of species, especially if human interference allows fishes to move over natural barriers. Thus, cutthroat trout have been eliminated from much of their native range in the Great Basin of the United States by hybridizing with introduced rainbow trout. For a variety of reasons, the rainbow trout phenotype usually becomes dominant after a number of generations. When the sheepshead minnow *(Cyprinodon variegatus)* was introduced into the range of the Pecos pupfish *(C. pecosensis)* in Texas, within 5 years the two species had hybridized over several hundred kilometers of river, making the Pecos pupfish an endangered species (Echelle and Connor 1989).

Although hybrids between many species are known, the mere fact that members of two species can mate with each other and produce offspring does not mean they belong to the same population—or even that they are not reproductively isolated from each other. The important test of the significance of hybridization between two populations is whether, under ***natural conditions***, significant gene flow between them occurs, thereby resulting in the mixing of gene pools and, eventually, a continuum of forms from one population to the next. Many species of fishes will freely interbreed in captivity, producing fertile hybrids that will backcross with either parent species; many varieties of aquarium fishes are the result of such hybridization between species. Some of the best examples of this are in swordtails (Poeciliidae), which rarely hybridize in the wild. Even when hybridization between fish species does occur in the wild, reproductive isolation usually is maintained, because the hybrids either die before reaching sexual maturity, are sterile, or have low survival and/or fertility.

Hybrid sterility, in which hybrids develop to adulthood but then prove to be sterile, seems to be fairly common in fishes. The best-known examples include crosses among species of the genus *Lepomis* (Centrarchidae), especially the bluegill–green sunfish cross *(L. macrochirus × cyanellus)*. This hybrid is common where the two species occur together; it presumably results mostly from male green sunfish sneaking into bluegill nests and releasing sperm when a bluegill pair is spawning. The hybrids that result are all male and exhibit hybrid vigor (***heterosis***), in that they typically grow faster than either parent species. During the spawning season, they build nests and defend them with great vigor, dominating nonhybrid fish. Such fish are sterile, however, so no introgression occurs between the two species.

Fertile hybrids typically have low survival rates and poor reproductive success. The reasons for this are complex but, presumably, are related in part to their intermediate nature. In most cases, the two parent species have a competitive advantage over the hybrids in occupying their respective niches. Hybrids are likely to be successful only when environmental conditions are intermediate. Thus, hybrids between two species of darters (Percidae) are abundant only in stretches of stream that have been altered by humans, thereby creating an intermediate type of habitat (Loos and Woolcott 1969). This hybridization seems to result from deliberate matings between members of two species. It is much more common among fishes, however, for the matings to be accidental, the result of the mixing of sex products when two species spawn in the same area at the same time. For example, Tsai and Zeisel (1969) found that hybrids among three species of cyprinids were common in a small stream in the eastern United States, because all three species spawned simultaneously on the gravel nest mound of yet another species.

12.5   NOMENCLATURE

One of the most frustrating aspects of ichthyology for anyone who is not a taxonomist is the constant changing of the scientific names applied to fishes at all levels, from subspecies to kingdom. The changes in names throughout the four editions of this text alone (starting in 1982) have been substantial, and names the authors learned as students often seem quaintly archaic. The changes are nevertheless necessary, reflecting that fish systematics is a dynamic field; the rate at which names are changing reflects the rate at which we are obtaining new information about fish and their interrelationships. To ease the problem for journeyman biologists of fish and fisheries in the United States and Canada, every 10 years the American Fisheries Society issues a standardized list of the common and scientific names of fishes (Robins et al. 1991). This has allowed the development of nearly universal, constant common names for fishes while keeping interested individuals informed of changes in the scientific nomenclature. The rainbow trout, for example, has been called the rainbow trout for approximately 50 years, but its scientific name has changed completely (Box 12.2). In this section, we discuss scientific names for levels higher than species and for species.

Higher-Level Names

Textbooks like this one, and most discussions of systematics, use a hierarchical system of names to designate how species and higher-level groups are related. This is largely the result of a tradition going back to Linneaus, which fits a very human need to put things into categories for purposes of learning. Thus, we have the basic Kingdom—Phylum—Class—Order—Family—Species structure of taxonomic systems, with various subdivisions (e.g., superclass, subclass, infraclass) to make finer distinctions. The basic assumption behind all this is that all classes, orders, and families are equivalent levels for all fishes, even though no really standardized way of doing this exists. A class in one taxonomic system may be a superclass in another one or an order in yet another. The fact remains, however, that each name usually does enclose a group of fishes with a recognizable set of characters. The elegant analyses that are now possible through cladistics demonstrate that these groups do exist in reality, but where one draws a line on a cladogram to separate, for example, a class from an order is still arbitrary. Cleverly, names on cladograms generally are not preceded by the level labels. Despite this problem, the hierarchical system of names does help to organize knowledge of fishes in human brains (at least in those of ichthyologists, who are reputed to be human) and so will continue to be used.

Species Names

Each species is assigned a unique, two-part scientific name: the genus name, and the species name. This scientific name is often followed by the name of the person who first described it and the year in which it was described. For example, the longnose dace is designated *Rhinichthys cataractae* (Valenciennes 1842). Valenciennes' name is in parentheses because he assigned the fish the species name, but the genus name was

BOX 12.2

The Complex History of Rainbow Trout

The rainbow trout [*Oncorhynchus mykiss* (Walbaum 1792)] is, thanks to introductions, one of the most widely distributed fish in the world today. It also has a fascinatingly complex history behind its scientific and common names, involving an international set of biologists. For more than 150 years, it was listed as *Salmo gairdneri* Richardson 1836 (Smith and Stearley 1989), because John Richardson, a British physician, was thought to be the first person to describe it (in 1836), using specimens he had collected from the Columbia River. Johann Walbaum, however, had described a similar species, *Salmo mykiss*, from the Kamchatka Peninsula of Asia in 1792. The similarity of the two forms was recognized for a long time, but Russian and Japanese workers did not present unambiguous evidence that they belonged to the same species until the 1970s and 1980s. By the strict rules of scientific nomenclature, the earlier Russian name had precedence, but resistance of American workers to recognize Russian science delayed application of the name *Salmo mykiss* for North American populations. Then, studies during the 1980s showed that rainbow trout were much more closely related to Pacific salmon of the genus *Oncorhynchus* than they were to Atlantic salmon and trout of the genus *Salmo*. Thus, rainbow trout were placed in *Oncorhynchus*, and because the Russian species name was finally accepted at the same time, rainbow trout became *Oncorhynchus mykiss* (Walbaum 1792) (Smith and Stearley 1989).

Oncorhynchus means "hook-snout" (referring to the hooked snout that develops in spawning salmon), and *mykiss* is derived from the common name for this fish as given by the natives of the Kamchatka Peninsula in Russia. *Mykiss* probably bears very little resemblance to the original common name because of its history. The common name was written down by German naturalist Georg W. Steller in Russian and was applied to specimens he collected as part of a Russian exploration of the Pacific Coast during 1733 to 1744. Steller died before he made it back to Russia, so his notes and specimens languished until 1784, when an Englishman, Thomas Pennant, described the species (in English) using Steller's common names but not the Linnaean system of scientific nomenclature. In 1792, Johann Walbaum, a German naturalist in the Russian court, took the common name from Pennant's account and rendered it into Latin, resulting in *mykiss*! Unfortunately, the names of other Pacific salmon have gone through this same tortuous transformation process—the culmination of which is *O. tshawytscha*, the unpronounceable name for chinook salmon.

The complicated history of rainbow trout names does not end with the scientific names. During the nineteenth century, rainbow trout, native to the North American west coast, were known as brook trout and were usually referred to as *S. irideus*, following the description of W. Gibbons in 1855. Gibbons and others thought that resident rainbow trout were a species separate from the anadromous rainbow trout (steelhead) described by Richardson. When brook trout (*Salvelinus fontinalis*) from the eastern United States were introduced into western streams, the confusion in common names lead to the application of a translation of the scientific name (*irideus*) to the common name, rainbow trout. Once it was recognized that resident and anadromous rainbow trout were the same species, *irideus* disappeared as a species name but remained in the common name!

subsequently changed. If his name was not contained in parentheses, it would mean that the genus name had remained unchanged since the time he described the species (a rare occurrence). As long as the longnose dace is recognized as a distinct species, its species name will be *cataractae*; however, the generic name can change depending on how ichthyologists perceive its relation to other species. The genus *Rhinichthys* now contains seven or eight (depending on who is counting) rather similar-looking species.

The scientific names for fish are created from a mixture of Latin and Greek root words, a tradition that dates back to the time when entire species descriptions were in Latin. It has been retained as a useful tradition. The names of the fishes generally are descriptive of the fish in some way or tell something of its history or habits. Figuring out what the names mean is a bit like working a crossword puzzle—a very satisfying, if esoteric, exercise. For example, for the longnose dace, *Rhinichthys* translates as "nose-fish" (referring to the long snout), and *cataractae* refers to the fast water in which it lives. Fishes, however, are often named after people and places as well. Thus, the coelacanth (*Latimeria chalumnae*) was named after the person who found the first specimen (Marjorie Courtenay-Latimer) and a river in South Africa (Chalumna) that empties into the ocean near the spot where the first speciman was caught.

12.6   LESSONS FROM FISH SYSTEMATICS, GENETICS, AND SPECIATION

The phenotypic and genotypic responses of fishes to changing conditions are impressive, a constant reminder that evolution is an ongoing process. Species can show surprisingly high morphological divergence with little apparent genetic divergence, as is demonstrated by the charrs of Iceland and the cichlids of the Great Lakes of Africa. On the other hand, species can appear to be fairly uniform morphologically over a wide geographic range while showing high genetic diversity, as in Pacific salmon. Such diversity makes fishes fascinating subjects for students of evolution, and it creates nightmares for purveyors of Linnean taxonomy. This apparent conflict has long made ichthyologists leaders in adopting the techniques and attitudes of phylogenetic systematics, in which knowledge of the evolutionary history of fish is essential to their classification. The study of the genetics of fish has also lead to the realization that genetic diversity is essential for allowing fish to adapt to environmental change. Thus, protecting genetic diversity and variability is important to fish conservation, including the conservation of economically important species.

SUPPLEMENTAL READINGS

Bell 1976a; Cailliet et al. 1986; Carvalho 1993; Hey 2001; Lundberg and McDade 1990; Mayden 1992a.

WEB CONNECTIONS

PHYLOGENETIC SYSTEMATICS
www.ucmp.berkeley.edu/clad/clad4.html
darwin.eeb.uconn.edu/systematics.html

CHAPTER 13
Evolution

The evolutionary history of fish is a complex and fascinating topic. The frequent lack of a good fossil record for critical periods of fish evolution, however, coupled with the diversity and specialized nature of both modern and fossil fish groups, makes controversies about relationships among major groups common. Even as the fossil record becomes better known, the controversies continue and increase, because as the paleontologist Alfred S. Romer has noted, the increasing knowledge of fossil forms often leads to a "triumphant loss of clarity." Partly for this reason, recent studies of the evolutionary relationships among fishes rely heavily on comparisons among living forms and on the use of cladistics (Patterson 1977). The results of the new inquiries into the evolutionary relationships of fishes, however, still produce many controversies. Thus, the relationships described in this chapter, which are presented in approximate evolutionary sequence, should be regarded as tentative and are necessarily oversimplified.

13.1   THE EARLIEST VERTEBRATES

The chordate ancestors of fish and, of course, of vertebrates in general are assumed to be attached, bottom-dwelling (sessile) forms that had free-swimming larvae for dispersal. The basic vertebrate organization presumably evolved as the larval stage became more active, perhaps as a means of increasing the probability that the larvae would find a suitable site for the adult stage. Eventually, however, larvae capable of reproduction evolved, and finally, the sessile adult stage was lost altogether. It is usually hypothesized that the typical vertebrate organization evolved as these neotenous forms assumed an increasingly planktonic existence in the oceans. Alternatively, the organization may have developed among forms adapted for bottom feeding on organic particles, much like modern lamprey larvae do (Mallatt 1984).

The switch to a vertebrate organization seems to have occurred fairly early during the history of metazoan life. The earliest known fossils of probable vertebrates are from the marine Chengjiang formation in Kunming, China, dating to the early Cambrian (530 Before Present (BP)). These tiny, soft-bodied fossils are a small part of a remarkable assemblage of invertebrates (Shu et al. 2001, Smith et al. 2001). Two species have been described, *Myllokumingia fengjiaoa* and *Haikouichthys ercaicunensis,* although *Haikouichthys* appears to be the more derived of the two forms, perhaps allied to lampreys. Both are fusiform and have V-shaped myomeres, a distinct head, a probable notochord, a dorsal fin, and ventral fin folds (possibly paired). They lack any biomineralized tissue (dentine, bone) otherwise considered to be characteristic of vertebrates (at least those for which only hard parts are preserved).

The next vertebrates known from the fossil record are found in deposits from the late Cambrian (490 million years BP), which are mainly ***conodonts*** (Smith et al. 2001). The conodonts (Conodonta) are known almost entirely from tiny (<2 mm), tooth-like structures that are abundant in marine sediments. Although they are now extinct, they were common for more than 300 million years (Fig. 13.1)! The nature of the organism containing the "teeth" was long unknown, although analysis of the chemical

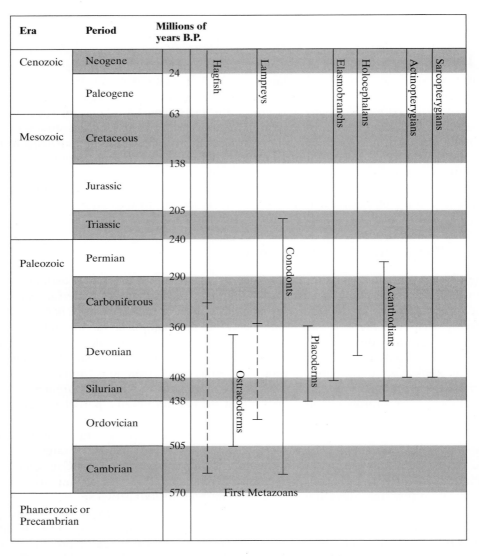

FIGURE 13.1 Geologic periods and time lines for major groups of fish.

composition of the teeth showed an outer coating of enamel over a layer of bone-like material, much like vertebrate teeth (Sansom et al. 1992). Placing conodonts firmly among the vertebrates required the discovery of fossils containing impressions of soft tissues (Donoghue et al. 2000). These fossils show that conodonts had large eyes and eel-like bodies supported by a notochord and that they were probably free-swimming (Briggs 1992) (Fig. 13.2). The presence of mineralized tissue suggests they are closer phylogenetically to the jawed vertebrates (Gnathostomata) than to lampreys and hagfishes, although how the tooth-bearing structures in the mouth relate to jaws is uncertain. It is nevertheless curious that the earliest known bone-like material is associated with a feeding apparatus (Donague and Aldridge 2001). The exact relationship of the condonts to

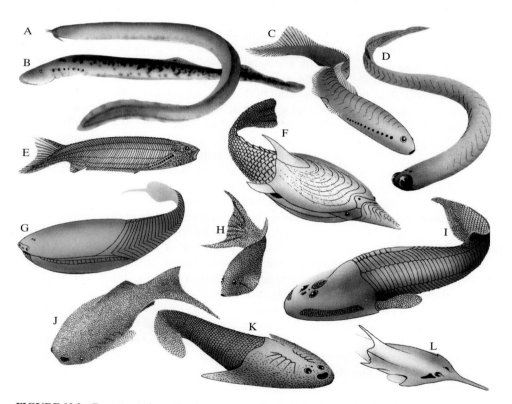

FIGURE 13.2   Representatives of early vertebrate lineages. **A.** *Eptatretus* (600 mm), Myxiniformes, a modern hagfish. **B.** *Petromyzon* (800 mm), Petromyzontiformes, a modern lamprey. **C.** *Jaymoytius* (130 mm), a soft-bodied ostracoderm from the Lower Silurian. **D.** *Clydagagnathus* (60 mm), Conodonta, a conodont from the Lower Carboniferous. **E.** *Pharyngolepis* (150 mm), Anapsida, an ostracoderm from the Upper Silurian. **F.** *Errivaspis* (150 mm), Heterostraci, an ostracoderm from the Lower Devonian. **G.** *Sacabambaspis* (300 mm), Arandaspida, an ostracoderm from the Upper Ordovician. **H.** *Furcacauda* (35 mm), Thelodonta, a fork-tailed thelodont from the Lower Devonian. **I.** *Hemicyclapsis* (150 mm), Osteostraci, an ostracoderm with paired fins from the Lower Devonian. **J.** *Loganellia* (120 mm), Thelodonta, a thelodont from the Lower Devonian. **K.** *Geraspis* (150 mm), Galeaspida, an ostracoderm from the Middle Silurian. **L.** *Pituriaspis* (head shield 45 mm long), Pituraspida, from the Middle Devonian. From Donaghue et al. (2000). Used with permission of the Cambridge Philosophical Society.

other vertebrates remains speculative, but their being a sister group to the Gnathostomata seems likely (Fig. 13.1).

13.2 OSTRACODERMS

The ***ostracoderms***[1] are the first major vertebrate radiation, aside from the conodonts, to be widespread in the fossil record (Fig. 13.2). The earliest fossils are found in marine deposits from the Ordovician period, approximately 460 million years ago, and they were abundant and diverse for approximately 100 million years (Forey and Janvier 1994). Approximately 600 species have been described, representing about seven evolutionary lines that seem as different from one another as the Chondrichthyes and Osteichthyes are today.

The ostracoderms were characterized by the lack of jaws and of paired fins (with some possible exceptions) and the presence of bony armor, an internal cartilaginous skeleton, and a heterocercal tail. The armor tended to be fairly massive around the head and more scale-like on the body. Most ostracoderms were quite small (<15 cm long) and dorsoventrally flattened. Their overall body structure indicates that they were mostly bottom dwellers that lived by sucking up organic ooze and small invertebrates. One curious group (the Thelodonta or fork-tailed thelodonts), however, was deep-bodied and large-eyed, suggesting that free-swimming modes of existence were possible (Wilson and Caldwell 1993). The gills of ostracoderms, which were quite large, presumably served as food filters as well as respiratory organs, much like those of the modern fish-like chordate, amphioxus. Ostracoderms presumably moved about much like tadpoles do, because most lacked the control of movement granted by paired fins. Later ostracoderms did develop bony, fin-like structures behind the gill openings that undoubtedly increased their stability and maneuverability. In one group (Osteostraci), the fins appear to be true pectoral fins, indicating a sister relationship of this group to jawed fishes (Forey and Janvier 1994). The habits of these advanced forms are a bit of a mystery, however, because most of them also had reverse heterocercal tails (Fig. 13.2). Such a tail would tend to push the animals off the bottom.

The armor is one of the most interesting features of the ostracoderms, because it represents the first undoubted bone. It is generally assumed that the bony armor evolved for protection against predators, although its use as an osmotic barrier in freshwater forms and as a phosphate-storage mechanism have also been suggested (Halstead 1968). Thanks to the work of Swedish paleontologist E. A. Stensiö and his students, who carefully sectioned and analyzed numerous ostracoderm fossils, many other details of ostracoderm structure are also known, including a surprising amount concerning the structure of the nervous system. Although a great deal of uncertainty remains as to how the ostracoderms relate to the jawed vertebrates or to the lampreys and hagfishes, ostracoderms

---

[1]The ostracoderms (meaning shell-skinned) were once regarded as a unified taxon of fossil jawless fishes distinct from the modern lampreys and hagfishes and placed together in the Cyclostomata. All the jawless fishes were then placed within the Agnatha, with the idea that a number of distinct evolutionary lines were within the ostracoderms, one of which gave rise to the lampreys and another to the hagfishes. Because recent research shows that the Agnatha represent several major evolutionary lines rather than one, the term is not used here, although the term ***agnathans*** is often used to refer to jawless vertebrates in general.

most likely are more closely related (i.e., are a sister group) to jawed vertebrates than they are to lampreys (Forey and Janvier 1993, 1994). In turn, lampreys are more closely related to ostracoderms and jawed vertebrates than are hagfishes. The unarmored, bone-less condition of lampreys and hagfishes thus represents the ancestral condition rather than the result of secondary loss of ostracoderm armor (as had generally been thought).

Although the ostracoderms were quite varied in morphology, their basic body plan seemed to limit their ability to become as ecologically diverse and active as later fish groups, which may explain in part why they eventually became extinct. Purnell (2001) points out that this widely accepted hypothesis assumes that competition and predation from jawed fishes was responsible for the demise of ostracoderms, which does not nec-essarily fit with the fossil record. Ostracoderms coexisted with jawed fishes for millions of years. In addition, the conodonts, also now extinct, were probably active predators on small invertebrates (Purnell 2001). We clearly have much to learn about the early history of the vertebrates.

13.3   PLACODERMS

The earliest major radiation of jawed fishes was the placoderms (class *Placodermi*, meaning plate-skinned). They are well represented in the fossil record because of their heavy, bony skeletons. Another group, the Acanthodii, actually appeared earlier in the fossil record and are important as the earliest known jawed fishes. They remained, how-ever, only a minor element of the fish fauna for millions of years. The placoderms domi-nated the waters of the late Devonian period and died out completely during the lower Carboniferous period (≈ 350 million years ago). They were a highly diverse group of fishes, often very bizarre in appearance, but they shared a common ancestry based on the structure of their armor, especially a ring of interlocking plates around the shoulder region (Young 1986). The large bony plate on the head region of most placoderms had a joint-like attachment to the bony plates covering the trunk. Besides their distinctive cov-ering of dermal bony plates, placoderms had an internal body skeleton, paired fins, jaws, and a dorsoventrally compressed body (Fig. 13.3).

The development of this combination of characteristics permitted the placoderms to achieve a much greater ecological diversity than the ostracoderms. Indeed, the jaws, paired fins, and internal skeleton are fundamental characteristics of the fishes that even-tually replaced them. At the same time, their bony armor, flattened bodies, and other specializations seem to have limited them to the role of bottom dwellers (with a few ex-ceptions). Their success as bottom dwellers was considerable; this is perhaps best attrib-uted to the development of jaws (from the first gill arch), which enabled them to become predators (Box 13.1). During their Devonian heyday, large-jawed predators such as the arthrodires achieved lengths of 6 m or more. The failure of the placoderms to evolve more advanced types of jaw suspension and true teeth, both of which are characteristics of the Osteichthyes and Chondrichthyes, may have been partially responsible for their replacement by these two groups.

What is the significance of the placoderms in the evolution of modern fishes? They appear in the fossil record at a time when we would expect to find ancestors of the modern groups, yet they are so bizarre and specialized that it hardly seems possible they

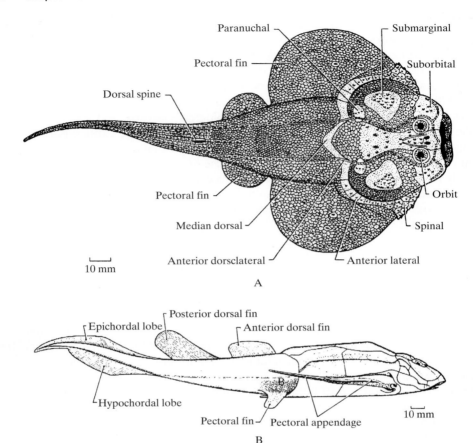

FIGURE 13.3 Placoderms. **A.** *Gemuendina*, dorsal view (Rhenaniformes). **B.** *Bothriolepis* (Antiarchiformes). From Moy-Thomas and Miles 1971.

could have served as such. Seven distinct evolutionary lines within the placoderms are recognized (Young 1986), but none shows any special affinities with modern fishes. Nevertheless, they are most likely the sister group to the combined lineages of Acanthodii, Chondrichthyes, and Osteichthyes, because they share some fundamental characteristics: (1) jaws with a common structural pattern, (2) two pairs of paired fins with bony girdles for support, and (3) three semicircular canals in the inner ear (ostracoderms, lampreys, and hagfishes possess two) (Goujet 2001). In turn, this suggests that further investigations of placoderm structure can provide clues regarding the origins of jawed vertebrates (Goujet 2001).

13.4 CHONDRICHTHYES

The cartilaginous fishes (class Chondrichthyes) are an easier group to define than the bony fishes (class Osteichthyes), partly because there are only approximately 850

BOX 13.1

The Origin of Jaws

Jaws are regarded as a key evolutionary development leading to the diversification of the vertebrates. As the ambiguous nature of the feeding apparatus of the conodonts suggests, their evolution was a long process. Anatomical and embryological evidence shows that jaws originally developed from the first gill arch. In ostracoderms, the gills were probably used for both feeding and respiration, so a gradual separation of these two functions is not hard to imagine. This separation would enable processing of larger, more high-energy food sources. The first gill arch was close to the mouth and essentially moved forward to provide greater support for it (Maisey 1996). The earliest jaws were entirely derived from the first gill arch (*autostylic suspension* from the cranium) and were relatively simple, hard-biting structures (much like our own jaws). To take advantage of larger and more active prey (probably other fish!), the jaws gradually incorporated the second gill arch (and associated muscles) into their structure, thereby allowing increased strength, mobility, and gape. The earliest versions of these more derived jaws had *amphistylic suspension*, in which elements of both arches attached the lower jaw to the cranium. The final step of this general process was to divorce elements of the first gill arch from the suspension apparatus and to allow only elements of the second gill arch (hyoid arch) to be involved (*hyostylic suspension*), mainly as a way to increase mouth gape. This resulted in the flexible arrangement of jaws that is characteristic of all modern fishes. In sharks and ancestral bony fishes, the remnants of the second gill arch are the spiracles, which are small gill openings that still perform a minor respiratory function. This neat picture of evolutionary progression has been presented in textbooks for decades, but the actual progression quite likely was much less straightforward.

species and partly because the fossil groups (with a few exceptions) are so poorly known that the characteristics of the living forms suffice for the entire class. Distinctive anatomical characteristics include: (1) cartilaginous skeleton; (2) teeth that are not fused to the jaw and that are replaced in rows; (3) presence of unsegmented, epidermal fin rays (*ceratotrichia*); (4) single, ventral nostrils on each side of head; (5) spiral valve intestine; and (6) claspers on pelvic fins of males, indicating internal fertilization. They also lack features of bony fish (described later in this chapter). Fossil forms may lack some of the features just listed; for example, well-preserved sharks from upper Devonian deposits lack claspers. The Chondrichthyes first appeared in the fossil record more than 400 million years ago, and they were common by the Devonian period. Most of the deposits from which they are known are marine. This is in marked contrast to the bony fishes, which accomplished much of their early evolution in fresh water. Within the Chondrichthyes are two distinct evolutionary lines going back independently as far as the Devonian: the subclass Elasmobranchii (sharks and rays), and the subclass Holocephali (chimaeras).

Elasmobranchii

The elasmobranchs have been important predators in the oceans ever since the first shark-like forms appeared in the middle Devonian. Unfortunately, their lack of a bony skeleton means that good fossils are rare, although teeth and spines are often abundant as fossils. They seem to have achieved a successful combination of characteristics early during their evolution, and the greatest deviation from the basic body plan occurred when the rays (superorder Batoidea) appeared during the Jurassic period. The characteristics that distinguish the elasmobranchs are five to seven gill openings, plus a spiracle (secondarily lost in some forms); placoid scales; upper jaw (palatoquadrate cartilage) not fused to the cranium, and lower jaw attached with either amphistylic or hyostylic suspension (Box 13.1); and teeth that are numerous and rapidly replaced. With a few exceptions, elasmobranchs are top-level carnivores that are not particularly dependent on sight for capturing prey (in contrast to most bony fishes).

The evolution of elasmobranchs is difficult to discuss, because it requires that leaps of imagination be made between the few well-preserved fossil forms and modern forms. The overall trend, however, is toward steady improvements of the basic Devonian shark design, with numerous side excursions into bottom-dwelling, invertebrate-feeding forms and a fairly wild explosion of diverse forms during the Carboniferous. The Devonian *cladoselachian sharks* (infraclass Cladoselachii, order Cladoselachiformes) had ancestral characteristics such as the lack of claspers, an elongate skull, amphistylic jaw suspension, and no anal fin; they also had broad-based, triangular, paired fins (Fig. 13.4). These early sharks had sharp, multicusped teeth (termed *cladodont*) that show signs of wear and so were probably replaced slowly. Cladoselachian sharks were undoubtedly

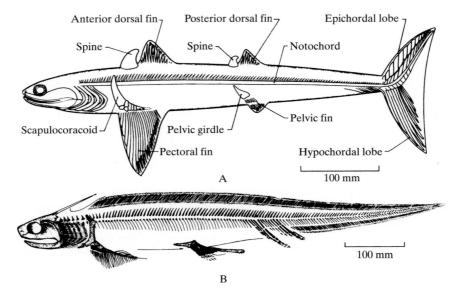

FIGURE 13.4   Devonian Chondrichthyes. **A.** *Cladoselache*. **B.** *Xenacanthus*, a freshwater form. From Moy-Thomas and Miles 1971.

predators on other fishes. When cephalopods and other molluscs radiated in the seas of the Carboniferous period, however, forms with flat, pavement-like teeth for crushing hard-shelled invertebrates became common. The evolution of diverse elasmobranch body forms in this period (as often reflected in peculiarly shaped spines and teeth) also may have been a reflection of a relatively low diversity of bony fish compared to present times. Curiously, whereas the cladoselachians were diversifying in salt water, another group—the *xenacanth sharks* (infraclass Xenacanthii, order Xenacanthiformes)—were common in fresh water (Fig. 13.4). They disappeared during the Triassic period.

During the Permian period, cladoselachians were replaced by *hybodont sharks* (infraclass Neoselachii—in part, order Hybodontiformes), some of which are usually considered to be ancestral to the modern sharks, skates, and rays (Shirai 1996). The hybodont sharks seem to have been adapted for feeding primarily on large, active invertebrates (e.g., squid), because they had sharp teeth for biting in the front of their jaws but blunt teeth for crushing in the rear. During the Jurassic period, the hybodonts gave rise to the modern elasmobranchs (infraclass Neoselachii) with a diversity of feeding mechanisms. Modern sharks are largely adapted for preying on bony fishes (albeit with many exceptions to this rule), while skates and rays are largely adapted for feeding on benthic invertebrates (also with many exceptions!). Modern neoselachians are characterized by a distinctive structure of pelvic fin cartileges, by the teeth and placoid scales, and by calcified vertebrae (see Chapter 15). Ideas concerning the relationships among major lineages of modern elasmobranchs are in a state of flux (Nelson 1994, Carvalho 1996, Shirai 1996).

Holocephali

Chimaeras (subclass Holocephali, order Chimaeriformes) are a strange group of bottom-dwelling, invertebrate-feeding fishes that have been present in small numbers apparently since the upper Devonian period, with a brief flowering during the Carboniferous. Among their more distinctive characteristics are a single gill cover over four gill openings, no spiracle; an upper jaw fused to the cranium; teeth consisting of only a few large, flat plates (with very slow replacement when lost); and no scales. The males have a clasper on the head in addition to ones on the pelvic fins. Modern-type chimaeras first appeared during the Jurassic period.

13.5   ACANTHODII

The so-called spiny sharks (Acanthodii) are the oldest-known jawed vertebrates (having lived approximately 440 million years ago), so they have long excited the interests of students of vertebrate evolution. Despite their early appearance in the fossil record, they were somewhat specialized forms. Most were small (<20 cm long), with large eyes and flexible, streamlined bodies covered with bony, dentine-tipped scales.

Their most distinctive feature, however, was the stout, ornamented spines in front of all fins. The earliest acanthodians also had two rows of ventral paired fins, one row on each side, and each fin was preceded by a spine (Fig. 13.5). The number of fins in a row was variable, and the presence of the row has been frequently cited as evidence of the

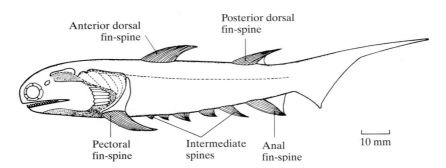

FIGURE 13.5 Typical acanthodian, *Climatius reticulatus.* From Moy-Thomas and Miles 1971.

fin-fold theory of the origin of paired fins (and limbs). According to this theory, the predecessors of fins were lateral folds on the sides of the presumably tadpole-like early fishes. The folds would have been used for stabilizing while swimming. With an evolutionary premium on control of movement, the fin-fold, by this theory, became subdivided into controllable segments, and eventually, the intermediate segments were lost, leaving the standard two sets of paired fins. The multiple paired fins of early acanthodians presumably represent the intermediate stage of the progression. Goujet (2001) suggests that this classic theory may not fit new observations regarding the development of paired fins both in modern fish embryos and in fossil placoderms. These observations suggest that fin-folds were not necessary as an intermediate stage and that pectoral and pelvic fins may have evolved independently. Even most acanthodians lacked the multiple paired fins. Acanthodians persisted as specialized plankton feeders in freshwater and marine environments well into the Permian period, long after the much more abundant and diverse placoderms had become extinct.

The relationship of the acanthodians to other fishes has been controversial. Early workers usually placed them with the placoderms, as just one more peculiar group. During the past 30 years or so, various workers have considered them to be a sister group of the Osteichthyes, whereas others have found them to be more closely related to the Chondrichthyes. Jarvik (1977) thought they had a surprising number of features (e.g., structure of the jaws, gill arches, and pectoral fins, etc.) in common with modern sharks, but the presence of such features as bony opercula, branchiostegal rays, and three otoliths argue in favor of bony fish affinities (Nelson 1994). They most likely are a sister group to all other bony fishes.

13.6 OSTEICHTHYES

The bony fishes (class Osteichthyes) are a large, diverse group with a fairly rich fossil record. For this reason, they are a hard group to define precisely. No one feature distinguishes modern forms; rather, a common structural pattern, combined with the absence of the features characterizing the Chondrichthyes, distinguishes them. The most distinctive elements of the osteichthyan structural pattern are the presence of lungs, bone, bony scales, and lepidotrichia (Long 1995).

Lungs (represented as swimbladders in derived forms) presumably indicate that the early steps of bony fish evolution took place in fresh water in a tropical region, where the ability to breathe air would be advantageous as protection against periods of stagnation. The history of the development of the lungs into a swimbladder that acts primarily as a hydrostatic (buoyancy) organ is an interesting one, because it reflects the overall structural changes that have taken place during the course of bony fish evolution. The first lungs were probably ventral outpouchings of the gut, which evolved in fishes that survived in stagnant waters by swallowing bubbles of air and then exchanging oxygen and carbon dioxide through the gut. Ventral lungs still characterize the South American and African lungfishes (Lepidosireniformes) and tetrapods. Such lungs, however, present problems to fish that swim about actively, because a pocket of air under the gut tends to make a fish top-heavy and subject to rolling over. The earliest solution to this problem seems to have been the development of a dorsal lung with a ventral opening. This situation is still found in the Australian lungfish (Ceratodontidae). The next step, found in most modern bony fishes, is to have a dorsal "lung" with a dorsal connection. Such *physostomous swimbladders* function primarily to keep the fish buoyant in the water. In many forms, the connection to the gut is lost altogether, and gases enter and leave the swimbladder entirely through the circulatory system (*physoclistous swimbladder*). In addition, a number of teleosts, such as the sculpins (Cottidae), have secondarily lost the swimbladder entirely, largely as an adaptation to a bottom-dwelling existence.

Bone, or at least some ossification, is present in most Osteichthyes, but it may be secondarily lost in a few forms such as the sturgeons and paddlefishes (Acipenseriformes) and the lungfishes. The skeletal structure of the head, although quite variable, also seems to have a general osteichthyan pattern, particularly in the cranium and jaws.

Bony scales in ancestral bony fishes contain a layer of dermal bone that is overlaid by enamel, forming very solid but flexible armor. In more derived fishes, the bony layer has been lost, and the scales consist of a thinly structured material called *isopedine* (Sire 1990). Scales, like bone and the swimbladder, may be secondarily lost in some groups of fishes (e.g., catfishes).

Lepidotrichia, like scales, are of dermal origin and probably are derived from scales. They form the soft rays of the fins, which are segmented and dumbbell-shaped in cross-section. True bony fishes first appear in the fossil record of the lower Devonian period. Most of these early bony fish are from freshwater deposits, which indicates that bony fishes in general originally evolved in fresh water. The two major evolutionary lines of the modern Osteichthyes appear in the fossil record nearly simultaneously: Sarcopterygii (lobe-finned fishes), and Actinopterygii (ray-finned fishes).

13.7   SARCOPTERYGII

The relationships among the lobe-finned fishes and their relationships to other groups is a subject of heated (well, at least warm) debate and no doubt will continue to be. The reasons for this are: (1) Modern forms are highly specialized, (2) smashed fossils that characterize the ancestral forms of a group are always subject to reinterpretation, (3) tetrapods (including humans) are a sister group to the Sarcopterygii, and (4) the modern fishes within the group are few and highly specialized. Among the distinctive features of

---

BOX 13.2

Picturing Ancient Fishes

This book is a textbook, so the focus is on words—often sleepy ones. It is often hard, therefore, to really picture what many of these ancient fishes looked like. Fortunately, two fascinating books are available that focus on color pictures of fish fossils and imaginative paintings of ancient fishes in their habitats: *Discovering Fossil Fishes* (1996) by John G. Maisey of the American Museum of Natural History, and *The Rise of Fishes* (1995) by John A. Long of the Western Australian Museum. The two books cover the same basic ground with surprisingly little overlap in their pictures. Their words are good as well. The books make a wonderful supplement to this chapter.

"We are fishes, whether we like it or not. It is more than a semantic issue, for our anatomy bears the indelible traces of our fish ancestry: the vestiges of the incompressible notochord, our spinal cord with all its myelin-coated nerves, our internal and external bony skeletons, our highly modified pectoral and pelvic fins—the list is much longer, but the point is made" (Maisey 1996, p. 217).

---

this group are fins with bony, leg-like supports. Following Nelson (1994) as well as Clouthier and Ahlberg (1996), we present the lobe-finned fishes as six presumed, separate evolutionary lines or clades—(1) Actinistia (Coelacanthimorpha), (2) Porolepimorpha, (3) Dipnomorpha (Dipnoi), (4) Rhizodontimorpha, (5) Osteolepimorpha, and (6) Tetrapoda—while recognizing that other fossil groups are as yet difficult to place within the grand scheme. The porolepimorphs and rhizondontimorphs are obscure fossil groups that will not be discussed further, whereas the tetrapods are outside the study of ichthyology—even if well represented by ichthyologists.

Actinistia

The coelacanths are a distinctive group of predatory fishes with a long fossil record and two living forms. They are easily recognized by their three-lobed (*diphycercal*) tail (with the vertebral column entering the middle lobe), forward-placed dorsal fin, and large cosmoid scales (see Chapter 16). They also have external nares (as opposed to having *choana* or *separate internal openings to the lungs*), no branchiostegal rays, no maxillary bone in the upper jaw, a large swimbladder, and unbranched lepidotrichia (fin rays). One of the more fascinating aspects of their biology is that they evolved initially (Middle Devonian period) in fresh water, became fairly diverse inhabitants of shallow-water marine environments by the late Triassic period, and then disappeared from the fossil record during the Cretaceous period. It was thus a major ichthyological event when, in 1938, coelacanths *(Latimeria chalumnae)* were discovered to be still living in the deep waters off the African coast, in the Indian Ocean (Thomson 1991). At the time, they were thought to be the closest living relatives to the tetrapods, a sort of living "missing link," but more recent work has cast doubt on this theory (see Chapter 16).

Dipnomorpha

The lungfishes have been a very conservative group as far as their evolution, and they have never achieved much diversity of form. Their evolutionary history, however, is well known, because they have consistently lived all but the earliest stages of their history in fresh waters that are prone to drying up and, hence, are conducive to making fossils. On the basis of these fossils, the amount of change from the earliest-known forms (lower Devonian) to the present-day forms has been plotted (Fig. 13.6). Lungfish evolution seems to have been most rapid early in their history, initially reflecting a switch from

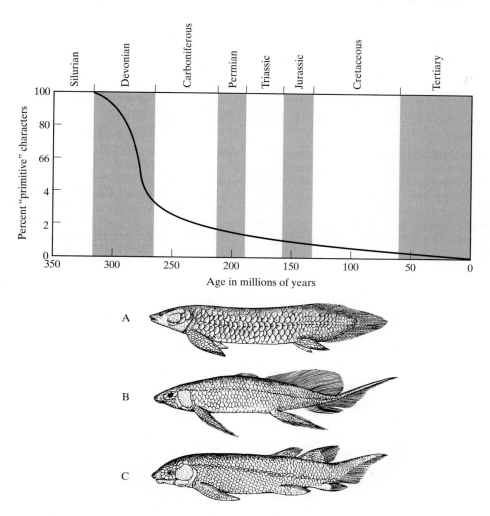

FIGURE 13.6 Rate of evolutionary change in lungfishes, from ancestral characteristics (defined as those possessed by the Devonian genus *Dipnorhynchus*) to derived characteristics (those of the modern genus *Protopterus*). The lungfishes illustrated are (**A**) *Neoceratodus* (recent), (**B**) *Scanmenacia* (late Devonian), and (**C**) *Dipterus* (middle Devonian). After Westoll 1949.

shallow-water marine to freshwater environments, which included the development of lungs. The rate of change has been extremely slow, however, for the past 200 million years or so.

This change in lungfish during their history has largely been one of reduction, including reduction in the amount of bone and the number of skull bones, the loss of various hard layers from the scales, and the loss of the separation between the dorsal, caudal, and anal fins. Besides lungs and a largely cartilaginous skeleton, modern lungfishes are characterized by choana, plate-like teeth, a spiral valve intestine, and many other features (Rosen et al. 1981). Because they seem to be intermediate in so many of their characteristics, during the nineteenth century lungfishes were considered to be the fishes most closely allied to terrestrial vertebrates. Subsequently, the similarities were regarded to be largely the result of convergent evolution, making the lungfishes only an interesting evolutionary sideshow. Rosen et al. (1981), however, examined some fossil lungfishes closely and revived the idea that the lungfishes are, in fact, the closest piscine relatives of the tetrapods. Recent morphological studies support this conclusion (Coates 2000), as does analysis of mitochondrial DNA (Meyer and Wilson 1990, Zardoya and Meyer 1997) and other molecular data (Tohyama et al. 2000).

Osteolepimorpha

For a long time, osteolepids and coelacanths were placed together as one evolutionary line. They differ, however, in a number of fundamental ways. For example, osteolepids have choana, branchiostegal rays, and branched lepidotrichia. Their relationship to other fishes is still subject to debate, but no question remains that they are intermediate between the first amphibians (Labyrinthodontia) and fishes. Indeed, some authorities prefer to place the osteolepids and the labyrinthodont amphibians together in one group. The most dramatic characteristics that link the osteolepids and the early amphibians are the lobed fins, autostylic jaw suspension, and structure of the teeth (Fig. 13.7). The fins are lobed, because they contain a series of bony elements that link them to the pelvic and pectoral girdles, much like tetrapod legs. The autostylic jaw suspension is like that of tetrapods. The teeth have the complex foldings of the enamel—visible as grooves on the worn tips of each tooth—that are also found in labyrinthodont ("labyrinth tooth")

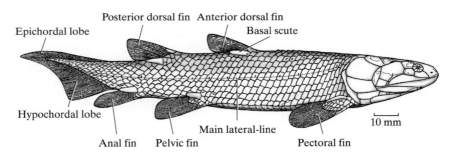

FIGURE 13.7 An osteolepid sarcopterygian, *Osteolepis macrolepidotus*. From Moy-Thomas and Miles 1971.

amphibians. The osteolepids were mostly armored, air-breathing predators of tropical freshwater environments. They became extinct by the lower Permian period, coincident with the rise of the amphibians (Maisey 1996).

13.8   ACTINOPTERYGII

The subclass Actinopterygii (ray-finned fishes) contains most of the bony fish species that exist today. When they first appeared in the fossil record in Devonian freshwater deposits, however, they were uncommon. By the beginning of the Carboniferous period some 360 million years ago, they had become the dominant freshwater fishes and had begun their invasion of the seas. The history of this group is one of constant change, with continuous "improvements" being made on the basic fish design that culminated in the modern teleosts. The ancestral features of the Actinopterygii include lepidotrichia, heavy ganoid scales, structurally distinctive pelvic and pectoral girdles, fins attached to the body by the fin rays (rather than with a fleshy lobe), branchiostegal rays, and no internal nares (choana). The Actinopterygii subclass is usually divided into two major evolutionary groups; the infraclasses Chondrostei, and Neopterygii. The odd African bichirs (with virtually no fossil record) are sometimes placed as a separate subclass (Brachiopterygii) because they are so distinctive structurally (Bjerring 1985). It is more likely, however, that they are an early, specialized derivative of some chondrostean line (Nelson 1994).

Chondrostei

The Chondrostei are the original ray-finned fishes, which achieved their greatest abundance and diversity in the Carboniferous period. There is little reason to think the group is monophyletic; instead, it is more or less a "garbage can" grouping of ancient fishes with multiple origins (***polyphyletic***) (Grande and Bemis 1996). At least one of the ancient chondrostean lines gave rise to the Neopterygii, which includes most modern fishes. All modern chondrosteans (25 species) and many fossil forms belong to the order Acipenseriformes, represented in modern fauna by two highly specialized families; Acipenseridae (sturgeons), and Polyodontidae (paddlefishes) (see Chapter 16).

The ancient chondrosteans developed many of the body shapes and, presumably, ways of living that also characterize the more derived groups (Fig. 13.8). They also, however, had many structural features considered to be ancestral, because the fishes that replaced the chondrosteans had either lost or modified them. Among these features are: (1) heavy ganoid scales (so called because the outer enamel layer is made up of a distinct substance, ganoine), (2) a spiracle, (3) a heterocercal tail, (4) a cranium consisting of three strongly fused units of bone, (5) bones of upper jaw (maxilla and premaxilla) fused to cranium, and (6) no interopercular bone. In more derived fishes, the structure of the skull is more complex and flexible, especially in relation to the jaws.

Neopterygii

This group contains most of the modern bony fishes and seems to be derived from one line of chondrosteans. Until recently, the neopterygians ("new fins") were generally

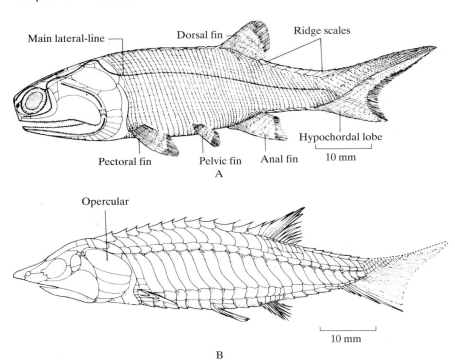

FIGURE 13.8   Fossil Chondrostei. **A.** *Moythomasia* (Devonian).
B. *Phanerorhynchus* (Pennsylvanian). From Moy-Thomas and Miles 1971.

divided into two infraclasses: Holostei, and Teleostei. The holosteans were considered to be an intermediate evolutionary stage, containing many fossil forms as well as modern gars (Lepisosteiformes) and bowfin (Amiiformes)—two groups whose exact relationships to the teleosts remain uncertain (Olsen and McCune 1991). It appears, however, that the bowfin is more closely related to teleosts than to the gars, justifying the breakup of the Holostei as a grouping (Gardiner et al. 1996). Although this classification system was convenient, it did not reflect the more complex reality of evolutionary trends among the fishes (see Fig. 1.1 for relationships). Early neopterygian fishes (Fig. 13.9) shared the seas and fresh waters with the chondrosteans through most of the Triassic, Jurassic, and Cretaceous periods but gradually became the dominant group. They presumably became dominant because of gradual—but significant—changes in structure that gave them a competitive edge over the chondrosteans. These changes culminated in the subdivision Teleostei, the dominant bony fishes of today. The teleosts are a highly diverse group of fish with a number of distinctive (but not necessarily exclusive) characteristics in common:

- The operculum consists of four bones, including the interopercular bone (absent from chondrosteans), which is derived from a branchiostegal ray.

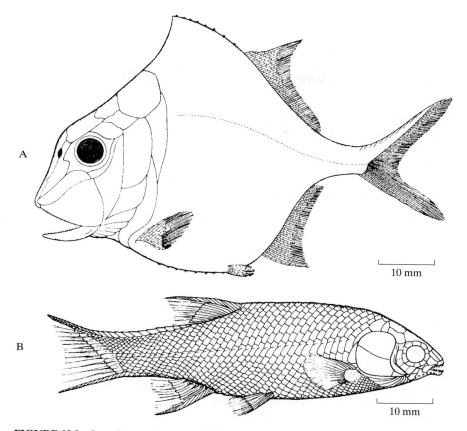

FIGURE 13.9 Jurassic neopterygians. **A.** *Platysomus* **B.** *Acentrophorus*. From Moy-Thomas and Miles 1971.

- The tail usually is homocercal and is supported dorsally by uroneural bones (modified from neural arches of vertebrae) and centrally by seven or fewer hypural bones, which may fuse to form a hypural plate. Even when fused, the hypurals form dorsal and ventral groups, associated with the dorsal and ventral division of the caudal fin rays, respectively.

- The scales, referred to as ***elasmoid scales***, are made of two thin layers: an external fibrous layer, and an internal bony layer called isopedine.

- The vertebrae are completely ossified and reduced in number from those of chondrosteans. The vertebrate are not solid structures but are cylinders of bone supported by bridge-like struts. Because of their structure, they are lighter and stronger as well.

- The swimbladder is reduced in size and functions primarily as a hydrostatic organ.

- The jaw structure has many distinctive features (dePinna 1996). For example, the premaxillary bones are the principal bones of the upper jaws and are highly moveable.
- The fins are highly maneuverable, giving teleosts good control over their movements. This function is associated with the distinctive structure of the pectoral girdle (e.g., two or more postcleithra, four radials) (dePinna 1996).
- The body shape is amazing in its variety, and a large number of the species are quite small (<30 cm long), thus enabling the teleosts to occupy niches and habitats not previously available to fishes.

13.9   STRUCTURAL CHANGES IN ACTINOPTERYGIANS

The characteristics of teleosts are the culmination of general trends in the evolution of actinopterygian fishes toward fishes that are capable of rapid and complex movements, have efficient respiratory systems, and can use a wide variety of foods. This can be seen by examining the evolutionary changes within the Actinopterygii in some of the more obvious structures.

Scales

The general trend in scales has been a reduction in heaviness and complexity with an increase in flexibility. The heavy scales of chondrosteans must have served as a moderately flexible suit of armor. Such armor was a definite improvement over the solid sheets of bone used by the placoderms but remained a heavy burden to carry. It is probably significant that the surviving chondrosteans (sturgeons and paddlefishes) have no scales at all over most of their bodies. The early neopterygian **ganoid** scale, as found on modern gars, still had three layers, but the general trend in the group was to reduce the heaviness of these layers. This culminated in the thin, two-layered **elasmoid** scale of teleosts, which is light and flexible. Many teleosts have gone a step further and eliminated scales altogether.

Branchiostegal Rays

In fossil chondrosteans, the floor of the branchial cavity was rather rigid, so water had to move across the gills by movements of the operculum, through the spiracle, or simply by being rammed across during swimming. The branchiostegal rays developed from bones in the floor of the branchial cavity. This increased the efficiency of active pumping of water across the gills. The efficiency of the "two-pump" respiratory system was further increased by the modification of one branchiostegal ray on each side to become part of the gill cover, as the **interopercular bone**. This bone, a definitive feature of teleosts, increased the size of the opercular cavity and, hence, the amount of water that could flow across the gills as part of the two-pump respiratory system. A secondary—but equally important—function of the branchiostegal rays has been to permit feeding methods based on suction rather than on grabbing. Thus, the evolution of branchiostegal rays has been an important factor in the development of teleost diversity. Within the neopterygian fishes, a general trend is toward decreasing the number of branchiostegal rays, although a reverse trend is present in some groups.

Swimbladder

Presumably, the swimbladder in the earliest actinopterygians functioned primarily as a lung, as it did for other bony fish groups. The general trend in actinopterygian evolution has been to increase the importance of the swimbladder's hydrostatic (buoyancy) function and to decrease its importance in respiration. In many teleosts, the swimbladder remains connected to the gut by a tube (physostomous), but this tube functions primarily as a means of regulating the volume of gas in the bladder to adjust buoyancy. In derived teleosts and in deepsea forms, even this connection is lost (physoclistous). In teleosts, the swimbladder is also reduced in size. A final trend in swimbladder evolution has been for it to assume other functions in addition to being a hydrostatic organ. In many teleosts, it serves as an amplifier of sound waves for hearing, and in some, it is used for sound production as well. So specialized has the swimbladder become for nonrespiratory functions that most teleosts adapted for stagnant waters have developed other means for breathing air.

Jaws

During the course of actinopterygian evolution, the jaws changed from being rigid, toothed structures adapted for biting and grabbing to much more flexible structures, often without teeth, adapted for feeding by a variety of methods. This is best illustrated by following the changes in the two principal bones of the upper jaw; the *maxilla*, and the *premaxilla*. In chondrosteans, these bones are firmly united to the skull. The maxilla is the main bone of the upper jaw and possesses many sharp teeth. In ancestral neopterygian fishes, the maxilla is reduced in size and is firmly attached to the skull only by its anterior portions. The premaxilla is greatly increased in size, and both bones still carry teeth. This arrangement developed in conjunction with branchiostegal rays and seems to be an intermediate step in the development of suction feeding. In most derived teleosts, the premaxilla is the dominant bone of the upper jaw and is largely free of strong attachments to other bones, so it can be easily extended. The maxilla in these fishes has become a sort of lever to increase the protrusibility of the premaxilla. The maxilla is usually without teeth, and teeth are frequently absent from the premaxilla as well. One result of having this type of mouth structure is that food processing (e.g., chewing) must occur at some location other than the rim of the mouth. Thus, predatory fish with protrusible premaxillae more often than not have pharyngeal teeth for this purpose. The flexible mouth also permits other types of feeding specializations to develop, such as plankton straining. In this case, the O-shaped mouth opening is very efficient, because it provides the maximum opening with the minimum perimeter. Suction feeding is most characteristic of the smaller, invertebrate-feeding teleosts; large piscivorous forms tend to reinvent the firm, toothed, biting mouth, which is elliptical in cross-section.

Tail

The original actinopterygian tail was heterocercal, but there seems to have been an inexorable trend in actinopterygian evolution toward the symmetrical, homocercal tail—even though the classic homocercal tail seems to have appeared rather abruptly in bony fish evolution (Metscher and Ahlberg 2001). The development of the homocercal tail is related, in large part, to the development of the swimbladder and neutral buoyancy, which eliminated the need for a heterocercal tail to provide lift. The homocercal tail is also advantageous for fast-swimming, pelagic forms (because it delivers uniform thrust) and for small, maneuverable forms that require a tail in which each ray can be controlled (for precise movements).

Fins

Two of the most important trends in fin structure have been the addition of spines and the changes in the positions of the pelvic and pectoral fins. True spines are characteristic of the most derived teleosts. Spines are antipredator devices that usually are best developed in fishes that cannot rely on speed to escape. Their development followed the loss of heavy, bony scales and coincided with changes in the relative positions of the pelvic and pectoral fins. Basically, in more derived teleosts, the pelvic fins are located immediately below, or even slightly anterior to, the pectoral fins, whereas in most other actinopterygians, the pelvic fins are well behind the pectorals. The anterior positioning of the pelvic fins is associated with increased maneuverability of the fish, because they are used for assisting the pectoral fins in controlling movement rather than primarily as stabilizers. The anterior positioning of the pelvic fins is also associated with the development of deep-bodied forms; however, even in deep-bodied chondrosteans, the pelvic fins occupied a more anterior position than they did in elongate forms.

13.10 EVOLUTIONARY TRENDS WITHIN THE TELEOSTS

Modern teleosts seem to represent four or five distinct lineages (see Fig. 13.10), although the relationships within and between the lineages are a matter of considerable debate. The present lineages were first recognized by Greenwood et al. (1966), who developed a provisional classification system that stirred the modern debate regarding how teleosts are related to one another, a debate which was fueled by Lauder and Liem (1983). The first line (subdivision Osteoglossomorpha) is a very distinctive group of fishes, the bonytongues (Osteoglossiformes) and mooneyes (Hiodontiformes), represented by a few peculiar species scattered through the world fauna and by the 190 species of African electric fishes. The second line (subdivision Elopomorpha) includes four rather disparate orders—Elopiformes (tarpons), Albuliformes (ladyfishes), Anguilliformes (true eels), and Notacanthiformes (spiny eels)—that are united by the presence of a distinctive larval form. The third line (subdivision Clupeomorpha) is a highly specialized group of plankton feeders, the Clupeiformes (herrings, anchovies, etc.). The fourth line is the Ostariophysi (the freshwater minnows, catfishes, and characins), which is very distinctive but seems to share some characters with the Clupeomorpha (Lecointre and Nelson 1996).

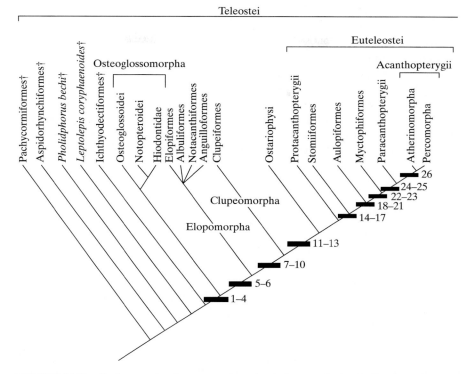

FIGURE 13.10   Cladogram showing one version of a phylogeny of teleosts. The characters numbered at each dark bar are ancestral characters that can be found in all groups above the bar (although they may be lost secondarily) but not in the groups below the bar. Groups known from fossils only are indicated by a dagger. The characters indicated by each number are as follows: **1.** Presence of an endoskeletal basihyal bone. **2.** Four pharyngobranchials present. **3.** Three hypobranchials present. **4.** Basibranchial and basihyal cartilages overlain by medium tooth plate. **5.** Two uroneurals extend anteriorly over the second ural centrum in support structure for caudal fin. **6.** Intermuscular bones present in abdominal and caudal regions of epipleural muscles. **7.** Retroarticular bone excluded from quadratomandibular joint surface. **8.** Toothplates fused with bones of gill arch. **9.** Neural arch on centrum of first ural vertebra reduced or absent. **10.** Articular bone of lower jaw fused with angular bone. **11.** Adipose fin present. **12.** Nuptial (breeding) tubercles present. **13.** A membranous outgrowth of first uroneural bone in tail present. **14.** Retractor dorsalis muscle present. **15.** Rostral cartilage present. **16.** Tooth attachment to bone is distinctive. **17.** Portion of adductor mandibulae muscle of jaw has a tendon inserting on the quadrate, preopercular, or opercular bone. **18.** Neural spine on second preural vertebrae reduced. **19.** Tendon or retractor dorsalis muscle inserts on third pharyngealbranchial bone. **20.** Interoperculohyoid ligament present. **21.** Third epibranchia bone has toothplate fused to it. **22.** Pharyngohyoideus muscle inserts on urohyal bone. **23.** Fourth pharyngealbranchial reduced or absent. **24.** Ctenoid scales present. **25.** Premaxillary bone of upper jaw with expanded ascending and articular processes. **26.** Retractor dorsalis muscle inserts on third pharyngobranchial only. † = extinct lineages. Figure and information modified from Lauder and Liem 1983.

The Euteleostei ("true teleosts"), with 22,000 species, can be divided into a number of separate evolutionary lineages as well. A lineage with many generalized or ancestral euteleost characteristics is the Protacanthopterygii, containing the smelts, salmonids, and pikes. In contrast, the euteleost lineage with the most derived features is the superorder Acanthopterygii ("spiny fins"), which contains numerous orders of the diverse spiny-rayed fishes, ranging from perch to seahorses to flounders. Between these two groups are other lineages whose relationships to each other are fairly fluid, although the cladogram developed by Lauder and Liem (1983) (Fig. 13.10) is still largely valid, whether the groups are divided into four superorders, as in Lauder and Liem (1983), or into seven, as in Nelson (1994) and as is followed in this text. The relationships of the fishes within and between the lineages currently recognized are complex, so the classification system presented here will no doubt change as more is learned about them.

13.11 LESSONS FROM FISH EVOLUTION

The evolutionary history of fishes is a dynamic field of study, because the fishes are not only diverse today but were diverse in the past, often in ways that are hard to understand. Throughout their long history, major groups have developed, flourished, and then died out: ostracoderms, placoderms, and major lineages of Chondrichthyes and Osteichthyes. Why some groups died out and others flourished is not as clear as it might seem. We are tempted to attribute the disappearance of "primitive" groups to competition and predation from "advanced" groups that had evolved improvements in fish design. Reality is not so simple. For one thing, representatives of many ancient fishes persist in modern times, and many others coexisted for millions of years with "advanced" forms. For another, evidence is growing that catostrophic events of global proportions (e.g., meteors colliding with the Earth) caused major and sudden changes to oceans and continents that would certainly have affected the course of fish evolution, perhaps in a nearly random manner. It is curious, for example, that teleost fishes, the dominant group today, are apparently monophyletic when good evidence exists for many potential ancestral groups among the ancient chondrosteans.

Although random events of global proportion likely did shift the course of fish evolution on occasion, it is also clear that evolutionary processes have resulted in a steady improvement of fish design, in the sense that fish have become increasingly effective in exploiting the watery world, thus resulting in increased specialization and, presumably, increased species richness. It is particularly fascinating to note that the Chondrichthyes and the Osteichthyes represent two independent evolutionary ways of being fish, convergent in some characteristics and divergent in others. Even within these major lines, some "minor" lines developed some of their own solutions to the problems of being a fish, such as the ratfishes (Holocephali) in the Chondrichthyes and the sturgeons (Chondrostei) in the Osteichthyes. Finally, it is worth noting that fish did not evolve independently of their environments but were both shaped by it and shaped it through their activities. For example, the structure of coral reefs is very much a result of the evolutionary interactions of teleost fishes and reef-building invertebrates, in some respects an "arms race" in which the invertebrates are continually evolving defenses against fish predation, whereas the fish evolve mechanisms to overcome those defenses (see Chapter 33).

SUPPLEMENTAL READINGS

Ahlberg 2001; Forey and Janvier 1993, 1994; Lauder and Liem 1983; Long 1995; Maisey 1996; Moy-Thomas and Miles 1971; Stiassney et al. 1996; Thomson, 1972.

WEB CONNECTIONS

PHYLOGENIES
www.tolweb/tree/phylogeny.html

EVOLUTION, GENERAL
www.ucmp.Berkeley.edu/history/evolution.html
www.pbs.org/wgbh/evolution

C H A P T E R 14
Hagfishes and Lampreys

PHYLUM CHORDATA
  Subphylum Myxini
    Class Myxini
      Order Myxiniformes
        Family Myxinidae (hagfishes)
  Subphylum Vertebrata
    Class Cephalaspidomorphi (?)
      Order Petromyzontiformes
        Family Petromyzontidae (lampreys)

The lampreys and hagfishes are slimy, eel-like, jawless, and without paired fins. They have long fascinated biologists as the only living representatives of the ancient creatures that gave rise to both fish and humans. They clearly possess many ancestral features, but they also have many specializations that make them fascinating in their own right. As we have learned more about the two groups, it has become clear that lampreys and hagfishes are as different from each other morphologically as either is from the jawed fishes (Table 14.1). In fact, the differences between the two groups are so profound that hagfishes can be excluded from the vertebrates altogether, as is done here, while lampreys remain firmly in the vertebrate column (see Chapter 13) (Fig. 14.1). This makes hagfishes the living sister group of all vertebrates, and lampreys become the sister group of all jawed vertebrates. Mallatt et al. (2001), however, found that the gene sequences in ribosomal DNA indicate that lampreys and hagfishes are closer to each other than lampreys are to the jawed vertebrates. These authors admit, however, that evidence derived from "a single gene family . . . is not entirely conclusive" (p. 115).

14.1 HAGFISHES

Having been neglected because of their alien appearance and habits, hagfishes are just beginning to be understood and appreciated (Jorgensen et al. 1998). Hagfishes are remarkable for their lack of striking external features (Fig. 14.2B). Linnaeus even classified them as worms! Although they are worm-like, they live in part by feeding on invertebrates buried in the bottom and in part by scavenging dead and dying fishes.

TABLE 14.1   **DIFFERENCES BETWEEN ADULT LAMPREYS AND HAGFISH**

Characteristics	Lampreys	Hagfishes
Dorsal fin	1–2	None
Preanal fin	Absent	Present
Eyes	Well developed	Rudimentary
Extrinsic eye muscles	Present	Absent
Oral disk	Present	Absent
Lateral-line system	Well developed	Absent
Barbels	Absent	3 Pairs
Intestine	Ciliated	Unciliated
Spiral valve intestine	Present	Absent
Buccal glands	Present	Absent
Nostril location	On top of head	In front of head
Nasohypophysial sac	Does not open into pharynx	Opens into pharynx
External gill openings	7	1–14
Internal gill openings	United	Separate
Cranium	Clearly cartilaginous	Poorly developed
Branchial skeleton	Well developed	Rudimentary
Vertebrae	Present	Absent
Spinal nerve pairs per body segment	2	1
Ducts of Cuvier to heart	Right only	Left
Pronephric kidney	Absent	Present
Osmoregulation	Hyper- or hypoosmotic	Isosmotic
Eggs	Small, without hooks	Very large, with hooks
Cleavage of embryos	Holoblastic	Meroblastic

Based on information in Hubbs and Potter (1971), Janvier (1981), and papers in Foreman et al. (1985).

Because of their lack of jaws, hagfish have a hard time biting through tough skin, so they enter a dead or dying fish through the gills, mouth, or anus—eating it from the inside out. Consequently, it is not uncommon for marine commercial fishermen who use setlines or gill nets to find in their nets fish that are occupied by feeding hagfish (Box 14.1). Once on deck, hagfish secrete incredible amounts of slime that stick to both the deck and the fishermen (the stem *myxin* means "slime").

Structure

The most conspicuous structures on hagfish are the three pairs of barbels around the nostril and mouth, which are used as tactile organs. The single nostril connects to the pharynx and functions as the water intake for both respiration and the sense of smell. The eyes are rudimentary, visible mainly as shallow depressions on the top of the head. The ventral mouth contains a tongue with four rows of keratinized teeth. These teeth work against each other and a cartilaginous dental plate to form a jaw-like structure that is capable of tearing out pieces of flesh from dead fish or marine worms. Along the sides of the body are from 1 to 14 gill openings and a long series of pores that are the openings for the slime glands. Hagfish also have such unusual features as a circulatory system with four "hearts" and a simple (pronephric) kidney, reflecting that their internal salt concentrations are the same as that of sea water. The kidney structure provides evidence that ancestors of the hagfish inhabited fresh water. Many features of hagfish can be regarded

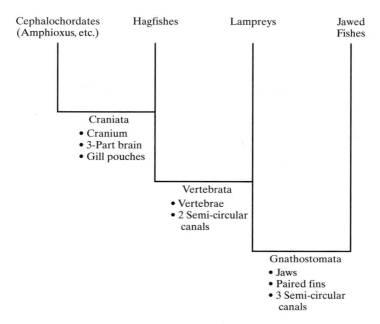

FIGURE 14.1   Phylogenetic relationships of the hagfishes and lampreys.

as adaptations for life in burrows (Martini 1998a), including the smooth and eel-like body with its coating of slime, the "degenerate" eyes and lateral-line system, the tidal (rather than continuous-flow) respiratory system, and the physiology that allows them to tolerate low oxygen levels.

Life History

All hagfishes are marine, and they are found mainly in temperate seas (Martini 1998a). They occur in the intertidal zone in some areas and have been observed as deep as 5,000 m, but most species occur at moderate depths (25–1,500 m). Most hagfish are found in association with soft mud and sand bottoms, into which they burrow and search for soft-bodied invertebrates for food, but some species also forage or live on hard-bottomed areas or among rocks. In some soft-bottomed areas, they can be among the most abundant large benthic organisms, both in terms of numbers and biomass. Because they live in burrows and "dig" into the bottom in search of their invertebrate prey, they may have a significant impact on the structure of benthic habitats in some areas. Although they prey extensively on invertebrates, hagfish are also important scavengers on the ocean floor. They can quickly congregate in large numbers on dead whales and fish, covering a carcass in a "writhing swarm" with enough action stir up the surrounding sediments and to cover the dead animal with slime (Martini 1998a). Modern commercial fishing techniques, which result in thousands of tons of discarded fish (bycatch), may have produced an increase in hagfish numbers as a result of an increase in their food supply.

BOX 14.1

Hagfish Leather

Although despised by many fishermen, hagfish are captured in large numbers in special fisheries, usually using traps baited with dead fish. During one 5-year period (1991–1996), more than 50 million hagfish were caught off New England alone, where they were processed and shipped to South Korea (Martini 1998b). The value of hagfish is for the attractive leather that can be made from the skin, which easily peels off the animal in one long strip. The dermal (subsurface) layer of hagfish skin consists of layers of densely packed collagen fibers, which provide a tough coating for live hagfish and tans well for designer handbags and briefcases. The leather is usually sold as "eel skin," under the assumption that few people would buy something labeled *hagfish leather* or *slime-hag hide* (Martini 1998b). Not surprisingly, some populations of hagfish are showing signs of being heavily overfished.

Hagfish are eaten by both predatory fish and mammals. Their main defense mechanisms appear to be their burrowing behavior and their slime, with which they can quickly coat themselves in large quantities. The slime is also used to coat dead fish, thereby making it unpalatable to other scavengers. To clear their gill openings and body surface of a heavy load of slime, hagfish have developed the remarkable ability to tie themselves into a knot, which passes down the body and pushes the slime away. This knotting behavior is also useful in giving hagfish extra leverage when taking a bite out of large fish.

Little is known about reproduction in hagfish, except that each female lays only a small number (20–30) of large (2–3 cm), leathery eggs. The eggs have small, hook-like structures on their ends for attaching to the bottom and to each other, although only a few have been collected. The early life history of hagfishes is poorly known, but they do not undergo the dramatic metamorphosis of lampreys. Curiously, juvenile hagfish (<17 cm long) have never been collected or observed (Martini 1998b).

14.2 LAMPREYS

Lampreys have acquired an evil reputation in modern times, because some species feed on fish that are also favored by humans, especially salmon and trout. Their habit of latching onto the sides of fish and sucking their blood and other body juices has caused them to be placed in the "nasty creature" category, along with vampires and leeches. Their prey often survives the attack, however, which is not true for the prey of more conventional predators. Lampreys also have some value as a gourmet food item themselves. Indeed, Henry I of England died after overindulging in a dish of cooked lampreys. Lampreys are also useful as experimental animals, particularly useful for neurobiological studies because of their large nerves (Rovainen 1979).

Structure

The most distinctive feature of adult lampreys (Table 14.1) is their oral disk, with its numerous, tooth-like plates of keratin that cover the disk and tongue in species-specific patterns (Fig. 14.2A). These plates are used for grasping the prey and then rasping a hole, through which fluids and tissues are sucked. Other features of lampreys, such as their large eyes and ability to pump water in and out of each gill opening when attached to a fish, also reflect the efficient means by which they have taken advantage of the abundant bony fishes. The sexes of spawning lampreys may be distinguished by the

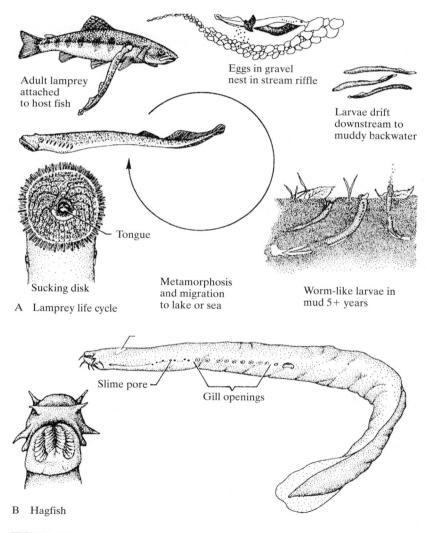

FIGURE 14.2   A. Life cycle of the sea lamprey (*Petromyzon marinus*). **B.** A typical hagfish (*Myxine*). From Moyle (1993); copyright © 1993 by Chris Mari Van Dyck.

presence of an anal fin (ventral fin-fold) on the female and a penis-like structure (genital papilla) on the male.

The larval lamprey (***ammocoete***), in contrast to the adult, is adapted for a secretive life buried in the mud of river backwaters, where it filter feeds on algae and detritus. Structurally, the ammocoete has the most ancestral features of any living vertebrate and is so different from the adult that the two were not connected until 1856. The eyes of ammocoetes are located beneath the skin of the head and are barely functional. In fact, the most important light-sensing organs are photosensitive cells in the tail. The fins are also barely visible and consist mostly of a low dorsal fold with a notch in the end to form the caudal fin. The most conspicuous feature is the expanded pharynx, which is used for both respiration and feeding. Water is drawn into the pharynx by the movements of a special muscular structure (***velum***) in the anterior portion and by expansion and contraction of the posterior portion (***branchial cavity***). The water departs through seven gill slits on each side. In the gill filaments and in the walls of the pharynx are ***goblet cells*** that secrete mucus, which traps particulate matter drawn into the pharynx and is swallowed in a continuous stream. This mechanism is very efficient but slow, so it requires dense concentrations of food to be effective (Mallatt 1982). Some mucus is also secreted to help stabilize the walls of the burrows in which ammocoetes live.

Distribution

Of the approximately 40 or so species of lamprey, 4 are found in temperate areas of the southern hemisphere, in South America, New Zealand, and Australia. The rest of the lampreys are found throughout the northern hemisphere, including six widespread anadromous species.

Life History

One of the most fascinating aspects of lamprey biology is the presence of two distinct life-history patterns. The most notorious lamprey species are those that migrate as adults to the ocean or large bodies of fresh water, where they prey on fish. Many of these predatory forms have given rise to nonmigratory "satellite species" that do not feed as adults (Vladykov and Kott 1977). These nonpredatory adults have poorly developed teeth, are small in size (usually <20 cm total length), and often inhabit streams that are small or are located a great distance from the ocean or other sources of large fish. The adult stage generally lasts less than 6 months. Predatory species are most abundant as spawners in short coastal streams, but they may migrate as much as 250 to 300 km up large streams. The adult stage of these species may last as long as 2 years, although 6 to 8 months of this time may be taken up with migrations, during which they do not feed. When feeding, they prey on a wide variety of fishes, but they concentrate on larger species. Vision seems to be the sense most involved in locating prey, although some evidence suggests that smell may be used as well. The lampreys grow rapidly on their diet of fish. Species that spend the predatory stage in the ocean reach a total length of 30 to 80 cm; those that spend this stage in fresh water seldom exceed 30 cm. The differences in the sizes achieved by adults with different life-history strategies is well illustrated by the lampreys of the Klamath River in Oregon and California (Moyle 2002). The anadromous Pacific lamprey (*Lampetra tridentata*) may reach lengths of 60 to 70 cm in this river. Landlocked versions, however, which prey on fish in the lakes and reservoirs of the upper river, typically reach only 20 to 30 cm. The nonpredaceous species from this area seldom exceed 20 cm as adults.

Regardless of the size and feeding habits of the adults, the life cycle of lampreys from the time of spawning through the ammocoetes stage through the period of metamorphosis shows the same basic pattern from species to species. Spawning adults generally choose a shallow, gravel-bottomed riffle as a spawning site, and spawning takes place during the late winter or early spring. Males tend to arrive first at the spawning site, and each begins to construct a nest depression by latching onto the larger stones and then pulling them downstream. Apparently attracted by olfactory cues released by the male, one or more females move into the nest site and help move the stones. The eggs fall into the interstices of the gravel and then adhere to the gravel. The number of eggs per female varies with the species. Females of large, anadromous lampreys produce 124,000 to 260,000 eggs, whereas most females of nonpredatory lampreys produce only 1,000 to 2,000 eggs. Most lampreys die after spawning, but some anadromous forms may survive to spawn a second time (Moyle 2002).

Newly hatched ammocoetes swim out of the nest and drift with the currents until they are swept into an area with a muddy bottom. There, they burrow into the mud and feed with only their oral hoods sticking out above the surface. The currents created by their feeding apparatus bring a steady stream of diatoms, desmids, detritus, and small invertebrates to the animal, although local depletions of digestible material apparently cause ammocoetes to change positions frequently. This stage usually lasts from 3 to 7 years, and ammocoetes may grow to a total length in excess of 10 cm. Ammocoetes of the nonpredatory forms seem to spend 1 to 2 years longer as larvae than do those of the predatory forms.

Lampreys as Predators

Lampreys are often labeled as parasites, but they do not really fit the image of a typical parasite, which has a single host for its lifetime and lives in intimate association with it. Instead, lampreys behave more like predators whose prey survives the experience, much like mosquitoes or leeches. An appropriate name for such animals is ***micropredator*** (Lafferty and Kuris 2002). In fact, under normal conditions, lampreys are "prudent" micropredators that do not seriously deplete the populations of their prey species (Box 14.2). In some instances, prey species may have a very high incidence of scars from lamprey attacks, with some individuals surviving repeated attacks. The attacks may seriously reduce growth of a prey fish, but at least it may survive to reproduce.

Conservation

Although some anadromous lampreys are abundant enough to be harvested for food, others are in serious decline. Because of the lengthy ammocoetes stage, lampreys are exceptionally vulnerable to human-caused changes in streams, especially pollution of backwater areas and siltation of spawning grounds. In the northern hemisphere, 60% of the species are in trouble: Ten are in danger of extinction, and nine are declining rapidly (Renaud 1997).

BOX 14.2
The Great Sea Lamprey Invasion

The bad reputation of lampreys stems in good part from the near-extinction of many fish populations following the invasion of the Great Lakes by the sea lamprey (*Petromyzon marinus*). The lamprey, an anadromous species native to the Atlantic Coast, invaded Lake Erie in about 1921. By 1936, they were established in Lake Huron and Lake Michigan, and 10 years later they were established in Lake Superior. Following the invasion of each lake, the populations of the large fish species—such as lake trout, burbot, and lake whitefish—collapsed, as did the fisheries for them. The only exception was Lake Erie, which is too warm for the lampreys and has few streams that are suitable for spawning, so the lamprey populations remained small. It is widely assumed that lampreys were able to invade the Great Lakes because the Welland Canal, opened in 1833, allowed them to move around Niagara Falls. Daniels (2001) points out that the canal was a largely unsuitable habitat for lampreys (stagnant, polluted, dried up annually) and so an unlikely route of invasion. An equally likely scenario is that they came in through bait-bucket introductions (ammocoetes were a popular live bait during the nineteenth century) or even a deliberate introduction as a food fish (Daniels 2001).

An intensive research program after the collapse of the Great Lakes fisheries made the life cycle of the sea lamprey one of the best known of any aquatic vertebrate (Applegate 1950) and, eventually, resulted in development of a poison specific for ammocoetes. Through the use of this lampricide and other control measures, the lamprey populations have been reduced, and most of the prey species have partially recovered. Sea lampreys, however, will never be eradicated from the Great Lakes, so continued control measures will be required to maintain the fish populations. It is interesting to note that in Cayuga Lake, New York, the sea lamprey coexists with many of the same species that were depleted by lamprey predation in the Great Lakes. This seems to be a situation in which predator and prey occurred together naturally and have coadapted so that both can survive. Given enough time, presumably the same thing would happen in the Great Lakes. It is also worth noting that the control programs in the Great Lakes streams may be having a detrimental effect on the native, nonpredaceous lampreys, which are also killed by the lampricide.

14.3   LESSONS FROM HAGFISHES AND LAMPREYS

Hagfishes and lampreys are continually teaching us new things about the evolution and biology of early vertebrates as we learn more about their anatomy and physiology. Recognition is also growing that they are highly successful creatures in their own way, preying or scavenging on the jawed fishes that replaced their presumed ancestors. Hagfishes have extraordinary specializations for burrowing and scavenging that make them major players in the ecology of the marine benthos. Likewise, lampreys show wonderful adaptations as both sophisticated predators and as larval detrital feeders, demonstrating the complete range of vertebrate feeding habits. In the Great Lakes, sea lampreys have exhibited an extraordinary capacity to colonize new habitats and to generate large populations, putting them in direct competition with human fishers/predators. In less-disturbed systems, lampreys are prudent predators that often do not kill their prey, indicating the importance of coevolution in predator–prey systems. Unfortunately, both lampreys and hagfishes, despite their survival abilities, are suffering from the negative effects of human activities worldwide.

SUPPLEMENTAL READINGS

Applegate 1950; Brodal and Fänge 1963; Foreman et al. 1985; Forey and Janvier 1994; Hardisty and Potter 1971; Martini 1998a, 1998b; Vladykov and Kott 1977.

WEB CONNECTIONS

GENERAL
www.calacademy.org/research/ichthyology
www.fishbase.org
www2.biology.ualberta/jackson.hp/IWR
www.flmnh.ufl.edu/fish/default.htm
www.people.clemson.edu/~jwfoltz/wfb418

EVOLUTION
tolweb/tree/phylogeny.html

LAMPREYS
www.glsc.usgs.gov/information/factsheets/sea_lamprey00

HAGFISHES
oceanlink.island.net/oinfo/hagfish

CHAPTER 15
Sharks, Rays, and Chimaeras

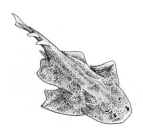

CLASS CHONDRICHTHYES
  Subclass Elasmobranchii
    Superorder Galea
      Order Heterodontiformes
        Family Heterodontidae (horn sharks)
      Order Orectolobiformes
        Families Parascyllidae (collared carpet sharks), Brachaeluridae (blind sharks),
          Orectolobidae (wobbegons), Hemiscyllidae (bamboo sharks),
          Ginglystomatidae (nurse sharks), Stegostomatidae (zebra shark),
          Rhincodontidae (whale sharks)
      Order Lamniformes
        Families Ondontaspididae (sand tigers), Mitsukurinidae (goblin shark),
          Pseudocarchariidae (crocodile shark), Lamnidae (mackerel sharks),
          Megachasmidae (megamouth shark), Alopidae (thresher sharks),
          Cetorhinidae (basking shark)
      Order Carcharhiniformes
        Families Scyliorhinidae (cat sharks), Proscyllidae (finback cat sharks),
          Pseudotriakidae (false cat shark), Leptochariidae (barbeled houndshark),
          Triakidae (houndsharks), Hemigaleidae (weasel sharks), Carcharhinidae
          (requiem sharks)
    Superorder Squalea
      Order Hexanchiformes
        Families Chlamydoselachidae (frill sharks), Hexanchidae (cow sharks),
          Notorynchidae
      Order Echinorhiniformes
        Family Echinorhinidae (bramble sharks)
      Order Squaliformes
        Families, Dalatiiae (sleeper sharks), Centrophoridae, Squalidae (dogfish
          sharks)
      Order Squatiniformes
        Family Squatinidae (angel sharks)
      Order Pristiophoriformes
        Family Pristiophoridae (saw sharks)

Order Rajiformes
Families Rhinobatidae (guitarfishes), Rajidae (skates)
Order Torpediniformes
Families Torpedinidae (electric rays), Hypnidae, Narcinidae (electric rays), Narkidae
Order Pristiformes
Family Pristidae (sawfishes)
Order Rhiniformes
Family Rhinidae
Order Rhynchobatiformes
Family Rhynchobatidae
Order Myliobatiformes
Families Platyrhinidae, Zanobatidae, Plesiobatidae (deepwater stingray), Hexatrygonidae (longsnout stingray), Urotrygonidae, Potamotrygonidae (river stingrays), Dasayatidae (stingrays), Gymnuridae (butterfly rays), Myliobatidae (eagle rays)
Subclass Holocephali
Order Chimaeriformes
Families Callorhynchidae (plownose chimaeras), Chimaeridae (shortnose chimaeras), Rhinochimaeridae (longnose chimaeras)

The sharks, rays, and chimaeras have long been held in low esteem by humans. As a group, they are generally considered to be vicious, inedible, and primitive. With a few notable exceptions, however, most members of this surprisingly diverse class possess none of these attributes. The common application of the word *primitive* to the group is particularly inappropriate, because they are as specialized in their own way as the teleosts are among the bony fishes. The main group, the elasmobranchs, *are* less diverse than the teleosts in the sense of having fewer species and lower genetic diversity (Heist 1999). Most of the species are predators on large invertebrates and fish, so extreme diversity in ways of "making a living" is also lacking. Nevertheless, the diversity that does exist within the elasmobranchs has been largely unappreciated until recently. Compagno (1973, 1977) stimulated interest in this diversity when he broke modern elasmobranchs into four evolutionary lines instead of the usual two (sharks and rays). Current classification systems restore the idea of two major evolutionary lines, but the members of the two groups have shifted, mainly because many sharks are related more closely to skates and rays than they are to other sharks. The details of the classification system remain in dispute, however, as indicated by the polite contradictions in three definitive papers on elasmobranch classification published in the same volume (Carvalho 1996, McEachran et al. 1996, Shirai 1996). The classification presented here follows that of Carvalho in most respects but that of McEachran et al. for classification of skates and rays. The skates, rays, and guitarfishes (the last seven orders listed here, under the superorder Squalea) are a monophyletic group and probably should be listed in a separate superorder (Batoidea) or as just one order (Rajiformes), but it is important to note their relationship to the dogfish and other sharks.

15.1 CHARACTERISTICS

The modern cartilaginous fishes are clearly a monophyletic group, although the two major branches diverged early (see Chapter 13) (Fig. 15.1). Their common origin, distinct from bony fishes, is shown by the many fundamental characteristics, beyond the cartilaginous skeleton, that the species have in common (Compagno 1999):

1. The cranium is a simple, box-like container for the brain and sense organs of the head.

2. The upper jaws (***palatoquadrate cartilage***) are not fused to the cranium but are separate and capable of being protruded using elements of the former second gill arch (hyoid arch). The lower jaw is a single element (***Meckel's cartilage***) on each side. The jaw teeth are in rows of replicate teeth, which replace one another in a conveyer belt–like fashion.

3. There are four to seven internal and external gill openings, although they are covered by one opercular flap in the chimaeras.

4. The vertebral column is essentially the notochord supported by calcified vertebrae.

5. The pectoral and pelvic fins are supported internally by a girdle skeleton and externally by rays (***lepidotrichia***) of flexible connective tissue. In males, the basal skeleton of the anal fins extend backward into the ***claspers***, which are paired copulatory organs.

6. Most have a covering of tiny ***placoid scales*** (dermal denticles).

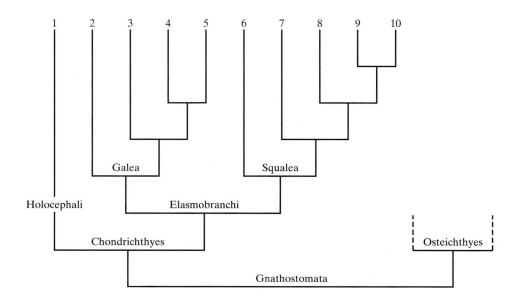

FIGURE 15.1 Phylogenetic relationships of the Chondrichthyes. **1.** Chimaeriformes. **2.** Hetereodontiformes. **3.** Orectolobiformes. **4.** Carcharhiniformes. **5.** Lamniformes. **6.** Hexanchiformes. **7.** Squaliformes. **8.** Squatiniformes. **9.** Pristiophoriformes. **10.** Skates and rays (six orders).

7. The snout usually protrudes above the mouth, with nostrils on the ventral surface. The nostrils are always single openings but are subdivided internally by folds of skin (Compagno 1999).

15.2 ADAPTATIONS

Regardless of how we classify them, the success of the sharks, skates, and rays as marine predators is undisputed. This success seems to result from the particular combination of adaptive characteristics, including those related to (1) buoyancy, (2) respiration, (3) external covering, (4) feeding, (5) movement, (6) sensory systems, (7) osmoregulation, and (8) reproduction. Because of their long, separate evolution from the Osteichthyes, the Chondrichthyes have developed different solutions to the problems of being a fish.

Buoyancy

Many chondrichthyan fishes spend their entire lives in close contact with the bottom (e.g., skates and rays), but many others spend most of their life in the water column. To do this without expending huge amounts of energy swimming, the ideal is to achieve neutral buoyancy, which bony fishes accomplish with a swimbladder. Chondrichthyan fishes instead use a combination of methods to reduce their average density. The primary method is the cartilaginous skeleton, which is much less dense than bone (1.1 vs. 2.0; water is 1.0) but still very strong. This is frequently augmented by a large, oil-filled liver (oil density, 0.8) and/or the hydrodynamic lift created by the heterocercal tail and pectoral fins while swimming (see Chapter 3).

Respiration

The Chondrichthyes have three basic—but not mutually exclusive—means of respiration. First, slow-moving, bottom-oriented sharks have developed a "two-pump" respiratory system that works on the same principles as that of teleosts (see Chapter 4) and provides for a steady flow of oxygenated water across the gills. Second, fast-moving sharks use "ram" ventilation, pushing the water across the gills during the process of swimming. Such sharks will die if they stop swimming. Third, the spiracles can be used to bring water to the gills, which is a technique especially favored by rays. The spiracles are small, round openings that immediately precede the principal gill openings on each side. In rays, they are located on top of the head and thus can be used to draw water into the gill chambers while the fish is lying motionless on the bottom, waiting to ambush prey or avoiding predators. Not surprisingly, the spiracles are greatly reduced—or even absent—in pelagic sharks that ram water across the gills. In some pelagic sharks, however, the spiracles may operate as an independent respiratory system for supplying oxygen to the large muscles of the eyes.

External Covering

All members of the Chondrichthyes have placoid scales in one form or another on the outside of their bodies. In skates and rays, these scales are typically found only as a few rows of large denticles on the back, which are sometimes modified into spines (as in stingrays). In sharks, the skin is filled with tiny, overlapping placoid scales, giving it a sandpaper-like feel. These scales form a lightweight, protective coat over the sides that seems to be particularly important for increasing hydrodynamic (swimming) efficiency (see Chapter 2). In fast-swimming sharks, such as the mako sharks (*Isurus* spp.), channels are found between the scales that absorb turbulent flow, thereby reducing the drag of the passing water (Moss 1984). Such placoid scales are characteristic of the oldest-known sharks, yet the bony fishes took several hundred million years to develop a scale covering (ctenoid scales) that apparently works as well as placoid scales for increasing the efficiency of swimming.

In more sluggish sharks, the placoid scales can be stout and projecting, often with sharp points, so they work well as lightweight, defensive armor. Placoid scales have been modified into a number of other structures as well, most conspicuously the "sting" of stingrays, the dorsal spines of dogfish sharks, and the defensive spines and denticles found on many skates. The teeth of sharks and rays are also modified placoid scales.

Feeding

Most members of the Chondrichthyes are specialized predators, and their dentition reflects this. Sharks that prey on large fish and marine mammals have triangular, blade-like teeth to grab their prey and saw or snap off large chunks. Because sharks cannot move their jaws back and forth to chew, a large shark must shake its head back and forth to get the "chewing" motion necessary when it takes a bite out of a large prey. Each species of shark or ray has distinctive teeth that reflect the way it makes a living. Sharks that swallow their prey (usually fish) whole have teeth that come to long, thin points for holding onto the prey so it can be swallowed. Many small inshore sharks have rows of small, sharp teeth for feeding on small invertebrates. Some species will "graze" on the protruding siphons of clams, resulting in stomachs full of mysterious O-shaped rings of muscle. Many skates and rays have flattened, pavement-like teeth for crushing hard-shelled invertebrates. Some sharks, such as the horn sharks (Heterodontidae), have pointed teeth in the front of the jaw for grasping prey but crushing teeth in the back for processing it.

One of the more remarkable aspects of Chondrichthyan teeth is that they are continually being shed and replaced. Consequently, some sharks may lose 30,000 teeth during their lifespan, dropping them either singly or in rows. Examination of the jaws of a shark reveals row after row of teeth behind the front biting row, with the next row ready to pop up as the outer one is lost. This results in a mouth full of ever-sharp teeth. Aside from the teeth, the main reason that sharks can ingest large prey is that their jaw suspension is hyostylic in most forms, which allows the gape to be maximized (see Chapter 13). The jaws are also rather loosely attached to the cranium, which again increases their gape and flexibility. In rays, the loose connection has allowed the development of an ability to protrude the jaws a short distance for picking or pulling organisms off the bottom. This results in a structure very much like the suction mouth of teleost fishes.

Once the prey is ingested, it quickly reaches the large stomach, where the initial stages of digestion take place. From the stomach, the food passes to the spiral valve intestine, a distinctive feature often considered to be primitive (presumably because only the most ancestral bony fishes possess one). In fact, the spiral valve seems to be an efficient way to increase the digestive surface area without increasing the length of the intestine.

Movement

How sharks and rays swim has already been discussed (see Chapter 2), but it is worth emphasizing that the characteristic heterocercal tail of modern sharks is a complex organ that serves the fish well for propulsion, steering, and hydrodynamic stability. Not surprisingly, sharks have complex movements even at fairly fine scales (Klimley et al. 2002). The most active pelagic sharks (Lamnidae) have countercurrent exchangers in their circulatory system that permit them to be "warm-blooded" and increase the efficiency of their swimming (see Chapter 3). This and other adaptations are convergent on the characteristics of tunas (Scombridae), which are among the most derived of bony fishes (Bernal et al. 2001). In contrast, pelagic rays (Myliobatidae) use their pectoral fins for propulsion and literally fly through the water with their "wings."

Sensory Systems

Sharks and rays have a number of remarkably well-developed sensory systems, which act in concert to help them locate prey and find their way about the environment (Box 15.1). This is unlike the teleosts, which tend to rely heavily on vision. Probably the first sensory cues that sharks and rays pick up when in search of prey are odor, particularly from injured prey, and low-frequency sounds. Their well-developed olfactory organs attest to the keenness of their sense of smell. Low-frequency vibrations and the turbulence created by struggling fish (or even by fish swimming in schools) may also be detected at distances of a kilometer or more, either with the inner ear (vibrations) or with the lateralis system (turbulence).

At closer range, vision comes into play. The visual systems of sharks and rays are well developed and used both during the day and when light levels are low (Gruber 1977). Many sharks possess both rods and cones, so color vision may be present. When sharks are almost ready to seize their prey, nictitating membranes often cover the eyes, and other sensory systems take over. One of these is the electroreception system, which can detect the tiny electrical fields created by muscular movement. The **ampullae of Lorenzini**, located on the snout, function in the latter capacity, as do sensory receptors on the lower jaw. The ampullae are also important in detecting hidden prey (e.g., fish hiding in soft substrate) by sensing their electrical fields (Kajiura and Holland 2002). The ampullae may also play a role in intraspecific communication, because many elasmobranchs, particularly among the skates and rays, possess weak organs (mainly in the tail) that generate electricity and whose function is otherwise unknown. After the prey is seized, it may be rejected because of taste. The Moses sole is a bony fish that is distasteful to sharks and will be rejected once bitten. It is therefore a possible source of chemical shark repellents.

Osmoregulation

The osmoregulatory system of the Chondrichthyes is extremely efficient. The concentration of solutes in the body is actually close to or higher than that of sea water—largely

---

BOX 15.1
What Is the Hammer Head for?

One of the most distinctive groups of sharks are the hammerhead sharks (*Sphyrna* spp.), with their broad, hammer-shaped head. With eyes on each end, the head is clearly not used as a hammer—or at least not often! Hammerhead sharks have been observed holding down stingrays with their heads, however, while taking bites out of them with their subterminal mouths (Strong et al. 1990). Their heads frequently contain stingray spines, but this does not seem to deter the predation. Another function of the head seems to be for increased flexibility of movement. Nakaya (1995) notes that the head acts as a sort of anterior rudder, allowing the fish to turn more quickly and sharply than ordinary sharks as well as to move its head up and down quickly, an advantage when feeding on benthic prey. Perhaps most importantly, the broad head provides lots of room for widely spaced sensory organs. Hammerhead sharks feed mainly at night, a good reason for having well-developed senses other than vision. The wide separation of nostrils may allow for more precise following of odor trails from prey in the water, by turning toward the side where the smell is strongest. The large number of ampullae of Lorenzini on the underside of the head allows them to detect the electrical fields of hiding or buried fish, such as stingrays, and hunting hammerheads will swing their heads back and forth over the bottom like a person with a metal detector searching for coins in the sand. Hammerheads are not more sensitive to electrical fields than other sharks, but they do have much more area available for detection and so have an increased ability to find hidden prey (Kajiura and Holland 2002). An acute ability to detect electrical fields may also be handy for navigation using the Earth's electrical fields. Thus, Klimley (1995) observed a large shoal (>200) of scalloped hammerheads (*S. lewini*) hanging out during the day around a seamount in the Gulf of California (off Mexico). After dark, sharks marked with ultrasonic tags swam 15 to 20 km on a fixed course, over deep water, to another, deeper seamount where prey (large squid) were abundant. The tagged fish engaged in a "yo-yo" movement while in transit, suggesting that they were periodically diving downward to get a better fix on the magnetic ridge-lines they were following. They would return to the home seamount at dawn in a similar fashion.

Why were the sharks returning each day to the one seamount in such numbers, thereby forming shoals? Given the large size of the fish (1–4 m), defense from predators is an unlikely explanation, nor is a hydrodynamic advantage because the fish were of mixed sizes (see Chapter 11). Klimley (1985, 1995) found that the shoaling was tied to reproduction. He observed that the females in particular circled the seamount slowly, with the largest females in the center of the circle and actively chasing smaller females away from central locations. Males could then easily find females ready to mate by dashing into the center of the circle.

---

because of the retention of nitrogenous wastes, mostly urea and trimethylamine oxide, to which the gills are nearly impermeable. A large rectal gland is used primarily for the excretion of sodium and chloride ions, whereas the kidney excretes divalent ions (see Chapter 6). This osmoregulatory system allows them to adapt readily to fresh water, as demonstrated by the abundance of stingrays in tropical rivers, including one family

(Potamotrygonidae) that is found *only* in fresh water. Some marine sharks and rays also enter fresh water on a regular basis (see Box 6.1).

15.3   REPRODUCTION

The osmoregulatory and reproductive systems of Chondrichthyes likely evolved simultaneously, because the long gestation periods of the embryos, either in egg cases or in body cavities of females, would hardly be possible without the ability of the embryos to withstand high concentrations of their own waste products. In some sharks, the urea surrounding embryos is periodically reduced by the ability of the female to flush the uterus with sea water. The cartilaginous fishes, unlike most bony fishes, expend most of the energy used for reproduction in producing a relatively small number of large, active young. They have a wide variety of ways of doing this, from egg laying (oviparity) to bearing (viviparity) and all stages in between (Wourms 1977, Compagno 1990, Wourms and Demski 1993).

Oviparity, or egg laying, is considered to be the ancestral means of reproduction in the Chondrichthyes and is still characteristic of 43% of them, including many sharks, all skates (Rajidae), and probably all chimaeras. Most of these fish produce large eggs covered with tough, leathery cases that are deposited in the environment, where they may take as long as 15 months to hatch (*extended oviparity*). Many of the cases have tendrils at the corners, apparently for attachment to rocks and corals off the bottom (Fig. 15.2). The tendrils also have tiny slits that permit water to flow through the case, sometimes assisted by movements of the tail of a developing skate (Zimmer 1999). Empty capsules are frequently washed up on beaches as "mermaids' purses." The eggs of the horn shark (Heterodontidae) are surrounded by a spiral shelf, which makes the eggs very difficult to remove from the rocky crevices in which they are wedged. The closest thing to parental care observed in sharks and rays has been female horn sharks carrying their eggs and nudging them into crevices. The young that emerge from the cases are essentially miniature adults, and they quickly disperse into the nearby cover of shallow nursery areas. For a few sharks (e.g., some cat sharks, Scyliorhinidae), the encased embryos spend little time in the environment, because they are retained by the female until they are nearly ready to hatch. This *retained oviparity* clearly represents an intermediate step in the process that led to viviparity.

Viviparity, or live bearing, in elasmobranchs takes a number of forms: (1) yolk-sac viviparity, (2) uterine viviparity, (3) cannibal viviparity, and (4) placental viviparity (Compagno 1990). Fish with *yolk-sac viviparity* (ovoviviparity) produce eggs. The shells are thin, however, and the eggs are retained in the uterus of the female. The shell soon disappears, and the young are retained in the uterus until fully developed. Like the young of oviparous forms, these young obtain their nutrition from a yolk sac. A modification of this method is found in stingrays and eagle rays (Myliobatiformes, 19% of all Chondrichthyes), which rely on *uterine viviparity*. In these rays, the mother secretes a nutrient-rich fluid that is taken up by the developing young through the skin or epithelia in the gut. Another modification of yolk-sac viviparity is *cannibal viviparity*, in which the young in each of the two oviducts consume unfertilized eggs produced

FIGURE 15.2   Egg capsule of a swell shark (*Cephaloscyllium ventriosum*) showing an embryo inside. After Springer (1979).

by the mother. It is characteristic mainly of large sharks in the Lamniformes (Gilmore 1993). The most extreme case of cannibal viviparity is the sand tiger *(Carcharias taurus)* and, perhaps, the thresher sharks *(Alops),* in which the first two young in the oviducts proceed to eat the other embryos in the oviduct with them. They have well-developed teeth to do this. The mother shark continues to produce unfertilized eggs for the nutrition of the two young, which are born at large sizes ($\approx$90 cm) and as experienced predators.

The most derived form of viviparity is ***placental viviparity***, which is found mainly in carcharhiniform sharks, in which most of the nutrients for the developing embryos are provided through a placenta-like arrangement, including an umbilical cord. The placenta in these sharks is derived from a wall of the embryonic yolk sac, which fuses with the uterine wall. Because the yolk-sac placenta evolved independently many times within the carcharhiniform sharks (in five families), a number of different arrangements now exist. For example, in the Atlantic sharpnose shark *(Rhizopriondon terranovae)*, early nutrition of the young is through egg yolk, and the placental connection does not develop until the yolk is largely consumed (Hamlett 1993). In the spadenose shark *(Scoliodan laticaudus)*, the embryo has virtually no yolk, and the yolk-sac placenta develops while the embryo is still tiny (3 mm). The placental arrangement in the spadenose shark, with the complete dependence of the embryo on direct nutrition from the mother, is remarkably like that of mammals (Wourms 1993).

Viviparity is characteristic of 57% of all Chondrichthyes and is especially characteristic of large, active forms, such as the requiem sharks and the eagle rays. Its presumed evolutionary advantage is that it allows the young to be delivered into the environment at a large size, so lots of prey are available to them and few predators are around to prey on them. Females usually move into shallow, protected areas to give birth and do not feed during the birthing period, perhaps as a mechanism to reduce cannibalism. The reproductive output of live-bearing sharks and rays is generally low (few young, long gestation period), but the spadenose shark, with the most highly developed form of placental viviparity, has a gestation period of only 6 months and produces 10 to 15 young that are approximately 15 cm long. This reproductive rate is about twice that of other placental sharks (Wourms 1993).

Behavior

All cartilaginous fishes engage in internal fertilization, which requires fairly elaborate courtship to carry off. This is well illustrated by Klimley's (1995) description of courtship in scalloped hammerhead sharks, in which a large number of females swim in slow circles above a seamount: "Females that win a central position in the school become the most desirable to males. A confident, sexually mature male will dash into the central cluster while performing a 'torso thrust' to advertise his readiness to mate with the 'boss' female. By beating his tail to one side, he propels his midsection to the opposite side, revealing his clasper. If a female accepts him, the couple leave the school and swim to the bottom of the seamount, where they may mate" (p. 35). During copulation, the male twists himself around the female and inserts one of his claspers into her cloaca, flushing in the sperm with an infusion of seawater.

Life Histories

The reproductive strategy of sharks and rays is to produce a relatively small number of young during a lifetime that necessarily have high survival rates. To produce these young, live-bearing forms have gestation periods of 6 to 22 months. Compared to teleosts, most sharks and rays are large. Recently developed methods for determing the age of sharks and rays indicate that most are slow-growing, long-lived, and have late ages of maturity. The best-studied shark is the spiny dogfish *(Squalus acanthias),* which rarely grows larger than 1.2 m but regularly lives for 70 to 80 years; females usually become mature at approximately 35 years of age and can theoretically produce 70 to 80 young in a lifetime (Saunders and McFarlane 1993). The similar-sized leopard shark *(Triakis semifasciatus)* lives 25 to 30 years, and females become mature at approximately age 17 (Cailliet 1992). The bat ray *(Myliobatus californica)* lives for approximately 23 years (Martin and Cailliet 1983). On the other hand, the Atlantic sharpnose shark *(Rhizoprionodon terraenovae)* reaches a maximum length of 1.1 m at 9 to 10 years of age, and females become mature at age 5 (Cortes 1995). Given that fecundities are low and gestation periods and reproductive intervals are long, such sharks and rays can produce only a small number of young during a lifetime. This "slow" life-history pattern has clearly served sharks and rays well through the millennia, but it is rapidly becoming their undoing. Fisheries for many species, but especially for the larger ones, have expanded greatly in recent years. It is already abundantly clear that these species can be easily overfished and that recovery of depleted populations likely will be slow at best (Klimley 2000). Species such as the white shark (*Carcharodon carcharias*) are rare enough that intensive fisheries could easily drive them to extinction. It has been estimated that for every human eaten by a shark, 10 million sharks are eaten by people—a clearly unbalanced ratio.

15.4 DIVERSITY

The preceding discussion indicates that the Chondrichthyes are a cohesive group taxonomically in that they share a common structural pattern, but an examination of adaptations by the various groups within the Chondrichthyes demonstrates their diversity and has resulted in a rather split taxonomy. The number of orders and families, considering the number of species, is high compared to teleosts, for example. Of the two main lines of evolution with the Chondrichthyes, the Holocephali (chimaeras) are a small, fairly uniform group, whereas the Elasmobranchii (sharks and rays) are highly diverse. The number of living species within the Chondrichthyes is usually given as approximately 850, but Compagno (1999) suggests the number is actually somewhere between 929 and 1164. Clearly, many species are still waiting to be discovered and described.

15.5 ELASMOBRANCHII

Galeid Sharks

The shark superorder Galea contains not only the sharks with the classic body shape that usually comes to mind but also many that diverge markedly from this shape (Fig. 15.3).

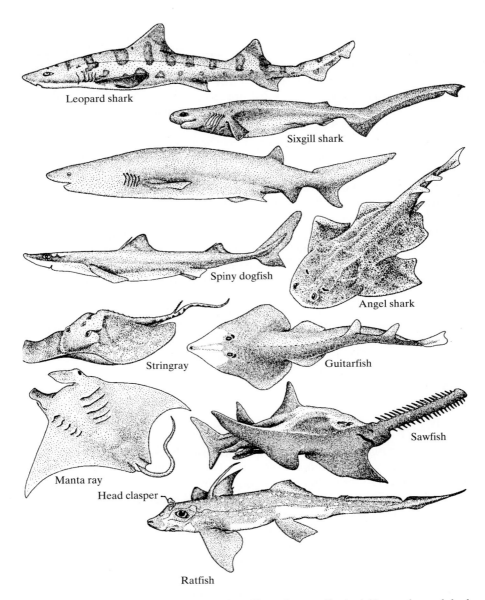

FIGURE 15.3 Representative Chondrichthyes. **Top to bottom:** Carchariniformes: leopard shark *(Triakis semifasciata)*; Hexanchiformes: juvenile sixgill shark *(Hexanchus griseus)*; adult sixgill shark; Squatiniformes: angel shark *(Squatina californica)*; Squaliformes: spiny dogfish *(Squalus acanthias)*; Myliobatiformes: stingray *(Dasyatis* sp.); Rajiformes: guitarfish *(Rhinobatos* sp.); Myliobatiformes: manta *(Manta birostris)*; Pristiformes: sawfish *(Pristis perotteti)*; Chimaeriformes: chimaera *(Hydrolagus collei)*. From Moyle (1993); © 1993 by Chris Mari Van Dyck.

The "typical" sharks are mostly in the families Lamnidae (mackerel sharks) and Carcharhinidae (requiem sharks). These sharks are mostly large, pelagic forms with blade-like teeth. They are efficient predators of large fish, cephalopods, and marine mammals and are also responsible for most attacks on humans. The sharks that perhaps most deserve the reputation of "man-eater" are the white shark *(Carcharodon carcharias)*, the oceanic whitetip shark *(Carcharinus longimanus)*, the bull shark *(C. leucas)*, and the tiger shark *(Galeocerdo cuvieri)*, although almost any shark will attack a human if provoked. Studies of shark behavior (e.g., Johnson and Nelson 1973) indicate that many attacks on humans may not be feeding attacks but actually defensive attacks on swimmers behaving in a way the shark interprets as threatening. Likewise, attacks on small submersible vehicles appear to be defensive responses to what the sharks perceive as a predator (Nelson et al. 1986). Klimley (1994) has documented that white sharks are very selective in what they eat and seem to prefer prey with high fat content, such as seals and sea lions; this may explain why they tend to spit out prey that are mostly muscle, such as sea otters, birds, and lean, rubber-covered humans.

Three sharks in this superorder that definitely are not capable of biting attacks on humans are the whale shark, the basking shark *(Cetorhinus maximus)*, and the megamouth shark *(Megachasma pelagios)*, although on occasion, male whale sharks apparently mistake boats for competitors and ram them. The whale shark, which may reach 18 m (Box 15.2), is the world's largest fish, whereas the basking shark runs a close second, with lengths approaching 15 m. Whale sharks and basking sharks are only distantly related to each other and have independently evolved mechanisms for straining plankton. Both have fine, elongate gill rakers for straining, but those of the basking shark have developed from placoid scales and those of the whale shark are cartilaginous rods with spongy material between that acts as a filter. The basking shark is largely a passive ram feeder, collecting zooplankton while swimming through the water with it gigantic mouth open wide, but the whale shark can also engage in more active pumping of prey into the maw. The megamouth shark is a large (4–5 m), deepwater shark discovered in 1976. It has an extraordinarily large mouth, through which it filters small shrimp, that may be lined with light-producing organs (Taylor et al. 1983). Because these large sharks swim through the dark water, they likely are largely invisible—except for a giant, Cheshire cat–like grin.

Among the more peculiar sharks in this superorder are the thresher sharks, with their extraordinarily long upper lobe of the caudal fin. The long lobe of the tail is used for stunning prey, which can be anything from fish to squid to birds. Thresher sharks have been observed overtaking prey and then whipping their tail around to stun it, after which the prey is easily eaten. One bit of evidence for this is that most thresher sharks caught on hooks in long-line fisheries are caught by their tails, presumably as they attempt to stun the bait (Gruber and Compagno 1981). Another odd galeid group is the horn sharks (Heterodontidae), which have been considered to be among the most ancestral groups of living elasmobranchs because they *look* primitive. As Compagno (1973) points out, however, they seem to have as many derived as ancestral characteristics and so are best treated as part of the "advanced" galeid sharks. Because they are rather sluggish, shallow-water bottom dwellers, horn sharks are comparatively well studied (McLaughlin and O'Gower 1971).

BOX 15.2

World's Largest Fish

The whale shark (*Rhiniodon typus*) is unquestionably the world's largest fish, although its maximum size is in dispute. The largest one accurately measured was 12 m in total length, but 14- to 18-m individuals have been reported (Colman 1997). Their size and small numbers make whale sharks difficult to study and easy to harvest to extinction, although they are protected in part by their wide distribution. They are found throughout the tropical and warm temperate seas of the world, concentrating in areas of upwelling where food is abundant (Colman 1997). They feed on a wide variety of small fishes and invertebrates, which they actively pump into their wide, terminal mouths, filtering out the organisms with dense filter screens on the gill arches. They can also feed passively by swimming through dense aggregations of prey with their mouths open, but the pumping appara-

tus allows them to capture faster-swimming prey, even small tuna (Colman 1997). How long whale sharks live is speculative, but the best guesses have them becoming mature at approximately 30 years (lengths >9 m) and living for more than 100 years (Colman 1997). Whale sharks are ovoviviparous, and females can contain more than 300 embryos. Once the young are born, they largely disappear from human view until they reach 3 to 4 m. When they aggregate close to shore, such as along Ningaloo Reef off western Australia, they often attract ecotourists to swim with the placid giants (Colman 1997). Divers occasionally grab a fin for a ride, mimicking the behavior of sharksuckers (Echeneidae). These bony fishes, specialized for attaching themselves to sharks and whales, are sometimes so abundant on whale sharks that they form a living beard.

Much more typical of the galeid sharks are the many species of cat sharks (Scyliorhinidae), hound sharks (Triakidae), and bamboo sharks (Hermiscyllidae). These are elongate, medium-sized (50–200 cm) sharks that have flattened heads, small teeth, small livers, and a highly asymmetrical heterocercal tail, with the lower lobe being small. These bottom-oriented sharks are abundant in shallow areas throughout the world. They cruise about feeding on small bony fish, squid, and various benthic invertebrates.

Squaleid Sharks

The 11 orders in the superorder Squalea represent much of the range of body plans within the elasmobranchs, from big sharks to small skates and rays. Despite differences among these orders, modern cladistic analyses (using multiple anatomical and biochemical characters) have found that the members are more closely related to each other than any is to the sharks placed in Galea. Only the first four orders, however, include traditional sharks, such as spiny dogfish, although even within these groups some peculiar forms exist, such as the frill shark (Chlamydoselachiformes, *Chlamydoselachus anguineus*). This deepwater fish has an eel-like body and a terminal mouth. The first gill extends across the throat, from one side to the other. This shark has been considered to be related to the ancient cladodont sharks because of its three-cusped teeth, but Compagno (1973) argues that similarities to ancient sharks are only superficial and that, in most respects, the frill shark is a modern form. Its peculiarities are derived from its specialization for feeding on squid (Kubota et al. 1991).

The order Hexanchiformes contains only five species of sharks, with six or seven gill openings, that also are largely inhabitants of deep water. The cow sharks (Hexanchidae) are rather flabby, bottom-oriented sharks with weak jaws and small teeth. They apparently live mostly by scavenging. The teeth in the upper jaw are sharply pointed, whereas the teeth in the lower jaw are saw-like, suggesting that they feed by anchoring their head onto large, dead animals (like whales) with their upper teeth and then sawing off chunks with their lower teeth.

The 74 sharks of the order Squaliformes are abundant and widely distributed. All have two dorsal fins, and many have spines preceding one or both fins. Although sleeper sharks (Dalatiidae) may exceed 6 m in length, most of the squaliform sharks seldom exceed 2 m. In fact, the smallest sharks known are from this order. The tsuranagako-bitozame[1] (Squalidae, *Squaloides laticaudus*) is mature at 11 to 15 cm. Another contender for smallest shark is the dwarf dogshark (*Etmopterus perryi*) from the deep water (300 m) of the Caribbean Sea, which matures at 16 to 20 cm (Springer and Gold 1989). All these small sharks are apparently deepwater dwellers, and some are luminescent, with photophores (Box 15.3).

BOX 15.3
The Cookie-Cutter Shark, Ambush Predator

Perhaps the most extraordinary of the small deepwater sharks is the cookie-cutter shark *(Isistius brasiliensis)*, so named because of its ability to bite cookie-shaped pieces of flesh from large, fast-swimming fish, such as swordfish and tuna (Jones 1971). The sharks take these bites by latching onto the fish with suctorial lips, biting down with needle-like upper teeth, and then slashing out the piece of flesh with a set of strong, fused teeth in the lower jaw (Fig. 15.4). The jaws are highly unusual, because they are heavy and highly calcified and so must be balanced—for buoyancy—by an extremely large, oil-filled liver. Cookie-cutter sharks are obviously not great swimmers, although apparently they do show vertical migrations. Thus, how they take bites out of big fish or capture large squid, an alternate prey, is a mystery. Widder (1998) presents evidence that the sharks make themselves look like small prey fish through patterns of biolumines-cence. Their ventral surface is covered with tiny photophores, which presumably allow the shark to blend in with light produced by the myriad small fishes and shrimp that also live in this habitat. The sharks also have a black band in the throat, however, that when surrounded by luminescence appears to be a small fish to a predator looking upward. All the shark has to do at that point is make a quick evasive maneuver and latch onto the passing fish! Widder (1998) speculates that the tendency of these sharks to shoal may improve the illusion.

---

[1]A Japanese name meaning "dwarf shark with a long face."

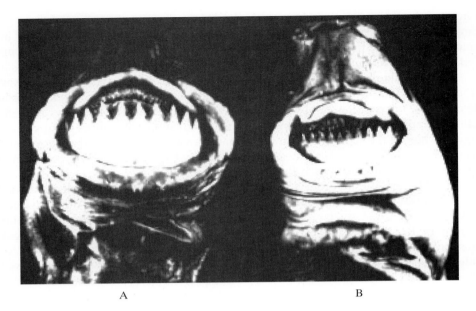

A   B

FIGURE 15.4   Cookie-cutter sharks: (**A**) *Isistius plutodus*, and (**B**) *I. brasiliensis*. These small sharks grab onto the side of a larger fish, hold on with their strong suctional lips, and then cut along a cookie-shaped plug of flesh with their lower teeth. Photo by L. J. V. Compagno.

Perhaps the best-known and most abundant squaliform shark is the spiny dogfish (Squalidae, *Squalus acanthias*), which commonly inhabits anatomy and ichthyology-teaching laboratories. They are also found worldwide in temperate to subpolar waters, mostly along coastlines. They form huge schools that contain both sexes when the dogfish are immature but only one sex or the other when they are mature (Jensen 1966). They may migrate considerable distances. A dogfish tagged off the coast of Washington has been caught off Japan, for example, and one tagged off Newfoundland was subsequently caught in the North Sea (Templeman 1976). In many parts of the world, they are an important commercial fish, but off North America, they are still considered to be mostly a nuisance that preys on more valuable fish. In contrast to the abundant dogfish, bramble sharks (Echinorhiniformes) are considered to be uncommon, although this is probably a reflection of their deepwater habitat. Despite their rather weak jaws and teeth, they are bottom-oriented predators that apparently feed by suction. Their rather flabby bodies are covered with a scattering of spine-like denticles, presumably for protection.

Angel Sharks

The 13 species of angel sharks (Squatinidae) appear to be intermediates between squaliform sharks and rays. They are flattened like rays, yet the large pectoral fins are not attached to the head. They have large spiracles on top of the head, but the five gill openings are as much lateral as ventral. In addition, the mouth is terminal rather than ventral.

Like rays, they lack anal fins and have two small dorsal fins on the caudal region of the body. In most other details, however, they bear little resemblance to the rays and have many characteristics in common with "standard" sharks as well as many characteristics of their own. The flattened profile and blunt heads of angel sharks qualify them as bottom lurkers that ambush large prey (Compagno 1990). Their mouths, armed with spike-like teeth, are highly protractible and expandable, so they can suddenly inhale unsuspecting passing fish of surprisingly large size.

Saw Sharks and Sawfishes

Another squaleid order characterized by modified external denticles is the Pristiophoriformes, the saw sharks. These fish have teeth attached to their snout, which is extended as a long, flat blade. In this respect, they are like the true sawfishes (Pristiformes), but the peculiar snout seems to have evolved independently in the two groups. In the saw sharks, the teeth on the blade are unequal in size and rather weakly attached; in the sawfishes, the teeth are all about the same size and are firmly held by sockets. Both forms, however, have flattened heads, although the gill openings are lateral on the saw sharks and ventral on the sawfishes. In both saw sharks and sawfishes, the "saw" is used for slashing through schools of fish, after which they return to devour the pieces and other incapacitated fishes. The present classification system reflects recent thinking that similarities between the two forms are more than just coincidental and that saw sharks are closely related to the skates and rays (Nelson 1994).

Skates and Rays

The 450 to 500 species of skates and rays are clearly related to one another and can be placed in a taxonomic subdivision (McEachran et al. 1996, Gill and Mooi 2002). The groups called orders in this textbook are also generally recognized as natural groupings within the larger group. What unites all skates and rays is their basic design for bottom living, even if a number of species have developed the ability to "fly" through the water with their enlarged pectoral fins and have become nektonic. Even the bottom-dwelling forms are surprisingly diverse, although all skates and rays are characterized by ventral gill openings, enlarged pectoral fins that attach to the side of the head, lack of an anal fin, eyes and spiracles located on top of the head, and pavement-like teeth. The skates (family Rajidae) are the most species-rich family and are often extremely abundant, especially in water less than 1,000 m deep. The tail of skates is very slender, usually without a caudal fin, and contains weak electric organs. It is common to find several morphologically similar species of skates, especially of the genus *Raja*, occurring together. These species segregate by subtle ecological factors, such as temperatures and depth preferences, and by feeding habits, which are reflected in specialized dentition. Closely related to the skates are the guitarfishes (family Rhinobatidae), which seem to have reinvented a shark-like body. Most of the guitarfishes are small, shallow-water feeders on small crustaceans, although one species may exceed 3 m in length.

Perhaps the most remarkable of the skates and rays are the electric rays (Torpediniformes), which can deliver as much as 200 V from electric organs in their heads. Although electric rays are rather flabby, they use electric discharges to stun the active fish that make up most of their prey (Bray and Hixon 1978). The eyes of all electric rays are quite small, however, and a number of narcid rays are blind. Therefore, it is quite possible that they also use electricity for navigation. Another group of rays with formidable weapons are stingrays and bat rays (Myliobatiformes). Their sting is a spine modified from a placoid scale with a venom gland at its base. The spine is whipped about by the tail, and because stingrays feed mostly on crustaceans and other invertebrates, it must be used solely for defense against large predators, including other sharks. The eagle rays (Myliobatidae) are closely related to the classic stingrays (Dasyatidae), and a number of species have venomous spines. The eagle rays fly through the water in schools, flapping their powerful pectoral fins. They have powerful jaws and teeth for crushing mollusks and appear to move around in search of concentrations of invertebrates. A group of eagle rays that are adapted for plankton feeding are the manta and devil rays. These rays are distinguished by large, scoop-like appendages on the head that direct the plankton into the mouth. Like whale and basking sharks, manta and devil rays have fine gill rakers for straining out the plankton. The mantas are also the largest of the skates and rays, with some achieving a pectoral fin span in excess of 6 m and weights of more than 1,360 kg.

15.6   HOLOCEPHALI

The chimaeras (subclass Holocephali) have had a long evolutionary history independent of the elasmobranchs, and their anatomy reflects this (see Chapter 13). The chimaeras are commonly known as ratfish because of their long, slender tail and large, pointed head (Fig. 15.3). This body shape is convergent to that of teleosts (e.g., Macrouridae) that also live on the bottom in deep water. Approximately 30 species of chimaeras are known, most of them in the family Chimaeridae. All apparently feed on bottom-dwelling crustaceans and mollusks and, consequently, have well-developed, pavement-like teeth for crushing (which are not replaced on a regular basis). For defense, all have a spine in front of the dorsal fin with an associated venom gland. All lay large-yolked eggs with leathery coverings.

The chimaeras have many anatomical peculiarities (Lund and Grogan 1997). For example: (1) The gills are nested together under a gill flap resembling the operculum of bony fishes; (2) most water for respiration is taken in through the nostrils; (3) there is just one mobile spine, in front of the dorsal fin; (4) the upper jaws are fused to the cranium; and (5) the placoid scales occur in only a few patches. Males have complex, spiny claspers on the pelvic fins and an additional, highly mobile clasper on the head. How these structures are used in reproduction is not known.

15.7   LESSONS FROM THE CHONDRICHTHYES

This chapter contains three big lessons. First, the Chondrichthyes, when contrasted with the Osteichthyes, represent a clear alternative in fish design. They have hit on many independent solutions to individual problems such as swimming, buoyancy, sensory perception, capturing prey, or being a flatfish. The modern Chondrichthyes are highly derived species that are primarily top carnivores in the oceans. Second, the presence of a few small species of sharks and rays and of a few freshwater species indicates that the Chondrichthyes have broad morphological and ecological limits, yet the fact that most species are fairly large and marine suggests their diversity is somehow limited by interactions with bony fishes. Third, the Chondrichthyes have a highly successful lifestyle based on long life, low reproductive rates, and high survival rates. This lifestyle makes them highly vulnerable to human predation and other activities, so their abundance and diversity are rapidly being diminished.

SUPPLEMENTAL READINGS

Castro 1983; Compagno 1973, 1990; Hamlett 1999; Klimley and Chinley 1996; Klimley 2003; Moss 1984; Springer and Gold 1989; Wourms and Demski 1993.

WEB CONNECTIONS

GENERAL
www.calacademy.org/research/ichthyology
www.fishbase.org
www2.biology.ualberta/jackson.hp/IWR
www.flmnh.ufl.edu/fish/default.htm
www.people.clemson.edu/~jwfoltz/wfb418

CHONDRICHTHYES
www.ncf.carleton.ca/~b2050/homepage
www.flmnh.ufl.edu/fish/sharks

EGG CAPSULES
www.rajidae.ftmfweb.nl

CHIMAERAS
www.reefquest.com/shark_profiles/chimaera.htm

CHAPTER 16
Relict Bony Fishes

CLASS OSTEICHTHYES
  Subclass Sarcopterygii
    Infraclass Actinistia (Coelacanthimorpha)
      Order Coelacanthiformes
        Family Latimeriidae (gombessa or coelacanth)
    Infraclass Dipnomorpha (Dipnoi)
      Order Ceratodontiformes
        Family Ceratodontidae (Australian lungfishes)
      Order Lepidosireniformes
        **Families Lepidosirenidae (South American lungfishes), Protopteridae
          (African lungfishes)**
  Subclass Actinopterygii
    Infraclass Chondrostei
      Order Polypteriformes
        Family Polypteridae (bichirs)
      Order Acipenseriformes
        Families Acipenseridae (sturgeons), Polyodontidae (paddlefishes)
    Infraclass Neopterygii
      Division Ginglymodi
        Order Lepisosteiformes
          Family Lepisosteidae (gars)
      Division Halecomorphi
        Order Amiiformes
          Family Amiidae (bowfin)
      Division Teleostei

The bony fishes have a long and complex evolutionary history, yet the world's fish are dominated by one group, the teleosts (Teleostei). Other groups of bony fishes have had similar heydays in the past, but only a few scattered descendants ($\approx$50 species) of these past groups persist in the modern fauna. Fortunately for students of evolution, the relict fishes represent many of the major lines of bony fish evolution. The relationships among the groups are highly controversial, as are their relationships to fossil forms and to tetrapods (see Chapter 14) (Fig. 16.1). One of the most interesting questions to ask, however, is how did these relict fishes manage to survive after most of their relatives perished?

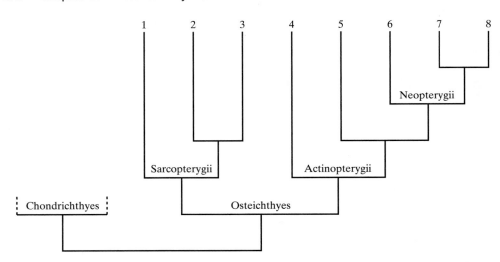

FIGURE 16.1   Phylogenetic relationships among relict bony fishes. **1.** Actinistia. **2.** Dipnomorpha. **3.** Tetrapoda. **4.** Polypteriformes. **5.** Acipenseriformes. **6.** Lepisosteiformes. **7.** Amiiformes. **8.** Teleostei.

16.1   SARCOPTERYGII

Modern lobe-finned (Sarcopterygii) fishes belong to two very different groups: the freshwater lungfishes (Dipnomorpha), and the marine coelacanths (Actinistia). The lobe-finned fishes, both living and fossil, are the fishes most closely related to the tetrapods, and modern classification systems usually place tetrapods as a subgroup of the Sarcopterygii (Clouthier and Ahlberg 1996). Within the Sarcopterygii, however, modern forms are so derived and so evolutionarily distant from the tetrapods that arguments will probably continue indefinitely about which forms are our closest living piscine relatives. Gorr and Kleinschmidt (1993), for example, analyzed the structure of the hemoglobin molecule in various vertebrates and concluded that "there is no reason to doubt that the coelacanth is the closest living relative of the tetrapods" (p. 81). On the other hand, Clouthier and Ahlberg (1996) argue that, based on fossil evidence, lungfishes and tetrapods form a natural group. This position, which is now widely accepted (e.g., Maisey 1996, Lévêque 1997, Gill and Mooi 2001), is supported by DNA studies as well (Meyer and Wilson 1990). The most obvious characteristics that unite sarcopterygians with tetrapod ancestors are the presence of lobed, paired fins, and lungs. The lobed fins resemble the limbs of tetrapods; they are supported by internal skeletal elements that attach to pelvic and/or pectoral girdles.

16.2   DIPNOMORPHA (DIPNOI)

Modern lungfishes are in three distinct families: Ceratodontidae of Australia, Lepidosirenidae of South America, and Protopteridae of Africa. Lungfishes are readily recognized by their elongate bodies and continuous rear fins that encompass dorsal, caudal, and anal fins (Bemis et al. 1986). The Australian lungfish *(Neoceratodus forsteri)* most resembles the fossil forms, with its flipper-like fins, large scales, unpaired lung, and compressed body. In contrast, the South American *(Lepidosiren paradoxa)* and African

(*Protopterus,* four species) lungfishes have fins that are reduced to filaments, small scales, paired lungs, and eel-like bodies. The Australian lungfish is native only to the Mary and Burnett rivers in Queensland but has been introduced into four other rivers as well (Berra 2001). These rivers have low flows during the summer, and the lungfish live in the deep pools that remain. Although their lung is functional, even during the dry season the Australian lungfish relies primarily on gill respiration; their lungs supplement the gills during times of stress. The South American and African lungfishes live in swamps that are likely to dry up, and they breathe air to survive. As the waters recede in their swamps, they dig burrows into the mud in which they can survive until the rains come. The African lungfishes undergo true estivation, which includes drastic metabolic changes. They secrete large amounts of mucus into their water-filled burrow, which then dries to form a lining and results in a hard, impermeable cocoon. Each burrow is connected to the surface by a narrow tube plugged with mud except for a small hole. When the rains come, the plug dissolves, water enters the lung of the fish, and it awakens by coughing. They can estivate in these burrows for as long as 4 years (Berra 2001).

During estivation, oxygen consumption by African lungfish is reduced to extraordinarily low levels, and what metabolic energy is needed is derived entirely from the breakdown of protein. Estivating lungfish live by metabolizing their muscles. During this process, carbohydrate (glycogen) reserves actually build up and, consequently, are available for immediate use when the lungfish emerges from the cocoon. Another substance that builds up during estivation is urea (see Chapter 7). Urea may concentrate in tissues to seven times its normal level, far beyond the levels that are toxic to most other bony fishes. Once the lungfish leave the cocoon, the urea is rapidly excreted, and ammonia becomes the main nitrogenous waste product.

The African and South American lungfishes hollow out nests in mud banks or bottoms. The female then lays her eggs in the nest, and the male guards them. The newly hatched lungfish resemble salamander tadpoles in that they have external gills. Such gills, as well as the nesting behavior, are not characteristic of Australian lungfish, which lay their eggs on aquatic plants.

All lungfishes have dermal tooth plates. These tend to be flattened in fossil forms, apparently for crushing and grinding, but are more blade-like in living lungfishes. The South American and Australian lungfishes are omnivorous, and plant material is an important part of their diet. African lungfishes are more carnivorous, preying on mollusks, crustaceans, and fish. On this diet, some species may exceed 2 m in length, although fish longer than 1 m are seldom encountered. Australian lungfish grow to more than 1.5 m and 40 kg (Berra 2001).

16.3 ACTINISTIA: COELACANTHS

Probably no single event in the history of ichthyology has received more public attention than the discovery of the first living coelacanth *(Latimeria chalumnae)* in 1938. The coelacanth excited the public imagination because coelacanths were previously known only from fossils (the most recent being ≈70 million years old), and the new fish would supposedly reveal much about the evolutionary transition between fish and tetrapods. It was also decidedly primitive looking, with its large size (to 2 m), gaping mouth, limb-like fins, and large scales. The coelacanth brought instant fame to the

South African ichthyologist J. L. B. Smith, who described it, and a certain celebrity—in scientific circles anyway—to Ms. M. Courtenay-Latimer (hence *Latimeria*). Latimer was the curator of a small, local museum who recognized the significance of the fish brought to her by a local fisherman and so notified Smith. Unfortunately, a second specimen was not obtained until 1952. It had been caught nearly 3,000 km north of the first collection locality, in the deep waters around the Comores Islands (Box 16.1). It is common enough that the local fishermen have a name for it (gombessa), although the total population in the area is small enough that a sustained fishery for specimens may be seriously depleting the population (Fricke 2001).

Since 1952, considerable research has been conducted on the Comores coelacanth, although new discoveries about its biology are still being made (Musick et al. 1991). For example, the coelacanth is ovoviviparous (i.e., retains the eggs in the body cavity, where the young hatch and develop), which was not discovered until 1975 (C. L. Smith et al. 1975). Initially, the developing young depend on nutrients in the large yolk sac. When the yolk material becomes depleted, however, the yolk sac apparently becomes a primitive, unattached umbilicus, because it is surrounded by uterine tissue of the female, thereby making transfer of nutrients possible (Wourms et al. 1991). Evidence also indicates that the late embryos feed in the uterus on unfertilized eggs produced by the female (Wourms et al. 1991). In 1987, the first observations on living coelacanths in their natural habitat were made using a small submarine. These observations revealed that coelacanths do not "walk" on the bottom with their lobed fins, as had been commonly supposed (Fricke et al. 1991). Instead, when feeding (at night), they drift in currents

---

BOX 16.1
The Second Coelacanth

The ability of coelacanths to capture headlines was demonstrated in 1998, when the world press carried stories about the unexpected discovery of a second species in Indonesia. This species was first spotted (and photographed) in a fish market on the island of Manado in 1997 by Mark Erdmann, his wife, and friends, but Erdmann had to wait another 10 months before acquiring a specimen from fishermen (Erdmann 2000). He discovered that coelacanths were caught only by fishermen who targeted deepwater fishes and who called it Raja Laut, "King of the Sea." Erdmann decided to delay announcing his discovery and its formal description until the Indonesian government instituted conservation measures, which it did with surprising rapidity (Erdmann 2000). After the initial announcement (Erdmann et al. 1998), genetic studies indicated the Indonesian coelacanth had long been isolated from the Comores Islands species and deserved species status. Unfortunately, before Erdmann and his Indonesian colleagues could publish the formal description, a French scientist unethically published a quick description with the name, *Latimeria menadoensis*. Despite the sloppy and incomplete description, under the strict rules of scientific nomenclature the name *L. menadoensis* sticks to the fish (Erdmann 2000). Today, the Manado coelacanth is fully protected under Indonesian law and has inspired a trade on the island involving souvenirs featuring coelacanths.

close to the bottom, with the head pointing downward at an angle and the paired fins used for stabilizers. During the day, they aggregate in caves, resting quietly. Most coelacanths live at depths of 150 to 400 m, and the small number of resting caves at such depths may be a factor limiting their populations (Fricke et al. 1991, Fricke 2001). The total population around the Comores Islands is probably less than 400 individuals (Fricke 2001). They feed on large bottom fishes, including small sharks, which they apparently detect with a special electric organ in the snout.

Coelacanths have a number of interesting anatomical adaptations. The bones of the paired fins articulate with the pelvic and pectoral girdles, much like those of tetrapods. The skeleton is made of both bone and cartilage, but the vertebral column is essentially a notochord constricted by cartilaginous neural and haemal arches (this is true of lungfishes as well). The fin "spines" are also cartilaginous and hollow, hence the name coelacanth (*coel*, "hollow"; *acanth*, "spine"). Bone, however, is present in the large, cosmoid scales that cover the body. The skull and jaws are also partially bony and allow the ingestion of large, active prey. Unlike other living fishes, the coelacanth has an intracranial joint in which the notocord functions as a hinge that allows the cranium to flex dorsally when the jaws are opened. In addition, the gape of the mouth is very wide, and the head can be moved somewhat independently from the trunk (Thomson 1973). Other unusual aspects of coelacanth biology include its osmoregulatory strategy and its fat-filled swimbladder. Coelacanths are nearly isosmotic with sea water, maintaining high concentrations of urea in the blood much like sharks and rays. The fat-filled swimbladder, combined with the reduced skeleton, presumably allows the coelacanth to be neutrally buoyant without the limitations of vertical movement imposed by a gas-filled swimbladder.

16.4   ACTINOPTERYGII

The ray-finned (Actinopterygii) fishes are the dominant fishes on the planet, with 42 orders, 431 families, and nearly 24,000 species (Nelson 1994). They have evolved into so many specialized forms that common structural characters are difficult to find; thus, the group is easiest to define by the absence of key characters of other major groups. The vast majority of actinopterygians are teleosts, but the structure and ecology of the few other species provide some clues as to what the ancestral fish were like. The forms with the most ancestral characteristics are placed in the Chondrostei, the name reflecting the largely cartilaginous skeletons of modern forms: the bichirs (Polypteriformes), and the sturgeons and paddlefishes (Acipenseriformes). The rest of the modern actinopterygians are placed in the Neopterygii ("new-fins"). Within the neopterygians are two groups with many ancestral characters and distant relationships to the dominant teleosts: the gars (Lepisosteiformes), and the bowfin (Amiiformes).

16.5   POLYPTERIFORMES

The bichirs (*Polyterus* spp.) and reedfish (*Calamoichthys calabaricus*) share a common ancestry with the early chondrosteans, although they superficially resemble lobe-finned fishes. For example, in common with lungfishes and coelacanths, they possess lobed pectoral fins, although the distinctive supporting structures evolved independently. In common with chondrosteans, bichirs possess spiracles, maxillary (upper jaw) bones firm-ly united to the skull, and ganoid scales. In common with both groups, bichirs possess such ancestral features as a spiral valve intestine, paired lungs that attach ventrally to the gut, and a heterocercal tail. In most features, however, the bichirs are quite distinct. The body is elongate, flexible, but covered with a heavy armor of scales. The dorsal fin con-sists of 5 to 18 separate finlets, each supported by a single spine. The bichirs normally respire through gills but can breathe air when the oxygen content of the water becomes low. The larvae are also adapted for low oxygen levels, possessing external gills. Most features of the bichirs are adaptations for being predators of fishes and invertebrates in the swamps and rivers of tropical Africa (see Box 1.1). Although locally they may be abundant enough to be collected for the aquarium trade, only 10 species exist (Fig. 16.2).

16.6 ACIPENSERIFORMES

The sturgeons and paddlefishes are large fish with cartilaginous skeletons, which is a cu-rious (and secondary) condition for our best representatives of ancient bony fishes. They also have a heterocercal tail, a spiral valve intestine, an upper jaw that does not articu-late with the cranium, only one branchiostegal ray, fin rays that are more numerous than the supporting basal elements, and a notochord that persists in adults. Otherwise, the sturgeons (Acipenseridae) and paddlefishes (Polyodontidae) are morphologically quite different from each other and probably are products of long-separate evolutionary lines. Fossil fishes of this order are known from the Triassic period of Europe, so the distribu-tion of the 27 species (in scattered large temperate and arctic rivers of Eurasia and North America) presumably reflects their Mesozoic distribution on the ancient northern continent (Bemis and Kynard 1997).

   Sturgeons (25 species) are distinguished by five rows of bony scutes on the body that represent the remnants of ganoid scales. They have a highly protrusible, bottom-oriented mouth that is preceded by four barbels (Fig. 16.3). The scutes are presumably protective in nature, because they are proportionally larger and sharper on small sturgeon than they are on adult sturgeon (Findeis 1997). The protrusible, toothless mouth was "in-vented" independently from the analogous mouth of the teleosts and, apparently, devel-oped along with the cartilaginous skeleton. The loss of bone allowed elements of the hard mouth of their chondrostean ancestors to become free to support a sucking mouth (Bemis et al. 1997). Retention and processing of food are accomplished by a group of flat cartilages (*palatal complex*) on the roof of the mouth that move across a tongue pad with sharp ridges on it, breaking up the food. When actively engaged in feeding or when processing large prey, sturgeons have the unique ability to pump water for respiration

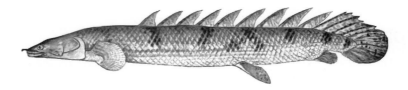

FIGURE 16.2 A bichir (*Polypterus*). After Boulenger (1907).

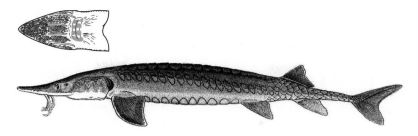

FIGURE 16.3 Shovelnose sturgeon (*Scaphirhynchus platorynchus*) from the Missouri River, showing the head with opercular-opening exposed gills, protrusible mouth, and barbels that enable it to feed effectively on benthic invertebrates. Drawing by Paul Vesci.

downward across the gills from an opening at the top of the operculum. Otherwise, they use a "two-pump" system like that of teleosts, with water taken in through the mouth.

With the combination of a strong and rounded body, flattened snout, and specialized mouth, sturgeons can be characterized as "benthic cruisers" (Findeis 1997). They feed by finding patches of prey and then sucking up the organisms from the bottom. Their prey is detected by their highly sensitive barbels and ampullary (electric) organs located on the underside of the flat-bottomed snout. Although various types of garbage found in their stomachs indicate that sturgeons feed, in part, by scavenging, most of their diet is benthic invertebrates, with fish becoming increasingly important prey as sturgeon grow larger. Most of the fish they eat are probably captured at night, when it is more difficult for the prey to avoid being sucked up by a large sturgeon. The shovelnose sturgeons *(Scaphirhynchus)* use their flat snouts for stirring up the bottom to expose the invertebrates on which they feed. Shovelnose sturgeons also have hard, curved "spines" on the leading edge of their pectoral fins that they can use like legs to drag themselves over the bottom while foraging. In other sturgeons, the spines are flat and apparently serve as stabilizers for depth control in swift currents.

Sturgeons (and paddlefishes) all make long upstream migrations for spawning—even species that never leave fresh water (Bemis et al. 1997). The longest migrations known are for beluga sturgeon *(Huso huso)*, which once moved from salt water 2,500 km up the Danube and Volga rivers. Those sturgeon that escape fishermen spawn in swift, gravel-bottomed areas of large rivers. There is no parental care.

Sturgeons are the largest fish found in fresh water, although the largest species are anadromous. The beluga sturgeon, the main source of true Russian caviar, has been found to achieve lengths of 8.5 m and weights of nearly 1,300 kg. In North America, apparently authentic records exist of white sturgeon *(Acipenser transmontanus)* growing to lengths of approximately 4 m and weights of 590 kg. Sturgeons have presumably been able to persist by adopting a life-history strategy combining the large size, slow growth, and long life of sharks with the high fecundity and small egg size of teleosts. Because of their extreme value, large size, and vulnerability to both pollution and overfishing, all sturgeon species are threatened with extinction in at least part of their range, and most are threatened in *all* of their range. Their value and reproductive habits, however, do make it possible to culture them successfully, and sturgeon farms now exist in several places around the world (Billard and Lecointre 2001).

Paddlefishes possess a long, paddle-like snout covered with thousands of passive electroreceptors (like the ampullae of Lorenzini in sharks) that detect patches of zooplankton and other prey by sensing electrical fields (Russell et al. 1999). This allows them to feed both at night and in turbid water on small prey. Their skin is smooth and scaleless, except for a small patch on the caudal peduncle. Like some sturgeons, they possess a small spiracle, which is covered with a greatly elongated operculum. Two species of paddlefish are known. One lives in the Mississippi River system and achieves lengths of approximately 2 m; the other lives in the Yangtze River of China and may reach 5 to 7 m in length. North American paddlefish feed on zooplankton, which they capture by swimming through the water with the mouth agape and filtering out the plankton with their many fine, elongate gill rakers. These paddlefish have a rigid mouth and obtain oxygen through ram ventilation of the gills; they must keep swimming to stay alive. In contrast, Chinese paddlefish feed on fish and benthic invertebrates and have a much more flexible jaw. Both species are declining in numbers because of habitat changes and overexploitation, but the Chinese paddlefish is close to extinction as the result of changes to the river from dams (Berra 2001).

16.7 LEPISOSTEIFORMES

The seven species of gars are among the most distinctive freshwater fishes. Their cylindrical bodies are covered with hard, nonoverlapping, diamond-shaped ganoid scales, and their hard, bony heads have long snouts with sharp, conspicuous teeth. The long body and snout, coupled with the placement of the dorsal and anal fins near the tail, make gars superb lie-in-wait predators (Fig. 16.4). Passing fish are seized sideways during a sudden dash in ambush by the gar, held firmly with the teeth, and eventually, turned around and swallowed whole. The heavy armor that gars carry with them as protection against predators is made possible by large swimbladders, which make gars neutrally buoyant. The swimbladders are also used for air breathing. Gars are not obligate air breathers but become increasingly dependent on the lung for oxygen as water temperatures increase and the dissolved oxygen level in the water decreases (Renfro and Hill 1970). Aside from the large, well-vascularized swimbladders, the most distinctive internal features of gars are the spiral valve intestine and the vertebral column with unique **opisthocoelous** vertebra; each vertebral centrum has a convex anterior surface and a concave posterior surface, much like that of reptiles.

The embryos of gars are often laid on aquatic plants, to which they adhere. Gar embryos have long been thought to be toxic to protect against predation; in fact, potential fish predators readily eat the embryos and will grow on an exclusive diet of gar eggs (Ostrand et al. 1996). The young cling to the plant stems with an adhesive disk on the head until the yolk sac is absorbed and they can swim actively. Most of the seven species reach a maximum length of 1 to 2 m, but the alligator gar *(Atractosteus spatula)* can exceed 3 m. Gars are relict species that are largely confined to the Mississippi River drainage. One species is found in coastal drainages down to Costa Rica, however, and another is found in Cuba.

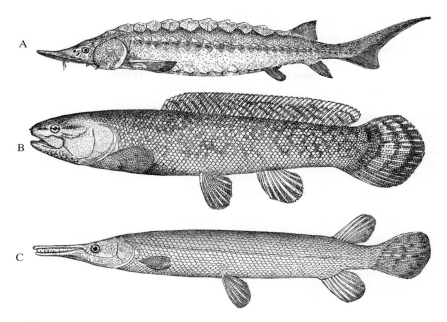

FIGURE 16.4 Three relict fishes **A.** Atlantic sturgeon, (*Acipenser oxyrhynchus*); **B.** Bowfin (*Amia calva*); and **C.** Alligator gar (*Atractosteus spatula*). From Jordan and Evermann (1900).

16.8 AMIIFORMES

The bowfin *(Amia calva)* is yet another ancient fish of the Mississippi River drainage, the last survivor of a once abundant group that coexisted with the dinosaurs (Fig. 16.4). With its cycloid scales and functionally homocercal tail, it is similar in appearance to teleosts. Its many sharp teeth, 10 to 13 branchiostegal rays, vestigial spiral valve intestine, stout body, solid jaws and head, and large lung, however, also make it seem distinctly archaic. One of its distinguishing features is the bony gular plate under the lower jaw. Like teleosts, the bowfin has vertebrae that are concave at both ends *(amphicoelous)*. The skeleton is bone. The large lung is definitely advantageous to the bowfin in the warm summer waters where it normally occurs. It is not an obligate air breather, but once the water temperatures exceed approximately 10°C, it supplements oxygen obtained through the gills with oxygen absorbed through the lung. The rate of air breathing increases with increasing temperature (Horn and Riggs 1973). On occasion, bowfin can apparently survive entirely by air breathing while lying torpid in muddy burrows during periods of drought (Ross 2000). Some evidence indicates that during such periods, they, like lungfish, actually may metabolize muscle rather than fat. Although they cannot accumulate urea in their body tissues, they can reduce water demand by excreting urea and uric acid.

Bowfin grow to more than a meter in length, and adults are piscivorous, making them a good game fish (although rarely sought). Males build nests and aggressively guard the embryos and young (Ross 2000). The juveniles have a distinct eye-spot (*ocellus*) on the caudal peduncle, which may inhibit predation by other bowfin; spawning males also have a distinct ocellus.

16.9   LESSONS FROM RELICT BONY FISHES

These holdovers from fish faunas of earlier eras have persisted because of specializations that allow them to coexist with—and frequently to prey upon—the more derived teleosts. The sturgeons in particular are specialized bottom cruisers that have independently developed characters regarded as being important to the success of teleost fishes. Most relict fishes are large in size. Many have heavy armor, and many are air breathers. That they have more than held their own is indicated by the abundance of gars, bowfin, paddlefish, and sturgeons in the Mississippi River drainage, a river system where more "modern" fishes are abundant and diverse. These forms have persisted for millions of years; however, the survival of some of them for the next hundred years is probematical.

SUPPLEMENTAL READINGS

Bemis et al. 1986; Billard and Lecointre 2001; Birstein et al. 1997; McCosker and Lagios 1979; Musick et al. 1991; Schaeffer 1973; J. L. B. Smith 1956; Thomson 1973.

WEB CONNECTIONS

GENERAL
www.calacademy.org/research/ichthyology
www.fishbase.org
www2.biology.ualberta/jackson.hp/IWR
www.flmnh.ufl.edu/fish/default.htm
www.people.clemson.edu/~jwfoltz/wfb418

RELICT FISHES
www.palaeos.com/Vertebrates/Units/090Teleostomi/
www.gen.umn.edu/faculty_staff/hatch/fishes/natural_history.html

COELACANTH
www.salab.ru.ac.za/coelacanth

PADDLEFISH
www.umesc.usgs.gov/aquatic/fish/paddlefish/main.html
www.palaeos.com/Vertebrates/Units/090Teleostomi/

CHAPTER 17
Bonytongues, Eels, and Herrings

SUBDIVISION TELEOSTEI
  Infradivision Osteoglossomorpha
    Order Osteoglossiformes
      Families Osteoglossidae (bonytongues), Pantodontidae (butterflyfishes),
        Hiodontidae (mooneyes), Notopteridae (knifefishes), Mormyridae (elephant-
        fishes), Gymnarchidae
  Infradivision Elopomorpha
    Order Elopiformes
      Families Elopidae (tenpounders), Megalopidae (tarpons)
    Order Albuliformes
      Families Albulidae (bonefishes), Halosauridae (halosaurs), Notacanthidae
        (spiny eels)
    Order Anguilliformes
      Families Anguillidae (freshwater eels), Heterenchelyidae, Moringuidae (spaghet-
        ti eels), Chlopsidae (false morays), Myrocongridae, Muraenidae (moray eels),
        Synaphobranchidae (cutthroat eels), Ophichthidae (snake eels), Colocongri-
        dae, Derichthyidae (longneck eels), Muraenesocidae (pike congers), Ne-
        michthyidae (snipe eels), Congridae (conger eels), Nettastomidae (duckbill
        eels), Serrivomeridae (sawtooth eels)
    Order Saccopharyngiformes
      Families Cyematidae (bobtail snipe eels), Saccopharyngidae (swallowers),
        Eurypharyngidae (gulpers), Monognathidae (one jaw eels)
  Infradivision Clupeomorpha
    Order Clupeiformes
      Families Denticipitidae (denticle herring), Engraulidae (anchovies), Pristigasteri-
        dae, Chirocentridae (wolf herring), Clupeidae (herrings), Sundasalangidae
        (Sundaland noodlefishes)

17.1   TELEOSTEI

Based on sheer numbers of species and individuals, the Teleostei are the main line of fish evolution. According to Nelson (1994), this group contains 42 orders, 431 families, and 23,681 species. It is a successful group because of a package of adaptations that have improved respiration, feeding, buoyancy, swimming, and reproduction (see Chapter 13).

The group is defined by the homocercal tail; *Teleostei* means "end-bone," referring to the specialized bones (uroneurals) at the end of the vertebral column that support the symmetrical caudal fin. The success of the teleosts overall has depended on the integrated design package, but the success of evolutionary lines within the teleosts has depended on specializations of various sorts. These include specializations that have resulted in the abandonment of what seem to be basic teleost traits, such as the "two-pump" respiratory system or the homocercal tail. The importance of specialization is reflected in the three major lines of teleost evolution that are covered in this chapter: (1) the bonytongues (Osteoglossomorpha); (2) the eels and tarpons (Elopomorpha); and (3) the herrings (Clupeomorpha). Together, these three groups contain approximately 1,400 species (roughly 6% of all teleosts) and include some of the most abundant fish in the world (herrings and anchovies); however, they are considered to be relict or specialized offshoots of the main line of teleost evolution (the Euteleostei) (Fig. 17.1).

17.2   OSTEOGLOSSOMORPHA

The osteoglossomorphs are an unusual group of approximately 220 species of freshwater fish, most of which (≈200 species) are in one African family, the Mormyridae. The rest of these species are scattered about the continents and generally are considered to be relics of a once much more abundant group. The relationships of the osteoglossomorphs to other fish groups are poorly understood, presumably because the osteoglossomorphs are so ancient. Their osteology is quite distinctive and ties the members of the Osteoglossomorpha together, if somewhat tenuously. Osteoglossomorphs *(osteoglosso,* "bony-tongue") have most of their teeth located on the "tongue" (actually basihyal and

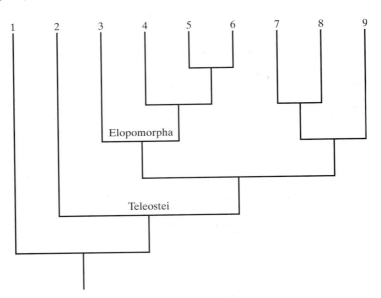

Figure 17.1   Phylogenetic relationships of Osteoglossomorpha (**2**), Elopomorpha (**3–6**) and Clupeomorpa (**7**). **1.** Amiiformes. **2.** Osteoglossiformes. **3.** Elopiformes. **4.** Albuliformes. **5.** Anguilliformes. **6.** Saccopharyngiformes. **7.** Clupeiformes. **8.** Ostariophysi. **9.** Euteleostei.

glossohyal bones) and on the roof of the mouth (parasphenoid bone). Together, these sets of teeth are used for biting, like the jaw teeth of other teleosts. They also have a caudal fin with 16 or fewer branched rays (most bony fishes have more), an elongate body (60–100 vertebrae), no intermuscular bones on the back (epipleurals), cycloid scales with ornate microsculpturing, and an intestine that curls around to the left side of the esophagus rather than to the right as in most other bony fishes (Nelson 1994). Six living families are contained in this group, all in the order Osteoglossiformes.

Osteoglossidae

Only seven species are in this family, yet they are found in tropical areas of Africa, South America, Asia, and Australia. This distribution pattern is frequently cited as evidence for plate tectonics, because fossils for this family are found as far back as the Paleocene epoch. Bonytongues are conspicuous members of their local faunas, with heavy, elongate bodies that are covered with large scales (Fig. 17.2). The dorsal and anal fins are long and on the rear half of the body. All species apparently can breathe air with their lung-like swimbladders. Most of the species make their living as predators in tropical rivers. The arapaima *(Arapaima gigas)* of the Amazon River is one of the largest freshwater fish species and may reach 4.5 m (Berra 2001). The silvery arawana *(Osteoglossum bicirrosum)* occurs with the arapaima but reaches lengths of only approximately 1 m. It is characterized by two chin barbels and a remarkably large, angled mouth which can capture other fishes and small animals (including birds and bats) that fall into the water from the overhanging vegetation. It incubates its young in a special pouch in its mouth, as do the bonytongues *(Scleropages* spp.) of Australia and Asia.

The only member of the Osteoglossidae that is not predatory is *Heterotis niloticus,* of western Africa (Fig. 17.2). This large fish (to 1 m) has its fourth gill arch modified into a spiral-shaped filtering apparatus. This organ secretes mucus in which phytoplankton and small bits of organic matter are trapped and then swallowed.

Hiodontidae

The two species in this family, the mooneye *(Hiodon tergisus)* and the goldeye *(H. alosoides),* are the most "normal"-looking fish in their entire superorder, because they superficially resemble clupeid fishes (Fig. 17.2). Their most distinctive external features are their large eyes, which have bright gold irises in the goldeye and gold-silver irises in the mooneye. The gold color in both species is from a tapetum lucidum that increases their ability to see in low light (see Chapter 10). Goldeye feed mostly at night and have only rods in their retinas (no cones). The goldeye and mooneye are both found in the backwaters of the large rivers and lakes throughout the Mississippi River drainage system. They are also found in the Hudson Bay drainages and Arctic drainages (goldeye only) of Canada, where the goldeye is abundant enough to be an important commercial species. Both feed on a wide variety of prey, but they are largely piscivorous as adults (Scott and Crossman 1973).

Notopteridae

The eight species of knifefishes (featherbacks) have long, strongly compressed bodies that taper to a point and a small, feather-like dorsal fin (Roberts 1992). They swim mainly through the rhythmic movements of the long anal fin that extends from just behind the

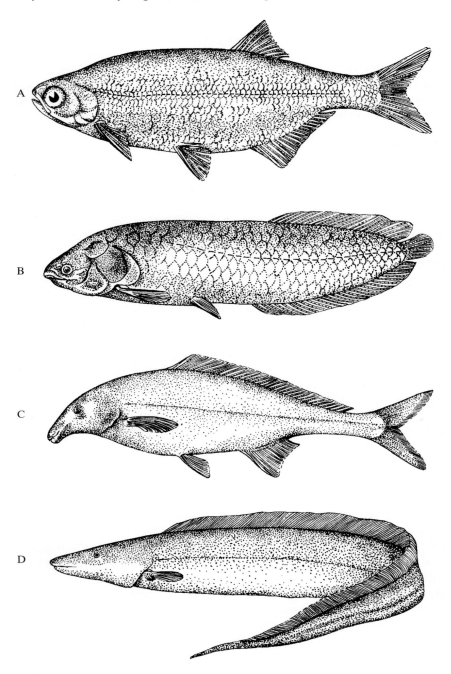

Figure 17.2 Representative osteoglossomorph fishes. **A.** Goldeye (Hiodontidae). **B.** African bonytongue (Osteoglossidae). **C.** Elephant fish (Mormyridae). **D.** Aba (Gymnarchidae). **A** from Jordan and Evermann (1900); **B–D** after Boulenger (1907).

head to the tiny caudal fin, which it joins. They appear to be able to swim forward or back-ward equally well. The swimbladder is connected to the gut and is used for air breathing and sound production. Thus, the knifefishes are extremely well adapted for living among submerged and emergent vegetation in stagnant backwaters and ponds of tropical Asia and Africa. They generally remain quietly in cover during the day, and come out to prey on invertebrates and small fish during the evening. The large species, which may approach 1 m in length, are favored food fishes in Southeast Asia. A fish with habits and morphol-ogy similar to the knifefishes is the aba (*Gymnarchus niloticus,* Gymnarchidae), a close relative of the mormyrids (Fig. 17.2). This species, however, uses its dorsal fin rather than the anal fin for propulsion. Also, like the mormyrids, it uses electricity to find its way.

Mormyridae

The elephantfishes include approximately 200 species of odd-looking fishes, including those with a trunk-like appendage on the chin or snout (Fig. 17.2). They are adapted for nocturnal living in muddy rivers and lakes in tropical Africa and are abundant in many areas. They have also successfully adapted to the reservoirs that now exist on many African rivers and contribute significantly to reservoir fisheries. A principal reason for their success is the use of electric organs to find their way, to detect prey, and to commu-nicate with each other (Möller 1995). Weak electric signals are produced by modified muscles in the caudal peduncle, and an electrical field is set up around each fish. Anything that disrupts this field can be detected and identified with a surprising (to us) degree of precision. The electric signals can be modified by the fish to communicate with other fish by changing the shape of the field, the form of the waves, the discharge frequencies, the timing of the discharges, and the pattern of stopping and starting the discharges (Hopkins 1974). Thus, the electric signals can be used in courtship, aggressive behavior, and other intraspecific encounters. Because each species has its own set of electrical patterns, recog-nition and avoidance of other species is possible. For example, African streams contain five co-occurring species of *Brienomyrus,* which are virtually identical morphologically yet can be readily distinguished by their electrical signals (Alves-Gomes 2001). The sig-nals also reveal that each species occupies a distinct microhabitat within each stream.

This capability for acute discrimination between different types of electrical sig-nals requires an extremely well-developed nervous system. The mormyrid cerebellum in particular is so large that, relative to body size, it is roughly the same size as that of hu-mans. The complexity of the mormyrid brain is also reflected in their complex behavior patterns. Aquarists have long admired mormyrids for their learning abilities and the fact that many species engage in apparent "play" behavior, which is considered to be a sign of intelligence. The play usually consists of batting around a small object with the head.

The morphology of mormyrid fishes reflects the importance of electricity as a sen-sory modality. The eyes are small, and the skin is thick. All species usually swim slowly, with their bodies rigid to avoid distorting the electrical field they are generating. They rely primarily on the tail and/or dorsal fin for propulsion. As a consequence, the tail is often deeply forked, and the caudal peduncle is very narrow. Despite the importance of the electrical system, they also possess a well-developed sense of hearing, which depends on the swimbladder for amplifying sound (Werns and Howland 1976). The swimbladder is quite small and is located mostly in the head region, where it is in contact with the inner ear. Another peculiar feature of the Mormyridae is that many of the species have proboscis-like snouts or long extensions of the lower jaw. These species presumably feed by probing the bottom mud for small invertebrates. Species without the protruding snout usually feed on plankton or small fish.

A predator on mormyrids is the large (to 1.5 m), eel-like electric fish, the aba (Fig. 17.2). The aba can turn its electric field off fairly readily, a good trick when stalking other electrical fishes. It is widely distributed across central Africa (Berra 2001).

17.3   ELOPOMORPHA

The fishes of the Elopomorpha are morphologically extremely diverse. They include the tarpons (which look like giant herrings), the streamlined bonefishes, the true eels, and the spiny eels (which look like a cross between a bonefish and a true eel). What links these diverse groups is the presence of leptocephalus larvae (Greenwood et al. 1966). These planktonic larvae are thin, transparent, and leaf-like, drifting passively with currents. The leaf-like body shape apparently allows the larvae to obtain a significant portion of their nutrition through direct absorption of dissolved organic matter (Pfeiler 1986), although eel larvae will actively feed and may be specialized for feeding on gelatinous larvaceans, *Oikopleura* spp. (Mochioka and Iwamizu 1996). Despite this seemingly basic link between the groups, the classification system used here for the Elopomorpha is far from being universally accepted by experts.

17.4   ELOPIFORMES

This order contains eight species (two families) of streamlined, predatory fish with large, cycloid scales and deeply forked tails (Fig. 17.3). The presence of leptocephalus larvae seems to link them with the eels, but the larvae are quite distinctive. Elopiform larvae have a forked tail and a distinct dorsal fin; eel larvae have a rounded "tail" that merges with an elongate dorsal fin. The members of this order are all shallow-water, tropical or subtropical marine forms that occasionally enter fresh or brackish water. Although not esteemed for eating, species such as the Atlantic tarpon *(Megalops atlanticus)* and the tenpounder *(Elops saurus)* are favorite sport fishes because of the spectacular, leaping fights they put up once hooked.

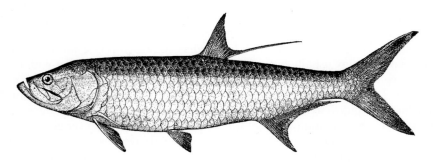

Figure 17.3   An elopiform fish, the tarpon (Megalopidae). From Jordan and Evermann (1900).

17.5   ALBULIFORMES

Only 29 species are recognized in this order, but they represent two very different groups: the bonefishes (Albulidae), and the halosaurs (Halosauridae) and spiny eels (Notacanthidae). The union of these two groups is based on obscure details in structure, because they look extremely different. The bonefishes are tarpon-like predators of shallow, tropical marine waters, whereas the 25 species of halosaurs and spiny eels are eel-like forms from the deep sea (Fig 17.4). The subterminal mouths and body structure of halosaurs and spiny eels indicate that they probably make a living by pulling small invertebrates from muddy bottoms. Although superficially similar to true eels, they differ in many important characteristics, such

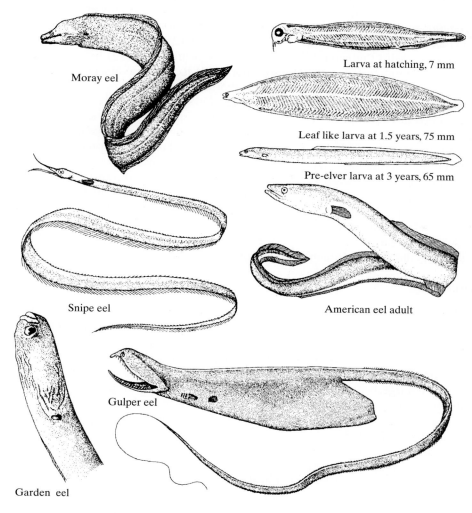

Larva at hatching, 7 mm

Leaf like larva at 1.5 years, 75 mm

Pre-elver larva at 3 years, 65 mm

Moray eel

Snipe eel

American eel adult

Gulper eel

Garden eel

Figure 17.4   Top right. Stages in the development of *anguillid* eels (Anguillidae). **Left, top to bottom.** Moray eel (Muraenidae), snipe eel (Nemichthyidae), garden eel (Congridae), and gulper eel (Eurypharyngidae). From Moyle (1993); © 1993, Chris Mari Van Dyck.

as the presence of both pelvic and pectoral fins, the flexible jaw structure, and the elongate anal fin, which merges with the caudal fin (if one is present) (Fig 17.4). One of the most striking aspects of spiny eels is that the leptocephalus larvae may be quite large (to 180 cm) but then metamorphose into much smaller juveniles (D. G. Smith 1970).

17.6   ANGUILLIFORMES

The diversity of eels is not widely appreciated. Their public image has been formed largely by the culinary and sporting qualities of freshwater eels and by lurid descriptions of fierce moray eels attacking divers on coral reefs. Their image is not helped by the fact that most species are quite secretive in their habits. In addition, many are quite rare, and superficially, not much variation is found in eel morphology. Eels are elongate fishes that: (1) lack pelvic fins (and often pectorals as well), (2) have dorsal and anal fins that are continuous with the caudal fin (and, hence, a pointed tail), (3) have cycloid scales that are deeply imbedded in the skin or absent, (4) lack gill rakers, and (5) have a reduced skeleton. They also share many other characteristics.

Despite this seeming uniformity, more than 730 species of eels, belonging to 15 families, are known. These species are found in a wide variety of habitats, from freshwater lakes and streams to coral reefs to the deep sea, although most eels live in shallow tropical or subtropical marine habitats. All, however, have planktonic leptocephalus larvae, and in temperate regions, eels of many species make extensive migrations to lay their eggs in places where oceanic currents will favor the growth, survival, and return "home" of their larvae. McCleave (1993), for example, documents the remarkable ability of different species of eels to spawn in different areas of the Sargasso Sea to take advantage of the various currents generated in the Subtropical Convergence Zone.

An idea of the diversity of eels can be obtained by briefly examining four of the families: (1) Anguillidae, (2) Muraenidae, (3) Congridae, and (4) Ophichthidae (Fig. 17.4).

Anguillidae

Although only 15 species are in this family, it is the best known of the various eel families, because most of the species live in fresh water and spawn in the ocean (Box 17.1). They are "typical" elongate eels, however, in that each has a wedge-shaped head with a hard-rimmed mouth that lacks the maxilla and premaxilla. The head also has gill openings that are far back and are covered with a small, flexible operculum. The branchial cavity is quite large, because the eels breathe mostly by "swallowing" water and moving it past the gills in pulses. The shape of the head greatly restricts the gape of an eel's mouth, and hence, the size of prey it can swallow whole. To eat a larger prey item, an eel grabs onto it and then spins around rapidly, until a chunk breaks off that can be swallowed (Helfman and Clark 1986). All these features make eels admirably suited for living secretive lives in crevices or burrows or among stalks of aquatic plants in which they can seek out equally secretive prey or ambush passing fishes.

In the eastern United States, the American eel *(Anguilla rostrata)* is an important (if declining) predator in many lakes and streams, and the closely related *A. anguilla* plays a similar role in European waters. After spending 6 to 12 years in these habitats and growing to approximately 35 to 150 cm, both species transform from being cryptically

BOX 17.1

Are All Anguillid Eels Catadromous?

Whenever catadromy is discussed, as in Chapter 11, anguillid eels are used as the standard example. All these eels live in fresh water and spawn in the ocean, but this life-history pattern is not as fixed as was once thought. Tsukamoto et al. (1998) discovered that some adults of known catadromous species from both the Pacific and Atlantic oceans apparently never enter fresh water. This was revealed through an analysis of the chemistry of otoliths; Eels rearing in fresh water have low levels of strontium in their otoliths, whereas those rearing in salt water have high levels. The failure of some eels to enter fresh water is just one of the fascinating discoveries coming out of the studies of Katsumi Tsukamoto, Jun Aoyama, and their colleagues at the University of Tokyo. Their work has revealed the different spawning areas of six Pacific *Anguilla* species that are much like the Sargasso Sea—areas where currents can carry leptocephalus larvae to appropriate coastal areas (Aoyama et al. 1999). Their molecular studies on the systematics of the genus show that the eel species native to Borneo (*A. borneensis*) is a sister group to all other members of the genus, suggesting a western Pacific origin of the genus (Aoyama et al. 2001). The line that colonized the Atlantic Ocean apparently originated from leptocephali drifting in the ancestral ocean between Asia and Africa, before the two continents collided 20 to 30 million years ago.

colored green and yellow to being silvery. The silver eels then migrate out to sea and, apparently, seek deepwater currents, with which they swim to their spawning grounds in the Sargasso Sea, a distance of as much as 5,600 km. Once there, they apparently spawn at great depths and then die after spawning. Schmidt (1922) located the spawning grounds of the eels by plotting the distribution of leptocephalus larvae by size on a map of the Atlantic Ocean. The smallest larvae were found in the vicinity of the Sargasso Sea. To reach Europe, the larvae must drift with the currents for approximately 3 years; to reach North America, they must drift for approximately 1 year. Once the leptocephali reach coastal waters, they metamorphose into elvers, which proceed to migrate into streams and estuaries. The habitats in which the eels rear is tied to their sex as adults: Males seem to rear mainly in riverine or estuarine habitats, whereas females tend to rear in lakes (Oliveria et al. 2001).

Muraenidae

Moray eels are efficient predators of fish and invertebrates of reefs and rocky shores in tropical and temperate regions. They lack both paired fins and scales. They have hard, pointed heads with many teeth; small, round gill openings; and a posterior set of nostrils set high on the head, usually between the small eyes. With such adaptations, they penetrate easily into the deep crevices of reefs to consume night-active fishes that are in hiding during the day and day-active fishes that are in hiding during the night. Most species have dagger-like teeth for grabbing their prey, but some have flattened teeth for breaking up hard-shelled invertebrates. Some morays grow as long as 3 m, but most do not exceed 1 m. Their colors range from drab yellow-green to bright yellow or red with spots or rings of white. Many morays are considered to be dangerous, because they can give severe bites to an unwary diver. They are probably most dangerous, however, as a cause of ciguatera fish poisoning. The flesh of tropical species of morays can become poisonous when they consume other fishes that have fed on the algae that produce the toxins (Halstead 1967). The fish have adapted to the toxins, but the humans eating the fish have not!

Congridae

Most of the 150 or more species of conger eels resemble moray eels, except that they usually have pectoral fins and their teeth, rather than being sharp fangs, are stout and cone-shaped. They feed on a wide variety of invertebrates. In temperate regions, most conger eels are associated with shallow, rocky areas, but in tropical regions, many of them instead construct burrows in soft bottoms. Most remarkable of the latter group are the garden eels *(Gorgasia, Heteroconger)*. Garden eels are small, extremely elongate eels with moderately large eyes, rounded heads, and small mouths; they form colonies on sandy bottoms in areas of moderate currents. Each eel constructs a burrow, from which it can extend its long (to 1 m) body and feed during the day on zooplankton carried by the current (Fricke 1970). When all the eels in a colony are feeding and waving about in the current, the colony has the appearance of a field of sea grass more than an aggregation of fish.

Ophichthidae

The family of snake and worm eels is the largest in the Anguilliformes, with approximately 250 species. Most are found in shallow water in tropical and subtropical areas, are small in size (<1 m), and are brightly colored. Despite their abundance, bright colors, and shallow-water habitat, they are rarely seen and are difficult to collect, because they burrow into soft bottoms and are active mostly at night. Unlike the garden eels, they do not have permanent burrows; instead, they use them for a day and then abandon them. The "true" snake eels (subfamily Ophichthinae) have a finless, spike-like tail, with which they can quickly penetrate sand and mud bottoms. The worm eels (Myrophinae) possess a caudal fin but are mostly found in soft mud bottoms. When in their burrows, snake and worm eels leave just their heads exposed. Consequently, they have exceptionally well-developed branchial pumping apparatus for moving water across the gills, and both sets of nostrils are located on the tip of the snout.

17.7 SACCOPHARYNGIFORMES

This group of deep sea eels has pushed teleost morphology to its extreme, in good part by abandoning structures essential to most other teleosts, such as opercular bones, branchiostegal rays, scales, pelvic fins, ribs, and swimbladders (Fig. 17.5). The loss of so many structures is, presumably, an energy-saving measure for fish that live under extreme conditions with infrequent meals. The least modified members of the three families in this

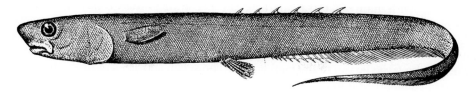

Figure 17.5   Spiny eel (Notacanthidae). From Goode and Bean (1895).

order are the bobtail snipe eels (Cyematidae). They still possess a caudal fin and traces of a maxillary bone in the mouth. They also have the most abbreviated bodies of any eels and elongate, snipe-like snouts, which are apparently advantageous for the life of a mid-water planktivore in the deep ocean. The other three families in this order have abandoned even more features (e.g., caudal fin, upper jaw bones) to focus on their role as predators capable of swallowing prey larger than themselves. The mouth and pharynx are extremely large and distensible, with small teeth for holding onto the prey. The gill openings are small and placed far back on the body, closer to the anus than to the tip of the snout in the gulper eels, whereas the tiny eyes are set near the tip of the snout. These fishes are among the largest (or at least the longest, up to 180 cm) bathypelagic fishes. Their watery flesh and poorly developed fins indicate that they do not actively pursue prey; instead, they hang suspended in the water column, waiting for their prey to come to them and be swallowed in slow motion. It seems likely that they attract their prey by using a light-producing organ at the tip of the tail. At present, nine species of swallower eels (Saccopharyngidae) and one species of gulper eel (Eurypharyngidae) are recognized. In addition, six species of small, gulper-type eels are placed in a separate family (Monognathidae), because they lack an upper jaw and pectoral fins.

17.8   CLUPEOMORPHA

The herrings and their kin form one of the most well-defined teleost orders (Clupeiformes, the only living order in the Clupeomorpha). They are distantly linked to the Ostariophysi (see chapter 18), and the two groups are more closely related to the rest of the modern teleosts (Euteleostei) than they are to the preceding groups in this chapter (Lecointre and Nelson 1996).

All clupeiform fishes are adapted for living in well-lighted surface water, where most species school and feed on plankton. Their most conspicuous adaptations for this way of life are silvery scales, a compressed and often keeled body, a flexible mouth, and fine gill rakers. The silvery scales and compressed body reduce the visibility of the fish; The scales scatter the light coming from above, and the compressed body reduces the profile visible from below. The mouth and gill-raker structure are used for plankton feeding. More definitive are their caudal skeleton characteristics, the structure of their lateral-line system, and the connection of the swimbladder with the inner ear by means of a narrow diverticulum of the bladder. The latter structure gives clupeids a sense of hearing that is among the most acute in fishes. Most clupeiform fishes belong to the families Clupeidae and Engraulidae, which occur in both salt and fresh water. Curiously, a relict species with many ancestral characteristics (Denticipitidae, *Denticeps clupeoides*) occurs in only a few streams in west Africa (Berra 2001). Perhaps the most curious of all clupeiforms, however, are the Sundaland noodlefishes (Sundasalangidae), which at an adult size of 15 to 22 mm total length are among the smallest of all vertebrates. They look more like large larvae than small adults! The seven species live in the fresh waters of Borneo, Thailand, and Malaysia (Berra 2001).

Clupeidae

The herrings, shads, sardines, menhaden, and other clupeid fishes ($\approx$215 species) can be readily recognized by their keeled (sawtooth) bellies and silvery, deciduous scales (Fig. 17.6). The lateral line is absent, as are teeth (usually). They play key roles in many food webs because of their abundance and their ability to feed on both zooplankton and phytoplankton. They tend to concentrate in coastal waters (but see Box 17.2), in areas of upwelling, so they have long been important to humans, either as the object of commercial fisheries or as food for predacious fishes harvested by humans. Because they often have boom-or-bust population cycles, especially when under pressure from commercial fisheries (see Chapter 34), they may have considerable impact on human affairs. During the fourteenth century, for example, the power of the Hanseatic League declined when herring populations in the Baltic Sea, on which the League depended for a steady source of fish for trade, collapsed. This shifted the trade to the Dutch, who could fish the North Sea, and it ultimately resulted in the paintings of Rembrandt, Vermeer, and others (Moyle and Moyle 1991).

Anadromous clupeids are found in many parts of the world: shad (*Alosa* spp.) are characteristic of coastal streams of eastern North America and Europe, and hilsa (*Hilsa*

BOX 17.2
Freshwater Herrings

Clupeids are usually thought of as marine fishes, but many species are found in inland waters around the world. In ancient Lake Tanganyika (eastern Africa), for example, two species (*Stolothrissa tanganicae* and *Limnothrissa miodon*) dominate the pelagic zone and play a major role in the food web of the lake, making it one of the most productive freshwater lakes in the world (Lévêque 1997). Related species are abundant in the fresh waters of India, Southeast Asia, and Australia. In North America, freshwater clupeids (especially the gizzard shads, *Dorosoma* spp.) are valued as forage for game fish; for this purpose, they have been planted in lakes and reservoirs far outside their native ranges, with mixed results. In such situations, they frequently have population explosions, taking advantage of the lightly exploited zooplankton populations. When this happens, growth rates of adult game fishes may increase dramatically. Unfortunately, the introduced shad are often too successful and, through competition for zooplankton, may decrease the growth and survival rates of juvenile game fish (which typically feed on zooplankton) and of native planktivorous fish.

kelee) spawn in streams that flow into the Indian Ocean, from China to Africa. These fishes typically move upstream during periods of high water to spawn. Following spawning, the embryos and/or larvae are quickly washed downstream, and the young live initially in estuaries or nutrient-rich coastal waters. It is worth noting that spawning migrations are characteristic of clupeids in general. Most of the marine species congregate for spawning in areas away from the adult feeding grounds. Herrings *(Clupea)* seek inshore areas with submerged plants or rocks to which they can attach their eggs, attracting large aggregations of predators (including fishermen).

Engraulidae

The 139 species of anchovies are distinguished from clupeids by their overhanging snouts and long upper jaws, which extend behind the eye (Fig. 17.6). As a consequence of this mouth structure, anchovies can open their mouths to an incredible extent, producing round openings that are efficient for filter feeding on plankton. The actual filtering apparatus consists of densely packed gill rakers on the first gill arch. Anchovies are usually small (<15 cm), translucent fish that inhabit inshore areas of the oceans where plankton densities are high, although a few species inhabit fresh water. The freshwater species include some of the curious rattail anchovies (*Coilia* spp.) of Asia, which have

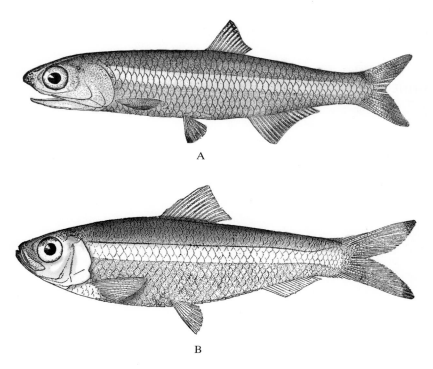

Figure 17.6   Clupeiform fishes. **A.** Anchovy (Engraulidae). **B.** Sardine (Clupeidae). **A** from Jordan and Evermann (1900); **B** from Jordan (1895).

tapering bodies, pointed tails, and long pectoral fins with independent rays (Berra 2001). Many anchovies are important commercial species; their populations, like those of the clupeids, fluctuate considerably in response to changing oceanographic conditions and collapse dramatically when overfished (see Chapter 34).

17.9   LESSONS FROM BONYTONGUES, EELS, AND HERRINGS

Clearly, there are many ways to be a successful teleost. The bonytongues can be regarded as early successes that faded as other teleost groups evolved. They managed, however, to leave specialized remnant species on all continents, and the African electric fishes found a way to maintain high numbers and diversity through specialized adaptations to murky waters. The eels have taken one particular morphological pattern and diversified in astonishing ways, occupying a wide array of habitats from streams and lakes to the open ocean to coral reefs to the deep sea. Likewise, the herrings and anchovies developed into specialized plankton feeders with relatively few species but extraordinarily high numbers, especially in coastal areas. One key to the success of both eels and herrings is their ability to make long migrations, including those into fresh water, that separate life-history stages.

SUPPLEMENTAL READINGS

Alves-Gomes 2001; Blaxter 1985; Bullock 1973; McCleave 1993; Möller 1995; Sinha and Jones 1975; Tesch 1977.

WEB CONNECTIONS

GENERAL
www.calacademy.org/research/ichthyology
www.fishbase.org
www2.biology.ualberta/jackson.hp/IWR
www.flmnh.ufl.edu/fish/default.htm
www.people.clemson.edu/~jwfoltz/wfb418

CHAPTER 18
Minnows, Characins, and Catfishes

SUBDIVISION TELEOSTEI
  Infradivision Euteleostei
    Superorder Ostariophysi
      Series Anotophysi
        Order Gonorynchiformes
          Families Chanidae (milkfish), Gonorynchidae (beaked sandfishes), Kneri-
            idae, Phractolaemidae (snake mudhead)
      Series Otophysi
        Order Cypriniformes
          Families Cyprinidae (minnows and carps), Gyrinocheilidae (algae eaters),
            Catostomidae (suckers), Cobitidae (loaches), Balitoridae (river loaches)
        Order Characiformes
          Families Distichodontidae, Citharinidae, Paraodontidae, Curimatidae,
            Prochilodontidae, Anostomidae (headstanders), Chilodontidae
            (headstanders), Crenuchidae (South American darters), Hemiodontidae,
            Alestidae (African tetras), Gasteropelecidae (freshwater hatchetfishes),
            Characidae (characids), Acestrorhynchidae, Cynodontidae, Erythrinidae
            (trahiras), Lebiasinidae, Ctenoluciidae (pike-characins), Hepsetidae
            (Kafue pike).
        Order Siluriformes
          Families Diplomystidae (velvet catfishes), Ictaluridae (bullhead catfishes),
            Bagridae (bagrid catfishes), Claroteidae, Australoglanidae, Olyridae
            (longtail catfishes), Cranoglanididae (armorhead catfishes), Siluridae
            (sheatfishes), Schilbidae (schilbid catfishes), Pangasiidae (pangasid cat-
            fishes), Amphiliidae (loach catfishes), Sisoridae (sucker catfishes), Ereth-
            sidae, Amblycipitidae (torrent catfishes), Akysidae (shortfin catfishes),
            Parakysidae, Chacidae (squarehead catfishes), Clariidae (labyrinth cat-
            fishes), Heteropneustidae (airsac catfishes), Malapteruridae (electric cat-
            fishes), Ariidae (sea catfishes), Plotosidae (eeltail catfishes), Mochokidae
            (squeakers), Doradidae (thorny catfishes), Auchenipteridae (driftwood
            catfishes), Pimelodidae (longwhisker catfishes), Cetopsidae (whale cat-
            fishes), Aspredinidae (banjo catfishes), Trichomycteridae (pencil catfish-
            es), Callichthyidae (plated catfishes), Scoloplacidae (dwarf catfishes),
            Loricariidae (suckermouth armored catfishes), Astroblepidae (climbing
            catfishes)

Order Gymnotiformes
Families Sternoptygidae (glass knifefishes), Rhamphichthyidae (sand knife-fishes), Hypopomidae, Apteronotidae (ghost knifefishes), Gymnotidae (knife eels), Electrophoridae (electric eel)

The superorder Ostariophysi contains more than 6,500 species—over one-quarter of the known fish species. More importantly, nearly three-fourths of all freshwater fish species belong in this superorder. They are the dominant freshwater fish on all continents except Australia and Antarctica. Thus, to appreciate the adaptations of fish in this group is to appreciate what it takes to live in freshwater environments, which frequently are turbulent, turbid, and subject to extreme fluctuations in both temperature and water chemistry. Such environments also are often small, isolated, and fragmented, making them subject to extreme and rapid changes from both human and natural causes. The fact that this superorder contains some of the most abundant species of freshwater fish and some of the most important species for aquaculture is a tribute to their adaptability. As this chapter demonstrates, the ancestors of the minnows, characins, and catfish developed a suite of adaptations that has allowed them to dominate fresh water and to diverge early from the evolutionary stream of other euteleostean fishes.

Some of the characteristics that fishes in this group share include:

- Fright substance (***Schreckstoff***) is present. Schreckstoff is released into the water when a fish is injured and causes a fright reaction in members of the same or closely related species.
- A swimbladder is present, usually with two chambers.
- Unculi are present (Roberts 1982). Unculi are small, unicellular projections on various parts of the body that may provide a rough surface for clinging or scraping.
- Breeding tubercles are well developed, including a keratinous cap.
- The upper jaw (premaxilla) is easily extended, for suction feeding.
- The pelvic fins are abdominal in position.

Although obviously a distinct evolutionary group, ostariophysans are tied phylogenetically to the herrings and their kin (Clupeomorpha) as a separate lineage from the rest of the modern teleosts (Lecointre and Nelson 1996) (Fig. 18.1).

18.1   GONORYNCHIFORMES

Before the studies of Greenwood et al. (1966), the Ostariophysi did not include the Gonorynchiformes, only the fishes placed here under the Otophysi. This is because the series Otophysi is the most cohesive of all the major teleost taxa; its members are easily recognized by a number of distinctive features. The Gonorynchiformes lack most of these distinctive features, but the members of this order do have enough in common with the otophysan fishes to be a sister group of the Otophysi (Rosen and Greenwood 1970).

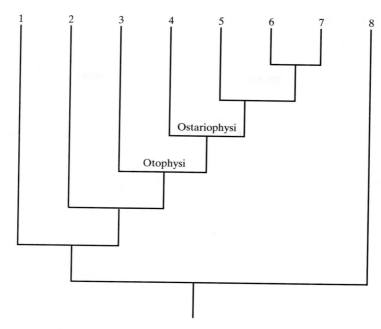

FIGURE 18.1 Phylogenetic relationships of the Otophysi. **1.** Elopomorpha. **2.** Clupeiformes. **3.** Gonorhynchiformes. **4.** Cypriniformes. **5.** Characiformes. **6.** Siluriformes. **7.** Gymnotiformes. **8.** Euteleostei.

This order contains only four families and approximately 35 species. Aside from the various family specializations, they are not a particularly distinctive group. They have small, usually toothless mouths and epibranchial organs (modified gill rakers) for breaking up the particles of food they ingest. Their first three vertebrae are modified in such a way as to suggest the condition that permitted development of the distinctive Weberian ossicles of the Otophysi. The best-known member of the Gonorynchiformes is the milkfish *(Chanos chanos)*, which is the sole member of the family Chanidae (Fig. 18.3). This marine and brackish-water species is one of the most important food fishes of Southeast Asia. Adults are large (to 1.8 m), active fish, with silvery sides and deeply forked tails. They feed mostly on planktonic algae. They are extremely tolerant of a wide range of temperature and salinity, so they are successfully raised in brackish and freshwater ponds. The fish used for pond culture are collected as fry in estuaries and other inshore areas.

The gonorynchiform family with the most species (27) is the Kneriidae, a group of small, loach-like freshwater fishes in Africa. These fish are well adapted for living in swift streams. They have fine (or no) scales, subterminal and protractile mouths, elongate bodies, and large pectoral fins that, when placed together, can form a sucker enabling the fish to cling to rocks in strong currents. Some species are also air breathers, enabling them to survive in stagnant pools. The kneriids are of special interest, because they provide some idea of what the early ancestors of the otophysan fishes may have been like (Berra 2001).

18.2   OTOPHYSI

The Otophysi group is the main line of ostariophysan evolution, with 6,500 species. The taxonomic arrangement of the Otophysi presented here is that of Fink and Fink (1981), who regard the Cypriniformes (minnows, etc.) as the sister group for the rest of the orders, with the Characiformes (Characins) being derived early in the evolutionary history of the Otophysi. The catfishes (Siluriformes) were derived somewhat later and gave rise to the South American electric fishes (Gymnotiformes). The Otophysi have a tremendous range of body forms and sizes, yet they are united by a number of distinctive skeletal features. The most unusual feature is the ***Weberian apparatus***, a chain of bones that connects the swimbladder to the inner ear and, thus, gives the fishes a sensitive sound-reception system similar in function to that of the inner ear in mammals. The connecting bones are modified portions of the first four or five vertebrae (see Fig. 10.3). The acute sense of hearing from the Weberian ossicles is presumed to be particularly useful in turbid water or at night, when most catfish are active. Another major feature is the protractile upper jaw (absent in most characins), which in combination with pharyngeal teeth allows for a diversity of feeding techniques. The pharyngeal teeth permit separation of the grabbing and chewing functions of the mouth, which is important for fish that feed mostly on active prey. The many specializations in pharyngeal dentition that are present in the Otophysi, from molariform grinding teeth to comb-like teeth for breaking up fine materials to sharp teeth for piercing prey, have undoubtedly contributed to the success of these fish in much the same way that the specialized jaw teeth of mammals have contributed to their success on land.

Other features that seem to be particularly important contributors to the success of the Otophysi include their fear scent, their generally small size, and their reproductive strategies. Like the keen sense of hearing, fear scent likely is particularly useful under conditions, common in fresh water, in which a predator is hard to see. Once the scent has been released, the fish may flee the area, hide, or school more closely—even though the predator is not visible—so that any additional attacks by the predator are likely to be less successful. The use of fear scent by these fishes reflects the fact that most of these species are highly social and shoal during at least one stage of their life history. It also reflects the fact that most otophysan fishes are quite small, even as adults. Some species reach 1 to 2 m in length, but they are decidedly a small part of the total. Most of these large forms are either piscivores or detritivore/herbivores. Small size seems to be advantageous for feeding on the myriad of small aquatic invertebrates as well as on the terrestrial invertebrates that fall into the water. Small size is also advantageous for occupying the numerous microhabitats in fresh water, such as between rocks in fast streams or among the aquatic vegetation in lakes, or for occupying small or intermittent waterways in which low oxygen levels and shallow water likely discriminate against large fish. Yet another advantage of small size lies in the ability of the fish to reach maturity quickly. Freshwater environments, but particularly streams, fluctuate from season to season and from year to year; wet years and droughts may randomly alternate with one another. Thus, the availability of conditions necessary for successful reproduction may vary considerably from year to year and from place to place. Mobile, small fish with early maturity can quickly take advantage of favorable conditions and flood the environment with young, thus maintaining large populations as long as the favorable conditions exist.

18.3   CYPRINIFORMES

The Cypriniformes order contains the Cyprinidae and specialized cyprinid relatives: the Catostomidae, and the four families of loaches. Together, these fish dominate the fresh waters (especially streams) of North America and Eurasia and, to a lesser extent, Africa. The nearly 2,700 species in the order are diverse in their external appearances, but most possess specialized pharyngeal teeth and protractile mouths without teeth. Their heads are scaleless, and with the exception of a few loaches, all lack an adipose fin.

Cyprinidae

The minnow or carp family is the largest family of fishes, with more than 2,000 species. Members of this family are distinguished by their pharyngeal teeth (one to three rows, but never more than eight teeth per row) and thin lips (upper jaw usually bordered only by the premaxilla). Although most have only soft rays in the fins, in some forms rays that have been modified into spines are present, most notably in common carp *(Cyprinus carpio)* and goldfish *(Carassius auratus)*. Remarkably, considering the number of species involved, cyprinids all have body plans that are mainly variations on the "classic" fish theme, with fusiform to moderately deep bodies, large eyes, conspicuous scales, abdominal pelvic fins, and small terminal or subterminal mouths (Fig. 18.2). Nevertheless, the cyprinids show considerable morphological, physiological, and behavioral diversity (Winfield and Nelson 1991). Most are diurnal predators of small invertebrates. Some are piscivorous, however, and others feed on algae, higher plants, and organic ooze. Cyprinids include one of the smallest freshwater fish known, *Danionella translucida,* which is mature at 10 to 11 mm in length (Roberts 1986), as well as a number of large species that reach 2 to 3 m in length. Many species exhibit various forms of parental care of the embryos and young, but many also simply scatter their eggs in selected places in the environment, sometimes making long migrations to do so.

In North America, approximately 286 species of cyprinids are known, more than 40% of them small, silvery "shiners" of the genera *Notropis* and *Cyprinella* (Mayden 1991). The shiners are often the most abundant, or at least the most conspicuous, fishes in small streams and lakes of eastern North America, where they form large shoals in shallow water. It is not unusual to find 5 to 10 species of small cyprinids living together in these habitats, and studies to determine how groups of such similar species can coexist have contributed much to our understanding of stream ecology. One of the most interesting and abundant of the eastern stream minnows is the central stoneroller, *Campostoma anomalum* (Jenkins and Burkhead 1993). The stoneroller is an herbivore with a hard plate on the lower jaw for scraping algae from rocks and a long intestine (eight times its body length) for digesting it. Schools of stonerollers are so abundant in places that they can keep the rocks clean of algae; thus, algae grow well only in pools where predatory fish keep the stoneroller populations low (Power and Matthews 1983). Stonerollers are also well known for the many large tubercles that cover the stout heads of large (10–28 cm), multicolored breeding males. The males build nests on the bottom of riffles, which are defended from other males by way of the breeding tubercles. During the spawning season, it is not unusual to find males missing eyes or dying from encounters with other males. Like antlers, the breeding tubercles are shed by surviving males at the end of the breeding season (see also Box 28.1).

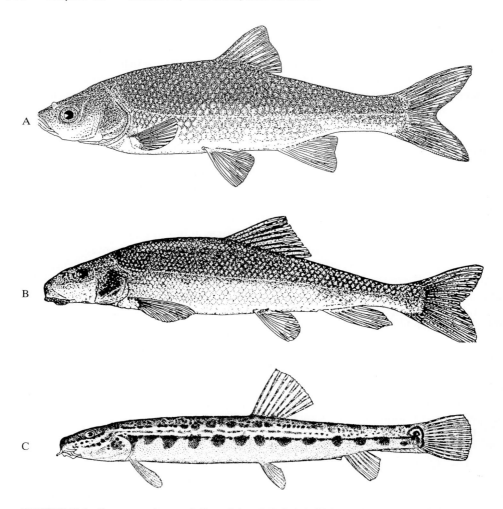

FIGURE 18.2   Representative cypriniform fishes. **A.** Tui chub (*Gila bicolor*, Cyprinidae). **B.** Sucker (*Catostomus occidentalis*, Catostomidae). **C.** Loach (*Cobitus* sp., Cobitidae). **C** from Nichols (1943).

Bright breeding colors, nuptial tubercles, and some type of nest building are actually characteristic of many North American cyprinids, so for a few weeks each year, they can rival their tropical relatives in brilliance. Some of these cyprinids have developed the curious habit of using the nests of other fishes, mostly those of larger cyprinid species or centrarchid basses, for spawning. Nests of the large cyprinids, fallfish *(Semotilus corporalis)* and hornyhead chub *(Nocomis biguttatus)*, may be used for spawning simultaneously by as many as three species of small cyprinids, resulting in occasional hybrids among the species.

In eastern North America, a large majority of cyprinids have a maximum length of less than 10 cm, although some species may reach 25 to 30 cm. In the major drainages of western North America, however, a majority of the species are large, commonly exceeding 25 cm as adults (Moyle and Herbold 1987). The largest of these western species are the four piscivorous pikeminnow (*Ptychocheilus* spp.). The Colorado pikeminnow *(P. lucius)*, now an endangered species, is reported to have reached nearly 2 m in length and a weight of 45 kg. Presumably, these pike-like cyprinids evolved in western rivers, because the true pikes (Esocidae) and centrarchid basses, the main large piscivores of the Mississippi and associated drainages, were absent. In altered habitats into which the basses and pike have been introduced, pikeminnows often gradually disappear.

Southeast Asia is generally considered to be the center of cyprinid evolution, because the cyprinids are extraordinarily numerous and diverse there as well as on the Indian subcontinent. Most African cyprinids belong to genera also found in Asia and, apparently, are derived from them. Many of these species are the small, bright, colorful, and highly active forms favored by aquarists, usually known as barbs *(Puntius, Barbus)*, rasboras *(Rasbora)*, and "sharks" *(Labeo)*. These species show a remarkably high degree of ecological segregation in streams (see Chapter 30). Many Eurasian cyprinids also grow to respectable sizes and are important sport and commercial fish. The largest of these is the giant barb (*Catlocarpio siamensis*) of the Mekong River, which reaches more than 3 m in total length. In many areas, large cyprinids, including common carp, are much sought after by anglers. Because of their ability to grow rapidly to harvestable sizes in ponds on diets of organic ooze, plants, and small invertebrates, a number of the Asiatic carps—such as common carp, goldfish, grass carp (*Ctenopharyngodon idella*), and silver carp *(Hypophthalmichthys molitrix)*—are among the most important aquaculture animals in the world (see Box 18.1).

Catostomidae

The suckers are a small (68 species) but very successful family of mainly North American fishes (Fig. 18.2). One species (*Myxocyprinus asicaticus*) with many ancestral characteristics is located in two Chinese Rivers, and one North American species, the longnose sucker (*Catostomus catostomus*), has managed to invade Asia across the Bering Strait (Berra 2001). Throughout North America, suckers are among the most abundant fishes, particularly in streams and rivers. Their success can be attributed to the tolerance most species have for a wide variety of environmental conditions and to their feeding habits. Most species are bottom browsers, sucking up organic ooze, algae, and small invertebrates through their subterminal mouths with fleshy lips. The food ingested is then broken up by comb-like pharyngeal teeth (one row of 16 or more teeth) and is digested in a long, winding intestine. A majority of the suckers are moderate-sized (40–80 cm), streamlined forms belonging to the genera *Catostomus* and *Moxostoma*. Species of these genera are most characteristic of streams, where they are often abundant. Other species have become adapted for life in large, fast-moving rivers. Most extreme in this regard is the razorback sucker *(Xyrauchen texanus)* of the Colorado River. The back of the razorback sucker rises steeply behind the head to form an inverted keel. An evolutionary trend in the opposite direction is toward lake living, and suckers adapted for living in quiet water, such as carpsuckers *(Carpoides)* and buffalo suckers *(Ictiobus)*, are large, deep-bodied forms. Many of these forms are still bottom feeders, but some have developed terminal mouths and fine gill rakers enabling them to feed on plankton.

BOX 18.1

Common Carp: Super Fish

In 1653, Isaak Walton wrote that, in England, the common carp was "Queen of the Rivers," but he also recognized that it was not native to Britain. In fact, it had already a long history of spread. The Romans transported carp from the Danube River to Italy. From there, carp slowly spread across Europe, mainly because they were raised in monastic ponds in the Middle Ages. Carp may even have been introduced into the Danube: Genetic evidence suggests that a single introduction from Asia founded both the Danube population and all introduced populations, including koi (Froufe et al. 2002). Regardless, common carp are now abundant around the world in habitats ranging from cold temperate to tropical (Lever 1996). They support commercial and sport fisheries and are widely cultured as both a food fish and an ornamental fish (koi).

Why have carp been so successful, aside from their desirability as a food fish? They are incredibly hardy, capable of surviving in water that is depleted of oxygen (<1 mg/L), very warm (to 36°C), very cold (near freezing), fairly saline (to 16 ppt), and polluted with various wastes (Moyle 2002). They grow to large sizes (to 37.9 kg), live long (up to 47 years in captivity), and are very fecund. Females can produce as many as 2 million eggs in a season. The eggs are scattered and fertilized on flooded vegetation, in which the young hide during the first weeks of life. Once loose in a river, carp spread widely and rapidly. They are omnivores but prefer aquatic insects buried in bottom mud, which they root out with vigor. On top of all this, they possess stout spines on their dorsal and anal fins and grow rapidly during the first years of life, thereby reducing predation risk. With high physiological tolerances, long life, high fecundity, and high dispersal abilities, these fish do not have to spawn successfully very often to become abundant, taking advantage of the good times when they come. Despite their considerable benefits to humans as a food fish, their spread has come at an environmental cost. Their destructive feeding behavior can make clear waters turbid, and their spread is frequently followed by declines in desirable native fishes and invertebrates. Unfortunately, they are still spreading, often through the release of ornamental koi into natural waterways. Koi, like goldfish, can produce wild-type young that quickly become the dominant form.

Cobitidae

This family (160 species) of small fish favored by aquarists is largely adapted for living in the streams of Eurasia. The loaches are most diverse in Southeast Asia, but one species has managed to invade North Africa. In addition, the oriental weatherfish *(Misgurnus anguillicaudatus)* has become established through introductions in a number of places in the United States and Europe. Most loaches are small (to 30 cm) fishes that range in body shape from worm-like to chunky (with a flattened belly). All have subterminal mouths with three or more pairs of barbels (Fig. 18.2). As might be expected, most are secretive bottom dwellers that feed on small invertebrates and plant matter. A closely related family is the ***Balitoridae*** (Homalopteridae), the river loaches, with approximately

485 species. These loaches are adapted for life in fast-flowing streams of Eurasia. Most of the species (mainly in the genus *Nemacheilus*) are small, dace-like fish that have subterminal mouths with three sets of barbels and large pectoral fins, and they actively forage among the rocks. One species (*Yunnanilus*) lives in hot springs at 5,200 m elevation in Tibet—the highest known elevation for any fish species (Kottelat and Chu 1988). Others are highly specialized for browsing on algae that grows on rocks in torrential streams and have a ventral sucking disk. The disk is created by the pelvic and pectoral fins joining on a flattened belly. The mouth is subterminal and designed for scraping algae. This arrangement allows them to live in faster-flowing water than any other fishes.

18.4   CHARACIFORMES

The characins are confined to Mexico, South and Central America ($\approx$1,350 species), and Africa ($\approx$200 species), with the exception of the Mexican tetra *(Astyanax mexicanus)*, which has managed to invade North America. In South America, characins, together with catfish, totally dominate the freshwater fish fauna. Although the many species possess a wide array of specialized features, they are readily recognizable as a group. In the past, this has resulted in all species being placed in just one family, the Characidae. Recent workers, however, have split this group in recognition of its diversity, but the 18 families listed here, following the classification of Buckup (1998), may change further as our understanding of the characiform fishes increases. The interrelationships among the families are also interesting, because some evidence indicates that much of the radiation of the characiform fishes took place before Africa and South America became separate continents (Orti and Meyer 1997).

Most characins are diurnal predators, with large eyes (absent in some cave-dwelling forms), no barbels, and numerous teeth set in the jaws. Their jaws are not very protractile, and most possess rather unspecialized pharyngeal teeth. They tend to be small and bright (often silvery) in color, and they have fusiform or laterally compressed bodies that are completely covered with (usually) cycloid scales. Most possess an adipose fin, a short dorsal fin located midway on the body, abdominal pelvic fins, and a caudal fin, usually with 19 principal rays.

African Characins

Four families of characins are endemic to Africa: Distichodontidae, Citharinidae, Alestidae, and Hepsetidae. These families contain more than 200 species. The first two families are regarded as forming a sister group to all the rest of the characins. Some of the 90 species of distichontids are elongate fishes, with jaws specialized for snipping off chunks of fins of other fishes, but most are bottom-oriented herbivores and invertivores (Berra 2001). The African tetras (Alestidae) were formerly lumped with the South American Characidae. The family (>100 species) is widely distributed across Africa and mostly contains small omnivores that are often sold as aquarium fishes. Most recognizable of the alestids, however, are the piscivorous African tigerfishes (*Hydrocynus* spp.), with their sharp-toothed, wide-gaped jaws (Fig. 18.2), which make them prized game fishes. Giant tigerfish (*H. goliath*) from the Congo River can reach 1.5 m and

50 kg, and they require steel leaders for catching. The sole member of the Hepsetidae, the Kafue pike (*Hepsetus odoe*), is also a toothsome piscivore. It closest relatives seem to be the pike characins (Ctenoluciidae) of South America.

Characidae

This is the largest and most varied of the characin families (≈700 species). Its members are found in South and Central America, with one species reaching North America (the Mexican tetra). The family Characidae essentially contains the various characins that lack the features used to characterize the other families (Fig. 18.1). All have good sets of jaw teeth, but a wide variety of feeding habits can be found, including species that feed largely on the scales of other fishes. Many characids are predatory, including the fearsome-looking piranhas (*Serrasalmus*). Undoubtedly, piranhas are well-developed predators, but their predation on humans and large mammals seems to have been exaggerated. Most human "victims" of piranhas may have drowned first and then been scavenged by the fish. A number of close relatives of the piranhas are vegetarians, and some have piranha-like jaws, with strong teeth, that enable them to invade flooded Amazon jungles and feed on fruit and seeds falling from trees (e.g., pacus, *Colossoma*, *Mylossoma*). Such species grow to large sizes, have strong interactions with floodplain forests, and support important indigenous fisheries (Goulding 1980). Other smaller species, such as the tetras, support fisheries and aquaculture operations to provide brilliantly colored fish for the aquarium trade.

Other South American Families

The 13 other characin families are small groups of rather specialized species (Roberts 1969, Berra 2001). Two families (Acestrohynchidae, Ctenoluciidae) are pike-like, lie-in-wait predators of other fishes. The pike characins are among the largest South American characins (>70 cm), rivaling the equally predatory Cynodontidae. The cynodontids are best known for the long, dagger-like canines on their lower jaws. The trahiras (Erythrinidae) are also largely piscivorous as adults and are the only charcins without adipose fins (Berra 2001). In contrast, many members of the Curimatidae are deep-bodied, bottom-feeding fishes of moderate size with subterminal, sucker-like mouths. Perhaps the most remarkable fishes are the freshwater hatchetfishes (Gasteropelecidae), which have a compressed head and body and large breast muscles, thus giving them a deep-bodied appearance (Fig 18.3). These small (3–10 cm) fish are capable of making long leaps from the water and were thought to have powered flight using rapid beats of their pectoral fins. It appears, however, that the fins are used mainly to push the fish out of the water when becoming airborne, assisting the tail, and are folded along their sides during leaps (Wiest 1995).

18.5   SILURIFORMES

The catfishes are one of the most distinctive groups of fish. Their Weberian apparatus and fear scents link them to other otophysan fishes, but their distinctive morphology also indicates a long-independent evolutionary history. In general, it appears that the

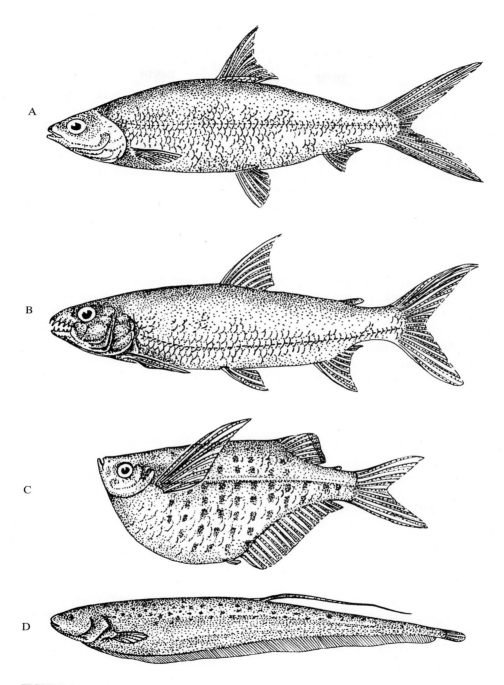

FIGURE 18.3   Representative ostariophysan fishes. **A.** Milkfish (*Chanos*, Chanidae, Gonorynchi-formes). **B.** Tigerfish (*Hydrocynus*, Alestidae, Characiformes) **C.** Freshwater hatchetfish (*Gasteropelecus*, Gasteropelecidae, Characiformes). **D.** Electric knifefish (*Sternarchus*, Apteronoti-dae, Gymnotiformes).

catfishes have diversified for being active after dark or in turbid water, thereby reducing interactions with the more vision-oriented cypriniforms and characiforms. This diversification has produced 31 families and approximately 2,600 species of catfishes, of which 13 families and 64% of species reside in South America. Although any group of fish with so many species is bound to have its share of bizarre forms, most catfish are readily recognizable by their whiskery snout, containing one to four pairs of barbels (invariably one of the pairs is supported by the maxilla), as well as by their small eyes, a head that is usually flattened, a mouth without big sharp teeth, and a body that is either without scales or covered with heavy, bony plates. Most also have a stout spine leading each pectoral and dorsal fin, which can lock into place. Some species have venom glands associated with the spines. This basic morphology is highly adaptable; catfishes show the greatest range of adult sizes of any group of bony fishes, from 12 mm in total length (blind banjo catfishes, Box 23.3) to more than 5 m (wels, *Silurus glanis*) (Bruton 1996). Most species, however, have adults in the 10- to 30-cm range. Although catfish are readily recognizable by their external anatomy, their most definitive features, which separate them from other otophysans, are osteological:

- The cranium is a solid box, somewhat flattened, with a reduced number of bones (e.g., the symplectic, basihyal, and subopercular bones are missing). Presumably, this is support for the shovel-like lower jaw of most catfishes (Bruton 1996).
- They lack intermuscular bones in the body. This feature makes them especially desirable as food fish, in contrast to the carps.
- The premaxilla is usually covered with small teeth, but these are absent from the maxilla, which is reduced to a rod to support the barbels. The only exception is found in *Diplomystes*, a South American genus considered to have the most ancestral characters of all catfish. In general, this means the catfishes do not have protrusible mouths.
- Small teeth are usually present on the vomer, on the roof of the mouth (except in the Ictaluridae).
- The Weberian apparatus is generally more complex than those of other otophysans.

Not surprisingly, many catfishes are superb nocturnal predators. The wels, for example, can follow the wake of a swimming fish through a mixture of hydrodynamic and chemical cures, allowing it to grab the fish from behind (Pohlmann et al. 2001). Many other catfishes are scavengers, omnivores, and herbivores, often consuming large amounts of food and growing very rapidly. Catfishes also show a wide variety of reproductive behaviors, although most are nonguarders (Bruton 1996). The nonguarders typically spawn in groups and scatter their adhesive embryos on plants or gravel, although they can be highly selective as to where and when they spawn. The most peculiar of the nonguarders is the bronze corydoras, *Corydoras aeneus*. In this species, the female swallows the sperm of the male, which then passes through her gut. The sperm is discharged along with a batch of eggs into a cup created by the pelvic fins, where the two products are mixed for fertilization (Kohda et al. 1995). All bullhead catfishes (Ictaluridae) and suckermouth catfishes (Loricariidae) are guarders, nesting in caves or under rocks, after which the male cares for the young for varying lengths of time.

The systematics of the catfishes is just beginning to be understood (Teugals 1996). The problem of determining interrelationships among the families is somewhat simplified by the fact that each continent has been an independent center of catfish radiation, so in most cases, each family is confined to just one continent. Thus, South America has 13 families of its own, Eurasia 10, Africa 5, and North America 1. Three other families are found in both Eurasia and Africa, whereas the remaining two (Ariidae and Plotosidae) are mostly marine. The 14 species of catfish found in the fresh waters of Australia are members of the two marine families.

Ictaluridae

The bullhead catfish family contains approximately 45 species of rather unspecialized catfish (Fig. 18.4) endemic to North America. All are scaleless and dark in color (except for cave forms), with large, flattened heads that support eight barbels (two on the

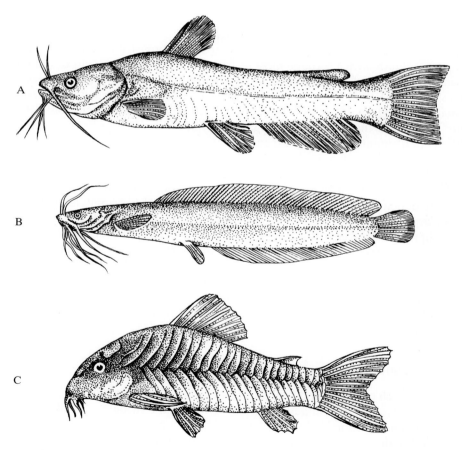

FIGURE 18.4   Representative siluriform fishes. **A.** Bullhead (*Ameiurus*, Ictaluridae). **B.** Labyrinth catfish (*Clarias*, Clariidae). **C.** Armored catfish (*Callichthys*, Callichthyidae).

snout, two on the maxillae, and four on the chin). These catfishes can be divided into three distinct groups: (1) the large species sought by anglers *(Ameiurus, Ictalurus, Pylodictis)*; (2) the small, secretive madtom species *(Noturus)*; and (3) the blind cave species *(Satan, Trogloglanis, Prietella)*. The species in the first group were originally native to warm waters east of the Rocky Mountains, with the exception of one species in Mexico. Thanks to their popularity as food fish, humans spread them to suitable waters throughout North America as well as to Europe, South America, and Australia. The second group, the madtoms, are a widely distributed group of 25 to 30 small (<10 cm) catfishes (Jenkins and Burkhead 1994). Most live in small- to moderate-sized streams, where they hide under rocks, logs, and trash (e.g., old automobile parts and beer cans) during the day and come out to forage on invertebrates at night. Besides their secretive habits, their main protection against predators seems to be stout spines, whose effectiveness is increased by the mild venom associated with them. The species in the third group, the cave catfishes, are small, white, blind, and known from only a few specimens, some collected from wells.

Clariidae

The labyrinth catfishes (150 species) are the most widely distributed catfish family, occurring from South Africa through most of Africa and the southern half of Asia over to Java and the Phillipine Islands. Their bodies are rather elongate (some species are remarkably eel-like), their heads flattened with four pairs of long barbels, and their dorsal and anal fins long (Fig. 18.4). The adipose fin is frequently absent. Their most distinctive feature, however, is the air-breathing organ (labyrinth or suprabranchial organ) made of modified gill filaments on the second and fourth gill arches. These gill filaments are supported by a tree-like structure with a cartilaginous "trunk," so they do not collapse (as do normal gill filaments) when exposed to air (see Chapter 3). The posterior portion of the gills is normal and can be used for aquatic respiration, although most species of clariids seem to depend, at least partially, on aerial respiration. Surprisingly, many clariids are found in large lakes and rivers, where oxygen levels are normally high. A number of such species support important fisheries. Many other species, however, are found in low-oxygen waters, and some are capable of moving for short distances across land to colonize new areas. The latter forms "walk" on the tips of their stout pectoral spines by using them as pivots as they shove themselves along by flexing their bodies. The species best known for this activity is the so-called walking catfish *(Clarias batrachus)*, which has become established, presumably, by escapees from aquaria and tropical fish farms in such places as Guam, Hawaii, and Florida (Courtenay et al. 1974).

Other Eurasian and African Families

Throughout much of Africa and southern and eastern Asia (to Japan), the "typical" catfishes belong to the closely related families Bagridae and Claroteidae. Ecologically, morphologically, and economically, they are similar to the ictalurid catfishes of North America as are the Asian Pungasiidae (Box 18.2). In Europe and northern Asia, the typical catfish role is played by 60 or so members of the family Siluridae. Members of this family diverge markedly, however, from the classic catfish shape of the bagrids by having

BOX 18.2

The Remarkable Pangasiid Catfishes of the Mekong River

The family Pangasiidae only contains approximately 21 species of rather ordinary-looking catfish in India and southeastern Asia. Twelve species, however, are among the most important fishes in the immense Mekong River for fisheries and aquaculture, with many people depending on them for sustenance (Hogan 2003). Most famous of these catfishes is the giant catfish (*Pangasianodon gigas*), one of the largest freshwater fish in the world, reaching approximately 3 m and 300 kg. Supposedly, this fish is an herbivore that grows extremely rapidly, so that large fish may be only 3 to 5 years old (Roberts and Vidthayanon 1991). Regardless of whether this rapid growth is true, the giant catfish is on the verge of extinction, in good part because of capture during its long migrations up the river to spawn. Long migrations (up to 1,500 km) are also known for *Pangasius krempfi*, which is one of the few (only?) catfish known to be anadromous; it rears in the South China Sea and spawns in the Mekong River in Thailand and Laos (Hogan 2003). Other species migrate up the river as juveniles after rearing in the Great Lake (a floodplain system) in Cambodia. The shortbarbel pangasius (*P. macronemus*) apparently makes long migrations up the river, where it supports important seasonal fisheries in Laos even though it is captured at an average length of approximately 10 cm (Baird et al. 2001). Apparently, these fish eventually work their way upstream to spawn. Despite the importance of these catfishes in regional fisheries and evidence that local traditions allow at least some species to be managed on a sustainable basis, the fisheries are in danger of being lost. Large dams are proposed for the river, which will block migrations and flood the spawning and rearing areas (Roberts 1993, Baird et al. 2001).

a long anal fin, a short (or no) dorsal fin, no adipose fin, a strongly compressed trunk, weak spines, and (for catfish) large eyes. Alexander (1966a) interprets these features and others as being adaptations for a more pelagic mode of existence than is true of most catfishes. The small glass catfishes *(Kryptopterus)* are, in fact, completely pelagic, as evidenced by their diurnal activity and semitransparency, which should offer them camouflage in open waters. The 45 species of the African and Indian family Schilbeidae have a morphology similar to that of the silurids, including "glass catfish" species *(Physalia)*. The remaining families in this group have various specializations. For example, five families (Amblycipitidae, Amphiliidae, Akysidae, Olyridae, Sisoridae) contain mostly small species adapted for life in fast-moving streams. The Heteropneustidae have lung-like extensions of the gill cavity they use for air breathing, and the electric catfishes (Malapteruridae) have powerful (450 V) electric organs they use for stunning prey.

South American Catfishes

The 13 families and more than 1,300 species of catfish in South America rival the characins in their diversity (Fig. 18.4). They include the poorly known velvet catfishes

(Diplomystidae), a small family that is the "primitive" sister group to all other catfishes. The largest of the families is the Loricariidae (650 species, discussed below). Two other important families are the armored catfishes (Callichthydae, 175 species) and the thorny catfishes (Doradidae, 90 species). Members of both families tend to be moderately deep bodied (but flattened ventrally) and protected with large, bony plates and spines. The callichthyids are characteristic of waters that frequently become stagnant, so they swallow bubbles of air, which are absorbed in a highly vascularized portion of the hind gut. Like the clariid catfishes of Africa and Asia, many of these catfish are capable of moving overland on their stout pectoral spines. Another family of armored catfish is the spiny dwarf catfishes (Scoloplacidae, six species), which are distinguished by their tiny adult size of 14 to 20 mm (Berra 2001). This puts them among the smallest of all vertebrates.

The remaining families of South American catfishes all have smooth bodies. The Pimelodidae, the long-whiskered catfishes, fill the "typical" catfish role, being widely distributed (including in Central America and Cuba), abundant (>300 species), and large enough to support fisheries for them (Berra 2001). The bodies of pimelodids, however, tend to be much more flattened than those of the ictalurid and bagrid catfishes, and they possess extremely long barbels (three pairs). One of the larger species (to 2 m) is the dourada, which migrates from the estuary of the Amazon River more than 3,300 km upstream to spawning areas, supporting major fisheries as they go. They feed on the characin fishes that eat the fruits and seeds of the flooded rain forest, demonstrating a connection between fisheries and rain forests (Barthem and Goulding 1997). Eight other families of smooth catfish are mostly distinguished by their minor morphological variations on the typical catfish theme.

Loricariidae

The suckermouth armored catfishes, with more than 650 species, account for approximately 25% of all catfishes (Berra 2001). These catfishes have taken the basic catfish morphology to some interesting extremes, including forms that are extremely attenuated and woody in apperance (stick catfishes, *Farlowella*) and forms with elaborate, plant-like bristles on their snouts (bristlenose catfishes, *Ancistris*). With their armor and elaborate disguises, such catfishes are able to thrive in predator-rich environments. Most are adapted for scraping or sucking algae from the bottom in streams and can use their sucker-like mouths for holding onto rocks in fast water. A few species not only live on and in logs but also feed on the wood, although how they digest wood is a mystery. Curiously, many species can breathe air when the water stagnates and have reinvented the original bony fish method for doing this, vascularized sections of gut. Loricariid catfishes are familiar to aquarists as "plecos," after species of algae scrapers that were once in a now-defunct genus *Plecostomus* (Berra 2001). One of the commonly kept species, however, is the suckermouth catfish, *Hypostomus plecostomus*.

Trichomycteridae

The more than 150 species of pencil catfishes are small, elongate catfishes that lack the adipose fin and have tiny eyes. Most live secretive lives in streams, under rocks in swift water, in dense beds of vegetation along the edges, or in sand, mud, or leaf litter. They are often known as parasitic catfishes, because a few tiny species are known to enter the urinary tracts of humans by mistake (Berra 2001). These fishes are apparently attracted to a urine stream, much as they would normally be attracted to the flow of water coming from the opercular cavity of large fishes. They then enter the gill cavities and feed on gill filaments and blood. Other species specialize in ambushing passing fish to snatch scales or chunks of fin.

Marine Catfishes

The only families in the Otophysi whose member species are primarily found in salt water are the Ariidae (sea catfishes) and the Plotosidae (eeltail catfishes). The sea catfishes are unspecialized-looking catfishes found in tropical and subtropical waters over much of the world but particularly in estuarine waters. On the east coast of the United States, the gafftopsail catfish *(Bagre marinus)*, so named because of its large dorsal fin, and the sea catfish *(Arius felis)* are common fish that move in and out of the estuaries on a seasonal basis. They swim about, feeding on benthic invertebrates, in noisy schools, with the noise being created by the clicking of pectoral spines and the vibration of the swimbladder. The males of these two species, as well as of other species in the family, incubate the eggs in their mouths. The eeltail catfishes look as if they have the head of a catfish welded onto the body of a stout eel. They are widely distributed in the Indian and Pacific oceans, where some species have even managed to become part of complex tropical reef faunas. In Australia, the freshwater catfishes evolved from marine eeltail catfish ancestors. Their spines are particularly toxic.

18.6 GYMNOTIFORMES

The more than 60 species of electric knifefishes are found entirely in the fresh waters of South America, where they presumably share a common ancestor with modern catfishes (Fink and Fink 1981). All have electric organs that are used for navigation and for detection and capture of prey at night or in murky water (Box 18.3). As a result, they are remarkably similar to the mormyrids of Africa, although they evolved independently (Fig. 18.3d). The distinctive morphology of the Gymnotiformes can be largely explained as an accommodation for efficient use of their electric organs (Möller 1995). Their bodies are eel-like, and much of their muscle mass has been converted to electric organs, which can generate large electric fields. Because of the large electric organs and the need to keep the body fairly rigid to create a uniform electric field, the fish propel themselves, slowly and gracefully, with wave-like movements of their long (>140 rays) anal fin. Using this fin, they can move either forward or backward. Each fin ray actually moves in a circular pattern, made possible by the way the fin rays connect to the internal supporting bony element (pterygiophore). Other fins on these fish are minimal; the pelvic fins are absent, the caudal absent or very small, the pectorals small, and the dorsal absent or reduced to a few filaments. The viscera of the gymnotiforms are confined to a relatively small anterior part of the body. The gills are small and have small openings, reflecting both the leisurely activity patterns of the Gymnotiformes and the ability of many of them to breathe air. The dependence on electric organs for sensory input is indicated by the small eyes of gymnotiform fishes, although some sand knifefishes

BOX 18.3

Electric Eels: Shocking Predators

Perhaps the best known of the gymnotiform fishes is the electric eel *(Electrophorus electricus)*, the only member of the family Electrophoridae. It inhabits the backwaters and shallow streams of the Amazon region of South America, where it may grow to 2.5 m in length. It is an obligate air breather and obtains oxygen by holding a bubble of air in its mouth and absorbing the oxygen through vascular tissue on the roof of the mouth (Berra 2001). Approximately half the body musculature of this fish has been converted into electric organs, which together can produce 350 to 650 V, depending on the length of the fish, although the amperage is low (<1 A). These extraordinary electrical powers are used both for defense and for stunning the fish on which it feeds. The eels also have two other, much smaller sets of electrical organs, one of which is used for navigation. Apparently, the main prey of electric eels are other electric fish, which they detect using their electrical sense organs (Westby 1988). The eels make themselves "invisible" to their prey by turning off their own electrical systems and waiting in ambush. Prey fish that detect a waiting eel attempt to confuse them by shutting down their own electrical activity for a few seconds at irregular intervals.

(Rhamphichthyidae) have an attenuated, finger-like caudal region that apparently enables them to feel their way into protective cover, which they frequently enter backward.

18.7 LESSONS FROM MINNOWS, CHARACINS, AND CATFISHES

The otophysans demonstrate how the "right" combination of behavioral, physiological, morphological, and life-history traits has allowed one major evolutionary line of fish to dominate freshwater environments. Members of this group are not only abundant and widely distributed but also are some of the smallest (<15 mm) and largest (>5 m) freshwater fishes. One factor contributing to their high diversity is the coexistence in all areas of forms that are active mainly during the day (Cypriniformes, Characiformes) with forms that are active mainly at night (Siluriformes, Gymnotiformes). Most remarkably, the gymnotiform fishes evolved electrical sensory systems to handle the highly turbid conditions of tropical rivers, paralleling the evolution of mormyrid fishes in Africa. Whereas minnows and characins frequently are visual predators, their superb sense of hearing and well-developed sense of taste make clear that visual cues are often of secondary importance to them. For catfishes, vision is clearly a secondary sense. All this reflects the frequent lack of water clarity and the turbulence of stream environments, the principal freshwater habitat. It is interesting to note that the ostariophysans became the dominant freshwater fishes after a meteorite hit the Yucatan Peninsula 65 million years ago, which caused worldwide changes to the environment, including massive erosion from devastated land masses. Presumably, a premium for survival would have been placed on fishes that could live in warm, turbid water.

SUPPLEMENTAL READINGS

Alexander 1966a, 1966b; Barthem and Goulding 1997; Berra 2001; Bruton 1996; Fink and Fink 1981; Sawada 1982; Teugals 1996; Winfield and Nelson 1991.

WEB CONNECTIONS

GENERAL
www.fishbase.org/home.htm
www.calacademy.org/research/ichthyology
www2.biology.ualberta/jackson.hp/IWR
www.flmnh.ufl.edu/fish/default.htm
www.people.clemson.edu/~jwfoltz/wfb418

NORTH AMERICAN FORMS
www.nativefish.org/articles/

CATFISH
www.planetcatfish.com

CHAPTER 19
Smelt, Salmon, and Pike

SUBDIVISION TELEOSTEI
  Infradivision Euteleostei
    Superorder Protacanthopterygii
      Order Argentiniformes
        Families Argentinidae (herring smelts), Microstomatidae, Bathylagidae
          (deepsea smelts), Opisthoproctidae (barreleyes), Leptochilichthyidae,
          Alepocephalidae (slickheads), Platytroctidae (tubeshoulders)
      Order Salmoniformes
        Suborder Salmonoidei
          Families Salmonidae (salmon, trout, whitefish)
        Suborder Osmeroidei
          Families Osmeridae (smelts), Salangidae (noodlefishes), Retropinnidae (New
            Zealand smelts), Lepidogalaxiidae (salamander fishes), Galaxiidae
            (galaxiids)
      Order Esociformes
        Families Esocidae (pikes), Umbridae (mudminnows)

A nyone who wishes to understand how modern fish systematics progresses should explore the history of the Protacanthopterygii, the fishes grouped together in this chapter (Fig. 19.1). A good place to start is with Nelson (1994), who notes that the large number of recent analyses "might suggest that we now have a strong basis for classifying this taxon; however, much disagreement exists . . . " (p. 175). Then, check out Johnson and Patterson (1996), who start with a quotation from a popular song: "We all make mistakes; then we're sorry," reflecting how careless work in systematics can *really* cause confusion. Johnson and Patterson's effort is the most recent definitive work on this mixed group of fishes, so we follow their suggestions for classification. The high level of interest in this group is reflected in its name: *Prot* (original)-*acanth* (spine)-*pterygii* (fin). This name indicates that the group is supposed to contain the fishes most similar to the presumed ancestors of the spiny-rayed fishes (Acanthopterygii) that dominate the world's oceans. The work of Johnson and Patterson indicates that the argentiniform and salmoniform fishes form one evolutionary line, whereas the esociform fishes form another, with the latter being closest to the rest of the "higher" teleosts.

The superorder as constituted here contains more than 310 species divided into three orders. The fish in these orders lack spines, but many possess an adipose fin. Their

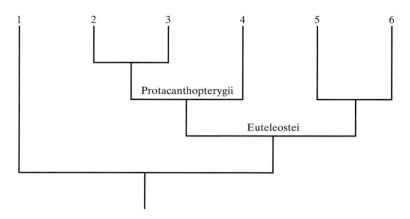

FIGURE 19.1 Phylogenetic relationships of the Protacanthopterygii. **1.** Ostario-physi. **2.** Argentiformes. **3.** Salmoniformes. **4.** Esociformes. **5.** Stenopterygii. **6.** Other Euteleostei.

pelvic fins are abdominal in position, widely separated from the pectoral fins (the pelvic and pectoral girdles are not connected), and placed low on the body. The scales are cycloid, and the upper jaw contains both the maxilla and the premaxilla but the latter is not protractile. The swimbladder is physostomous, connected to the gut with a duct. From an ecological perspective, the superorder contains three rather different kinds of fish: (1) freshwater predators, (2) diadromous species, and (3) deep sea species. The diadromous fishes are among the most studied of fishes, because they include the valuable and much-favored salmonids and smelts plus the curious galaxiids of the southern hemisphere.

19.1 ARGENTINIFORMES

The more than 160 species, from seven families, in this order are all small, smelt-like fishes with large or tubular eyes that live in deep sea environments. They are either silvery in color (Argentinidae, Microstomatidae, Bathylagidae, Opisthoproctidae) or black (Leptochilichthyidae, Alepocephalidae, Platytroctidae). Many lack the adipose fin, and only a few have photophores. All, however, possess an ***epibranchial (crumenal) organ*** for grinding up small prey. This organ consists of a small diverticulum (pouch) just behind the fourth gill arch. The gill rakers from both sides can fit into this pouch, where they interdigitate with one another to break up the food particles. One of the more peculiar adaptations in some members of this order, including the 10 known species of barreleyes or spookfishes (Opisthoproctidae), is tubular eyes. Such eyes are pointed upward and contain exceptionally large numbers of rods. The fishes that possess such eyes are apparently capable of binocular vision under extremely low-light conditions. Fishes with tubular eyes typically prey on small, active invertebrates toward the lower limits of light penetration; the upward-pointing eyes enable them to locate prey silhouetted against the light above.

19.2   SALMONIFORMES

The smelts and the salmonids, with a few interesting exceptions, are largely cold-water fishes with adipose fins that show considerable capability for moving between fresh and salt water. Generally, however, they have been regarded as being only distantly related. Johnson and Patterson (1996) bring the two groups together in the Salmoniformes on the basis of 11 shared characters, including characteristics of the caudal skeleton, absence of radii on the scales, and presence of nuptial tubercles on at least some members of each group. The distinctness of the two groups, however, is reflected in the existence of two suborders; Osmeroidei, and Salmonoidei.

19.3   OSMEROIDEI

The suborder Osmeroidei contains small, elongate fishes that prey on small invertebrates. They are frequently found in large numbers and are favored as food fishes. Three distinct groups are known: (1) the northern smelts, (2) the noodlefishes, and (3) the southern smelts.

Northern Smelts

Within this group are "true" smelts of the northern hemisphere (Osmeridae). Within the Osmeridae (Fig. 19.2C) are only approximately 13 species, but these species are often enormously abundant in coastal areas of the northern hemisphere. Some species are entirely marine (although confined to inshore areas). Some are entirely freshwater or brackish-water dwellers, and some are anadromous. All make excellent eating, but only about half the species are regularly sought as food by humans. Smelts are generally small (usually <20 cm, although some species may reach 40 cm), but they are still voracious pelagic carnivores, consuming both zooplankton and small fish. The mouth is well equipped with small teeth, including those on the maxilla and premaxilla. As in salmonids, the adipose fin is present, but unlike salmonids, the axillary process (a projecting modified scale) on the base of the pelvic fin is absent and the number of pyloric cecae ranges from 0 to 11. Smelts are silvery in color and, as McAllister (1963) notes, have "a curious cucumber odor" that is caused by a specific chemical but that has an unknown function. Smelts generally seek out gravelly areas, either beaches or riverine riffles, for spawning, and they lay adhesive demersal eggs. When they aggregate for spawning, smelts are very vulnerable to capture by humans. Spawning periods are also when whales, predatory fish, and sea birds move into the inshore areas to feast on the concentrated fish.

Noodlefishes

The noodlefishes or icefishes (Salangidae) are also the subject of coastal fisheries in Japan, China, and Southeast Asia. Like the true smelts, these 11 species are small (to 16 cm) and often very abundant, especially in brackish-water lakes and bays (Sarawatari and Okiyama 1992), although some species are found in fresh water. These fish are elongate, scaleless, and nearly transparent because of a poorly ossified

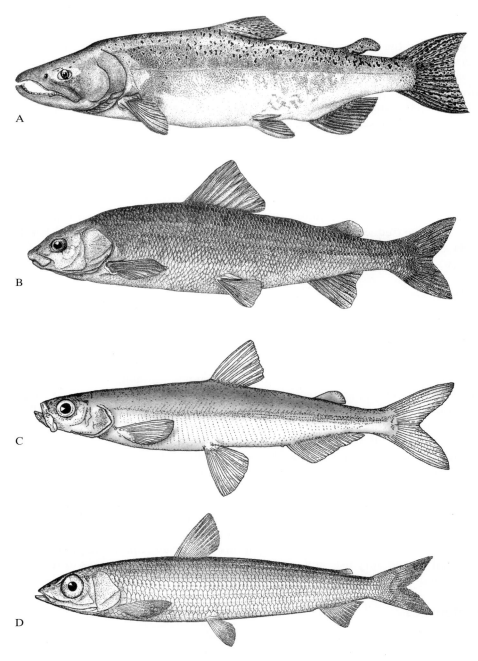

FIGURE 19.2 Representative protacanthopterygian fishes. **A.** Chinook salmon (*Oncorhynchus tshawytscha*, 64 cm, Salmonidae). **B.** Mountain whitefish (*Prosopium williamsoni*, 28 cm, Salmonidae). **C.** Delta smelt (*Hypomesus transpacificus*, 8 cm, Osmeridae). **D.** Atlantic argentine (*Argentina silus*, 36 cm, Argentinidae). **A–C** from Moyle (1976a); **D** from Goode and Bean (1895).

skeleton; their transparent flabbyness is the reason they are called noodlefishes. These and other characteristics can be regarded as being typical of larvae, so noodlefishes have long been regarded as neotenic (adult fish possessing larval traits) (Roberts 1984).

Southern Smelts

Throughout the coastal and fresh waters of Australia and New Zealand are a group of small, often trout-like fishes that fall into three (or more) families: (1) Galaxiidae, (2) Retropinnidae, and (3) Lepidogalaxiidae. The galaxiids are a widely distributed and readily recognizable group. They are small (usually <15 cm), elongate fishes lacking an adipose fin but having a dorsal fin placed far back on the body, above the anal fin. The caudal fin is usually rounded, the pectoral fins large, and the pelvic fins small or absent in some species. Scales and maxillary teeth are lacking. Most of the 50 or so species spend their entire life cycle in fresh water, but a few spawn in estuaries and have marine larvae. Inanga (*Galaxius maculatus*) move downstream to spawn in response to lunar cycles so they can lay their eggs among grasses flooded by high spring tides. When the tides ebb, the embryos are stranded among the grasses, but they hatch when the next high tide covers the grasses, approximately 2 weeks later (McDowall 1990). In streams, galaxiids (Box 19.1) are the approximate ecological equivalents of salmonids in the northern hemisphere. They are either amphidromous or catadromous, however, rather than being anadromous (see Chapter 11 for definitions). The migratory nature of New Zealand

---

BOX 19.1
Galaxiids: Small Fish in Remote Places

The galaxiids are an especially fascinating group for zoogeographers because of their occurrence in fresh water on all the southern continents except Antarctica. They are found in Australia, Tasmania, New Zealand, southern South America (including Tierra del Fuego and other islands), South Africa, and New Caledonia. Most of these areas have endemic genera and species, but the inanga is present in all regions, except South Africa, where galaxiids are found. The distribution pattern of the galaxiids as a family can be explained by the former connection of the continents and their subsequent separation through continental drift (see Chapter 25). On the other hand, McDowall and Robertson (1975) argue that the distribution of inanga is best explained by the fact that larvae of this species are planktonic and marine and, thus, are capable of being carried to distant shores by oceanic currents. Recent genetic evidence also suggests that the separation of species on different islands took place, at least in part, after the continents split up, indicating that marine dispersal is a potential mechanism for galaxiids to colonize distant places (Waters and Wallis 2001). Curiously, many species in streams become nonmigratory and can only disperse through freshwater routes. Thus, on the South Island of New Zealand are approximately 10 species of galaxiids whose evolution in isolation from a common ancestor is best explained by rivers becoming connected and then separated during periods of geologic upheaval (Waters et al. 2001).

stream fishes like the galaxiids means that stream fish assemblages become progressively less rich in an upstream direction because of the differing abilities of the fish to overcome barriers (McDowall 1998).

One of the more curious purported galaxiid relatives is the salamander fish (*Lepidogalaxias salamandroides*), a tiny (<60 mm) fish discovered in 1961 in southeastern Australia, where it inhabits acidic, ephemeral streams and ponds (Berra and Allen 1989). It is eel-like and has a wedge-shaped head, which allows it to burrow into damp sand or litter when the water disappears (Fig. 19.3). After rains, it emerges quickly and preys on insects with its surprisingly formidable teeth. This fish is so peculiar that its relationships to other fishes are in doubt. It does not seem to be related to the galaxiids (although it is now placed with them in the Osmeroidei), and it may be more closely related to the pikes, Esociformes (Waters et al. 2000). If it is a relative of the pikes, then it would be the only member of this group in the southern hemisphere.

19.4 SALMONOIDEI

The Salmonoidei suborder contains just one family, the Salmonidae, with approximately 70 species. The Salmonidae can be divided into three readily recognizable groups, usually treated as subfamilies (but often as families): (1) Salmoninae (salmon and trout), (2) Coregoninae (whitefishes), and (3) Thymallinae (graylings). Salmon and trout have fine scales (>110 in the lateral line), a short dorsal fin, and teeth on the maxillary bone of the upper jaw. Whitefishes have coarse scales (<110 along the lateral line), a short dorsal fin, and no teeth on the maxillary bone. Graylings have moderately large scales (70–110 in the lateral line); a long, sail-like dorsal fin; and teeth on the maxillary bone.

Salmonids are the dominant fishes in the cold-water streams and lakes of North America and Eurasia, where they support major sport and commercial fisheries. Because of the favor they find with anglers and their ease of propagation, salmonids now dominate the cold fresh waters of the southern hemisphere as well. A member of the family is readily recognized by its streamlined body, forked tail, adipose fin, axillary process, and large number of pyloric ceca (11–210) and branchiostegal rays (7–20) (Fig. 19.2A and B). Most species grow to at least 20 cm, and the largest species is the chinook salmon, which may grow to nearly 1.5 m (57 kg).

Perhaps no group of fishes has been more studied than the Salmonidae, as their presence in numerous examples of this textbook attests. The reasons for this high interest are their mystique as a sport fish combined with their value in commercial fisheries and aquaculture. Because of their abundance and their occurrence in streams and hatcheries near research centers, they also are convenient to study. As a group, salmon and trout are highly adapted physiologically for survival in the cold and fluctuating waters of the northern hemisphere. Their migratory behavior also enables them to persist in geologically and climatically unstable regions. Most species are anadromous or are derived from anadromous forms. Nonanadromous forms have evolved repeatedly when populations of anadromous fish become trapped above new barriers or otherwise find it evolutionarily advantageous not to go to sea.

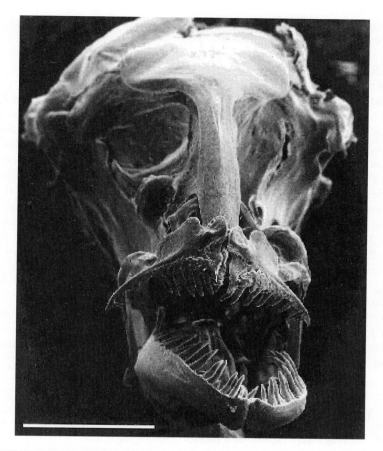

FIGURE 19.3   Scanning electromicrograph of the head (**top**) of a 3-cm Australian salamanderfish, *Lepidogalaxias salamandroides* (**bottom**). The salamanderfish is a small species that lives in acidic, ephemeral streams in south Australia and uses its teeth for the capture of large prey. From Berra and Allen (1989). (Photos by Clay Bryce [top] and Tim M. Berra [bottom].)

The advantages of anadromy to salmon and trout are considerable. First, it provides a superb dispersal mechanism that has allowed rapid recolonization of regions suddenly made available by retreating glaciers or other events. Second, the abundance of small fish and shrimp in the ocean provides food for rapid growth. In the ocean, salmon and trout grow extremely rapidly and, consequently, produce more (and larger) eggs than fish that spend their entire life cycle in fresh water. A 500-g rainbow trout that has lived in a stream for 4 years typically produces fewer than 1,000 eggs, whereas a 4-kg anadromous trout (steelhead) of the same age produces 4,000 or more. The advantage of continuing to use streams for spawning is, presumably, the protection that streams afford the embryos and young from marine predators, including other salmon and trout. The embryos are particularly well protected, because they are buried within riffles in gravel beds that permit penetration of oxygen-rich water but few predators. By the time the young fish emerge from the gravel, they are large (for newly hatched fish) and quite active, so they can avoid most invertebrate predators and feed on a wide range of prey themselves. Despite the advantages of anadromy, however, it is a variable trait among salmonids (Box 19.2). It is obligatory in some species of Pacific salmon and in Atlantic salmon (*Salmo salar*), optional for most species of trout, and not present at all in charr (*Salvelinus*), most species of whitefish, and grayling. Interestingly, some species of charr and trout have populations that move into salt water as juveniles, but they do not move far, often not even leaving the estuaries and coming back to rear some more in fresh water after a few months (Stearley 1992).

A question that has long fascinated people is what evolutionary advantage is provided to Pacific salmon by dying after spawning (see Chapter 9)? Crespi and Teo (2002)

BOX 19.2
Special Names for Special Fishes

Anadromy is not a universal feature of the salmonids, but most species do undertake spawning migrations and possess a series of distinctive life-history stages. Thanks to interest by salmon and trout anglers, a special set of terminology has arisen in association with these stages. The eggs are laid in a depression dug in the gravel, called a **redd**. The embryos develop into **alevins** or sac fry, small fish that still possess a yolk sac. As the yolk sac is absorbed and the alevins emerge from the gravel, they become **fry**. The small, active fry develop a series of bars on their sides (parr marks) and are then called **parr**, a stage that may last a few months or years. Stream-dwelling trout frequently retain the parr marks throughout their life. In anadromous populations, parr transform into silvery **smolts** and migrate to the sea. Because the parr–smolt transformation is in preparation for a dramatic change in habitat, it includes profound changes in morphology, physiology, and behavior (Hoar 1976). In the ocean (or large lakes), the smolts gradually become mature adults and return to their home streams for spawning. Some males, called **jacks** or **grilse**, return to spawn at an early age and a small size. Small females are called **jills**. Spawned-out fish are termed **kelts**. Finally, people who spend their lives fishing for (or studying) trout and salmon are said to be afflicted with **adipose fin disease**.

present evidence that semelparity is advantageous if adults have a low probability of survival but juveniles have a fairly high probability of survival. Because salmon must make long migrations in the ocean in search of food, their mortality rates are high. Under such conditions, it "pays" to invest as much energy as possible in the eggs, so females of semelparous species tend to produce more and larger eggs than females of iteroparous species. Larger embryos tend to have higher survival rates than smaller embryos, thereby providing an additional advantage to semelparity.

In both inland and coastal drainages, the migratory tendencies of salmonid fishes tend to promote the creation of isolated populations that become quite distinct in their morphology and color patterns. High-mountain populations of trout and charr often develop into distinctive, spectacularly colored forms that can be described as subspecies (e.g., Behnke 1992). Good examples include the brilliant green and red Colorado cutthroat trout and the three subspecies of golden trout in California. Less appreciated perhaps is the tendency of whitefish to develop distinctive morphological types when isolated in lakes. Before the invasion of non-native fishes caused extinctions of several species of *Coregonus*, the Laurentian Great Lakes had a "flock" of species/forms that segregated by temperature and depth (S. H. Smith 1968).

Despite the enormous interest in salmonids, including whitefish and grayling, many species, runs, and populations are either already extinct or in danger of extinction. Many factors are working together to produce this situation. Foremost are the destruction of habitats in which the fish spawn and rear their young and the construction of dams that deny the fish access to spawning areas or that change the nature of the streams. Salmon and trout require cold water of high quality, preferably flowing through relatively undisturbed watersheds. Increasingly, however, such habitats are not available. Salmonids are relatively easy to culture, so many large salmon and trout hatcheries have been developed to allow fisheries to continue despite massive habitat loss. Though obviously successful, hatcheries can have negative effects on wild salmonids. First, they have historically allowed habitat destruction to be justified because of promises for no net loss of fish to fishermen. Second, hatchery fish can be harvested at much higher rates than wild fish; hatcheries require many fewer spawners to replace their stocks each year than do wild populations. Unfortunately, most salmon fisheries on mixed stocks do not distinguish between fish of natural and hatchery origin, so wild-fish stocks may decline because of excessive exploitation. Third, hatchery fish can spread diseases to wild fish, including diseases not native to a region. Fourth, traits that improve survival in a hatchery often are not the same as those that improve survival in the wild. Hatchery fish tend be less wary, may produce smaller eggs, be less aggressive, or have other traits that make it difficult to use them to reestablish populations in the wild. More important, fish released from a hatchery do interact with the wild fish already in the system. The hatchery fish may be less fit for long-term survival in the wild, but their sheer numbers can disrupt the breeding and feeding behavior of wild salmonids, especially in streams. In some instances, the larger fish released from a hatchery may actually prey on their smaller cousins, resulting in further declines in wild populations.

The increased documentation of these problems is gradually changing the practices of hatcheries run by governmental agencies and increasing the focus on restoring wild populations of salmon and trout by restoring habitats. Unfortunately, at the same time, massive private aquaculture operations are developing for salmonids, especially

Atlantic salmon, and are resulting in "inadvertent" (but inevitable) releases of millions of hatchery fish throughout various parts of the world (see Chapter 37.7).

19.5   ESOCIFORMES

The order Esociformes contains just 10 species, all confined to fresh water: five species of pike (Esocidae), and five species of mudminnow (Umbridae). The pikes and mudminnows are widespread in North America and northern Eurasia. Their body morphology is that of a lie-in-wait predator, with dorsal and anal fins placed far back on the body, about equal in size, and aligned with each other. The adipose fin is lacking, as are pyloric cecae, teeth on the maxillary bone, and the mesocoracoid bone in the shoulder girdle. With their elongate snouts and forked tails, the pikes are important piscivores in weedy lakes and other slow-moving bodies of water (Craig 1996). In North America and Europe, they are favorite sport fishes, especially the circumpolar northern pike (*Esox lucius*), which may reach 24 to 26 kg (1.3 m), and the muskellunge (*E. masquinongy*) of eastern North America, which may reach 32 kg (1.6 m). Pike generally capture their prey during a sudden rush from cover. They grab the prey sideways, impaling it on sharp teeth before retreating to cover to turn the prey around and swallow it head first. Pike require flooded vegetation for spawning and nursery areas for their young, so stabilization of lake levels and stream flows by humans is frequently detrimental to their populations.

In contrast to the pikes, the mudminnows are all small, rarely reaching 20 cm, and have blunt snouts and rounded tails. All, however, are voracious lie-in-wait predators of invertebrates. They are most characteristic of weedy ponds, lakes, and backwaters, where temperatures may be very warm and oxygen levels low. At least three species, including the Alaska blackfish (*Dallia pectoralis*), have limited abilities to breathe air. The mudminnows are of particular interest to zoogeographers, because they are obligatory freshwater fishes with a distinct family distribution pattern. The three species of *Umbra* are found in Atlantic coast drainages of North America, the Mississippi River system, and Europe, respectively. The Olympic mudminnow (*Novumbra hubbsi*) is confined to the Olympic peninsula of Washington, whereas the Alaska blackfish is widely distributed in arctic North America and Siberia (see Chapter 25). Blackfish are extraordinarily abundant in some arctic lakes and ponds and have been used by the native people of these areas as food for themselves and their dogs.

19.6   LESSONS FROM SMELT, SALMON, AND PIKE

Three major lessons can be drawn from this chapter:

1. The pikes, smelts, and salmonids may have many characteristics that make them reasonable representatives of the ancestors of more derived groups of fishes, but they also are highly specialized and successful fishes in their own right. Salmonids, for example, are superbly adapted in their morphology, physiology, and behavior as cold water predators in streams, lakes, and the ocean.

2. Life cycles that entail movement between fresh and salt water can be very successful in terms of numbers of individuals produced by a species, but the total number of

species that can adopt such strategies seems to be limited. As the galaxiids illustrate, migratory behavior can lead to widely distributed species and to localized species that develop in isolation when migratory behavior becomes disadvantageous.

3. For fishes, having a high value to humans can be a mixed blessing. Salmonids are among the most sought-after fishes in the wild and support high-value aquaculture operations, yet they also have more than their share of endangered species.

SUPPLEMENTAL READINGS

Berra 2001; Craig 1996; Groot and Margolis 1991; McDowall 1990; Nelson 1994; Stearley 1992.

WEB CONNECTIONS

GENERAL
www.calacademy.org/research/ichthyology
www.fishbase.org
www2.biology.ualberta/jackson.hp/IWR
www.flmnh.ufl.edu/fish/default.htm
www.people.clemson.edu/~jwfoltz/wfb418

SALMONIFORMES
www.natureserve.org/explorer/speciesIndex/Order_Salmoniformes_100077_1.htm

PACIFIC SALMON, SMELT
www.psmfc.org

ATLANTIC SALMON
www.asf.ca/Overall/atlsalm.html
www.nefsc.noaa.gov/sos/spsyn/af/salmon

CHAPTER 20
Anglerfish, Barracudinas, Cods, and Dragonfishes

SUBDIVISION TELEOSTEI
  Infradivision Euteleostei
    Superorder Stenopterygii
      Order Stomiiformes
        Families Gonostomatidae (bristlemouths), Sternoptychidae (marine hatchet-
          fishes), Photichthyidae (lightfishes), Stomiidae (barbeled dragonfishes)
      Order Ateleopodiformes
        Family Ateleopodidae (jellynose fishes)
    Superorder Cyclosquamata
      Order Aulopiformes
        Families Aulopodidae (aulopus), Pseudotrichonotidae, Synodontidae (lizard-
          fishes), Chloropthalmidae (greeneyes), Notosudidae (waryfishes), Ipnopi-
          dae (spiderfishes), Alepisauridae (lancetfishes), Paralepididae
          (barracudinas), Evermannellidae (sabertooth fishes), Scopelarchidae
          (pearleyes), Omosudidae, Bathysauridae, Giganturidae (telescope fishes)
    Superorder Scopelomorpha
      Order Myctophiformes
        Families Neoscopelidae, Myctophidae (lanternfishes)
    Superorder Lampridiomorpha
      Order Lampridiformes
        Families Veliferidae, Lampridae (opah), Stylephoridae (tube-eye), Lophotidae
          (crestfishes), Radiicephalidae, Trachipteridae (ribbonfishes), Regalecidae
          (oarfishes)
    Superorder Polymixiomorpha
      Order Polymyxiformes
        Family Polymixiidae (beardfishes)
    Superorder Paracanthopterygii
      Order Percopsiformes
        Families Percopsidae (troutperches), Aphredoderidae (pirate perch), Amblyop-
          sidae (cavefishes)
      Order Ophidiiformes
        Families Carapidae (pearlfishes), Ophidiidae (cuskeels), Bythitidae (viviparous
          brotulas), Aphyonidae, Parabrotulidae (false brotulas)
      Order Gadiformes
        Families Ranicipitidae (tadpole cod), Euclichthyidae (eucla cod), Macrouridae
          (grenadiers), Steindachneriidae (luminous hake), Moridae (flatnose cods),

> Melanonidae (pelagic cods), Macruronidae (southern hakes), Bregmaceroti-
> dae (codlets), Muraenolepidae (eel cods), Phycidae (phycid hakes), Mer-
> lucciidae (merluccid hakes), Gadidae (cods)
> **Order Batrachoidiformes**
>   **Family Batrachoididae (toadfishes)**
> **Order Lophiiformes**
>   **Families Lophiidae (goosefishes), Antennariidae (frogfishes), Brachionichthyi-
>   dae (handfishes), Chaunacidae (sea toads), Ogcocephalidae (batfishes),
>   Caulophrynidae (fanfins), Neoceratiidae, Melanocetidae (black devils),
>   Diceratiidae, Himantolophidae (football fishes), Oneirodidae (dreamers),
>   Thaumatichthyidae, Gigantactinidae (whipnoses), Centrophrynidae
>   (deepsea anglerfish), Ceratiidae (seadevils), Linophrynidae (netdevils)**

In this chapter, we enter the freeway leading to the most complex metropolis of teleost systematics, sometimes called Neoteleostei (new teleosts). The exits eventually lead to between 15,000 and 16,000 species. A major problem (from a human point of view) is that the freeway is still under construction, and in many cases, the locations of exits, both major and minor, have not yet been decided. Considering the complexity of the task before the systematists, who must find ways to fit an incredibly diverse array of fishes into a common plan, this is not surprising. Most of the fishes ($\approx$13,500 species) are located in Acanthopterygii, the dense metropolis of spiny-rayed fishes that dominate the shallow, inshore waters of the oceans and are treated in the following chapters. This, however, still leaves approximately 2,000 species that exited early from the main line and have successfully exploited a wide variety of environments—especially the deepsea, soft-bottomed environments on continental shelves and caves in fresh water. Their specializations obscure their relationships with the Acanthopterygii and with each other, but they also make these fishes some of the most curious known. Five major groups are discussed together in this chapter, as much for convenience as for any other reason (Fig. 20.1).

20.1   STENOPTERYGII

The Stenopterygii ("narrow-fins") have a mixture of ancestral and derived teleost characteristics and, seemingly, a long-independent evolutionary history. They are tropical-to-temperate deepsea fishes, with photophores and large mouths and teeth on both the maxilla and premaxilla (Fig. 20.2). Many species have an adipose fin. The scales, if present, are cycloid. Most species are black in color, but a few are silvery. This group includes the most abundant fishes in the world (bristlemouths) as well as some of the most fearsome-looking small predators (dragonfishes).

Gonostomatidae

The bristlemouths are extraordinarily abundant in many parts of the world's oceans; in fact, members of the genus *Cyclothone* may be more numerous than any other fishes. All 75 known species are quite small (most <5 cm long), however, and live at great depths, so they cannot be exploited. Bristlemouths have large, horizontal mouths with numerous

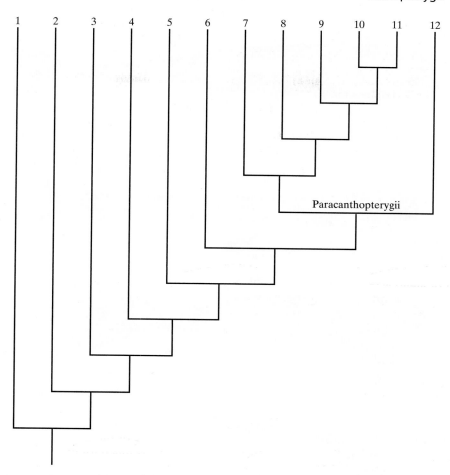

FIGURE 20.1   Probable phylogenetic relationships of mostly deepsea groups of neoteleostean fishes. **1.** Protacanthopterygii (see Chapter 19). **2.** Stomiiformes. **3.** Aulopiformes. **4.** Myctophiformes. **5.** Lampridiformes. **6.** Polymixiformes. **7.** Percopsiformes. **8.** Ophidiiformes. **9.** Gadiformes. **10.** Batrachiformes. **11.** Lophiiformes. **12.** Acanthopterygii (see Chapter 21). The Ateleopodiformes is not shown because of uncertainties in its relationships with the other groups.

small teeth, which is typical of fish that feed on large prey, yet they also have fine gill rakers, which is typical of fish that feed on small prey (Fig. 20.2A). This arrangement may partially explain their success, because it enables them to feed on whatever prey comes along, regardless of its size. The bodies of bristlemouths are elongate, somewhat rounded, and lined with light organs (photophores).

Sternoptychidae

In many deep areas of the oceans, the marine hatchetfishes ($\approx$50 species) are second in abundance only to bristlemouths. These small (to 10 cm) fish are, indeed, hatchet-shaped,

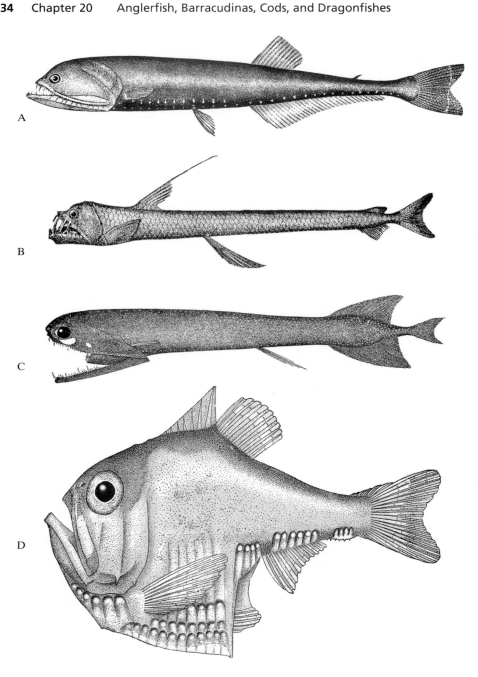

FIGURE 20.2   Representative stomiiform fishes. **A.** Bristlemouth (*Cyclothone elongata*, 8 cm). **B.** Viperfish (*Chauliodus sloani*, 35 cm). **C.** Loosejaw (*Malacosteus niger*, 12 cm). **D.** Hatchetfish (*Argyropelecus olfersi*, 3 cm). **A–C** from Goode and Bean (1895); **D** from Jordan and Evermann (1900).

with deep and extremely compressed bodies (Fig. 20.2D). Not only does the belly taper to a thin edge, but the top of each fish has a "blade" made of fused pterygiophores (the internal bony supports of fin rays). The sides of the body are silvery and contain numerous large photophores. The eyes and mouth are all turned upward, and the eyes of some species are tubular. All these features indicate that most hatchetfishes move up and down the water column, preying on copepods and other migrating zooplankton. The slender profile of these fish when seen from above or below, coupled with the light-scattering abilities of their silvery sides and photophores, helps to conceal them from both predators and prey.

Stomiidae

The barbeled dragonfishes are a diverse group of dark-colored, deepsea predators equipped with a chin barbel and a large, fang-rich mouth that allows some species to capture prey larger than themselves. The viperfishes (*Chauliodus* spp.) have extraordinary fangs and an ability to swing open their mouths to swallow very large prey (Fig. 20.2B). The scaleless dragonfishes (subfamily Melanostomiinae, >160 species) and loosejaws (subfamily Malacosteinae, 15 species) are elongate, lie-in-wait, deepsea predators with rows of tiny photophores (Fig. 20.2C). Their jaws are large, with many sharp teeth, and the long chin barbel of most species has a luminescent "lure" at the end, presumably for attracting prey. The maximum size of these fishes is 10 to 25 cm. The black dragonfishes (*Idiacanthus* spp.) are similar in many respects to the preceding families, except that their bodies are nearly eel-like, with elongate dorsal fins. The black dragonfishes also possess extremely peculiar larvae that are colorless and have eyes on the tips of long stalks. These fishes can achieve lengths between 35 and 40 cm (but only 25–30 g in weight), yet only the females reach such sizes. The males are much smaller and, apparently, serve only for reproduction, because they lack teeth, the chin barbel, and a functional digestive tract.

20.2   CYCLOSQUAMATA

The more that we understand about the anatomical details of deepsea fishes, the more they become subdivided into lines with long-separate evolutionary histories. The fish in the Cyclosquamata ("circle-scales") have been lumped with trout and smelt in the Salmoniformes and then with the lanternfishes in the Myctophiformes (next section). Nelson (1994) places the diverse fishes in one order, Aulopiformes, with 12 to 13 families.

This order is a mixed bag of odd fishes, including one family of inshore fishes—the lizardfishes, Synodontidae—that are all united by the unique structure of the gill arches. Most of the deepsea forms are small (<50 cm) predators capable of devouring a wide range of prey, including other fishes of nearly the same size. The more than 55 species of barracudinas (Paralepididae) look like miniature barracudas, complete with sharp teeth (Fig. 20.3B). They have large eyes but no swimbladder. Although typically less than 15 cm long, barracudinas seem to be important in marine food chains, because they prey on smaller deepsea fishes and are themselves extensively preyed on by such predators as salmon, tuna, and swordfish. The daggertooth *(Anotopterus pharao)* is a barracudina

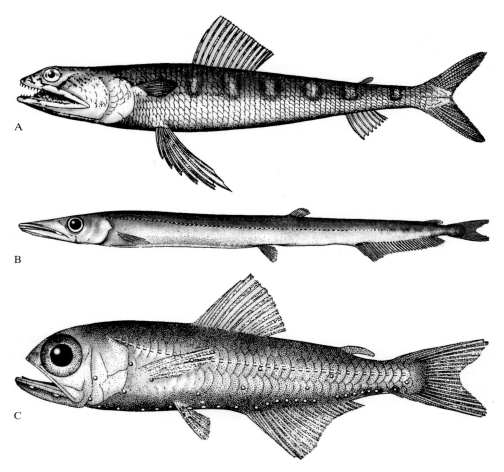

FIGURE 20.3   Scopelomorph fishes. **A.** Lizardfish (*Synodus*). **B.** Barracudina (*Lestidium*).
C. Lanternfish (*Myctophum*, Myctophidae). From Jordan and Evermann (1900).

that has enlarged jaws, with numerous blade-like teeth. Extremely large jaws and teeth
are also characteristic of other families in the order, and the sabertooth fishes (Ever-
mannellidae) are capable of swallowing fish larger than themselves. Both the sabertooth
fishes and the pearleyes (Scopelarchidae) have telescopic eyes that point upward, thereby enabling them to see their prey from below. While these fishes are spectacular predators, they are less than 50 cm long and rarely receive much public attention. This is not
true, however, of the lancetfishes (Alepisauridae), which may exceed 2 m in length and
occasionally wash up on beaches. Besides having long fangs and elongate bodies, these
fish have an enormous, sail-like dorsal fin. They seem to be among the largest fish that
spend all their time in the mesopelagic region of the world's oceans, feeding on all other
fish that are present. A number of rare deepsea fishes are known mainly from specimens
obtained from lancetfish stomachs.

Most aulopiform fishes inhabit the water column of the deep sea, but some of the most curious members are deepsea bottom-dwellers, especially the greeneyes (Chloropthalmidae) and the spiderfishes and grideye fishes (Ipnopidae). The greeneyes are small, slender fish with exceptionally large, pale-green eyes that permit them to see in the dim light that exists just above the bottom at the edge of the continental shelf. There, they feed on small, pelagic invertebrates. In deeper water, the spiderfishes (18 species) play a similar role, except that they "stand" on the bottom with the extraordinarily long rays on their pelvic and caudal fins (see Fig. 35.2). These fishes have tiny eyes, and they likely detect the small invertebrates on which they prey with the long, independent rays of their pectoral fins (Sulak 1977).

The exception to the rule that aulopiform fishes inhabit the deep sea is the family Synodontidae, but even two of these species are deepwater dwellers. The lizardfishes are moderate-sized (25–50 cm) benthic fish that live mainly in the shallow waters of tropical and subtropical areas. Some species, such as the Atlantic lizardfish *(Synodus foetans)*, inhabit temperate waters as far north as Cape Cod. Lizardfish have a lizard-like appearance, with their flattened bony heads, large scales on the body and head, and large mouths in pointed snouts (Fig. 20.3A). The mouth contains numerous long, pointed teeth that are useful for holding onto their prey. The behavior of tropical forms is also lizard-like, in that they will scurry along the surface of a reef in search of prey. They will also conceal themselves to ambush prey by resting quietly on their large pelvic fins or by partially covering themselves with sand in soft-bottomed areas.

20.3 SCOPELOMORPHA

Fishes of the superorder Scopelomorpha are, in many respects, similar to the salmoniform fishes, and the two groups have previously been lumped together in the Salmoniformes (Greenwood et al. 1966). Both have many ancestral features, such as soft-rayed fins, the presence (usually) of an adipose fin, abdominal pelvic fins, and various skeletal features. The Scopelomorpha differ from the Salmoniformes in many important respects, such as the exclusion of the maxillary bone from the gape of the mouth (although the upper jaw is not protrusible) and the presence of a physoclistous (closed) swimbladder. This superorder contains 14 families with approximately 560 species, most of which inhabit deep sea environments where they prey on fish and invertebrates. The diversity of the group is reflected in the fact that 8 of the 15 families contain fewer than 10 species each.

Myctophiformes

Most (235) of the species of myctophiform fishes are lanternfish (Myctophidae) (Nafpaktitis 1978). They are a remarkably abundant group that can be found in all the oceans of the world. They generally live between 200 and 1,000 m in the open water, but some have been observed at depths greater than 2,000 m and others at the surface at night. They are often an important part of the deep scattering layer, so-called because sound produced by sonar units of ships bounces off the thousands of swimbladders in a school, at times giving the impression of a false bottom. The lanternfishes are small, with blunt heads, large eyes, and rows of photophores on the body and head (Fig. 20.3C). The photophore patterns are different for each species, and they are also different for the sexes within each species, which led some early investigators to describe the males and females as separate species. It is not unusual for several species of lanternfish to be caught together in sampling devices. How they manage to coexist is an interesting question, because most are generalists in their feeding, apparently taking whatever prey of suitable size that happens to be available (Tyler and Pearcy 1975). There may be some segregation by depth preferences and by patterns of vertical migration.

20.4   LAMPRIDIOMORPHA

The Lampridiformes (*Lamprid-*, "brilliant") are a bizarre mixture of large, often brightly colored pelagic fishes. Most species apparently live at depths between 100 and 1,000 m, so they are rarely seen despite broad, often worldwide distribution patterns. Their diversity of form is indicated by the fact that seven families, but only 19 species, are recognized. Their relationship to other fishes is not clear, but they have sufficient ancestral features to group them with the rest of the fishes in this chapter. The mouth is highly protrusible, because not only the premaxilla but also the maxilla can be pushed outward. Protrusibility is especially characteristic of the tube-eye *(Stylephorus chordatus)*, which has a membranous pouch connecting the mouth to the cranium. The pouch can be expanded to a volume nearly 38 times that of the closed buccal cavity, thereby creating the tremendous suction pressure for drawing in planktonic organisms (Pietsch 1978b). Most have an elongate dorsal fin.

In body shape, Lampridiformes range from extremely deep-bodied (Lampridae, Veliferidae) to extremely elongate (Lophotidae, Regalecidae). Most spectacular of the deep-bodied forms are the two species of opahs (Lampridae). Not only are these species large (up to 1.5 m), disk-shaped, and laterally compressed, they are also brilliantly colored: The body is dark blue on top, shading to a silver-flecked green and iridescent purple on the sides and then to pink on the belly; the fins and jaws are vermillion. How these colors function is not known, because opahs are found mostly below 100 m, where they feed on other fish, squid, octopus, and crustaceans. In marked contrast to opahs are the two species of oarfish (Regalecidae). These fish are extremely attenuated, reaching lengths of 8 m. The anterior rays of the dorsal fin are modified into a peculiar red "cockscomb" over the head, and the pelvic rays are modified into long filaments. Their mouth and eyes are quite small, and they seem to feed mainly on pelagic crustaceans. Apparently, oarfish normally maintain themselves in the water column in a vertical position, which enables them to see their prey silhouetted against the downstreaming light (Pietsch 1978b). Because oarfish are rarely seen on the surface, the appearance of a "monster" oarfish in a coastal area is likely to be reported as a "sea serpent".

20.5   POLYMIXIOMORPHA

The polymixiomorphs ("many mixed types"), despite their long name, contain just one order, one family, and five species of beardfishes. Just where these fish belong in the great highway of teleost phylogeny remains uncertain, and they are moved about freely in different classification schemes (Nelson 1994). The beardfishes are moderate-sized

(30–40 cm), deep-bodied, and large-eyed fishes found in tropical oceans. They are named for the pair of long barbels that dangles from their chins and are defined by a mixture of derived (e.g., well-developed spines in their fins), ancestral (e.g., two sets of intermuscular bones along the back), and unique (e.g., long hyoid barbels) characteristics.

20.6   PARACANTHOPTERYGII

The Paracanthopterygii (*para*, "like") are a major evolutionary line of predominantly marine fishes (>1,200 species) with many ancestral features in their bone and muscle structure but also many of the derived features of the Acanthopterygii. Among the derived features that characterize most species are: (1) an elaborate and protractile premaxilla; (2) spines on the dorsal, anal, and pelvic fins; (3) reduced numbers of pelvic and caudal fin rays; (4) ctenoid scales; and (5) pelvic fins in the thoracic or jugular position (Fig. 20.4). There is good reason to doubt, however, that the Paracanthopterygii represent a single, unified evolutionary line (Nelson 1994).

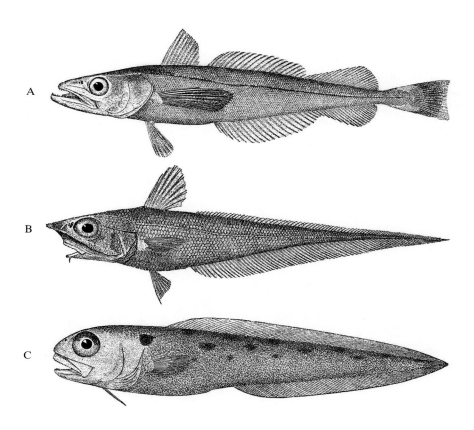

FIGURE 20.4 Paracanthopterygian fishes. **A.** Hake (*Merluccius*). **B.** Grenadier (*Coelorhynchus*). **C.** Cusk eel (*Otophidium*, Ophidiidae). From Goode and Bean (1895).

20.7 PERCOPSIFORMES

The order Percopsiformes ("perch-like form") contains just three families and nine species of peculiar North American freshwater fishes that are all small (<20 cm) in size. These species have an interesting mixture of ancestral and derived characteristics. The premaxilla forms the entire margin of the upper jaw but is not protractile. The pelvic fins (if present) are subthoracic in position, in that they are close to but slightly behind the pectorals. Spines are weakly developed. Scales are either ctenoid or cycloid.

The two species of troutperch (Percopsidae) (Fig. 20.5), with their adipose fins, ctenoid scales, and weak dorsal, pelvic, and anal fin spines, look like a cross between trout and perch. The pirate perch (*Aphredoderus sayanus*), the sole member of the Aphredoderidae, is more perch-like with its ctenoid scales, its spines in the dorsal and anal fins, subthoracic pelvic fins, and most distinctively, its anus in its throat. In contrast to the pirate perch, the eight species of Amblyopsidae have cycloid scales, spines on the anal and dorsal fins (optional), and no pelvic fins (except in one species). They are adapted for living mainly in caves and springs of the limestone regions of the south eastern United States (Fig. 20.6).

20.8 OPHIDIIFORMES

Most members of the order Ophidiiformes ("snake form") are distinctly eel-like, with long, tapering tails (Fig. 20.4C, Box 20.1). Their pelvic fins, when present, are located on the chin, where they are reduced to just one or two fin rays. Many cuskeels and brotulas (families Ophidiidae, Bythitidae, and Aphyonidae) are also inhabitants of the deepsea bottom and may show extreme adaptations for living there, such as poorly developed eyes. Some of the deepest-living fishes are brotulas (see Fig. 35.1). Curiously, some brotulas have managed to adapt to the water of caves, including brackish and fresh water.

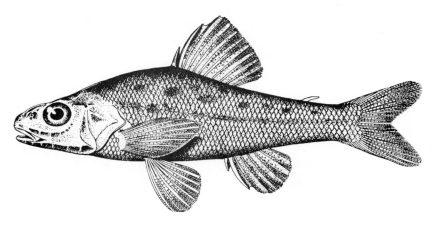

FIGURE 20.5   Sandroller (*Percopsis*, Percopsidae). From Jordan and Evermann (1900).

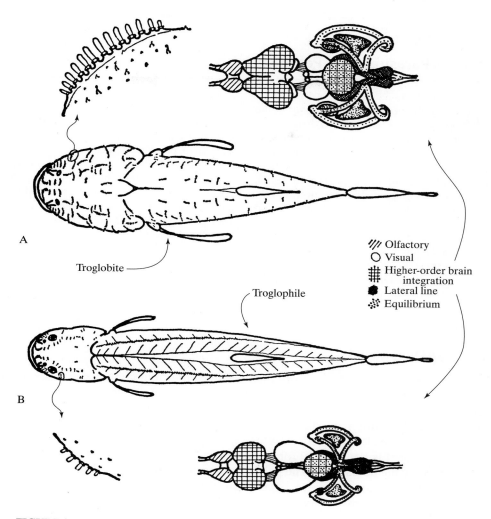

Olfactory
Visual
Higher-order brain integration
Lateral line
Equilibrium

FIGURE 20.6   Adaptations of two species of Amblyopsidae from different cave habitats. The northern cavefish **A**; (*Amblyopsis spelaea*) lives in total darkness and has a well-developed lateral-line system. The sections of the brain dealing with the lateral line are also well developed. The head is larger and has more sensory canals for detection of obstacles than that of the spring cave-fish (**B**; *Forbesichthys agassizi*), which lives in caves that have some light. The longer fins of the northern cavefish allow it to "row" slowly and search for prey efficiently. From Poulson and White (1969); © 1969, American Association for the Advancement of Science.

These forms are typically blind or have reduced eyes. The cave forms occur in a wide range of localities, from the Bahamas to Cuba to the Yucatan Peninsula of Mexico to the Galapagos Islands. The largest family in the order, the cuskeels (Ophidiidae, >160 species) also occurs in a wide range of habitats, from tide pools to the deepest parts of the ocean. Most species are small, however, and not particularly common.

BOX 20.1

Fish as Parasites: The Pearlfishes

Among the more remarkable fishes in the Ophidiiformes are the 27 species of pearlfish (Carapidae), because some live in close association with marine invertebrates, such as sea cucumbers, clams, sea urchins, and starfish (Smith et al. 1981, Trott 1981). Their relationships with the invertebrates range from commensalism to parasitism. In the latter case, certain species live inside the body cavity of sea cucumbers, where they feed on the internal organs (Fig. 20.7). Pearlfish are small (to 30 cm), with elongate, slender bodies that taper to a fine point. The pelvic, the caudal, and sometimes, the pectoral fins are absent, and the scales are small or absent. This morphology allows a pearlfish to back into the body cavity of its host species through the mouth or anus. The anus of a pearlfish is located close to the head, allowing the fish to void wastes without emerging far from the host. Besides the peculiar way of life of the adults, pearlfish are also noted for their complex life cycle. They pass through two larval stages: a vexillifer stage, which is planktonic; and a tenuis stage, which is benthic. The larval forms are so different from the parents that the first ones known were described as distinct species.

20.9   GADIFORMES

The order Gadiformes contains more than 480 bottom-oriented marine fishes, some of which play major ecological roles in their respective communities (but, particularly, in the benthic communities of deep water). They are a fairly uniform group from a taxonomic perspective, with many distinctive anatomical features (Markle 1989). Gadiform fishes typically have elongate bodies (often tapering to a point at the tail), with long dorsal and anal fins. It is not unusual for these fins to be broken up into two or three sections (especially the dorsal fins). The pelvic fins are either thoracic or jugular in position or are absent. There are no spines. All of these fish have small, cycloid scales. Because they prey on fish and invertebrates, most have large terminal or subterminal mouths, often with barbels.

Macrouridae

The grenadiers or rattails are an important group (285 species) of benthic fish found in deep water throughout the world, although they are most abundant in tropical and subtropical areas. All have large heads and long, tapering bodies. The dorsal and anal fins are continuous with the sharply pointed caudal fin (Fig. 20.4B). The grenadiers are major predators and scavengers of the deepsea bottom. They can be photographed in large numbers by suspending a camera over a bucket of dead fish lowered to the bottom. Although their preference for deep water makes them difficult to catch, interest is developing in commercial fisheries for grenadiers, because populations of more accessible fish are becoming depleted.

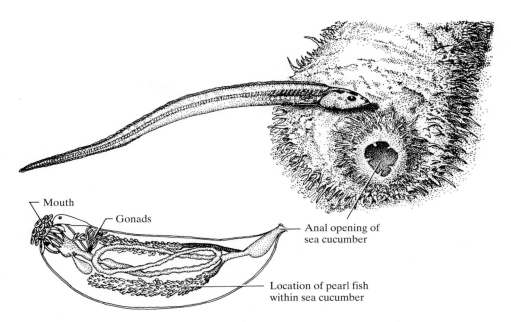

Mouth

Gonads

Anal opening of
sea cucumber

Location of pearl fish
within sea cucumber

FIGURE 20.7   A pearlfish (Carapidae) and its host, a sea cucumber. From Moyle (1993); © 1993 by
Chris Mari Van Dyck.

Moridae

The flatnose cods are moderate-sized (50–70 cm), deepsea fish that apparently are
very abundant on the upper continental slope throughout the Atlantic, Pacific, and
Indian oceans. The approximately 100 species resemble hake in many respects but
can be distinguished from them by their pointed snout, which is quite flat over the
mouth, and by their swimbladder, which connects to the inner ear. Two closely related species of *Antimora* are abundant off the Atlantic and Pacific coasts of North
America.

Merlucciidae

The merluccid hakes are often placed in the Gadidae but differ from them by having
one or two dorsal fins, no chin barbel, and the first ray of the first dorsal fin being
hardened into a "spine" (Fig. 20.4A). A few species have a long, pointed tail in which
the anal, caudal, and dorsal fins are united. All have terminal mouths with sharp
teeth, indicating their tendency to be piscivorous. The true hakes (*Merluccius*, seven
species) are found near coastlines at moderate depths over much of the world. Although their flesh is not as prized as that of the cods, the hakes now rank among the
most important commercial fishes of the world because of their abundance and ease
of capture with trawls.

Gadidae

The "true" cods are distinguished from other cod-like fish by their separate caudal fin and by the dorsal fin, which is divided into two or three sections. None of the fins has spines. The swimbladder is physoclistous, and a chin barbel is usually present. As might be expected from their morphology, the cods are bottom-oriented species, feeding on such animals as crabs, fish, and mollusks. Some species make extensive seasonal migrations for spawning and feeding. Besides the classic cods (*Gadus*), the commercially important members of this family are haddock (*Melanogrammus aegelfinus*) and pollock (*Pollachius, Theragra*). In the Arctic, some cold-adapted species of cod are the dominant predators (see Chapter 36) and have some importance as commercial fish. One species of cod, the burbot (*Lota lota*), is confined to cold, freshwater lakes and rivers of North America and Eurasia.

The cod family only contains approximately 55 species, but most of them are highly desired as food fish and, therefore, have long played an important economic and nutritional role for Western peoples. The Atlantic cod (*Gadus morhua*) has been particularly important because of its size, abundance, and ease of capture. At least 100 years before the formal colonization of North America, European (particularly Portuguese) fishermen were capturing cod by hook and line off the coast of North America and then landing in Newfoundland to dry and salt their catches. One indication for the importance of cod fisheries to the early development of America is that since 1784, the effigy of a cod has been prominently displayed in the Massachusetts statehouse. This, of course, has not kept the cod populations from collapsing because of overfishing. At one time, it was thought that the Grand Banks and other fishing areas had an inexhaustible supply of cod, haddock, and similar fishes. Cod in particular have such high reproductive potential (e.g., a 6-year-old female can produce 5 million eggs) that it was often assumed they could replace themselves under almost any fishing pressure. This has proven to be incorrect (see Chapter 37).

20.10   BATRACHOIDIFORMES

The order Batrachoidiformes ("frog-like forms") includes just the toadfish family (Batrachoididae), which only contains approximately 69 species. What the order lacks in species, however, it makes up for in conspicuousness, because toadfishes are commonly encountered in shallow marine waters, especially off North America. As the common name suggests, toadfish have a flat, toad-like head with protruding, dorsally located eyes and a large mouth. The body is squat and usually without scales. The pelvic fins are jugular in position, and spines are present in the dorsal fin. In some South American forms, the spines are hollow and associated with venom glands. Perhaps the best-known toadfish is the oyster toadfish (*Opsanus tau*), which can make incredibly loud sounds by vibrating its swimbladder and has the habit of laying its eggs (which it defends vigorously) in old cans and other trash. Midshipmen (*Porichthys*) are also well-known members of this family, because they are one of the few inshore fishes that have numerous photophores. They also make loud noises when courting—to the annoyance of people who live in boats anchored near a spawning area, because the hulls of the boats can pick up and even amplify the loud, buzzing sound.

20.11   LOPHIIFORMES

All 280 to 300 members of the order Lophiiformes (*Lophi-*, "crest") are called angler-fishes, because most possess a fishing pole (***illicium***) on the head, complete with an artificial lure (***esca***). The illicium and esca develop from the first ray of the spinous dorsal fin. Anglerfish are inactive forms with large heads and mouths (Fig. 20.8), cryptic coloration, small or absent scales, small gills, jugular pelvic fins if present, no ribs, and physoclistic swimbladders if present (Pietsch and Grobecker 1987). Anglerfish fall into two ecological groups: those that live on the bottom or attached to drifting seaweed, usually in shallow water (families Lophiidae, Brachionichthyidae, Antennariidae, Chaunacidae, Ogcocephalidae); and those that live in the Bathypelagic zone (the remaining 11 families). The fishes in the first group lie hidden on the bottom and wave the illicium and esca to attract potential prey. The esca mimics particular types of organisms; at least one species of anglerfish (*Antennarius* sp.) has an esca that looks remarkably like a small fish (Pietsch and Grobecker 1978). Most species in the latter four families of the first group have muscular pectoral fins that look like arms and are used either for "walking" across the bottom or for clinging onto floating seaweed. Most of the shallow-water anglerfishes are small, but some goosefishes (Lophiidae) may reach 1.3 m and weigh nearly 32 kg. Despite their ugly appearance (to us), such fish are quite edible and sometimes are referred to as the "poor man's lobster."

The shallow-water anglerfishes are extremely peculiar in appearance, but many of the deepsea anglers can only be described as bizarre. By and large, they are small, soft-bodied forms without scales or pelvic fins but with large mouths and teeth, dark-colored bodies, and a variety of illicia and esca. The angling organs, however, are present only on females; males exist primarily for reproduction and, in many species, live attached to the females (Fig. 20.8A) (see Chapter 35 for a discussion of reproductive strategies). As in the shallow-water forms, the esca seems to mimic other organisms to attract them to the anglerfish. Because most potential prey in the deep sea are luminous, the esca contains an organ in which light is produced by symbiotic bacteria (O'Day 1974). This light attracts not only prey but also (because the esca is different for each species of anglerfish) males. As might be expected of an organ that is important for both prey capture and reproduction and that likely becomes damaged by animals attracted to it, the illicium is apparently capable of regeneration (Pietsch 1974). In addition to the esca, some deepsea anglerfish have barbels that also produce light. Curiously, these barbels do not contain light-producing bacteria but instead have an intrinsic light-producing system (Hansen and Herring 1977).

20.12   LESSONS FROM ODDS AND CODS

The many bizarre (to us) shapes of the deepsea fishes make perfect sense for the extraordinary environment in which they live (see Chapter 35) and demonstrate the extreme flexibility of the basic teleost body plan. Primarily, teleosts are a vision-oriented group, but vision as a major sensory mode has been abandoned repeatedly, whether to live in caves or the deep sea. Another response to unlighted conditions has been to evolve means of producing light—adaptations that presumably have developed independently in the three major evolutionary lines discussed here. In short, the process of natural

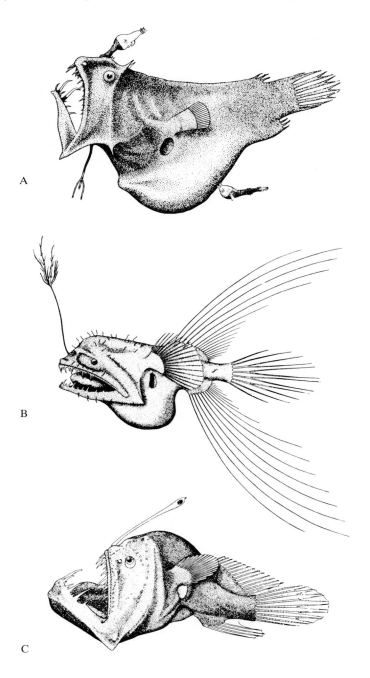

FIGURE 20.8 Deepsea anglerfishes (Lophiiformes). Life size of these fish is 5 to 10 cm. **A.** Netdevil *(Linophryne)*. Note the parasitic male attached to female. **B.** Fanfin anglerfish *(Caulophryne)*, and **C.** Black devil *(Melanocetus)*. From Regan and Trewavas (1932).

selection has produced teleosts that have found extremely efficient ways to use energy in resource-poor environments.

In contrast, the cods and their relatives have found ways to become extremely abundant in resource-rich benthic environments of northern seas, maximizing the possibilities of the small-egg/high-fecundity life-history strategy. When conditions are good, these fish can flood the pelagic environment with larvae, and when conditions are bad, they can live long enough to persist. Humans, of course, have found ways to overcome the astonishing ability of these fish to maintain high populations.

SUPPLEMENTAL READINGS

Fitch and Lavenberg 1968; Gill and Mooi 2001; Kurlansky 1997; Nelson 1994; Pietsch and Grobecker 1987.

WEB CONNECTIONS

Myctophidae

www.museums.org.za/sam/resource/marine/lantern.htm

CHAPTER 21
Mullets, Silversides, Flying Fish, and Killifish

SUBDIVISION TELEOSTEI
Infradivision Euteleostei
Superorder Acanthopterygii
Series Mugilomorpha
Order Mugiliformes
Family Mugilidae (mullets)
Series Atherinomorpha
Order Atheriniformes
Families Bedotiidae (Madagascar rainbow fishes), Melanotaeniidae (rainbow fishes), Pseudomugilidae (blue eyes), Atherinidae (Old World silversides), Atherinopsidae (New World silversides), Notocheiridae (surf sardines), Telmatherinidae (sailfin silversides), Dentatherinidae, Phallostethidae
Order Beloniformes
Families Adrianichthyidae, Belonidae (needlefishes), Scomberesocidae (sauries), Exocoetidae (flying fishes), Hemiramphidae (halfbeaks)
Order Cyprinodontiformes
Families Aplocheilidae (panchaxes), Rivulidae (rivulines), Profundulidae (Middle American killifishes), Fundulidae (killifishes), Valenciidae, Anablepidae, Poeciliidae (livebearers), Goodeidae (splitfins), Cyprinodontidae (pupfishes)

This is the first chapter devoted to the Acanthopterygii (spiny-fins); with more than 250 families and 13,500 species, this is the most diverse group of fishes. As might be expected, the taxonomic relationships of these fishes are complex, and our concepts about them are changing as we learn more about various species and groups. This is particularly true of the higher categories of classification, which are groups with presumed common ancestry (Fig. 21.1). In this text, we divide the acanthoptergyians into three main presumptive evolutionary groups—Mugilomorpha, Atherinomorpha (this chapter), and Percomorpha (next three chapters)—even though good arguments can be made for creating different groupings of orders within these groups (Nelson 1994). What all acanthopterygian fishes have in common are structural anatomical details

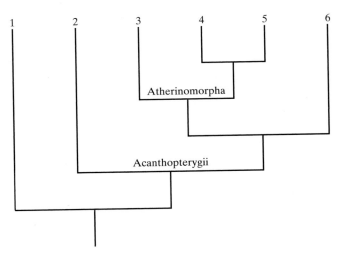

FIGURE 21.1 Phylogenetic relationships of the Mugilomorpha and Atherinomorpha. **1.** Paracanthopterygii (see Chapter 20). **2.** Mugiliformes. **3.** Atheriniformes. **4.** Beloniformes. **5.** Cyprinodontiformes. **6.** Percomorpha (see Chapter 22). The relationships shown here are very tentative (Nelson 1994).

of the mouth, pharyngeal jaws, and paired fins that permit increased efficiency of feeding and movement.

21.1 MUGILOMORPHA

The mullets (Mugilidae, where *Mugil-* is Latin for "mullet") are the only fishes in this evolutionary line, but they are often lumped together with many other groups, especially the Percomorpha. The 80 species of mullets are readily recognizable with their thick-yet-streamlined bodies, forked tails, hard-angled mouths, large cycloid or faintly ctenoid scales, subabdominal pelvic fins, and two widely separated dorsal fins, with the first containing just four spines (Fig. 21.2). The streamlined body is necessary both to avoid the numerous predators that attack as their schools move through shallow inshore waters and to speed them along in their spawning migrations, which closely follow the coastlines. The main reason that mullets favor the inshore environment is for feeding. They subsist largely on organic detritus and small algal cells, which the fish scoop up when swimming at an angle to the bottom, running their mouths through the sediment. The larger particles are retained by their fine gill rakers and then ground up in their gizzard-like stomachs. Digestion takes place in an extraordinarily long intestine (five to eight times the body length), which is necessary because much of the ingested material is sand and other indigestible matter.

Mullets are important food fishes, because they are both abundant and accessible. The striped mullet (*Mugil cephalus*) is particularly important as a food fish, because it is found in subtropical waters around the world. It also enters fresh water frequently, as do

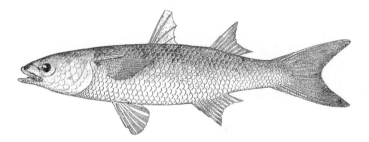

FIGURE 21.2   Mullet (*Mugil,* Mugilidae). From Jordan and Starks (1895).

other species of mullets. In Australia and New Guinea, several species of mullets can apparently spend their entire life history in fresh water (Berra 2001).

21.2 ATHERINOMORPHA

The members of the diverse (≈1,300 species) Atherinomorpha ("arrow-body") are usually recognizable by their generally surface-oriented body shapes; most species have large eyes, flattened heads, upturned mouths, and dorsal fins placed far back on the body. Anatomical details used to link the three orders include the absence of spines or serrations on the opercular bones, the rarity of ctenoid scales, the lack of an orbitosphenoid bone, and the connection of the pectoral girdle to the skull with Baudelot's ligament. Among the most distinctive characteristics, however, are those associated with reproduction (Parenti 1993). Males have testes that produce sperm only in restricted areas and pass them into a common duct, thereby enhancing internal fertilization. Females are either livebearers and/or produce eggs with long filaments emerging from the chorion (outer coat); the filaments attach the fertilized eggs to the spawning substrate. Spawning is almost always done in pairs, with the male and female "head to head, with bodies close together and bent in a S-curve" (Parenti 1993, p. 177). Associated with this behavior is a relatively long development period from the embryo to hatching, typically 1 to 2 weeks (as opposed to 1–2 days for most fishes). The long development essentially allows the fish to skip the larval stage, and the newly emerged young are adult-like in appearance. This development also seems to be associated with the rapid maturation of many species. The general reproductive strategy is well adapted to the environments in which atherinomorph fishes live. Some, such as flying fish and needlefish, are abundant marine fish, but most live in estuarine or inland environments, including some of the harshest environments that support fish of any type.

21.3 ATHERINIFORMES

The more than 285 atheriniform fishes are mostly elongate species, with terminal mouths, large eyes, and two dorsal fins. The first dorsal fin is made of weak spines, and a similar spine usually precedes the anal fin as well. The lateral line is weak or absent, whereas deciduous cycloid scales are present. Approximately 160 of the species are silversides (Atherinidae, Atherinopsidae), which are slender, silvery fishes with abdominal pelvic fins and large scales (Fig. 21.3). Silversides are often extremely abundant in the inshore regions of lakes, estuaries, and various shallow marine environments. They are schooling, diurnal planktivores. Perhaps the most famous of the atherinids is the grunion (*Leuresthes tenuis*), which deposits its eggs in the sands of southern California beaches

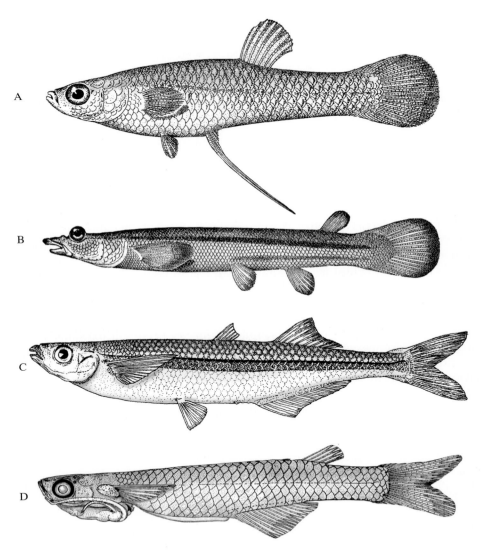

FIGURE 21.3   Atherinimorph fishes. **A.** Male mosquitofish (*Gambusia affinis*, Poeciliidae). **B.** Four-eyed fish. (*Anableps anableps*, Anablepidae). **C.** Topsmelt (*Atherinops affinis*, Atherinidae). **D.** *Ceratostethus* male (Phallostethidae). **A** and **B** from Jordan and Evermann (1900); **C** from Moyle (1976a); **D** from Roberts (1971).

during high spring tides (see Chapter 32). In lakes, silversides are frequently important forage fishes for predatory fishes. This and their supposed potential for the control of nuisance insects have resulted in the introduction of the inland silverside *(Menidia beryllina)* to lakes and reservoirs in California, Oklahoma, and elsewhere. Such introductions, however, often seem to create more problems than they solve. In California, for example, the silverside spread rapidly from the northern California lake into which it was introduced (for gnat control) and colonized much of the state to the south, including the San Francisco Estuary, where it may be reducing populations of native fishes by preying on eggs and larvae (Moyle 2002).

Two families of moderately deep-bodied, freshwater atheriniforms are the rainbow fishes (Melanotaeniidae) and the blue eyes (Pseudomugilidae). The rainbow fishes (>68 species) and the blue eyes (16 species) are small (5–12 cm), brightly colored fish with compressed bodies and subthoracic pelvic fins, which are characteristics that make them favorite aquarium fishes. They are confined to the fresh waters of Australia, New Guinea, and adjacent islands. A number of blue eyes are confined to isolated desert springs or similar habitats.

Perhaps the most curious family in this group is the Phallostethidae (21 species), which reside in brackish and fresh waters of Southeast Asia (Fig. 21.3D). These small, elongate, translucent fishes are distinguished by an extremely complex copulatory organ (priapium) located on the throat of the males, part of which is used to hold the female during copulation; the priapium is derived from bones of the pelvic girdle (Parenti 1989). Despite internal fertilization, the females are egg layers.

21.4 BELONIFORMES

The order Beloniformes *(Belon-,* "needle") unites the killifish-like medakas (Adrianichthyidae, 18 species), with the flying fishes (Exocoetidae, >50 species), halfbeaks (Hemiramphidae, >85 species), needlefishes (Belonidae, >30 species), and sauries (Scomberesocidae, 4 species). The medakas are abundant inhabitants of fresh and brackish waters of tropical Asia. The ricefishes *(Oryzias)* are important for laboratory studies because of their ease of rearing in captivity and high reproductive rates. The other four families consist of small to moderate-sized epipelagic fish (Fig. 21.4). Most are marine, but approximately one-third of halfbeaks and needlefishes live in fresh water (Berra 2001). The four families are all characterized by the single dorsal and anal fins that are approximately equal in size and located close to the tail. The pelvics are abdominal in position, and none of the fins have spines. The lateral line is well developed and low on the body, which typically is long and narrow.

Members of four families can readily be distinguished by their unique specializations. Flying fishes have long pectoral fins. Halfbeaks have an elongated lower jaw. The needlefishes and sauries have extremely terete (pointed and elongated) bodies. Adult flying fish (Box 21.1) have blunt snouts and small, unspecialized mouths but feed on zooplankton. Curiously, the juveniles of some species have elongated lower jaws, such as those possessed by halfbeaks (as do juveniles of needlefishes). In return, some halfbeaks have enlarged pectoral fins for making longer leaps from the water, which indicates the close relationship between the two groups. The elongate lower jaw of halfbeaks apparently functions in directing small, surface-oriented organisms into the

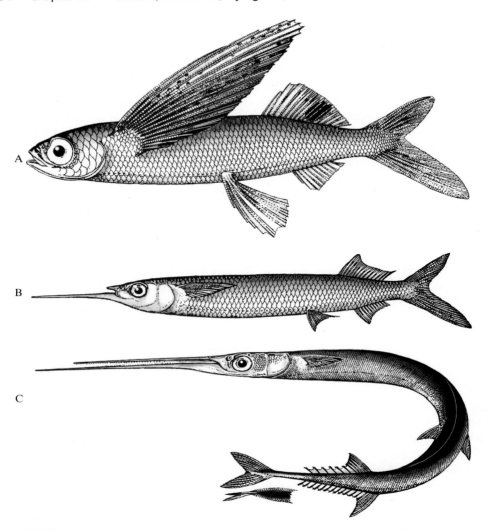

FIGURE 21.4   Beloniform fishes. **A.** Flying fish (*Cypselurus*, Exocoetidae). **B.** Halfbeak (*Hemiramphus*, Hemiramphidae). **C.** Needlefish (*Belone*, Belonidae). From Jordan and Evermann (1900).

mouth, because halfbeaks capture such prey while swimming in schools close to the surface. Most halfbeaks are largely herbivorous, however, and how the beak functions in gathering algae or pieces of eelgrass is a mystery. In contrast to halfbeaks, needlefish and sauries prey on small fishes, as their long jaws with pointed teeth attest. The needlefishes are elongate and look like unarmored gars, whereas the sauries are even more terete. Both groups are speedy, pelagic fishes. The speed of the sauries is enhanced by the five to seven finlets behind both the dorsal and the anal fins, much like those possessed by tuna. Despite their speed, these fish (but particularly the sauries) are important prey for

BOX 21.1

Do Flying Fishes Really Fly?

Flying fishes got their name from sailors who were surprised—and pleased—when these edible fish with long, wing-like pectoral fins landed on their decks, clearly flying out of the water. Biologists, however, tend to define flight as being powered flight, during which the time in the air is sustained by rapid beating of wings. Flying fish instead use their "wings" for gliding after they jump out of the water. The length of the glide is increased by sculling the lower lobe of the tail on the water's surface. This lobe is larger and longer than the upper lobe, and it possesses extra, stiffened rays (Davenport 1994). Some species have enlarged pelvic fins as well, which assist in the gliding. The "flying" of these fish appears to have developed as a means of escaping predators. Larger individuals can gain nearly a meter in altitude and glide as far as 100 m, although most flights are lower and shorter. It was also once thought that the freshwater hatchetfishes (Characiformes, Gasteropelecidae) were capable of powered flight, but this myth has also been debunked (Berra 2001).

larger predators, such as tuna and swordfish. Because of their difficulty of capture, the stocks of sauries in many parts of the oceans are relatively unexploited by humans. This situation, however, is changing rapidly.

21.5 CYPRINODONTIFORMES

Throughout the tropical and temperate world, wherever water is too saline, too warm, too small, too isolated, or too extreme in its fluctuations in quality for most fish, chances are reasonably good that water will be inhabited by one or more cyprinodontiform species. By and large, these species are small (<15 cm) omnivores, although many are such efficient predators on insects that they are used for mosquito control and other species are capable of digesting blue-green cyanobacteria. Their omnivory is one feature that allows them to live in harsh environments. Another is their surface-oriented morphology, which enables them to pump the thin, oxygen-rich layer of water at the air–water interface across the gills, an especially useful ability for life in warm, stagnant waters. Their ability to colonize unusual places is also enhanced by their diverse reproductive strategies. Despite their hardiness and versatility, however, cyprinodontiform fishes include more than their share of threatened species, because many of their habitats, ranging from desert springs to coastal salt marshes, have been destroyed, modified, or polluted by humans. This is particularly true for the many endemic species that have adapted to peculiar local conditions.

As a result of their surface orientation, these fishes are frequently called topminnows. They are distinguished from similar true minnows (Cyprinidae) by the presence of small teeth in the jaws, and they are sometimes referred to as the "tooth-carps" (*Cyprino-don*, "carp-tooth"). The pelvic fins are reduced or absent in many species, and fin spines are absent. Scales are generally cycloid, but the males of a few species develop ctenoid scales. The lateral line is usually well developed on the head but is absent from the body. The nostrils have double openings, and the upper jaw is bordered by a protractile premaxilla. The caudal fin is symmetrical and often rounded. The evolutionary diversity of this order has been hidden by the superficial similarity of the many species around the world (Parenti 1981, 1984).

Aplocheilidae

The 185 species of panchaxes or lyretails from Africa, India, and Southeast Asia are usually lumped together with rivulines from the New World, which makes little sense from a zoogeographic perspective. They are small, surface-oriented fishes that often play a strong role in mosquito control in shallow waters. Some of the most remarkable members of this family, however, are the nothos *(Nothobranchius)*. These fish complete their life cycles in a year. Eggs are deposited in the mud of the pond bottoms, where they can survive after the pond dries. When the rains come again, the eggs hatch quickly, and the pond is soon populated with fish once more. At least one species, the redtail notho *(Nothobranchius guentheri)*, grows from egg to spawning adult in 4 weeks—the fastest such growth known in fish (Haas 1976a, 1976b).

Rivulidae

The more than 200 rivuline species also includes fishes with some remarkable reproductive strategies. Individuals of *Rivulus marmoratus*, for example, contain both ovaries and testes and engage in internal self-fertilization, the only known case in fishes of a synchronous, self-fertilizing hermaphrodite (Harrington 1961). This results in localized homozygous clones, although some cross-fertilization is also possible (Soto and Noakes (1994). The annual lifecycle of the nothos also evolved independently in South America, in the 100 or so species of rivuline pearlfishes (mainly *Cynolebias*).

Fundulidae

The approximately 50 species of killifish are scattered throughout North America (into southeastern Canada and the Mexican Yucatan). Perhaps the most widely distributed species (Atlantic coast from Canada to Florida) is the mummichog (*Fundulus heteroclitus*), which is important in estuarine food webs. Morphologically, the killifishes are the least specialized of the cyprinodontiform fishes, and most other families in the group are defined according to how they diverge from the standard killifish pattern. Many of them are brightly colored and are favored as aquarium fishes. All lay eggs, and many exhibit territorial behavior at breeding time. Killifish are successful in a wide range of habitats, including lakes, streams, salt marshes, and estuaries.

Cyprinodontidae

More than 100 species of pupfish are known. They are renowned for their ability to live in conditions of extreme isolation, temperature, and salinity in North and South America and the Mediterranean region of Europe and Africa (Box 21.2). Pupfish got their

---

BOX 21.2

Devils Hole: Smallest Range of Any Species

A remarkable fish even among the pupfishes is the Devils Hole pupfish *(Cyprinodon diabolis)*, which occupies the smallest known range of any animal species. Devils Hole is a pocket of water on a desert hillside in Nevada that is fed by a deep aquifer; the pupfish, however, live mainly on a shallow shelf approximately 20 m² in area. The fish subsist on the sparse growth of algae on the shelf and on the small invertebrates associated with it. Their population fluctuates between 200 and 700 individuals, reflecting their small body size (15–20 mm). They have evolved a number of energy-saving features to survive, including a lack of pelvic fins, reduced pigmentation, and elimination of the vigorous courtship behavior found in other pupfishes. When reared in a more energy-rich environment, Devils Hole pupfish become larger and may sprout pelvic fins.

---

name from the mistaken notion that the brightly colored males busily engaged in the serious business of reproduction were "playing" like puppies. This name has contributed to their popularity in the United States, however, and made it easier to protect obscure species and their unusual habitats, especially desert springs. An example of an extreme environment for pupfish is the floor of Death Valley, California, where the Cottonball Marsh pupfish *(Cyprinodon salinus milleri)* regularly experiences salinities up to 4.6 times that of seawater, and water temperatures from nearly freezing in winter to 40°C in summer, with daily fluctuations of 14°C to 15°C. The sheepshead minnow *(C. variegatus)* survives in hypersaline lagoons along the Texas coast in conditions nearly as harsh. Despite their remarkable survival abilities, however, pupfishes in many areas are endangered because of diversion of the meager water sources that feed the habitats in which they live (Minckley and Deacon 1991).

A major contrast to the pupfishes are the Andean pupfishes (*Orestias* spp.), which consist of 43 species inhabiting high-elevation lakes and streams of the Andes Mountains of South America. Twenty-three of these species inhabit Lake Titicaca, which is large (8,100 km²), cool (11°C to 16°C), and located at 3,803 m elevation. In the absence of competing species from other groups (except one catfish species), the Andean pupfish species have taken the basic cyprinodont morphology to its limits. In Lake Titicaca, species have evolved into a rather peculiar-looking array of piscivores, planktivores, insectivores, and grazers. All are clearly recognizable as cyprinodonts, yet some are deep-bodied (inshore forms), others slender-bodied (offshore forms), and others large and with big mouths (piscivores) (Parenti 1984).

Poeciliidae

The family name Poeciliidae was formerly used to encompass only approximately 190 species of small, livebearing fishes now relegated to the subfamily Poeciliinae. Mode of reproduction, however, has proven not to be entirely reliable as a family level character, so the family now includes 110 additional species that were formerly lumped together with the killifishes and pupfishes. In this account, we discuss only the livebearers, in which the anal fin of the male is highly modified into a copulatory organ, the gonopodium (Meffe and Snelson 1989) (Fig. 21.3A). The livebearers are small. Most rarely exceed 10 cm in length, and some species have a maximum length of only 2 to 3 cm. They are found in warm, fresh, or brackish water at low elevations from northeastern Argentina to the southeastern United States and in the West Indies as well. They are most numerous in Central America, perhaps because this region is so unstable geologically. Poeciliids in general are well suited for maintaining populations in such areas, because:

- They have a broad tolerance of extremes in temperature, salinity, and dissolved oxygen.
- They are small in size, so a large population can exist in a small area.
- They will eat whatever food is available.
- Livebearing frees them from the need to have special substrates for spawning and assures high survival rates of the young, even under adverse conditions.
- Many species are unisexual, giving them unusual adaptive abilities (see Chapter 9).
- Their quick attainment of maturity results in high reproductive rates that can produce rapid population expansion under favorable conditions.

These same factors help to explain the success of livebearers such as guppies *(Poecilia reticulata)*, mollies *(Poecilia* spp.), and swordtails *(Xiphophorus* spp.) as aquarium fish, as well as the ability of guppies to establish populations in sewage treatment plants (into which they frequently are flushed by aquarists who tire of their prolific charges).

These factors are also important reasons why mosquitofish *(Gambusia affinis* and *G. holbrooki)* have been widely used as mosquito-control agents, resulting in their introduction to warm, shallow, mosquito-breeding habitats throughout the world (Fig. 21.5). In some situations (e.g., rice paddies and urban ponds), they have been a successful alternative to the use of pesticides (Swanson et al. 1996a). In the process of becoming established around the world, however, mosquitofish have probably displaced similar native fishes, although this has been documented mainly in the American Southwest. Their success as mosquito-control agents has also led to reduced consideration, in many parts of the world, of native fishes that might be equally useful (e.g., annual rivulines).

Other Families

Other families of cyprinodontiform fishes contain few species but a number of interesting adaptations. The Goodeidae (40 species) includes 36 livebearing species largely confined to the Rio Lerma basin in the highlands of Central Mexico, where the various species are specialized for a wide variety of niches (Fitzsimmons 1972). The male copulatory organ is a muscular "pseudophallus," because the anal fin is only slightly modified. The embryos obtain nutrients from a placenta-like affair called a trophotaenia. The four-eyed fishes (Anablepidae, three species) are also viviparous, but they are distinguished by their remarkable eyes, each of which is divided into two parts with separate pupils (Fig. 21.3B). With this system, they can see simultaneously both above and beneath the water. These fish forage in shallow water in both freshwater and brackish-water environments.

FIGURE 21.5 Female western mosquitofish (*Gambusia affinis*, Poeciliidae) with mosquito larvae. This fish shows some of the traits that have made cyprinodontiform fishes so successful in environments that most fish find inhospitable. These traits include small size and a body shape that allows it to feed and respire at the air–water interface. Photo by Jack K. Clark, University of California, Davis.

21.6 LESSONS FROM THE ATHERINOMORPHA

The atherinomorphs demonstrate how a basic body pattern—large eyes, upturned mouth, flattened head, abdominal pelvic, and rearward dorsal fin—can be modified for life under extreme conditions or for specialized lifestyles. The Atheriniformes have classically streamlined bodies, which are advantageous for living in the large shoals where they are often found. In the Beloniformes, the fish retain the terete body (or stretch it out) and develop elongate jaws, pectoral fins, or both. In the Cyprinodontiformes, the body is compacted, small, and often chunky, which is ideal for living in shallow, isolated habitats. This body form is also present in some atheriniforms (rainbow fish and blue eyes) that live in similar environments. The atherinomorphs' success seems to depend on their ability to live and reproduce under conditions that most fish find difficult. Silversides, for example, are abundant in estuaries and salt marshes. Killifish and their kin thrive in water that is near their physiological limits of temperature and salinity. Flying fishes and halfbeaks live in the well-lighted surface waters, where they would seem to be exceptionally vulnerable to predation. Finally, the atherinomorphs show a complete array of reproductive strategies, from the small pelagic eggs of the sauries to the live-bearing poeciliids to the all-female "species" found among silversides and livebearers. The adaptability of many species within this group is demonstrated by the explosive success of introductions of mosquitofish, mollies, guppies, and various silversides into human-altered environments. On the other hand, this group has more than its share of endangered species, usually those that cannot escape when humans dry up or pollute their isolated habitats or introduce predators into them.

SUPPLEMENTAL READINGS

Berra 2001; Meffe and Snelson 1989; Minckley and Deacon 1991; Naiman and Soltz 1981; Parenti 1981, 1984, 1989, 1993; Rosen 1964.

WEB CONNECTIONS

GENERAL
www.calacademy.org/research/ichthyology
www.fishbase.org
www2.biology.ualberta/jackson.hp/IWR
www.flmnh.ufl.edu/fish/default/htm
www.people.clemson.edu/~jwfoltz/wfb418

CYPRINODONTIFORMES, NORTH AMERICA
www.desertfishes.org/na/index.shhtml

RAINBOW FISHES, ETC., AUSTRALIA
www.nativefish.asn.au

CHAPTER 22

Opahs, Squirrelfish, Dories, Pipefish, and Sculpins

SUBDIVISION TELEOSTEI
Infradivision Eutelostei
 Superorder Acanthopterygii
  Series Percomorpha
   Order Stephanoberyciformes
    Families Melamphaidae (bigscales), Gibberichthyidae (gibberfishes), Stephanoberycidae (pricklefishes), Hispidoberycidae, Rondeletiidae (redmouth whalefishes), Barbourisiidae, Cetomimidae (flabby whale-fishes), Mirapinnidae, Megalomycteridae (largenose fishes)
   Order Beryciformes
    Families Anoplogastridae (fangtooth), Diretmidae (spinyfins), Anomalopidae (lanterneyes), Monocentridae (pinecone fishes), Trachichthyidae (roughies), Berycidae (alfonsinos), Holocentridae (squirrelfishes)
   Order Zeiformes
    Families Parazenidae (parazen), Macrurocyttidae, Zeidae (dories), Oreosomatidae (oreos), Grammicolepididae, Caproidae (boarfishes)
   Order Gasterosteiformes
    Families Hypoptychidae (sand eel), Aulorhynchidae (tubesnouts), Gasterosteidae (sticklebacks), Pegasidae (seamoths), Solenostomidae (ghost pipefishes), Syngnathidae (pipefishes and seahorses), Indostomidae, Aulostomidae (trumpetfishes), Fistulariidae (cornetfishes), Macrorhamphosidae (snipefishes), Centriscidae (shrimpfishes)
   Order Synbranchiformes
    Families Synbranchidae (swamp eels), Chaudridae, Mastacembelidae (spiny eels)
   Order Dactylopteriformes
    Family Dactylopteridae (flying gurnards)
   Order Scorpaeniformes
    Families Scorpaenidae (rockfishes), Caracanthidae (orbicular velvetfishes), Aploactinidae (velvetfishes), Pataecidae (prowfishes), Gnathanacanthidae (red velvetfish), Congiopodidae (racehorses), Triglidae (searobins), Bembridae (deepwater flatheads), Platycephalidae (flatheads), Hoplichthyidae (ghost flatheads), Anoplopomatidae (sablefishes),

Hexagrammidae (greenlings), Normanichthyidae, Rhamphocottidae (grunt sculpin), Ereuniidae, Cottidae (sculpins), Cottocomephoridae (Baikal sculpins), Comephoridae (Baikal oilfishes), Abyssocottidae (deepwater Baikal sculpins), Hemitripteridae, Agonidae (poachers), Psychrolutidae (flathead sculpins), Bathylutichthyidae, Cyclopteridae (lumpfishes), Liparidae (snailfishes)

22.1 PERCOMORPHA

The percomorph fishes are an immensely variable and successful group, and account for approximately half of all known fish species. Representatives can be found in most major habitats, and their food and feeding mechanisms are as varied as the group itself. Most members, however, are easy to place in this series because of their common structural plan. One reason for these structural similarities is that, even though some species live in habitats as diverse as the deep sea and swift streams, most live in shallow-water marine habitats or in lakes, especially in tropical regions. Most also live in, on, or closely associated with the bottom. The general percomorph plan includes:

1. Pelvic fins that, if present, are thoracic or jugular in position;
2. Pectoral fins that are placed high on the body;
3. Fin spines, although often reduced;
4. Protractile premaxilla that is part of a very flexible skull and jaw;
5. Upper and lower sets of pharyngeal teeth for processing of food;
6. Swimbladder that is either physoclistous or absent;
7. Pelvic fins usually, with one spine and five rays each;
8. Caudal fin, usually with 15 branched rays;
9. Spinal column, usually with 24 vertebrae;
10. Small spines, usually on bones of head and/or operculum;
11. Pleural ribs;
12. Ctenoid scales (usually); and
13. Well-developed eyes.

Many of the groups listed under the Percomorpha lack one or more of these characteristics, but whether the condition is ancestral (and, therefore, reason to exclude the group from the Percomorpha) or secondarily lost is often not certain. The extreme specialization of many forms makes schemes purporting to demonstrate phylogenetic relationships among the various lines very tentative (Fig. 22.1).

This mosaic of characteristics (and others) made the percomorph fishes successful because the characters evolved, in concert with one another, into an extraordinary array of specializations. Most of these specializations allow the fishes to take advantage of the complexity of benthic inshore habitats in both salt and fresh water, and they fall into three general categories: (1) predator avoidance, (2) feeding, and (3) social (especially reproductive) behavior.

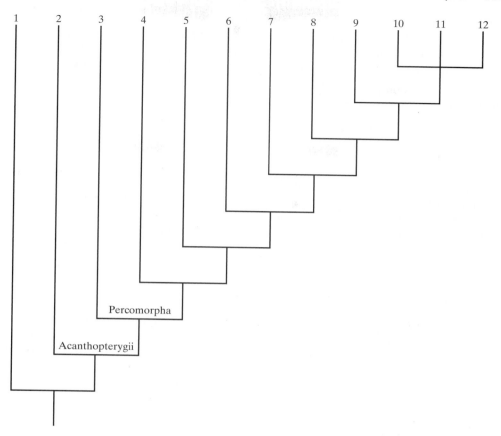

FIGURE 22.1 Possible phylogenetic relationships of the Acanthopterygii and Percomorpha. **1.** Para-canthopterygii. **2.** Atherinomorpha. **3.** Stephanoberyciformes. **4.** Beryciformes. **5.** Zeiformes. **6.** Gasterosteiformes. **7.** Synbranchiformes. **8.** Dactylopteriformes. **9.** Scorpaeniformes. **10.** Perciformes. **11.** Pleuronectiformes. **12.** Tetraodontiformes. For an alternative arrangement, see Gill and Mooi (2002).

Predator Avoidance

Perhaps the most important percomorph characteristic that reduces predation pressure is spines. Spines increase the effective diameter of small fish and, thereby, increase the minimum size of their predators. Big predators are fewer in number than small predators, and even they are likely to prefer prey with an edible body size close to the effective diameter (for easy swallowing). Spines thus give fish the type of protection that otherwise is available mainly through the use of heavy armor, which greatly reduces mobility. With spines, a fish can afford to be small, slow-moving, and day-active and to have a body shape advantageous for something other than a speedy escape. The maneuverability of percomorph fishes, which results from the placement of the paired fins and from the fine control they possess over buoyancy, provides further protection from predators. These fish can turn quickly, dive into cover, and in some species, swim backward.

Of course, many percomorph fishes have abandoned both stout spines and swimbladders (and, in many cases, pelvic fins) in favor of a completely benthic existence, in which small size, cryptic coloration, secretive habits, and unusual body shapes offer protection from predation; examples of these include blennies, flounders, and sculpins.

Feeding Specializations

The maneuverability of percomorphs has also been extraordinarily important in their diversification of feeding habits. Being able to hover, twist, and turn, they can capture a wide variety of prey and feed in a wide variety of microhabitats. Most live by capturing small invertebrates of all shapes and sizes, but specializations exist for feeding on everything from organic ooze to plant material to large fish. These specializations are reflected in their flexible mouths for sucking in or grabbing food items combined with pharyngeal dentition for breaking up, grinding, or holding them. The acute vision of most percomorph fishes (resulting from the tight pattern of cones in the retina) allows, in addition, precise location of prey, whether snapping at zooplankton, pursuing small fish, or grazing on the substrate.

Behavior

As might be expected with such varied and complex feeding habits, the behavior of these fishes in general is quite complex. Many are highly social, exhibiting either schooling or territorial behavior. In addition, many interact symbiotically with other organisms. To accomplish these various inter- and intraspecific interactions, the percomorph fishes have developed varying means of communication, including rapid color changes, sound production, and stereotyped movements (fin and body language).

Because the series Percomorpha contains nine orders, more than 230 families, and roughly 12,000 species, only the largest or most representative families in each order are discussed here. Most of the families excluded are small and poorly known, although they invariably have fascinating specialized features of their own. This chapter covers six relatively small percomorph orders, plus the large Scorpaeniformes. Chapter 23 covers only the immense order Perciformes, whereas Chapter 24 covers the highly derived Pleuronectiformes and Tetraodontiformes.

22.2 STEPHANOBERYCIFORMES AND BERYCIFORMES

The fish in these two orders are a mixed bag of mostly deepsea or nocturnal species possessing a number of ancestral osteological features. Many of these 16 families and more than 200 species are poorly known, so the composition and status of both orders is fluid. Some of the families are well represented in fossil deposits from the Cretaceous period, and later in that period, they may have played the dominant role in the inshore waters that now is played by perciform fishes.

The Stephanoberyciformes are soft, rounded, toothless, deepsea predators that include the gibberfishes, pricklefishes, and whalefishes. The three families of whalefishes have a vaguely whale-like appearance but only range from 10 to 30 cm long. The deepsea forms in the Beryciformes are mostly small and chunky, with large (or no) scales and, usually, conspicuous spines on the body and/or fins. A few species have photophores. The deepsea species make up all the beryciform families except the Holocentridae, Moncentridae, and a few species in the Trachichthyidae, Anomalopidae, and Berycidae. The shallow-water fishes are characterized by large eyes, bright-red coloration, and rather conventional perch-like bodies. One of the Trachichthyidae, the orange roughy (*Hoplostethus atlanticus*), has become an important commercial fish species off New Zealand and Australia during recent decades. Because the roughy is slow-growing and late-maturing, these fisheries are mining the populations of large individuals and have been sustained mainly by the discovery of new, unexploited populations.

The bright red squirrelfishes (Holocentridae, >60 species) are the largest and most widely distributed family in the Beryciformes and are found throughout the tropical and subtropical marine regions (Fig. 22.2B). Most species inhabit coral reefs or shallow, rocky areas, in which they hide during the day; they come out at night, often in amazing numbers, to feed on zooplankton in the water column. As a group, they are rather noisy fish, communicating with each other through a variety of clicks, croaks, and grunts. Some species of lanterneye fishes (Anomalopidae) are also active at night over reefs. These fish have a large, light-emitting organ beneath each eye that apparently enables the fish to see prey, confuses predators (it can be quickly covered with a lid), and also relates to intraspecific communication (Morin et al. 1975).

22.3 ZEIFORMES

The zeiform fishes are a curious mixture of six families and 39 species (Fig. 22.2A). Most live in the deep sea, but several better-known forms occur at only moderate depths, in midwater, and may actually support small fisheries. Most zeiform fishes have deep, compressed bodies, with ctenoid scales and a few well-developed fin spines. Their heads are large, with extremely distensible jaws that enable them to capture the fishes and large crustaceans on which they feed. Anatomically, these fishes have a mixture of ancestral, derived, and specialized percomorph features, so they are usually linked with the Beryciformes. Perhaps the best-known fishes in this order are the elegant dories (Zeidae), which are easily recognized by the large, upward-angled mouth, silvery color, and long filaments on the dorsal fin.

22.4 GASTEROSTEIFORMES

Although small (≈260 species), the Gasterosteiformes contains some of the most unusual and best-known teleost fishes, such as the seahorses, pipefishes, and sticklebacks. It is divided into two suborders, that are so different from each other they often are treated as separate orders: the Gasterosteoidei (Hypoptychidae, Aulorhynchidae, Gasterosteidae), and the Syngnathoidei (the remaining eight families). The two groups are treated together here, following the suggestion of Pietsch (1978a), who has demonstrated that the seamoths (Pegasidae) are intermediate in structure between the two groups. Previously, the seamoths had been placed in an order by themselves. The problem with defining this order, however, even with the seamoths included, is that few anatomical features are possessed by all members. Each group has its own bizarre specializations, but most have

FIGURE 22.2 Percomorph fishes with ancestral characteristics. **A.** Boarfish (*Antigonia*, Zeiformes). **B.** Squirrelfish (*Myripristis*, Beryciformes). From Jordan and Evermann (1900).

armor plates (of one sort or another) and small mouths. Also, enough overlap exists between groups to link all the families listed, although the development of these characteristics shows considerable variation, as the following accounts should demonstrate.

Gasterosteidae

The sticklebacks are the one family in this order with body shapes approaching that of "normal" fish (Fig. 22.3). They are small fishes that live in either fresh or salt water (or both) and are found throughout the northern hemisphere, usually in close association with coastlines. They are readily recognized by the presence of 3 to 16 isolated spines on the back that precede the dorsal fin; large eyes; small, upturned mouths; and narrow caudal peduncles. Most possess a row of bony plates on each side. Sticklebacks are territorial nest-builders, and their elaborate reproductive behavior is perhaps the best documented of any animal species (Wooton 1977, 1984) thanks to their abundance in the wild and the ease with which they reproduce under artificial conditions. Their physiology (mostly as related to reproduction) has also been well documented. Only seven or eight species of stickleback are usually recognized, but two of these—the threespine stickleback (*Gasterosteus aculeatus*) and the ninespine stickleback (*Pungitius pungitius*)—are widespread in northern Asia, North America, and Europe. Haglund et al. (1992) note that ninespine sticklebacks on the three continents form distinct evolutionary lineages that can be recognized as separate species. A similar situation may exist for threespine sticklebacks. These "species," however, are in fact complexes of hundreds of divergent populations. Many of these populations may have the characteristics of "good" species, reflecting the ability of sticklebacks to adapt to local conditions, such as food supply, substrates, and abundance of predators. This nightmare for traditional taxonomists has been a dream for students of evolution and genetics, however, providing many fascinating insights regarding general problems in these fields (Bell 1976a, 1976b).

Pegasidae

The five species of seamoths have excited the interest of biologists since the first dried specimens were imported to Europe from China by early explorers. They are small (to 15 cm) marine fishes that are completely encased in bony plates and have wing-like pectoral fins and long, bony snouts. Rather than being at the end of the snout, the mouth is underneath it. A uniquely complicated structure, the mouth can be folded when not in use into a cavity beneath the snout. When the fish is feeding, the mouth unfolds, and the buccal cavity expands, creating a powerful suction device (Pietsch 1978a).

Syngnathidae

The pipefishes (200 species) and seahorses (25 species) are a family that has sacrificed streamlining and speed for armor, cryptic coloration, and secretive behavior (Box 22.1). They are nevertheless very successful and are found in shallow marine waters throughout the world and, occasionally, in fresh water as well. All are long and thin, are encased in bony rings, have tube-like snouts with the mouth at the end, and lack pelvic fins (Fig. 22.3). Pipefishes propel themselves with their tails. Seahorses employ the dorsal and pectoral

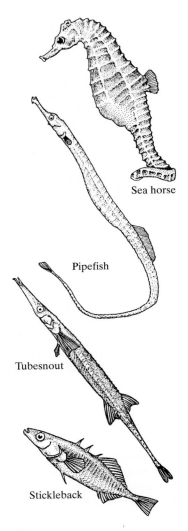

Sea horse

Pipefish

Tubesnout

Stickleback

FIGURE 22.3 Representative Gasterosteiformes.
Top to bottom. Seahorse (Syngnathidae), pipefish
(Syngnathidae), tubesnout (Aulorhynchidae), and
stickleback (Gasterosteidae). The fish are arranged to
represent steps in the hypothetical evolution of the
seahorse from a stickleback-like ancestor. The modern
fishes shown, of course, are not directly ancestral to
the seahorse. From Moyle (1993); ©1993 by Chris
Mari Van Dyck.

> ## BOX 22.1
> ### Pregnant Males
>
> One of the most fascinating aspects of syngnathid biology is how they care for their eggs and young. Each male seahorse has a sealed brood pouch on the underside of the tail with a tiny opening on the top. The female lays her eggs in this brood pouch, where they are fertilized, incubated, and hatched. The young are expelled once they are capable of swimming on their own. The pipefishes have brood pouches as well, but these pouches are open down the middle or sealed with overlapping flaps. Male pipefish carry the developing young for several weeks, and large individuals can be pregnant with 900 or more young. In the most ancestral-like species, the brood pouch is completely lacking, and the female merely attaches the eggs to a bare spot on the belly of the male. Because this mating system requires a high energy investment by the males, they can be very choosy about their mates, preferring larger females, which produce larger eggs (Berglund 1995). Some species are monogamous, but others will allow multiple females to lay eggs in their pouches (Jones and Avise 1997)

fins, having converted the caudal peduncle into a prehensile organ for holding onto plant stems; consequently, seahorses swim upright in a very slow, unfish-like manner. To compensate in part for their slow swimming, the tube-shaped mouth of syngnathids allows them to suck in small crustaceans from some distance after locating the prey precisely with their binocular vision.

Aulostomidae and Fistularidae

The trumpetfishes (three species) and cornetfishes (four species) are long (0.8–1.8 m), extremely slender fishes that are generally associated with tropical reefs. The snout is especially long but is tipped with a surprisingly large mouth, which enables them to capture other fishes. As their body shape and fin placement indicate, these are lie-in-wait predators that conceal themselves by hanging in the water column above the reef, hanging close to the reef itself (often at an angle to it), or associating closely with large, browsing fish (e.g., parrotfishes).

22.5 SYNBRANCHIFORMES

Fish in the Synbranchiformes order are convergent on true eels (Anguilliformes), with elongate bodies, no pelvic fins, small gill openings, and hard mouths. The two main groups—swamp eels and spiny eels—are otherwise quite different from each other.

Synbranchidae

Swamp eels (15 species) enjoy a wide distribution in the fresh and brackish waters of tropical Africa, Asia, Australia, and South America. Pectoral fins are present only during the larval stages. Dorsal and anal fins are reduced to small folds, without rays, in the caudal region. The caudal fin is reduced or absent, and scales are absent or confined to the caudal region. The eyes are small or absent in cave-dwelling species. The gill membranes are united and continuous, so only one, continuous ventral gill opening is found. The gills themselves are often quite small, because swamp eels are capable of breathing air, either through the vascularized lining of the gill pouch or through a pair of lung-like sacs off the gill pouch. Some species are capable of moving across land for short distances; others may burrow into the mud and survive even if the covering water evaporates. Swamp eels are nocturnal predators of small fishes. They also are the only freshwater fishes known to be protogynous hermaphrodites (capable of changing sex with increasing size and age) and are often large enough (up to 1.5 m) and abundant enough to support fisheries for them. Unfortunately, at least two species of *Monopterus* have been successfully introduced into Florida, Georgia, and Hawaii in the United States (as well into at least three other countries), where their potential for spreading and doing harm to native fishes is considerable (Collins et al. 2002).

Mastacembelidae

Spiny eels (60 species) live in the fresh waters of tropical Africa and Asia. They have soft dorsal and anal fins placed far back on the body, often joining with the tail (Fig. 22.4A).

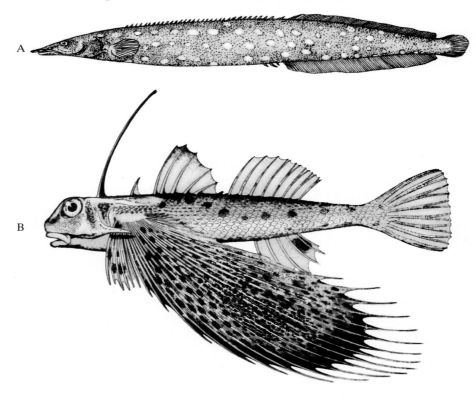

FIGURE 22.4   A. Spiny eel (*Mastacembelus*, Synbranchiformes). **B.** Flying gurnard (*Dactyloptena*, Dactylopteriformes).

Each soft dorsal fin is preceded by a row of stout, isolated spines that can inflict painful wounds on anyone unfortunate enough to grab a spiny eel around the middle. Their most peculiar feature is their pointed snout, which has a fleshy appendage at the end. This appendage has three lobes, with the two side lobes containing the anterior nostrils. Apparently, the appendage is used for finding invertebrates on and in the soft bottoms of the backwaters commonly inhabited by these eels. The only other family in this order (Chaudhuriidae) contains just three species, which lack the appendage on the snout as well as the dorsal spines and scales.

22.6   DACTYLOPTERIFORMES

Superficially, the seven species of flying gurnards (Dactylopteridae) that make up this order resemble searobins (Triglidae). Like the searobins, they have an armored head; large, dorsal eyes; bony plates on the body; large pectoral fins with free inner rays that can be used to "walk" over the bottom; and the ability to make loud sounds (Fig. 22.4B). The flying gurnards differ from the searobins, however, by the arrangement of the pectoral fins, their use of hyomandibular stridulations rather than "twanging" the swimbladder to produce sound, and the presence of two long, free spines just before the dorsal fin. Although the large pectorals have given rise to the supposition that these small fish (30–50 cm) can fly, it appears that the pectorals are used mainly to frighten potential predators. When a gurnard is startled, the brightly patterned pectorals are flashed open, thereby startling the predator.

22.7   SCORPAENIFORMES

With nearly 1,300 species in 24 families, the Scorpaeniformes are one of the largest teleost orders. Exactly where they should be placed among percomorph fishes, however, is a matter of dispute, although most likely they are closely related to—or even part of—the Perciformes (Nelson 1994). Gill and Mooi (2002), for example, place them within the Perciformes, but they also point out that the Perciformes is not monophyletic. Most of the Scorpaeniformes are bottom-oriented and so possess large, rounded pectoral fins, large heads, rounded caudal fins, and bodies and heads with many spines and/or bony plates. All possess a *suborbital stay* on each cheek, which is a ridge of bone that runs across the operculum below the eye. As a result, the Scorpaeniformes are frequently referred to as the mail-cheeked fishes. Most members of this group are found in marine environments at depths of less than 100 m, but a number of species are freshwater forms.

Scorpaenidae

The rockfishes (Box 22.2) and scorpionfishes, with nearly 400 species, make up the largest family in the order. They have large heads, mouths, and eyes as well as stout, laterally compressed bodies with large pectoral fins (Fig. 22.5A). Scales are either absent or very fine and ctenoid. Most scorpaenids are bottom-oriented predators, although many species enter the water column to feed on shoals of fish and squid.

BOX 22.2
World's Longest-Lived Fish?

The development of sophisticated techniques for determining the age of fish using rings on otoliths has resulted in the discovery that some rockfishes are extremely old. So far, the record-holder is a meter-long rougheye rockfish (*S. aleutianus*) from Alaska—at 205 years of age. As stated by Love et al. (2002) " ... some rockfish holed up in a deepwater outcrop may have been swimming around as a youngster when Thomas Jefferson was our president (1801–1809)" (p. 59). Other species have been aged at between 100 and 160 years. Most individuals or species, of course, do not live so long, although 20- to 50-year-old adults (50–70 cm) are not unusual. Love et al. (2002) show that the oldest fish come from the coldest water, either down deep or far north. They also belong to species

that are among the largest in the genus (provided they live where it is cold). Slow growth is fairly typical of scorpaenids, which is one reason why the commercially important species, such as Pacific ocean perch (*S. aleutus*) and Atlantic redfish (*S. marinus*), have been so easy to overexploit. The largest fish not only are old but also have a disproportionately high fecundity. As a result, fish taken by fisheries are not rapidly replaced. Many species that were once enormously abundant in their respective northern oceans now support only modest fisheries—or the fisheries have collapsed entirely. Recovery is possible only if the fisheries are banned for long periods of time and/or refuges are set up that allow populations of large, old fish to keep sending their young into the surrounding areas.

Scorpaenids are mostly found in the Indian and Pacific oceans; only approximately 60 species are known from the Atlantic. In the northern Pacific Ocean alone are more than 96 species of rockfish (mostly *Sebastes*), many of them not only morphologically similar to each other but also found together, apparently in mixed aggregations (Love et al. 2002). This complex of species presents some interesting problems in evolution and ecology. For example, why are so many species in the north Pacific but only four in the north Atlantic? One way that rockfish species segregate is by depth; Those found in the deepest water typically are bright red or orange in color, because at that depth, such coloration serves as camouflage. The color patterns of many inshore forms are also quite bright, however, which may serve to advertise their toxic spines in the dorsal, anal, and pelvic fins. The spines not only are toxic but also are stout and numerous; when combined with the head and operculum spines, they give the fishes a very prickly appearance. Even so, rockfish males accomplish internal fertilization, and rockfish females carry the embryos, giving birth to tiny larvae. As livebearers, the females have not sacrificed much in the way of total egg production; Their eggs are tiny and retained in the body cavity after fertilization, but the larvae are released immediately after hatching. Some species may carry more than 2.7 *million* embryos, presumably the record for a viviparous fish (Love et al. 2002).

Scorpaenids generally are known as rockfishes in northern waters, where they support valuable sport and commercial fisheries, but in tropical waters, they usually are known as scorpionfishes—because of the extreme toxicity of the spines. Most spectacular

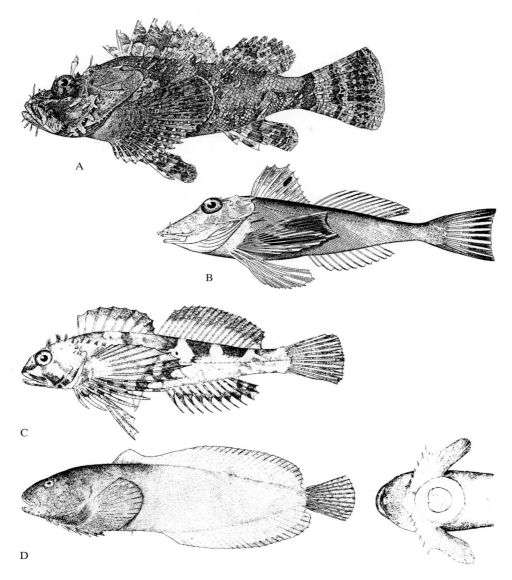

FIGURE 22.5 Scorpaeniformes. **A.** Scorpionfish (*Scorpaena*, Scorpaenidae). **B.** Searobin (*Trigla*, Triglidae). **C.** Sculpin (*Oligocottus,* Cottidae). **D.** Snailfish, with ventral view to show sucker (*Liparis,* Cyclopteridae). **A** from Jordan and Everman (1900); **B** from Goode and Bean (1985); **C** and **D** from Jordan and Starks (1895).

are the turkeyfishes and tigerfishes of coral reefs (*Pterois* spp.), which have bright bodies crossed with white and black stripes that extend onto the extremely long rays of the pectoral and dorsal fins. These species are toxic, and they act as if they are well aware of the fact; They cruise the reefs during the day and have been known to swim aggressively toward human swimmers with their toxic spines angled forward. In contrast to the turkeyfishes, the stonefishes *(Synanceia)* are rather dull in color. They sit quietly on reefs and are extremely well camouflaged; however, they possess the most deadly fish venom known. This venom is a neurotoxin that can actually be injected by a stonefish into the foot of a hapless wader by venom glands at the base of hypodermic-like, hollow dorsal spines.

Triglidae

The searobins or gurnards (100 species) are a widely distributed group of marine fish adapted for living on soft ocean bottoms at moderate depths. Typically, they are red in color, with large eyes set toward the top of a head that is covered with heavy, bony plates (Fig. 22.5B). The body may be covered with such plates as well. The mouth is subterminal and quite protrusible. The benthic invertebrates on which searobins feed apparently are located, in part, through the use of two or three finger-like, independent rays that are part of each pectoral fin. These rays are used to probe the bottom and also to rest on. The swimbladder of searobins is large and muscular and is used for sound production. They are perhaps the noisiest fish on the Atlantic coast of North America.

Cottidae

The sculpins are a large (>300 species) family of bottom-dwelling fishes. With the exception of approximately five species, all are confined to the marine coastal waters and fresh waters of the northern hemisphere. They are characterized by a broad, flattened head that usually has conspicuous spines and large, dorsal eyes; a smooth body (scales embedded, absent, or modified into tiny prickles); lack of a swimbladder; and large pectoral fins (Fig. 22.5C). Many species of sculpin live in the turbulent waters of the intertidal zone of oceans, swift streams, or the wave zone of lakes. They maintain themselves in these habitats by hiding beneath objects or by taking advantage of their negative buoyancy and hydrodynamic body shape. Presumably, when their pectoral fins are spread and the fish face into a current, the current flows over them, actually pressing them to the bottom.

Perhaps the best-known members of this family are the freshwater sculpins of the genus *Cottus*, which are called bullheads in Europe. They inhabit coldwater streams and lakes and, consequently, are frequently accused of competing with or preying on trout and salmon, charges that usually are not justified (Moyle 1977). Like other members of the family, however, they are efficient predators of active benthic invertebrates.

Two other scorpaeniform families that are confined to fresh water, the Abyssocottidae (37 species) and the Comephoridae (2 species), are derived from cottid ancestors. Most of these species are found only in ancient Lake Baikal in Russia. They (and some cottid species) have radiated into a number of unsculpin-like niches, including that of pelagic planktivores (see Box 25.2).

Agonidae

The poachers and alligator fishes make up a curious family of approximately 44 species with a bipolar distribution pattern. Most of these species, however, occur in the northern Pacific. They are small (25–30 cm), elongate fishes that are covered with bony plates, so many species look somewhat like a cross between a sculpin and a pipefish. They are bottom-dwellers (no swimbladders) and frequently are common in shallow marine waters, but their biology is poorly known.

Cyclopteridae

The lumpsuckers (Cyclopteridae, >26 species) are found in the temperate and arctic portions of the northern hemisphere, whereas the snailfishes have a bipolar distribution pattern. The members of this family tend to have rather flabby bodies that are globular in shape (Fig. 22.5D). The pelvic fins are modified into a large sucking disk or are absent altogether, and the gill openings are small. Both features indicate the inactive life these fishes live. The lumpsuckers are found attached to the bottom and to drifting objects, but they are often caught in midwater as well. Despite the repulsive appearance created by their lumpy build, complete with wart-like tubercles, lumpsuckers are favored foodfishes (they reach >60 cm) not only of humans but also of sperm whales, seals, and sleeper sharks. The eggs are frequently sold as lumpfish caviar.

Liparidae

The snailfishes (>195 species) are somewhat more fish-like in appearance than the related lumpsuckers, but their skin is without scales and is jelly-like in texture. They also possess a sucking disk. They favor cold waters and are common in both the Arctic and the Antarctic as well as in deep water (down to 7 km). Some species lay their eggs in the gill cavities of crabs and other crustaceans.

22.8   LESSONS FROM PERCOMORPHS

The first four orders discussed in this chapter are—or have been—regarded as most resembling the ancestral condition of percomorph fishes. It is probably significant that most members of these orders are deep sea or open-sea fishes that are adapted for life away from the shallow-water habitats dominated by the Perciformes. The deepsea environment in particular is shared with a wide variety of other fishes of ancient lineages. In contrast, the sticklebacks, seahorses, and their kin usually are found with perciform fishes and seem to defend themselves with heavy armor and good disguises. The synbranchiform eels are also very cryptic species. Within both groups, species may be few, but population numbers are often high.

The scorpionfishes are most abundant in shallow-water marine environments. They have evolved a very successful strategy as bottom-dwelling ambush predators, protecting themselves from predators by both concealment and poisonous spines. Smaller species, especially the cottids, can live in turbulent environments where food is abundant. Perhaps because of the abundance of active perciform predators, most scorpianfishes engage in some sort of parental care, from nest-brooding to livebearing. Rockfish have successfully combined livebearing with the teleost small-egg/high-fecundity strategy, with each female giving birth to thousands of larvae. Rockfish (and many other of the larger species in the percomorph orders discussed in this chapter) have opted for life in the slow lane. Under natural conditions, the cryptic, toxic-spined adults have high survival rates, so it pays for them to delay reproduction even though growth rates are slow. Individuals can live long, grow large, and thereby produce lots of young when conditions are good, even if those conditions are infrequent. This life-history strategy, of course, makes them extremely vulnerable to fisheries, which select for the largest individuals and can quickly push populations to collapse.

SUPPLEMENTAL READINGS

Gill and Mooi 2002; Johnson 1993; Love et al. 2002; Mayden 1992a; Nelson 1994; Wooton 1990.

WEB CONNECTIONS

GENERAL
www.calacademy.org/research/ichthyology
www.fishbase.org
www2.biology.ualberta/jackson.hp/IWR
www.flmnh.ufl.edu/fish/default.htm
www.people.clemson.edu/~jwfoltz/wfb418

SCORPAENIDAE
www.id.ucsb.edu/lovelab/

SYNGNATHIDAE
www.aqua.org/seahorses/

CHAPTER 23
Perciformes: Snooks to Snakeheads

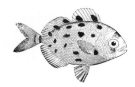

SUBDIVISION TELEOSTEI
  Infradivision Eutelostei
    Superorder Acanthopterygii
      Series Percomorpha
        Order Perciformes
          Suborder Percoidei
            Families Centropomidae (snooks), Chandidae (Asiatic glassfishes), Moronidae (temperate basses), Percichthyidae (temperate perches), Acropomatidae (ocean basses), Serranidae (sea basses), Ostracoberycidae, Callanthiiidae, Pseudochromidae (dottybacks), Grammatidae (basslets), Plesiopidae (roundheads), Notograptidae, Opistognathidae (jawfishes), Dinopercidae, Banjosidae, Centrarchidae (sunfishes), Percidae (perches), Priacanthidae (bigeyes), Apogonidae (cardinalfishes), Epigonidae (deepwater cardinalfishes), Sillaginidae (smelt-whitings), Malacanthidae (tilefishes), Lactariidae (false trevallies), Dinolestidae (longfinned pike), Pomatomidae (bluefishes), Nematistiidae (roosterfish), Echeneididae (remoras), Rachycentridae (cobia), Coryphaenidae (dolphins), Carangidae (jacks), Menidae (moonfish), Leiognathidae (ponyfishes), Bramidae (pomfrets), Caristiidae (manefishes), Emmelichthyidae (rovers), Lutjanidae (snappers), Lobotidae (triple tails), Gerreidae (mojarras), Haemulidae (grunts), Inermiidae (bonnetmouths), Sparidae (porgies), Centracanthidae, Lethrinidae (emperors), Nemipteridae (threadfin breams), Polynemidae (threadfins), Sciaenidae (croakers), Mullidae (goatfishes), Pempheridae (sweepers), Glaucosomatidae (pearl perches), Leptobramidae (beachsalmon), Bathyclupeidae, Monodactylidae (moonfishes), Toxotidae (archerfishes), Coracinidae (galjoens), Drepanidae, Chaetodontidae (butterflyfishes), Pomacanthidae (angelfishes), Enoplosidae (oldwife), Pentacerotidae (boarfishes), Nandidae (leaffishes), Kyphosidae (seachubs), Arripidae (Australasian salmon), Teraponidae (grunters), Kuhliidae (flagfishes), Oplegnathidae (knifejaws), Cirrhitidae (hawkfishes), Chironemidae (kelpfishes), Aplodactylidae, Cheilodactylidae (morwongs), Latrididae (trumpeters), Cepolidae (bandfishes)
          Suborder Elassomatoidei
            Family Elassomatidae (pygmy sunfishes)

Suborder Labroidei
  Families Cichlidae (cichlids), Embiotocidae (surfperches), Pomacentridae (damselfishes), Labridae (wrasses), Odacidae, Scaridae (parrotfishes)
Suborder Zoarcoidei
  Families Bathymasteridae (ronquils), Zoarcidae (eelpouts), Stichaeidae (pricklebacks), Cryptacanthodidae (wrymouths), Pholidae (gunnels), Anarhichadidae (wolffishes), Ptilichthyidae (quillfish), Zaproridae (prowfish), Scytalinidae (graveldiver)
Suborder Notothenioidei
  Families Bovichthyidae, Nototheniidae (cod icefishes), Harpagiferidae (plunder fishes), Bathydraconidae (Antarctic dragonfishes), Channichthyidae (crocodile icefishes)
Suborder Trachinoidei
  Families Chiasmodontidae, Champsodontidae, Pholidichthyidae (convict blenny), Trichodontidae (sandfishes), Pinguipedidae (sandperches), Cheimarrhichthyidae, Trichonotidae (sanddivers), Creediidae (sand-burrowers), Percophidae (duckbills), Leptoscopidae (southern sand-fishes), Ammodytidae (sand lances), Trachinidae (weeverfishes), Uranoscopidae (stargazers)
Suborder Blennioidei
  Families Tripterygiidae (threefin blennies), Dactyloscopidae (sand stargazers), Labrisomidae, Clinidae (clinids), Chaenopsidae (pike-blennies), Blenniidae (combtooth blennies)
Suborder Icosteoidei
  Family Icosteidae (ragfish)
Suborder Gobiesocoidei
  Family Gobiesocidae (clingfishes)
Suborder Callionymoidei
  Families Callionymidae (dragonets), Draconettidae
Suborder Gobioidei
  Families Rhyacichthyidae (loach gobies), Odontobutidae, Eleotridae (sleepers), Gobiidae (gobies), Kraemeriidae (sand gobies), Xenisthmidae, Microdesmidae (wormfishes), Schindleriidae
Suborder Kurtoidei
  Family Kurtidae (nurseryfishes)
Suborder Acanthuroidei
  Families Ephippidae (spadefishes), Scatophagidae (scats), Siganidae (rabbitfishes), Luvaridae (louvar), Acanthuridae (surgeonfishes)
Suborder Scombrolabracoidei
  Family Scombrolabracidae
Suborder Scombroidei
  Families Sphyraenidae (barracudas), Gempylidae (snake mackerels), Trichiuridae (cutlass fishes), Scombridae (mackerels and tunas), Xiphiidae (billfishes)
Suborder Stromateoidei
  Families Amarsipidae, Centrolophidae (medusafishes), Nomeidae (drift-fishes), Ariommatidae, Tetragonuridae (squaretails), Stromateidae (butterfishes)
Suborder Anabantoidei
  Families Luciocephalidae (pikehead), Anabantidae (climbing gouramies),

Helostomatidae (kissing gourami), Belontiidae (gouramis), Osphrone-
midae (giant gourami)
Suborder Channoidei
Family Channidae (snakeheads)

The Perciformes, with more than 9,200 species, is the largest order of vertebrates. These species are extremely diverse, but most are adapted for life as predators in the shallow or surface waters of the oceans or for life in lakes. Clearly, this is not a monophyletic group; rather, the order serves as a catchall for fishes that do not easily fit into other orders (Gill and Mooi 2002). Perciform fishes usually have the following characteristics, but none of these is unique to the group, either individually or in combination: (1) fin spines; (2) dorsal fins either double or made of two distinct parts, with the lead part being spiny; (3) no adipose fin; (4) pelvic fins thoracic or jugular in position, if present; (5) pelvic fins with one spine and five or fewer rays; (6) pectoral fins on the side of the body, with a vertical insertion; (7) 17 or fewer principal caudal fin rays; (8) scales ctenoid or absent (but cycloid in a few forms); (9) premaxilla as the only bone bordering the upper jaw; (10) no orbitosphenoid, mesocoracoid, and intermuscular bones; and (11) swimbladder physoclistous or absent. The forms that are recognized most readily as perciform range from classic deep-bodied fish to rover-predator types, all with stout spines on the lead part of the dorsal fin.

Another important (but less obvious) adaptation of perciform fishes—and of teleosts in general—is the dominance of small, free embryos (eggs) in reproduction, usually with pelagic larvae for dispersal. Thus, many perciform fish have high fecundity rates; the record is probably the louvar (*Luvarus imperialis*, Luvaridae), in which a 1.7-m individual can produce more than 47 million eggs (Nelson 1994). The advantages of this strategy include wide dispersal of the young in ocean currents and large numbers of young put into the environment when conditions are good. This strategy is the complete opposite of the Chondrichthyes', which emphasize producing a few large young, usually by livebearing. It is interesting, however, that a few perciform fishes (e.g., Embiotocidae) have adopted the chondrichthyan strategy.

Within the Perciformes, general lines of evolution are indicated by the 19 suborders, although nearly 57% of these species are found in just eight families (Gobiidae, Cichlidae, Labridae, Serranidae, Blenniidae, Pomacentridae, Sciaenidae, and Apogonidae). In contrast, many of the 149 families recognized here contain fewer than 20 species. The relationships of the Perciformes to other acanthopterygian orders are not well understood, nor are the interrelationships of the various groups within the Perciformes. In this chapter, we discuss every suborder (-oidei) as a major evolutionary line. We recognize, however, that many may not be monophyletic, whereas others will probably be raised to orders in future classification systems.

23.1 PERCOIDEI

The suborder Percoidei contains more than 2,850 species, but it is the "trash can" suborder of a "trash can" order (Gill and Mooi 2002). By and large, these have the classic perciform appearance: deep to moderately elongate body, two dorsal fins, large mouth and eyes, ctenoid scales, and thoracic pelvic fins. Most are inshore, diurnal (or crepuscular) predators, and those of any size are usually harvested by humans for food. Some of the most popular saltwater and freshwater game fishes, such as striped bass, snook, bluefish, jacks, snappers, croakers, perches, and sunfishes, belong to this suborder.

Serranidae

The 450 species of sea basses are large, piscivorous fishes that are associated with tropical and temperate reefs as well as inshore environments. They can be distinguished from most similar species by the following; (1) three spines on the operculum, (2) complete lateral line, (3) caudal fin that is often rounded, (4) no scaly process at the base of the pelvic fins, and (5) a long, continuous dorsal fin, with from 7 to 13 spines. Most species are hermaphroditic, a trait first noticed by Aristotle around 300 B.C. The hermaphrodites typically are female when small and convert to males when large, although in some species, male and female gonads develop simultaneously.

In recent years, three "new" families have been split from this family: (1) Moronidae (six species), (2) Percichthyidae (22 species), and (3) Acropomatidae (40 species). These basses can be best distinguished from the sea basses by the presence of only two spines on the operculum, a tail that is usually forked, and the absence of hermaphroditism (Fig. 23.1A). In North America, the four species of the Moronidae are found in fresh water and estuaries and are much sought after as sport and commercial fishes. The best known of these species is the striped bass *(Morone saxatilis)*, the "rockfish" of the East Coast; this species was introduced into California during the 1870s, and quickly became extraordinarily abundant. During the 1970s and 1980s, populations on both coasts showed large-scale declines in abundance because of the deterioration of environmental conditions in estuaries. By the beginning of the twenty-first century, both populations showed substantial recovery, resulting from a combination of improved environments, managed fisheries, and other factors. In Australia, several endemic freshwater game fishes are in the Percichthyidae, including the legendary Murray cod *(Maccullochella peelii)*. The Murray cod once grew to more than 113.5 kg and 1.8 m, but dams, overfishing, and alien species—the usual cast of characters—have reduced it (and other Australian percichthyids as well) in both numbers and size (Berra 2001).

Centrarchidae

The sunfish are a family of 29 freshwater species native only to North America. They are most characteristic of warm-water lakes and sluggish streams. Because the black basses *(Micropterus)*, sunfishes *(Lepomis)*, and crappies *(Pomoxis)* are very popular sport fishes, many populations have been established far outside their native ranges. The largemouth bass *(M. salmoides)* in particular has become established in ponds and lakes throughout the world. In several localities, such as Lake Atitlan, Guatemala, the establishment of largemouth bass has been responsible for the disastrous decline of native fishes and fisheries. In the Central Valley of California, introduction of various

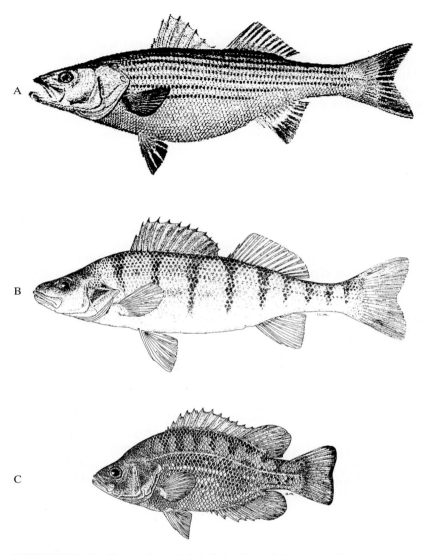

FIGURE 23.1 Perciformes, Percoidei. **A.** Striped bass (*Morone saxatilis*, Moronidae). **B.** Yellow perch (*Perca flavescens*, Percidae). **C.** Sacramento perch (*Archoplites interruptus*, Centrarchidae). **A** by H. L. Todd and **B** and **C** by M. Marciochi, from Moyle (2002).

centrarchids from the eastern United States was accompanied by the near disappearance of the one native centrarchid, the Sacramento perch *(Archoplites interruptus)*, from its native habitats. The centrarchids are recognizable as fishes that are deep bodied (or moderately deep bodied), with the spinous and soft dorsal fins joined together and moderately forked tails (Fig. 23.1C). They are nest builders and show well-developed parental behavior.

Percidae

The Percidae family is widely distributed in fresh waters of the northern hemisphere, but approximately 90% of the more than 190 species are found in North America east of the Rocky Mountains. The reason for this is the presence of 155 species of darters (*Etheostoma, Percina, Crystallaria, Ammocrypta*) in the streams and lakes. These are small, elongate, bottom-dwelling fishes, with small mouths and conspicuous eyes and the swimbladder either reduced or absent. Although they are secretive fish that pick small invertebrates from the bottom, the males of many of these species are brightly colored, especially when spawning. In small, warm streams, darters are usually second in abundance only to cyprinids and have complex ecological interactions. Curiously, only four species of darter-like percids (*Zingel, Romanichthys*) are found in Eurasia, mostly in the Danube River, although they are generally larger (to 48 cm) than American darters (Berra 2001). In addition to the darters, the North American percids contain two other groups of fishes: the perches *(Perca)*, and the pike-perches *(Stizostedion)* (Fig. 23.1B), which also have representatives in Europe. They mostly inhabit lakes and large rivers and are highly prized as food fish. The most favored of these fish is the walleye *(S. vitreum)*, a species that prefers deep, cool water and has large, reflective eyes that help it to detect and capture other fishes in dim light.

Apogonidae

The cardinalfishes are superficially similar to the beryciform squirrelfishes, but they usually are smaller (5–15 cm as adults) and possess two short, well-separated dorsal fins. Like the squirrelfishes, most of the more than 195 species are red in color, have large eyes, and are active mainly at night. The majority are associated with tropical reefs, where they hide in crevices and caves (often with squirrelfishes) during the day and emerge in large schools to feed on plankton at night. A few species have managed to invade the freshwater streams and mangrove swamps of the Pacific islands; a few others inhabit deep (to 1,200 m) water. They are unusual among coral reef fishes in that many—and perhaps most—of them are mouth brooders. A number of species live in fresh water in New Guinea and Australia, where they have earned the wonderful name "mouth-almighty" for their large gape (Berra 2001).

Pomatomidae

A text on fish would be remiss if it did not mention the nearly cosmopolitan bluefish *(Pomatomus saltatrix)*, which is famous for its voracious appetite and sharp teeth and was celebrated by John Hershey in his 1987 book *Blues*:

> It has such an elegant hull. It is so wicked and wild when it's hooked. It has such a cruel eye and such passionate jaws. ... It's so full of life! And it's true to life; there is nothing fake or soft about it; life is harsh. I've said it before and I'll say it again: I'm deeply in awe of the bluefish. (p. 186)

Echeneididae

The remoras, or sharksuckers, are among the most specialized of perciform fishes. The specializations of these eight species center around the remarkable sucking disk on top of their head. This disk is formed from the spiny dorsal fin and contains 10 to 28 slat-like, transverse ridges, which are modified spines. When a remora presses this disk against a large fish, turtle, or whale, the ridges are erected to create a powerful suction, and the remora becomes very difficult to dislodge. The disk is fused to the upper jaw and, consequently, forms part of the snout, beyond which projects the lower jaw. The body of a remoras is smooth (small, cycloid scales) and fusiform. The fish are good swimmers, but their dependence on rides from other fish is indicated by the absence of a swimbladder. Some remoras will pick parasites and diseased tissues from their hosts, but such activity is unlikely to compensate for the energy expended by the host to carry one or more remoras. On the other hand, the remoras get a free ride, enjoy the protection of a large fish, and make short forays away from their host to capture small fish.

Carangidae

The body shapes of the more than 140 species of jacks and pompanos range from torpedo-like to nearly plate-like, but all are fast-swimming predators. This is reflected in their deeply forked tails and narrow caudal peduncles; the fine, cycloid scales (often absent or modified into scutes along the lateral line); the deeply sloping heads (with large eyes and mouth); and usually, the laterally compressed bodies. Carangids range in color from silvery to metallic blue or green to bright yellow or gold. Adults range in size from approximately 25 cm to 2 m. Most jacks are schooling fish that feed by making rapid, slashing attacks at shoals of smaller fish, especially herrings and anchovies.

Lutjanidae

The snappers are among the most important food fishes in tropical and subtropical waters. Nothing is remarkable about their morphology, which may, in part, explain their success (125 species); they are generalized, bottom-oriented predators. The typical snapper is heavy bodied, with a continuous dorsal fin, a slightly forked tail, a fully scaled body, and a triangular head with a large mouth located at the apex of the triangle. The mouth is protractile and equipped with many teeth, usually including some large canines. Most are brightly colored, ranging from bright red to yellow to iridescent blue, often with contrasting stripes and bars. Most species are associated with tropical reefs, submerged banks, or inshore areas. They have shown themselves to be quite adaptable; for example, the red snapper *(Lutjanus campechanus)*, an important commercial species in the South Atlantic and Gulf of Mexico, is one of the more abundant fish around oil platforms, shipwrecks, and artificial reefs.

Haemulidae

The grunts (>150 species) are a much more attractive family than their name suggests; shoals of grunts with iridescent yellow, blue, and orange stripes are commonly seen over tropical reefs. At night, these color patterns are altered, and the grunts of many species leave the reefs to forage on hard-shelled invertebrates that live in the surrounding flats. The invertebrates are crushed by the grunt's formidable pharyngeal teeth. These species are called grunts because they also use their pharyngeal teeth for making rude sounds, which are amplified by the swimbladder.

Sciaenidae

The croakers or drums are another group of noisy, bottom-oriented, invertebrate crunchers. Despite many similarities, however, they can be distinguished from the four families previously discussed by their deeply notched dorsal fins, rounded caudal fin (rather than forked or straight), and one or two anal spines (rather than three). In addition, many possess one or more chin barbels. Internally, croakers are notable for their multibranched swimbladder and huge otoliths. Both features presumably relate to the fact that croakers produce loud sounds, especially during spawning season, by vibrating muscles associated with the swimbladder. Because the 270 species are often inhabitants of turbid estuaries, bays, and rivers, their elaborate sound-producing and -receiving systems may assist them in finding their way about and in communicating with one another. The well-developed lateral line (extending into the rays of the tail) is also indicative of their murky habitats. In North America, approximately 34 species of croakers are found, many of which are important as sport and commercial fishes, including the famous "blackened redfish" of gourmet Cajun restaurants (red drum, *Sciaenops ocellatus*). The freshwater drum *(Aplodinotus grunniens)* is an important commercial species throughout much of the Mississippi River drainage.

Chaetodontidae

It is unusual to see a photograph of a reef that does not include at least one individual of the more than 110 species of butterflyfish or the closely related angelfishes (Pomacanthidae, 74 species) (Fig. 23.2A). All are extremely colorful, with striking patterns of stripes and spots on multihued backgrounds that include most colors of the visible spectrum. Yellow, however, seems to be the most common. The colors are well displayed, because butterflyfishes are thin in cross-section but nearly circular in side view, with large dorsal and anal fins. The spectacular patterns are made even more noticeable by the constant activity of these fishes during the day.

Butterflyfishes and angelfishes are remarkable not only for their coloration but also for their feeding specializations. Most have small, protractile mouths, with many tiny teeth. Different species are specialized for feeding on invertebrates at different depths in crevices, as indicated by the length of the snout. Others specialize on prey such as zooplankton, coral polyps, sponges, and polychaete worms, because each food type requires particular morphological and behavioral specializations to be fed on efficiently (Motta 1985).

23.2   ELASSOMATOIDEI

The pygmy sunfishes (Elassomatidae), formerly placed in the Centrarchidae, are now regarded as quite distinct (e.g., they lack a lateral line, have cycloid scales, have only five branchiostegal rays, and spawn on vegetation). Following Nelson (1994), they are placed here in a separate suborder. They may, however, belong in the Gasterosteiformes (Gill and Mooi 2002). The pygmy sunfishes are six species of small (25 cm), attractive fish from the southeastern United States (Fig. 23.2C).

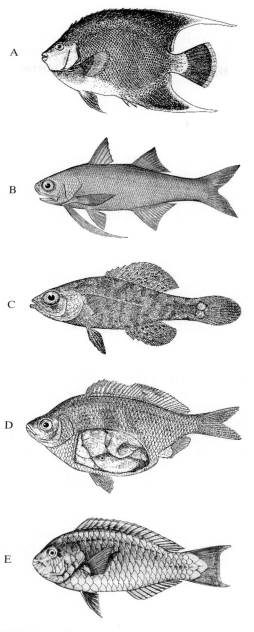

FIGURE 23.2   Representative perciform fishes.
A. Angelfish (*Holacanthus*, Pomacanthidae, Percoidei).
B. Threadfin (*Polydactylus*, Polydactylidae, Percoidei). **C.** Pygmy
sunfish, (*Elassoma*, Elassomatoidei). **D.** Pregnant shiner perch
(*Cymatogaster aggregate*, Embiotocidae, Labroidei). **E.** Parrotfish
(*Scarus*, Scaridae, Labroidei). From Jordan and Evermann (1900).

23.3 LABROIDEI

The suborder Labroidei contains six fascinating families and more than 2,200 species, all of which possess highly specialized pharyngeal jaws that have allowed them to diversify and prey on a wide variety of food items. Three of the families (Pomacentridae, Labridae, Scaridae) contain large numbers of species that are especially important on tropical reefs. This suborder also contains the cichlids, which are extraordinarily diverse in tropical lakes, and the surfperches (Embioticidae), which are famous for their spiny viviparity.

Cichlidae

The cichlids include more than 1,300 (probably closer to 2,000) species of freshwater and brackish-water fishes that inhabit Africa, South America, Central America, and parts of Asia and North America. From an evolutionary perspective (Box 23.1), their most amazing characteristic is a tendency to form flocks of extremely specialized species in large tropical lakes, especially in Africa (see Chapter 30). Lakes Malawi, Victoria, and Tanganyika together contain approximately 900 endemic species of cichlids (Fryer and Iles 1972, Ribbink 1990). That number is actually conservative, because new

---

BOX 23.1

Nature's Grand Experiment in Evolution

The title of this box is the subtitle of a book by George Barlow called *The Cichlid Fishes* (2000), which enthusiastically describes the diversity of cichlid forms and behaviors. Barlow has spent a lifetime studying their behavior in aquaria and in the wild. He provides many insights into the diverse adaptations of this group, but especially into their behavior: " . . . what captivates me is the complexity of their social lives and their devotion to family, not to mention the sheer beauty of many species" (p. xiii). For example, in a section called "Munching on Mom and Dad," Barlow describes how the fry of many cichlids browse on the slime covering their parent's bodies as part of the parental care strategy. The behavior is facultative in some species but mandatory in others. For species in which browsing on the parents is well developed, the nibbling actually stimulates the parents to increase production of mucus cells, which Barlow argues is analogous to mammals producing milk. Apparently, the mucus contains not only nutrients but also hormones and other chemicals that promote growth and survival of the fry. This parental feeding is best developed in the pancake-shaped discus cichlids (*Symphysodon*) of South America. In discus, larval browsing on the adult skin is mandatory, and the adults develop specialized skin cells that "are released intact as free cells on the surface of the parent where fry graze on them like tiny hors d'oeuvres" (p. 197). This is just one example of cichlid adaptations that seem to push the evolutionary envelope of fishes. Not only are such traits unique among fishes, they can often evolve quite rapidly (hence the subtitle of Barlow's book). For further discussion of cichlid evolution, see Chapter 30.

species from these lakes are continually being described. Sadly, however, as many as 200 species have been lost from Lake Victoria in recent years after introduction of the predatory Nile perch (*Lates* sp., Centropomidae) in combination with pollution, overfishing, and other factors. This has been one of the greatest mass extinction events in recent times.

A major factor promoting speciation in the cichlids is that most species have a strong feeding orientation toward the bottom, a heterogeneous environment with a wide variety of invertebrates and plants on which to feed. The tendency in lake-dwelling cichlids is to specialize in feeding on just one type of food, such as filamentous algae growing on rocks or fly larvae living in sand. According to Liem (1974), the extraordinary ability of the cichlids to specialize results from the unique and complex structure of the pharyngeal "jaw" apparatus, which permits efficient grinding and chewing of ingested food using highly specialized teeth. The pharyngeal jaws free the mouth to develop specializations just for food gathering, such as long, forceps-like teeth for picking up small invertebrates from plants or pulling scales from other fish.

As might be expected of substrate-feeding fishes, most cichlids are deep bodied, with large and spiny fins, rounded tails, protractile mouths, and well-developed eyes (see Fig. 11.3). The "signature" look of cichlids is produced by their long dorsal and anal fins, which often come to graceful points over the base of the tail, providing a pleasing symmetry in form. Many are brightly colored, although many also have color patterns to match the substrate. Many of the more colorful forms, such as discus fishes *(Symphysodon)*, angelfishes *(Pterophyllum)*, South American cichlids (multiple genera formerly lumped together in *Cichlasoma*), and jewelfishes *(Haplochromis)*, are favorite aquarium fishes; this preference is increased by the ease with which they breed in captivity. Reproductive behavior is often elaborate, and many species brood the developing embryos in the mouth of the female (and a few in the mouth of the male).

Embiotocidae

The surfperches form a small (23 species) but abundant family that is found only along the northern Pacific coasts of North America and Asia. One species, the tule perch *(Hysterocarpus traski)*, is confined to fresh water in California. The surfperches are not particularly remarkable in appearance, being moderately deep-bodied percoid fishes, with cycloid scales and small mouths that are handy for picking up small invertebrates (DeMartini 1969). When discovered during the early 1800s, however, the surfperches greatly aroused the interest of biologists, including Louis Agassiz, because they are viviparous (Fig. 23.2D). The young develop in uterus-like sacs in the ovary and obtain most of their nutrition from the body fluids of the mother. This is possible because the dorsal, pelvic, and anal fins of the embryo are greatly enlarged, highly vascularized, and in close contact with the tissues of the mother, thereby allowing transfer of nutrients to take place. When born, the young are 3 to 5 cm long, and they may become sexually mature a few weeks after birth. Females, however, typically give themselves time to grow by storing sperm for several months before fertilizing their eggs. The number of young produced per female varies among the species; Those living in the most variable environments produce the most young, whereas those living in more predictable environments produce fewer young (Baltz 1984).

Pomacentridae

The damselfishes are another group of deep-bodied, conspicuous fishes that are characteristic of tropical reefs. The conspicuousness stems not just from their bright color patterns, because many are, in fact, rather plain, but also from their abundance and activity. Many of the more than 315 species in the family actively defend territories on reefs, not only against members of the same species but against many other species as well (see Chapter 33). The functions of these territories are multiple. One of the most important, however, appears to be in providing food for the fish, because defended areas typically have much heavier growths of algae than do undefended areas and the main food of many damselfishes is filamentous algae.

Damselfishes that do not defend garden territories mostly fall into two general categories: shoaling fish that feed on plankton above the reef, and fish that are commensal with anemones. The plankton-feeding forms are similar in body shape to the territorial forms; both types seem to represent a compromise between the convenience of being deep bodied for hovering and picking out small food items and the ability to flee quickly to cover when a predator approaches. The latter "need" places a limit on just how deep bodied they become, and it results in a forked tail. The plankton-feeding and territorial forms also have large eyes and small, protractile mouths. The teeth of the territorial forms are incisor-like, for nipping off pieces of algae, whereas the teeth of the plankton-feeders are conical or villiform, for holding onto small prey. The group of 27 damselfish species that live in close association with anemones generally are small, brightly colored forms, many with rounded tails. These fish actually live among the tentacles of the invertebrates and have a mucus layer that seems to inhibit the anemones from stinging them (see Box 33.1).

Labridae

With more than 500 species, the wrasses make up one of the largest perciform families. Variation in body shape and size is considerable among the wrasses, but most can be quickly recognized based on the following: (1) the pointed snout; (2) the small to moderate-sized mouth containing conspicuous, outward-pointing teeth; and (3) the solid, semicylindrical body with cycloid scales, long dorsal fin, and square or slightly rounded tail. Most are less than 12 cm long, but members of the genus *Cheilinus* may reach 2.3 m. Whether they occur in tropical or temperate regions, wrasses usually are brightly colored, diurnal fishes that live by picking up small invertebrates. Perhaps the best-known trait of wrasses is that a number of them are cleaners that pick parasites from other fishes. Such fish usually are solitary or live in pairs, and they may behave very aggressively toward other wrasses. Most noncleaner wrasses behave similarly, although a few species, especially wrasses that inhabit temperate regions, are shoaling fishes. On the Atlantic coast of North America, two such species are the cunner *(Tautogolabrus adspersus)* and the tautog *(Tautoga onitis)*, both of which have some value as commercial and sport fish.

Scaridae

The more than 80 species of parrotfish look like large, heavy-bodied wrasses that have their jaw teeth fused into a solid, parrot-like beak and have some of the most complicated sexual systems among vertebrates (Box 23.2). They also have heavy pharyngeal teeth consisting of solid units of bone (Fig. 23.2E). These structures permit parrotfish to scrape algae and invertebrates from hard surfaces of the reef and then to crush the ingested material. They move about the reefs in conspicuous, small schools during the day, and they hide in caves and crevices at night. A few species secrete a mucus cocoon about themselves when at rest, presumably to foil night-feeding predators such as moray eels.

23.4   ZOARCOIDEI

All 320 species in the suborder Zoarcoidei are elongate fishes, with long dorsal and anal fins and small (or absent) pelvic fins. Most live in temperate to arctic waters of the northern hemisphere, but curiously, a few species of eelpout (Zoarcidae) live in Antarctic waters. Although many of the species are offshore or deepwater bottom-dwellers, many pricklebacks (Stichaeidae) (Fig. 23.3) and gunnels (Pholidae) are common in tide pools. The rockweed gunnel *(Xererpes fucorum)*, like a number of other intertidal fishes, has red, green, and brown forms; The green forms are found among beds of green algae

---

BOX 23.2

Color and Sex Change in Parrotfishes and Wrasses

A confusing—but conspicuous—feature of parrotfishes and wrasses is their color patterns, which change dramatically with age and sex (which may also change), reflecting delightfully complicated mating systems. The colors are dramatic enough that early taxonomists sometimes described different color phases as different species. Robertson and Warner (1978) recognized three basic color phases in parrotfishes: (1) a *juvenile phase*, (2) an *initial phase* (characteristic of young adults), and (3) a *terminal phase* (characteristic of large, dominant males). Depending on the species, males and females may or may not have different color patterns, and color patterns may or may not be reversible. To complicate matters further, parrotfish may contain two patterns of sexual development: fishes that have their sex genetically fixed (*gonochorists*), and fishes that can change sex from female to male (*protogynous hermaphrodites*). In many species, much of the mating is controlled by large, terminal-phase males (often protogynous hermaphrodites) who defend permanent territories, each of which contains a harem of females. In these situations, the smaller, initial-phase males manage to spawn largely by *sneaking* and *streaking* (Robertson and Warner 1978). Sneaking occurs when a small male manages to get a member of a harem to spawn with it within the territory of the terminal male; streaking occurs when a male dashes into a territory and spawns with a female just as the terminal male is also spawning. Not all species have permanent territories. Some resort instead to either group spawning or lek spawning. In the latter case, the terminal males defend territories only temporarily and mate with females, either singly or in groups, that enter the territory. One advantage of being a terminal male is that its gonads are smaller than those of other males or of females, so it can put more energy into growth. As a result, terminal males typically are the largest individuals in a population—and often the oldest as well (up to 20 years) (Choat et al. 1996).

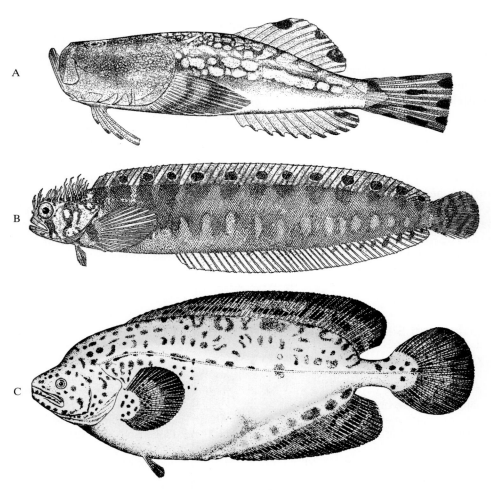

FIGURE 23.3 Representative perciform fishes. **A.** Stargazer (*Astroscopus*, Uranoscopidae, Trachinoidei). **B.** Mosshead warbonnet (*Chirolophis nugator*, Stichaeidae, Zoarcoidei). **C.** Ragfish (*Icosteus aenigmaticus*, Icosteoidei). **A** from Jordan and Evermann (1900); **B** from Jordan and Starks (1895); **C** from Goode and Bean (1895).

in the upper intertidal zone, the brown forms among brown algae in the lower intertidal zone, and the red forms in deep water among red algae (Burgess 1978). The largest members of this suborder are the wolffishes (Anarhichadidae), which reach 2.5 m and are famous for their ferocious appearance and tendency to lunge at divers who get too close, especially when the fish are guarding a nest of embryos.

23.5 NOTOTHENIOIDEI

The more than 120 species of notothenioid fishes are the dominant fish of the Antarctic, although a few species are also found along the coasts of southern South America,

Australia, and New Zealand. The remarkable morphological and physiological adaptations of these fishes are discussed in Chapter 36.

23.6 TRACHINOIDEI

Thirteen families and approximately 210 species make up the Trachinoidei suborder. Most of these species are dwellers in the sandy bottoms of shallow waters. They bury themselves in sand to ambush prey, which they suddenly engulf in their large, upward-pointing mouths. To protect themselves, weeverfishes (Trachinidae) and stargazers (Uranoscopidae) have poisonous spines that can inflict painful injuries on waders and other predators (Fig. 23.3). The stargazers also have electric organs that are formed from modified eye muscles, which they probably use to detect prey. Some species, however, can deliver up to 50 V, so these organs may also function in stunning the prey. When buried, stargazers can inhale water through their nostrils, which open into the branchial cavity. Some species also have a worm-like lure growing from the floor of the mouth that can be moved about to attract prey.

A recent addition to this suborder are the sand lances (Ammodytidae, 18 species), which are small, shoaling fishes. They are abundant enough in the inshore waters of the northern hemisphere to be of major importance in marine food chains leading to larger, commercially exploited fishes. They are elongate, rarely exceeding 30 cm, with deeply forked tails, long dorsal fins (without spines), reduced or absent pelvic fins, cycloid scales, and no swimbladder (Fig. 23.4). Their heads are pointed, with projecting lower jaws. Apparently, this latter characteristic is an adaptation that allows them to burrow quickly, head first, into sand when attacked by predators. Despite this important defense mechanism, shimmering schools of sand lances frequently are found feeding on zooplankton offshore.

23.7 BLENNIOIDEI

The blennioid fishes range from eel-like to sculpin-like in body shape, and they include nearly every conceivable combination of these two conditions. The pelvic fins are either absent or small and located in front of the pectoral fins. Most have long dorsal and anal fins, which are fused with the caudal fin in eel-like forms. The suborder contains more than 730 species in six families.

Clinidae

The clinids are an enormously variable group of approximately 75 species that mostly inhabit the intertidal zone. They are perhaps the most common group of intertidal fishes in Australia and South Africa, but they are not unusual elsewhere (particularly along the Pacific coast of North America). Many species have the elongate body and blunt head so typical of blennies, but they are distinguished from the Blenniidae by the predominance of spines in the dorsal fin (many more than rays) and patches of fixed, conical teeth in the mouth. Some species have pointed snouts and laterally compressed heads and bodies. Most species have small cycloid scales, although scales are absent from one species. Their reproductive habits vary considerably. Several species are viviparous, so the males

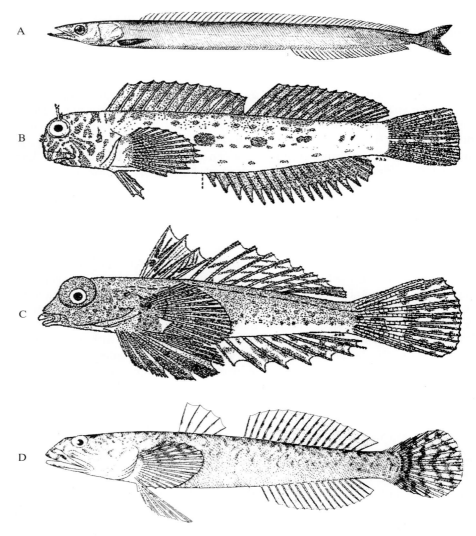

FIGURE 23.4   Representative perciform fishes. **A.** Sand lance (*Ammodytes*, Ammodytidae, Trachinoidei). **B.** Blenny, *Entomacrodus*, Blennoidei). **C.** Dragonet (*Synchiropus*, Callionymoidei). **D.** Goby (*Clevelandia ios*, Gobioidei). **A** from Jordan and Evermann (1900); **B** and **C** from Schultz et al. (1960); **D** from Jordan and Starks (1895).

have intromittent organs. Some of the most distinctive of the clinids are the fringeheads *(Neoclinus)* of the Pacific coast, which are 20-cm, blenny-like fishes with long, fleshy cirri over the eyes and extraordinarily large mouths. The common name of one species is the sarcastic fringehead.

Blenniidae

Combtooth blennies are an important part of the rocky inshore marine fauna over much of the world, but especially in tropical and subtropical regions. The common name stems from the numerous, closely packed teeth on their jaws, although a few species also possess dagger-like teeth on each side of the lower jaw. Most of the 345 species have blunt heads, usually topped by cirri, and moderately long, scaleless bodies that include dorsal fins containing more rays than spines (Fig. 23.4B). Many species have pelagic larvae or postlarvae, so they can quickly colonize disturbed areas of coastline, such as new breakwaters. Wherever they are, blennies tend to be secretive in their behavior but eclectic in feeding, consuming both invertebrates and algae. One of the more famous exceptions to this characterization is the sabertooth blenny *(Aspidontus taeniatus),* which mimics cleaner wrasses so that it can sneak up on unsuspecting fishes to take bites out of their fins and tails. As the name implies, these are among the blennies with large jaw teeth; the large teeth are not used for feeding, however, but for aggressive displays to members of the same species.

23.8 ICOSTEOIDEI

Only one family (Icosteidae) and one species are in the Icosteoidei. The ragfish *(Icosteus aenigmaticus)* of the deep waters off the Pacific coast of North America is well described by its scientific name, which means "puzzling fish with soft bones." The adults can exceed 2 m and are extremely flabby, with a largely cartilaginous skeleton and no spines, scales, or pelvic fins. The body shape of the adult is somewhat trout-like, with a blunt head and chocolate-brown color. The juvenile fish, in contrast, are deep bodied, with rounded tails, pelvic fins, and color pattern of blotched brown and yellow (Fig. 23.3C). It is not surprising, therefore, that the juveniles were originally described as a different genus and species from the adults.

23.9 GOBIESOCOIDEI

The 120 species of clingfish have been moved back and forth between the Paracanthopterygii and the Acanthopterygii, their true nature being well hidden by their specializations. Another proposal is to put them in their own order along with the Callionymoidei (discussed later) (Gill and Mooi 2002). Clingfish are designed for clinging to rocks in the pounding surf of the intertidal zone or to swaying fronds of kelp. They are mostly small, tadpole-like fish, with their pelvic fins united into a large, sucking disk. They lack scales, spines, and a swimbladder.

23.10 CALLIONYMOIDEI

The dragonets (Callionymidae, 130 species; Draconettidae, 7 species) are small, colorful, goby-like inhabitants of tropical reef areas (Fig. 23.4C). The callionymids have the principal opening of each operculum as a nearly dorsal notch, which permits the fish to remain partially buried on the bottom but still capable of taking in water for respiration. A number of dragonets are covered with evil-smelling mucus, which apparently decreases their desirability to predators. They are best known, however, for the brightness of the males (in contrast to the females) and the readiness with which they court and spawn in aquaria.

23.11 GOBIOIDEI

The gobioid fishes form a distinctive group of bottom-dwellers on the basis of both osteology and external morphology, so perhaps they should be given the status of a separate order. They usually lack both lateral line and swimbladder. They have gill membranes that are joined to the isthmus (i.e., have opercula that are not free at the bottom), and they have a short, spiny dorsal fin (one to eight spines) and pelvic fins that are either united to form a sucking disc or at least close together. This suborder contains more than 2,100 species in eight families, representing 8% to 9% of all fishes. The authority on this group is Akihito, the Emperor of Japan (e.g., Akihito et al. 2000).

Gobiidae

The number of species in the Gobiidae (1,875) is only slightly smaller than that of the Cyprinidae. Gobies are found worldwide in both fresh and salt water, with the majority of these species being associated with shallow tropical and subtropical environments. Despite their wide distribution and richness of species, however, the gobies are a readily recognizable group (Fig. 23.4D). They are characterized by: (1) a sucking disk created by the pelvic fins; (2) two dorsal fins, with the first consisting of two to eight flexible spines; (3) a rounded caudal fin; (4) a blunt head, dominated by relatively large eyes; (5) visible scales, either cycloid or ctenoid; and (6) small size (few reach 20 cm, and most are <10 cm). Their success seems to be best explained by their remarkable ability to adapt to habitats or microhabitats most other fishes, find inaccessible, such as cracks and crevices in coral reefs, burrows of invertebrates, mud flats, mangrove swamps, fresh water on oceanic islands, and inland seas and estuaries. The sucking disk has enabled some species to climb over waterfalls and may serve as an anchor in loose substrates (Akihito et al. 2000).

The cracks and crevices of coral reefs and other inshore habitats provide homes for perhaps most goby species. Many of these gobies are very small (Box 23.3), and many of these reef gobies are brightly colored and live symbiotically with other reef animals. Some species are found only in the "chimneys" of sponges; others are found in the arms of branched corals. These probably are commensal relationships. Some gobies associated with sea urchins, however, seemingly behave as parasites, because they browse on the tube feet of the urchins (Teylaud 1971). In Atlantic reefs, *Gobisoma* gobies leave their crevices to pick ectoparasites from other fishes.

When gobies are found on soft bottoms, they typically live in the burrows of invertebrates, ranging from polychaete worms to shrimp to clams. On tide flats, some gobies manage to survive by breathing air and "walking" across the mud. The longjaw mudsucker *(Gillichthys mirabilis)* absorbs oxygen from the air in its large, highly vascularized mouth cavity. It is also capable of making "sojourns" from one muddy pool to another (Todd 1968). The most extreme examples of a goby adapted to the terrestrial environment, however, are the mudskippers *(Periopthalamus),* which inhabit the mangrove swamps and mud flats of the tropical Indian Ocean. These fishes actually climb out of the water, along mangrove roots, and seek terrestrial insects (Nursall 1981). They "walk" on their pectoral fins, propelled in part by the tail, and breathe air trapped in highly vascular opercular cavities. Another characteristic of the mudskippers is their

BOX 23.3
World's Smallest Fishes

Just how small can an adult vertebrate be? The answer seems to lie among the gobies. The current record-holder for smallest vertebrate is the dwarf goby *(Trimmatom nanus)* from reefs along the Chagos Islands in the Indian Ocean. Dwarf gobies are mature at 8 to 10 mm (Winterbottom and Emery 1981). Former record-holders include the dwarf pygmy goby (*Pendaka pygmaea*) from rivers of the Phillipines, in which males are mature at 9 to 11 mm and females at approximately 15 mm. Another freshwater goby of about the same size from Luzon (the sinarapan, *Mistichthys luzonensis*) is the smallest fish subject to commercial fisheries. The sinarapan was once scooped up in small nets by the thousands; it is now regarded as an endangered species. Nearly as small (16 mm) is the peculiar phallostethid fish (*Phenacostethus smithi*) of Southeast Asia. Various otophysan fishes can also be quite small. Two species of Sundasalangidae, which are larva-like fishes from the freshwaters of southeast Asia, mature at 15 to 22 mm. The Schindleriidae are similar. The females of the dwarf, blind banjo catfish (Aspredinidae, *Micromyzon akamai*) can be mature at 12 mm (Friel and Lundberg 1996), whereas other South American catfishes and characins are adults at approximately 18 to 30 mm. So far, 109 such "minature" species have been described from South America (Berra 2001). Small fishes seem to thrive either by being cryptic in color and hiding in tiny places or by being transparent and pelagic. Either way, with short life cycles, they can quickly take advantage of abundant resources or new habitats.

tolerance of low salinities. Indeed, this seems to be a characteristic of the entire family, because gobies have invaded freshwater habitats throughout the world, but particularly on oceanic islands and in Asia. On the islands, they frequently are the principal native freshwater fishes, and many are amphidromous, spending part of their life cycle in the ocean (see Chapter 11). On the Hawaiian islands, a widespread goby is the alamo'o *(Lentipes concolor),* in which the rear half of males is bright yellow and the front half is black. The males attract females by sitting on top of boulders in fast water and letting their rear end wave back and forth in the current. In Asia, gobies are common in inland waters, from estuaries to rivers to brackish inland seas. In the Black and Caspian seas, they are—(or were)—abundant enough to be exploited commercially, despite their small size. Some of these gobies have been spread to the Laurentian Great Lakes through the ballast water of ships, as have Asian gobies to western North America and Australia.

Eleotridae

The sleepers, gudgeons, and bullies (150 species) are essentially chunky gobies without the sucking disk (and, as a consequence, are sometimes placed in the Gobiidae). They have a circumtropical distribution in marine, brackish, and fresh waters, and they often are important members of island freshwater fish faunas, including those of New Guinea, Australia and New Zealand. Most freshwater forms seem to be amphidromous, although some may be either catadromous or anadromous (Lucas and Barras 2001). Most are also benthic ambush predators, lying quietly on the bottom (hence the name "sleeper") until a small fish or crab comes along. Eleotrids are gudgeons in Australia and bullies in New Zealand.

Schindleriidae

The two species of *Schindleria* are very peculiar indeed. They are tiny (2–3 cm), surface-dwelling fishes of the tropical Pacific. They would be classified as larvae, with the adults unknown, except that they become sexually mature. Like larval fishes, they are elongate, transparent, and have a poorly developed skeleton, except for the bones of the jaw, which bear fine teeth. They are extremely abundant in many areas, especially around Hawaii.

23.12   KURTOIDEI

Only two species (*Kurtus*, Kurtidae) are placed in the suborder Kurtoidei, both of them rather peculiar but abundant fishes in the estuaries and streams of Southeast Asia, New Guinea, and northern Australia. They are unmistakable fishes, with their deep and compressed bodies, long anal fins, deeply forked tails, small cycloid scales, large and upward-oriented mouths, and—most of all—depressed foreheads topped by a hook-like structure of modified dorsal fin spines (Fig. 23.5A). The latter feature is found only on males and is used to carry eggs. This has earned them the names of nurseryfish or humpheads.

23.13   ACANTHUROIDEI

Although the suborder Acanthuroidei is small, it contains some of the most conspicuous and colorful fishes on tropical reefs: the surgeonfishes and moorish idols (Acanthuridae, 72 species), and the rabbitfishes (Siganidae, 27 species). Surgeonfishes are deep-bodied fishes of moderate size (20–30 cm) that are extremely compressed laterally (Fig. 23.5B). They tend to have beak-like snouts, with small mouths and incisor-like teeth that are used for scraping algae from rocks and coral. Their name comes from the scalpel-like spines on the caudal peduncle, which generally point forward when erected. A few species that pick ectoparasites from other fishes are called doctorfishes. Moorish idols (*Zanclus canescens*) resemble surgeonfishes in body shape and mouth structure, but they lack the "scalpels" on the tail. They are readily recognized by their strong, black-and-white striped pattern and symmetrical dorsal and anal fins. In fact, moorish idols probably are the most familiar of all reef fishes; stylized versions of them are standard decorations on shower curtains, tiles, towels, bathroom wallpaper, and other water-related items.

23.14   SCOMBROIDEI

The scombroid fishes are only approximately 135 species distributed among five families, but these species usually are characterized by superlatives for their value as food

FIGURE 23.5 **A.** Nurseryfish (*Kurtus*, Kurtoidei). **B.** Surgeonfish (*Acanthurus*, Acanthuroidei). **A** from Day (1878); **B** from Jordan and Evermann (1900).

fishes, their speed, their voraciousness as predators, their long migrations, the sport they provide anglers, and their peculiarities of anatomy and physiology. Curiously, they have lost the protractile upper jaw so characteristic of teleosts and, instead, have a solidly fused unit to support a biting mouth, much like more-ancestral bony fishes.

Sphyraenidae

Barracudas are fearsome predators on other fishes, with elongate and pike-like bodies, protruding lower jaws, formidable pointed teeth in the jaws, two widely separated dorsal fins, forked tails, and tiny or nonexistent gill rakers. Their body plan is an interesting compromise between that of a lie-in-wait predator and that of an active predator (Fig. 23.6D). Barracudas swim about actively in clear water, often in schools, searching for prey. Their narrow head-on profile and silvery color reduce their visibility to potential prey; their long caudal region, assisted by posteriorly placed dorsal and anal fins, allows a quick burst of speed to close the gap between them and their prey. This strategy is particularly effective when light levels are low. The great barracuda *(Sphyraena barracuda)* of the Atlantic may reach lengths of nearly 2 m and has been known to attack divers, but such attacks are rare and virtually unknown in the other 19 species. They are excellent food and favored sport fish, however, so as usual, they have much more to fear from humans than humans do from barracudas.

Scombridae

Only approximately 50 species of mackerels and tunas are known, and they make up less than 10% of the total catch of marine fish by weight. Their monetary value, however, is extremely high because of the favor they find as a food fish, particularly with Americans, Europeans, and Japanese. Tuna and mackerel are the top carnivores of the epipelagic zone of the tropical and subtropical seas. Their prey consists of small schooling fish and squid, which they locate and capture through high-speed swimming, also in schools. Thus, most of the features that characterize the scombrids are adaptations for fast and continuous swimming. Their bodies are beautifully streamlined: spindle-shaped and oval to round in cross-section (Fig. 23.6A). The skin is smooth, with tiny cycloid scales; its coloration tends to be shades of iridescent blue or green on top of the fish, countershading to silver and white below. The first dorsal fin, although made up of stout spines, depresses into a groove on the back while the fish is swimming rapidly. The soft dorsal and anal fins are short, identical in shape and size, and located opposite one another just in front of rows of small finlets that run down the caudal peduncle to the tail. The tail fin is deeply forked to lunate (high aspect ratio) and provides the tremendous thrust that is needed to maintain high speeds. The caudal peduncle is very narrow and mainly contains the bony keel on each side (created by the flattening of the caudal vertebrate) and the tendons running to the tail from both red and white muscle. Because the tendons run over the keel, the keel acts as a pulley to increase the amount of pull exerted by the muscles. In many species, the swimbladder is reduced or absent, thereby enabling the fish to move throughout the water column in search of deepwater prey as well as surface prey.

The respiratory pump system in scombrids is reduced, because constant rapid swimming forces plenty of water across the gills. In the larger scombrids, the circulatory system has been modified with countercurrent exchangers to reduce the loss of the heat generated by the tremendous muscular activity. In turn, this increases the muscle efficiency (see Chapter 4).

Xiphiidae

The 12 species of billfish roam the surface waters of tropical and subtropical seas and are readily recognizable as swordfish *(Xiphias gladius)*, sailfish *(Istiophorus* spp.), spearfish *(Tetrapturus* spp.), and marlin *(Makaira* spp.). These fish all have a long bill

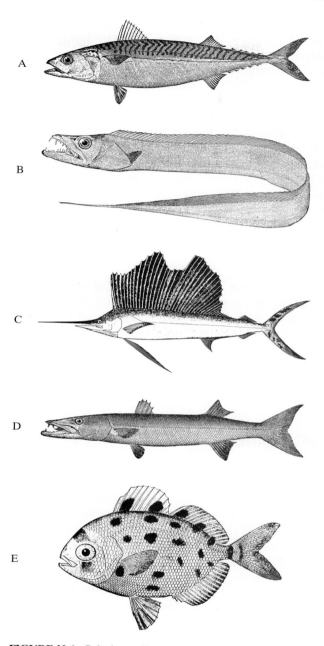

FIGURE 23.6   Pelagic perciform fishes. **A.** Tuna (*Thunnus*, Scombridae). **B.** Cutlass fish (*Trichiurus*, Trichiuridae). **C.** Sailfish (*Istiophorus*, Xiphiidae). **D.** Barracuda (*Sphyraena*). (**A–D** are all Scombroidei). **E.** Driftfish (*Psenes*, Nomeidae, Stromatoidei). **A**, **C**, and **D** from Jordan and Evermann (1900); **B** and **E** from Goode and Bean (1895).

derived from the premaxilla, two anal fins, no true spines in the dorsal fin, and reduced or absent pelvic fins (Fig. 23.6C). The sword of the swordfish is flat and, at its maximum extent, is approximately one-third the length of the fish (which may reach 4.5 m). The sword of other species is rounded in cross-section and generally less than one-quarter of the total length (up to 4 m). Billfish capture other fish for food by swimming through schools and slashing the bill back and forth, thereby stunning or injuring the prey. Swordfish apparently can also spear large fishes with their snouts. Occasionally, billfish will attack boats and submersibles, usually when provoked. In one instance, a commercial fisherman was killed when fighting a hooked swordfish. The swordfish charged the man's dory, causing the line to go slack and the man to fall backward into the boat; the fish then rammed the boat, stabbing through both boat and man (Wilcox 1887). The swordfish also died, of course. Most such contests do not wind up with such even results, and swordfish are now among the species with populations that are declining because of the effects of longline and gillnet fisheries.

Trichiuridae

The cutlassfishes, ribbonfishes, and snake mackerels (32 species) are widely distributed marine predators, although many are mesopelagic and poorly known. They are recognized by their elongate, eel-like bodies and pointed head with large, sharp teeth (Fig. 23.6B). The dorsal fin runs much of the length of the body, and the caudal and pelvic fins are reduced or absent. How they capture prey is a bit of a mystery, but they are well known by anglers for latching onto smaller fish that have been hooked and allowing themselves to be dragged in (Hoese and Moore 1998). Large (1–2 m) individuals that wash up beaches are sometimes reported in newspapers as "sea serpents."

23.15 STROMATEOIDEI

In this curious group of six families and 65 species, the juveniles are, for the most part, better known than the adults. This is because, with the exception of members of the Ariommatidae (which are bottom-oriented, deepsea fishes), juvenile stromateoid fishes are associated with jellyfish or drifting objects in the epipelagic region. The juveniles found with jellyfish swim among the tentacles, dodging the stinging nematocysts, to which they are not immune. With this association, they obtain not only protection from large piscine predators but also food, because they nibble on the tentacles. These juveniles, in contrast to the adults, have a well-developed swimbladder and a color pattern of vertical stripes. The rather drab and unremarkable-looking adults (Fig. 23.6E) are mostly pelagic as well, and they may compensate, in part, for the reduced or absent swimbladder by having a high lipid content of the body and a reduced firmness of muscle and bone. Exceptions, however, include the 13 species of deep-bodied, iridescent butterfishes (Stromateidae), which have a high lipid content but nevertheless are negatively buoyant. The butterfishes are active, inshore species that apparently move freely up and down the water column, feeding on coelenterate medusae, ctenophores, and other soft-bodied animals. Like other stromateoid fishes, which also feed on such organisms, they possess tooth-covered pharyngeal sacs just behind the last gill arch (Horn 1984).

23.16 ANABANTOIDEI

The gouramis are a group of five freshwater families that are well known to aquarists through such species as the climbing gourami (Anabantidae, *Anabas testudineus*), the Siamese fighting fish (Belontiidae, *Betta splendens*), the kissing gourami (Helostomatidae, *Helostoma temmincki*), and the giant gourami (Osphronemidae, *Osphronemus goramy*). The more than 80 species in these families are mostly small (<10 cm), surface-oriented fishes, with moderately deep bodies, rounded tails, and long anal fins (Fig. 23.7A). Internally, they have two particularly distinctive features: a long body cavity, so that the swimbladder extends into the caudal region; and a labyrinthine (suprabranchial) organ (Liem 1963). The labyrinthine organs permit the gouramis to breathe air. They are made of elaborately folded layers of highly vascularized skin that cover modified lamellae from the first functional gill arch. The organs occupy much of the gill chamber as well as an extra chamber on its roof. The gills have become reduced in size as a consequence, and gouramis suffocate if deprived of access to the surface. Gouramis breathe by swallowing bubbles of air, holding them in the labyrinthine organs for extended periods of time, and then forcing them out through the gill covers. The air bubbles taken in by the gouramis are also important for hearing and reproduction. When in the chambers of the labyrinthine organ, these air bubbles are next to the sides of the cranium containing the inner ear. In gouramis, membranous "windows" are present at this point, so vibrations picked up by the air bubbles are easily transmitted to the inner ears (Alexander 1967). For reproduction, male gouramis of some species expel air bubbles through their mouths

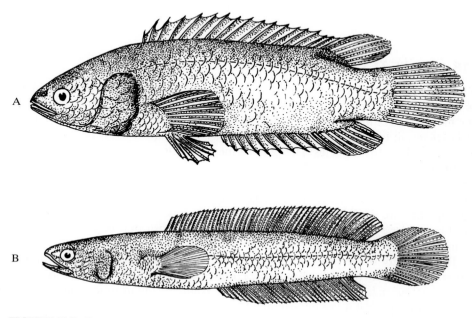

FIGURE 23.7 Representatives of tropical freshwater perciform suborders. **A.** Climbing perch (*Anabas*, Anabantidae, Anabantoidei). **B.** Snakehead (*Channa*, Channidae, Channoidei).

and build a nest of froth on the surface of the water, into which eggs are laid by the female and subsequently tended by the male. The froth nests and labyrinthine organs of the gouramis are obviously adaptations for life in the stagnant waters of tropical Asia and Africa in which they live.

A member of this group so distinctive that it is placed in a family by itself is the pikehead (*Luciocephalus pulcher*, Luciocephalidae), a small (to 18 cm), freshwater fish from Indonesia. It essentially resembles a minature pike, with classic lie-in-wait morphology. A major difference from "true" pikes, however, is that the mouth is extremely protrusible, with premaxillaries capable of extending to one-third the length of the head (Berra 2001).

23.17   CHANNOIDEI

The 21 known species of snakeheads (Channidae) are famous for their voraciousness as predators. They have a large mouth, with an underslung and well-toothed jaw, large eyes, and a thick and elongate body having long dorsal and anal fins (Fig. 23.7B). Like the gouramis, with which they typically occur, snakeheads have a labyrinthine organ in the gill chamber, but the structure is quite different from that of the gouramis. Other adaptations of snakeheads for life in stagnant waters are eggs that float (by means of oil droplets) at the surface, where they are guarded by the males, and the ability of some species to move overland by "swimming" in an eel-like fashion. They are highly regarded as food fish, in part because they are easy to keep fresh by transporting alive. This has resulted in their establishment (and eradication) in a few ponds the eastern United States, with newspaper stories making the most of their voracious nature. Some species can reach 1 m in length.

23.18   MORE LESSONS FROM THE PERCOMORPHA

The extraordinary success of the perciformes clearly relates to three factors: (1) a basic suite of adaptive characters that has allowed them to take advantage of complex, shallow-water habitats; (2) an evolutionary "willingness" to abandon key general adaptations, such as spines, in favor of specializations; and (3) complex behavior patterns, including interactions with other species. The perciformes also show amazing variability in their reproductive strategies. Many of the species are small and cryptic, but many others are brightly colored, day-active fishes that are protected by their spines and behavior. Others are large and predatory. If any two external features unite the diverse members of this group, they are the large, maneuverable pectoral fins, placed near the center of gravity, and the large eyes. These two features characterize fish that rely on precise movement and vision to capture prey.

The diversity of percomorph fishes is particularly threatened by human activity, because they thrive in clear, shallow waters of the oceans and in clear lakes and streams. Their marvelous adaptations, fine-tuned to a complex environment in high-quality water, are of little avail if the environment is clouded and simplified by pollution, habitat destruction, and overfishing.

SUPPLEMENTAL READINGS

Barlow 2000; Gill and Mooi 2002; Nelson 1994; Pitcher 1993; Ribbink 1990; Wooton 1990.

WEB CONNECTIONS

INTRODUCTION TO PERCOMORPHS
people.clemson.edu/~jwfoltz/wfb300/subjects/perco

GENERAL
www.flmnh.ufl.edu/fish
www2.biology.ualberta.ca/jacksonhp/IWR

CICHLIDAE
cichlidresearch.com
cichlidae.come
malawicichlids.com

C H A P T E R 24
Flounders, Puffers, and Molas

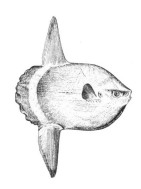

SUBDIVISION TELEOSTEI
  Infradivision Euteleostei
    Superorder Acanthopterygii
      Order Pleuronectiformes
        Families Psettodidae, Citharidae, Bothidae (left-eye flounders), Achiropsetti-
        dae (southern flounders), Scopthalmidae, Paralichthyidae (sand dabs),
        Pleuronectidae (right-eye flounders), Samaridae, Achiridae (American
        soles), Soleidae (soles), Cynoglossidae (tonguefishes)
      Order Tetraodontiformes
        Families Triacanthodidae (spikefishes), Triacanthidae (triplespines), Balistidae
        (triggerfishes), Monacanthidae (filefishes), Ostraciidae (boxfishes), Tri-
        odontidae (threetooth puffers), Tetraodontidae (puffers), Diodontidae
        (porcupinefishes), Molidae (ocean sunfishes)

The two orders considered in this chapter contain fishes that are so modified it is difficult to figure out with whom they share common ancestry among the Acanthopterygii. It is certainly not each other! The flatfishes and puffers are usually placed at the end of fish classification schemes, on the assumption that they are the most-derived groups of fishes and are descended, somehow, from highly derived, perciform-like ancestors.

24.1 PLEURONECTIFORMES

There are few problems in recognizing a member of the flatfish order, because it is the only group of fishes in which the adults are not bilaterally symmetrical. One side of the body is white and eyeless, but the other side is dark-colored and has both eyes (Fig. 24.1). Flatfish lie on the bottom, with the pale side down and the dark side up. Because they are very flat, secretive in their behavior, and able to change color to match the substrate, flatfish are masters at hiding from both predators and prey. Most flatfish remain close to the bottom even while swimming, which they accomplish by undulating motions of the body. Presumably, the efficiency of this type of swimming is improved by the long dorsal and anal fins, which, together with the caudal fins, may almost completely encircle the body. These

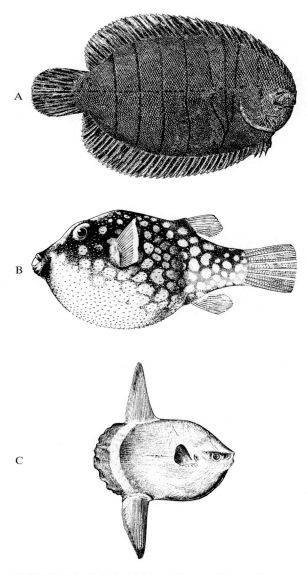

FIGURE 24.1 **A.** Sole (*Achirus*, Pleuronectiformes).
B. Puffer (*Canthigaster*, Tetraodontiformes). **C.** Ocean sunfish
(*Mola mola*, Tetraodontiformes). From Jordan and Evermann
(1900).

fins are usually without spines and are very flexible. As might be expected, the swimbladder is absent in adults, and the body cavity is very small. Within the limits of their peculiar morphology, flatfish are very diverse, with more than 570 species in 11 families.

One of the most fascinating—and phylogenetically defining—aspects of flatfish biology is the change that occurs as the pelagic, bilaterally symmetrical larvae become benthic, asymmetrical juveniles (Fig. 24.2). The most visible parts of this process are the change in pigmentation patterns and the migration of one eye across the top of the head to its final resting place close to the other eye. This involves not only movement of the eye but also changes in nerves, blood vessels, skull bones, and muscles. Pectoral and pelvic fins on the blind side also tend to be reduced in size, and the blind side is flat,

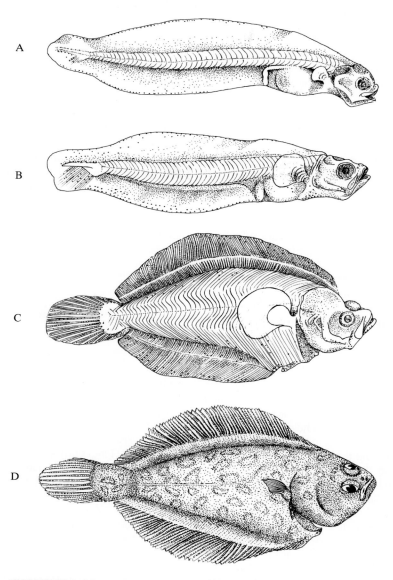

FIGURE 24.2 Metamorphosis of halibut (*Hippoglossus stenolepis*) from a pelagic **(A)** larva to a bottom fish **(D).** Note the migration of the eye from one side to the other **(B, C).** After Thompson and Van Cleve (1936).

while the "eyed" side more rounded. One of the features used to distinguish various flatfish groups is whether they have both eyes on the left or the right side of the body. Left-eyed flatfishes rest on the right side of the body, and vice versa. Some species, however, such as the starry flounder (*Platichthys stellatus*), have both right- and left-eyed forms. Molecular evidence now suggests that the "sidedness" of flatfish is not particularly useful in their classification (Berendzen and Dimmick 2002).

Flatfish are found mostly on soft bottoms of continental shelves. A few species inhabit the continental slope, and others may invade fresh water. On continental shelves, they have an abundant, worldwide distribution, often supporting important fisheries. The large number of flatfish species means that many species coexist with each other. Ecological segregation is reflected in differential depth distributions (see Chapter 32) and in different feeding habits, as reflected in the structure of the mouth and pharyngeal teeth. Piscivorous species, such as halibut (*Hippoglossus*), have large, symmetrical mouths that are lined with tiny teeth for grasping and also have sharply pointed pharyngeal teeth. Species that feed by taking chunks out of the siphons of clams and other buried invertebrates have asymmetrical mouths (favoring the down side), with chisel-like teeth for biting and molariform pharyngeal teeth for grinding. Most flounders are ambush feeders, lying partially buried on the bottom with their colors closely matching those of the substrate. When a prey organism comes near, it is grabbed during a sudden, lunging charge.

Flatfish are favored food fish (Box 24.1), so a great deal is known about some species, especially European plaice (*Pleuronectes platessa*) and halibut (*Hippoglossus* spp.). Fisheries often catch a mixture of flounders, soles, sanddabs, turbots, and tonguefish. Because Pacific halibut (*H. stenolepis*) are so valuable and competition for them among Alaskan

---

BOX 24.1
Floundering for Names

The unusual appearance of flounders has resulted in many unusual names. The commonly fished flounders of Europe had English names, like sole, plaice, whiff, dab, brill, turbot, witch, and halibut—all referring to single species. As more species were described around the world, these names were applied to the new species without regard for the relationship to the original bearer of the name (e.g., New Zealand sole, Queensland halibut). The original common names, however, were descriptive. *Sole* is from the Latin word for *sandal* or *shoe*, whereas *plaice* is from the Greek word for *flat* (same root as *plate*). The origins of *whiff* and *dab* are obscure, but these names probably refer to the small size of these flounder. *Witch* is a dark-colored flounder usually caught in the winter, and so is associated with evil. Halibut, in contrast, was supposedly a desired fish to be eaten on holy days—hence, *holibutt*, where *butt* is an old Germanic word for *flounder*. Elsewhere, equally creative names were coined. Tonguefish is the name given to a whole family (Cynoglossidae) of flounders with small eyes and tapering tails; they do, indeed, resemble tongues. The hogchoker (*Trinectes maculates*) is a small, rough-scaled flounder of the Atlantic Coast of North America that must have choked a hog (or greedy person) at some point!

Most of the information in this box came from the Oxford English Dictionary (1971).

fishermen was so severe, at one point the fishing season was reduced to a single frantic, 24-hour period! In contrast, in the fishery in British Columbia, where the number of fisher-men is restricted, the season lasts for months. In 1995, the United States, working through the International Pacific Halibut Commission, instituted similar regulations.

24.2   TETRAODONTIFORMES

If the degree of alteration of the basic acanthopterygian body plan is the key to deter-mining how "derived" a fish group is, then the Tetraodontiformes order includes the most derived of all teleosts. Perhaps the best indications of this are found internally, in modifications of the skeleton (e.g., they lack parietals, nasals, infraorbitals, and lower ribs), but the tetraodontiform fishes are also visibly different from other fishes. To begin with, they come in a variety of body shapes that are the antithesis of the fusiform bodies of "typical" fish, ranging from globular to triangular to extremely compressed. All are slow swimmers, propelled by a rounded caudal fin, by sculling with the pectoral fins, or by movements of the dorsal and anal fins, which typically are symmetrical and placed far back on the body. Presumably because of the vulnerability of such fish to predators, most possess some form of protection, such as inflatable and spine-covered bodies, body armor (often of scales), stout fin spines that can be locked erect, tough leathery skin, poi-sonous flesh, or even some combination of these. Gill openings are small, usually just holes on the side of the fish, in front of the pectoral fins. The food of tetraodontiform fishes is mostly invertebrates that other fishes will not eat, mostly because of their heavy shells, spines, or other armor. As a result, tetraodontiform fish have stout teeth in the jaw or tooth-like bones fused into a beak. The maxillae and premaxillae are also fused to each other. This hard jaw is powered by large muscles. The pharyngeal teeth are also stout and provide additional crushing abilities. Most of the 340 or so species in this order are associated with tropical reefs, which are dominated by hard-shelled invertebrates, although a few species live in fresh water.

Balistidae

The triggerfishes are typified by the 20-cm humuhumunukunukuapuaa (*Rhinecanthus aculeatus*) of the Hawaiian reefs. Most members of this family (40 species) and of the re-lated Monacanthidae (filefishes, 95 species) are slow-moving, solitary, and often brightly colored reef-dwellers; they have laterally compressed bodies, tiny mouths, and no pelvic fins. Triggerfish are so named because the stout first dorsal spine can be locked rigidly in place by depressing the second spine (the "trigger").

Triodontidae, Tetraodontidae, and Diodontidae

The more than 140 species in these three families have the outer bones of their jaws modified into strong, beak-like structures for shearing off corals and other such invertebrates (Fig. 24.1). The beaks are divided by sutures, which give them the appear-ance of teeth. The number of "teeth" is distinctive, however, and handy for distinguish-ing the three families: Diodontidae means two-toothed; Triodontidae, three-toothed; and Tetraodontidae, four-toothed. The body shape of these fishes is the antithesis of sleek streamlining; they are globular to pyramidal in shape, with tiny gill openings, small and paddle-like fins, and a stiff and rounded tail. They are slow but maneuverable, hovering over their stationary prey and biting with precision. For protection, these slow-moving fishes rely on heavy skin, prickles, and armor, as well as on their ability to puff themselves up and their toxicity. To inflate themselves, the puffers and porcupinefishes suck water into a ventral diverticulum of the stomach. This greatly increases their diameter, especially in the species that possess spines or prickles on the sides. Fishes unfortunate enough to be caught by humans will inflate themselves with air. The neurotoxin possessed by many of these fishes, called *tetraodotoxin*, is found mostly in the internal organs and can be fatal to humans (Box 24.2). Tetraodontid puffers are common in tropical rivers in Asia, Africa, and South America, and some of the smaller species are popular in the aquarium trade (Berra 2001).

Molidae

The three species of molas are unlike any other members of their order. They are adapted for life as sluggish, pelagic predators of jellyfish or other large invertebrates that come close enough to be sucked in. They possess the fused "teeth" in the jaws that are typical of the order, but in most respects, they are unique. They are large, flattened fishes that lack a caudal peduncle but that do have a fin, of sorts, running along the posterior edge of the body, immediately behind the tall, short-based dorsal and anal fins (Fig. 24.1). Pelvic fins and swimbladder are absent, as are spines. The body, however, is covered with a tough, leathery skin. Despite these disadvantages, molas are commonly observed jumping out of the water, although how they do this is a mystery. The mouth is small and has sharp-edged plates of fused teeth. These fish feed by grazing on various kinds of jellyfish, although

---

BOX 24.2
Fugu Fame

The word *fugu* is associated with dangerous eating and high-tech science. The dangerous eating comes from people snacking on filets of poisonous species (genus *Takifugu*), called fugu in the restaurant trade in Japan. Because safe eating is possible only if the filets are not contaminated by toxins from the internal organs, the fish must be cleaned very carefully by chefs with special training. Presumably, the risk involved in eating fugu enhances the flavor, although trace amounts of the toxin in the muscle tissue may provide a narcotic effect. High-tech science comes into play in an effort to sequence the genome of the Japanese pufferfish (*Fugu rubripes*). This genome was chosen for sequencing because it is the shortest known for any vertebrate, with 365 megabases (Aparico et al. 2002). It is short because repetitive, "junk" DNA occupies only approximately one-sixth of the sequence and gene loci approximately 33% of the genome. Approximately 75% of the genome is similar to that of humans, and it is thought that it will provide useful comparisons with the human genome, including identifying common gene sequences and functions. It is also fascinating to consider how much we egotistical humans actually have in common with the pufferfish.

fish (e.g., the leptocephalus larvae of eels) are occasionally found in their stomachs. Two of the species (*Mola mola* and *Masturus lanceolatus*) may reach 3 m in length and weigh 1,500 kg. Because such large fish are typically observed basking on the surface (and are often called, as a result, ocean sunfish), they can be a hazard at times to boats. A collision of a large sharptail mola with a submersible vehicle at 670 m in depth led Harbison (1987) to suggest that this species may be more an inhabitant of the mesopelagic zone than of surface waters, especially because large jellyfish are most abundant at depths well below the surface.

The molas are considered to be among the most fecund of all vertebrates, producing more than 300 million eggs. They are not considered to be highly desirable as food, however, because the flesh is flabby and often parasite-ridden. Nevertheless, they are being fished commercially to supply the sashimi market in Japan. Despite their large size, the yield of edible flesh in a mola is low, roughly 20% of the body weight, because of the large, cartilaginous skeleton and thick skin.

24.3 LESSONS FROM FLATFISH AND PUFFERS
Flatfish

Two major, independent evolutionary lines of fish are flat bottom-huggers: flatfish, and rays. A majority of both groups are cryptic in behavior, modest in size, and feed on bottom-dwelling invertebrates, which they catch with many tiny, sharp teeth. A few larger species use their flatness to lie in ambush for other fishes. Both groups are very diverse (570 species of flatfish vs. 460 species of rays), abundant, and widespread, although rays have a slight advantage in being considered less desirable to eat by humans. In a way, it is curious that flatfish are so successful, because rays were already common when flatfish were still figuring out how to distort their bodies into flatness. Arguably, rays hit on a much more elegant solution to the problem by simply flattening out a basic morphology already headed in that direction. The original shark morphology, in turn, reflects less reliance on vision for finding prey than seems to be true of bony fishes, resulting in subterminal mouths and elongate snouts. The ancestral condition in bony fishes seems to be large eyes and terminal mouths. Perhaps this is also why some rays have reversed the trend and become pelagic, flying through the water with a grace virtually unmatched by other fish. Flatfish will often enter the water column to feed or to move, but their movements would rarely be called elegant or graceful! Still, the migration of the eye across the head in flatfish must be regarded as one of the developmental wonders of the animal world.

Puffers

A general trend in bony fish evolution has been a reduction in armor in favor of increased maneuverability. In the evolutionary arms race that is taking place on tropical reefs, however, the puffers and their kin have reinvented the concept of armor to survive as slow-moving crunchers of hard-shelled invertebrates. Those without heavy bony plates typically are covered with spines, which can be erected when the body is inflated with water. In addition, many of the species are highly toxic, which also discourages predators. The fact that even the toxic species are taken in fisheries, however, is a tribute to human omnivory.

SUPPLEMENTAL READINGS

Fitch and Lavenberg 1968; Gordon 1977; Nelson 1994; Wilson and Wilson 1985.

WEB CONNECTIONS

INTRODUCTION TO PERCOMORPHS
people.clemson.edu/~jwfoltz/wfb300/subjects/perco

GENERAL
www.flmnh.ufl.edu/fish
www2.biology.ualberta.ca/jacksonhp/IWR

PUFFER GENOME
www.fugu.sq.org

HALIBUT
www.pac.dfo-mpo.gc.ca/ops/fm/Groundfish/Halibut

C H A P T E R 25
Zoogeography of Freshwater Fishes

The study of fish zoogeography is alternately one of the most fascinating and one the of most frustrating areas of ichthyology. It is fascinating because the explanation of the world patterns of fish distribution requires putting together knowledge from many other areas of ichthyology, such as ecology, physiology, systematics, and paleontology, as well as from other disciplines such as geology and biogeography. It is frustrating, however, because so much of our knowledge in these areas is fragmented or incomplete, so any attempt to explain fish distribution patterns, particularly over large areas, is bound to contain gaps that must be bridged with guesswork until better information becomes available. Nevertheless, the development of plate tectonics as a unifying theory to explain the configuration of the Earth's surface has improved understanding of the historic distribution of organisms. Likewise, the rise of phylogenetic systematics has provided new tools for exploring zoogeographic hypotheses by seeing how distribution patterns of species and higher taxa reflect how they are related to one another. The use of new information and tools has revolutionized the study of fish zoogeography, in part because ichthyologists have been among the leaders in developing these new approaches (Mooi and Gill 2002). Our understanding of patterns has improved for freshwater fishes in particular (Matthews 1998, Berra 2001). Therefore, in this chapter, we summarize the following: (1) the broad zoogeographic patterns of freshwater fishes, (2) the fish faunas and drainage patterns of the major continental areas, and (3) the zoogeographic history of freshwater fishes, especially the minnows, characins, and catfishes (Otophysi).

25.1 ZOOGEOGRAPHIC TYPES

Two basic types of fish are found in fresh water; euryhaline marine fish, and obligatory freshwater fish. *Euryhaline marine fish* are those that primarily are marine but are capable of entering fresh water for extended periods of time. They frequently are characteristic of the lower reaches of coastal streams. Most commonly, juveniles inhabit the fresh water, perhaps to avoid predation. Adult forms of such fish as the large bull sharks (*Carcharinus leucus*) and sawfishes (*Pristis* sp.), however, which move in and out of Lake Nicaragua are important in some areas.

   Obligatory freshwater fish are those that must spend at least part of their life cycle in fresh water. For zoogeographic studies, two basic types can be recognized: freshwater dispersants, and saltwater dispersants. ***Freshwater dispersants*** belong to fish families or species whose members by and large are incapable of dispersing for long distances through salt water (defined as water with salinity usually >25–30 ppt). The distribution patterns of these families and species are best explained through a combination of active movement (***dispersal***) by following water courses and of being carried to new places (***vicariance***) by plate tectonics and other geological means. Freshwater dispersants include the primary and secondary freshwater fishes of Myers (1951). As Rosen (1975a, 1975b) points out, the distribution patterns of most secondary freshwater fishes (fish that are characteristic of fresh water but that have distribution patterns seeming to indicate saltwater dispersal) can be better explained on the basis of vicariance than on the basis of long-distance movements through salt water. In addition, many "primary" freshwater fish (those that supposedly are incapable of entering salt water) have been shown to be tolerant of a wide range of salinities. Typical examples of fish families that are freshwater dispersants include most of the Otophysi and families such as the Esocidae (pikes), Percidae (freshwater perches), Mormyridae (African electric fishes), Poeciliidae (livebearers), Lepisosteidae (gars), and Cichlidae (cichlids). Many species or groups of species with distribution patterns that reflect strict overland dispersal, however, also are members of families whose members largely are saltwater dispersants. Freshwater dispersants dominate the fresh waters of the world—even in Australia, where the freshwater fishes have strong taxonomic ties to marine fishes (Fig. 25.1).

Saltwater dispersants are those fishes whose distribution patterns can be explained, in large part, by movements through salt water. Even in this group, however,

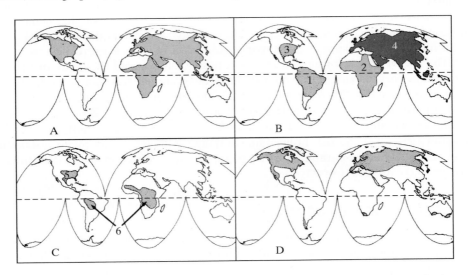

FIGURE 25.1 Representative distribution patterns of freshwater fish families. **A.** Distribution of the Cyprinidae, a widely distributed family of freshwater dispersants that is absent from South America and Australia. **B.** Distribution of four families of freshwater dispersants largely confined to one continent (*1*—Gymnotidae, *2*—Mormyridae, *3*—Centrarchidae, *4*—Cobitidae); **C.** Distribution of two ancient groups of freshwater dispersant fishes (*5*—Lepisosteidae, *6*—Lepidosireniformes); **D.** Distribution of two saltwater dispersant families; Salmonidae (northern hemisphere), and Galaxiidae (southern hemisphere). From Rosen (1975a).

many families have broad distribution patterns in freshwater environments that are related to continental movements. Saltwater dispersant fishes are of two basic types: diadromous fishes, and freshwater representatives of marine families. Diadromous fishes are those that regularly move between fresh water and salt water, spending different parts of their life cycle in each environment (see Chapter 11 for further definitions). Diadromous fishes frequently are important members of the faunas of coastal streams, especially if the region is geologically young. In most areas where they occur, diadromous species have given rise to numerous nonmigratory forms, which if isolated from the parent stock may evolve into distinct species. This has happened repeatedly in families such as the Galaxiidae, Salmonidae (Fig. 25.1), and Petromyzontidae (lampreys).

If a region lacks a well-developed fish fauna of freshwater dispersants, the fauna likely will be made up of a mixture of diadromous species and species that are members of primarily marine families. An example of the latter group are the sculpins of the genus *Cottus* (Cottidae), which have a circumarctic distribution, mostly in streams that are dominated by salmonids. Although some species of *Cottus* live in estuaries, it is mostly a freshwater genus. Like sculpins, many species in "marine" families are quite successful members of well-developed freshwater fish faunas, such as the many freshwater herrings (Clupeidae) and freshwater puffers (Tetraodontidae). Most of these species, despite their origins, are freshwater dispersants.

The zoogeographic types presented here reflect the general switch away from the use of dispersal explanations for broad distribution patterns and toward vicariance explanations (Mooi and Gill 2002). With dispersal explanations, the assumption is that species evolved in limited geographic regions and then spread outward from that area, with new species developing in isolation after a few colonists crossed some nearly impassible barrier (e.g., an oceanic strait). For vicariance explanations, the land moves more than the fish. Land masses split apart, or mountain ranges rise up. Either way, populations of species are separated into new evolutionary lineages. Vicariance explanations work especially well to explain broad distribution patterns, usually above the species level, whereas dispersal explanations often work especially well to explain local or regional patterns. The obvious explanation of distribution patterns, however, should never be taken for granted without multiple lines of evidence. For example, the Galaxiidae, although containing a number of diadromous species, have a broad distribution pattern (Fig. 25.1) that seems to be explained best by vicariance, because it reflects the ancient connections of the southern continents to each other. McDowall (2002), using multiple lines of evidence, suggests that this pattern could be explained just as easily by dispersal through salt water, perhaps with ancient vicariant events contributing as well. Dramatically, one species (inanga, *Galaxius maculatus*) has dispersed through salt water to several continents. Likewise, local distribution patterns of Galaxiidae may be explained by both changes in the landscape (vicariance) and local movements (dispersal) (see Box 19.1).

25.2 ZOOGEOGRAPHIC REGIONS

For the study of fish zoogeography, it is convenient to divide the world into six zoogeographic regions (Darlington 1957): (1) the African Region, consisting of the African continent; (2) the Neotropical Region, consisting of South and Central America; (3) the

Oriental Region, which includes the Indian subcontinent, Southeast Asia, most of Indonesia, and the Phillipines; (4) the Palaearctic Region, which includes Europe and Asia north of the Himalaya Mountains and the Yangtze River; (5) the Nearctic Region, which is North America down to central Mexico; and (6) the Australian Region, which consists of Australia, New Zealand, New Guinea, and the many smaller islands in the same region. The fish fauna of each region has distinctive elements that reflect its isolation from (as well as its connections to) other regions. In fact, using a scheme based on the present position of the continents may do an injustice both to broader distribution patterns and to areas of endemism within the regions (Mooi and Gill 2002).

25.3 AFRICAN REGION

The *African Region* possesses an extremely diverse fish fauna and is of special interest to students of fish evolution and zoogeography because of its wealth of ancient forms, such as lungfishes (Protopteridae), bichirs (Polypteridae), and osteoglossiforms (five families). The region contains approximately 3,000 species of freshwater fishes, which belong to approximately 340 genera and 76 families (Lévêque 1997, Lundberg et al. 2000). The high diversity of the African fish fauna relates in good part to the fact that much of the continent has been in place as a stable landmass for more than 600 million years, resulting in many families that are both endemic to Africa and ancient in origin (Lévêque 1997). More than 95% of the species are freshwater dispersants. More than one-third of these belong to the Otophysi ($\approx$475 cyprinid species, 210 characoid species, and 450 catfish species). The otophysans belong to 18 families, which include 6 endemic families and 2 families of marine catfishes. The number of otophysan species likely is greatly underestimated. For example, the electric catfish family (Malapteruridae) has only one described species but probably contains 20 or so species, including some dwarf forms (Lundberg et al. 2000).

The nonotophysan freshwater fishes belong to a variety of endemic and nonendemic families. One of the most numerous (>200 species) of the endemic families is the Mormyridae, an osteoglossiform family possessing many unusual adaptations, including electric organs, for living in muddy tropical waters. Two other endemic Osteoglossiformes, with single species, are the Pantodontidae (butterflyfish) and the Gymnarchidae (aba). Six other endemic families of teleosts, with just a handful of species, may date back to the Mesozoic (Lévêque 1997). Even more ancient are the Protopteridae (African lungishes, related to the Lepidosirenidae of South America) and the Polypteridae (bichirs and reedfish, 10 species), which are lobe-finned fishes related to the ancient Chondrostei. A number of families are shared with other zoogeographic regions, but especially with the Oriental region. They include Osteoglossidae (zone species, bony-tongues, which is shared with South America, Southeast Asia, and Australia), Anabantidae (gouramis, 28 species), Mastacembelidae (spiny eels, 46 species), and Aplocheilidae (panchaxes and annual killifishes, 150 species), all shared with the Oriental Region, and Cichlidae and Cyprinodontidae (cichlids and killifishes, respectively, shared with South America, North America, and the Oriental Regions).

The cichlids are the single most species-rich family in Africa, with at least 870 species, mostly in the Great Lakes of eastern Africa. Because cichlids are diverse in

South America as well as in Africa, it usually is assumed that cichlids arose as a family before the splitting of the continents during the early Cretaceous period and then speciated in isolation from one another (e.g., Mooi and Gill 2002). Fossil and other evidence, however, suggests the cichlids are, in fact, a much younger group, appearing approximately 65 million years ago (Murray 2001). In turn, this indicates that the dispersal of the group probably required marine routes, which seems to be possible given the physiological tolerances of many species. Interestingly enough, the cichlids with the most ancestral characteristics are found in Madagascar and are related more closely to cichlids in Asia than to those on the African continent (Mooi and Gill 2002) (Box 25.1).

Interesting additional components of the fauna are freshwater members of marine families, such as Tetraodontidae (pufferfishes), Syngnathidae (pipefishes), Mugilidae (mullets), Clupeidae (herrings), and Centropomidae (snooks). The latter family is important ecologically, because the seven freshwater species in Africa, including the Nile perch *(Lates niloticus)*, are major predators in many lakes and rivers. Most of the species in these groups occur only in fresh water and can no longer disperse through marine environments.

One curious aspect of the African freshwater fish fauna is that each of the dominant groups tends to be characteristic of a broad habitat type (Roberts 1976). Thus, the cichlids exist mostly as members of the spectacular species flocks that characterize the Great Rift Lakes of East Africa. The characoid fishes (e.g., Hepsetidae, Characidae, and

BOX 25.1

Madagascar: A Refuge for Relict Fishes

Madagascar broke off from Africa when the entire Gondwanaland continent was splitting up into the present southern continents, sending India crashing into Asia. Isolated for 150 million years, the large island has long been recognized as a home to relict plants, birds, and mammals, but the fishes largely were regarded as both depauperate and closely related to mainland forms. Neither of these assumptions is true. The island contains at least 127 species of freshwater fish, 60% of them endemic (Mooi and Gill 2002). The most species-rich families are the Cichlidae, Bedotiidae, and Gobiidae, although other families, such as the Aplocheilidae, Eleotridae, and Ariidae, are well represented. The cichlids and aplocheilids (panchaxes) seem to have the most ancestral traits of members of their families and are related more closely to forms in India and Southeast Asia than to forms in mainland Africa. This suggests that these families (and others) may have been transported to Asia on the Indian landmass and then spread from there (see Section 25.9). The Bedotiidae is an atheriniform family endemic to Madagascar and appears to be related most closely to the Melanotaeniidae (rainbowfishes) of Australia (Mooi and Gill 2002). This suggests a very long history in fresh water for this combined group, thereby lending credence to the idea that much of Australia's freshwater fish fauna may be more ancient than is generally supposed. Unfortunately, the fascinating fish fauna of Madagascar is rapidly disappearing under the pressure of an exploding human population (Stiassny 1996).

Citharinidae) are found almost exclusively in sluggish, lowland streams, as are most mormyrids. A number of minor families that have air-breathing abilities (e.g., Protopteridae, Polypteridae, and Anabantidae) also are confined to the lowland habitats. The cyprinids are most characteristic of flowing waters at middle to high elevations. The catfishes are found in most habitats, although with different types of specializations for the different environments. Their success presumably results primarily from their nocturnal activity patterns, because most of the other large families (except the mormyrids) are diurnal.

Africa contains an extraordinary mixture of endemic (40%) and nonendemic families (60%) with different ages of origin (Lévêque 1997). These families have produced a large number of endemic species, because Africa contains a number of ancient, isolated drainage basins. Within those basins, speciation, especially of the cichlid fishes, may be extremely rapid. (Owen et al. 1990). Thus, Africa can be divided into at least 10 ichthyological provinces, each with its own characteristic fish fauna and complicated history of connections to other basins (Roberts 1976, Lévêque 1997).

25.4   NEOTROPICAL REGION

The *Neotropical Region* has two distinct zoogeographic subregions; mainland South America, and Central America. The fish faunas of the two subregions reflect the fact that South America was isolated from the other continents for a long period of time, whereas geologically, Central America is of recent origin. The region has a wide diversity of climates, but the vast majority of the fishes are associated with tropical rivers and lakes.

The diversity of fish fauna in *South America* is just beginning to be fully appreciated, and new species are being described at a rapid pace (Lundberg et al. 2000). This area has the most species of obligatory freshwater fish of any continent ($>$3,600), many of them undescribed; there may be as many as 8,000 species (Mooi and Gill 2002). One of the most striking aspects of the South American fish fauna is the complete absence of the family Cyprinidae, which dominates the freshwater fish fauna in all other zoogeographic regions (except the Australian region). Instead, the dominant fishes are characins (order Characiformes, eight families, seven of them endemic, with $\approx$1,800 species) and catfishes (order Siluriformes, 13 endemic families, with $\approx$1,400 species). The characins show an extraordinary variety of ecological and morphological adaptations, yet the basic similarities in body plans of the many forms seem to reflect a common ancestry. Only the family Characidae occurs outside the Neotropical Region, in Africa. The South American catfishes are as diverse as the characins and include the most ancestral family (Diplomystidae) of the order, consisting of three large families distinguished by bony plates on their bodies (Doradidae, Callichthyidae, and Loricariidae), and a family (Trichomycteridae, $>$200 species) of small catfishes, of which some species live in the gill cavities of larger fishes. Apparently sharing a common ancestry with the catfishes is the order Gymnotiformes. This order consists of six families of electric fishes ($\approx$100 species) that are strikingly convergent on the unrelated mormyrids of Africa.

In contrast to Africa, nonotophysan freshwater fishes in South America are a relatively minor component ($\approx$10%) of the fish fauna, although they often play important roles in South American lakes and streams. Cichlids (Cichlidae, $\approx$450 species) are widespread in South American streams, whereas killifishes, livebearers, and their relatives

(Cyprinodontiformes, 375 species) are most abundant in habitats not utilized by larger fishes (e.g., stagnant backwaters). In Lake Titicaca, high in the Andes, a species flock of 23 endemic killifishes (Cyprinodontidae) is present, presumably derived from one or two ancestor species that were uplifted with the mountains. Other lakes and streams in the Andes also contain these unusual cyprinodonts. The two most ancient families of freshwater fishes in South America are the Lepidosirenidae (lungfishes, one species) and the Osteoglossidae (bonytongues, two species). Both families also have relatives in Africa, and osteoglossids occur in Australia as well.

The final elements of the South American fish fauna are fishes that are assigned to marine families and diadromous fishes. The lower Amazon alone has 20 to 30 endemic freshwater representatives of marine families and an approximately equal number of euryhaline marine species. Among the freshwater species are members of the families Dasyatidae (stingrays), Soleidae (soles), Clupeidae (herrings), Engraulidae (anchovies), and Sciaenidae (drums). The diadromous species (mainly galaxiids) are confined to coastal Chile, Patagonia, Tierra del Fuego, and the Falkland Islands.

South America can be divided into eight provinces that are characterized by distinctive fish faunas (Gery 1969). Because many of the major geological features of South America are comparatively recent in origin (e.g., the Andes Mountains), however, and because connections between drainage systems are common, the regions share many species and genera.

Another portion of the Neotropical Region with a fish fauna resembling that of South America is *Central America*. The Central American fish fauna, which includes species in part of Cuba and other Antillean islands, also shows affinities to that of North America but also has many unique characteristics. The distributional history of the fishes of the region is confusing, because the geological history of the region is extraordinarily complex (Rosen 1975b). Miller (1966) lists 456 species from the fresh waters of mainland Central America. Of these, 269 (59%) are freshwater dispersants, 57 (13%) are freshwater representatives of marine families, and the rest (28%) are euryhaline marine fishes. Most species in the latter group are only sporadic in their occurrence in the lower reaches of the streams. In contrast to both North and South America, only 38% of the freshwater dispersant species are otophysans. Most of these are found in the southernmost part of Central America and belong to families and species that otherwise are found only in South America. Many species are endemic, however, with highly localized distribution patterns. In Costa Rica, the best-documented part of Central America, Bussing (1998) recorded 139 species (33 families) in freshwater, of which 65% were from freshwater dispersant families (mainly cichlids, characins, and poeciliids).

In the northern part of Central America, three otophysan species belong to the North American families Ictaluridae (catfishes) and Catostomidae (suckers). Even here, however, most of the few otophysans that are present have South American affinities (e.g., Pimelodidae and Characidae). Still, the otophysans do show a high degree of endemism, indicating long isolation from South America.

The most curious aspect of the Central American fish fauna is not the otophysan element but the fact that the most widely distributed—and often most abundant—fishes are cichlids (Cichlidae, >78 species), livebearers (Poeciliidae, >60 species), and killifishes (Cyprinodontidae, Fundulidae, Rivulidae, and Profundulidae; >23 species). The success of these families most likely is related to adaptations that have enabled them to survive in an area that is geologically very unstable. These adaptations include an ability to live under fluctuating temperature and salinity regimes, viviparity (Poeciliidae), and small body sizes, which permit large populations to live in small, isolated streams and lakes. In addition, the success of the cichlids is related, at least in part, to their ability to do well in lakes. For example, the species "flocks" of the great lakes of Central America have many similarities with the species flocks of the Great Lakes of Africa, representing similar phenomena of speciation (Echelle and Kornfield 1984) (see Box 25.2).

Bussing (1985) notes that the killifishes, poeciliids, and cichlids, along with a few other fishes, may represent the original fish fauna of Gondwanaland (the great southern continent). They became isolated in Central America as early as the late Cretaceous period, when the region presumably was a series of large islands. Following the reconnection of North and South America during the Pliocene (or later), fish were able to invade from South America (various characins and catfish) and from North America (a gar, Lepisosteidae; three shads, Clupeidae; two suckers, Catostomidae; and a catfish, Ictaluridae). The invasions have been slow, however, so that South American forms, for the most part, have penetrated only as far north as mid–Central America.

25.5   ORIENTAL REGION

The *Oriental Region* has two subregions: Peninsular India, including Sri Lanka; and Southeast Asia, including Sumatra, Java, Borneo, and Mindanao. Although the subregions have distinctive aspects to their fish faunas, they have more similarities than differences. In particular, they are all dominated by cyprinids (>1,000 species), river loaches (Balitoridae, 300 species), loaches (Cobitidae, 100 species), bagrid catfishes (100 species), and gobies (Gobiidae, 300 species) (Mooi and Gill 2002). The fish fauna, however, is poorly described. Therefore, these numbers are low (Yap 2002), and the total number of species probably is approximately 3,000 (Lundberg et al. 2000). Perhaps the best indication of the diversity of fishes is the high number (121) of fish families that are found in fresh water (other regions have <55 families). Most of these families have both marine and freshwater species.

Peninsular India

Peninsular India was an independent island-continent for perhaps 100 million years before it collided with Asia, creating the Himalaya Mountains. It is somewhat surprising, therefore, to find that its freshwater fish fauna seems to contain no truly ancient forms but instead is made up mostly of derived otophysans of Asiatic or African origin. Recent evidence, however, suggests that some groups, such as the cichlids and panchaxes (Aplocheilidae), may have arrived in the region on the Indian landmass and then spread from there (Mooi and Gill 2002) (see Box 25.1). Fossil evidence indicates that once invasion routes were open, the Asiatic otophysans must have overwhelmed the original fish fauna, a decline that may have been hastened by the climatic change that presumably took place as the Himalaya Mountains rose. More than 700 species of freshwater dispersant fishes, from 27 families, are known from Peninsular India (Jayaram 1974). Of these, 373 species (≈53%) belong to the family Cyprinidae, and 176 species (≈25%) belong to 12 families of catfishes. Because the presence of endemic genera usually indicates long isolation from surrounding areas, it is significant that no endemic genera of cyprinids or catfishes are found in Peninsular India. The genera that are present are found in Southeast Asia, in Africa, or in both (Menon 1973). Those shared with Africa also are found in Southwest Asia (the Middle East). Other otophysan families found in India are the Cobitidae (loaches) and Balitoridae (river loaches), which also are found in the Palearctic Region. The nonotophysan freshwater fish families found here include the Channidae (snakeheads), Mastacembelidae (spiny eels), Anabantidae (climbing gouramis), Belontiidae (gouramis), Notopteridae (knifefishes), Cichlidae, Aplocheilidae (panchaxes), Nandidae (leaf-fishes), and Synbranchidae (synbranchid eels). All these families have representatives in Africa as well as elsewhere in the Oriental Region. The Notopteridae are particularly noteworthy, because they are highly specialized members of the order Osteoglossiformes, the members of which usually are regarded as relict species. Saltwater dispersant fishes play a relatively small role in the fresh waters of Peninsular India.

Southeast Asia

Southeast Asia contains more than 2,000 species of fish, but distribution patterns are complex and not well understood (Zakaria-Ismail 1994, Lundberg et al. 2000). An idea of the composition of the fish fauna of this area is provided by the fish fauna of Thailand (Smith 1945). Of the 549 fish species listed for Thailand, from 48 families, 72% are from freshwater dispersant families, principally Cyprinidae (39%) and various catfish families (18%). Likewise, Roberts (1989), in a survey of the fishes of western Borneo, found 290 species (40 families), of which 84% came from freshwater dispersant families, especially Cyprinidae (35%). Even the species from marine or estuarine families, however, were mostly freshwater dispersants. The marine families found in Southeast Asia's fresh waters include stingrays (Dasyatidae), anchovies (Engraulidae), threadfins (Polynemidae), gobies (Gobiidae), pipefishes (Syngnathidae), tonguefish (Cynoglossidae), and puffers (Tetraodontidae) (Roberts 1989, 1993). A few representatives of ancient fish families (Notopteridae and Osteoglossidae) also are found.

Zakaria-Ismail (1994) divides Southeast Asia into five zoogeographic regions:

1. The Salween River Basin (Nyanmao), with close ties to the Indian Subcontinent and low endemism ($\approx$3% of 150 species).

2. The Indo-Chinese Peninsula (including the Mekong River), a high-diversity region (>500 species) with high endemism (48%).

3. The Malay Peninsula, with 260 species and low ($\approx$2%) endemism.

4. The Indo-Malayan Archipelago (including Sumatra, Borneo, and Java), with high diversity (>400 species) and many endemic species, especially gouramis (Belontiidae).

5. The Phillipines, where fishes from freshwater dispersant families are found only on Mindanao. The Mindanao fauna is depauperate but contains an endemic flock (15 species) of cyprinids in Lake Lanao. Most of the freshwater species are gobies (Gobiidae).

The fish fauna of the Indo-Malayan Archipelago is very similar to that of the mainland, because the islands all protrude from a shallow continental shelf area that was dry during much of the Pleistocene because of lower sea levels. At that time, most of the river systems of the islands and those of the mainland drained into a large, central system, thereby giving the fishes ready access to all the areas (Myers 1951). The boundary between this region and the Australian region is hard to define because of its complex geologic history (Mooi and Gill 2002).

25.6   PALAEARCTIC REGION

The *Palaearctic Region* consists of all of Eurasia that is not included in the Oriental Region. Berg (1949) divides the area into six subregions, but a general idea regarding the composition of the fish fauna in this region can be obtained by examining the fauna of Europe outside the former Soviet Union (Lundberg et al. 2000). In this area, 36 families of fish, with at least 350 species, are found in fresh water. Species from eight families make up 80% of the fauna, mostly Cyprinidae (minnows, 36%), Salmonidae (trouts and whitefishes, 24%), Gobiidae (gobies, 9%), and Cobitidae (loaches, 6%). The Otophysi are otherwise represented in the entire palaearctic region by catfishes from just three families (Siluridae, Bagridae, and Sisoridae) and one sucker (Catostomidae). The sucker, which is found in Siberia, is a recent invader from North America, although the loaches and silurid and sisorid catfishes apparently evolved in Asia. Other freshwater dispersant families found in the area are the Percidae, Channidae, Cyprinodontidae, Esocidae, and Umbridae. All families, except the Channidae, are found in North America as well. The Salmonidae usually are considered to be saltwater dispersants, but most of the diverse members of this family are freshwater dispersants, especially the small flocks of endemic species of char (*Salvelinus*) and whitefishes (*Coregonus*) that are characteristic of many lakes. Anadromous species are most important in Arctic drainages, where they make up approximately 51% of the fishes, as opposed to 30% for freshwater dispersants. Freshwater representatives of eight marine families make up another 11% of the total fauna if the endemic families of sculpins from Lake Baikal (Box 25.2) are counted, although all members of the Lake Baikal families live in fresh water.

A region with high fish endemism that was relatively unappreciated until recent times is the north Mediterranean region of Europe. At least 132 species are endemic to the fresh waters of this region, representing 13 families (Crivelli and Maitland 1995). More than 80% of the species are freshwater dispersants (mainly Cyprinidae and Cobitidae), and many are highly localized in their distribution. In Spain, for example, 86% of the 29 freshwater dispersant fishes are endemic species or subspecies (Elvira 1995). In Anatolia, Turkey, 24% of the 86 native species are endemic to the region, and most of the others have limited distributions (Balik 1995).

25.7   NORTH AMERICAN REGION

The *North American (Nearctic) Region* consists of North America down to the southern edge of the Mexican plateau (Fig. 25.2). It can be divided into three broad ichthyological subregions: (1) the Arctic–Atlantic Subregion, consisting of all drainages into the Arctic and Atlantic oceans as well as the Gulf of Mexico down to the Rio Panuco; (2) the Pacific Subregion, consisting of all drainages along the Pacific coast down to the Mexican border (including Baja California but excluding the Yukon River drainage in Alaska) and all interior drainages west of the Rocky Mountains; and (3) the Mexican

BOX 25.2
Lake Baikal: Center of Speciation

Lake Baikal in central Siberia is largest lake in the world, covering 31,500 km^2 and containing 20% of the world's fresh water. It has a watershed of approximately 534,000 km^2 and drains into the Arctic Ocean via the Angara River. It occupies a huge rift in the continent's crust, so it averages approximately 730 m in depth, with a maximum depth of more than 1,740 m. Because it is also the oldest known lake (25 million years), it has been a center of regional evolution, supporting more than 1,500 species of algae, plants, and animals, most of them endemic to the lake and/or its watershed and adapted to its cold, transparent waters. The lake supports 56 species of fish, most of them (30 species) in three endemic families of sculpin-like fishes: (1) Comphoridae, (2) Cottocomphoridae, and (3) Abyssocottidae (Smith and Todd 1984). The Comphoridae contains just two species, called Baikal oilfishes, which are pelagic, that live in the deep (>1,000 m) waters of the lake, and use oil for buoyancy (Berra 2001). The abyssocottids (20 species) and the cottocomphorids (eight species) are mostly bottom-dwelling fishes, but some species are pelagic. One pelagic form, *Cottocomephorus grewingki*, is extremely abundant and a major link in the lake's food webs. The three families represent a classic **species flock**, like those found in other large lakes of the world, such as Lake Titicaca in South America (Cyprinodontidae flock) and the Great Rift Lakes of East Africa (cichlid flocks). Species flocks form when a lake is created and then invaded by a limited number of stream-dwelling species. In the presumed absence of other fishes that are adapted for lake living, the new invaders radiate into more specialized species. In the case of Lake Baikal, the invader was one or more species of sculpin (Cottidae). Species flocks that develop from a common ancestor in lakes most likely are the result of sympatric (ecological) speciation, as is discussed for African cichlids in Chapter 30. Curiously, the rest of the lake's fish fauna are widespread species of salmonids, cyprinids, cobitids, and other families, although some have forms that are endemic to the lake (Kozhov 1963). Given the variety of fishes with access to the lake, the physiological adaptations of the cottoids to the extremely cold water of the depths likely have played a role in speciation.

Subregion, consisting of most of Mexico except the drainages flowing into the upper Gulf of Mexico and related interior drainages, down to the imperceptible upper limits of Central America. These broad regions are called *ecoregions* by Abell et al. (2000) and are broken down further into 76 subregions for the purposes of aquatic conservation, a finer breakdown than is used here.

The zoogeography of North American fishes is the best documented of any continent thanks to two monumental works (Lee et al. 1980, Hocutt and Wiley 1986) and to the general fascination of North American ichthyologists with the subject. Approximately 1,061 species of fishes are native to the three subregions, belonging to more than 200 genera and 56 families (Burr and Mayden 1992, Lundberg et al. 2000). If the Mexican Subregion is excluded because of its transitional nature, then approximately 916 species are in the two other subregions combined. Freshwater dispersant fishes make up approximately 93% of the fauna. They represent 30 families, although most of the species are from the families Cyprinidae (minnows, 34%), Percidae (perches and

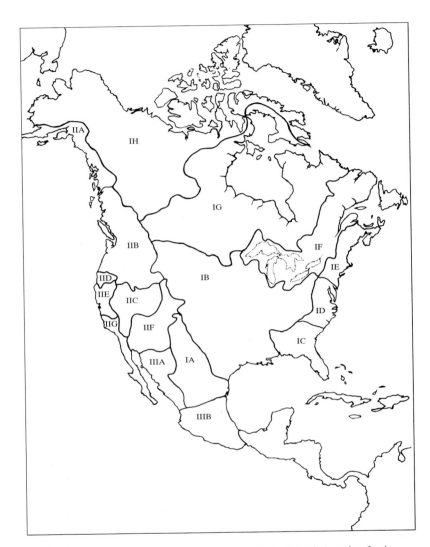

FIGURE 25.2 Zoogeographic subdivisions for fishes of North America: *I*—Arctic–Atlantic subregion, *II*—Pacific subregion, *III*—Mexican subregion, *IA*—Rio Grande province, *IB*—Mississippi Province, *IC*—Southeastern Province, *ID*—Central Appalachian Province, *IE*—Northern Appalachian Province, *IF*—Great Lakes–St. Lawrence Province, *IG*—Hudson Bay Province, *IH*—Arctic Province, *IIA*—Alaska Coastal Province, *IIB*—Columbia Province, *IIC*—Great Basin Province, *IID*—Klamath Province, *IIE*—Sacramento Province, *IIF*—Colorado Province, *IIG*—South Coastal Province, *IIIA*—Sonoran–Sinaloan Province, *IIIB*—Central Mexico Province.

darters, 18%), Catostomidae (suckers, 8%), Poeciliidae (8%), Goodeidae (5%), Fundulidae (4%), Ictaluridae (bullhead catfishes, 5%), and Centrarchidae (sunfishes, 3%). Approximately 6% of the fishes belong to diadromous families, principally the Salmonidae (salmon, trout, and whitefishes), Osmeridae (smelts), and Petromyzontidae (lampreys), but many of the species within these families are now freshwater dispersants (even if their recent ancestors were not). Freshwater representatives of marine families, including the Cottidae (sculpins, 3%), and of euryhaline marine species from 15 families make up the remainder of the fauna.

Some of the general patterns emerge from examining how the fishes are distributed among the zoogeographic subregions:

- Species richness and diversity is highest in the southeastern United States and Central Mexico.
- Regions with the highest rates of endemism are the Tennessee/Cumberland basins, the south Atlantic coastal plain, the Ozark and Ouachita Mountains, California/southern Oregon, and the Rio Grande/Rio Pecos basins.
- Species richness declines toward the north, especially in recently glaciated areas.
- Species richness declines toward the west, mainly because so many western basins are arid.
- The regions of highest species richness north of Mexico are dominated by small cyprinids, bullhead catfishes (Ictaluridae), and darters (Percidae).
- Cyprinodontiform fishes (Poeciliidae, Cyprinodontidae, and Goodeiidae) tend to dominate in areas where aquatic environments are most hostile.
- Diadromous species are important components of coastal faunas, contributing in terms of both diversity and volume of individuals.
- The Mississippi River is simultaneously the great refuge for fishes, the great highway for dispersal, and the great divider among regions (Box 25.3).

Arctic–Atlantic Subregion

The *Arctic–Atlantic Subregion* is the largest and most species-rich of the Nearctic subregions. It can be divided into eight ichthyological provinces that largely reflect the major drainage systems or collections of watersheds: (1) the Mississippi, (2) the Rio Grande, (3) the Northern Appalachian, (4) the Central Appalachian, (5) the Southeastern, (6) the Great Lakes–St. Lawrence, (7) the Hudson Bay, and (8) the Arctic.

The *Mississippi Province* consists of all the United States and Canada that is drained by the great Mississippi–Missouri river system as well as many western Gulf Coast drainages, Burr and Mayden, (1992) treated as a separate province. Several tributary systems have large numbers of endemic forms and so, perhaps, should be treated as separate provinces. For example, the Tennessee and Cumberland rivers alone contain more than 240 fish species (Starnes and Etnier 1986), so a relatively small state like Tennessee can list 297 fishes (Etnier and Starnes 1993). Other nearby states are nearly as rich in freshwater fishes: Alabama (257 species), Kentucky (220 species), Georgia (219 species) and Mississippi (204 species) (Ross 2001). The reasons for the high diversity lie in the southeastern United States' diverse topography, its abundant water, and its long and complex geologic history. There have been both lots of opportunities—and lots of time—for speciation in isolation and for dispersal into new areas.

The Mississippi Province, partly because of these species-rich areas, is by far the richest in North America, with 375 species (40% endemic) in 31 families. Freshwater dispersants make up most of these species, predominantly Cyprinidae (32%), Percidae (30%), Centrarchidae (5%), Catostomidae (6%), and Ictaluridae (6%). Diadromous fishes and freshwater representatives of marine families are poorly represented. In addition, a number of euryhaline marine forms occur in the Mississippi Delta area; Ross (2001) lists 69 from Mississippi alone. As the high percentage of endemic freshwater dispersants attests, the Mississippi–Missouri drainage system is an ancient one (Box 25.3).

The *Rio Grande Province* is essentially the Rio Grande River and its tributaries in Texas, Mexico, and New Mexico as well as its nearby, allied basins. Species known from

---

BOX 25.3
Mississippi River as a Refuge for Fishes

The Mississippi River is uniquely important in North America as a center of fish evolution. It was a refuge during times of glaciation, so that species have been able to reoccupy waters once covered by continental glaciers, and it is a refuge for representatives of past fish faunas. Its role in the development of the North American fish fauna is indicated by the fact that the families Centrarchidae, Percidae (Etheostominae), Ictaluridae, Cyprinidae, and Esocidae are most abundant and diverse in its waters and those of its tributaries. Its role as a glacial refuge is indicated by the fact that most of the species found in the Great Lakes–St. Lawrence, Hudson Bay, and Arctic provinces, which once were largely covered by ice sheets, also are characteristic of the northern parts of the Mississippi Province. Its role as a refuge, however, has not been confined to just the Pleistocene period, because it remains a refuge for many relict Chondrostei and Neopterygii, including teleosts. The Chondrosteans consist of three species of sturgeons (Acipenseridae), including two species in the endemic genus *Scaphirhynchus* and one species of paddlefish (Polyodontidae).

Only one other species of paddlefish exists, in the Yangtze River of China, whereas the closest relatives of the *Scaphirhynchus* sturgeons live in the Palaearctic Region. Ancient neopterygians are represented by six species of gar (Lepisosteidae) and by bowfin (Amiidae). Aside from one additional species of gar found in Central America, these fishes are the only representatives on Earth today of the fishes that were abundant in fresh water during much of the Mesozoic. Among the relict teleosts are the mooneye and goldeye (Hiodontidae, order Osteoglossiformes), a sucker (*Cycleptus*, which seems to be closest to the relict sucker *Myxocyprinus* of the Yangtze River in China), and the peculiar families Elassomatidae (pygmy sunfishes), Percopsidae (trout-perches), Amblyopsidae (cavefishes), and Aphredoderidae (pirate perch), which are confined to North America.

It is worth noting that the huge, muddy river and its floodplain in the southern United States also has served as a barrier to the dispersal of small stream fishes, thereby promoting speciation in drainages on opposite sides of the river.

this province total 154, but 22 are euryhaline marine species that are found in lower reaches of the main river or coastal streams (Smith and Miller 1986). The province includes 125 species (81%) that are members of freshwater dispersant families, mainly Cyprinidae (41 species), Cyprinodontidae (23 species), Poeciliidae (20 species), Catostomidae (11 species), Ictaluridae (10 species), and Percidae (6 species). Approximately half the fish are endemic to the province, including many of the cyprinids and most of the cyprinodontids and poeciliids (mostly in isolated spring systems). Almost all fish in the Rio Grande Province are related to species that also are abundant in the Mississippi River drainage. The principal exceptions are a characin (Mexican tetra, *Astyanax mexicanus*) and a cichlid (Rio Grande perch, *Cichlasoma cyanoguttatum*), whose closest relatives are in the Mexican Subregion.

The **Northern Appalachian Province** is an artificial assemblage of independent drainage systems that flow into the Atlantic Ocean along the coast from the Gaspé Peninsula, New Brunswick, to the Delaware River drainage. The streams are dominated by anadromous fishes and by minnows and suckers that can colonize new waters through headwater connections (e.g., blacknose dace, *Rhinichthys atratulus*; creek chub, *Semotilus atromaculatus*; and white sucker, *Catostomus commersoni*). This province was covered by glaciers during the Pleistocene; consequently, little endemism is found. The one nominal endemic species is a whitefish (Salmonidae), such as are found in the Great Lakes–St. Lawrence Province (Burr and Mayden 1992). The Northern Appalachian Province most likely should be lumped into the one of the adjacent provinces, despite its independent watersheds.

The **Central Appalachian Province** is made up of many separate drainage systems with former headwater connections to the Mississippi River drainage, through the low ridges of the Appalachian Mountains. The province is bounded on the north by the Susquehanna River and other Chesapeake Bay tributaries and on the south by the Edisto River. The streams in this province have both rich faunas and numerous endemic species; 177 species of fish, including 28 endemic species, spend at least part of their life cycle in fresh water (Jenkins et al. 1972). Each watershed has its own endemics. For example, the Roanoke River of Virginia contains 82 freshwater dispersant species, 6 of them endemic (Jenkins and Burkhead 1994). Curiously, the nearby ancient and ironically named New River, above Kanawha Falls, contains only 46 freshwater dispersants, 8 of them endemic.

The **Southeastern Province** is another collection of independent watersheds with some ichthyological affinities for each other. It starts with the Savannah River and ends in the Gulf Coast (Lake Pontchartrain) drainages of Louisiana. With 268 species, this province is exceeded in diversity only by the huge Mississippi River drainage, and approximately 29% of the species are endemic. The total number of species encompasses several euryhaline marine species (Burr and Mayden 1992). Because of local endemism, this province can be subdivided into at least 10 subprovinces. The richest of these is the Mobile Basin, which drains most of Alabama and by itself contains 236 species, 42 of them endemic, 25 of which are darters (Percidae) and 14 of which are minnows (Cyprinidae) (Mettee et al. 1996).

The **Great Lakes–St. Lawrence Province** had major connections to the Mississippi River system in the late Pleistocene, so it is not surprising that its fauna essentially is that of the northern part of the Mississippi Province. Most of this province was glaciated at one time or another during the Pleistocene. It is treated as a separate province here, both because it is a distinct, large drainage system and because the Great Lakes do (or did) have several endemic forms, such as the members of the whitefish (*Coregonus*) species complex, the siscowet lake trout (*Salvelinus namaycush siscowet*), and the blue walleye (*Stizostedion vitreum glaucum*). Of the 173 native obligatory freshwater fishes listed for this province by Underhill (1986), 73% belong to freshwater dispersant families (28% of which are Cyprinidae), 17% to diadromous families, and 10% to marine families.

The ***Hudson Bay Province*** includes most of central Canada and the portion of the United States that is drained by the Red River of the North. Although all its waters drain into Hudson Bay, this province's fish fauna shows strong affinities with that of the Mississippi Basin, particularly in its freshwater dispersants. Of the 101 native species listed for this drainage (Crossman and McAllister 1986), 73% belong to freshwater dispersant families (31% of which are Cyprinidae), 21% to diadromous, and 6% to marine. These figures are deceptive, however, because most of the freshwater dispersant fishes occur at the southern edges of the basin. The more northern lakes and streams are dominated by salmonids, cottids, and a few cold-hardy cyprinids, catostomids, esocids, and percids.

The ***Arctic Province*** includes all those streams of Canada and Alaska that flow into the Arctic Ocean, up to and including the Yukon River and Mackenzie River drainages. Sixty-five species of obligatory freshwater fish are native to this region (Lindsey and McPhail 1986). Of these, 54% belong to diadromous families, 35% to freshwater dispersant families, and 11% to marine families. As in the Hudson Bay Province, most of the "true" freshwater fishes occur at the southern edge of the province. These species have invaded from both the Mississippi Province and the Pacific Subregion. The salmonids in this province also are characteristic of both areas. No unique forms are found, although the Alaskan blackfish (*Dallia pectoralis*), the inconnu (*Stenodus leucichthys*), the Arctic cisco (*Coregonus autumnalis*), the arctic lamprey (*Lampetra japonica*), and the least cisco (*C. sardinella*) are otherwise found mainly in Siberia. Thus, this province should be regarded as a transitional region between the Palaearctic and Nearctic Regions as well as between the Pacific and Arctic–Atlantic subregions.

Pacific Subregion

The ***Pacific Subregion*** contains many fewer species than the Arctic–Atlantic Subregion, both because it is smaller in size and, more importantly, because much of it is arid or semiarid. The fish faunas of its seven provinces, however, are much more distinct from each other than are the faunas of the provinces of the Arctic–Atlantic Subregion. The presence of mountain ranges and deserts in this subregion has created many barriers to fish movements, thereby resulting in many fish species that are endemic to small regions (Minckley et al. 1986). Many of these species appear to be relicts of the nearctic fish fauna that partly gave rise to—and were displaced by—the dominant fishes of the Arctic–Atlantic Subregion. The zoogeographic provinces of the Pacific Subregion are as follows: (1) the Alaska Coastal, (2) the Cascadia, (3) the Great Basin, (4) the Klamath, (5) the Sacramento, (6) the Colorado, and (7) the South Coastal.

The *Alaska Coastal Province* consists of all the minor coastal drainages from the Aleutian peninsula down to the Canadian border of the Alaskan panhandle. Only 34 freshwater species are found in this province (excluding euryhaline marine forms), 79% of them in diadromous families, 12% in marine families, and 8% in freshwater dispersant families. The three freshwater dispersant species (*Esox lucius, Catostomus catostomus*, and *Dallia pectoralis*) also are found in Siberia. This province could be regarded as part of the Arctic Province (Burr and Mayden 1992).

The *Cascadia Province* includes the great Columbia River drainage, which penetrates into the interior as far as Montana and Idaho, and its associated coastal drainages of British Columbia, Washington, and Oregon. It contains the greatest number of obligatory freshwater fishes species (61) of the provinces of the Pacific subregion. Only 38% of these are freshwater dispersant fishes (21% of which are Cyprinidae), but these are the most distinctive elements of the fauna, because 58% of these species are endemic to the province (Miller 1958). They include a relict mudminnow (Umbridae) and a trout-perch (Percopsidae). Of the remaining species, 41% are from diadromous families (23% from the Salmonidae alone), and the rest are freshwater representatives of marine families, especially the Cottidae (sculpins, 18%).

The *Great Basin Province*, which is the site of the classic studies of fish zoogeography and speciation by R. R. Miller and C. L. Hubbs, is a collection of 150 internal drainage systems roughly bounded by the Sierra Nevada on the west and south, the Wasatch Mountains on the east, and the Columbia plateau on the north. All the basins within the area are quite arid, although they were filled with large lakes during various periods of the Pleistocene. A high degree of endemism is found in the desert basins, but the species also tend to show a common faunal ancestry. The basins with the most fish species within the Great Basin are Death Valley (nine species); the Lahontan drainage, which includes Lake Tahoe and Pyramid Lake (13 species); and the Bonneville drainage, including Bear Lake and Utah Lake (20 species). The Great Basin Province contains a total of approximately 50 species. The exact number is uncertain, because several undescribed and/or controversial species are present. Nearly 60% of the species are either cyprinids or catostomids. Six species of pupfish (Cyprinodontidae and Goodeidae) make up most of the Death Valley fish fauna, however, and four species of whitefish (Salmonidae) are found in Bear Lake, Utah. Approximately 80% of the species found in the Great Basin are endemic (Miller 1958).

The *Klamath Province*, consisting of the Klamath and Rogue drainages on the California–Oregon border, contains approximately 30 obligatory freshwater fish species, with only 8 freshwater dispersant species (3 cyprinids and 5 suckers). It is characterized instead by its representatives of diadromous families (53% of fishes, including four lamprey and eight salmonid species) and its six species of sculpin (Cottidae). Thirty-seven percent of the species, including all the suckers, are endemic to the system.

The *Sacramento Province* contains most of the water that flows in California, because it is dominated by the great Sacramento–San Joaquin drainage system. It also includes numerous coastal drainage systems from Monterey Bay north to the mouth of the Klamath River. It contains at least 43 native obligatory freshwater species, of which 42% are endemic (Moyle 2002). Among these endemic forms is a unique complex of 10 minnow species and the Sacramento perch (*Archoplites interruptus*), which is the only native centrarchid west of the Rocky Mountains. Only 28% of the obligatory freshwater fishes in the province are freshwater dispersant species, whereas 47% of the fishes belong to anadromous families and the rest are freshwater representatives of marine families. Of particular interest is the tule perch (*Hysterocarpus traski*), which is the only freshwater species in the marine family Embiotocidae (surfperches).

The **Colorado Province** is drained by the Colorado River system. This province is home mostly to freshwater dispersant fishes, which make up approximately 81% of its approximately 32 obligatory freshwater species. Among these fishes are six species of the endemic cyprinid group of spinedaces (Plagopterini), at least three species in a complex of the cyprinid genus *Gila* that live in the main Colorado River, and the Gila topminnow (Poeciliidae, *Poecilopsis occidentalis*), which apparently entered the system from the Yaqui River of Mexico (Miller 1958). The rest are representatives of anadromous or marine families.

The **South Coastal Province** is an artificial collection of small, coastal drainage systems that enter the ocean between Monterey Bay on the north and the tip of Baja California on the south. Only 12 species of obligatory freshwater fishes are found in this province, 4 of them freshwater dispersants. Minckley et al. (1986) show that at least one of these species, the arroyo chub *(Gila orcutti)*, arrived in the region "riding" on a piece of continental plate that shifted from its original position in Mexico.

Mexican Subregion

The **Mexican Subregion** contains two major provinces; the Sonoran–Sinaloan Province, and the Central Mexico Province. This subregion has a fish fauna that is related to fishes in both the Arctic–Atlantic and the Pacific subregions as well as to Central American fishes. Thus, it is transitional in its fish fauna between North America and Central America. Seven families of North American fishes, for example, reach their southern limits in this subregion, whereas one family of South American catfishes (Pimelodidae) reaches its northern limits here (Miller and Smith 1986).

The **Sonoran–Sinaloan Province** includes the watersheds of northwestern Mexico that drain into the Gulf of California. The largest of these is the Rio Yaqui, which contains approximately 45 species, including at least 16 endemic forms, a number of them undescribed. Approximately 125 euryhaline species have been collected from the lower reaches of the streams, however (Minckley et al. 1986). The streams support such "western" fishes as minnows of the genus *Gila* and the Mexican golden trout *(Oncorhynchus chrysogaster)* as well as such "eastern" fishes as minnows of the genus *Notropis* and the Yaqui catfish *(Ictalurus pricei)*.

The **Central Mexico Province** contains the lowlands of both coasts of Mexico, which seems to have been the dispersal route for fishes coming from both the north and the south. In the south, the most species-rich groups are the characins, cichlids, and poeciliids, but in more northern drainages, fishes with North American affinities become more common, especially minnows, suckers, and ictalurid catfishes. Physiologically hardy freshwater dispersants, however, dominate the freshwater fauna of 205 species: Goodeidae (39 species), Poeciliidae (43 species), and Atherinidae (30 species). Marine fishes or fishes of marine families are common in the lower reaches of these streams

($\approx$170 species). The central plateau of this province is an isolated area of particularly high endemism. It contains, for example, 17 species of the odd viviparous goodeids and at least 16 species of the endemic atherinid genus *Chirostoma*. Extinction rates are high in this province.

25.8   AUSTRALIAN REGION

The ***Australian Region*** has long been cut off from the major evolutionary events taking place on the other continents. This is reflected dramatically in its fish fauna, which contains only four species from three families of ancient lineage that are exclusively freshwater dispersants: Australian lungfish (*Neoceratodus forsteri*, Ceratodontidae), two species of saratoga (*Scleropages* spp., Osteoglossidae), and probably, salamanderfish (*Lepidogalaxius salamandroides*, Lepidogalaxiidae). The rest of this region's fish fauna usually is characterized as being representatives of diadromous or marine-derived families. The Melanotaeniidae (rainbow fishes) and Pseudomugilidae (blue eyes), however, can be regarded as freshwater dispersant families that are endemic to the region, even though they are related to the marine/estuarine atheriniform groups. In addition, representatives of many marine familes, such as the Terapontidae and Percichthyidae, probably were present when the continent split from the other southern continents (Unmack 2001). Because these lineages have been evolving in Australia's waters for millions of years, they now are effectively freshwater dispersants. A few species of fishes also are capable of moving through brackish water (e.g., swamp eels, Synbranchidae; archerfishes, Toxotidae) and are represented both within this region and in the adjacent Oriental region. The Australian Region can be divided into three subregions in terms of its fish fauna: (1) Australia including Tasmania; (2) New Zealand; and (3) New Guinea, including nearby islands.

Australia

Australia contains more than 200 native species that spend all (167 species) or most of their life cycles in fresh water, plus approximately 150 species of marine fishes that occasionally are found in fresh water or estuaries (Allen 1989, Unmack 2001, Mooi and Gill 2002). The number of species is growing, however, as more are being described. The two endemic freshwater dispersant families, Melanotaeniidae (15 species) and Pseudmugilidae (five species), are widely distributed on the continent (Unmack 2001). Other fishes are diadromous species or their freshwater derivatives, including Geotriidae (lampreys, three species), Anguillidae (eels, four species), Galaxiidae (17 species), and Retropinnidae (southern smelts, two species). A dozen or so families of marine fishes have freshwater representatives in Australia. Of particular interest are the marine catfishes (Ariidae and Plotosidae, 18 species), gudgeons (Eleotridae, 32 species), gobies (Gobiidae, 16 species), hardyheads (Atherinidae, 13 species, all in the endemic genus *Caterocephalus*), flagtails (Kuhliidae, two species), glassfishes (Ambassidae, eight species), grunters (Terapontidae, 25 species), and temperate perches (Percichthyidae, 17 species). Members of the latter two families are the typical perciform predators of Australia's lakes and rivers, including the Murray cod *(Maccullochella peeli)*, which reaches weights of more than 110 kg.

The freshwater fish fauna shows a high degree of endemism. Allen (1989) found that 69% of the species were endemic to Australia, whereas 20% were shared only with New Guinea and 2% with New Zealand. The rest have wider distributions in Southeast Asia. Based on patterns of endemism, Australia can be divided into 10 fish provinces (Unmack 2001). Fish species are most numerous on a per-area basis in coastal areas, where rainfall is highest and the most permanent water exists. It is interesting to note that 30 species manage to exist in the sparse desert waters of the Lake Eyre Province of the interior, but that only 29 freshwater species inhabit the extensive Murray-Darling Province of southern Australia, with the 1,900-km-long Murray-Darling River (Unmack 2001).

New Zealand

New Zealand has long been isolated from other land masses, even Australia, so its 27 recognized freshwater species are wholly of marine or diadromous origin (McDowall 1990). Most abundant are species from the diadromous families, including Geotriidae (lampreys, one species), Retropinnidae (southern smelts, two species), Galaxiidae (>14 species), and Anguillidae (eels, two species). Despite their ability to enter salt water, ancestral fishes in the four families possibly already were present in New Zealand when it separated from Antarctica during the late Cretaceous, although dispersal from other areas cannot be ruled out (McDowall 2002). Marine families with freshwater representatives are the Pleuronectidae (flounders, one species), Eleotridae (gudgeons, six species), and Cheimarrichthyidae (freshwater sandperch, one species). All but four of the species found in New Zealand are endemic to it.

New Guinea

The fish fauna of *New Guinea* is very similar to that of Australia, reflecting a presumed connection of the two land masses (because of falling sea levels) less than 8,000 years ago, although Unmack (2001) suggests even at this time, little or no faunal exchange took place. New Guinea is divided in half by a mountain range, and the fish fauna reflects this division. The fauna of the southern half is similar to that of northern Australia (32 of 130 species are shared), whereas that of the northern half has a large number of species of its own. The two share just three native upland species (Allen and Coates 1990). Northern New Guinea has approximately 75 freshwater species ($\approx$60% of them endemic). In both regions, most of the fishes are gobies (Gobiidae), gudgeons (Eleotridae), rainbow fishes (Malanotaeniidae), or catfishes (Plotosidae, Ariidae) (Allen and Coates 1990).

25.9 PLATE TECTONICS AND THE DISTRIBUTION OF FRESHWATER FISHES

The advances in geophysics that led to acceptance of the concept of plate tectonics have permitted the solution of many perplexing zoogeographic problems. Many of the families and species distribution patterns, however, remain puzzling and await solution.

Therefore, the following discussion, based in part on the papers of Cracraft (1974) and of Novacek and Marshall (1976), should be considered hypothetical.

Approximately 200 million years ago, during the Triassic, the continents were united as one land mass, **Pangaea** (Fig. 25.3). The freshwater fish fauna at this time presumably was made up of a mixture of chondrostean fishes (represented today by the paddlefishes and sturgeons), early neopterygian fishes (represented today by bowfin and gars), and lobed-finned fishes (e.g., lungfish). Some early teleosts also possibly were present at this time (represented today by the osteoglossids), although teleost radiation usually is considered to have taken place in the Cretaceous. By the end of the Triassic (180 million years ago), Pangaea had split into two continents: **Laurasia** (the future North America and Eurasia), and **Gondwanaland** (the future South America, Africa, Antarctica, India, and Australia). In both areas, the Pangaean fish fauna probably remained dominant until the Cretaceous, when various modern freshwater fish groups started to develop. The modern descendants of the first of these groups to develop in Laurasia are the Esocidae (pikes), Umbridae (mudminnows), and Salmonidae (salmon and trout), as well as the exclusively North American Percopsidae (trout-perches), Aphredoderidae (pirate perch), and Amblyopsidae (cavefishes). Most of whatever distinctive teleost fauna developed in Laurasia, however, appears to have been swamped

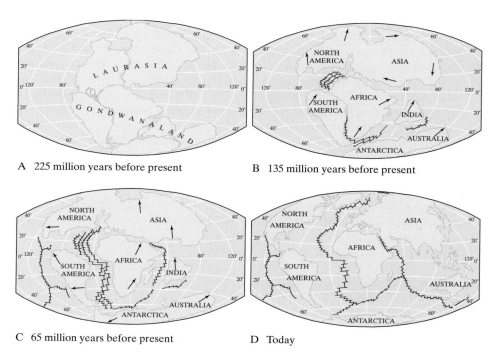

A   225 million years before present

B   135 million years before present

C   65 million years before present

D   Today

FIGURE 25.3   Probable relationships of the continents following the breakup of the large landmass Pangaea into Laurasia and Gondwanaland and then into the modern continents. From McKnight (1998).

by the otophysans invading from Gondwanaland and by the perciform groups that evolved somewhat later.

The Otophysi (minnows, characins, and catfish) seem to have developed in Gondwanaland during the early Cretaceous period, when the African and South American plates were well on their way to separation. Antarctica, Australia, India, and Madagascar had separated from Gondwanaland soon after Laurasia split off, so they played no role in the early development of the Otophysi. Because the most ancestor-like otophysans are known from South America, the early stages of their evolution likely took place on the South American half of Gondwanaland (see Briggs 1979 for an alternative point of view). Representatives managed to invade western Africa before the two continents finally separated in the middle Cretaceous. South America then became an island-continent, and its characoid fishes and catfishes diversified enormously into numerous endemic families. Africa, in contrast, made contact with the Eurasian portion of Laurasia at about the end of the Cretaceous, thereby permitting ostariophysans to enter Laurasia. Then, in the early Paleocene, the Cyprinidae evolved either in North Africa or in Europe. This family, together with the catfishes, spread throughout Laurasia and most of Africa. The derived characins that had evolved in Africa managed to hold their own and diversify, however, as did the African catfishes, mormyrids, and other endemic groups.

During the early Cenozoic, as the North American and Eurasian continents gradually separated from each other, the modern representatives of the holarctic fish fauna developed and spread: Cyprinidae, Catostomidae (suckers), and Percidae (perches). Following (and during) the final separation of North America and Eurasia, endemic groups of fishes developed on both continents. In North America, the sunfishes (Centrarchidae), the catfishes (Ictaluridae), and the darters (subfamily Etheostominae of the Percidae) evolved. In Asia, the highly derived subfamilies of cyprinids developed, probably in the Oriental Region, along with the Cobitidae (loaches), Siluridae, and other otophysan families with more restricted distributions. Many families of freshwater dispersants are endemic to the Oriental Region, which gradually separated from the rest of Asia following the rise of the Himalaya Mountains, which in turn seems to have been related to the collision of India with Asia during the late Eocene. The Oriental fauna subsequently invaded India, in exchange for cichlids and aplocheilids that had rafted on the subcontinent. There appear to have been some exchanges of fishes with Africa somewhat later, as indicated by the broad distributional patterns of catfishes of the families Bagridae and Clariidae and the perciform families, such as the Channidae and Mastacembelidae.

25.10   PLEISTOCENE GLACIATION

Plate tectonics explains many of the broad patterns of fish distribution, but explaining continental patterns requires an understanding of the events of the Pleistocene as well. This period began approximately 2.5 million years ago and ended (presumably) approximately 10,000 years ago. During that time, as much as one-third of North America and Eurasia was covered with continental glaciers, which advanced and retreated at least four times during the period. As the glaciers advanced and more and more of the Earth's water became ice, sea level dropped, thereby connecting many land areas currently separated by salt water. The ice wiped out all fishes in the land it covered, and it changed the water temperatures in nearby areas. As it advanced and retreated, the glacial ice scoured lakes out of bedrock or created them by damming streams with moraines or by leaving behind melting blocks of ice buried in the gravels. The melting ice then created large rivers and enormous lakes. Some of the lakes were behind ice dams that burst, resulting in catastrophic floods, as shown by the "scablands" of the Columbia Plateau that were sculptured by such a flood (with unknown effects on the fishes). The general climatic conditions that favored the buildup of glaciers in northern or high-elevation areas caused increased precipitation in other areas. In the intermountain basins of the American West, the cooling and rainfall caused huge lakes to develop, which have now dried to small remnants that often are too salty to support fish (e.g., the Great Salt Lake). Much of Nevada was once covered with huge lakes (mainly Lake Lahontan), and the abundant fish fauna of those lakes now survives in isolated springs, streams, and lake remnants (e.g., Pyramid Lake).

The Pleistocene events explain such patterns as the presence of freshwater dispersant fishes on many of the islands of Indonesia, which once were uplands surrounding a large river system that existed when sea levels were lower. The paucity of native fishes in the western United States and Europe is, at least in part, due to the lack of refuges from Pleistocene events, whereas the richness of the fish fauna of glaciated areas of eastern North America is due to the ease with which the areas could be recolonized from the lower Mississippi River basin and other refuges (Moyle and Herbold 1987). Indeed, the last retreat of the glaciers was so recent that that many of the glaciated areas likely are still subject to natural invasions by fishes from unglaciated areas.

25.11 LESSONS FROM FRESHWATER ZOOGEOGRAPHY

The discovery that the Earth's surface is constantly in slow motion—and that fish can be rafted to new areas on continental plates—created a new understanding of the zoogeography of freshwater fishes. Darlington (1957), in his classic book on zoogeography, had to postulate chains of sunken islands and long ocean voyages by freshwater fishes to get some species to remote locations on a stable Earth. Following Myers (1951), he used a zoogeographic classification system that gave major groups of fishes (e.g., killifishes and poeciliids) the capacity to colonize islands through saltwater dispersal, even though such events were difficult to imagine. We now realize that whereas dispersal is important, the larger patterns of freshwater fish distribution usually are the result of vicariant events in which the Earth moves and the fish remain in place. As the examples of galaxiids and cichlids in this chapter suggest, however, it is important to realize that some fishes can, in fact, disperse over long distances—even between continents. Australia provides a large example of another phenomenon; that species derived from ancient invasions of marine fishes become "true" freshwater fishes, in that they lose their ability to disperse through salt water. Not surprisingly, the fish faunas of all regions of the world have complex histories, with each family and species passing through multiple "filters" to wind up with their particular distribution pattern.

SUPPLEMENTAL READINGS

Berra 2001; Hocutt and Wiley 1986; Lévêque 1997; Lundberg et al. 2002; Matthews 1998; Mayden 1992a, 1992b; McDowall 1978; Mooi and Gill 2002; Rosen 1975a, 1975b; Unmack 2001.

WEB CONNECTIONS

NEOTROPICAL FISHES
www.neodat.org

AUSTRALIA
www.peter.unmack.net/biogeog
www.nativefish.asn.au

CHAPTER 26
Zoogeography of Marine Fishes

The zoogeography of marine fishes is much less well understood than that of fresh-water fishes. Not only are many more marine fishes spread over a much greater area, but fewer dramatic, seemingly permanent barriers to movement, such as mountain ranges, are found than on land. The distribution patterns of freshwater fishes can be broadly related to plate tectonics (continental drift) and drainage systems. For marine fishes, broad distribution patterns seem to be related to oceanographic features, to the positions of the continents, and to the historic effects of Pleistocene events. Because most marine fishes are shallow-water species located on the rims of land masses, plate tectonics undoubtedly has played an important role in the distribution and evolution of marine fishes, that is, in vicariance (Springer 1982, Mooi and Gill 2002, Santini and Winterbottom 2002). Recent studies suggest that many groups of closely related species have distribution patterns that are best explained by a combination of vicariant events and the interaction of dispersal abilities and ecological factors, such as ocean currents (e.g., Bowen et al. 2001, Grant and Leslie 2001, Muss et al. 2001)

Most discussions of marine zoogeography are still largely descriptive in nature, dividing the marine world into zoogeographic regions whose boundaries are rather vague. Typically, these boundaries are bands where oceanographic conditions and, con-sequently, the fish fauna change rapidly as the result of current patterns or the position of continental shelves and islands (Bailey 1998). Rarely are such boundaries complete barriers to fish movement, nor are they necessarily permanent features. Thus, the desig-nated zoogeographic regions are even more arbitrary than those found in fresh water. The dynamic nature of marine fish faunas has generally frustrated the search for broad rules to explain worldwide patterns. Two rules, however, hold up: (1) that the richness of fish species increases as the latitude decreases (i.e., there are more fishes in the tropics), and (2) that the richness of fish species decreases with depth. For the latitude–diversity relationship in the eastern Atlantic, Macpherson and Duarte (1994) calculated that species richness declines northward by approximately 1% for each degree of latitude (in 5° bands), a number that seems to be reasonable for much of the world. For the depth–diversity relationship, Stevens (1996) indicates that deepwater fishes also have broader depth ranges than shallow-water fishes, an application of the much-contested Rapoport's rule for terrestrial vertebrates.

This chapter briefly describes the major zoogeographic regions of the world's oceans, and it discusses the kinds of fishes that occur in them. Following Briggs (1974), we first discuss the fishes of the continental shelves and ocean reefs and then the fishes of the open sea and of the deep sea.

26.1   CONTINENTAL SHELVES

The distribution of fishes on continental shelves can be related not only to the presence of continents and islands and the extent of the shelf surrounding them but also to annual temperature regimes and oceanic currents (Fig. 26.1). On the basis of temperature regimes, the shelves can be divided into a wide tropical region around the equator, northern and southern temperate regions, and two polar regions. Of course, broad transition areas exist between adjacent regions, but particularly between the tropical and temperate waters (which, as a consequence, frequently are divided into warm temperate waters and cold temperate waters).

26.2   TROPICAL REGIONS

Nearly 40% of all known fish species occur in the shallow waters of the tropics. Most of these species are associated with tropical reefs (see Chapter 33). Largely on the basis of the reef fishes, the tropical oceans can be divided into four large regions: (1) the Indo-Pacific Region, (2) the Eastern Pacific Region, (3) the Western Atlantic Region, and (4) the Eastern Atlantic Region. These regions are separated by continents or by vast

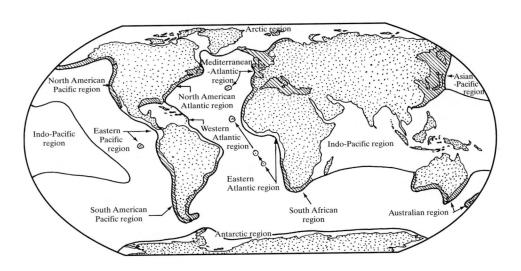

Temperate regions

FIGURE 26.1   Major zoogeographic regions associated with continental shelves and islands. Boundaries between regions are not firm but are wide, shift seasonally, and contain elements from both adjoining fauna.

areas of open ocean. The northern and southern boundaries of these regions are roughly the 20°C isotherm for the coldest month of the year.

The ***Indo-Pacific Region*** contains by far the most diverse fish fauna (≈4,500 species) of the four regions, presumably because its sheer size has provided many opportunities for speciation. Most of these fishes belong to the same families that are characteristic of the other three regions as well, such as Muraenidae, Holocentridae, Scorpaenidae, Serranidae, Apogonidae, Carangidae, Lutjanidae, Mullidae, Gobiidae, Chaetodontidae, Pomacentridae, Labridae, Scaridae, Blenniidae, Acanthuridae, and Tetraodontidae. A few small families, however, are endemic to the region (i.e., Pegasidae, Sillaginidae, Kraemeriidae, Siganidae, and Plesiopidae). Despite this region's size, the presence of many apparently wide-ranging species makes it fairly homogeneous, although some are cryptic species that show more genetic differences than morphological ones (e.g., Colborn et al. 2001). Springer (1982) noted that the distribution of many species is coincident with an enormous piece of the Earth's crust, the Pacific lithospheric plate, thereby demonstrating the importance of plate tectonics to marine fish distribution. Within the region, however, are areas with fish faunas distinctive enough that Briggs (1974) divided the region into provinces. Four of these provinces are associated with groups of isolated oceanic islands (i.e., Lord Howe-Norfork, Hawaiian, Easter, and Marquesas) that have a fairly high degree of endemism in the fishes (15%–25%). Other designated provinces are sections of the coastline containing endemic species, but because the continental shelf in this region is essentially continuous—from Australia to the Malay Peninsula, to India and down along the east coast of Africa (including Madagascar and the Comoro Islands) to South Africa—much of the fish fauna is continuous as well. The fish fauna of the African coast, however, is quite different from that of the Malay Peninsula, although no really abrupt breaks separate one fauna from the other.

An area that has always been of special interest because of the extraordinary richness of its fauna (≈2,800 species) is the ***Indo-Malayan Subregion*** (Mooi and Gill 2002). This area forms a triangle roughly bounded by the Phillipines, the Malay Peninsula, and New Guinea. Although few species are unique to this area, the total number of species is higher than that in any similar-sized area of the world, and many families seem to have the most species here. The standard explanation for this diversity (e.g., Briggs 2000) has been twofold: (1) The region is the place where most of the taxa evolved and from which they dispersed to other regions (i.e., it is a ***center of origin***), and (2) the complex topography of the area, with its many island arcs and deep trenches, provided many opportunities for allopatric, microallopatric, or ecological speciation (see Chapter 12). Santini and Winterbottom (2002), however, provide evidence showing that rather than being a center of origin, the region may instead be where species originating elsewhere have collected. Their main tool is cladograms that show the relatedness of various species groups with wide distributions. These cladograms suggest that in many groups, the related species that co-occur in the Indo-Malayan subregion are related more closely to more distant members of the same group than they are to each other. Their analysis also suggests that the subregion can be divided into at least 18 areas with distinct faunal elements, each having a distinct geologic history related to the breakup of the southern continents and the rise of arcs of volcanic islands. Santini and Winterbottom suggest there was plenty of time for speciation to take place in each of

BOX 26.1
The Suez Canal: An Experiment in Zoogeography

The Red Sea is of special interest to zoogeographers because it is a semi-isolated part of the Indo-Pacific Region without a particularly high number of endemics (10%–15% of the fish fauna). It is also of interest because it is a participant in a great zoogeographic experiment. The other participant is the Mediterranean Sea, which has been connected to the Red Sea by the Suez Canal since 1869. The high salinity of the Bitter Lakes reach of the canal originally was a barrier to the movement of fish between the two basins, but deepening the canal has made the salinity of the entire canal close to that of the water at both ends (≈41 ppt). This high salinity at the Red Sea end is natural. At the Mediterrean end, however, it results from the diversion of most of the flow of the Nile River, so the local sea water is no longer diluted (Boudouresque 1999). Thus, fish can move through the canal in either direction, although the length of the canal (163 km), and its soft bottom and turbidity undoubtedly have acted as a selective filter of species. The native fish faunas of the two basins are quite distinct, with that of the eastern Mediterranean (≈400 species) being largely the temperate fish fauna of the rest of the Mediterranean, and that of the Red Sea being the typical diverse Indo-Pacific fauna. As a result of the canal, at least 45 species of Red Sea fishes have immigrated to the Mediterranean Sea, where some of them have become established in large enough numbers to support fisheries (Boudoresque 1999). Only three fish species from the Mediterranean have managed to move into the Red Sea, but all three are established mainly near the canal entrance. This mostly one-way dispersal seems to relate to the species-poor nature of the regional Mediterranean fauna compared to that of the Red Sea, combined with the presumption that the temperate origin of the eastern Mediterranean fauna means they are less well adapted to tropical conditions and, thus, are unable to compete successfully with Red Sea fishes (Ben-Tuvia 1978). The net flow of water down the canal also tends to be toward the Mediterranean most of the year, and Red Sea fishes in general seem to be better adapted for canal conditions. The movement of the fauna is continuing, and no doubt, new species will keep appearing in the Mediterranean Sea as long as the canal is open (Boudouresque 1999).

The movement of fishes through the Suez Canal is in marked contrast to the lack of movement of fishes through the Panama Canal, which also links disparate zoogeographic regions. This seems to be the result of a combination of the freshwater middle reaches of the canal combined with strong differences in marine habitats at each end.

these areas, followed by the mixing of faunas with the breakdown of barriers and natural dispersal.

The **Eastern Pacific Region** is often referred to as the Panamanian Region, because the Pacific coast of Panama is central to this region, which extends from Bahia Magdalena near the tip of Baja California (but including the Gulf of California as well) down to the Gulf of Guayquil on the coast of South America. The southern boundary of this region is only approximately 3° of latitude from the equator, because the cold

Humboldt Current from the Antarctic keeps tropical conditions at bay (most of the time). The northern boundary is not very definite, and considerable mixing of "temperate" and "tropical" species occurs along the coasts of Baja California and southern California and in the Gulf of California. The Eastern Pacific Region is separated from the Indo-Pacific Region by a wide stretch of deep water and from the Western Atlantic Region by the isthmus of Panama. Most of the 750 species therefore are confined to the region, although at least 62 species are common with the Indo-Pacific Region and at least 12 are common with the Western Atlantic region (Briggs 1974). Before the closing of the Panamanian isthmus in the Pliocene, the eastern Pacific and the western Atlantic presumably had a common fish fauna (Rosen 1975b, White 1986). Few offshore islands exist in the Eastern Pacific Region, but those that do include the fascinating Galapagos. These islands' fish fauna is fairly rich (>300 species), with 23% of the species being endemic, 54% shared with the coast of Central America, and 12% of Indo-Pacific origin (Grove and Lavenberg 1997). Dana (1975) presents evidence that coral reefs were wiped out in this region as the result of colder sea temperatures during the Pleistocene and that the present reef systems may be only approximately 5,000 years old. If this is true, then the endemic species of this region evolved very rapidly from the reinvasions of Indo-Pacific fishes.

The **Western Atlantic Region** includes the Caribbean and Gulf coasts of Central America and Mexico, the coast of South America to Cape Frio, the multitudinous islands of the West Indies, Bermuda, and the tip of Florida. It contains at least 1,500 species of fishes, most of them endemic to the region (Mooi and Gill 2002). Much of the fish fauna is associated with coral reefs and is similar to those of coral reefs elsewhere, at least superficially. Many differences do exist, however. For example, in the Indo-Pacific Region, the principal cleaner fishes are wrasses (Labridae); in the Western Atlantic Region, they are gobies (Gobiidae). Within the region, most species are mostly widely distributed, although some areas have enough endemism for Briggs (1974) to divide the region into provinces. The origins of this fauna are complex (Joyeux et al. 2001, Muss et al. 2001). Before the closure of the Isthmus of Panama approximately 5 million years ago, this region was connected to the Indo-Pacific Region, which accounts for the great similarity of the faunas at the level of family and genera. Regular cross-Atlantic currents, however, have infrequently carried fishes from (and to) the west coast of Africa and permitted colonization of the extremely remote islands of the central southern Atlantic (e.g., St. Helena, Ascension, and St. Paul's Rocks). Thus, a combination of major vicariant events, long-distance dispersal, and local speciation in isolation shaped the fauna of this region.

Within the Western Atlantic Region, two areas are of particular interest to fish zoogeographers: Bermuda, and the Brazilian Coast. A cluster of approximately 360 small islands, Bermuda is the northernmost (32° latitude) bastion of tropical fishes. The tropical nature of its fauna (365 species) is maintained by the warm Gulf Stream flowing up from Florida and the Gulf of Mexico. The Gulf Stream also helps to maintain a genetic connection of the fishes to their counterparts in the Caribbean—despite a separation of more than 1,400 km. Bermuda is a fairly ancient group of islands, but its fauna shows little sign of differentiation. Presumably, this is because of the present influence of the Gulf Stream as well as its lack of influence during the Pleistocene (11,000 years ago), when it is believed that most of the tropical fauna of Bermuda disappeared as a result of lower water temperatures (Briggs 1974).

Along the coast of Brazil, the shallow-water coral reef fauna is sporadic for nearly 2,900 km because of the influence of the Amazon, Orinoco, and other rivers, which decrease salinity, increase turbidity, and deposit large amounts of silt over a large area. Reef fishes are found primarily in association with sponges growing on hard bottoms in areas of oceanic salinity. Otherwise, typical reef fishes in this stretch are replaced by fishes that are more characteristic of soft bottoms, such as sea catfishes (Ariidae) and croakers (Sciaenidae). The shallow-water reef fauna resumes as the influence of the rivers diminishes. This separation has resulted in a certain amount of endemism in the southern Brazilian fauna. The endemism, however, seems to be most characteristic of shallow-water forms, because deepwater reefs apparently are still present along the entire coast. Before the rise of the Andes, the influence of the rivers likely was not as strong in Atlantic coastal waters, so some dispersal between regions was possible (Joyeux et al. 2001).

The **Eastern Atlantic Region** consists of the continental shelf along the west coast of Africa from Cape Verde to central Angola in the south plus three islands off the coast (i.e., Cape Verde, St. Helena, and Ascension). The northern boundary is rather arbitrary, but it does mark a region south of which temperate fishes are rare. It is the smallest and most isolated of the tropical regions, and it contains the fewest fish species. Presumably, one reason for its comparative lack of faunal richness is the near absence of coral reefs. Briggs (1974) lists 434 shore fishes from the region—compared with 1,500 on the opposite side of the Atlantic. The main affinities of the fishes are to those of the Western Atlantic Region. Not only do they have approximately 120 species in common, many of the genera in the remaining species are the same as well. Approximately 40% of the species are endemic, however, reflecting the area's isolation. The Eastern and Western Atlantic regions presumably had a common fauna at the time Africa and South America were drifting apart and a much smaller Proto-Atlantic Ocean existed, which was less of a barrier to movement than the present ocean. A fascinating aspect of the Eastern Atlantic fish fauna, however, is that it includes approximately 32 species also found in the Indo-Pacific region. Sixteen of these species are found off the African coast as well as the three islands. Eight are also found in the western Atlantic, four only off West Africa, and four only around the island of St. Helena. Such fishes made it around the Cape of Good Hope at one time or another; the presence of four such species at St. Helena is particularly indicative of this (Box 26.2).

The fish fauna of St. Helena points out nicely the interconnectedness of the fauna of the four regions. They share most of their families, many of their genera, and even a few species. The similarities of the faunas largely reflect the closeness of the continents during the time that tropical fish fauna was evolving; the differences reflect the considerable isolation of the faunas that now exist—and that have for some time—as well as the limited dispersal abilities of most of the fishes.

26.3   NORTH TEMPERATE REGIONS

Although the North Temperate Regions of the Atlantic and Pacific oceans are marked on maps and generally broken into a number of provinces or subregions, the boundaries

BOX 26.2

St. Helena: Evidence of Long-Distance Dispersal in Fishes?

St. Helena, one of the most isolated tropical islands in the world, is located 1,870 km from the coast of Africa and 1,270 km from Ascension Island to the north (Joyeux et al. 2001). Its reef fish fauna, although not rich (55 species), consists of a mixture of endemic species (18, including 6 shared with Ascension Island), circumtropical species, and species from all four major tropical regions (Briggs 1974). This fauna reflects the island's ancient origin as a volcano and its strategic position in the central southern Atlantic, where it can intercept fishes (and emperors) coming from various directions. Because it was never connected to a continent itself, St. Helena would seem to provide a good demonstration that tropical reef fishes are capable of dispersing, presumably as larvae, over extremely long distances. Bowen at al. (2001), however, noted that the trumpetfish (*Aulostomus strigosus*) is divergent enough genetically from a related Indian Ocean species that it must have colonized the island before a cold-water upwelling barrier existed around the horn of Africa ($\approx$2.5 million years ago). At this time, a single species would have been present in both oceans, and the hop from Africa to St. Helena (and other mid-Atlantic islands) would have been less formidable. In fairly recent times, *A. strigosus* has managed to colonize the coast of Brazil from the islands, a distance of more than 2,500 km via the South Equatorial Current (Bowen et al. 2001).

between them and the Arctic Region on the north and the Tropical Regions on the south are very hard to fix. The reason is that the change from one fauna to the next is gradual, and the ranges of many species fluctuate from year to year, depending on the vagaries of coastal currents. Some coastal features, such as Point Conception of the California coast, do mark rather sharp changes in the fish fauna, yet even in such situations, typically more similarities than differences are found on the two sides of the boundary. Distinctive faunas of temperate-water fishes, however, do exist along long stretches of coast on both sides of the Atlantic and both sides of the Pacific. Therefore, these four areas are treated as separate zoogeographic regions: (1) the Mediterranean-Atlantic Region, (2) the North American Atlantic Region, (3) the North American Pacific Region, and (4) the Asian-Pacific Region.

The *Mediterranean-Atlantic Region* consists of the Atlantic coast of Europe and North Africa, together with the Mediterranean, Black, and Caspian seas. In the north, it merges with the Arctic Region; in the south, it merges with the tropical Eastern Atlantic Region. Because the Gulf Stream swings south along the coast of Europe, forming the North Atlantic Current, comparatively warm water usually is present as far north as the British Isles. As a result, representatives of "tropical" families, such as the Gobiidae (gobies), Labridae (wrasses), Mugilidae (mullets), Sparidae (porgies), Mullidae (goatfishes), and Scaridae (parrotfishes), are common in the more southern portions of this region, including the Mediterranean Sea (Lythgoe and Lythgoe 1991). Representatives of typically cold-water marine families, however, such as Gadidae (cods), Agonidae (poachers),

Cottidae (sculpins), Pleuronectidae (right-eye flounders), and Cyclopteridae (snail-fishes), are common throughout much of the region. Many of the fish species and genera, but none of the families, found in this region are endemic. Many of the species that are characteristic of the more northern areas are shared with the Arctic Region and with the North American Atlantic Region, whereas a number of the southern species occur in tropical areas on both sides of the Atlantic. The gradual faunal changes that are characteristic of this region make provinces probematic, but it is worth examining some of the more isolated components of the region: (1) the Mediterranean Sea, (2) the Black Sea, and (3) the Baltic Sea.

The Mediterranean Sea, despite its apparent isolation, contains a fauna of approximately 600 fish species that differs little from that of the Atlantic coasts of northern Africa and southern Europe. Indeed, 35% of the fish species found north of the Arctic Circle in Europe are also found in the Mediterranean (Lythgoe and Lythgoe 1991). The Mediterranean, however, does possess a number of local endemic fishes, especially in the families Blenniidae, Gobiidae, and Labridae. It also possesses a distinctly tropical fauna in the eastern portion, a fauna currently being augmented by Red Sea fishes emigrating through the Suez Canal (see Box 26.1).

The Black Sea has a complex history of connections with the Mediterranean and Caspian seas, presumably with considerable fluctuations in salinity. At present, its narrow connection with the Mediterranean, coupled with freshwater inflow from the Danube and other large rivers, gives it a salinity of 17 to 18 ppt over much of its surface. This fish fauna (140 species) is mostly a depauperate Mediterranean fauna, with a few (33) endemic species and a few species (mostly gobies) shared with the Caspian Sea. Unfortunately, even this diversity is being lost as the result of human-induced changes, as is also happening in the Baltic Sea (Lappäkoski and Mihnea 1996).

The Baltic Sea is the largest estuary in the world, with salinities that gradually increase from the upper to the lower end. The salinity gradient, the shallowness of the sea, and the extreme temperature fluctuations that it experiences greatly limit the Baltic Sea's capacity to support a diverse fish fauna, although it does contain a mixture of freshwater and marine species. The marine fishes are the same as those found in the North Atlantic, although some (e.g., the cod [*Gadus morhua*]) are considered to have distinct Baltic "races."

The *North American Atlantic Region* consists of the Atlantic coast of North America (excluding the tip of Florida, with its distinctly tropical fauna) and the northern Gulf of Mexico. Because of the high diversity (1200 species) and the presence of "tropical" fishes from the Caribbean along the Gulf and southern Atlantic coasts, there have been many attempts to subdivide this region (Briggs 1974), but the general lack of agreement among people who work with different animal groups indicates the gradual nature of the change along the coasts (Mooi and Gill 2001). Usually, however, a boundary of some sort is made in the Cape Hatteras (North Carolina) area, because the Gulf Stream tends to swing away from the coast in this area, thereby permitting the cooler waters from the north to have greater influence. Seasonal temperature fluctuations and vagaries in the flow of the Gulf Stream, however, make even this boundary a tenuous one, especially for mobile fishes. Thus, the region between Cape Hatteras and Cape Cod contains many typically "southern" species in the summer, when water temperatures are high, whereas in the winter, the dominant species are typical fishes of northern waters.

These northern fishes include representatives of the same families as those found off northern Europe, such as the Gadidae (cods), Pleuronectidae (right-eye flounders), Cottidae (sculpins), and Cyclopteridae (snail fishes), as well as diadromous members of the Anguillidae, Clupeidae, and Salmonidae. The fishes that are more characteristic of southern (warmer) waters include members of such families as Sciaenidae (croakers), Sparidae (porgies), Serranidae (sea basses), Haemulidae (grunts), and Labridae (wrasses). Even these families, however, are widely distributed up and down the Atlantic coast, with some species being more characteristic of northern than southern faunas.

One factor complicating any discussion of the distribution patterns is the general decrease in the number of species toward the north. Approximately 400 species are found along the Gulf Coast of Texas, perhaps 350 along the coast of South Carolina, approximately 250 in the area south of Cape Cod, approximately 225 (130 shore fishes) in the Gulf of Maine, 61 along the Labrador coast, and 34 from Greenland (Briggs 1974). The figures for Cape Cod and the Gulf of Maine are somewhat deceptive, because 50% to 60% of the species recorded are warm-water species of sporadic or rare occurrence.

The *North American Pacific Region* extends north along the Pacific coast from Baja California to the end of the Aleutian Island chain. As in the North American Atlantic region, a very gradual change occurs from a tropical fauna to a mixed tropical-temperate fauna to a largely temperate fauna to a mixed temperate-arctic fauna. The region is extremely rich in fishes. At its southern end, most of the 800 or so species that are known from the Gulf of California (Thomson et al. 2000) can be expected at one time or another, this is especially the case in years when the California Current is not flowing strongly, so that the cool water from the Gulf of Alaska does not extend so far south. For the California coast, Miller and Lea (1972) list 554 species, of which 439 should perhaps be considered continental shelf species. For the coast of Alaska, Mecklenburg et al. (2002) report 501 marine and andromous fishes, with perhaps another 100 species being present but not confirmed. Presumably, this list would also fit most of the Canadian coast to the south as well. Farther north, the fauna merges with that of the Arctic region and, to the west, with the Asian-Pacific Region.

The fauna in this region have many fascinating aspects. It is extremely diverse, with 144 families represented off California alone. Several of these families are found only in this region and in the Asian-Pacific Region, most prominently the Embiotocidae (surfperches, 23 species), Hexagrammidae (greenlings, 11 species), and Anoplopomatidae (sablefish and skilfish). A number of other families have attained great diversity here. These include the Scorpaenidae, with more than 65 species in the genus *Sebastes* alone; the Salmonidae, which includes seven species of Pacific salmon and trout (*Oncorhynchus*); the Cottidae (sculpins), with approximately 85 species; the Pleuronectidae (flounders), with 20 to 25 species; the Agonidae (poachers), with 22 species; and 35 to 40 species in various blennioid families (Stichaeidae, Blenniidae, Clinidae, etc.).

The *Asian-Pacific Region* extends roughly from Hong Kong north past the Kamchatka Peninsula. These boundaries are very arbitrary, because the warm Kuroshio Current brings many Indo-Pacific Region species north to Japan, occasionally as far as the island of Hokkaido. Hong Kong marks the southern limit of many temperate species. The fishes of the coast of China (up to the Yellow Sea), of Formosa, and of southern Japan, however, are predominantly tropical. For example, only 10% of the 924 species known from the southern Japanese island of Shikoku are also found in northern Japan, although another 29% are more or less endemic to the area and many of the tropical species do not occur on a regular basis (Kamohara 1964). The Yellow Sea, although at approximately the same latitude as Shikoku, contains a more characteristically cool-water fauna, because the sea is not heavily influenced by the Kuroshio Current. Thus, the main resident fishes are approximately 40 to 50 species of typical northern Pacific fishes (Cottidae, Hexagrammidae, Pleuronectidae, etc.), although at least 174 tropical species make sporadic appearances (Briggs 1974). Farther to the north, the fauna of the Okhotsk Sea is very similar to that of Canada and Alaska in the families and genera that are present (Schmidt 1950). Only a few of the 300 or so species in the Okhotsk Sea are shared with the North American Pacific Region. The entire northern Pacific coast of Asia is also considered to have been an important center for salmonid evolution.

To the north, the fauna of this region merges with that of the Arctic Region in the Bering Sea. Presumably, the Bering Sea serves as a partial barrier to the exchange of inshore species between North America and Asia because of its essentially arctic temperature regime, although it has also been dry when sea levels were lower during the Pleistocene.

26.4 ARCTIC REGION

The *Arctic Region* includes all of the Arctic Ocean, the waters around Greenland, and the Bering Sea. Its fishes are discussed in some detail in Chapter 36 and only briefly described here. Probably fewer than 110 species occur in the Arctic on a regular basis, most of them cold-tolerant species that also occur in the northern Atlantic or Pacific and in such families as the Cottidae, Cyclopteridae, Zoarcidae, and Pleuronectidae. Approximately half the Arctic species are largely confined to the Arctic Region, and most of these species are circumpolar, providing connecting links to the faunas of the northern Atlantic and northern Pacific.

26.5 SOUTH TEMPERATE REGIONS

Only three South Temperate Regions are considered here: (1) the South American, (2) the South African, and (3) the Australian. Briggs (1974) and others divide these regions into provinces, mostly on the basis of differences in the faunas of the coasts that border on different oceans and Antarctic waters. Despite differences in the faunas of the various coasts of the three continents, however, a great deal of faunal interconnectedness is found on each continent, thereby making it possible to treat each as a unit. Also, many similarities in the faunas of the three continents are found, presumably the result of their connections by ocean currents and their ancient existence as just one continent.

The *South American Region* extends from the coast of Peru around the tip of South America to about Rio de Janeiro on the Brazilian coast. The northern ends of this region merge gradually with the tropical faunas of their respective coasts, whereas the fauna of the tip of South America shows strong affinities with that of the Antarctic. The strong prevailing winds along the west coast are responsible for the far-northward extent of this region along the coast; they also create an immense upwelling along the coasts of Peru and Chile, which once made the anchoveta (*Engraulis ringens*) the most abundant harvestable fish in the world (see Chapter 34). Although many purely tropical fish families are represented along the west coast of South America, especially in the more northern areas, families that are typical of the southern areas of the North Temperate Regions are also abundant here: Engraulidae (anchovies), Clupeidae (herrings), Serranidae (sea basses), Carangidae (jacks), Pomadasyidae (grunts), Sciaenidae (croakers), and Scorpaenidae (rockfishes). On the more southern portions of the coast, families such as the Gadidae (cods), Zoarcidae (eelpouts), Bovichthyidae, and Nototheniidae (plunderfishes) become well represented. A few small, peculiar families, such as the Normanichthyidae (with a single species apparently related to the Cottidae), the Aplodactylidae (otherwise found only in the Australian Region), the Psychrolutidae (cottid-like fishes with species in the North Temperate Regions as well as the South African Region), Cheilodactylidae (found in the three South Temperate Regions and in the Asian Pacific Region), Latridae (otherwise known only from the Australian Region), and Congiopodidae (scorpaeniform fishes confined to these three regions and the Antarctic), are also found. The fish fauna of the tip of South America contains many notothenioid fishes, many of them endemic, and so bears a great resemblance to that of the Antarctic Region.

The ***South African Region,*** because of the combination of its more northward location and the presence of warm currents (the South Equatorial Current on the Pacific side, the Agulhas Current around the Cape of Good Hope, and the Benguela Current on the Atlantic Coast), lacks most of the cold-water fish families that are characteristic of the South American Region, such as the Nototheniidae, Bovichthyidae (except one island species), Cyclopteridae, and Zoarcidae (Smith and Heemstra 1986). Species in the latter two families, however, do occur in deep water off the South African coast. The region contains more than 2,200 species, from more than 80% of the world's fish families. Most of the region's fishes belong to widespread tropical families, although a large number (13% of the fauna) are endemic species, especially in the families Clinidae ("klipfishes," the most species-rich inshore family), Gobiidae, Blenniidae, Gobiesocidae, and Batrachoididae (Smith and Heemstra 1986). This region shares with the Australian Region a number of distinct families, such as the Pentacerotidae, Oplegnathidae (also found off Japan, Hawaii, and the Galapagos Islands), Congiopodidae (racehorses), and Callorhynchidae (plownose chimaeras). The latter two families are also found in South America. Many of the species and genera are endemic to the region, however, reflecting its long isolation.

The ***Australian Region*** is inhabited by a temperate fish fauna (1,500 species) with a strong tropical element. This is because warm currents from the Indian and Pacific oceans moderate the water temperatures along the southern half of Australia, from Shark Bay on the west to about Hervy Bay on the east. Around Tasmania and the southern tip of Australia, as well as around much of New Zealand, the tropical element tends to drop out, and a more cold-water element comes in. Southern Australia, New Zealand, and small islands around New Zealand constitute the Australian region, although New Zealand, with more than 1,000 species (11% endemic) could easily be treated as a separate region. New Zealand has 80 species (in 40 families) that otherwise are found only along the Australian coast.

Along the coast of southern Australia are approximately 600 species, most of which (85%) are endemic to the region (Mooi and Gill 2002). Scott (1962) found that 101 of the 253 inshore species were characteristic of the entire Australian region (and that many lived in the Indo-Pacific Tropical Region as well). Another 114 species were endemic to the region between Shark Bay and Tasmania, although 72 of these were not found in the colder waters around Tasmania. Thirty-eight species have a break in their distribution created by the cold water around the tip of Australia. Apparently, this break has also resulted in the evolution of several pairs of species, one on each side of the cold-water area. Tasmania supports the wide-ranging species but also has a number of endemic forms, especially in the families Rajidae, Clinidae, Bovichthyidae, and Ostraciidae (Briggs 1974). Curiously, the fish fauna of Tasmania bears many resemblances to that of southern New Zealand, including the presence of the notothenioid fish family, Bovichthyidae.

26.6 ANTARCTIC REGION

The *Antarctic Region*, thanks to the stability of the Antarctic convergence, is one of the most isolated marine regions. Most of the species (≈300, including 200 inshore species) are endemic, and many (>120) belong to six families in the suborder Notothenioidei (see Chapter 36). The remaining inshore fishes are divided among five families, one of which (Muraenolepididae) is an Antarctic family with representation on the tip of South America and two of which, the Liparidae (≈40 species) and the Zoarcidae (≈25 species), are widely distributed cold-water families that are also present in the Arctic. Unlike the Arctic Region, however, the Antarctic Region does not connect the faunas of the surrounding continents, and it shares few species with them. Nevertheless, the fact that the South Temperate Regions share most of their fish families, many of their genera, and even a few species is a strong indication of their past connection through the ancient continent of Gondwanaland. Presumably, the more temperate elements died out in the Antarctic Region as the continent drifted into its present position, whereas the cold-adapted notothenioid fishes (represented in the South Temperate Regions by the Bovichthyidae) evolved into many species.

26.7 PELAGIC REGIONS

The Mesopelagic Zone and the offshore portions of the Epipelagic Zone (see Chapters 34 and 35) are inhabited by fishes that are, for the most part, not bound to the continents in their distribution. Many of these species have a worldwide distribution, but others are more restricted. Because the distribution patterns of species from many groups, both vertebrate and invertebrate, that inhabit the open ocean often coincide, it is possible to divide the world ocean into pelagic regions. The boundaries between these regions are areas where oceanographic conditions and, particularly, temperature change more rapidly than elsewhere. Such regions of change, although they tend to fluctuate seasonally, often are coincident with isotherms and, thus, have latitudinal limits. For mesopelagic fishes, temperature boundaries may be less important determinants of distribution patterns than the location of oxygen minimum layers or regions of low productivity (e.g., Johnson and Glodek 1975). Even so, mesopelagic regions usually are largely coincident with epipelagic regions. Most of the regions used here are widely recognized (Parin 1968, Backus et al. 1977): Arctic, Subarctic, North Temperate, North Subtropical, Tropical, South Subtropical, South Temperate, and Antarctic (Fig. 26.2).

The ***Arctic Region*** seems to contain no true pelagic fishes, although the arctic cods *(Arctogadus, Boreogadus)* are associated with pack ice on a regular basis. Mesopelagic fishes apparently are of only irregular occurrence (Backus et al. 1977).

The ***Subarctic Region*** is, roughly, the area between the Arctic Region and the 8°C to 10°C isotherm. Its fish fauna differs considerably in the Atlantic and Pacific oceans. In the Pacific, the dominant epipelagic piscivores are Pacific salmon, with a few odd forms such as pomfret (Bramidae: *Brama japonica*), skilfish (Anoplopomatidae: *Erilepis zonifer*), and ragfish (Icosteidae: *Icosteus aenigmaticus*) joining in. When temperatures warm up a bit in the summer, many large, predatory fishes from the North Subtropical Region may move into this region. This movement is perhaps even more marked in the Atlantic than in the Pacific because of the lack of a well-developed endemic epipelagic fish fauna, although Atlantic salmon *(Salmo salar)* and pomfret *(Brama brama)* are widely distributed in the region. Backus et al. (1977) note that the mesopelagic community in this region is dominated by just one species of myctophid.

The **North Temperate Region** is bordered on the south by the 14°C to 16°C isotherm. The dominant epipelagic fishes are such fishes as tuna (Scombridae), swordfish *(Xiphias gladius)*, basking shark *(Cetorhinus maximus)*, sauries (Scomberesocidae), and opah *(Lampris guttatus)*. They are joined during the summer months by many species, especially other scombrids, from the North Subtropical Region.

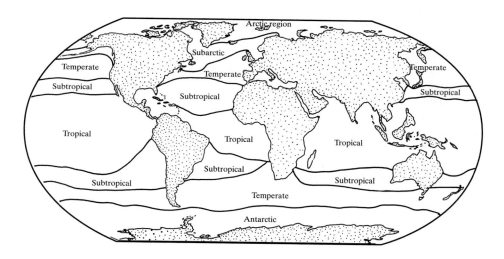

FIGURE 26.2   Major zoogeographic regions for pelagic fishes. Based on Briggs (1974) and Backus et al. (1977).

The **North Subtropical Region** and its counterpart, the *South Subtropical Region,* are bounded by the 14°C to 16°C isotherm on one side and by the 18°C to 20°C isotherm on the other. These regions are characterized by tunas, sauries, flying fish (Exocoetidae), louvar *(Luvarus imperialis)*, and marlins (*Makaira* spp.). Backus et al. (1977) note that in the Atlantic, 13 species of mesopelagic myctophids have their center of abundance in the two subtropical regions, and 11 of these species are present in the two regions but are missing from the Tropical Region in between. A similar *antitropical pattern* with many of the same species presumably exists in the Pacific as well. A unique feature of the North Subtropical Zone of the Atlantic is the Sargasso Sea, with its distinct set of fishes associated with drifting sargassum weed (see Chapter 34). During the summer, virtually all the epipelagic fishes that are characteristic of the Tropical Region occur here as well, in both subtropical regions.

The *Tropical Region* consists of a broad, worldwide belt of equatorial water that rarely drops below 20°C. Its limits in the various oceans correspond roughly to the limits of the tropical shelf regions. Much of the epipelagic fish fauna in this region has a worldwide distribution, although distinct differences exist between the those of the Atlantic and the Pacific/Indian oceans. Because the epipelagic fishes move into the subtropical regions during the summer, this region is perhaps better defined by the absence of many subtropical species of fishes than by its year-round residents. The characteristic species of the region are tropical flying fishes, scombrids, pelagic sharks, and billfishes (Xiphiidae). The mesopelagic fish fauna is most abundant and diverse in this region, especially in the Pacific and Indian oceans, although the distribution patterns of most species are not completely known.

The *South Temperate Region* is marked in the north by the 14°C to 15°C summer isotherm (subtropical convergence) and in the south by the 3°C to 5°C isotherm (Antarctic convergence). Its fish fauna is not as well studied as those of the other pelagic regions, but it shares many species and genera with the North Temperate Region. The lamnid sharks, the basking shark, and the sauries, pomfrets, and scombrids belong to the same or closely related species. This region is a relatively narrow one, so it shares many fishes seasonally with the Subtropical Region.

The *Antarctic Region* consists of the shallow waters that circle the Antarctic continent. It is remarkable for its lack of true epipelagic fishes, except for a few coast-bound forms such as the antarctic herring (Nototheniidae: *Pleurogramma antarcticum*). The mesopelagic fauna is better developed, with a number of species in the families Myctophidae, Bathylagidae, Gonstomatidae, Paralepididae, and Scopelarchidae largely confined to the region (Eastman 1993).

26.8   DEEPSEA REGIONS

In the deep sea, two ecological groups of fishes tend to be confined to zoogeographic regions: the bathypelagic fishes, and the deepsea benthic fishes (see Chapter 35). For both groups, two factors are of great importance in their distribution patterns: that the greatest number of individuals and species are located in tropical and subtropical areas; and that in all areas where they are found, their abundance tends to be greatest near the continents. By and large, bathypelagic fishes tend to be located in regions approximately equivalent to those described for epipelagic fishes. Relatively few species have world-wide distribution patterns, however, and most tend to be associated with specific ocean basins or water masses with distinct salinities, temperatures, and dissolved oxygen levels.

For benthic species, those inhabiting the upper continental slope tend to fall into zoogeographic patterns like those of fishes inhabiting the continental shelf, although the greater uniformity of temperatures at great depths makes the fishes even less respectful of zoogeographic boundaries. In deeper water, the dominant fish families are the Ophidiidae and Macrouridae, but in more polar waters, the Zoarcidae and Cyclopteridae usually are more common. Despite their commonness, however, most species in these families are confined to single ocean basins; undersea mountain ranges and trenches seem to be barriers to dispersal. A few species, however, do seem to have wide distributions. For example, the flatnose cod *(Antimora rostrata)* is found in deep water in the Atlantic, Pacific, and Indian oceans (although these populations are often listed as separate species). In the deepwater family Chlorophthalmidae are 18 species of spiderfishes *(Bathypterois)*. Most of the species have fairly restricted distributions, but one is found in the tropical and subtropical regions throughout the world, except for the northeastern Atlantic (Sulak 1977). Consequently, the distribution patterns of deepsea fishes can be related to those of the pelagic regions (using distribution patterns of families and genera) or can be divided on a basin-by-basin basis (using the distribution patterns of species).

26.9 ANTITROPICAL DISTRIBUTION PATTERNS

One of the recurring patterns in marine zoogeography is the presence of species, genera, and families that are absent from the tropics but are present in the regions on both sides. These antitropical (or bipolar) patterns have been noted for entire families, such as hagfishes (Myxinidae), eelpouts (Zoarcidae), snailfishes (Cyclopteridae), and cods (Gadidae); for genera within families, such as Scorpaenidae (rockfishes), Clupeidae (herrings), Engraulidae (anchovies), Merluccidae (hakes), Rajidae (skates), and Squalidae (dogfish sharks); and for species such as Atlantic pomfret *(Brama brama)*, flying fishes *(Hirundichthys speculiger, Exocoetus volitans)*, basking shark *(Cetorhinus maximus)*, and saury *(Scomberesox saurus)*. Almost all species with antitropical distributions are pelagic, so their distribution patterns likely are the result of movements across the tropical regions. Many of the families or genera are bottom-oriented or shallow-water forms, however, which are not likely to move over long distances as pelagic forms. Furthermore, the general lack of antitropical species of this sort indicates that the events leading to the present distribution pattern were fairly ancient.

Four theories have been advanced to explain the antitropical distribution patterns:

1. In the past, lower sea temperatures in the tropics may have permitted more continuous distributions of cool-water fishes.
2. Cool-water fishes may have moved through deep water of suitable temperatures, thus avoiding the warm surface waters.
3. Fishes with antitropical distributions may once have been present in the tropics as well but were displaced through competition with more "advanced" groups.
4. The distribution pattern may be the result of the movement of continental plates.

The first theory was proposed by Hubbs (1974) and is supported by White (1986), who provides evidence of climatic change during the Miocene that would have caused sea temperatures to rise at low latitudes, effectively eliminating cool-water species from the tropics. Evidence for the disappearance of tropical conditions in the eastern Pacific during the Pleistocene indicates that in at least some parts of the ocean, continuous areas of cool surface waters may have connected the two hemispheres for short periods of time as a result of generally cooler conditions and long stretches of cool coastal waters from upwelling. This could explain the antitropical distribution of the pelagic species, which seem not to have diverged much in the two hemispheres. It also helps to explain the distribution of genera like *Sebastes* (rockfishes), in which genetic evidence suggests that the southern hemisphere was colonized by fish from the Pacific Ocean between 80,000 and 140,000 years ago (Love et al. 2002).

The second theory may explain the patterns of some antitropical genera of the subtropical and temperate regions, but it is applied with difficulty to explain the presence of families such as the Zoarcidae and Cyclopteridae at the poles. The difficulty arises from the lack of family representatives in the deep waters of intermediate regions, because presumably, if such water was suitable for migration, it should also be suitable as a permanent place in which to live. Dramatic evidence that fish can move between regions, however, was the capture off Greenland of a Patagonian toothfish (*Dissoichthus eleginoides*) from the southern hemisphere (Moller et al. 2003).

The third theory is favored by Briggs (1974), although no evidence supports such drastic effects of competitive interactions among species that presumably coevolved (White 1986). If there were once species of eelpouts and snailfishes adapted to oceanographic conditions of the tropics, it is not unreasonable to expect relict populations in a few localities.

In regard to the fourth theory, and as indicated at the beginning of this chapter, the role of continental drift in determining the present distribution of marine fishes is poorly understood. It seems very likely, however, given the fact that families such as the Zoarcidae and Cyclopteridae are closely tied to continents, that this theory will prove to be of major importance and help to explain many of the distribution anomalies that now exist (Mooi and Gill 2002).

26.10   LESSONS FROM MARINE ZOOGEOGRAPHY

Marine fishes present in local or regional assemblages are the result of biotic dispersion, long-distance dispersal, and vicariance (Cracraft 1994). *Dispersion* is the gradual spread of a species from its center of origin as the population expands into areas with suitable conditions. In marine environments, dispersion can be fairly rapid when changes in ocean currents and temperatures allow larval fishes to settle in areas that previously had not been suitable for them; likewise, ranges of marine fish can shrink rapidly as oceanographic conditions change. *Long-distance dispersal*, rather than being a population phenomenon, is the result of a small number of individuals somehow being carried over a natural barrier to settle in distant, suitable habitat. The fish fauna of the Hawaiian Islands and St. Helena Island, for example, are largely the result of long-distance dispersal from other reef areas, presumably by long-lived larvae being carried long distances by unusual events. ***Vicariance*** implies that most elements of fish faunas develop in place, as the result of isolation caused by plate tectonics or changes in current patterns and then by local biotic processes (e.g., speciation and extinction). Recent genetic evidence indicates that most faunas are the result of all three general mechanisms working together, with colonization by new species superimposed on structures originally created by vicariance events. By and large, however, the history of local and regional faunas is still poorly understood—as this chapter reflects. What processes, for example, led to the development of 65 similar species of rockfish (*Sebastes*) along the Pacific Coast of North America and so few in other oceans? Clearly, to understand such patterns, we need a better appreciation of distributional, speciation, and extinction processes in the ocean and a better understanding of how plate tectonics has affected fish distribution. In particular, applying what is being learned about plate tectonics and seafloor spreading to the distribution of marine organisms is a major challenge for the marine zoogeographer, as is application of the growing knowledge of the mechanisms of dispersal.

SUPPLEMENTAL READINGS

Briggs 1974; Cracraft 1994; Mooi and Gill 2002; Springer 1982; White 1986.

WEB CONNECTIONS

SUEZ CANAL
www.yale.edu/environment/publications/bulletin/103pdfs/103golani.pdf

PLATE TECTONICS
www.scotese.com/earth.htm

CHAPTER 27
Introduction to Ecology

Fish ecology is one of the most dynamic areas of ichthyology. Understanding how environmental factors influence the distribution and abundance of fishes is important for managing fisheries, protecting aquatic ecosystems, and setting water-quality standards. Increasingly, fish are being studied to answer fundamental questions of interest to all ecologists, such as how complex communities or food webs are structured. These studies have contributed to the "paradigm shift" that ecology has undergone in recent decades. In a clamshell, this paradigm shift has been away from the "balance of nature" and toward "the only constant is change." Historically, biologists tended to view each ecological system as having an underlying, predictable structure to which it would bounce back if perturbed. Today, a general realization exists that the underlying structure of ecological systems is a moving target, constantly changing through time, so it becomes important to understand how internal and external forces drive the change. The problem is that change takes place not only on local and regional scales, in response to environmental fluctuations and biological events, but also on much broader scales, in response to climatic changes and evolutionary events. Today, of course, humans are often the main drivers of change, at least in the short term, and it has become increasingly important to understand how human perturbations of the biosphere interact with natural changes.

27.1 FILTERS

The dynamic nature of fish faunas can be revealed by thinking of each local assemblage of species as having passed through a series of filters to arrive at the particular assemblage observed (Smith and Powell 1971). In each marine or freshwater region, a large group of fishes could potentially be part of the local fish fauna. The first filter this broad fauna has to pass through is the response of individual species to vicariant events (see Chapter 26) or other landscape-shaping events, especially those of the Pleistocene period 10,000 or more years ago (Fig. 27.1). The Earth's biota is still recovering from the massive glaciers that covered good chunks of North America and Eurasia, accompanied by lower sea levels worldwide and climatic changes in nonglaciated areas. For example, the upper Mississippi River basin and many lakes and rivers in Canada may lack many species now present in areas that never were glaciated, simply because the species have not had the time or the opportunity to invade (or reinvade). Moyle and Herbold (1987) argue that the superficial resemblance

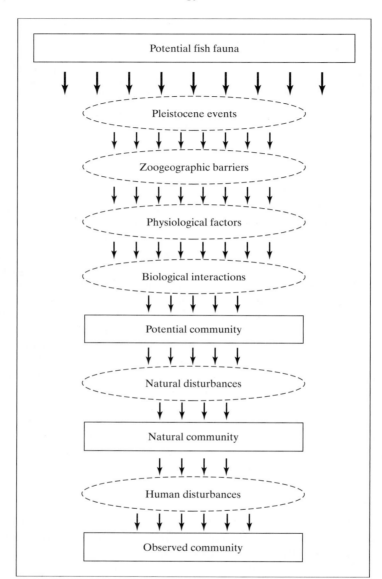

FIGURE 27.1 Filters (dashed ovals) through which fishes have passed that result in particular fish assemblages.

between the freshwater fish faunas of Europe and of western North America (e.g., dominance by large cyprinids) is the result of Pleistocene events in those regions that selected for particular life-history patterns.

The next filter is a zoogeographic filter, which sorts out fish of continental or oceanic faunas into sets of regional faunas through natural barriers to dispersal, such as mountain ranges, deserts, or ocean current systems. The importance of zoogeographic barriers in separating marine systems is indicated by the massive invasion of organisms into the Mediterranean Sea from the Red Sea after the Suez Canal was built (see Chapter 26).

Once a fish species is in a body of water, its distribution is often restricted by a physiological filter; its tolerances of such environmental factors as temperature, salinity, dissolved oxygen, and pH. A species will be "filtered out" of an area if the environment regularly exceeds its physiological limits for one or more of these factors. For example, codfish cannot move into water colder than 2°C in the North Sea, because the cold water inhibits the chloride-secreting mechanism of their gills (Cushing 1968). These abiotic filters determine the *fundamental niche* of a species, which is a concept developed by G. E. Hutchinson to describe the potential range of environmental conditions a species can inhabit in the absence of predators and competitors.

The next filter is a biotic filter, representing interactions with other species. A species can be excluded from an otherwise suitable area by competitors, predators, parasites, or lack of suitable prey. When these factors are considered, the range of conditions in which a species lives is its *realized niche*. An assemblage of fishes found in a common set of environmental conditions or habitat usually is referred to as a *fish assemblage or community*, implying that it is structured by the interlocking realized niches of the species.

The idea of predictable fish communities neatly assembled through biotic interactions has been assaulted by the development of the everything-is-change paradigm in ecology. Yet, this idea still works well for predicting that fish will divide existing resources fairly effectively, at least for short periods of time, because fish have an astonishing ability to interact with other species and to establish what appear to be highly structured communities—even when some of the species are alien species introduced by humans. In fact, the frequent ability of alien species to invade established assemblages without causing extinctions has lead to speculation that most freshwater fish communities are undersaturated with species (Moyle and Light 1996). The flexibility of different fish species in their living arrangements with each other should not be surprising, because even under natural conditions, the environment is constantly changing through time, alternately favoring one species or group of species over another. For example, Strange et al. (1992) found that the fish community of a small alpine stream fluctuated between being dominated by nonnative brown trout and by a cluster of native species. The driving force in changing the composition of this community was the timing of flood events. In the winter, these events apparently scour out the embryos of the brown trout that are incubating in the gravel. The native species are less affected by winter spates, however, because they are all spring spawners. They need high-flow conditions at that time for successful spawning. If brown trout are abundant, they suppress the native fish populations through predation, regardless of the spawning success of the natives. During extended periods that favor brown trout, some native species of fish may be locally extirpated. Thus, understanding the composition of the local fish community often requires understanding the recent local environmental history, which is another filter (Fig. 27.1).

The final filter that every community of fishes passes through is the one imposed by human activity. Humans have drastically altered most watersheds, even those that appear to be natural. In marine systems, fishing typically favors some species over others, with cascading effects throughout the ecosystem. These "filters" have increased the dynamism—and unpredictability—of local fish assemblages.

Despite the strong interaction between biotic and abiotic factors in fish ecology, the rest of this chapter focuses on the biological factors that determine the realized niches of fishes. The influences of abiotic factors on fishes are explained in the chapters on physiology (see Chapters 2–10) as well as in the subsequent chapters on ecology (see Chapters 28–36). For convenience, biotic factors can be divided into interspecific and intraspecific interactions. The interspecific interactions in turn can be divided into (1) predation, (2) competition, (3) symbiosis, and (4) parasites and pathogens.

27.2   PREDATION

With few exceptions, fish are simultaneously predator and prey—especially if humans are counted as predators (Box 27.1). Relatively few species are herbivores or detritivores, and even these typically are predators on invertebrates during their early life-history

---

BOX 27.1

Cannibalism: A Special Kind of Predation

Cannibalism is common in fish, presumably because the young typically are so small compared to the adullts and because external fertilization and high fecundities provide plenty of opportunity for it. The fact that the life-history strategies of many fishes result in spatial segregation of adults and young suggests that reducing cannibalism is important. Thus, salmon and herring have breeding grounds that are far removed from the adult feeding areas. Many marine and lake-dwelling fish have planktonic larvae that spend the most vulnerable stage of their life history in a food-rich environment that is quite different from that of the adult (see Chapter 9). One reason for cannibalism might be that it allows one individual to increase its fitness at the expense of that of a neighboring individual by consuming their young, a nearly perfect food. For example, both sticklebacks and sunfish regularly raid the nests of conspecifics to consume developing embryos. Males may raid neighboring nests, or schools of nonbreeding fish may overwhelm a guarding male to consume its young. In bluegill sunfish, some females engage in "pseudocourtship" of males to gain access to a nest and consume its contents (Dominey and Blumer 1984). Where cannibalism is important, it may have a major impact on the population of a species. Thus, in Swedish lakes, large Eurasian perch (*Perca fluviatilis*) limit the abundance of juveniles (Persson et al. 2000). When a die-off of large perch occurs, young of year perch become much more abundant in open waters of the lake, and zooplankton populations decline substantially. Curiously, this decline in a major food source may cause yearling fish to starve. These fish, originally limited by adult predation, need fairly high densities of zooplankton to grow and survive. Yearling fish that survive to a large-enough size to become cannibals have accelerated growth from that point on, reaching to large size. This may start the cycle over again!

stages. As a result of their dual predator–prey role, fishes have evolved a wide array of feeding and defense mechanisms. These range from the strong beak and spiny covering of porcupinefishes (Tetraodontidae) to the sharp teeth and sleek bodies of barracudas (Sphyraenidae) to the pharyngeal teeth and fear scents of minnows (Cyprinidae) to the fine gill rakers and schooling behavior of herrings (Clupeidae). Such adaptations are a reflection of the ***coevolution*** of predators and prey through time. Presumably, defense mechanisms evolve because they confer a reproductive advantage on the possessors, at least until their predators evolve the means to overcome the defense. One of the best indications of how prey adapt to the presence of their predators is that predators introduced into new waters often devastate populations of their prey. In New Zealand, alien brown trout either eliminated native galaxiid fishes from streams or greatly changed their behavior patterns (Townsend 1996). In a chain of Spanish lakes, introduced northern pike eliminated native fishes not adapted to their style of predation; as a result, their diet now consists mostly of an introduced crayfish, supplemented with several other introduced fishes (Elvira et al. 1996).

A major question in ecology (and fisheries biology) is whether predators control populations of their prey. The answer is often (but not always) yes. For example, Möller (1984) demonstrated that herring populations in a fjord off the Baltic Sea were regulated, in good part, by predation on their larvae by jellyfish. Understanding such relationships is important, because such predation can regulate how ecosystems function. For example, when planktivorous fishes are introduced into lakes that previously lacked them, they can depress populations of larger species of zooplankton and cause small species to become dominant (Brooks and Dodson 1965). This results in a ***trophic cascade***, in which change occurs at all levels of the ecosystem (Carpenter and Kitchill 1993). For example, the explosive invasion of alewife (*Alosa pseudoharengus*) into Lake Michigan resulted in the near disappearance of two large species of zooplankton, the decline of five other species, and the increase in 10 species of small zooplankton (Wells 1970). When Pacific salmon were introduced into the lake, they greatly reduced the alewife populations through predation; as a result, large zooplankton species once again flourished in the lake. Because these species were more efficient grazers on phytoplankton than were the smaller species that could coexist with the alewife, phytoplankton densities decreased dramatically, causing the lake to become extremely clear (Scavia et al. 1986).

In some situations, a prey population may control that of a predator. In a pond, for example, bluegill may limit the population of piscivorous largemouth bass by preempting limited nesting sites and by preying on the bass eggs in the nests. This situation is particularly likely to develop if anglers remove the largest bass, because such bass can keep bluegill numbers under control. Despite examples such as this, however, most predator populations seem to fluctuate independently of that of any single prey species, because they usually are capable of switching among prey species when the population of one gets too low.

A dramatic illustration of the importance of alternate prey is the collapse of cod populations in the Barents Sea. The cod population collapsed after gross overfishing reduced the population of herring, their favorite prey. The cod then starved, because populations of alternative prey were also low as the result of natural conditions (Hamre 1994). In most situations, however, alternate prey are available for predators, but the decision to switch prey is a complex one. Under *optimal foraging theory*, it is assumed that maximizing the intake of energy is a major factor determining this choice. When it takes less effort for a bluegill to become satiated on zooplankton than to feed on invertebrates that live among aquatic plants, the bluegill will switch to feeding on zooplankton. This foraging strategy is tempered by other factors, however, such as the increased risk of predation by largemouth bass when the bluegill feeds too far away from the cover provided by aquatic plants.

A similar problem exists for reproductive behavior. Male fish must continually balance the need to attract females with the increased risk of predation that such behavior entails (Sih 1994). The commonness of red as a spawning color in sticklebacks, salmon, minnows, and other freshwater fish presumably reflects this conflict. Red is conspicuous in shallow, well-lighted waters but is hard to see in deep cover or deep water (see Chapter 2).

27.3   COMPETITION

Competition is "the demand, typically at the same time, of more than one organism for the same resources of the environment in excess of immediate supply" (Larkin 1956, p. 327). By definition, competition should not allow two species with identical ecological requirements to coexist if resources are limited. It works as a mechanism for structuring biotic communities, because two species rarely have identical ecological requirements. Take, for example, two species that evolved in different geographic regions and whose ecological requirements are similar but not identical. If these two species are brought together, then the differences between them presumably will become emphasized through competitive interactions in such a way that the resources under dispute will be divided between them; that is, they will show *resource partitioning*. Selection pressure eventually may cause the species to diverge permanently from each other through behavioral, morphological, and physiological specializations (*character displacement*). An extreme result of this process is the many highly specialized fishes in the ancient rift lakes of Africa (see Chapter 30). Character displacement can occur very rapidly. Crowder (1984) found that after the invasion of alewife into the Great Lakes, a native cisco shifted from plankton feeding to bottom feeding, with a concomitant reduction in the number of gill rakers in the ciscos.

At the opposite end of the spectrum are the species of trout that are characteristic of streams and lakes in recently glaciated areas. These trout have a behavioral plasticity that allows them to use interference competition to segregate quickly from other species of trout as well as from other kinds of fishes with which they come in contact (see Chapter 28). Because competition is such an ephemeral phenomenon, it often is difficult to demonstrate in natural situations (Fausch 1988). This does not prevent it, however, from frequently being invoked (usually without evidence) to explain patterns of distribution and abundance of many fish species, particularly if the species are economically important and declining in number while less-valuable species are on the increase.

Evidence for competition between fish species comes from two general sources: observational, and experimental. An example of observational evidence is the study of Nilsson (1963), who found that Arctic char (*Salvelinus alpinus*) and brown trout in

Swedish lakes occupy both inshore and offshore habitats and feed on everything from zooplankton to benthic and terrestrial insects when either one of these species is the sole fish occupant of a lake. When the two species occur together, however, the trout concentrate close to shore and the bottom, where they feed on both benthic and terrestrial insects. The char, on the other hand, are found almost entirely in the open waters of the lake, where they feed on zooplankton. In aquaria, trout are more aggressive than char and consistently dominate them; in a lake, this behavior presumably keeps char out of the inshore areas. The competition in this case is for the limited and productive inshore space, and segregation results directly from the behavioral interactions of the two morphologically similar salmonids. There *are*, however, morphological differences between the species that also have some bearing on the results of their interactions. Arctic char, for example, have more and longer gill rakers than brown trout, which give them an advantage in feeding on zooplankton, whereas brown trout are more effective predators on benthic invertebrates. In experimental situations, char preferred zooplankton when by themselves, whereas brown trout preferred benthos. This suggests that the extreme segregation observed when the species are together is based as much on intrinsic differences as on interactions (Jansen et al. 2001). This may be a fairly typical situation in fishes (Matthews 1998).

Experimental demonstrations of competition require controlled manipulations of fish populations or of the limiting resource, such as food or cover. For example, two species of surfperch (Embiotocidae) occur on rocky reefs off central California, but each occupies a different depth zone. The striped surfperch (*Embiotoca laterals*) lives in food-rich shallow areas; the black surfperch (*E. jacksoni*) lives in deeper areas, where food is relatively scarce. Hixon (1980) found that black surfperch moved into shallow areas after he removed all striped surfperch from a small reef, but that striped surfperch would not move into deep water when he removed black surfperch from another reef. Further experiments showed that striped surfperch were aggressively dominant over black surfperch and forced them to live in the less productive areas. This example indicates that several behavioral mechanisms may be acting simultaneously to enable fish to avoid competition by segregating them ecologically. Some of the most important of these mechanisms include differential exploitation, aggressive behavior, predation, and habitat imprinting.

Differential exploitation of resources in a given environment (exploitation competition) occurs when one species is more efficient than its competitors at using a limited resource (Nilsson 1967). Thus, when brown bullhead (*Ictalurus nebulosus*), a bottom-adapted species, and the deep-bodied pumpkinseed (*Lepomis gibbosus*) are kept in separate aquaria, both show a strong preference for benthic insect larvae over large zooplankton. When kept together, however, the pumpkinseed switches to zooplankton, presumably because the bullhead can exploit the insects at a very rapid rate but cannot feed effectively on the zooplankton (Ivlev 1961).

Interspecific ***aggressive behavior***, especially when expressed as territoriality, may be one of the most important mechanisms that initially forces two similar species into different habitats or different diets (interference competition). This is demonstrated by the interactions described for brown trout and arctic char and those for striped and black surfperch. This mechanism may also be important within species (Box 27.2).

BOX 27.2

Intraspecific Competition: An Important Regulatory Mechanism in Fishes

---

The interactions among members of the same species can be even stronger than the interactions among individuals of different species. Because most intraspecific interactions may alter the chances that particular combinations of genes will be passed on to the next generation, intense intraspecific competition is more the rule than the exception. Thus, the territorial behavior that causes salmon and trout species to segregate also is important within each species as a way to divide the limited space that is available for feeding and reproduction. In streams, trout that cannot find a suitable space to establish themselves must either colonize new areas or be eaten by predators. In some species, intraspecific competition may be one of the main mechanisms regulating population size. In pelagic sardines and anchovies, high populations reduce plankton densities, so growth rates of individual fish may be reduced. In turn, this reduces the ability of individual fish to produce eggs or sperm, thereby resulting in fewer young and a smaller population in the years that follow. In the absence of severe environmental fluctuations, intraspecific competition alone will produce regular population oscillations in such species.

Predation can reduce the numbers of potentially competing prey species in an area so that resources are not limiting for them. On the other hand, predators also can force species that otherwise might avoid competition into competitive situations. Thus, largemouth bass apparently can keep juvenile bluegill and pumpkinseed sunfish confined to beds of vegetation, where they compete for limited supplies of invertebrates (Mittbelach 1984). Competition sometimes may be reduced when one species preys on the eggs and larvae of a potential competitor (***intraguild predation***; Polis and Holt 1992). Thus, the rapid depletion of plankton-feeding whitefish in the Great Lakes following the alewife invasion may have been as much the result of alewife predation on whitefish eggs and larvae as of direct competition.

Habitat imprinting is a poorly understood phenomenon that may permit species, or even morphs of the same species, to coexist with a minimum of interaction. The basic idea is that young fish become imprinted on the particular type of habitat with which they are first associated; thereafter, they choose that type of habitat over others that might be equally suitable. For example, juveniles of the reef damselfish (*Dascyllus aruanus*) will selectively choose, in an experimental situation, the type of coral from which they had originally been collected (Sale 1971). It has been hypothesized that one of the factors accounting for the extraordinary diversity of cichlid fishes in the rift lakes of Africa is that because of the high degree of parental care, the juvenile cichlids become imprinted on the specific habitat of their parents, thereby permitting divergence to occur over even minor habitat differences.

Although many mechanisms exist to reduce or avoid competition, coexistence of competitors also often is possible, because the environment is so variable and resources are so "patchy" in their distribution. In general, potential competitors rarely reach population sizes at which resources become limiting. Some species actually have adopted life-history strategies that allow them to take advantage of environmental variability even though they are poor competitors. Thus, the bay blenny *(Hypsoblennius gentilis)* manages to coexist with two similar species by having larvae that disperse quickly and early in the season to settle in newly disturbed areas before they are colonized by the other species (Stephens et al. 1970).

Of course, usually no single factor is responsible for the pattern of species segregation that is observed in any association of fishes (Ross 1986). This is well illustrated in the studies by P. A. Larkin and his students of Paul Lake, British Columbia, which show how competitive and predator–prey relationships interact to produce an observed community structure. Originally fishless, Paul Lake was first planted with rainbow trout, which thrived. Later, redside shiner *(Richardsonius balteatus)* were introduced into the lake (Johannes and Larkin 1961). Before the introduction of the shiner, the trout fry lived in the shallow waters of the lake, where they fed on small amphipods associated with beds of aquatic plants. The adult trout lived mostly in the open waters of the lake, feeding on surface insects and zooplankton, a diet that limited their ability to grow larger than 40 cm. Immediately following the introduction of the shiners (apparently by anglers) and their subsequent population explosion, both shiners and trout fry were feeding mostly on amphipods. The shiners, however, were more efficient at capturing amphipods, because they could penetrate the plant beds to get them. The trout generally had to wait for the amphipods to emerge from the plants. In 15 years, the amphipod population had become so depleted by the shiners that they were no longer a major item in the diet of either shiners or trout. Instead, the two species had become segregated, with the trout fry feeding mostly on aquatic and terrestrial insects and the shiners feeding mostly on zooplankton. Because the shiners also had moved out into the open waters of the lake, they came in contact with large rainbow trout. The trout avoided competing for zooplankton with the shiners by preying on the shiners themselves. As a result of these interactions, the overall biomass of fish in the lake probably was increased, although the biomass of rainbow trout was lower than it had been previously. Adult rainbow trout, however, were able to grow to larger sizes because of the availability of shiners as prey.

27.4 SYMBIOSIS

Symbiosis means "living together," but it usually is considered to include just three special types of interactions: (1) mutualism, (2) commensalism, and (3) parasitism. Mimicry is another special type of symbiosis, but because examples in fishes primarily involve those from coral reefs, it will be discussed in Chapter 33.

Mutualism

Perhaps the most common instances of ***mutualism*** among fishes, in which two or more species form a close association for their mutual benefit, are shoals involving more than one species. In streams and lakes of eastern North America, for example, it is common to find several species of minnows shoaling together. Presumably, such behavior confers on all the species the advantages of large shoals, which the species could not achieve individually (Morse 1977).

Another common mutualistic behavior is cleaning, in which one species feeds on the external parasites and diseased tissues of other species. Both the fish that is cleaning and the fish being cleaned have special behavioral patterns that accompany the interactions. Cleaning behavior has been observed in many marine and freshwater fishes, but it has been studied most intensively in the cleaner wrasses (Labridae) of tropical reefs, which make their living largely by cleaning. Typically, a cleaner wrasse has a station on a reef, over which it hovers and displays its often brilliant color patterns. Large fish that approach the station to be cleaned assume a special relaxed posture and permit the cleaner wrasse to move over their bodies and even to enter their mouth cavities. Only rarely is a cleaner wrasse eaten. Although the advantages to both parties of cleaning behavior seem to be obvious, Losey (1978) points out that the relationship between the cleaner and the cleanee is not as clear-cut as once thought. When levels of parasitic infections are high, the fish being cleaned may gain from the cleaning, but their level of infection usually is low whether or not cleaner fish are available. Often, the main food of cleaner fishes is mucus, pieces of fin, and other healthy tissues, as well as the eggs of reef fishes, thus making the cleaners more parasites or predators than mutualists. It appears that a fish being cleaned tolerates the attentions of the cleaner because it has a positive response to the tactile stimulation provided by the cleaner. This response appears to have evolved for other purposes and is taken advantage of by cleaners (Gorlick et al. 1978). In a sense, the cleaner is a vice of the cleanee! Foster (1985), however, provides evidence that wounded fish attended by cleaners have higher rates of recovery than might be expected.

Clearer examples (perhaps) of mutualism are the interactions between certain species of burrow-dwelling shrimp and small gobies (Gobiidae). The shrimp construct burrows that are inhabited by both species, so the goby gains a home by being associated with the shrimp. The shrimp also benefits, however, because the goby, with its superior vision, can warn the shrimp of approaching predators. The shrimp uses its long antennae to remain in physical contact with the goby when at the mouth of the burrow or foraging nearby. A special series of signals between goby and shrimp have developed to enhance this relationship (Preston 1978).

Commensalism

Whereas the goby–shrimp relationship is mutualistic, many other instances occur in which gobies use the burrows of invertebrates and the invertebrate gains nothing from the relationship (but also is not harmed). These are examples of *commensalism*. The remoras (Echeneidae) are a whole family of fish adapted for commensal living with large fish, particularly sharks. Each remora possesses a large sucker on top of its head by which it attaches to a host that carries the remora to new sources of food. The presence of remoras may have a negative impact on the hydrodynamics of their hosts, but remoras also may benefit the host by keeping them clear of parasites.

Parasitism

Clear examples of fishes acting as parasites (**parasitism**) are few. The pearlfishes (Carapidae) are one such example. These small, elongate fishes live in association with sea cucumbers (Holothuroidea) and actually enter the gut through the anus. Once inside, they penetrate into the body cavity of the host and feed on the gonads. The sea cucumbers usually survive the experience, although the activities of the pearlfishes may interfere with their reproduction (see Fig. 20.5). Pearlfishes come closest to the definition of a true parasite, because the adults apparently spend most of their life in just one host. A number of fishes, however, lay their eggs in the gill cavities of large invertebrates for incubation. For example, bitterling (*Rhodeus* spp.) incubate their embryos in freshwater mussels, and some snailfishes (Cyclopteridae) lay their eggs in crabs. The effects of this habit on the invertebrates are not known, but it most likely is harmful.

Most fishes that are called parasites are better labeled as ***micropredators*** (Lafferty and Kuris 2002). Micropredators, like parasites, are smaller than their hosts and usually live on or in the host for a period of time without killing it. Like predators, however, they feed on multiple victims. For example, some members of the South American catfish family Trichomycteridae enter the gill cavities of large fish and feed on gill filaments and blood. One species (*Vandellia cirrhosa*) is renowned for entering, by mistake, the urogenital openings of bathing humans—with painful results for both species. Other micropredators include the many species of fish that feed largely by removing scales from other species or by taking bites out of fins. Likewise, lampreys feed by attaching to the sides of fish and then sucking out blood and other body fluids.

27.5 PARASITES AND PATHOGENS

The effects of parasites and pathogens on individual fish, on fish populations, and on fish communities are one of the least understood areas of aquatic ecology. All fish carry parasites and pathogens, at some cost to the fish, and outbreaks with negative impacts on individuals or populations of fish can be characterized as disease. Presumably, disease is a major source of mortality of most fishes, especially at the early life-history stages. Fish that can minimize the effects of parasites and pathogens through biochemical or behavioral means ultimately will have a reproductive advantage. Just what causes a low-cost infection to become a potentially fatal disease depends on a number of interacting factors (Fig. 27.2). Anything that increases stress on individual fish, from natural droughts to pollution to the presence of competitors, can cause an outbreak of disease. For example, in 1973 a bacterial disease eliminated approximately 80% of the spawning run of steelhead rainbow trout in the Snake River of Oregon and Idaho. This disease is virulent mainly when water temperatures are warm, such as occurred in 1973, a drought year (Becker and Fugihara 1978). During most years, however, water temperatures during the time of steelhead runs are too cold for the disease to have an effect. The resident native cyprinids, which regularly experience warm temperatures in the river, are much more resistant to the disease but may act as a source of infection for the salmonids. The steelhead running up the Snake River also may be more vulnerable to disease because of the additional stress they face while passing over dams and through reservoirs on their migration back to the home stream. Realization is growing that the absence of fish from some apparently suitable waters, previously attributed to zoogeographic barriers, may result, at times, from the presence of resident pathogens (Bayley and Li 1993).

Sometimes, the impacts of a disease can be considerable, leading to massive die-offs of fish. Wurtsbaugh and Tapia (1988) observed a major kill in Lake Titicaca as the

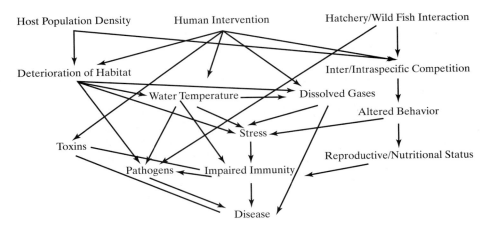

FIGURE 27.2   How factors interact to cause a disease outbreak in a wild population of fish. Pathways are both natural and human-related, and they often intersect. Thus, hatchery-raised salmon may introduce a pathogen into a population of wild salmon that has reduced immunity as the result of stress caused by an unfavorable habitat change (e.g., construction of a dam). The result is a disease outbreak in the wild salmon population. From Hedrick (1998).

result of an infestation of a parasitic protozoan. One of the most bizarre (and newly discovered) disease organisms that produces massive fish kills in shallow marine waters is a dinoflagellate (*Pfiesteria piscicida*) labeled as a "toxic ambush-predator" (Burkholder et al. 1995). When large numbers of fish are present in a confined situation (e.g., menhaden in shallow bays in North Carolina), the secretions from the fish stimulate the emergence of free-swiming zoospores from less-mobile, nontoxic forms of the dinoflagellate. Continued stimulation by the presence of fish causes the zoospores to secrete a neurotoxin, which then drugs the fish and causes them to slough off their skin and develop bleeding lesions. The dinoflagellates then consume the blood and bits of tissue in the water, and the affected fish die. Kills caused by this organism are common, and past kills attributed to other causes may have been caused by it as well (Burkholder et al. 1995). Such kills also could be regarded as a natural mechanism for the population regulation of superabundant fishes, but the frequency of the kills apparently is increased by organic pollution from hog farms and other types of agriculture (Magnien 2001).

Despite examples like these, parasites (and pathogens) usually are under evolutionary constraints not to kill off their hosts, because an extinct host population also means an extinct parasite population. They may, however, just depend, in an evolutionary sense, on the natural variability in the host's defensive strategies to keep the host available. For example, female threespine sticklebacks discriminate against males whose courtship dance is impaired by a parasitic infection, thereby suggesting a mechanism for strong selection of parasite resistance (Bronseth and Folstad 1997). Not all parasites seek to keep their hosts alive, however. In fact, parasites with multistage life cycles may go to some lengths to make sure their host is killed. Some tapeworms, for example, may change the behavior of threespine sticklebacks to ensure that the host will be eaten by a predatory bird, which is host to the next stage of the parasite's life cycle (Milinski 1985).

In some Alaskan lakes, a tapeworm (*Schistocephalus solidus*) causes infected stickle-backs to turn white and then increases their buoyancy so that they bounce along at the surface, becoming highly visible prey for a passing bird (Lobue and Bell 1993). Only a small percentage of the sticklebacks are infected so severely.

All fish carry parasites, but some carry more than others. This is partly related to habitat. Wilson et al. (1996) found that bluegill inhabiting the inshore waters of lakes contained more parasites than those living a more pelagic lifestyle. Presumably, this is because the inshore fish are always around snails, which are intermediate hosts for many of the parasites. Likewise, male sticklebacks typically carry heavier parasite loads than females, apparently because their nest-building behavior puts them in closer contact with snails (Reimchen and Nosil 2001). Females reduce parasitic attacks by remaining in the water column and feeding more on zooplankton. On a broader scale, pelagic fish in the swifter waters of African rivers tend to have lower parasite loads than other fishes, although the single biggest predictor of number of parasites is fish size: The bigger the fish, the more kinds of parasites (Guegan et al. 1992).

In a way, each fish can be regarded as the habitat for a community of parasites. The composition of this community is the result of many interacting, often stochastic factors, such as the physiological tolerances of both host and parasite, the presence of secondary hosts for parasites with complex life cycles, the ability of the host fish to resist or avoid infection, and the geographic distribution of the host. Thus, Simkova et al. (2001) cataloged the parasite communities of 39 species of fish from Central Europe but found it difficult to find the factors that predicted the richness of parasitic species in the fishes.

From a more selfish perspective, one of the most important fish–parasite relationships is the ability of some fishes to control the intermediate hosts of human parasites, thereby reducing prevalence of diseases. Thus, mosquitofish (*Gambusia affinis, G. holbrooki*) have been distributed worldwide to control mosquitoes, largely to reduce the incidence of malaria and other mosquitoborne parasitic diseases. In Lake Malawi in eastern Africa, the incidence of schistosomiasis seems to have increased after fishing reduced the abundance of mollusk-eating cichlids (Stauffer et al. 1997). Schistosomiasis is a parasitic disease for which snails are an intermediate host, so the control of snails by fish is the one of easiest ways to control the spread of the disease.

27.6 LESSONS FROM FISH ECOLOGY

Our views of how fish fit into their environment and about what regulates their numbers have undergone dramatic changes in recent decades. We now realize that populations and communities are much more dynamic than once thought. They change in response to environmental variability and in response to other species. In particular, the role of parasites and pathogens in ecological systems likely has been greatly underappreciated. Unfortunately, disease may become a more common factor in regulating fish abundance and diversity, because fish are increasingly under stress from changes that humans are causing to aquatic environments. Still, developing an understanding of the incredible complexity of aquatic systems containing fish is a wonderful challenge for anyone interested in ecology.

SUPPLEMENTAL READINGS

Diana 1995; Fausch 1988; Grossman et al. 1982; Matthews 1998; Ross 1986; Wooton 1990.

WEB CONNECTIONS

GENERAL
www.calacademy.org/research/ichthyology
www.fishbase.org
www2.biology.ualberta/jackson.hp/IWR
www.flmnh.ufl.edu/fish/default.htm
www.people.clemson.edu/~jwfoltz/wfb418
See also Web Connections for habitats, Chapters 28–36.

CHAPTER 28
Temperate Streams

Temperate streams are among the better-understood aquatic environments, both because of their accessibility and because they support important fisheries. Unfortunately for the fish, however, these streams also are among the most manipulated of aquatic habitats and are dammed, diverted, polluted, channelized, or otherwise altered for the supposed good of humankind. Despite the radical changes in the amount and quality of water flowing through temperate streams, so far few extinctions of temperate-stream fish species have occurred, although local faunal depletions are common and many species are very depleted. The ability of temperate-stream fishes to persist is the result of adaptations for living in an environment that naturally fluctuates—often considerably—on both a daily and a seasonal basis. Climatic changes also have created long-term environmental fluctuations. Thus, most temperate-stream fishes are able to maintain populations under a wide range of physical and chemical conditions. They are also capable of successfully interacting with many other fish species and are quick to colonize new areas of suitable habitat. Usually, however, each species tends to be found in a rather predictable set of conditions, as part of a distinct faunal assemblage. The purpose of this chapter, therefore, is to describe how environmental factors affect the distribution and abundance of stream fish and how interactions among these fishes affect the composition of fish assemblages over broad geographic areas as well as in specific habitats. Understanding the interactions among stream fishes and their environment is key to developing conservation strategies for stream ecosystems.

It is convenient to divide the factors that affect the distribution and abundance of stream fishes into four categories: physical, chemical, biological, and zoogeographic. Naturally, the distribution pattern of any one species is caused by an interaction of factors from all four categories.

28.1 PHYSICAL FACTORS

In the study of stream fish, four physical factors have been most useful in predicting patterns of distribution and abundance: (1) temperature regime, (2) gradient, (3) stream order, and (4) flow regime. Fishery managers have long recognized the importance of **temperature** in fish distribution, and they have divided the fluvial world into warm-water streams and cold-water streams. In North America, warm-water streams have temperatures that exceed 24°C to 26°C for extended periods of time and are characterized by species such as largemouth bass, green sunfish, bullhead catfish, and a diversity of small fishes, especially cyprinids and darters (Percidae). In Europe, the dominant species in warm-water streams tend to be various cyprinids and loaches (Cobitidae). In contrast, cold-water streams seldom exceed 24°C to 26°C and are characterized by trout and sculpins in much of the northern hemisphere. Many of the "problem" streams for fishery managers are those that do not easily fit into either category and, consequently, may not provide good fishing for any of the standard game fish. For example, when flows of a cold-water stream are reduced by a dam, suckers (Catostomidae) and minnows (Cyprinidae) may invade further upstream, even if the water temperatures are still suitable for trout. Then, a slight upward shift in water temperature may make conditions more suitable for the suckers and minnows, thereby resulting in a sudden increase in their abundance. Even in the face of presumed competition from such fishes, trout could maintain the upper hand as long as the water temperatures remain fairly cold and the fisheries do not eliminate large, aggressive individuals. At low temperatures, trout have comparatively high standard metabolic rates and are more active; therefore, they are able to use food resources more effectively than most other species. They are also effective predators on the young of other species, especially juvenile suckers.

Even in streams that fall clearly into the warm-water or cold-water category, considerable differences are found in the temperature tolerances and preferences of the species that are present. Brown trout, at least in Colorado, usually are found in streams that exceed 13°C for extended periods of time (Vincent and Miller 1969), whereas brook trout rarely are successful in such waters, especially when other species of trout are present. On the other hand, some warm-water species, such as the Rio Grande perch (*Cichlasoma cyanoguttatum*), cannot survive temperatures lower than 14°C to 18°C (Deacon and Minckley 1974). In some species, the temperature tolerances at different life-history stages may be different. In speckled dace (*Rhinichthys osculus*), the young can survive in water that is 2°C warmer than the maximum temperature adults can tolerate, which may permit them to survive periods of extremely low water in their streams (John 1964).

Whereas temperature is of great importance in determining the broad distribution patterns of stream fishes, ***gradient*** (the number of meters of drop per kilometer of stream) often is of equal importance. Gradient has a profound influence on the water velocity and substrates and on the number, size, and depth of pools. Where gradients are high, the streams tend to have little slow water. Their bottoms are predominantly bedrock, boulders, and cobbles, and there are few deep pools. More often than not, water temperatures also are cool, the water is saturated with oxygen, and the dominant fishes are trout and sculpins. In the Himalayas, however, the dominant native fishes are trout-like cyprinids and hillstream loaches (Homalopteridae), which are flattened fishes with enormous suckers for clinging to rocks.

At the opposite end of the spectrum are the sluggish, muddy-bottomed reaches of low-gradient streams with a fish fauna characterized by deep-bodied forms. In big rivers, however, the water velocities often are high despite low gradients. In general, as the gradient increases, the diversity of habitat for fish decreases, and fewer species of fish are present. Fish that feed on insects drifting in the water have an optimal range of velocity conditions for feeding associated with each species or size class within a species (Hill and Grossman 1993). Thus, streams with moderate gradients likely have the highest number of drift-feeding species. The importance of habitat diversity is demonstrated by the effects of stream *channelization*. In a channelized section of a stream, not only is the gradient increased (because the stream drops the same amount but over a shorter distance in the straightened channel) but the habitat variety (e.g., large pools) is reduced. As a result, the species, numbers, and biomass of fish are reduced as well. Thus, when sections of a mountain stream in California were channelized, speckled dace and sculpin became dominant, whereas populations of two species of trout decreased greatly— and the Modoc sucker (*Catostomus microps*, an endangered species) nearly disappeared altogether (Moyle 2002).

Stream gradient and water temperature both contribute to the impact of the third factor, *stream order*, which is a means of classifying streams according to their pattern of branching. The headwater streams are first-order streams and unite to form second-order streams, which in turn unite to form third-order streams, and so on, until the main river is reached. Essentially, stream order is a simplified way of presenting a complex of interacting physical factors. In most systems, first-order streams often are too high in gradient and too intermittent in flow to support fish. Where they do support fish, the streams are small, swift, and often dominated by boulders and coarse substrates, although in forested areas, they may contain large amounts of woody debris. As the stream order increases, habitat diversity, stream size, turbidity, and temperatures usually increase as well, and gradient and environmental fluctuations usually decrease. As a consequence, the number of species tends to increase with stream order. Thus, in eastern Kentucky, Kuehne (1962) found that first-, second-, third-, and fourth-order streams contained 1, 12, 20, and 27 species of fish, respectively. In most instances, as stream order increases, species are progressively added, and few drop out. The ones that do drop out are either fishes that are adapted to cold water, such as brook trout and sculpins, or species that are adapted for surviving in intermittent headwater streams, such as fathead minnow (*Pimephales promelas*) or green sunfish (*Lepomis cyanellus*). The latter species are capable of surviving under harsh physical and chemical conditions, but they tend to be eliminated from complex fish communities by predation or competition (although they may hang on in edge habitats). In some systems, the number of species actually decreases in the highest-order streams, usually because of pollution, water removal by humans, or floods.

As might be expected, the trophic structure (food webs) of the fish community in a stream system changes with stream order. In first-order streams, the dominant fishes usually feed largely either on insects that drop into the water from the overhanging vegetation (e.g., brook trout and creek chub) or on detritus (e.g., fathead minnow). In the downstream progression toward higher-order streams, predators on aquatic insects are added first, then piscivores, herbivores, and other specialists (Table 28.1).

One factor that greatly modifies the usefulness of the preceding three factors as predictors of distribution patterns is fluctuation in *flow*, either annually or over a longer period. Severe floods or extremely low flows eliminate or reduce populations of some fishes from sections of a stream where they would be expected to occur, whereas long periods of moderate fluctuations in flow may allow species that normally are found only in high-order streams to invade those of lower orders. The actual extirpation of species from an area by a flood is relatively rare, because most adult fishes that are native to a stream system seem to be able to find refuge during floods. Thus, a tremendous flood in the Salt River of Arizona virtually eliminated the alien fishes, but the native suckers survived (Deacon and Minckley 1974). Floods also may modify fish populations by washing

TABLE 28.1 OCCURRENCE OF FISH SPECIES IN FIRST-, SECOND-, AND THIRD-ORDER PORTIONS OF CLEMONS FORK, KENTUCKY

Species	Food[a]	Order First	Order Second	Order Third
Semotilus atromaculatus	T	X	X	X
Campostoma anomalum	A		X	X
Etheostoma sagitta	I		X	X
Etheostoma nigrum	I		X	X
Etheostoma flabellare	I		X	X
Etheostoma caeruleum	I		X	X
Hypentelium nigricans	I		X	X
Catostomus commersoni	A		X	X
Ericymba buccata	I			X
Lythrurus ardens	T			X
Luxilus chrysocephalus	T			X
Pimephales notatus	D			X
Ambloplites rupestris	V			X
Lepomis megalotis	V			X
Micropterus dolomieui	V			X

After Lotrich (1973).

[a] The principal food type of each species is indicated as follows: T—terrestrial insects; A—algae; I—aquatic invertebrates; D—detritus; V—vertebrates, principally fish.

away developing embryos or larvae. This is a common cause of missing year classes in stream fishes. Although floods have many negative effects on stream fishes, many species actually require them for reproduction, because they spawn on flooded vegetation and use flooded areas as nursery grounds for their young (Starrett 1951). Indeed, the loss of flood plains in large river systems is a major factor in the decline of riverine fishes, because the areas of forest and meadow that once flooded annually were major spawning and nursery areas for the fish as well as major sources of energy and nutrients for riverine ecosystems (Bayley 1995).

Fish also may be eliminated from streams by low flows, whether caused by drought or by human diversion of water. Smith (1971) lists 12 species whose ranges in Illinois were greatly reduced by long-term droughts. Even if a stream does not dry up completely, fishes may be eliminated by their inability to tolerate high temperatures, low oxygen levels, or even heavy growth of aquatic plants created by the low water. When more favorable conditions return, the "normal" fish fauna returns gradually. The order of appearance of the species depends on their relative abilities to colonize new waters or to withstand extreme conditions (Larimore et al. 1959). Examples of rapid colonizers are red shiner (*Cyprinella lutrensis*), fathead minnow, Rio Grande killifish (*Fundulus zebrinus*), western mosquitofish, and green sunfish. In some desert streams in Arizona, the dominant species may change depending on the amount of water that is flowing. During wet years, speckled dace and other species flourish; during dry years, longfin dace (*Agosia chrysogaster*) may be the only fish in abundance (Deacon and Minckley 1974). In general, the more variable the flows of a stream, the more variable the fish populations and the lower the number of species. For example, fish assemblages in the streams of Panama, with fairly constant flows, are richer and much more predictable than fish assemblages in the streams of Minnesota or Illinois that are of similar size but have much more variable flows (Angermeier and Schlosser 1989).

28.2   CHEMICAL FACTORS

Because they flow over varied substrates, streams tend to be well buffered chemically and also well oxygenated. Nevertheless, broad relationships occur between fish distributions and water chemistry. In Arkansas, for example, distinct clusters of species are found along a gradient of water-quality variables, such as total dissolved solids, hardness, conductivity, and dissolved oxygen, even though these variables are associated with different physical habitats as well (Matthews et al. 1992). Thus, fish found in streams in the Ozark uplands are associated with high dissolved-oxygen levels, high alkalinities, low conductivities, and other factors that typify "clean" water. Under more extreme conditions, the ability of chemical factors to limit fish distribution becomes more apparent. Thus, acid waters draining mines and road beds are a major limiting factor of stream fish in parts of the southeastern United States (e.g., Huckabee et al. 1975); these waters affect fish through a combination of low pH and heavy metal and sulfide poisoning. In Pennsylvania, sculpins (Cottidae) are absent from streams that are subject to episodic acidification caused by acid rain (Carline et al. 1994). In Oklahoma a definite group of fishes is associated with moderately saline streams. This group increases in abundance in streams that are polluted by brine from oil-drilling operations (Stevenson et al. 1974). Many of the chemicals that may limit fish distribution in both acid and saline waters are present in sewage (e.g., heavy metals and chlorine), and as a result, many fish species may be absent from the immediate vicinity of sewage outfalls (Tsai 1973). Equally important, however, is the depletion of oxygen in the water caused by the decay of organic matter in sewage. This favors species such as common carp, which are tolerant of low oxygen levels.

28.3   BIOLOGICAL FACTORS

Temperate stream fish are, for the most part, members of complex communities (e.g., Box 28.1). Yet, because the physical environment fluctuates so much, the particular set of organisms with which any species interacts likely varies considerably from place to place, from year to year, and even from season to season. As a result, most species are quite flexible in their interactions with other organisms, particularly other fishes, although the morphological, physiological, and behavioral characteristics of each species place definite limits on these interactions (Matthews 1998). The ecology of species of minnows (Cyprinidae) in North American streams, for example, can be predicted, to a certain extent, from their morphology (Douglas and Matthews 1992). Morphology, behavior, and other traits in fishes reflect adaptations to the hydraulic and geomorphic environment of the streams. As a result, fishes living in similar zones of temperate streams in different parts of the world may show a convergence in biological traits despite different evolutionary origins (Lomouroux et al. 2002). Often, however,

---

BOX 28.1

The Stoneroller: Mover and Shaker in Streams of the Eastern United States

---

The stoneroller (*Campostoma anomalum*) is an abundant and widely distributed minnow (Cyprinidae) that Matthews (1998) considers to be a **keystone species** in small streams of the Mississippi River watershed. A keystone species is one that significantly changes the nature of the ecosystem it inhabits by its interactions with other member of the biota. Stonerollers live by grazing on attached algae, and they grow to approximately 15 cm in total length. At times, thousands can be observed scraping the algae from rocks on the bottoms of stream pools. This grazing not only greatly reduces the amount of algae, but it also changes the kinds of algae that coat the rocks (to species that are more difficult to scrape). This was nicely illustrated by studies in Brier Creek, Oklahoma, where stonerollers are a favorite prey of largemouth bass. In pools that are large enough to contain large bass, stonerollers often are rare; as a consequence, algae carpet the rocks much more thickly than in pools without bass (Power and Matthews 1983). Matthews (1998) provides a long list of ways in which stonerollers affect the stream ecosystems they inhabit, including changing the abundance of invertebrates through direct and indirect means. Their grazing probably increases the abundances of certain snails by making the preferred kinds of algae more available, but it reduces the abundance of crayfish through direct competition for the algae (Matthews 1998).

---

distribution of stream fishes is limited by predator–prey, competitive, and symbiotic interactions with other fish.

Predator–Prey Interactions

Predation is probably the ultimate cause of death for most stream fishes, especially during early life-history stages. Thus, an increase in complex cover in a stream, such as fallen trees, undercut banks, and boulders, increases the numbers of fish by increasing survivorship as well as the complexity of the fish community. The presence of large, woody debris (e.g., fallen trees) in the stream channel greatly increases the production of coho salmon from Pacific coast streams and is a strong argument for limits on the removal of streamside trees by logging (Fausch and Northcote 1992). The cover provided by fallen trees helps small salmon to avoid predators, such as larger fish, birds, and snakes, and provides refuge from strong currents during high-flow periods. Often, small fish seek shallow water to avoid predatory fish, but they must leave the shallow water when they become big enough to be sought after by predatory wading birds (Harvey and Stewart 1991). Likewise, the abundance of large fish, such as smallmouth bass, in a stream depends on adequate cover, such as deep pools and deeply undercut banks, which the fish can use to avoid their own predators, such as herons, otters, and humans. These same pools also are good rearing areas for the larvae of stream minnows. The presence of bass keeps out medium-sized fishes that prey on the larvae, whereas the larvae are too small to serve as prey for large bass (Harvey 1991).

Not surprisingly, when piscivorous fish are introduced into a stream, the entire biota of that stream, from algae to invertebrates to fish, can be drastically altered. For example, Lemly (1985) showed that when green sunfish were introduced into piedmont streams of North Carolina, they eliminated or greatly reduced populations of minnows that were formerly the dominant species. Despite results like these, attempts by fishery managers to increase salmon and trout production through predator-control programs on streams generally have not worked well, because many other factors operate to counterbalance the reduced predation. Control of predatory birds, such as mergansers and kingfishers, on salmon streams may cause an increase in the numbers of young salmon; however, this may result in increased competition for food among the juvenile salmon, which in turn results in the salmon going to sea at smaller sizes. This makes them more vulnerable to marine predators, and the increased predation at sea may compensate for the decreased predation in the streams.

Competitive Interactions

Most temperate-stream fishes have a considerable degree of behavioral plasticity, which allows them to interact successfully with other species in ways that minimize competition for food and space. The plasticity of each species does have distinct limits, however, which are imposed by its morphology, physiology, and total behavioral repertoire (Matthews 1998). Much of the ecological segregation observed among temperate-stream fishes results from differences in morphology. For example, a typical North American cold-water stream may contain four species of fish: (1) a trout (Salmonidae), (2) a dace (Cyprinidae), (3) a sculpin (Cottidae), and (4) a sucker (Catostomidae) (Fig. 28.1). Body shape alone indicates that the trout is a fast-swimming predator, the sucker a bottom-oriented suction feeder, the sculpin a bottom-dwelling ambusher of large invertebrates, and the dace an active, bottom-oriented browser on small organisms. Similar assemblages are found in cold-water streams in Eurasia as well, only in this case, the sucker and dace may be by a loach (Cobitidae) and another cyprinid. In warm-water streams, such as those of the eastern United States, the number of species tends to be much higher and the morphological differences among the species not to be as sharp, but such differences nevertheless can account for much of the segregation that is observed (Gatz 1979).

Although morphological differences among species often can explain habitat differences, behavioral differences that lead to differences in feeding habits and microhabitats probably are the most important mechanisms of segregation among species that regularly occur together. Such species may segregate according to where in the water column they take their food, the time of day when they feed, the type of food that they eat, and the size of food items that they take. There also may be seasonal changes in distribution that reduce competition. Some of the best examples of these rather subtle kinds of segregation occur among the many species of small, silvery minnows, referred to as "shiners," in the streams of eastern North America. Many of these species are quite similar to one another, and typically, two or more species are found together, in different

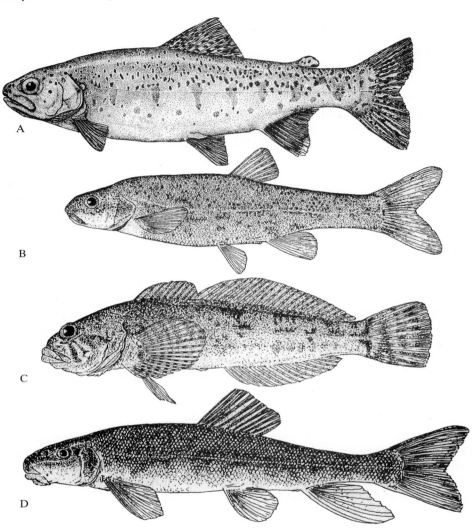

FIGURE 28.1 Typical fishes of cold-water streams of North America. **A.** Trout (*Oncorhynchus*). **B.** Dace (*Rhinichthys*). **C.** Sculpin (*Cottus*). **D.** Sucker (*Catostomus*). Drawings by Chris Mari Van Dyck (**A**) and A. M. Marciochi (**B–D**).

combinations. In a Wisconsin stream, Mendelson (1975) found that four species minimized competition for food by feeding at different places in the water column, in slightly different microhabitats (e.g., head of pool vs. bottom of pool), on different types of food (e.g., drifting vs. bottom-dwelling organisms), and on different sizes of food. The difference between any two species in any one category is not sharp, but the overall differences in all four categories result in an apparent subdivision of the food and spatial resources.

Among the most interesting aspects of temperate-stream fish ecology are the behavioral mechanisms that result in the observed segregation. It appears that the two main mechanisms are aggressive behavior and differential exploitation (see Chapter 27). Among salmonids, aggressive behavior can determine hierarchies within as well as between species. In streams of the Pacific Northwest, the use of different microhabitats by similar-sized juveniles of different species often is determined by the basic hierarchy: coho salmon > rainbow trout > cutthroat trout > chinook salmon (Bayley and Li 1993). Under this arrangement, coho salmon often occupy the best habitats for feeding and cover, although the order can change if members of one species are substantially larger than those of the other species. Likewise, juvenile Atlantic salmon (*Salmo salar*) aggressively defend territories against minnows and suckers as well as other salmon (Symons 1976). Such aggressive interactions, although common in salmonids, are less frequent in other families of stream fishes, which suggests that differential exploitation is the main mechanism that segregates similar species in streams. Presumably, the many species of shiners and darters in the species-rich streams of the southeastern United States have broader feeding habits and occupy a wider range of microhabitats when by themselves than when they occur with similar species. Just how each species specializes in the presence of another depends on slight differences in body morphology (e.g., subterminal vs. terminal mouths) that give each species advantages in exploiting slightly different parts of the environment.

Despite many demonstrations of ecological segregation among temperate-stream fishes, the frequent observation of large overlaps in the use of food and space indicates that such resources may not always be limiting fish populations. This is especially true in the complex fish assemblages of warm-water streams in eastern North America, particularly those that are heavily disturbed by humans. In such cases, unpredictable fluctuations in the environment may reduce the ability of each species to reproduce successfully every year or of their young to survive the first few months of life. Because such factors can affect each species differently, no one species can maintain dominance through competition indefinitely (Grossman et al. 1982; Schlosser 1982). Taking this a step further, studies of streams in areas as diverse as North Carolina and France have shown little evidence for interactions among the fishes in structuring the communities because of the combined effects of environmental variability and the morphological, behavioral, and physiological differences among the species (Grossman et al. 1998; Oberdorf et al. 1998).

Symbiotic Interactions

Clear examples of parasitism, commensalism, or mutualism affecting the distribution and abundance of stream fishes are hard to find. In the western United States, infestations of exotic parasites, brought in with exotic fishes, have been given some blame for the decline in native fish faunas (e.g., Vanicek and Kramer 1969). Under more natural conditions, stream fish that are confined to narrow habitats, such as riffles, may have fewer parasites than those that are found in broader or quieter habitats (Guegan et al. 1992). This may explain why the parasite load of fish sometimes increases when the stream flow is reduced (Heckmann et al. 1986). One of the most interesting fish parasites are the **glochidia** larvae of freshwater mussels of the family

Unionidae. These larvae attach to the gills of fish, and the fish then serve as both a source of nutrition and of dispersal. The larvae often are fairly host specific, and major declines of mussel populations in eastern North America are at least partly the result of declines of their fish hosts. The opposite relationship exists between European bitterlings (*Rhodeus*) and freshwater mussels, where the bitterling embryos develop in the gills of the mussels. At the mutualistic end of the symbiotic spectrum, some evidence suggests that the multispecies shoals of minnows that frequently are encountered in streams may give protection from predators for all species while still permitting members to retain their feeding specializations (Mendelson 1975).

Intraspecific Interactions

In streams, young fish commonly are segregated ecologically from the adults. Many stream fishes make spawning migrations up small tributary streams, where the young may spend the first year or so of their lives without having to compete with adults for food and space. In situations where juveniles and adults occur together, microhabitat segregation or segregation by feeding habits frequently is found (Moyle and Vondracek 1985). Gee and Northcote (1963) found that the young of two species of dace (*Rhinichthys*) shoaled together in shallow water, but as they matured, the two species moved into different habitats, one into pools and one into riffles. Among trout, which consume a wide variety of organisms, the segregation among different-sized individuals of the same species is based largely on the size of food items. Juvenile trout typically eat large numbers of small fly and mayfly larvae, whereas adults eat large stoneflies and caddisflies. Presumably, this difference in food is why large trout typically tolerate small trout within their dominance hierarchies, even though a main function of the hierarchies is to reduce competition for food among members of the same species.

28.4   ZOOGEOGRAPHIC FACTORS

One factor complicating any study of a local stream fauna is the presence of zoogeographic barriers that prevent fish from moving into habitats they currently do not occupy even though the habitats are suitable for them. Sections of stream with different temperatures, salinities, or flow may act as barriers to some specialized species, but zoogeographic barriers usually are waterfalls (or at least high-gradient areas). For example, St. Anthony Falls in Minnesota has acted as a major barrier for fishes to the upper Mississippi River (Eddy et al. 1963). Likewise, Kanawha Falls in West Virginia, although only 7.3 m high, has been a major factor separating the fishes of the upper and lower Kanawha River. The upper river contains 49 species, including six endemic forms; the river below the falls contains more than 92 native species. Perhaps because of the "undersaturated" fauna (but also because of reservoirs and other habitat changes), the upper river now supports an additional 37 introduced species, whereas the lower river supports only approximately 12 introduced species (Hocutt et al. 1986).

28.5 ZONATION

Despite the fact that distribution patterns of fish species largely are determined by interactions between the fish and various physical and chemical factors, most regions have groups of species that all respond to these factors in a similar fashion. Species tend to occur together in particular stream environments, thereby forming recognizable associations. Because such associations of species tend to replace one another as the stream environment changes from high-gradient headwaters to low-gradient rivers on the valley floor, they and the physical environment that they typify are together called *fish zones*. Generally, the species found together in a fish zone complement one another ecologically, thereby minimizing competition and, presumably, maximizing the utilization of the resources present. Fish zones are not sharply distinct from one another, however. Rather, they blend together as the environment changes. Indeed, many fish species may typify several zones, because frequently, downstream zones are created mainly by the addition of species without any deletions. For example, in most temperate-stream systems, the uppermost waters are dominated by trout and are referred to as the ***Trout Zone***. Typically, however, trout also occur in one or more downstream zones, usually becoming less and less abundant relative to other species as one moves downstream. Examples of fish zones can be found in many parts of the world, with varying degrees of overlap in the species making up the zones (Matthews 1998). In some streams, the downstream change in the fish fauna is so gradual that clearly recognizable fish zones cannot be distinguished. In streams where zones can be recognized, to do so greatly facilitates the descriptions of regional and local fish faunas and can even permit management of streams for assemblages of species rather than just for individual species of game fish.

An example of a region with a fish fauna that has been described as having fairly sharp fish zones is the Sacramento–San Joaquin drainage of central California (Moyle 2002), although even here, considerable overlap is found in the distribution of some species (Fig. 28.2). At highest elevations is a ***Rainbow Trout Zone***, where rainbow trout occur either by themselves or with riffle sculpin (*Cottus gulosus*). The fish in this zone extend downstream (the distance varies seasonally with the water temperatures) into a ***Foothill Fishes Zone***, in which California roach (*Lavinia symmetricus*) often occur by themselves in small streams, part of a larger assemblage in big streams, and with rainbow trout in cooler permanent streams. In the low-gradient streams and backwaters of the valley floor is a ***Valley Floor Fishes Zone***, containing a number of deep-bodied species as well as widely distributed species, such as the Sacramento pikeminnow (*Ptychocheilus grandis*) and Sacramento sucker (*Catostomus occidentalis*). Complicating the picture a bit is the presence of anadromous fishes, whose young often are part of the local fish assemblages for varying periods of time. Juveniles of fall-run chinook salmon, for example, may be present in the two lower zones for only 2 to 4 months (in spring), whereas juveniles of spring-run chinook may spend an entire year in the cool waters of the Rainbow Trout Zone. When dams constructed in the foothills block rivers flowing through the zones, the original zonation scheme often is broken up, but the zonation often is reestablished below the dam. Thus, the Rainbow Trout Zone may occur for a short distance below the dam, maintained by the clear, cold water of the tailwater discharge. In such a low-gradient situation, the water quickly warms, and a version of the Foothill Fishes

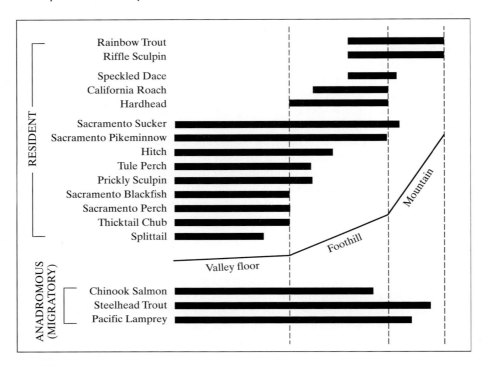

FIGURE 28.2 Generalized diagram of the historic (pre–dam construction) distribution patterns of native fishes in the Sacramento–San Joaquin drainage of California in relation to general topographic regions. On the valley floor, the native fishes have been largely replaced today by alien fishes.

Zone emerges, yielding quickly to the Valley Floor Fishes Zone. Under present conditions, most of the native fishes on the valley floor are replaced by introduced species, such as largemouth bass and bluegill (Moyle 2002).

Despite the fact that fish zones in different regions contain different species, the zones often have much in common in terms of environmental factors and fish morphology. Three general types of zones, similar to the zones just described, can be recognized in most stream systems; (1) an erosional zone, (2) an intermediate zone, and (3) a depositional zone. ***Erosional zones*** occur in high-gradient regions and are characterized by rocky bottoms and swift, usually cold water. Long riffles and small pools are the main habitat types. The fishes of the erosional zones tend to be streamlined, active forms, such as trout, and small, bottom-dwelling forms, such as sculpins and dace. In regions where gradients in headwater streams are not very high, such as the central United States, the classic cold-water erosional zone may be replaced by zones that are dominated by warm-water fishes (e.g., fathead minnow and green sunfish) that are adapted for fluctuating or extreme conditions, such as are found in the headwaters of Oklahoma streams or in streams that drain swamps.

Intermediate zones characterize the long middle reaches of tributary streams. They typically have moderate gradients, warm water, and approximately equal amounts of shallow riffles, deep and rock- or mud-bottomed pools, and runs that undercut the banks. They also merge imperceptibly with the erosional zones above and the depositional zones below, and they may be subdivided into two or more fish zones. In North America, the typical fishes of the intermediate zones are minnows, suckers, sunfishes (Centrarchidae), darters (Etheostomatinae), and madtom catfishes (*Noturus*). None of these forms is particularly deep bodied, and many specialize in particular habitats within the zone.

Depositional zones occur in the warm, turbid, and sluggish lower reaches of stream systems where the bottoms are muddy and beds of aquatic plants are common. A wide variety of fishes occur here, but most typical are deep-bodied forms, such as those adapted for bottom feeding (e.g., carpsuckers, *Carpiodes*), picking up small invertebrates from plants (e.g., sunfish), plankton feeding (shads, *Dorosoma*), or predation (centrarchid basses). The assemblages of fishes are nearly the same as those found in lakes in the regions. Where a river runs into the sea, the depositional zone will merge gradually with an estuarine zone that contains a mixture of euryhaline freshwater and saltwater fishes (see Chapter 31).

Fish zones often are descriptive tools of fish biologists as much as they are real entities. They typically merge into one another very gradually, and in low-gradient areas, one "zone" may occupy many miles of stream. In some streams, no apparent zones exist, because the distribution patterns of the fishes do not coincide enough to form easily identified associations of species. Neither is it rare for the sequence of zones described to be absent, either because the lower reaches of a stream have a higher gradient than the upper reaches (e.g., Hocutt and Stauffer 1975) or because high-gradient stretches are interspersed with low-gradient stretches (e.g., Barber and Minckley 1966).

28.6   FISH ASSEMBLAGES

The presence of fish zones indicates that distinct, recognizable assemblages of stream fishes exist. It generally is assumed that such assemblages are not just groups of fishes that co-occur because of common physiological responses to environmental conditions, but that they are coevolved assemblages that efficiently subdivide the resources that are available. Many stream assemblages clearly show such coevolved characteristics, but others are more ambiguous (Moyle and Vondracek 1985). This ambiguity may be the result of inadequate study of complex stream-fish assemblages; however, many indications also point toward environmental fluctuations that often prevent fish populations from reaching high-enough numbers so that resources are limiting and competitive segregation is necessary (Angermeier and Schlosser 1989). For example, a severe summer freshet washed nearly all of one species of darter (*Etheostoma*) out of a Minnesota stream, apparently because they were spawning in riffles at the time. Despite the absence of this once-abundant species, Coon (1982) could detect no expansion of the niches of two other darter species that were present in the stream but were not affected by the flood. Such drastic changes in the species composition of fish assemblages, as well as the lack of striking ecological segregation among similar species, seem to be common in streams of northeastern North America. Differing interpretations of such information can lead to drastically different conclusions about the nature of stream-fish assemblages

(Grossman et al. 1982, Herbold 1984, Rahel et al. 1984, Yant et al. 1984), but their dynamic nature is undisputed (Matthews 1998).

28.7   THE ROLE OF FISH IN STREAM ECOSYSTEMS

Typically, fish are important players in stream ecosystems, but they usually are not as important as fish ecologists like to think. The biomass and diversity of stream invertebrates, for example, suggests that they have a much greater role in local ecological processes than the fish, although fish predation and grazing can help to determine who the major players are (Matthews 1998). Often, impacts of fish can be modified by predation on the fish by birds or by the actions of ecosystem engineers, such as the beaver. It also is important to recognize that streams are narrow ribbons of water set in large terrestrial ecosystems, and the terrestrial ecosystems often are major sources of energy (e.g., leaves and insects) and cover (e.g., fallen trees, also known as *large woody debris*). In coastal ecosystems, anadromous fishes may provide major subsidies of energy and nutrients to both the stream and riparian ecosystems. This is especially true of Pacific salmon, which die after spawning; their carcasses serve as direct sources of food for many other creatures and as fertilizer for the riparian systems (Wilson et al. 1998). Thus, Henfield

---

BOX 28.2
Shigeru Nakano's Legacy of Stream Studies

The interdependent nature of the interactions between aquatic and riparian ecosystems, and the central role of fish in these interactions, has been brilliantly elucidated by intensive studies of Horonoai Stream, Hokkaido, Japan, by Shigeru Nakano and his colleagues (Fausch et al. 2002). The resident fishes of Hornoai Stream were four species of salmonids and a sculpin. Nakano's group first showed that production of trout and char was much higher in forested streams than in streams surrounded by grasslands because of the greater availability of terrestrial insects as high-quality food. They then demonstrated that without this subsidy of terrestrial insects, stream salmonids preyed more heavily on stream invertebrates, thereby reducing the effects of grazing by insects and causing blooms of algae on the rocks. Then, however, they realized that the birds and spiders living in the forest were also benefiting from the emergence of the adults of stream insects. Nakano and Murakami (2001) found that the emergence of aquatic insects peaked in the spring, when they were major items in the diets of the fishes and also of the riparian birds. In summer, however, aquatic insect availability was relatively low, so the fish and birds fed more on terrestrial insects. This *reciprocal subsidy* allowed both fish and birds to be more abundant than they otherwise might be (Nakano and Murakami 2001).

Tragically, Shigeru Nakano died in a boating accident in 2000 that also took the lives of four other Japanese and American ecologists. Although only 38 years old, he left more of an intellectual legacy than most scientists leave in a long lifetime of work (Fausch et al. 2002).

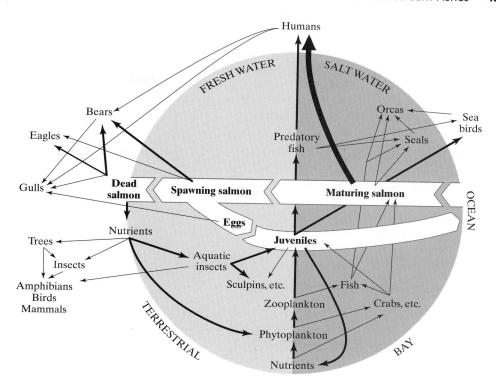

FIGURE 28.3 How spawning chum and pink salmon impact the saltwater, freshwater, and terrestrial ecosystems around a small Alaskan stream. The arrows suggest the general direction for the flow of energy and nutrients.

and Naiman (2002) found that riparian spruce trees were bigger and faster growing as a result of the uptake of salmon nutrients. Because such trees improve the quality of stream habitat for embryonic and juvenile salmon, this subsidy of the trees by salmon can be regarded as a positive-feedback loop: Salmon benefit the trees, which benefit the salmon. In fact, salmon provide a major source of energy for both aquatic and terrestrial ecosystems that benefit a wide array of plants and animals (Fig. 28.3).

28.8 LESSONS FROM STREAM FISHES

Even more than most fishes, those that live in temperate streams have lives that are regulated by environmental conditions that fluctuate daily, seasonally, annually, and longer. As a consequence, populations rarely reach densities where predation and competition are limiting factors, although this clearly happens on occasion. Nevertheless, a high degree of morphological, physiological, and behavioral diversity is found among stream fishes, indicating that interactive as well as historical processes play a role in the evolution of species and assemblages. Convergence of characters of fishes in distant regions, however, emphasizes the importance of physical factors. Each assemblage of stream fishes is the result of a unique combination of historical, zoogeographic, physical, chemical, and biological factors, and the structured systems that so frequently are observed are the result of the remarkable adaptability of most fish, both as individuals and as species. In recent years, we have seen a growing appreciation of the role of fish in stream ecosystems, including roles as a brokers of energy and as providers of nutrient subsidies to both aquatic and terrestrial ecosystems.

SUPPLEMENTAL READINGS

Allan 1995; Bayley and Li 1993; Matthews and Heins 1987; Matthews 1998.

WEB CONNECTIONS

GENERAL
www.calacademy.org/research/ichthyology
www.fishbase.org
www2.biology.ualberta/jackson.hp/IWR
www.flmnh.ufl.edu/fish/default.htm
www.people.clemson.edu/~jwfoltz/wfb418

CHAPTER 29
Temperate Lakes and Reservoirs

Most lakes are ephemeral, with ages ranging from fifty to several thousand years. They disappear as they fill with sediment, dry up in response to changes in climate, or are covered by advancing continental glaciers. They also are created rather easily by water collecting in depressions and excavations left from various natural processes, such as the scouring action of glaciers, by the blockage of rivers by landslides, or by oxbows pinched off from meandering rivers. In recent years, humans have been creating "lakes" by the thousands through impounding rivers. North America alone now has more than 1,500 reservoirs with a surface area of 200 hectares or greater as well as countless smaller ones. Reservoirs are even more ephemeral than lakes, with life spans of 100 to 200 years. In view of the impermanence of lakes and reservoirs, it is not surprising to find that comparatively few species of fish in the north temperate and Arctic regions are adapted only for life in lakes. Instead, most lake fishes are the same as those in nearby streams, especially large rivers. Many lake populations have made modest behavioral, physiological, and morphological adaptations to the lake environment, and a number of species are most abundant in lakes. Most, however, must still rely on streams for dispersal and long-term survival as species. Nevertheless, streams and lakes tend to have different subsets of the regional fish faunas, so the mouths of inflowing streams tend to have more species of fish, in unstable assemblages, than either the lake or the stream (Willis and Magnuson 2000). In very large or old lakes, distinctive fish faunas nevertheless may develop, such as the species complexes of salmonids found in the Great Lakes and other northern lakes or the unique fish fauna of Lake Baikal in Russia. Lake Baikal not only is one of the largest and deepest lakes in the world but also is one of the oldest (see Box 25.2).

Despite the paucity of fish species that are highly specialized for lake living, lakes have many special characteristics that restrict the distribution of fish both within and among them. The effects of physical, chemical, temporal, biological, and zoogeographic factors on fish result in distinctive regional fish faunas as well as partitioning by the fishes of the ecological resources of each lake.

29.1 PHYSICAL FACTORS

The principal physical factors that affect distribution and abundance of fish in lakes are temperature, light, water movements, water-level fluctuations, size (e.g., surface area and depth), and substrate.

Temperature

Fishery managers have long recognized that the basic fish fauna of lakes is determined, in large part, by temperature. Thus, they have tended to divide lakes into three basic types: (1) cold-water lakes, in which the dominant sport or commercial fishes are salmonids; (2) warm-water lakes, in which the dominant sport or commercial fishes are centrarchids (black basses and sunfishes), percids (perch and walleye), esocids (pike), and percichthyids (white, yellow, and striped basses); and (3) two-story lakes, with warm-water fishes in the warm upper layer (*epilimnion*) and cold-water fishes in the cold deep water (*hypolimnion*). Of course, many lakes do not fit nicely into any of these categories, but the system does demonstrate that in lakes, as in streams, temperature is a major determinant of distribution patterns. The presence of the two-story lake category indicates that temperature also is important in determining distribution patterns within lakes, at least during the summer months. When the thermocline forms during the spring in such lakes, separating the epilimnion from the hypolimnion, fish species quickly segregate by temperature preferences.

Even in lakes that are not easily categorized, temperature may be one of the main determinants of distribution. For example, in Lake Michigan, the native fishes show fairly narrow ranges of preferred temperatures and seem to have segregated within the lake partly on the basis of these preferences. One reason why the invasion of alewife (*Alosa pseudoharengus*) has been so disruptive to the other fish communities of this lake is that alewife occupy a broad range of temperatures and, hence, deplete zooplankton, on which the native fishes depend, both above and below the thermocline (Wells 1968).

Light

Light has many rather subtle effects on fish distribution patterns both within and between lakes. Most warm-water lakes are rather turbid, so the typical deepwater fishes are species that do not rely on vision for prey capture, such as catfish, carp, and suckers. In such lakes, most aquatic plant growth also is in shallow water. Because beds of aquatic plants increase both habitat diversity and abundance of invertebrates, fish species and numbers tend to be highest in shallow water. Pelagic plankton-feeding fishes tend to concentrate in well-lighted waters, where their prey is most visible. In lakes that are fairly clear, including most cold-water lakes, vision-oriented predators, such as lake trout (*Salvelinus namaycush*), can be found at considerable depths. One predatory species that is successful in both turbid and clear lakes is the walleye (*Stizostedion vitreum*), which has exceptionally large eyes that are equipped with a light-gathering layer, the tapetum lucidum. In turbid lakes, the walleye tends to feed during the day; in clear lakes, its feeding tends to be crepuscular (Ryder 1977).

Curiously, the clarity of lakes often is related to the abundance of plankton-feeding fish because of **top-down** control of the pelagic ecosystem structure. Where abundances of these fish are high, water clarity often is relatively low, because the fish feed on large zooplankton species, which in turn graze on the algae (phytoplankton). When zooplankton populations are reduced, phytoplankton densities are higher, so light does not penetrate as far into the lake. Thus, when a mass die-off of planktivores occurred in Lake Mendota, Wisconsin, the lake became substantially clearer (Vanni et al. 1990). In small lakes, the reduced penetration of light resulting from fish predation can substantially change other lake characteristics as well, such as decreasing the depth of the thermocline (Mazumder et al. 1990) and reducing the growth of aquatic plant beds (McQueen 1990). Despite these observations, abundance of phytoplankton and zooplankton in many lakes is tied to factors other than just top-down control by planktivorous fishes (Stein et al. 1995). In eutrophic lakes, the ecosystem structure tends to be controlled by nutrients and other factors affecting phytoplankton production, including the rooting activity of common carp (***bottom-up control***). In highly turbid lakes, the activities of visual-feeding fishes may be restricted as a consequence.

Water Movements

Steady and predictable currents, such as those in the oceans, seldom occur in natural lakes, yet the movement of water by winds can have considerable impact on fish distribution within a lake. A series of days with wind blowing from one direction can cause zooplankton to concentrate along the leeward shore. This may result in a concentration of zooplankton-feeding fishes as well. The distribution of larval fishes that drift in the surface waters of a lake may be quite patchy because of temporary gyres and currents created by winds, although these same currents also concentrate the zooplankton on which the larval fish feed. Most natural lakes have definite inlets and outlets, which create a general, if barely detectable, pattern of flow through the lake. Usually, such patterns are too slight to affect the distribution of fish within the lake. In Lake Memphremagog, straddling the border of Quebec and Vermont, however, such flow results in gradually decreasing amounts of nutrients from the inlet end of the lake to the outlet end as well as corresponding drops in the abundance and diversity of fish species (Gascon and Leggett 1977).

Water-Level Fluctuations

Water-level fluctuations are primarily a problem to fish in reservoirs and lakes with regulated water levels. In natural lakes, however, flooded shoreline marshes are important spring spawning sites for fish such as northern pike (*Esox lucius*). Later in the year, emergent vegetation serves as cover for the young of many species, such as largemouth bass and common carp. Drops in water level also may create barriers to migrations by lake-dwelling species that spawn in streams. In Pyramid Lake, Nevada, the near extinction of the cui-ui (*Chasmistes cujus*) and the extinction of cutthroat trout (*Oncorhynchus clarki*) were caused largely by reduced flows into the lake from the Truckee River, which had been diverted for irrigation. The drop in lake level exposed an extensive, shallow delta that was impassable to the spawning fish.

Size

Three aspects of lake size particularly affect both the number of species and the total fish production in lakes: (1) surface area, (2) mean depth, and (3) shoreline length. The relationship of surface area to number of species is not strong, because so many other factors, especially zoogeographic factors, can affect the number of species. In general, however, the greater the surface area, the more species likely are present (Barbour and Brown 1974).

One factor that obscures the relationship between surface area and fish numbers is the negative relationship between mean depth and species number as well as with standing crop (i.e., total fish biomass). Deep water in lakes often is cold, poorly lighted, and poorly oxygenated; hence, it supports few fish. The length of shoreline has the opposite effect of depth; usually, more shoreline means more species, in part because more shallow-water (littoral) areas are likely to exist. The littoral areas not only support more fishes, they tend to favor species that spawn and rear in shallow water, especially among aquatic plants. Thus, European reservoirs with extensive littoral zones tend to be dominated by cyprinids, whereas those with steep sides tend to be dominated by perciform fishes (Duncan and Kubecka 1995). All these size variables point toward the generality that within a zoogeographic region, as habitat variety increases, so does the number of species and the standing crop of fishes. Overall, the greatest variety of habitat suitable for fish is likely to occur in a large, shallow lake with many bays and other irregularities in its shoreline.

Substrate

It is quite difficult to relate the overall distribution pattern of any lake-dwelling species to the presence of a particular kind of substrate; nevertheless, substrate does have a strong influence on fish distribution within a lake, especially for spawning. For example, spawning sunfishes and black basses (Centrarchidae) require areas that are soft enough for digging a nest depression but firm enough that the embryos will not be smothered in mud. Sculpins (Cottidae), darters (Percidae), and some minnows (Cyprinidae) lay their eggs on the underside of overhanging rocks or logs. Yellow perch (*Perca flavescens*) and many minnows scatter adhesive eggs over beds of aquatic plants. Lake trout lay their eggs in deep water among concentrations of boulders.

Even when they are not spawning, most lake fishes show some preference for a general type of substrate. Particularly attractive to many fishes are beds of aquatic plants, which contain abundant food and cover. Such beds most often are located on soft and, usually, muddy bottoms. Other fish species are associated mainly with rocky bottoms. Thus, in many lakes of eastern North America, sculpins and some darters are found mainly in rocky, inshore areas. Cover provided by submerged objects, such as fallen trees and old boats, is also very attractive to fish, including many important game fish (e.g., catfish, bass, sunfish, etc.). This fact is used by fishery managers, who construct artificial reefs on sandy- or muddy-bottomed areas with little natural relief to improve fishing (Johnson and Stein 1979). Another type of cover that often is attractive to lake fish is shade created by overhanging trees, bushes, piers, and anchored boats. Although fish hovering in the shade may seem to be in a rather exposed position, they are less visible to predators; at the same time, it is easier for them to see predators approaching through sunlit areas (Helfman 1981a).

29.2   CHEMICAL FACTORS

Gases

The only dissolved gas in lake water that exerts a major, direct effect on fish distribution is oxygen, although carbon dioxide is indirectly important to fish through the carbonate–bicarbonate cycle and the relationship of this cycle to primary production.

Primary production is the main source of oxygen in lakes, although the action of wind and waves is an important secondary source. The absence of these two sources of oxygen, combined with the removal of oxygen through decay of organic matter, is the reason that major fish kills occur in shallow, ice-covered lakes during the winter.

Because some species (e.g., common carp) have a much higher tolerance for low oxygen levels than other species (e.g., northern pike), the composition of the fish fauna of many lakes in cold regions is determined, in part, by dissolved oxygen levels in winter, when they are lowest (J. Moyle and Clothier 1959, Tonn et al. 1990). Similarly, the absence of trout from the deep water of many two-story lakes is caused by low oxygen levels in the hypolimnion. Despite dramatic examples such as these, however, the influence of oxygen on fish distribution both in and between lakes is quite subtle and closely bound to temperature and ionic effects.

Ions

The influence of dissolved ions on the distribution of lake-dwelling fishes is complex, because the concentrations of the various ions in a lake affect not only the fish but also every other organism that is present. Moreover, the chemistry of a lake's water is related to that of the surrounding rocks, soils, and plant communities, so broad relationships between fish distribution and lake chemistry may be the result, at least in part, of zoogeographic coincidence. Usually, however, zoogeography does not go very far in explaining the high correlations that often are found between patterns of fish distribution and production and the following factors: (1) water chemistry patterns (expressed as pH), (2) nutrient concentrations, and/or (3) some measure of total dissolved inorganic and organic salts, such as salinity, total dissolved solids, conductivity, or alkalinity (Fig. 29.1).

Lakes with pH values above 8.5 (alkaline) or below 5 (acidic) support only a few, rather tolerant fish species or, if the values are extreme, none at all. Thus, one of the first signs that a region is being affected by acid rain is the disappearance of fish from its lakes. Within the normal pH range, alkaline lakes do have higher fish production than acidic lakes. Fish production in turn is determined through food webs by primary production, which is limited by the availability of nutrients, especially phosphorus and, occasionally, nitrogen. Thus, the fertilization of oligotrophic lakes in British Columbia and Alaska increases the growth and survival of juvenile sockeye salmon by increasing the supply of zooplankton on which the salmon feed (Hyatt and Stockner 1985). Unexploited oligotrophic lakes may support surprisingly high standing crops of fish (usually salmonids) despite low nutrient levels. Most of the fish biomass is tied up in large, old, slow-growing adults. Johnson (1976) compares such populations to "climax" communities of plants, because almost all the energy derived from primary production goes into maintenance and reproduction rather than growth.

The complexity of the interactions among dissolved salts, nutrients, and fish distribution and production is well illustrated by the broad distributional patterns of the four major fish associations in Minnesota lakes (Fig. 29.1). Not only can the presence or absence of the fish associations be related to water chemistry, but it also can be related to the surrounding plant communities and, to a lesser extent, climatic variables, such as temperature. These same factors are related to the potential production of fish from lakes, a subject of great interest to fishery managers. On a broad scale, fish production increases with

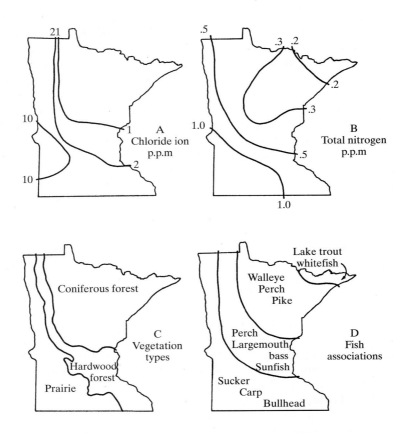

FIGURE 29.1   Generalized clines of salinity (**A**; as measured by chlorinity), dissolved nutrients (**B**; as indicated by total nitrogen), and (**C**) vegetation types in relation to (**D**) major fish associations in Minnesota lakes. From Moyle (1956).

increases in temperature, phosphorus concentration, phytoplankton production, and pH (Downing and Plante 1993). This suggests that total fish abundance largely is controlled by bottom-up factors. Indeed, predatory and plankton-feeding fish (top-down control) have much less effect on ecosystem structure in highly productive (eutrophic) lakes than they do in lakes of low productivity (oligotrophic) (Trippel and Beamish 1993). A factor adding complexity to the already naturally complex pattern of fish distribution among lakes is human activity, which changes landscapes surrounding lakes, increases eutrophication, and brings in new species—especially predatory game fishes (Allen et al. 1999).

29.3 TEMPORAL FACTORS

The composition of the fish fauna in most lakes rarely is constant. It is likely to change as a lake ages, as the climate fluctuates from year to year, as the abundance of species changes from season to season, and as fish engage in daily movements.

Long-Term Changes

Long-term changes in the fish faunas of lakes are poorly documented. They are, however, inevitable, because the chemistry of lakes changes with the accumulation of nutrients and other ions from the surrounding land and the physical environment changes with the accumulation of sediments. A gradual addition of species also is likely, as species enter lakes through outlets or inlets. Actually, the types of changes that are likely to occur naturally in lakes can be understood, more or less, through the study of lakes that have been strongly affected by humans through "cultural eutrophication" and the introduction of exotic species. For example, many lakes that are close to urban areas in North America and Europe formerly were dominated by salmonids. Under the influence of increased nutrients, species introductions, and overexploitation, they have become dominated by cyprinids and perch, with an overall increase in fish biomass (Loftus and Regier 1972).

Climatic Fluctuations

Changes in climate are likely to cause fluctuations in the composition of a lake's fish fauna, mainly as a result of their effect on water temperatures and lake levels. Most fish have a limited range of temperatures at which they will spawn. If these temperatures are not reached, spawning will not occur; if they are reached late in the season, the embryos and larvae may face larger populations of small, predatory fish. In either case, reproductive failure is likely, and the species—or at least the year class—is likely to be less abundant in the future. Reproductive failure also is likely in drought years for species that require flooded vegetation for spawning or emergent plants as cover for their young. The reproductive failure of one species also may affect the abundance of other species. In Oneida Lake, New York, young yellow perch are the principal prey of walleye. When the perch have poor reproductive success, walleye prey more heavily on their own young, resulting in a less-abundant year class of walleye (Forney 1974). Such complex interactions both among and between species and their environments indicate that even minor long-term climatic shifts can create major changes in fish faunas (Schindler et al. 1990). Thus, global warming will result in major changes in the many lake ecosystems.

Seasonal Changes

Seasonal changes in the fish fauna of lakes are related to a combination of reproductive and seasonal movements. The major fish species in lakes tend to reproduce at different times, which results in a definite succession of planktonic larvae in the open waters of lakes during the summer (e.g., Amundsrud et al. 1974) as well as a succession of abundance of juvenile fishes at the inshore areas. Many adult and juvenile fish make seasonal migrations to different depths or parts of lakes, thereby resulting in a seasonally changing species composition in some habitats, particularly inshore areas. The ability of alewife to move to different habitats as the seasons change has enabled it to disrupt the less-migratory inshore and deepwater populations of whitefish in Lake Michigan (Smith 1968).

Daily Changes

The activity patterns of most lake fish are cued to light (Helfman 1981b), so striking differences can be found in the distribution of fish within a lake according to the time of day. Vision-oriented predators are most active during the day, often with peaks of feeding in the early morning and evening hours, when invertebrates become more available: Zooplankton are moving upward, benthic insects start becoming active, and flights of terrestrial and aquatic insect adults increase. Dawn and dusk are also the times when piscivorous fishes move into shallow water, attracted by the feeding activities of the smaller fishes. The predators are less visible at low light intensities. After dark, most day-active fishes lie quietly on the bottom, beneath submerged objects, or among aquatic plants, although some species, such as the sunfishes, may remain in slowly circling schools. Night-active fishes, particularly catfish and suckers, then move into shallow water to forage.

Daily movements in relation to temperature also are common, often for energetic reasons. In Bear Lake, Utah, larval sculpins (*Cottus extensus*) forage on the bottom during the day, where water temperatures are 4°C to 5°C, but move up to the water column at night, where temperatures are 13°C to 16°C (Wurtsbaugh and Neverman 1988). The vertical migrations take place because sculpins digest their food faster at higher temperatures and, thus, grow three times faster. In contrast, kokanee salmon apparently move up into warm water to forage on plankton at night and then back into cold water to digest their prey in an energetically advantageous manner during the day (Bevelhimer and Adams 1993).

29.4   BIOLOGICAL FACTORS

Like the stream fish, lake fish are quite flexible in their living arrangements with other species of animals and plants, particularly other fish. These arrangements are best examined under the general headings of predation, competition, and symbiosis.

Predation

Predation by fish is a major factor influencing the composition of biotic communities in lakes, from plants through invertebrates to fish. Grazing and rooting about by fish, particularly common carp, can influence greatly the amount and species of aquatic plants growing on lake bottoms as well as the species and numbers of invertebrates that are associated with the plants. Most fish that browse on invertebrates associated with aquatic plants or the bottom are selective as to what species they prey on, and such selective predation affects the composition of the invertebrate community (Stein and Kitchell 1975). The impact of selective predation on the zooplankton community of lakes is even more pronounced—or at least is better understood. In lakes without specialized planktivores, the dominant zooplankton species tend to be relatively large, effective grazers on phytoplankton. In lakes with specialized planktivores, the dominant zooplankton are much smaller, may have protective spines, or may have smaller eyes, making them less visible (Brooks 1968, Zaret 1972). The impact of fish on zooplankton has been demonstrated repeatedly when zooplanktivorous fish have been introduced into lakes. For example, the development of alewife populations in Lake Michigan resulted in declines in the populations of seven species of large zooplankton, including the near disappearance of two species, and increases in the numbers of 10 small species (Wells 1970).

Predation also has a major impact on the structure of fish communities and populations in lakes (Box 29.1). Occasionally, the introduction of a new predator can drastically reduce the populations of many of the lake fishes already present, as the impact of the sea lamprey on the larger fish species of the Great Lakes so graphically illustrates. Interestingly enough, however, as the large, predatory fishes of the Great Lakes declined, populations of fish species that were too small to be preyed on by lampreys increased greatly, presumably as a result of the absence of predation and competition from the larger fish (Smith 1968) (Fig. 29.2). The complexity of predator–prey interactions in lakes is well illustrated by the long-term studies of northern pike, perch (*Perca fluviatilus*), and Artic charr (*Salvelinus alpinus*) in Lake Windermere, England (Le Cren et al. 1972). Before 1941, large pike apparently kept char numbers down by preying on the adults as they moved inshore to spawn. Although perch were the main food of the pike for the rest of the year, the number of adult pike apparently was too low to limit the size of the perch population. Between 1941 and 1964, adult perch were removed from the lake by intensive fishing, but the perch populations failed to recover when the fishing was stopped. The reason for this, apparently, was that an experimental pike fishery had removed most pike larger than 55 cm from the lake from 1944 onward, thereby resulting in optimum conditions for the growth and survival of small pike. The abundant small pike fed on small prey, particularly young perch, making it difficult for the perch populations to regain their former levels. Meanwhile, in the absence of predation from large pike, char numbers increased.

BOX 29.1

Fish Integrate Benthic and Pelagic Foodwebs in Lakes

Most studies of lake ecosystems focus on the pelagic ecosystem, which is dominated by plankton and pelagic fish, in part because this type of ecosystem is relatively easy to study. The concept that food webs were regulated by top-down or bottom-up processes (***trophic cascades***) in lakes also suggested that humans could control these processes and make lakes clearer or more productive as desired. The importance of food webs associated with the bottoms and beds of aquatic plants therefore often was ignored, even though studies of fishes often showed that aquatic insects, snails, and other benthic organisms were important in fish diets. Using both stomach data and stable-isotope analysis, a relatively new tool, Vander Zanden and Vadeboncouer (2002) examined the diets of 15 species of fish from more than 500 lakes, and they showed that on average, 65% of the food came from the benthos. They suggest that this results in a net transfer of energy (a subsidy) to the pelagic food webs, thereby changing their nature. Thus, small fish feeding on benthos may be important prey for pelagic predators, whereas inshore predators occasionally may feed heavily on planktivorous fishes. Thus, the next generation of models of lake food webs will need to take a more integrated view of the pathways involved.

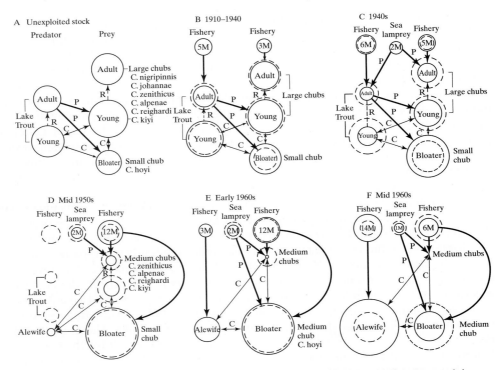

FIGURE 29.2   Interrelations of major deepwater fishes of Lake Michigan: (**A**) before exploitation; (**B**) during moderate exploitation; (**C**) during the period of increasing sea lamprey abundance; (**D**) during the peak of sea lamprey abundance, when large fish became rare and small chubs increased in numbers and size; (**E**) during the period when bloaters were at their peak abundance and the alewife was just becoming established; and (**F**) during the period of maximum abundance of the alewife. The nature of the interactions is indicated by *P* for predation, *C* for competition, and *R* for recruitment. From Smith (1968).

Obviously, predators and prey can affect each other's populations in lakes in many ways. Under more or less stable environmental conditions, self-regulating predator–prey systems can, theoretically, develop. Steady "cropping" of a prey species by a predator will reduce intraspecific competition and, consequently, increase growth rates of the prey. Such rapid growth may increase the reproductive potential of the prey by allowing it to mature at an earlier age and by increasing egg production, because the increase in fecundity of fishes tends to have an exponential relationship to length. If the increased reproduction results in a "surplus" of prey, two things may happen simultaneously: The predator population may increase, and intraspecific competition among the prey may increase, resulting in slower growth and lower reproduction. These two factors then cause a rapid decline in the prey population, followed by a decline in the predator population to the former, lower level of both species. The cycle then repeats itself. This system may, in fact, operate in lakes. It is likely to be highly modified, however, because predators will switch prey and because reproduction and growth also likely will be affected by changing environmental conditions. For example, evidence suggests that in some lakes used by sockeye salmon as nurseries, the feedback scheme described here plays a major role in determining the abundance of adult salmon (Hartman and Burgner 1972). In salmon nursery lakes with a number of alternative prey species for the predators, however, the system may not work, because the alternative prey either serve to maintain high populations of predators or as buffer species that reduce predation on the young salmon.

Competition

Fish in lakes, like those in streams, show a considerable degree of ecological segregation. Although this segregation can be explained, in part, by differences in body morphology, the coexistence of similar, closely related species in lakes is common. Fish populations in lakes probably are more constant in general than those in streams, but nevertheless, considerable year-to-year and season-to-season variation occurs in species composition. In addition, food and habitat (e.g., aquatic plant beds) show considerable variation in availability from season to season. All these factors indicate that competitive interactions likely are important in shaping the fish communities in lakes (Box 29.2). As the studies of trout and char in Scandinavian lakes show, such interactions can be expressed as direct aggression by one species on another, resulting in domination of the preferred habitat by the most aggressive species. Differential exploitation of resources as a cause of observed ecological differences, however, probably is more common. For example, when three species of sunfish (*Lepomis*) occur together, each is found in a different habitat, and each feeds on different invertebrates. In the absence of other species, however, any of the species will, at least partially, use the habitat and food that otherwise would be used by the other two species (Werner and Hall 1976). The least change in niche occurs with green sunfish (*L. cyanellus*), which feeds on large invertebrates among aquatic

---

BOX 29.2
Lake Ecosystem Changes Lead to Road-Killed Bald Eagles

When species segregate by diet or habitat, competition usually is inferred as the ultimate cause. For better or worse, introductions of alien species often result in direct demonstrations of competitive effects. For example, when a predatory mysid shrimp invaded Flathead Lake, Montana, it consumed most of the large zooplankton in the lake (Spencer et al. 1990). This resulted in a major decline in the population of kokanee salmon, which depended on the zooplankton (exploitation competition). Reduction in the spawning runs of kokanee salmon up tributary streams in turn affected black bears and bald eagles, which came to the streams annually to forage on the salmon. Several bald eagles apparently were killed by cars on local highways, because they began feeding on road-killed animals in the absence of their usual salmon. Curiously, the kokanee had been introduced into the lake themselves many years previously and undoubtedly had caused shifts in the habits of native fishes that are not appreciated (because they were not recorded).

plants. Green sunfish apparently protect these resources from the other species in part by aggressive behavior and in part by more efficient foraging for the prey. As a result of these interactions, bluegill (*L. macrochirus*) and pumpkinseed (*L. gibbosis*) prey on smaller organisms. Pumpkinseed become more bottom oriented, favoring mollusks as food, and bluegill become more oriented toward the water column and zooplankton. In the spring, when food is most abundant and fish populations are lower, more overlap occurs in food and habitat than in the fall, when resources likely are depleted (Seaburg and Moyle 1964).

Symbiosis

Examples of symbiotic effects on fish communities in lakes are few, possibly because they often have not been sought. It is common to find different species schooling with each other, presumably forming mutualistic or commensal associations. In Long Lake, Minnesota, small groups of bluntnose minnows (*Pimephales notatus*) were found in shallow water, outside weed beds, only when they were in association with large schools of mimic shiners (*Notropis volucellus*). In the mixed schools, the minnows were found below the main school of shiners, feeding on the bottom, whereas the shiners were feeding in midwater (Moyle 1973). Presumably, the minnows could take advantage of the protection from predators that was afforded by the large school of shiners to feed in a habitat that otherwise would be too hazardous for them.

29.5   ZOOGEOGRAPHIC FACTORS

Barriers to fish dispersal are major determinants of the fish faunas of lakes (Tonn et al. 1990). The closer a lake is (or was) to a major river system, the more species likely are present. Lakes with long stretches of stream, especially high-gradient stretches, between them and a major river or other lakes are likely to contain only the same fishes that are found in headwater streams or derivatives of them, such as trout, suckers, dace, and chubs. Indeed, most of the lakes of the high mountains in the American West were without fish until trout and other species were introduced by humans. In remote lakes already containing fish, the fact that a number of new species can be added indicates the importance of barriers in limiting fish faunas. For example, Clear Lake, California, originally contained 11 native fish species, even though it is one of the oldest lakes in North America. Since 1880, 17 exotic species have been added to the fauna, and six of the native species have become extirpated, leaving a total of 22 species. Although it can be argued that the present fish fauna is unstable, with a number of further extinctions likely to occur, the ultimate fish fauna undoubtedly will contain more than 11 species (Moyle 2002).

29.6   FISH ZONES

The designation of fish zones in lakes is even more arbitrary than in streams, because lake fish are comparatively free to move from one zone to another. Fish in lakes, however, do tend to sort themselves out along environmental gradients during the summer months, when most individual and population growth occurs. As a result, distinct clusters of species tend to be associated with the broad habitat types within lakes, and these associations can be described as fish zones. Within each fish zone, the species tend to segregate further by microhabitat and food preferences. As examples of these patterns, the zones in two lakes will be described here: a small, two-story lake (Long Lake); and a large, cold-water lake (Lake Tahoe).

Long Lake

This lake in north-central Minnesota is somewhat unusual for its bathtub-like morphology (deep and steep-sided) as well as for its clarity (Moyle 1969, 1973). These same factors, however, result in sharp environmental gradients and easily observable fish zones. The 19 species of fish that are common in the lake are typical of small lakes in areas that were subjected to Pleistocene glaciation, and the patterns of segregation also are typical of such lakes (Keast 1965, Gascon and Leggett 1977, Werner et al. 1977). Four zones are recognizable in the lake: (1) the shallow-water zone, (2) the aquatic plant zone, (3) the deepwater zone, and (4) the open-water zone (Fig. 29.3).

The ***shallow-water zone*** occurs in water less than 1 m deep and is strongly influenced by wave action. Aquatic plants are thinly distributed, and cobbles are the main substrate close to shore. The most conspicuous fish in the zone are schools of mimic shiners, which feed in midwater and at surface areas. Bluntnose minnows and common shiners often are associated with these schools as well. On the bottom are Iowa darters (*Etheostoma exile*) and, beneath the cobbles, mottled sculpin (*Cottus bairdi*). The main resident predators are green sunfish, which are found associated with whatever cover

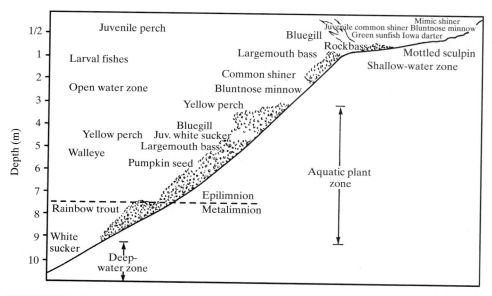

FIGURE 29.3   Typical summer locations of common fish species during the day in Long Lake, Minnesota. From Moyle (1969).

(e.g., logs, etc.) the zone affords, although piscivorous largemouth bass and walleye move into the zone to forage at dusk.

The ***aquatic plant zone*** is characterized by dense beds of mixed species of aquatic plants that are between 1 and 9 m deep. The beds are thickest and most diverse at depths between 2 and 3 m, and populations of the six most abundant species are the densest here. Bluntnose minnows are found in clearings among the plant beds, where they feed on detritus, diatoms, and small, benthic invertebrates. In contrast, large common shiners and yellow perch typically are associated with clumps of aquatic plants that grow higher than the main mass of vegetation. Both species are opportunistic foragers, with the perch consuming large invertebrates and small fish. Rockbass (*Ambloplites rupestris*) are the main ambush predators of large prey in this zone and, consequently, are associated with logs, large rocks, and other cover in the shallower areas. Largemouth bass are roving predators of crayfish and fish. Bluegill and pumpkinseed are, perhaps, the most morphologically similar species in the zone, and both feed by picking invertebrates from the aquatic plants. Pumpkinseed, however, feed largely on snails, whereas bluegill feed on aquatic insects.

The ***deepwater zone*** is found below the lower limits of the aquatic plant beds and below the epilimnion. Thus, it is cold, dark, and silty-bottomed. Only two species, johnny darter (*E. nigrum*) and white sucker (*Catostomus commersoni*), are regularly found here. The darter feeds on small, benthic invertebrates, whereas the sucker consumes detritus as well and may move into shallow water to forage at night.

The ***open-water zone***, which is away from the influence of the bottom, is characterized by large schools of juvenile perch and by all size classes of walleye. The walleye generally move inshore to feed during the evening. During the summer, there undoubtedly also is a succession of larval fishes, with each species occupying the zone on a temporary basis.

Lake Tahoe

This lake is a large (surface area, 304 km^2), deep (mean depth, 313 m), and clear (the bottom is visible at 20–30 m) mountain lake on the California–Nevada border. The six native fish species that now are characteristic of the lake also are found in the local streams. Originally, the main piscivore in the lake was cutthroat trout, but it has been replaced by lake trout, rainbow trout, and brown trout. In addition, kokanee salmon have been added to the fauna. The lake can be divided into three broadly overlapping zones: (1) the shallow-water zone, (2) the deepwater benthic zone, and (3) the midwater zone (Fig. 29.4).

29.7   RESERVOIRS

Reservoirs are treated separately in this chapter because they emphatically are *not* lakes, although they are often labeled as such. They share species with lakes and are big, open bodies of water, but they behave very differently. Those used mainly for power production often more are like giant riverine pools than lakes because of the high turnover rate of the water. Those used mainly for water supply typically have strong seasonal fluctuations in depth and area, which prevent the establishment of beds of aquatic

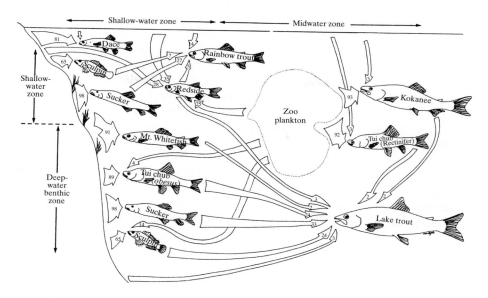

FIGURE 29.4   Zones and feeding relationships of Lake Tahoe fishes. The numbers represent the percentage of diet by volume of the most important food items for each species. The major food categories are benthic organisms, flying insects, zooplankton, and fish. Note that two forms of tui chub (*Gila bicolor*) are in the lake; a bottom-feeding form (*obesus*), and a zooplankton-feeding form (*pectinifer*). The importance of zooplankton now probably is less than is indicated, because the introduction of a mysid shrimp into the lake has resulted in the near elimination of the larger species. From Moyle (2002), modified from Miller (1951).

plants and other types of fish habitat. Overall, reservoirs are (1) different in their physical and chemical characteristics from natural lakes, (2) less predictable ecologically, and (3) more ephemeral on the landscape. The United States, however, has more surface acres of water in reservoirs than surface acres of natural lakes, excluding the Great Lakes, so they need to be discussed here. A similar pattern exists worldwide in temperate areas and, increasingly, in tropical areas as well.

Temperate reservoirs can be classified by temperature regime, like lakes, but most are dominated by fishes that are characteristic of warm-water lakes. Those in the United States tend to be remarkably similar in their fish faunas, both because extreme fluctuations in water levels and other factors that are typical of reservoirs select for certain species and because the reservoirs always are planted with a few preferred species of game fish. Many of these same species (e.g., largemouth bass and common carp) also are characteristic of reservoirs in Europe and elsewhere. The main variations in the fish faunas of reservoirs are the result of the persistence of species that are native to the river impounded, although most such species are eliminated early in a reservoir's history through predation and competition from introduced species (Paller et al. 1992). Thus, for the first few years, fish populations and species numbers are high in reservoirs, because both the lacustrine species (planted by humans) and the riverine species already present grow rapidly in the new environment. Reproduction also may be good initially, both because spawning areas are not yet silted over and because water-level fluctuations are minor. As the reservoir ages, however, and extreme fluctuations become the rule, the total fish biomass declines, as does the number of abundant species, particularly riverine species. The species that remain abundant are (1) those that are long-lived, so that one successful spawn will sustain their populations over a long period (e.g., common carp); (2) those that migrate upstream to spawn (e.g., many suckers); (3) those that spawn pelagically (e.g., shads, *Dorosoma*); or (4) those that spawn in water deep enough not to be affected by the fluctuations (e.g., channel catfish, *Ictalurus punctatus*). The habitat of larval fish also is important, because pelagic larvae may be washed downstream. Species that are most likely to decline from lack of reproductive success are piscivores, such as smallmouth bass. Angling and lack of suitable prey also may contribute to the decline of these species.

The fluctuating conditions on short and long time scales confer considerable instability on reservoir fish assemblages. Each species, however, generally is most characteristic of one of three zones: (1) the inshore zone, (2) the open-water zone, and (3) the deepwater benthic zone.

The *inshore zone* is associated with the sides of the reservoirs. Fish are most abundant where the bottoms are soft and the water less than 3 m deep, but adults of larger species also commonly are associated with steep, rocky areas. The typical species of this zone are sunfishes, black basses, catfish, crappie (*Pomoxis*), common carp, carpsuckers (*Carpiodes*), buffalofishes (*Ictiobus*), minnows (Cyprinidae), and silversides (Atherinidae). The exact composition of the inshore fauna will depend on the particular reservoir, the location within the reservoir, and the time of day. Carp, carpsuckers, and buffalofishes are mobile, bottom-feeding species that are most likely to be found close to shore at night. Predatory catfishes may move into shallow water at night to prey on the juvenile centrarchids and small minnows that are the most abundant fishes in the shallow areas. Fish assemblages in this zone show considerable variability, reflecting the fluctuating conditions of the reservoir (Gelwick and Matthews 1990).

The *open-water zone* is host to most reservoir species at some stage in their life cycle, but it is characterized consistently by plankton-feeding shad and their predators, typically white bass (*Morone chrysops*) but often striped bass (*M. saxatilis*) or walleye. Large crappie also may be found in this zone, feeding on both zooplankton and shad. If a reservoir stratifies every summer, trout and/or kokanee salmon may be found in the hypolimnion. More often than not, the salmonid populations are not self-sustaining and must be supplemented each year by plants of fish.

The *deepwater benthic zone* contains mainly predatory catfish and fish that subsist on detritus or detritus-feeding insect larvae. The detritus-oriented species typically are suckers (Catostomidae) and common carp. In most reservoirs, detritus-oriented fishes, including bullheads and other species that are more characteristic of the inshore zone, make up a majority of the fish biomass (Cherry and Guthrie 1975).

The apparent structure of the fish assemblages of reservoirs reflects the adaptability of the fish more than the stability of the habitat. In reservoirs, the effects of competitive interactions among species normally are minimized by the extreme fluctuations in the physical environment. The main exception to this occurs among the plankton-feeding fishes, in part because zooplankton populations are less likely to be affected by fluctuations in water level than are populations of benthic invertebrates. When a planktivore, such as threadfin shad (*Dorosoma petenense*), is introduced into a reservoir that has lacked such a species, it can reduce the zooplankton populations and, consequently, the populations of bluegill, largemouth bass, and other fishes with early life-history stages that depend on abundant zooplankton for survival (von Geldern and Mitchell 1975).

High turnover in reservoir water also can reduce populations of zooplankton and those of pelagic fish, including fish larvae. Thus, reservoirs frequently have conditions that are intermediate between those in rivers and those in natural lakes, because the demand for the water they store is likely to cause detectable currents at both their upper and lower ends. Consequently, the fish at the upper end of a reservoir may be somewhat different from those in the more lake-like middle areas. In many North American reservoirs, smallmouth bass are most abundant at the upper ends of the reservoirs and largemouth bass in other areas. When spotted bass (*M. punctulatus*) are present, they frequently occupy an intermediate position, often in deeper water.

The consistent problems that reservoirs create for fishery managers (usually reflected in disappointing catches) reflect the fact that they typically are constructed for purposes other than maintaining fish populations. In the long run, therefore, reservoirs cannot be relied on to provide reliable fish habitat and to support consistent fisheries.

29.8   LESSONS FROM LAKE FISHES

With a few notable exceptions, temperate lakes are short-term features of the landscape, with expected life spans measured in a few hundred or few thousand years. The original fish faunas of lakes were assemblages of regional stream fishes that had passed through major zoogeographic and physiological filters to become established. Introductions by humans have greatly increased the number of fish species in most lakes, even in regions where the fishes are native. Despite the highly variable composition of lake-fish assemblages, studies invariably find they have a high degree of organization. This structure is the result of inherent differences among the species working with complex biotic interactions and playing on a constantly changing environmental stage. It also is apparent that fish predation is a major factor affecting the biotic and abiotic characteristics of lakes, although this top-down effect lessens as the lake productivity increases. We probably have a better understanding of how the pelagic system in temperate lakes "works" than we do of any other aquatic system; Therefore, we have, in relation to other systems, even fewer excuses to continue letting lake ecosystems deteriorate to conditions of poor water quality and few or no fish species. Reservoirs have many characteristics of lakes, but they are poor substitutes for them.

SUPPLEMENTAL READINGS

Carpenter and Kitchill 1993; Matthews 1998.

WEB CONNECTIONS

GENERAL
www.calacademy.org/research/ichthyology
www.fishbase.org
www2.biology.ualberta/jackson.hp/IWR
www.flmnh.ufl.edu/fish/default.htm
www.people.clemson.edu/~jwfoltz/wfb418

LIMNOLOGY
wpw.nrri.umn.edu/wow

LAKE TAHOE
tahoe.usgs.gov

CHAPTER 30
Tropical Freshwater Lakes and Streams

Fish in tropical fresh waters are affected by the same factors as fish in temperate waters, so patterns of distribution and abundance often can by explained by making comparisons to the much-better-known temperate fish communities. On the other hand, the tropics contain an enormous number of fish species, often with extraordinary specializations for feeding and reproduction, and tropical lakes and streams have many distinctive characteristics. One of the generalities that has emerged from recent studies is that within lowland tropical lake or stream systems, biological factors often are more important than physical and chemical factors in determining fish distribution, abundance, and diversity. Physical and chemical factors, however, retain their importance for explaining broad distribution patterns. Why tropical freshwater fishes are so diverse and how they maintain their diversity are not as easy to explain as you might think. Tropical climates often are thought of as being benign, especially to fish, because the water is assumed to be highly productive, with little variation in temperature, during the entire year. Because temperatures are warm year-round, multiple generations of small fishes in a single year are possible, thereby allowing for rapid speciation, especially in rain-forest areas. In many tropical areas, however, the alternation of wet and dry seasons produces considerable environmental fluctuation, including major droughts and floods. In addition, the climates that we see today in tropical regions are not necessarily the same as those that existed even a few thousand years ago. Rain forests may have been fragmented, stream flows reduced, and lakes dried up. On the positive side, tropical rivers are among the largest in the world; the Amazon River alone drains 30% of South America and has an average flow of 2 to 3 million cubic feet per second at its mouth. This provides many opportunities for speciation in isolated subwatersheds. More important, tropical landmasses typically are very stable geologically, so many of the lakes and river systems of the tropics are quite old. This gives plenty of opportunity for many highly specialized species to evolve under localized physical, chemical, and trophic conditions (Lowe-McConnell 1975). Thus, the remarkably diverse fish fauna in tropical fresh water is the result of a combination of past climates, large river systems, ancient landscapes, warm temperatures, productive environments, and biotic interactions that all promote speciation. The present distribution of fishes within a tropical landscape is the result of species reacting to a combination of physical, chemical, and biological factors.

30.1   PHYSICAL FACTORS

The physical factors that seem to have the most noticeable impact on the distribution of tropical freshwater fishes are: (1) temperature, (2) water-level fluctuations, (3) gradient and stream order, and (4) turbidity.

Temperature

Fish in large, tropical lakes and rivers live in environments that are warm year-round and that have comparatively little temperature fluctuation. As a consequence, these fish tend to have rapid growth and short life cycles. Temperature is not a particularly important environmental cue for movements and reproduction. Seasonal low temperatures place broad distribution limits on tropical fishes (most do not seem to be able to survive temperatures below 15°C for extended periods of time), but local distribution patterns are much more likely to be affected by high temperatures. In Lake Victoria (Africa), the young of two groups of cichlids sorted themselves in shallow water on hot days according to their ability to tolerate high temperatures: Tilapiine cichlids were found in the hottest (to 38°C), shallowest water, whereas haplochromine cichlids were found in slightly cooler and deeper water (Lowe-McConnell 1975).

Water-Level Fluctuations

In contrast to temperature, annual fluctuations in water level are extremely important cues for many tropical fishes. With the advent of the rainy season, intermittent streams in the drier areas start flowing again, stagnant jungle pools and backwaters on the floodplains are flushed out, lake levels rise, and the flows of major rivers greatly increase. As a consequence of the flooding of land and the flushing of terrestrial nutrients into the rivers, habitats and food resources for fish expand greatly (Fig. 30.1). Thus, it is not surprising to find that reproduction and growth often are strongly related to rising water levels. Many tropical riverine fishes make extensive upstream spawning migrations at this time, or they move into the floodplain to spawn. In either case, the young hatch rapidly and find abundant food for rapid growth. Adult fish also find abundant food and experience most of their growth during this period. The seasonal differences in growth rates often are striking enough that annual rings are deposited on scales and other bony structures, thereby making it possible to determine the age of the fish (Lowe-McConnell 1975).

Tropical-stream fish not only grow rapidly during the wet season, many of the large species also accumulate substantial fat reserves to last them through the dry season. In the Amazon River basin, fish that move into the flooded forests to feed show considerable specialization in their feeding habits, including feeding on fruits falling from trees. During the dry season, when food is less available, the fish become more omnivorous, and dietary overlap is common, with the larger species living, in part, on their stored fat. In the Orinoco River floodplain in Venezuela, large numbers of fish are isolated in floodplain lakes during the dry season, representing 116 species from the riverine fauna. The fish assemblages in these turbid lakes show a predictable shift in structure as the season progresses, from a fauna dominated by small, visual-feeding

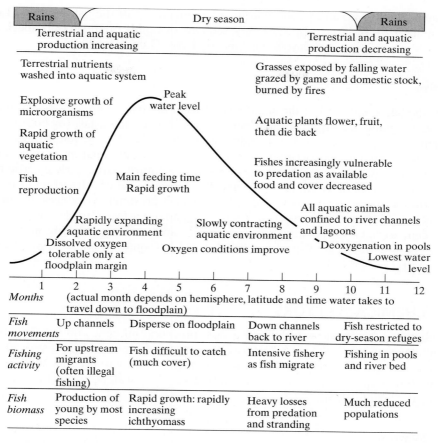

FIGURE 30.1   The seasonal cycle of events in a tropical floodplain river. From Lowe-McConnell (1975).

characins and cichlids to one dominated by catfishes and knifefishes, which have adaptations for finding prey at low light levels (Rodriguez and Lewis 1994). The most likely mechanism mediating this shift is predation.

Gradient and Stream Order

The gradient and order of streams factors usually are strongly interrelated in their effects on fish distribution, and they seem to have the same general effects on tropical as they do on temperate fishes. In high-gradient headwater streams, the typical fishes are streamlined cyprinids (Africa and Asia), highly specialized hillstream fishes (Balitoridae; Asia), and specialized catfishes (South America). As the gradient decreases and stream order increases, the variety of body shapes increases, as do the divesity of feeding specializations. In the headwater streams, most fishes feed either on terrestrial invertebrates or detritus. In higher-order streams, piscivorous forms become common, and they often seem to make up an extraordinarily large percentage of the fish biomass (Lowe-McConnell 1975). In large, tropical rivers, sections of rapids often support specialized fishes, such as tiny blind catfishes in the Amazon River (Lundberg 2001).

Turbidity

Usually, the lakes and most of the smaller streams of the tropics are quite clear, but the large rivers usually are turbid with suspended or dissolved materials (or both). Often, considerable seasonal variation in turbidity occurs, with the clearest water flowing during the dry season. One impact of high turbidity has been the radiation in rivers of fish groups that do not rely primarily on vision for prey capture, particularly electric fishes (the gymnotid fishes of South America and the Mormyridae of Africa) and catfishes. These fishes also are mostly nocturnal in habit, which presumably reduces the amount of interaction with day-active cyprinids, characins, and other fishes. In clear water, bright colors seem to have an advantage for communication among the fish; most of the brightly colored "tropical" fishes that are favored by aquarists come from lakes and clear streams (Roberts 1972).

30.2   CHEMICAL FACTORS

Three chemical factors seem to have the most profound influence on fish distribution and abundance in tropical waters: (1) dissolved oxygen, (2) pH, and (3) dissolved nutrients. These factors interact with each other and with turbidity to produce three general types of stream environments: (1) white waters, (2) clear waters, and (3) black waters (Roberts 1972, Sioli 1975).

White Waters

White waters are characterized by high turbidity caused by suspended material, pH values of around neutrality (i.e., 7), moderate levels of dissolved oxygen, and high nutrient levels. Because whitewater rivers tend to be the main rivers of most tropical systems, their flooding creates the floodplain lakes and pools. When the rivers recede, these lakes and pools may become more transparent as the suspended material settles. If the light is not blocked by trees, the high nutrient levels and clear water result in large blooms of plankton. The large amount of decaying organic matter, however, can create low oxygen levels. In this situation, many of the fish are either air breathers or small fishes that are capable of utilizing the oxygen that is present in a thin band of water at the surface. In addition, the commonness of species that are livebearers, oral incubators, or nest builders in this type of habitat suggests that such methods may have developed to protect eggs and young from low oxygen levels. In the floodplain lakes of the Orinoco River of Venezuela, however, oxygen is not a limiting factor. Instead, the predictable shift in each lake's fish fauna after the recession of the river is related to strong predation by large piscivorous fishes on smaller fishes (Rodriguez and Lewis 1994). Curiously, despite the low transparency of the water, the aquatic food webs in the Orinoco floodplain seem to be based mainly on algae rather than on the abundant detritus from terrestrial plants or the abundant aquatic plants (Lewis et al. 2001). High fish production nevertheless is maintained, because most fish feed directly on the algae or the consumers of algae.

Clear Waters

Clear waters are transparent jungle streams with moderate to low gradients that usually flow into large, whitewater rivers by way of a mouth bay. Clear waters range in pH from 4.5 to 7.8, but they typically are slightly acidic. Oxygen levels in clearwater streams are high enough to support abundant fish life, although the food chains are based primarily on organic matter of terrestrial origin. The frequent presence of abundant aquatic insect life in these streams increases the complexity of the food webs, and the variety of fishes, particularly small, brightly colored forms, often is considerable. Many of these streams also may be important as spawning areas for main-river and floodplain fishes during the wet seasons. One of the more curious aspects of South American clear waters is that the mouth bays are lower in dissolved nutrients and fish numbers than might be expected.

Black Waters

The water of these streams of the deep forest is dark brown but transparent; it appears black from a distance. Because of dissolved organic humic matter, black waters are extremely acidic, with pH values often being lower than 4.5, and they contain almost no dissolved nutrients. The decay of organic matter on the bottom of slow-moving stretches greatly reduces the amount of dissolved oxygen that is available. Such water typically is regarded as not being conducive to supporting large numbers of fish. Henderson (1990), however, argues that in fact, black waters can have a high diversity of fishes, often with dense local populations. These species live in the dense leaf litter, which supports large populations of invertebrates. Other species are concentrated around the edges, especially in flooded vegetation, where terrestrial invertebrates are likely to be found. In the Amazon Basin, a number of fishes are largely confined to black waters and seem to be able to spawn only in waters with extremely low pH values (Roberts 1972).

30.3 BIOLOGICAL FACTORS

Because the number of fish species is so high in tropical environments, the interactions among them are complex. Predator–prey, competitive, and symbiotic interactions often are expressed in extraordinary morphological and behavioral specializations. These are most extreme in the cichlid fishes of the Great Lakes of eastern Africa, although the specializations of many stream fishes are almost as extreme.

Predator–Prey Relationships

Food webs in tropical fish assemblages tend to be more complex than those in temperate environments, because detritus, algae, and plant matter are much more important in the diets of fish in tropical systems than in those of temperate fish (Winemiller 1990). In addition, predatory fish seem to exert a stronger influence on the community structure of tropical fishes. For example, fish have had a strong effect on the abundance of invertebrates in a Venezuelan stream, both through direct predation and through altering the environment by foraging on algae and detritus (Flecker 1992). As in temperate systems, however, the most dramatic examples of predation affecting aquatic communities occur where humans have introduced new piscivores into a lake or a stream. Thus, the addition of peacock cichlids (*Cichla ocellaris*) and largemouth bass to Central American lakes has greatly altered their trophic structure by the elimination, through predation, of a number of small, plankton-feeding fish species (Zaret and Paine 1973). In eastern Africa, introduction of piscivorous Nile perch (*Lates*) into Lake Victoria, in combination with eutrophication and other factors, has resulted in the elimination of perhaps 200 species of native cichlids (Goldschmidt et al. 1993). On the other hand, reduction in the populations of piscivorous fishes in tropical lakes can cause a great increase in plankton-feeding fishes. This was illustrated by the increase in biomass of native herrings (Clupeidae) in another eastern African lake, Lake Tanganyika, following the depletion, through fishing, of the populations of three species of predatory perch (Centropomidae); a heavy fishery for the herrings subsequently destabilized the entire pelagic fish community in the lake (Craig 1992).

In the absence of human disturbance, predator–prey relationships in tropical waters presumably are fairly stable. Nowhere, except on coral reefs, does one find fishes with more remarkable adaptations that reflect the coevolution of predators and prey. The abundant small catfishes of various families have a wide variety of antipredator features, including heavy bony armor, stout spines that lock in place, cryptic coloration, and nocturnal activity. Even these bony catfishes are not immune to predation by birds, however. Power (1983) showed that algae grew thickly in the shallows of Central American streams because herons, including night herons, prevented grazing catfish from entering such areas. Other fishes (e.g., small characins) seem to thrive by being small and living in environments that are too shallow, too low in oxygen, or too hot for large predators. The piscivorous predators have evolved many mechanisms to overcome the defenses of their prey, such as the cryptic coloration and behavior of the patient leaffishes (Nandidae), the stunning ability of the electric catfish (*Malapterurus electricus*) and electric eel (*Electrophorus electricus*), and the formidable teeth and jaws of piranhas (*Serrasalmus*) and African tigerfish (*Hydrocynus*).

In tropical waters, as in temperate waters, predation probably is the ultimate cause of death for most fish; this has led to the hypothesis that predation has been a major force in creating the large number of species in tropical waters, especially in lakes (Lowe-McConnell 1975). Predation could increase isolation by making it difficult for fish with specific habitat requirements to move from one patch of habitat to another and, thus, could promote speciation. The large number of piscivorous fishes that seems to be typical of tropical lakes and streams also may permit potentially competing species to coexist by keeping the populations of the competitors low enough that resources are never limited. Another way that predators may permit two competitors to coexist is to prey selectively on the one form that has a competitive advantage over the other in a different environment. For example, bright-gold males of the Midas cichlid (*Cichlasoma citrinellum*) in Lake Nicaragua are more successful at attracting females and defending nests, whereas dull-colored males apparently are more successful at avoiding predators. As a result, the bright-gold males breed mainly in deeper water, where they are less visible to predators, and the dull-colored males occupy the prime breeding sites in shallow water (Barlow 1976, 2000).

Competition

As is true for predator–prey interactions, the most dramatic examples of competitive interactions in tropical fishes occur with introduced species. When redbelly tilapia (*Tilapia zilli*) were introduced into Lake Victoria, their young displaced the young of a native tilapia species from crucial nursery areas, thereby causing a severe decline in the native species (Lowe-McConnell 1975). In undisturbed tropical streams and lakes, it usually is assumed that direct competition is no longer of much importance, because past competition has resulted in resource partitioning through morphological character displacement. The best evidence for this among tropical fishes is the extreme specializations in body shape and feeding habits that often is encountered, such as those found in the hundreds of coexisting cichlid species in the Great Lakes of Africa (discussed later in this chapter). Detailed studies of fish ecology in small, tropical streams may show a high degree of species segregation on the basis of habitat, diet, and microhabitat (Moyle and Senanayake 1984, Bussing 1993). Other studies, however, frequently find just the opposite. For example, most of the 50 species of fish collected from some streams and small lakes of the upper Amazon were feeding mainly on plant matter of terrestrial origin or on terrestrial insects, especially ants (Saul 1975). Even Goulding (1980), in his study of the extraordinary habits of the fishes that invade the flooded forests of the Amazon, noted that rarely was a food item used only by one species. Most species occurred with several potential competitors. Possible explanations for these differing observations include: (1) The studies were done when certain kinds of food were super-abundant, and segregation occurred at other seasons; (2) segregation occurred at other life-history stages; and (3) the fish populations were regulated by random environmental events or by predation (Rodriguez and Lewis 1994), so they rarely reached a point at which competition and/or segregation was likely.

Symbiosis

Few examples of symbiotic interactions are found among tropical freshwater fishes (Box 30.1), but this probably is mostly a reflection of how poorly known the habits of tropical fishes are in general. Mutualistic relationships among shoaling fishes probably are common, given the frequency with which small South American characins seem to be found together and the mixtures of cichlid species that are found in the Great Lakes of Africa. In Central America, cichlids guarding a school of their own young have been observed to "adopt" the young of other cichlid species in a manner that may reduce predation on their own young while giving the foreign young some protection as well as the opportunity for more rapid growth (McKaye 1977). At the opposite extreme, a number of species of small South American catfishes enter the gill chambers of larger fish, where they feed on the gill filaments and blood (Baskin et al. 1980). At least one of these catfishes feeds mainly on the scales of other fishes, a feeding adaptation (micropredation) that also is found in many South American characins and African cichlids. The scale-feeding behavior of some cichlids is so well developed that in some species, individual fish have their mouths bent either to the left or the right, to make grabbing the scales easier when a victim is approached from behind (Barlow 2000). Still other cichlids and characins feed primarily by taking bites out of the fins of other fishes. In a Brazilian reservoir, a high percentage of the fishes had their fins, especially their caudal fins,

---

BOX 30.1
Symbiotic Interactions Among Fishes in African Lakes

The diversity of symbiotic interactions that are possible among tropical fishes is illustrated by the interactions between various species of catfishes and cichlids in the Great Lakes of Africa. At the mutualistic end of the spectrum are the cichlids and catfish that brood their young together, with both species defending the nest (McKaye 1985). A presumed commensal interaction is the catfish that lays its eggs in the nests of another catfish species but then leaves the care to the host species. It is possible that this results in the juveniles of the intruding species preying on the eggs and young of the host species (Ochi et al. 2001). Cleaning behavior has been documented for a number of cichlids, but it appears to be used mainly to supplement diets consisting of algae (Barlow 2000). A more ambiguous example is the cichlid that cleans ectoparasites from catfish but then takes advantage of the relationship to steal eggs from the catfish nests (Ribbink and Lewis 1982). At the parasitic end is the catfish that somehow gets its embryos into the oral cavity of a mouth-brooding cichlid. The developing catfish young then proceed to consume the young cichlids that are developing along with them, until the young cichlids are gone (Sato 1986).

"cropped" by piranhas (*Serrasalmus spilopleura*). Northcote et al. (1987) characterized the fins as a "renewable resource."

30.4 FISH COMMUNITIES

Because the fish species of tropical fresh waters are so numerous and diverse, it is difficult to find a body of water whose fish communities can be said to be typical of anything but itself. Three examples are presented here, mostly to give an idea of the range of types of fish communities and the interactions that are possible in tropical fresh waters: (1) the rain-forest streams of Sri Lanka, (2) the Kainji Reservoir of Nigeria, and (3) Lake Malawi.

Sri Lanka Streams

The rain-forest streams of Sri Lanka contain few (10–20) species of fish compared with such streams elsewhere, but they have the advantage of having been well studied (Pethiyagoda 1991). The fishes show a remarkable degree of ecological segregation by habitat, microhabitat, and diet (Fig. 30.2). For example, two species of barbs (*Puntius*) can be observed grazing on the same rock, yet one will be feeding mainly on filamentous algae and the other on diatoms. Likewise, when the two species of gobies (Gobiidae) are encountered in the same reach of stream, one species will be found clinging to boulders and the other on patches of sand. A high diversity of body shapes and sizes occurs, and the most similar species usually occur in different habitats (Moyle and Senanayake

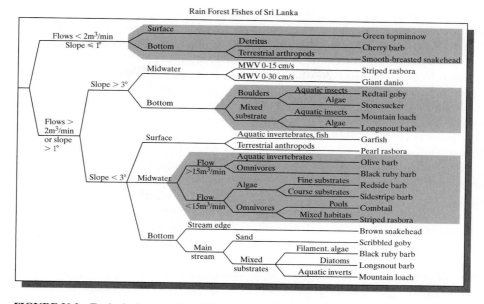

FIGURE 30.2   Ecological segregation of 19 species of fishes in rain-forest streams of Sri Lanka. *MWV* refers to mean water column velocity selected by the fishes. Categories such as *detritus* and *aquatic invertebrates* refer to principal foods; categories such as *boulders* and *sand* refer to substrates selected. Based on Moyle and Senanayake (1982).

1984). When four specialized species were introduced into a stream that naturally contained only a small subset of the rain-forest fish fauna, two of the introduced species retained their narrow niches, and the other two showed niche shifts (Wikramanayake and Moyle 1989). For example, the black ruby barb (*Puntius nigrofasciatus*) shifted from feeding primarily on diatoms to feeding on seeds and plant material. Species introduced into a stream with the fewest native species showed more rapid growth and larger sizes than they showed elsewhere. This suggests that despite the fact the fishes have evolved morphological means to reduce competition, competition still plays a role in structuring the communities. Another aspect of the Sri Lankan fish communities that seems to be fairly typical of tropical-stream fish communities, in contrast to temperate-stream fish communities, is the importance of algae and plant material in the diets of many fishes (e.g., Wootton and Oemke 1992).

Kainji Reservoir

In 1968, the closure of a dam turned 137 km of the Niger River into a 1,280-km^2 reservoir (Lowe-McConnell 1975). From 50 to 60 fish species occur in the reservoir (101 species have been recorded at various times), living in three fish zones. The inshore zone is shallow and contains considerable habitat diversity because of flooded trees and vegetation. Consequently, it also contains most of the fish species and much of the fish production. The open-water zone is dominated by pelagic clupeids and characins as well as by their predators, particularly the tigerfish. The deepwater benthic zone is not well developed because of the shallowness of the reservoir, but it seems to contain mostly omnivorous characins, cyprinids, and catfishes. Despite this apparent general similarity of fish zones to those of temperate reservoirs, tropical reservoirs such as Kainji differ greatly from temperate reservoirs in that the number of species is much larger, the environment is likely to be more constant in both temperature and water level, and piscivorous and omnivorous fish species seem to have a greater predominance.

One of the more interesting aspects of tropical reservoirs is the succession of fishes that occurs following the closure of the dams. The species that dominate such reservoirs mostly are native to the dammed rivers, yet the reservoir fish faunas are considerably different. The preimpoundment fish fauna of the Niger River was diverse but dominated by species that are typical of flowing water, mostly members of the families Mormyridae, Mochokidae (catfishes), and Citharinidae (moonfishes). As the reservoir filled, an initial population explosion of one species of moonfish occurred, followed by (1) increases in the predatory tigerfish and an omnivorous characin species; (2) increases in species of the catfish family Schilbeidae; (3) a slight increase in members of the catfish family Mochokidae, but with a dramatic change in species composition; and (4) a decline in the Mormyridae. As the reservoir aged further, (1) the moonfishes declined, (2) the characins continued to be abundant, (3) herrings (Clupeidae) became abundant in the open waters of the lake, (4) introduced cichlids became abundant, and (5) the mormyrids increased in abundance once again. In general, the species that became abundant in the reservoir were those that were relatively rare in the main river, being mostly characteristic of backwaters and swampy edges. The species that declined after the reservoir filled tended to be bottom feeders (on aquatic insects) or detritus feeders, whereas those that increased were piscivores, planktivores, and omnivores (Lelek 1973, Lewis 1974, Blake 1977).

One reason that such changes are so fascinating is that most of them were unexpected, reflecting our lack of knowledge regarding tropical fish ecology, especially that for large rivers. For example, in Kariba Reservoir, on the Zambezi River, a number of the dominant species now are fishes that formerly were known only from above Victoria Falls, 100 km upstream (Balon and Coche 1974). Previously, Victoria Falls was thought to be a barrier to fish movement, both upstream and downstream. As Balon (1978) points out, the changes brought about by the creation of such larger reservoirs as Kariba most often are more detrimental than beneficial, not only to the local fauna but to the local people as well. An increase in fish production may be balanced by a decrease in animal production from the terrestrial systems as well as by a decrease in agricultural production (from the flooding of vast valleys). Highly predictable systems become unpredictable as species replace one another in response to new perturbations of these systems. Given the rapidity with which tropical rivers now are being "developed," significant parts of the unique fish faunas endemic to these rivers may disappear in the near future.

Lake Malawi

Lake Malawi is one of the Great Lakes of the East African rift region. It usually is compared to the two other large rift lakes, Victoria and Tanganyika, which have their own rich assortment of fish species, especially cichlids. Lake Malawi is quite large (600 km long and up to 75 km wide, with a maximum depth of ≈785 m) and quite old (>2 million years). In part because of its size and age, this lake contains approximately 550 species of fish, all but 4 being endemic to it and 500 of them being haplochromine cichlids. The highly developed pharyngeal "jaws" of cichlids have permitted extraordinary feeding specializations to evolve (Liem 1974), as described below, which in part explains why so many species coexist. The species also segregate by habitat use. Lake Malawi can be divided into five broad habitat types, each with its characteristic fish community: (1) shallow bays and lagoons, (2) mud- or sand-bottomed shore areas of the main lake, (3) open water, (4) deep-bottom water, and (5) rocky-bottomed inshore areas of the main lake (Lowe-McConnell 1975).

The shallow bays and lagoons are a relatively small part of the lake, yet they are the most productive in terms of food fish. They contain a complex of tilapia species (Cichlidae) that feed on small invertebrates. Associated with these interesting cichlid species are two species of *Corematodus*, one of which mimics the tilapias so that it can feed on their scales and the other of which mimics another inshore cichlid species, on which it preys. Other species found in this habitat are large cyprinids and large, predatory catfishes.

The muddy- and sandy-bottomed areas have their own highly specialized cichlids as well as various cyprinids and catfishes (Mochokidae). Among the cichlids are species with shovel-like lower jaws and stout gill rakers for digging in the sand and filtering out invertebrates. The different species feed on different sizes and types of prey. Other species of cichlids feed on organic deposits on the bottom, scrape periphyton from aquatic plants, or feed on the plants themselves. Piscivorous cichlids also are abundant.

The open waters of the lake are dominated by a species flock of zooplankton-feeding haplochromines, which tend to be localized in their distribution near shore or rocky reefs, and the small, pelagic cyprinid *Engraulcypris sardella*. These fishes are preyed on by a number of large cichlids, catfishes, and cyprinids.

The deep-bottom water contains a poorly known group of fishes that live at depths as great as 100 m, below which the water is anoxic. These species are a mixture of catfishes, cyprinids, cichlids, and at least one electric fish (Mormyridae). Many of these species are piscivorous, and they may feed by moving into shallow water or up into the water column, perhaps after dark.

The rocky-bottomed inshore areas, which may have a steep profile, contain perhaps the most ecologically diverse group of fishes found anywhere in the world in such a limited habitat. Most of these fishes are the "mbuna" cichlids. There actually seem to be more fish species present than there are basic types of resources to exploit, so species with similar feeding habits segregate by depth, size of food, method of feeding, and other subtle ways. Cichlids also have developed many ways to feed on other cichlid species. Among the basic feeding types listed by Fryer and Iles (1972) (Fig. 30.3) are:

- Epilithic algae feeders that live by scraping algae from the rocks. A number of different species are in this group, with mouths modified variously for scraping diatoms, combing filamentous algae, or nipping off strands of algae.
- Periphyton collectors that feed mostly on algae and invertebrates scraped from the leaves of higher plants. These species typically have a pointed snout with scraping teeth on the sides of the jaws. They feed by grabbing the leaf and then moving down it sideways, scraping as they go. Such activity does not harm the leaf.

- Leaf choppers that feed by taking chunks out of leaves with their stout teeth.
- Mollusk feeders of two types: shell crushers, and foot grabbers. Shell crushers suck in small molluscs and crush them with their stout pharyngeal teeth. Foot grabbers grab the foot of a snail and, somehow, pull the snail out of its shell.
- Invertebrate pickers that pick small invertebrates from algae beds with long, forceps-like teeth. These species also have the large eyes necessary for precision

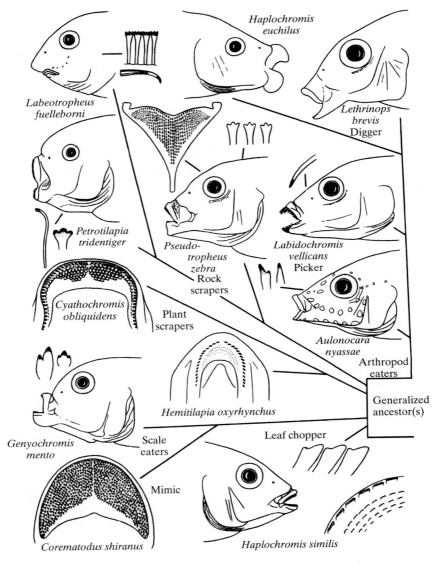

FIGURE 30.3   Feeding mechanisms of "mbuna" cichlids from Lake Malawi, showing adaptive radiation from a generalized ancestor. From Fryer and Iles (1972).

feeding. Other species are more generalized predators on large invertebrates, possessing protractile mouths and bands of inward-directed teeth inside their mouths.

• Zooplankton feeders that consist of a number of species with sucking mouths and long gill rakers. Although they feed in open water, they seldom venture far from rocky or sandy areas.

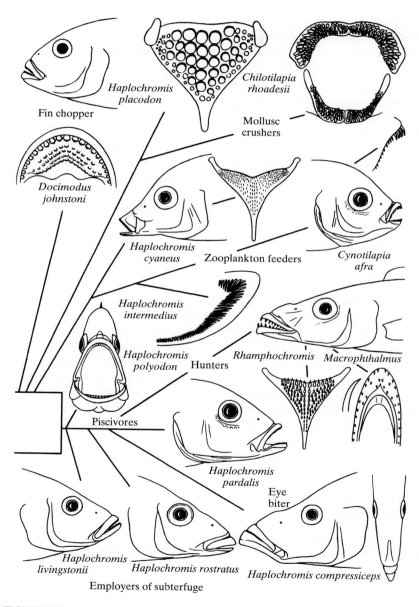

FIGURE 30.3 (*continued*)

- Scale eaters that have a scraping type of mouth similar to those of epilithic algae feeders, so that they can scrape small scales (usually from the caudal peduncle region) of other cichlids and also of cyprinids. Each species typically seems to specialize in a particular type of host or prey. Other scale eaters devour large scales that they obtain by sliding their lower jaw under a scale, clamping down with the upper jaw, and then jerking out the scale. Stealing the scales from other species requires some highly specialized behavior just to get close enough to the prey to be able to grab the scales (Nshombo 1994).
- Fin choppers that sneak up on their prey, presumably in part by mimicry, and then take bites out of their fins.
- Piscivorous cichlids that appear in a number of standard forms, from rover-predator to pike-like, lie-in-wait predator. One of the most bizarre forms is a species that lies on the bottom in plain sight and is colored to look like a dead fish. Any other fish that comes over to scavenge the "corpse" gets eaten (Barlow 2000).
- Egg, embryo, and larval fish eaters that seem to live mainly on the young of other species of cichlids, with different species apparently being specialized for feeding on different stages of development. Because most cichlids they prey on are mouth brooders, they get the brooding fish to jettison their young by ramming them (McKaye and Kocher 1983). Such predation possibly is an important means of population control, and the susceptibility of adults to releasing their eggs or young might depend on population density.

The picture of Lake Malawi that is presented here—as an environment whose resources are extraordinarily finely subdivided among its fishes—has been widely accepted, but it is based mostly on studies of cichlid morphology. Ongoing studies of the lake indicate that whereas cichlid fishes do, indeed, exhibit many amazing morphological and behavioral adaptations, their actual food often is much more diverse than the morphology indicates (Bouton et al. 1997). For example, the abundant zooplankton in the lake is an important food source for many of the cichlids, including those forms that are adapted for bottom feeding. It now seems to be likely that the extreme feeding specializations are an advantage mainly in times of food shortages, which, if they do not occur seasonally, may occur on an irregular basis over a longer period of time. An additional problem is that many pairs or even groups of species possess only small differences in their feeding morphologies. They differ from one another mainly in the color patterns of the adult males (see Section 30.6).

30.5 FISH AS MAJOR PLAYERS IN TROPICAL ECOSYSTEMS

The importance of fish in tropical aquatic ecosystems seems to be obvious when their extraordinary abundance and diversity in clear tropical lakes is noted. Only recently, however, have we really begun to appreciate how much they link the terrestrial and aquatic ecosystems of tropical rivers or how much they move energy and nutrients up and down rivers through long migrations. Goulding (1980) was the first to document in detail the way in which fishes, especially various characins, move in and out of flooded tropical rainforests, consuming seeds and fruits as they go. Their feeding presumably promotes dispersal of the seeds of the trees while bringing nutrients and energy into the main river ecosystems. Because annual flooding is a major event in all tropical rivers, such interactions most likely are the rule rather than the exception. Winemiller and Jepson (1998) demonstrated the complex movements of fish on and off a seasonal flood-plain in Venezuela that resulted in continuous changes in the fish assemblages in the river and the floodplain, with predation by fish being a major structuring force. This floodplain, like others, is a major rearing area for juvenile fishes because of its abundance of cover and invertebrate food. As the floodplains dry, the fishes move back into the rivers, during which time they are heavily preyed on by other predators and humans.

The importance of long-distance migrations of fishes in tropical rivers is another aspect of tropical rivers that is just beginning to be appreciated (Barthem and Goulding 1997, Winemiller and Jepson 1998). In the Amazon and the Mekong rivers, large cat-fishes migrate hundreds of kilometers to their spawning and feeding areas, as do several characiform fishes in the Orinoco and Amazon rivers. The characiform prochilodus (*Semiprochilodus kneri*) in the Orinoco River inhabits whitewater rivers during adult-hood but spawns in nutrient-poor blackwater rivers; Winemiller and Jepson (1998) suggest that this migration provides a major subsidy to the food webs of the blackwater streams. Likewise, pimelodid catfishes spawn in the upper reaches of the Amazon, but their new juveniles actually rear in the freshwater estuary 3300 km downstream (Barthem and Goulding 1997). Under high flows, they can be flushed into the estuary in only 13 days. The catfish rear in the estuary for several years, where they are abundant enough to support a fishery. They then move up into the main river until they reach adult size ($\approx$2 m). Thus, these abundant catfish alone have the potential to integrate food webs up and down the entire river! A similar situation seems to exist for pangasid catfish in the Mekong River (Hogan 2003). Unfortunately, construction of dams and overfishing are disrupting these movements in many rivers, with unknown long-term impacts on the river ecosystems.

30.6 SPECIATION IN AFRICAN CICHLIDS

The diversity of endemic cichlids in lakes Malawi (500 species), Victoria (300 species), and Tanganyika (165 species) is extraordinary by any standards; arguably, these fish constitute 4% of the world's fish fauna. The big question here is how did so many similar species come to exist in these lakes? The answer is made even more interesting by recent discoveries that the lakes have undergone major fluctuations in both size and depth during the past few thousand years (Stager et al. 1986; Owen et al. 1990). At least one area of Lake Malawi that was dry approximately 200 years ago now supports endemic species. The trench-like Lake Tanganyika apparently did not suffer so much during these droughts but no doubt still had major changes in depth and area. The differences among the lakes are reflected in the evolutionary history of their cichlid faunas. Lake Tanganyikan cichlids are derived from at least five distinct evolutionary lines that diverged from a common ancestor millions of years ago, although approximately half the fish are from just one of these lines (Nishida 1991). In contrast, the cichlids of lakes Victoria and Malawi appear to be derived from relatively recent invasions by a single species in each lake. The 300 species of cichlids in Lake Victoria are nearly identical genetically; less variation is found among them than within most *single species* of vertebrates (Meyer et al. 1990). This implies that the flock of species could be derived from a single female carrying young in her mouth.

Clearly, speciation in cichlids has taken place very rapidly in these lakes. Traditionally, the speciation has been explained as either ***allopatric speciation*** or as ***microallopatric speciation*** (Lowe-McConnell 1994), but more likely, it is an example of ***sympatric speciation***, perhaps in combination with microallopatric speciation. Allopatric speciation requires each species to have initially evolved independently, in isolation from the others, followed by invasion into the lake. Conceivably, during drought periods, each lake could break up into multiple basins, thereby allowing speciation to occur before refilling, with multiple events resulting in the large number of species. Alternately, parts of the lake could become temporarily isolated through geological processes. For example, a sandbar separated Lake Nabugabo from Lake Victoria approximately 4,000 years ago, and the lake has developed its own flock of five endemic cichlid species. Unfortunately, traditional allopatric speciation is unlikely to explain the sheer number of species and requires a rather unlikely series of events to have occurred in all the lakes. It also does not explain how rocky points in the lakes obtained their own endemic faunas.

Microallopatric speciation has been the explanation put forth by most workers on the cichlid species flocks (e.g., Fryer and Illes 1972). This explanation is based on the observation that the shorelines of the lake tend to be broken up into isolated islands of habitat; rocky points alternate with sandy beaches or swampy areas. The mbuna (rock-dwelling) cichlids in particular have been characterized as being very strongly tied to specific habitats, and because they are mouthbrooders, their young become imprinted on the same habitats. Such fish presumably are very reluctant to move across areas of different habitat, a reluctance that is enforced by predatory fish, birds, and mammals that pick off fish that look out of place. The best evidence for this theory is that most mbuna species are found only in small parts of the lake; no mbuna species has a lakewide distribution.

Microallopatric speciation may work as an explanation for the mbuna cichlids, but it does not explain the diversity of other, more widely distributed forms, such as the plankton feeders. It also does not explain the rapidity of the speciation events, with large numbers of species evolving from a single ancestor in only a few thousand years.

Sympatric (ecological) speciation has been proposed for cichlids in the Great Lakes and other lakes in Africa (Schliewen et al. 1994). The most plausible mechanism is sexual selection by females for distinctly colored males (Seehausen et al. 1997). The invocation of sympatric speciation requires the convergence of several of disparate facts in addition to the evidence of rapid speciation:

- In captivity, all the species can interbreed and produce fully fertile hybrids.
- Cichlids have color vision, with absorbance peaks in the eye pigments for the blue and red parts of the spectrum.
- Males are brightly colored, mostly reds, yellows, and blues, whereas females are drab.

- Being brightly colored has a cost: Bright males are more vulnerable to predators.
- Lakes with species flocks are very clear; turbid lakes have few cichlid species.
- Females prefer males with bright colors.
- Individual females have genetically based color preferences; they consistently seek males of one color, even in species that are polymorphic for color. Under experimental conditions where light is kept to a single color, females cannot distinguish males of their species from those of closely related species of a different color (Seehausen et al. 1997).
- The pharyngeal jaw apparatus of cichlids is relatively easy to modify, both phenotypically and genotypically (Liem 1980).

Sexual selection by females promotes speciation as follows: If a male is a bit different in color from its compatriots, it will be preferred by the females. Presumably, a male that looks different but that still can avoid been eaten by predators is a superior male; thus, it is worth selecting. Just as natural variation occurs in the color patterns of males on which female selection can act, natural variation also occurs among females as to which color of male they prefer. So, a given population will have strong selection for males of different colors, usually red/yellow and blue. Under this scenario, the color groups will very quickly sort themselves out, as generations of blue-loving females select males that are brighter and brighter blue and as red or yellow-loving females select brighter and brighter red/yellow males. The next stage is for the noninterbreeding groups to diverge in feeding habits and morphology to reduce competition for scarce resources. Specialization or extinction of one of the color morphs seems to be a likely alternative in this scenario. Here, too, cichlids are unusually plastic in their ability to modify their pharyngeal feeding apparatus, both in response to behavioral specialization on prey types and through rapid natural selection. Like the color pattern of males, the feeding apparatus of cichlids can show considerable variation in structure and, in some species, can be polymorphic (Meyer 1990). Thus, there appears to be the opportunity for rapid specialization in diet following isolation through sexual selection.

The hypothesis that sexual selection can promote speciation in cichlids needs further testing. It is very appealing, however, in that it ties together many bits of information. The actual events leading to the development of these large species flocks likely were a combination of allopatric, microallopatric, and sympatric speciation events. The rapidity of the speciation, however, is a strong indication that sexual selection was the driving force. Unfortunately, strong evidence indicates that reversal of the process is leading to a loss of diversity. In Lake Victoria, increasing eutrophication and runoff of silt from deforested areas are making the lake more opaque. In areas of low water clarity, usually just one mbuna species of a particular species group manages to persist; in areas of higher water clarity, multiple species persist, even in the face of predation from the introduced Nile perch (Seehausen et al. 1997). The high turbidity is apparently both causing species to hybridize and preventing further speciation, thereby putting a halt to the most extraordinary example of explosive speciation and adaptive radiation among vertebrates (Galis and Metz 1998).

30.7 LESSONS FROM TROPICAL FISHES

The remarkable fish communities in tropical waters are just beginning to be understood, however, they also are being severely disrupted by human activities, such as dam building, introduction of exotic species, pollution, and commercial fishing. Clearly, many of the communities are highly structured, often containing species whose narrow specializations are rivaled only by those of coral-reef fishes. The specializations and the intricate interactions both among the fishes and between the fishes and their environments can only be called awesome. Until recently, we had no suspicion that:

- There might be a dependent relationship between fruit-eating fish and trees in the Amazonian floodplain forest.
- Complex communities of fishes communicate mainly with electric signals.
- Communities of fishes live in the leaf litter of acid backwaters.
- The extraordinary assemblages of specialized fish in the Great Lakes of Africa may have evolved, at least in part, in less than 500 years.
- The single biggest human-caused extinction event in historic times was the elimination of hundreds of species of cichlids from Lake Victoria through a combination of the introduction of a predatory fish, eutrophication, and overfishing.

The last revelation indicates that tropical systems may be extremely fragile and need extraordinary conservation efforts to protect them.

SUPPLEMENTAL READINGS

Craig 1992; Echelle and Kornfield 1984; Goldschmidt et al. 1993; Goulding 1980; Lowe-McConnell 1975, 1987, 1994; Lundberg 2001; Payne 1986; Welcomme 1985; Winemiller and Jepson 1998.

WEB CONNECTIONS

GENERAL
www.calacademy.org/research/ichthyology
www.fishbase.org
www2.biology.ualberta/jackson.hp/IWR
www.flmnh.ufl.edu/fish/default.htm
www.people.clemson.edu/~jwfoltz/wfb418

NEOTROPICAL FISHES
www.neodat.org

CICHLIDAE
cichlidresearch.com
malawicichlids.com

C H A P T E R 31
Estuaries

E stuaries are transitional environments between fresh water and salt water. Essentially, they are bays whose waters are significantly diluted by water flowing in from rivers. Estuarine fish faunas consequently are a mixture of tolerant species from both marine and freshwater environments, species migrating from one environment to another, and a small number of resident species. The fluctuating physical and chemical environment limits the number of species in estuaries, but the abundance of nutrients allows estuaries to support large concentrations of individuals. These individuals often belong to economically important species or are major contributors to the food webs that support such species. An estimated 8% of the total commercial marine-fish catch comes from estuaries, which is high considering their relatively small area and tendency to have overexploited fisheries (Houde and Rutherford 1993). Unfortunately for fish, estuaries also typically are the focus of major urban areas and so are likely to be highly disturbed and polluted. Fortunately, estuarine fishes show remarkable resiliency and respond quickly to environmental improvements. Thus, they are worth considering in some detail, because they present many interesting challenges to the fish as well as to the biologists who study them. In this chapter, we consider (1) characteristics of estuaries, (2) types of estuarine fishes, (3) factors affecting the distribution and abundance of the fishes, (4) estuarine food webs, and (5) estuarine fish communities.

31.1 CHARACTERISTICS OF ESTUARIES

The characteristics of each estuary depend on its size, shape, geological history, location, amount and quality of inflowing fresh water, and nature of the surrounding land. All estuaries, however, have two important characteristics in common: the harshness of the physical and chemical environment, and high concentrations of nutrients. The harshness of the estuarine environment is caused by the mixing of fresh water and salt water. This creates not only salinity gradients but also simultaneous temperature gradients, because salt water and fresh water rarely are of the same temperature. The gradients created by the mixing are not stable phenomena but move up and down the estuary on a daily basis, according to the tidal cycles. They also are affected seasonally by the amount of fresh water entering the estuary and by oceanic storms that push in more salt water. The inflowing water carries suspended organic and inorganic matter, which may create gradients of turbidity and dissolved oxygen as well. Not only are there strong gradients from the head of the estuary to its mouth, there are gradients in the water column as well, because fresh water is less dense than salt water and tends to "float" on top.

The fluctuating nature of estuarine environments means that estuarine fish must expend considerable amounts of energy adjusting to the changing conditions, either metabolically or by moving about in search of less-stressful conditions. These energy costs usually are easy to pay. The same factors that create harsh conditions also cause nutrients to become concentrated, and the nutrients support large populations of food organisms. The nutrients mostly are associated with detritus that is washed in with fresh water or is created from the decay of plants in surrounding marshlands. Salt marshes are particularly important as sources of nutrients in many temperate estuaries, as are mangroves in tropical and subtropical estuaries (see Chapter 32). Such fringing vegetation also is important as cover and substrate for many estuarine organisms. The high turbidity of estuaries typically limits photosynthesis, but phytoplankton is another important source of energy input. Today, sewage often is one of the most important sources, although many efforts are underway to reduce this usually undesirable source of nutrients. The constant mixing that occurs in estuaries assures that most of the nutrients are recycled within them and that major losses occur only during times of flood.

At least as important as the mixing processes in the retention of nutrients within estuaries are the activities of the invertebrates. Estuaries have large concentrations of filter-feeding zooplankton, particularly copepods, that feed on detritus and phytoplankton in the water column and are preyed on by fish. The fecal pellets from the zooplankton and fish drop to the bottom, where they form part of the organic ooze that serves as food for the abundant benthic invertebrates, such as amphipods and nereid worms. Often, even more abundant on the bottom are clams and oysters, which filter-feed from the water column. Because estuarine currents typically concentrate nutrients and zooplankton in the upper, low-salinity (2–12 ppt) parts of the estuary, benthic invertebrates and fishes often show peak abundances in this region as well. The same processes that concentrate nutrients in estuaries also concentrate pollutants, including pesticides and heavy metals. Not only may this have direct, adverse effects on estuarine faunas (which often are naturally under stress), but it may render the surviving organisms toxic to humans.

31.2   ESTUARINE FISHES

Fish found in estuaries are of five broad types: (1) freshwater, (2) diadromous, (3) true estuarine, (4) nondependent marine, and (5) dependent marine. Typically, an estuary has representatives of all five types, although their relative abundance varies from season to season and from locality to locality. On a worldwide scale, however, major differences are found among estuaries. In tropical estuaries, anadromous fish are insignificant parts of the estuarine fish faunas compared with the situation in temperate estuaries, whereas species that spend their entire life in estuaries tend to be more important in tropical than in temperate estuaries (Potter et al. 1990, Blaber 2002).

Freshwater fish may complete their entire life cycle in the upper reaches of estuaries and invade the lower reaches in response to decreasing salinities. Most "true" freshwater fishes are not found at salinities higher than 3 to 5 ppt, and even the most tolerant species are not found at salinities much higher than 10–15 ppt. Examples of freshwater fishes that will live in the upper reaches of North American estuaries are white catfish (*Ictalurus catus*), mosquitofish, various sunfish species, and common carp (nonnative). Examples of species found in tropical estuaries include Mossambique tilapia (*Oreochromis mossambicus*, Africa), archerfishes (Toxotidae, Southeast Asia), livebearers (Poeciliidae, Central America), and pimelodid catfishes (South America).

Diadromous fish are found in large numbers within estuaries as they pass through on their way to either fresh or salt water. The estuaries frequently act as staging areas for anadromous fishes; for example, salmon and shad (Clupeidae, including tropical species) may remain in them for several days or weeks before finally moving upstream. For many anadromous species, estuaries also serve as nursery grounds for the young. American shad (*Alosa sapidissima*) typically spend the first few months to a year of life in estuaries. Some of the most important estuarine fishes can be labeled as *semianadromous*, because they spawn just above the head of the estuary, use the estuary as a nursery area, and may or may not go out to sea. Examples include striped bass and some sturgeons.

True estuarine fish, or those that usually spend their entire life cycle in estuaries, are represented by a relatively small number of abundant species worldwide. Whitfield (1994) speculates that the temporary and fluctuating nature of estuarine environments mitigates against the evolution of true estuarine species, although they often are among the most abundant fishes in estuaries, especially in the tropics (Blaber 2002). True estuarine fishes are particularly scarce on the Pacific coast of North America, where estuaries are geologically very young. In the San Francisco Estuary of California, for example, only delta smelt (*Hypomesus transpacificus*) falls into this category. In the older, Atlantic coast estuaries, true estuarine fishes are more abundant and include white perch (*Morone americana*) and spotted seatrout (*Cynoscion nebulosus*), both of which are of considerable importance to anglers. In tropical and subtropical estuaries, many gobies (Gobiidae), anchovies (Engraulidae), mullets (Mugilidae), croakers (Sciaenidae), and other resident fishes are major players in estuarine ecosystems.

Nondependent marine fish are commonly found in the lower reaches of estuaries but do not depend on them to complete their life cycles. Usually, they make up approximately half the species list for an estuary but typically are abundant only seasonally or in the saltiest part of the estuary (e.g., Thiel et al. 1995). In subtropical and tropical estuaries of Australia and South Africa, nondependent marine species can make up as much as 70% of the estuarine fish fauna (Potter et al. 1990). These species may be important parts of estuarine ecosystems, but they also are important in shallow-water marine environments in general. On the Pacific coast of North America, two abundant estuarine species are in this category; staghorn sculpin (*Leptocottus armatus*), and shiner perch (*Cymatogaster aggregata*). On the southern Atlantic coast, typical species include pinfish (*Lagodon rhomboides*), weakfish (*C. regalis*), harvest (*Peprilus alepidotus*), and pigfish (*Orthopristis chrysoptera*) (Rulifson 1985, Peterson and Ross 1991).

The species and abundances of nondependent marine fish have a strong relationship to the oceanic conditions. In the Thames River Estuary of Great Britain, herring and most flounders become less abundant when ocean waters become warmer, whereas a variety of more southern fishes, such as seabass (*Dicentrarchus labrax*), become more abundant (Attrill and Power 2002). Some of these fishes use estuaries on a facultative basis as nursery areas, thereby resulting in increased production of commercially important fishes.

Thus, understanding how climate shifts affect the use of estuaries by fishes can have major implications for the management of fisheries.

Dependent marine fish usually spend at least one stage of their life cycle in estuaries, using them as spawning grounds, as nurseries for the young, or as feeding grounds for the adults. Species such as the herrings that use the estuaries for spawning are relatively few, and they tend to attach their eggs to submerged plants or objects. The turbulence of estuaries likely creates problems for eggs that passively float in the water column. The main advantage of laying eggs in estuaries is that the young hatch close to abundant supplies of food. Most marine species whose young take advantage of estuaries spawn outside the estuaries and then the young migrate into them. Examples of this strategy are common in Atlantic and Gulf coast estuaries, with groups such as croakers (Sciaenidae) and menhadens (*Brevoortia* spp.). One of the few examples of this strategy on the Pacific coast seems to be the starry flounder (*Platichthys stellatus*); its abundance appears to be highest during years with a high freshwater outflow through estuaries. In tropical estuaries, it is thought that a major attraction for marine dependent fishes is mangroves, which serve as nursery areas for the young, but their importance in comparison to other factors has not been firmly established (Blaber 2002).

31.3   FACTORS THAT AFFECT DISTRIBUTION

The distribution and abundance of estuarine fishes are determined primarily by physical and chemical factors and only secondarily by biological factors. One of the main reasons for this is that most estuarine fishes are part-time residents. They move in when conditions are favorable to take advantage of abundant food, either for themselves or for their young, but then move out when the physical and chemical conditions become too severe. Dahlberg and Odum (1970) noted that only approximately half of the 70 fish species that they collected from a Georgia estuary could be found there year-round. Results from other estuaries are similar (Fig. 31.1). The seasonality of estuarine fish populations seems to be created primarily by their responses to temperature and salinity, but oxygen levels, marsh vegetation, predation, and interspecific competition also may play a role, as may, increasingly, invasions by nonnative species.

Temperature is probably the single most important factor affecting fish distribution both between and within estuaries seasonally, although the effects of temperature are closely tied to the effects of other variables. The often striking differences between the summer and winter fish faunas of temperate estuaries probably result, in large part, from the temperature tolerances and preferences of the different species, as does the gradual change in the composition of estuarine faunas from north to south along the Atlantic coast of North America. For example, gulf menhaden (*B. patronus*) in Gulf coast estuaries are most abundant at temperatures between 25°C and 35°C. The distribution of menhaden, however, also is strongly influenced by salinity and food supply; Young menhaden seek out the upper reaches of estuaries, where salinities are low and detritus is abundant (Copeland and Bechtel 1974). In contrast to temperate estuaries, tropical estuaries typically are warmer than 24°C year-round, and temperature rarely seems to be a limiting factor by itself.

Salinity exercises a strong influence over the distribution of fish within estuaries (Table 31.1). It has a particularly strong influence over the composition and structure of

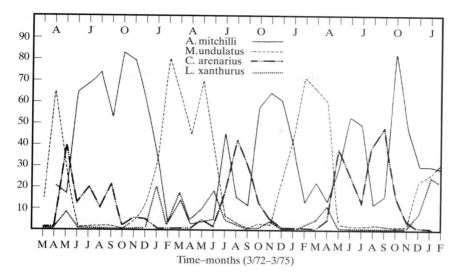

FIGURE 31.1   Relative importance (percentage of total caught) of four species of estuarine fishes at different times of year in the Apalachicola Bay estuarine system of Florida. From Livingston et al. (1975).

TABLE 31.1   DISTRIBUTION OF COMMON FISHES IN THE NAVARRO RIVER AND ITS ESTUARY, CALIFORNIA, IN RELATION TO SALINITY

Species	Classification[a]	O[b]	O[c]	1	3	9–10	23–25	30+
Sacramento sucker	1	X[d]	X	X				
California roach	1	X	X	X	X			
Prickly sculpin	2	X	X	X	X	X	X	X
Rainbow trout	2	X	X	X	X	X	X	R
Threespine stickleback	1, 2	X	X	X	X	X	X	X
Starry flounder	3		X*	X*	X*	X*	X	X
Shiner perch	3		R*	R*	X*	X	X	
Jacksmelt	3			R*	X*	X	X	
Bay pipefish	3					X	X	X
Plainfin midshipman	3[e]					X	X	X
Penpoint gunnel	3						X	X
Pacific herring	3[e]						X	X
Surf smelt	3						X*	X
Northern anchovy	3						X	X
Lingcod	3							X*

From unpublished data of D. H. Varoujean and P. B. Moyle, 1973–1976.
[a]1, Freshwater; 2, diadromous; 3, nondependent marine. No true estuarine or dependent marine fish are in this estuary.
[b]More than 1 km upstream from first riffle.
[c]Just above first riffle.
[d]X, the fish were present and common; R, the fish were present but rare; *, predominantly young-of-the-year fish.
[e]Spawning.

fish assmblages within an estuary (Thiel et al. 1995). The intermediate and fluctuating salinities that are typical of estuaries help to keep the number of species down, because they prevent stenohaline marine and freshwater fishes from penetrating far into estuaries. Most estuarine fish are capable of living in a wide range of salinities, and the most tolerant species, such as striped bass and mullets, can survive abrupt transfers from fresh water to sea water. Some estuarine fish tolerate abrupt changes in salinity caused by sudden increases in freshwater inflow and stay in one area, but other species, such as spotted seatrout, move to more saline regions. The life-history stages of a species also may differ in their ability to survive changes in salinity. Young fish that are unable to avoid low-salinity waters may, when confronted by sudden increases in freshwater inflow, suffer mass mortalities, either from osmotic shock or from being flushed into less-productive portions of the estuary, where they can starve.

A major problem for estuaries, particularly in the more arid parts of the world, is the diversion of rivers that flow into estuaries and that provide the salinity gradients on which various species depend. Estuaries deprived of their freshwater inflow not only become increasingly saline, they have less capacity to absorb and to flush out pollutants. They also have a greatly reduced capacity to support estuarine-dependent species, many of which are important to fisheries.

Oxygen levels usually are high in estuaries because of the constant inflow and mixing of both fresh water and salt water, although the naturally high levels of organic matter may reduce oxygen levels during times of low flow. In the Elbe River estuary of Germany, some species avoid areas with low oxygen, but eels (*Anguilla anguilla*) have their highest abundance in such areas (Thiel et al. 1995). The modern practice of dumping sewage into estuaries increases the amount of organic matter in estuarine systems and may reduce oxygen levels to the point where fish will not enter them or, if they are already present, will be highly stressed or killed. Reversal of this practice, as happened in the Thames River estuary, can result in recovery of fish populations to the point where fisheries can be reestablished (Colclough et al. 2002).

Other physical factors that can strongly influence estuarine fishes are redundant turbidity, tidal range, and current speeds in channels (from both tidal and riverine sources). These factors interact strongly with temperature, salinity, and dissolved oxygen to produce the complex patterns of distribution that often are observed in estuaries. Catastrophic events, such as major floods, also can cause long-lasting changes to estuaries (Box 31.1).

Marsh vegetation, from salt marshes to mangroves, has long been recognized as being an important source of organic matter to "power" estuarine ecosystems. It also, however, plays a major role as cover for the larvae and juveniles of many fishes. In a Louisiana estuary, various species, such as gulf menhaden, spot, and spotted seatrout, appear to rely, at least partially, on vegetation as cover during their early life-history stages (Baltz et al. 1993). In addition, naked goby (*Gobisoma bosci*), darter goby (*Gobionellus boleosoma*), and speckled worm eel (*Myrophis punctatus*) are species that appear to spend their entire life cycle in the vegetation. Submerged plants (e.g., eel grass) also are very important as cover for fish (Humphries et al. 1992). The influences of vegetation on growth and survival of juvenile estuarine fishes, however, appear to be secondary to the influences of physical and chemical variables, such as temperature, salinity, and dissolved oxygen (Baltz et al. 1998).

BOX 31.1

Life on the Edge: Fish in a Newfoundland Estuary

Fish diversity decreases and seasonal patterns become stronger in the south-to-north gradient of estuaries. In a study of the shallow waters of an estuary in Newfoundland (Trinity Bay), Methven et al. (2001) collected 22 species during 32 months of sampling. Threespine sticklebacks made up 85% of the fish collected, with another stickleback, a smelt, and two species of cod making up an additional 10%. Most of the fish (65%) collected were adults of relatively small species that spawn on the bottom in shallow water. Fish were most abundant and diverse in the summer, when temperatures approached 17°C, but catches became extremely low in the winter, when temperatures were in the 0°C range. Even the five common species largely disappeared in the winter, in part because they moved to deeper water. Curiously, Atlantic seasnails (*Liparis atlanticus*) appeared only in the depths of winter, apparently to breed. The conclusion of this study was that seasonal changes of temperature and light structure the fish assemblages, although the lack of much structure in the habitat sampled may have limited diversity. The study also suggested that unlike the situation in more southern estuaries, a majority of the fish using far northern estuaries are adults and benthic spawners (as opposed to juveniles and pelagic spawners).

Predation is an important process in estuaries, because most of the species are carnivores. The populations of the carnivorous fishes are large, and the impact of their predation on both invertebrates and fish populations seems to be considerable. Large concentrations of predators also may deplete local prey populations. For example, spot (*Leiostomus xanthurus*) may severely limit the density of benthic invertebrates on the soft bottoms of Atlantic coast estuaries (Virnstein 1977). Large schools of plankton-feeding menhaden may cause the water to be much clearer behind the school than ahead of it, presumably because the menhaden have filtered out most of the edible organisms (McHugh 1967). Plankton production, however, often is so high in estuaries that high densities typically are present despite intense predation. Benthic invertebrates similarly compensate for high losses because of predation through rapid growth and short generation times. Young fish also are subject to intense predation in estuaries, but their heavy mortality rates also appear to be offset, in part, by their rapid growth rates. Rapid growth enables them to more quickly reach sizes at which they are less vulnerable to predation and to achieve adult (i.e., reproductive) status more quickly than they could in areas where food is less available. In general, the physical and chemical fluctuations of the estuarine environment seem to have a much greater influence than predation on fish populations.

Competition, like predation, does not seem to be as important as environmental fluctuations in regulating the distribution and abundance of estuarine fishes. Most studies concerning the food habits of estuarine fishes show a high degree of overlap among all the species present. One reason for this is that the number of species of abundant invertebrates in an estuary typically is low. Nevertheless, it can be argued that the staggered use of estuaries by estuarine-dependent marine fishes may be, at least partially, a mechanism by which the species reduce interspecific competition for food by their young (Livingston 1976). In the York River estuary of Virginia, 10 species of croaker (Sciaenidae) manage to segregate partially on the basis of feeding habits (as reflected in body shape and mouth structure) and partially on the basis of differences in distribution within the estuary and timing of the use of the estuary (Chao and Musick 1977). In a tropical Australian estuary, there are guilds of fishes with similar diets, but the most abundant species change on a monthly basis. Thus, the trophic structure of the fish assemblages remain more constant than the species composition (Wilson and Sheaves 2001).

Invasions by nonnative species are a growing problem in estuaries worldwide, especially through the dumping of ballast water by ships (Carlton and Geller 1993). The most extreme example of an invaded estuary is the San Francisco Estuary of California, where several new species successfully invade every year and much of the biota is now alien species (Cohen and Carlton 1998). The result is a highly altered ecosystem with many endangered native species. For example, the shimofuri goby (*Tridentiger bifasciatus*) from Japan invaded during the 1980s and had a population explosion at the same time that the estuary's resident fauna was declining (Matern et al. 2002). Although abundant, the goby presumably was not a cause of the declines of potential competitors and predators but a beneficiary from them. A major contributing factor to the decline of the resident fauna, however, was the spectacular invasion of an Asian clam, whose filter-feeding has greatly reduced populations of the zooplankton that are required as food by larval fishes (Bennett and Moyle 1996). The clam has become a major item in the diet of the native white sturgeon. It is feared, however, that even this may ultimately be harmful, because the clams are concentrating selenium and other heavy metals, which are then concentrated further in the sturgeon. Other estuaries also are suffering more and more frequent invasions, so the biota of the world's estuaries is becoming increasingly cosmopolitan as new, hardy species invade and native species disappear.

31.4   FOOD WEBS

Food, particularly in the form of detritus, detritus-feeding invertebrates, and small fishes, is abundant in estuaries, but the availability of any particular type of food is likely to show considerable fluctuation even over short periods of time. Consequently, most estuarine fishes are not specialized feeders. Each species or life-history stage shows a preference for some general type of food, such as small fish or benthic invertebrates, yet sooner or later, almost every potential source of energy will appear in the diet. As a consequence of the flexibility of fish-feeding habits, at any given time it is unlikely that any source of food will not be used. The simplicity of estuarine food webs, compared to other inshore marine systems, is caused not only by the lack of extreme specialization of many species but also by the small number of species that are abundant at any one time. For example, Dalhberg and Odum (1970) found that only 12 of 70 species in a Georgia estuary made up more than 90% of the individuals and that, at any given time, only three to five of the common species were present in large numbers. In a subtropical estuary, 17 of 119 species made up most of the catch—with 45% being just two species (Kupschus and Tremain 2001). In many temperate estuaries, the number of species increases in summer, when temperatures are higher and outflows from rivers are low, but most of the additional species are relatively rare. In contrast, the number of species in tropical estuaries tends to remain consistently high (100–200 species) year–round (Blaber 2002).

31.5   FISH COMMUNITIES

The species that make up estuarine fish communities change constantly, yet the basic structure of the communities usually is fairly consistent, or at least predictable, in the absence of major disturbances (Livingston 1976). This stability is the result of (1) the regular distribution of species along gradients of salinity, temperature, and other variables; (2) the regular movements, often seasonal, of fishes in and out of estuaries; (3) the dominance of estuaries by relatively few but interchangeable (in terms of trophic position) species; and (4) the robust food webs. These aspects of estuarine fish communities can be illustrated by the fishes that are found in estuaries along the coast of the Gulf of Mexico, but they also seem to apply widely to estuaries in general (e.g. tropical estuaries [Wilson and Sheaves 2001] and California [Matern et al. 2002]). Long-term changes in estuarine fish faunas are likely, however, as ocean temperatures and sea levels rise in response to global warming (Box 31.2).

BOX 31.2

Long-Term Changes in a Portuguese Estuary

The Ria de Averio lagoon is an estuary on the Portuguese coast whose fish fauna has been studied since the early twentieth century, so it provides some idea of how estuarine fish faunas can change over time (Pombo et al. 2002). The lagoon is shallow, turbid, and highly productive, with a strong connection to the Atlantic Ocean and inflow from several rivers. Over the past century, it has become increasingly altered (e.g., dredging of ship channels), polluted (both organic and industrial pollutants), and overfished. Ninety-two species of fishes have been collected from the lagoon, and the overlap among species from the early 1900s to the present is approximately 75%. Since the early sampling, 13 species have disappeared, and 29 have been added (including 20 species that are now fairly common). During the early part of the century, most fishes were nondependent marine species that were benthic, oriented to soft bottoms, and had pelagic eggs. Today, benthic fishes remain important, but more resident species and more species that lay eggs on hard substrates are found. These fishes also generally are more tolerant of adverse water quality. Much of the shift in species took place during the last 25 years. These changes are attributed largely to human-caused changes to the environment, but overfishing has been a problem as well. For example, sea lampreys were eliminated by fisheries during the early twentieth century. A number of the recent arrivals are subtropical fishes, thereby suggesting that rising ocean temperatures are allowing these species to move further north. Overall, fish have remained abundant and diverse, and the basic trophic structure has remained similar through the years. Pombo et al (2002) therefore conclude that the fish community is resilient, but needs to be monitored as an indicator of water quality and other conditions. Without reversal of the trend of deterioration from human causes, however, it is hard to be optimistic about the future of the lagoon and its fishes.

Although the fishes of Gulf coast estuaries have definite preferred temperature and salinity ranges, these ranges tend to be broad, so the alignment of species along environmental gradients often is hard to detect. Nevertheless, distinct differences usually are found between the fish communities at the upper ends of the estuaries and those of the more saline and less thermally fluctuating lower ends. The most numerous fishes at the upper ends usually are small planktivores, such as the bay anchovy (*Anchoa mitchilli*, a year-round resident) and juvenile gulf menhaden (an estuarine-dependent marine species). Croakers, typically juveniles, also are abundant, however, along with some "true" estuarine species, such as the hogchoker (*Trinectes maculatus*). Freshwater fishes often are present at low numbers, as are anadromous species, such as shad and Atlantic sturgeon (*Acipenser oxyrhynchus*). As the water becomes more saline, nondependent marine fishes dominate the species list, but each species generally is uncommon. In terms of numbers of individuals, the dominant species are still true estuarine forms and estuarine-dependent marine species.

The exact composition of the fish community at any given place in a Gulf coast estuary depends strongly on the season of the year (Fig. 31.1). The dominant species, particularly the various croakers, have a tendency to peak in abundance at slightly different times of the year. These peaks are related mostly to the influx of young, although the exact timing and size of the peaks also are related to natural fluctuations in the estuarine environment as well as to human-caused disturbances. Both dependent and nondependent marine fishes show seasonal patterns of abundance in estuaries. In a Georgia estuary, sea catfish (Ariidae) were found almost exclusively in the summer, whereas hake (Gadidae, *Urophycis* spp.) were found only in the winter (Dahlberg and Odum 1970). The seasonal use of estuaries by fish species may show considerable variation from one estuary to another. For example, silver seatrout (*Cynoscion nothus*) are found in Georgia only in the summer but are most common in Texas estuaries in the winter.

The most numerous fishes in Gulf coast estuaries usually are juveniles that are feeding on plankton, but the biomass often is dominated by larger individuals that are feeding on fish, benthic invertebrates, detritus, or all three. Despite seasonal changes in the fish fauna, no major source of food is neglected. In Apalachicola Bay of Florida, bay anchovies usually are the dominant planktivore in the summer, whereas juvenile Atlantic croakers (*Micropogon undulatus*) are dominant in the winter (Livingston et al. 1975). Similarly, Perret and Caillouet (1974) indicate that at least one species of benthic-feeding croaker is always present in a Louisiana estuary.

31.6   LESSONS FROM ESTUARINE FISHES

Despite the high natural variability of estuarine conditions, estuarine fishes typically have predictable patterns of distribution and abundance. On a broad scale, the number of species using estuaries increases from temperate to tropical areas, as does the proportion of species that are year-round residents. Within estuaries, species succeed one another along complex environmental gradients both within and across each estuary as well as from season to season. Interactions among fish species typically seem to be less important in regulating fish numbers or assemblage composition than abiotic factors, although interactions between fish and invertebrates (especially mollusks) may be more important than is generally recognized. The importance to juvenile fishes of estuarine marshes and mangroves also is achieving increased recognition. Estuarine fish populations can change dramatically in response to severe disturbances, particularly those that result from human activities. Although estuarine fish and food webs are very hardy and flexible, the low diversity of fishes found in many urban estuaries today attests to the fact that even estuarine fish have limits to their tolerance of extreme environmental conditions. To add insult to injury, alien species of fish and invertebrates are being introduced into estuaries worldwide. Interest is growing in estuarine restoration (to restore diversity and productivity), but invasions of new species make this effort increasingly difficult.

SUPPLEMENTAL READINGS

Blaber 2000; Elliott and Hemingway 2002; Haedrich 1983; Houde and Rutherford 1993; Potter et al. 1990.

WEB CONNECTIONS

GENERAL
estuaries.gov/welcome.html
www.epa.gov/owow/estuaries/about1.htm

CHAPTER 32
Coastal Habitats

Most marine fish live on or near the edges of the continents, from the intertidal regions to the edge of the continental shelf. Within this region are a wide variety of habitat types, each of which is inhabited by a distinctive set of fishes. Because of the large numbers of individuals and species, the variety and complexity of habitats, and the open nature of coastal ecosystems, the interrelationships among the fishes and between the fish and the environment is complex and kaleidoscopic. Not surprisingly, the amount of information available on the fish increases in proportion to their closeness to shore, their closeness to institutions devoted to the study of marine biology, their attractiveness to divers, and their commercial value. When studying the ecology of coastal marine fishes, three critical factors must always be kept in mind. First, the fishes are closely tied to the bottom and are strongly influenced by the habitat structure, whether it be physical (e.g., rocks and sand) or biological (e.g., kelp forests and colonial invertebrates). Second, the habitat structure is surrounded by water that is constantly changing in its characteristics in response to both short- and long-term perturbations, such as changes in oceanographic conditions (e.g., El Niño events), big storms, or pollution (e.g., oil spills). Third, fish populations can be connected over long distances through the dispersal of larvae on currents or the active movements of adults in response to changed oceanographic conditions.

In this chapter, the fish communities that are associated with the following habitat types will be briefly discussed: (1) rocky intertidal areas, (2) exposed beaches, (3) mud-flats, (4) salt marshes, (5) mangrove swamps, (6) sea grass flats, (7) kelp beds, (8) near-shore rocky bottoms, (9) near-shore soft bottoms, and (10) hypersaline lagoons and salt ponds. The important inshore communities of tropical reefs and polar regions are discussed in separate chapters. The various coastal habitats grade into one another, and many fishes are found in several different habitats or move between them. The categories, however, do represent the most abundant and conspicuous habitat types in coastal regions, and each has distinct types of fishes associated with it.

32.1 ROCKY INTERTIDAL AREAS

Much of the coastline of the world is rocky and subject to the rise and fall of tides. The environment in these rocky areas is harsh, subject to crashing surf, strong currents, and daily exposure to the air. The faunas are quite diverse, however, because the intertidal environment also has high biological productivity (Leigh et al. 1987). Combined with the region's accessibility to observers, this faunal diversity has made rocky intertidal areas perhaps the most studied of temperate marine environments. The fish in particular present an interesting complex of adaptations to a harsh environment (Horn et al. 1998).

Intertidal rocky areas, as well as other intertidal environments, have four basic kinds of fish: (1) true residents, (2) partial residents, (3) tidal visitors, and (4) seasonal visitors.

True residents are found in the intertidal zone throughout the entire year and have the greatest degree of specialization for living in this environment. Typical representatives of this group in temperate areas are members of the families Cottidae (sculpins), Blenniidae (blennies), Gobiesocidae (clingfishes), Gobiidae (gobies), Stichaeidae (pricklebacks), Tripterygiidae (triplefin blennies), Clinidae (clinids), and Pholidae (gunnels). All these fishes are small and have reduced or absent swimbladders, compressed or depressed bodies, and some means of clinging to the substrate (Gibson 1993). The wide and flattened head, large pectoral fins, and smooth body of sculpins serve them just as well for a bottom-hugging existence in intertidal areas as they do for life in swift streams. Blennies tend to be laterally compressed and elongate, so they can easily fit into crevices and holes. Gobies have their pelvic fins modified into a sucker for holding onto the rocks. Clingfishes also have a pelvic sucker and are extremely flattened dorsally, which allows them to live in areas of high turbulence. Gunnels and pricklebacks, in contrast, are eel-like fish that avoid turbulence by squeezing into narrow crevices and cracks or by living underneath the rocks. Among the eel-like fishes, distinct differences exist among species that live at different places in the intertidal zone. Those species that live highest in the intertidal zone and are out of water for the longest periods are more cylindrical in cross-section than are those species that live lower down. The reason is that the lower amount of surface area in comparison to body volume reduces the amount of water that is lost by evaporation from the skin (Fig. 32.1).

Partial residents are fish that are consistently found in the intertidal areas but that also are found in deeper water. Typically, these fish are juveniles of species with large adults, and they often are representative of the same families that make up the true residents (e.g., Cottidae, Pholidae). Partial residents, however, also may be small, deep-bodied forms (e.g., Embiotocidae, Labridae).

Tidal visitors are fish that move into the intertidal region with the tide to feed. These fish can belong to almost any species that inhabits the more stable regions below the intertidal region.

Seasonal visitors often are fish that use rocky intertidal areas for spawning, but the species that use them in this manner are few. Tidal visitors often are seasonal visitors as well, because the inshore fish fauna in general shows seasonal changes in composition in temperate areas.

The dominant fishes of the rocky intertidal areas are residents and partial residents, and their distribution patterns strongly reflect the physical, chemical, and biological aspects of their environment. Because there generally are strong, fluctuating gradients of temperature and environmental severity from upper to lower intertidal areas, the distribution patterns of species tend to reflect their tolerance for these conditions. In the upper intertidal areas, the species must have special adaptations to survive the harsh environment, such as a tolerance for fluctuating salinities and the ability to breathe air (Yoshiyama

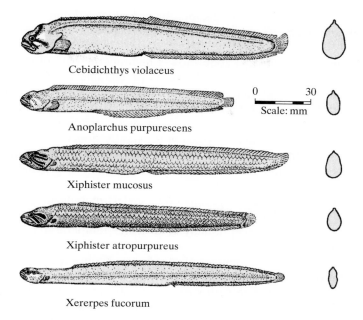

Cebidichthys violaceus

Anoplarchus purpurescens

0   30
Scale: mm

Xiphister mucosus

Xiphister atropurpureus

Xererpes fucorum

FIGURE 32.1   Five species of eel-like fishes from the rocky intertidal
zone of southern California, with cross-sections of their bodies. They are
placed, from top to bottom, in order of their ability to resist drying out and
surviving out of the water. The top two species are most likely to be found
under rocks in upper intertidal areas. From Horn and Riegle (1981).

and Cech 1994). In the lower intertidal areas, interactions among fish species may play
an important role in determining distribution patterns, but similar coexisting species
usually have strong preferences for particular environmental conditions that enable
them to avoid competition. For example, the tidepool sculpin *(Oligocottus maculosus)*
and the fluffy sculpin *(O. snyderi)* show segregation on the basis of habitat selection and
temperature tolerances. The fluffy sculpin selects habitats with plenty of cover (eel
grass) and low temperatures; the tidepool sculpin shows a preference for shallow water
and is more tolerant of a wide range of temperatures and salinities (Nakamura 1976).
Because intertidal species often tend to align themselves along various environmental
gradients, a succession of species is found from the upper to the lower intertidal regions
(Yoshiyama 1981). The strong vertical zonation so often demonstrated by intertidal invertebrates, however, is less apparent in fishes, probably because fish can move about
more readily.

Although much of the ecological segregation among intertidal fishes is based on
responses to the physical environment, fishes with overlapping distributions also tend to
segregate on the basis of feeding habits. Most are carnivores, but herbivores and omnivores are not uncommon. One of the more remarkable omnivores is the amphibious
pejesapo *(Sicyases sanquineus)*, a clingfish of the rocky intertidal areas of the Pacific
coast of South America. It lives where wave action is heavy, and it scrapes a wide variety

of invertebrates and algae from the rocks, thereby occupying a dominant grazer-predator role that normally is occupied by invertebrates in the upper intertidal zone (Paine and Palmer 1978).

Fishes of the rocky intertidal zone show a high degree of ecological segregation and local assemblages of fishes tend to be fairly predictable through space and time (Gibson and Yoshiyama 1999). In a way, this predictability is surprising, because the assemblages of intertidal invertebrates and algae seem to be much less predictable. A constantly changing community organization might, in fact, be expected given the dynamic nature of the intertidal environment, which is constantly subjected to disturbances on various scales, ranging from logs being bashed onto shore to heavy waves caused by storms to major shifts in temperature caused by changing oceanographic conditions. Intertidal fish communities, however, show remarkable powers of recovery from disasters. Tide pools from which fish have been removed by collectors typically recover their faunas in a few months (Grossman 1982). Longer recovery times are experienced following larger-scale disasters, such as exceptionally cold winter temperatures in warm coastal areas. Usually, a core of tolerant species survive such conditions with few ill effects; less tolerant species return at various degrees of speed (e.g., Thomson and Lehner 1976). An important feature contributing to the recovery of intertidal communities is that most intertidal fishes have pelagic larvae with strong substrate preferences when they settle. This is one reason why so many intertidal fishes are widely distributed in coastal areas. Once established in an area, however, many intertidal fish become territorial, thereby reducing the probability of additional colonization from larvae of their own and, perhaps, similar species. Such fish usually also have strong homing abilities and quickly return to their territories if displaced (Yoshiyama et al. 1992, Gibson 1993).

32.2   EXPOSED BEACHES

The fish associated with exposed coastal beaches live in the turbulent environment of the surf. The turbulence and currents would seem to require high energy expenditures by most of the fish that live there—even if only to maintain position. On the other hand, the turbulence also provides a constant source of small, disoriented invertebrates that are exceptionally vulnerable to capture. Thus, the surf zone is inhabited by a small but select group of fish, often in surprisingly large numbers and mostly of the following types: (1) small, active plankton-feeders; (2) roving substrate-feeders; (3) flatfishes; (4) migratory species; (5) beach spawners; and (6) piscivores. Most species that are found in the surf are widely distributed in coastal habitats; few are found primarily in the surf.

Small, silvery, streamlined planktivores, such as silversides (Atherinidae), anchovies (Engraulidae), and herrings (Clupeidae), often are the most numerous fish in the surf. Fish species that are specialized for feeding on the peculiar invertebrates that live in sand also may be abundant. Examples of roving substrate-feeders are the Atlantic threadfin *(Polydactylus octonemus)*, with its specialized pectoral fin rays used for poking into the sand to find prey, and the gulf kingfish *(Menticirrhus littoralis)*, a typical member of the bottom-feeding family Sciaenidae, with its sensitive barbels. The flatfishes in the surf zone also feed largely on benthic invertebrates. The flattened bodies of flounders and rays, particularly stingrays (Dasyatidae), enable them to avoid being swept about by the turbulence in the water column. Fishes that migrate along the coast in the surf zone and that spawn on beaches typically do not feed in the surf, so their distinctive behavioral patterns probably developed, in large part, to reduce their vulnerability to predation. The best-known examples of fish that migrate through the surf are mullets (Mugilidae), large schools of which often can be observed from shore. Fishes that spawn on beaches are few, but one, the California grunion *(Leuresthes tenuis)*, attracts hordes of people to witness the spawning and to collect the fish to eat. Their spawning is remarkably predictable, because the fish move inshore on high tides that follow a new or full moon at certain times of the year. The grunion allow themselves to be washed ashore by the waves. Between breakers, the females wriggle tail first into the sand, lay their eggs, and have them fertilized by males lying close by. The eggs hatch on the next series of high tides, and the larvae are washed out to sea.

Considering the variety and abundance of fish that are found in the surf, it is not surprising that piscivorous fish make excursions into the region to feed. Typical examples are bluefish *(Pomatomus saltatrix)* and jacks (Carangidae). On the New England coast, one such predatory fish is the silver hake *(Merluccius bilinearis)*, which is found in the surf during the fall and winter months. Because they frequently become stranded on beaches while pursuing their prey, they are known as "frost fish" to the people who go down to the beaches in season to pick them up. Occasionally, even large predators, such as white sharks, are present in the surf as they pursue other fish or marine mammals.

Fish abundance in the surf zone shows strong seasonal and diel patterns. In the Gulf of Mexico, fish are most abundant at night and in the summer, but the pattern is strongly influenced by the weather; Fewer fish are present when the surf is low (Ross et al. 1987). Many beach areas are heavily fished, and fishing can significantly reduce both fish numbers and fish sizes. With protection, however, the fish communities can recover rapidly (Bennett and Attwood 1991).

32.3   MUD FLATS

Mud flats that are exposed by the falling tides are associated with backwaters, bays, or other features that allow sediment to accumulate. When exposed, they appear to be barren of life, yet they are extremely productive of polychaete worms, clams, and other burrowing invertebrates. Consequently, as the tide moves in to cover the flats, large numbers of fish move in to feed. These species are mobile forms that are typical of bays, estuaries, and other inshore environments: (1) flounders from various families, (2) skates and sharks, (3) drums (Sciaenidae), (4) and deep-bodied forms, such as the surfperches (Embiotocidae). These species find their prey through a variety of mechanisms. Many bony fishes search the substrate carefully, either by sight or with barbels, to find individual prey, often pulling worms and small crustaceans from the mud. Rays often flap their "wings" rapidly to remove sediment that covers the clams they love to crunch. Some small sharks and flounders go about biting off the protruding siphons of clams. To keep sucking in water with food in it, the clams with shortened siphons have to move closer to the surface of the mud, which then makes them more vulnerable to fishes that feed on clams and to birds that feed on mudflats when the tide is low (Goesij et al. 2001).

Although most mudflat fish are tidal visitors, a small number of species are fulltime residents—aside from the few small tidal visitors that manage to survive in shallow drainage channels. These true tideflat residents are mostly small forms that can live in the burrows of marine invertebrates. Gobies (Gobiidae) are particularly well known for this trait. One species of goby, the longjaw mudsucker *(Gillichthys mirabilis)*, has developed the ability to breathe air using a highly vascularized buccal-pharyngeal chamber. If conditions in an invertebrate burrow or mudflat pool become too severe, the mudsucker is capable of "walking" with its pectoral fins across the muddy surface to a more favorable spot. Tropical mudskippers *(Periopthalamus spp.)*, on the other hand, apparently can find refuge from outside predators by staying in anoxic burrows and shutting down much of their metabolism (Graham 1997).

32.4 SALT MARSHES

When mudflats are present in sheltered coastal areas, particularly bays and estuaries, for long periods of time, they are invaded by salt-tolerant plants and become salt marshes. On temperate coasts, salt marshes and their associated networks of drainage channels and embayments are major coastal habitats. Like other intertidal environments, salt marshes are subject to extreme daily and seasonal changes of water levels, salinity, and temperature. Nevertheless, they are one of the most productive environments known and are important sources of nutrients for estuarine and bay systems. Salt marshes also support large fish and invertebrate populations of their own, and the standing crops (g/m^2) of these organisms may be among the highest of any coastal system.

As in other intertidal systems, the fishes of salt marshes are a mixture of true residents, partial residents, tidal visitors, and seasonal visitors. At any given time, the number of species in a salt marsh likely is small (typically <15). The *true residents* of North American salt marshes, those fish that complete their entire life cycle in the marshes, are mostly killifish *(Fundulus, Cyprinodon)*. Killifish can live at salinities ranging from nearly zero (fresh water) to several times that of sea water and at temperatures from near freezing to between 35°C and 40°C. If stranded by the falling tide, individuals of some species, such as the common mummichog *(F. heteroclitus)*, will bury themselves in the mud or flop toward the receding water. As the tide rises, the killifish typically penetrate into the marsh as far as they can to feed on the abundant invertebrates that are associated with the marsh vegetation.

Partial residents are fishes associated with the marshes only for part of their life cycle, usually as juveniles, taking advantage of abundant forage and structure for cover year-round. Worldwide, abundant partial residents of salt marshes are silversides (Atherinidae, Atherinopsidae, and related families). Adults tend to school along sandy or gravelly beaches or among beds of eel grass.

Tidal visitors typically are larger fishes, such as adult sciaenids, flounders, and halfbeaks (Hemiramphidae), that move into salt marshes at high tide to feed on the abundant juvenile fishes and invertebrates. The species that play this role vary from season to season (Table 32.1).

Seasonal visitors use salt marshes as spawning or nursery areas or as seasonal refuges from predators. In both the Atlantic and Pacific coasts of North America, the main salt-marsh spawners are sticklebacks (Gasterosteidae). During the spring, they build nests in the vegetation where the water is more or less permanent. The most

TABLE 32.1 ECOLOGICAL CHARACTERISTICS OF 16 COMMON FISHES IN NORTHERN FLORIDA SALT MARSHES

	Life-history stage[a]	Main food[b]	Type[c]	Months											
				J	F	M	A	M	J	J	A	S	O	N	D
Longnose killifish	A,J	S,D	TR	X	X	X	X	X	X	X	X	X	X	X	X
Gulf killifish	A,J	S,M,D	TR		X	X	X	X	X	X	X	X	X	X	X
Sheepshead minnow	A,J	D	TR		X	X		X	X	X	X	X	X	X	X
Diamond killifish	A,J	?	TR		X	X							X	X	X
Sailfin molly	A,J	D,S	TR?					X	X	X	X	X	X	X	X
Rainwater killifish	A,J	S,D	TR?		X	X	X	X	X	X	X		X	X	X
Tidewater silverside	J	S,D	PR		X	X	X	X	X	X	X	X	X	X	X
Spot	J	D	PR	X	X	X	X	X	X	X	X		X	X	
Pinfish	J	D,S	PR	X	X	X	X	X	X	X	X	X			
Pinfish	A	S,M,D	TV					X			X	X	X	X	X
Spotfin mojarra	J	S	SV					X		X		X	X	X	X
Bay anchovy	J	S	SV			X	X		X	X	X	X	X	X	X
Striped mullet	J	D	SV		X			X		X	X			X	
Halfbeak	A	S,D	TV								X		X	X	X
Atlantic needlefish	A	C,S	TV	X		X		X	X	X	X	X	X	X	
Sand seatrout	J	C,F	TV					X	X	X		X	X		
Atlantic threadfin	A,J	S	TV				X	X	X	X		X	X		

Information from Subrahmanyam and Drake (1975).

[a] A, adults; J, juveniles.

[b] S, small invertebrates; D, detritus; M, mollusks; C, crabs and other large invertebrates; F, fish.

[c] TR, true resident; PR, partial resident; SV, seasonal visitor; TV, tidal visitor.

important seasonal visitors to salt marshes of the southern Atlantic and Gulf coasts are juvenile drum, anchovies, mullet, and mojarras (Gerreidae). They spend several months in the marshes, taking advantage of abundant food and warm temperatures that promote rapid growth. The nursery function of salt marshes appears to be less important in northern Atlantic marshes, although Nixon and Oviatt (1973) noted that juvenile menhaden concentrated in a New England salt-marsh embayment during late summer, largely to avoid predators.

Although the fish communities of salt marshes often are dominated in terms of biomass by the true residents, they show considerable fluctuations in both species composition and species numbers as well as in total numbers of individuals. These fluctuations are caused not only by fish movements that are related to reproduction and tides but also by the differing responses of fishes to changes in salinity and temperature and to the abundance of both predators and prey. For example, Subrahmanyam and Drake (1975) found in a Florida salt marsh that the abundance of killifish was related to warm temperatures, whereas the abundance of pinfish was associated with high salinities. The abundance of spot was related to low temperatures and high salinities together. Although the fluctuating nature of the salt-marsh environment undoubtedly exacts high energy costs from the fish that live there, the abundance of food usually more than makes up for this. Fish that can take advantage of the food supply largely are detritus feeders, small invertebrate-pickers (from the bottom or from plants), and piscivores (Table 32.1). The most abundant species typically are invertebrate-pickers, but the detritus that they ingest while capturing their prey also may be a significant source of nutrition. Invertebrate-feeders, such as the common mummichog, may be so abundant and efficient at utilizing the invertebrates associated with salt marshes that they affect the species abundance and size distributions of the invertebrate populations (Vince et al. 1976). At the same time, they may achieve extremely high production of their own species. In turn, this high production of killifish allows a large number of predators (birds and fish) to exist by feeding on them. Because the predatory fish (e.g., bluefish and white perch) are important sport and commercial fish, the high productivity of salt marshes gives the marshes a high value to humans, even apart from their importance as nursery areas.

32.5   MANGROVE SWAMPS

On tropical coastlines, the sea–land interface often is covered with dense thickets of mangroves, which create rich swamps that may extend for miles inland along estuaries or form narrow belts along coastlines. In subtropical regions, such as Florida, they also may be important, but their distribution is less extensive. The shallow water beneath the mangroves ranges from hypersaline to fresh and from moving (by tidal currents or freshwater outflow) to stagnant. Intersecting the mangroves usually are muddy drainage channels a meter or so deep. Mangrove swamps thus provide a wide variety of aquatic habitats. Although variations in salinity, temperature, and dissolved oxygen levels can make this environment hostile, fish, particularly juveniles of marine fishes, abound wherever there is much exchange of the water. There are two main reasons for this. First, mangrove swamps are among the most productive ecosystems in the world. Second, the interconnecting root systems and shallow water have a complex structure that provides extensive cover from predation (Laedgdsgaard and Johnson 2001). The mangroves support large populations of aquatic animals, mainly through a detrital food chain, because few of the invertebrates feed on the mangroves directly. Detritus-feeding invertebrates form the center of food webs leading to fish. In Florida, many game fishes have young that may spend part of their life cycle among the mangroves. Some of the more important species are tarpon *(Megalops atlanticus)*, ladyfish *(Elops saurus)*, snook *(Centropomus undecimalis)*, and grey snapper *(Lutjanus griseus)*.

Mangrove swamps frequently show zonation in the vegetation, which is based on the different tolerances of the various mangrove species to factors such as salinity and water depth (Lugo and Snedaker 1974). Not surprisingly, a zonation of animals often follows that of the vegetation. The areas that are important as nurseries for marine fish are on the edges of the swamps or in the estuaries, where water flow prevents stagnation and limits temperature extremes. The more severe interior environments of mangrove swamps usually are occupied by fishes such as the mollies (Poeciliidae), killifish, and in tropical regions, by mudskippers (Gobiidae). Mudskippers often are extremely abundant in mangroves, especially along muddy drainage channels, because they are capable of breathing air, of "walking" across exposed mudflats, and even of climbing the exposed roots of the mangroves.

32.6 SEA GRASS FLATS

The intertidal mudflats, salt marshes, and mangrove swamps of bays, estuaries, and other shallow coastal areas often grade into another habitat that also is extremely productive of fish: the subtidal, muddy-bottomed flats on which beds of sea grass grow. From 35 to 50 species of grass can characterize these beds, but the dominant species in temperate regions is eelgrass *(Zostera marina)*. In tropical regions, the dominant grass is usually turtlegrass *(Thalassia testudinum)*, which often is associated with flats around coral reefs (see Chapter 33). Sea grass beds are among the most productive plant communities in the world, comparing favorably with corn and other intensively cultivated crops and with plankton blooms in areas of upwelling. As a consequence of this high productivity, sea grass beds support large populations of fish, which find both food and cover in the grass. These fish are principally either juveniles of large species or species with small (<200 mm) adults. Because sea grass beds are located in shallow coastal areas, bays, and estuaries, the water is subject to both seasonal and daily changes in temperature and salinity, although the salinity changes are seldom as dramatic as the temperature changes. As a consequence of these environmental fluctuations, the number of species in a bed likely is small, and the species composition resembles those of nearby or associated estuaries and salt marshes. Adams (1976a) collected only 39 species of fish in two North Carolina eelgrass beds over a 1-year period, and 15 of these species were collected only once. Of the biomass of the two beds, 45% and 67%, respectively, was from just one species, the pinfish. A similar pattern was found for New York eelgrass beds, except that two dominant species (rather than one) were found: Atlantic silverside *(Menidia menidia)* and fourspine stickleback *(Apeltes quadracus)* (Briggs and O'Connor 1971).

The number of fish in a sea grass bed fluctuates both diurnally and seasonally. Adams (1976a) found that densities of fish in eelgrass beds during the summer were highest at night, when temperatures were lowest. Because movement into the beds apparently was not related to feeding, it may well have been to avoid nocturnal predators. The movement out of the beds in the morning permitted the fish, particularly larger individuals, to avoid potentially stressful temperatures. On a seasonal basis, densities of fish are highest during the summer, when the waters are warm and the eelgrass beds thickest. Again, temperature seems to be the key factor regulating the movement of fish into and out of the beds. When the water cools down during the winter, many of the fish seek slightly warmer deep water, away from the beds, just as during the summer they may seek cooler water when the water around the beds becomes excessively warm. Regardless of the temperature of the beds, at least some fish always seem to be present. Usually, they are members of the same four or five resident species or of species using the beds as nursery grounds.

Sea grass beds are important as nursery areas, because they are a rich source of food for young fish. The grass itself rarely is eaten, but the detritus that it produces is consumed by fish either directly or indirectly through detritus-feeding invertebrates. Given the abundance of food in eelgrass beds, it is not surprising that the diets of the various species overlap considerably. Nevertheless, four basic feeding types can be found in the beds: (1) detritivores, (2) carnivores, (3) planktivores, and (4) omnivores. Usually, the most abundant fishes are *carnivores*, which feed on the abundant invertebrates and small fishes that are associated with the grass, although omnivorous species, such as pinfish, actually may be the most numerous fishes, particularly in the more southern eelgrass beds (Adams 1976b). The *omnivores* include not only invertebrates, fish, and algae in their diets but also substantial amounts of detritus (organic matter). There are few complete *detritivores*, such as spot, but detritus is, at times, the dominant item in the diet of juveniles of many species. Such juveniles, particularly during their youngest stages, also frequently are *planktivores*, feeding on zooplankton that thrives in the nutrient-rich waters around the eelgrass beds.

Overall, the structure of the fish communities of sea grass beds is very similar to that of salt marshes. Both environments are highly productive, and both are dominated by large populations of a few species of small, tolerant fishes that are capable of utilizing a wide spectrum of food resources. Like salt marshes, sea grass beds frequently are destroyed or altered by human activity. Fortunately, they can become established quickly on new substrates, such as dredging spoils (Brown-Peterson 1993).

32.7 KELP BEDS

Kelp beds are undersea forests of brown algae that typically are found in temperate waters between 6 and 30 m. The kelp are anchored to rocky or sandy bottoms with holdfasts, and they have fronds extending to the surface, thereby forming a dense canopy. The kelp forests provide a variety of habitats for invertebrates and fish and are located in naturally productive waters, so they support an abundant and diverse fish fauna. Relatively few species are exclusively kelp dwellers, however, although many are most abundant among the kelp. The most spectacular—and most studied—of the kelp-bed fish communities are those in the beds of giant kelp *(Macrocystis)* off the California coast. In this section, we deal with the ecology of these beds (Ebeling et al. 1980, Foster and Scheil 1985), recognizing that the basic ecology presumably is similar to those of fish communities in kelp beds elsewhere throughout the world. In all cases, kelp represents a constantly changing, but highly desirable, complex habitat for fishes. Fishes that inhabit these beds therefore have excellent dispersal abilities that allow them to take advantage of the dense forests as they grow.

Most fishes that are associated with kelp beds are acanthopterygians, particularly of the orders Perciformes and Scorpaeniformes, and the number of species can be quite high. Off California, probably most of the more than 150 inshore subtidal species can be found in kelp beds at one time or another. Only 50 to 60 of these are common, however, and only 15 to 20 are abundant and conspicuous enough to be easily observed by divers. The exact species composition of a kelp-bed fish community will vary from place to place and from time to time, depending on factors such as ambient water temperatures, depth, bottom type, season, water clarity, and the amount of turbulence created by wave action. At the same time, the variety of body shapes and sizes in kelp-bed fishes suggests that the species segregate on the basis of habitat preference and feeding habits. According to Ebeling et al. (1980), these fishes have five main habitat groupings: (1) a kelp-rock group, (2) a kelp-canopy group, (3) a bottom group, (4) an inner marginal group, and (5) a commuter group.

The ***kelp-rock group***, which is the most diverse group of kelp fishes, is found where there are high-relief rocky reefs and/or dense beds of kelp. Some of the species are members of families that also are abundant on tropical reefs (see Chapter 33): wrasses (Labridae), damselfishes (Pomacentridae), and rudderfishes (Kyphosidae). These fish always are found close to cover and show strong daily rhythms of behavior, much like their tropical relatives. They also have other complex behavior patterns that usually are associated with tropical reef fishes. For example, a large wrasse, the California sheepshead *(Semicossyphus pulcher)*, has a complex reproductive behavior that includes the ability of females to change into males. Likewise, the bright orange garibaldi (a damselfish, *Hypsypops rubicundus*) gardens algae on rock surfaces as nest sites. Another damselfish, the blacksmith *(Chromis punctipinnis)*, forages on zooplankton above the reef during the day. It then returns to the reefs to hide for the night and deposits there, as feces, large amounts of organic matter, which is a significant source of nutrients for benthic organisms (Bray et al. 1981). This group of tropical derivatives is absent from the kelp forests north of Point Conception, where the California coast makes an abrupt bend, thereby diverting the cold currents away from the shore. On the Atlantic coast, in contrast, fishes of tropical families can be found fairly far north.

The ***kelp-canopy group*** consists of species that largely are found beneath the canopy of kelp fronds. Two of this group, the señorita *(Oxyjulis californica)* and the kelp perch *(Brachyistius frenatus)*, pick small invertebrates from the kelp plants, although some individuals specialize in picking ectoparasites from other fish (Bray and Ebeling 1975). Other abundant fishes in this group include juvenile rockfishes (*Sebastes* spp.).

The ***bottom group*** comprises fishes that live in continuous contact with the substrate, either the bottom or the kelp plant. Many of these species are small, cryptic forms, such as blennies (Blenniidae), sculpins (Cottidae), kelpfish (Clinidae), clingfish (Gobiesocidae), and greenlings (Hexagrammidae), that feed on small invertebrates. Feeding on larger invertebrates and small fish are adult rockfishes and large ambush predators, such as lingcod *(Ophiodon elongatus)*. In more southern kelp forests, California moray eels *(Gymnothorax mordax)* occupy crevices in the rocky bottom. Within this group is considerable segregation on the basis of depth and feeding habits. For example, two similar species of rockfish *(S. chrysomelas, S. carnatus)* segregate by depth; the territorial behavior of one species excludes the other from the favored shallow-water habitat (Larsen 1980).

The *inner marginal group* consists primarily of adult surfperches (Embiotocidae) that inhabit the shoreward margin of the kelp beds, where the kelp forest thins and dense turfs of other algae often grow. These fishes segregate on the basis of depth and feeding methods. Some species are adapted for winnowing invertebrates from the turf; others pick individual invertebrates. Competition results in spatial segregation between at least two of the surfperch species (Hixon 1980).

The *commuter group* is a group of large, active fishes that move among habitats in search of prey. Most are surfperches and rockfishes, but the piscivorous kelp bass *(Paralabrax clathratus)* also is part of the group. Within this group is segregation on the basis of time of feeding and type of food. The pile perch *(Rhacochilus vacca)*, for example, feeds largely during the day on large, hard-shelled mollusks, whereas the rubberlip perch *(R. toxotes)* feeds on smaller, thin-shelled invertebrates that it finds at night with its thick, sensitive lips (Alevizon 1975).

Perhaps the most remarkable aspects of the biology of kelp forests are the complex biological interactions that maintain the forests themselves (Box 32.1). Occasionally, a severe storm will destroy a kelp forest, and regeneration may be prevented by the grazing of fishes or sea urchins on new kelp plants. Regeneration can occur, however, if rapidly growing green algae colonize the area first, thereby providing the new kelp plants with shelter from grazing fishes (Harris et al. 1984). One of these grazers is the señorita, which actually is seeking a small bryozoan that encrusts the blades of the plants. To eat the bryozoans, the señorita must take bites from the blades; this can seriously weaken the plant if it is small but has little effect when the plant is large. Another invertebrate, a herbivorous isopod, however, can become so abundant that it can destroy mature kelp plants. The señorita is the main predator on this isopod and keeps its populations small, thereby maintaining the kelp beds in the process (Bernstein and Jung 1979).

32.8 NEAR-SHORE ROCKY BOTTOMS

Shallow (<50 m), rocky-bottomed areas, from cliff faces to flat shelves, have fish communities that are similar to those of kelp beds. The reason for this is, in part, that kelp beds grow mostly on rocky bottoms and, in part, that to a fish, the two habitats have similar attributes. Both habitats contain a diversity of microhabitats in which small fish can hide and large fish can forage. Both provide substrate for the attachment of sessile invertebrates and cover for small, active forms, thereby providing a diverse and abundant supply of food. The differences between the two habitats are mostly in the relative abundances of the different species. On the Pacific coast, kelp beds are likely to contain larger numbers of plankton-feeding fish, such as blacksmith, that use the beds for shelter between forays into the water column, whereas rocky areas on the Pacific coast are likely to have a higher density of crevice-dwelling forms, such as sculpins, blennies, gunnels, and pricklebacks. The most conspicuous fish in both habitats are invertebrate-pickers that roam over the substrate in schools or loose aggregations, feeding on small invertebrates. Such fish mostly are members of the families Embiotocidae (surfperches), Scorpaenidae (mostly rockfishes),

BOX 32.1

How Offshore Fisheries Affect Kelp Bed Fishes

In the northern Pacific are large, industrial trawl fisheries that focus on the capture of abundant fishes, such as hake. Such fisheries are notorious for depleting the fishes that they harvest. In Alaska, evidence suggests that trawl fisheries in the northern Pacific may be linked to the decline of kelp forests and their associated fishes in coastal waters (Estes et al. 1998). The basic scenario is as follows: Trawl fisheries in offshore waters have depleted the fish that seals and sea lions depend on for food, thereby causing a drastic decline in the mammal populations. The seals and sea lions are major prey of orcas *(Orcinus orca)*, and their decline has forced some whales to switch to feeding on sea otters *(Enhydra lutra)*, thereby resulting in a major decline in otter populations.

Otters feed heavily on sea urchins, which graze on kelp; when sea urchin populations are high, kelp beds disappear. With the reduction in the sea otter populations, sea urchin numbers increase, kelp forests disappear, and kelp-bed fishes decline. All the links in this complex chain are not thoroughly documented, but it is plausible. It also is a good demonstration of the complexity of interactions within inshore habitats (Estes et al. 1998). Indeed, archaeological evidence from middens indicates that indigenous peoples periodically switched their diets from fish to inshore invertebrates (e.g., mussels, etc.). It is thought that this was the result of an overharvest of sea otters for fur, thereby resulting in demise of the kelp beds and reduction of the inshore fish populations.

Labridae (wrasses), Sparidae (porgies), and Sciaenidae (drums). Similar complex communities exist on rocky areas of the Atlantic coast (Sedberry and Van Dolah 1984).

Most fish that favor rocky bottoms have pelagic larvae (as do the invertebrates). As a result, new areas can be rapidly colonized. Off southern California, sewage outfall pipes that extend for several kilometers across muddy-bottomed areas have become spectacular, if narrow, reefs, with large populations of fishes that otherwise would be rare in the area (Allen et al. 1976). It is now common practice to deliberately create artificial reefs from concrete, old tires, automobile bodies, and old ships to increase the fish populations in muddy-bottomed bays. This practice has met with mixed success.

The fish fauna of rocky areas deeper than 50 m is less well known because of the difficulty of sampling the habitat, although use of submersibles is starting to provide better information (Yoklavich et al. 2000). Many species of fish now considered to be rare because of their infrequency in fish collections likely are actually abundant in such habitats, especially members of such families as the Cottidae (sculpins), Zoarcidae (eelpouts), and Ophidiidae (cusk-eels). Some of the better-known species associated with deep, rocky-bottomed areas are various species of rockfish, which feed in the water column and so are taken by fishermen. Among these species are two of high commercial value that have—or had, before overexploitation—enormous populations: Pacific ocean perch *(Sebastes alutus)*, and Atlantic redfish *(S. marinus)*. The Pacific ocean perch typically is associated with gullies and canyons at depths of 150 to 460 m. They generally stay on or close to the bottom during the day. At night, they move up the water column to feed on large planktonic crustaceans, squid, and small fish and so are caught by commercial fishermen. The standing crop of Pacific ocean perch in the Gulf of Alaska is estimated to have once been more than a *billion* kilograms. In just 6 years (1963–1968), this stock was reduced by 60% because of commercial fishing.

32.9   NEAR-SHORE SOFT BOTTOMS

On the continental shelf and slope, most of the sea bottom is soft and relatively featureless, covered with a layer of sand, silt, broken shells, and other fine materials in various proportions. The number of fish species in a given soft-bottomed area is usually low. From 40 to 60 species may be expected if the area is sampled for several years, with 15 to 20 occurring on a regular basis and only between 2 and 5 making up most of the fish biomass. The most abundant species, however, often are of major commercial importance. In particular, the trawl fisheries for various species of flatfish (Pleuronectiformes) and codfish (Gadidae) largely are associated with soft bottoms. Despite the importance of these fisheries, the actual standing crops of fish from these areas are not particularly high in comparison to such productive areas as salt marshes and estuaries. The ease of fishing over soft bottoms, however, and the large size of the areas that they cover along with the fact that many of the important species concentrate there at certain times of the year combine to make up for the low overall productivity. Furthermore, it often is possible to be fairly selective in a fishery over soft bottoms, because the different species show different patterns of distribution, depending on bottom type, depth, oceanographic conditions, pollution levels, and interactions with other fish species.

Bottom Type

Commercial fishermen have long known that the presence of a particular species of fish in an area often can be predicted on the basis of bottom type. As Bigelow and Schroeder (1953, p. 184) note, the bottoms where cod (*Gadus morhua*) and hake (*Urophycis* spp.) are found "are so distinct that a long line set from a hard patch out over the soft surrounding ground will often catch cod at one end, hake at the other." Because most fish can readily move from one bottom type to the next, the consistent association of species with a particular bottom type (e.g., clay, silt, or sand) likely is related to feeding habits. For example, bottoms with a high percentage of silt support a diverse "infauna" of benthic invertebrates, which in turn support fishes that are specialized for feeding on them. The association of species with bottom types, however, also may be coincidental, the result of their responses to other factors, such as depth or oceanographic conditions. Thus, Day and Pearcy (1968) found that the associations of fishes at different depths off the Oregon coast also were associated with distinct bottom types. Similar patterns have been found in areas as diverse as the coastal areas off South America and western Africa (Lowe-McConnell 1975).

Depth

Most species of bottom-oriented fishes have fairly narrow depth ranges (Fig. 32.2), and distinct associations of fish within depth intervals have been recognized all over the world. Usually, on soft bottoms between depths of 50 and 2,000 m is a gradual decrease

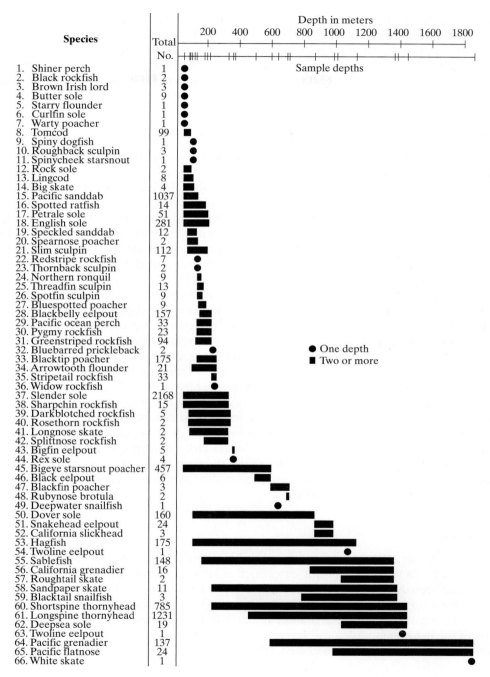

FIGURE 32.2 Depth ranges of benthic fishes collected off the central coast of Oregon. From Day and Pearcy (1968).

in the number of fish species, in the overall abundance of fish, and in fish biomass, although the average size of the fish tends to increase. At depths of less than 50 m, trends are less predictable. The causes of the observed depth distributions of fishes are complex, but the fact that positions of fishes frequently vary with the season indicates that responses to oceanographic conditions and changes in behavior related to spawning and food availability play important roles.

Oceanographic Conditions

Perhaps the most famous example of a benthic fish species with a distribution limited by oceanographic conditions (principally temperature) is the tilefish *(Lopholatilus chamaeleonticeps)* off the coast of New England, which is restricted to water with temperatures of between 8°C and 17°C and depths of between 81 and 540 m. This species dies off in large numbers when conditions in its habitat change (Bigelow and Schroeder 1953). Off southern California, upwelling, which brings close to shore deep water that is low in temperature and oxygen and high in salinity, causes dramatic changes in the distribution patterns of the inshore fishes (Fig. 32.3). Seasonal changes that are less dramatic than upwelling also have a profound effect on the soft-bottomed fish fauna. Tyler (1971) noted that in the shallow waters (<55 m) of the Atlantic coast are four types of fishes in terms of their seasonal occurrences: (1) year-round regulars, (2) summer periodics, (3) winter periodics, and (4) occasionals. The latter group consists of species that are found in low numbers on an irregular basis. The relative proportions of these components in a local fish fauna seem to depend on the amount of annual temperature fluctuation. Where the temperature fluctuation is low, the regular component is more abundant than the seasonal component; where the temperature fluctuation is high, the seasonal component (and the occasional component) is much more important than the regular component (Fig. 32.4). Seasonal movements of fish, presumably in response to oceanographic conditions, also have been noted in deep water. For example, witch flounder *(Glyptocephalus cynoglossus)* off Newfoundland aggregate to spawn in waters 500 to 700 m deep in the spring (the young use the continental slope as a nursery area), but they move into the Gulf of St. Lawrence in the summer and then back into deeper water in the winter (Bowering 1976).

Pollution

Increasingly, the effluent from cities is having a significant impact on the inshore fish communities. In the most extreme cases, such as areas where the wastes from New York City have been dumped repeatedly or where organic wastes carried down the Mississippi River wind up, vast areas of anoxic or near-anoxic "wastewaters" have been created that are barren of fish life (Rabalais et al. 2002). In less-extreme situations, the composition of the fish communities are changed, the abundance and diversity of fishes reduced, and the incidence of disease increased. For example, off Los Angeles, the fish community appears to have shifted from one that is associated with sand bottoms to one that is associated with silt bottoms as a result, at least in part, of the deposition of sewage on the bottom. In the immediate vicinity of wastewater discharge sites, the diversity of fishes is low, even though some species, such as white croaker *(Genyonemus lineatus)* and Dover

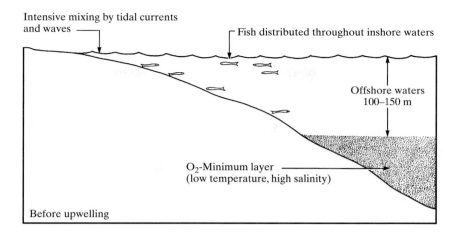

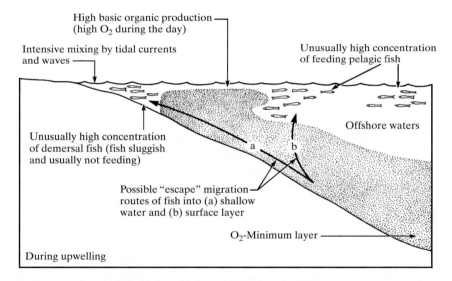

FIGURE 32.3 Impact of upwelling on the distribution of benthic and epipelagic fishes along the coast of Southern California. Used by permission of Southern California Water Research Project.

sole *(Microstomus pacificus)*, actually may be attracted to the discharges. In addition, there is a high incidence of disease—particularly fin erosion disease and various tumors— in the benthic fishes of the area, apparently as a result of contact of the fish with toxic materials on the bottom (Mearns 1973).

Species Interactions

Competition and predation play an important role in determining the structure of soft-bottomed fish communities. The succession of similar species with depth indicates that

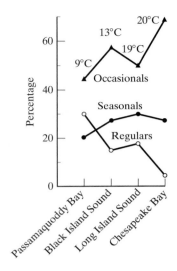

FIGURE 32.4 Percentages of fish species that are found on a year-round basis (regulars), seasonally, and irregularly (occasionals) in coastal waters at different latitudes of eastern North America. Annual temperature ranges are shown at the top of the graph. From Tyler (1971).

some level of competitive exclusion takes place, based in part on physiological specializations for different combinations of temperature and depth (i.e., pressure). Along the northern portions of both coasts of North America, for example, flounders of the genus *Glyptocephalus* tend to replace other species of flounders in water deeper than 150 m. The impact of overfishing on fish communities also indicates that the presence of one species may result in the exclusion of others. On the Georges Bank in the North Atlantic, overfishing of haddock *(Melanogrammus aeglefinus)* led to their replacement by yellowtail flounder *(Limanda ferruginea)*. Overfishing of flounder in turn led to their replacement by red hake *(Urophycis chuss)*. Before the fishery collapsed completely, the dominant fishes became the spiny dogfish *(Squalus acanthias)* and skates. Another line of evidence indicating the importance of competitive exclusion, at least over long periods of time, is that studies regarding the feeding habits of coexisting fishes consistently show partitioning of the available food resources among the species or among guilds of species (Tyler 1972, Garrison and Link 2000). Overall, it appears that fishes on soft bottoms are specialized in both habits and habitat requirements, but a number of them seem to be flexible enough to take advantage of situations created when a normally dominant species is absent.

Zonation

Although the fishes of the continental shelf and slope may show seasonal patterns of movement, especially in relation to reproduction, it usually is possible to detect distinct fish zones in relation to depth. Substrates and other habitat variables change with depth, so the role of each is hard to determine. As in freshwater streams, these zones appear to be broad areas in which the distributions of species with more or less similar habitat requirements overlap. Species frequently inhabit more than one zone (often at different times during their life cycles), but within zones, segregation by feeding habits seems to be the rule. Off most coasts, five basic habitat zones can be recognized: (1) shallow water, (2) inner continental shelf, (3) outer continental shelf, (4) upper continental slope, and (5) middle continental slope.

The *shallow-water zone* typically is found in water less than 50 m deep and includes bays and other sheltered areas. Within this zone, of course, are the other coastal ecosystems discussed previously.

The *inner shelf zone* typically is found between 50 and 100 m but also up to 150 m. The bottom often is sand (or mixed sand, silt, and broken shells) and is flat and feature-less (although interrupted by submarine canyons and islands or rocky reefs). The domi-nant fishes here are such forms as hagfish (Myxinidae), flounder, codfish (Gadidae), rockfish, and skates.

The *outer shelf zone*, usually between 150 and 400 m in depth, is similar in many respects to the previous zone, although the bottom tends to contain more silt and clay. The fishes also are similar but usually are of different species. In addition, members of families such as the Ophidiidae (cusk-eels) and Zoarcidae (eelpouts) are more common. It is worth noting that the species determined to be "dominant" are dominant in the catches of the sampling gear used and are not necessarily dominant on the bottom itself. Photographs taken by cameras lowered to the bottom in such areas often reveal quite different patterns of species abundance than those determined from sampling programs.

The *upper slope zone* generally shows an increase in gradient from the previous zone, dropping from approximately 440 to 1,200 m. The bottom is mixed clay and silt. In the Atlantic Ocean, the fishes are mostly true deepwater forms that rarely are seen in shallower water: grenadiers (Macrouridae), longnose eel (*Synaphobranchus kaupi*), a small hake (*Phycis chesteri*), and eelpouts. Although species tend to be year-round resi-dents in this zone, one other common species, the witch flounder, uses the zone as nursery grounds, so the young flounder eventually move up onto the continental shelf (Markle and Musick 1974, Haedrich et al. 1975). Off the Pacific coast, many distinct deepwater forms also are present in this zone, but the dominant species seem to be some of the same species of rockfish, flounders, and hake that are abundant on the shelf.

The *middle slope zone* is similar to the upper slope zone, but the number of species is much fewer and all are largely confined to the continental slope. The benthic fish fauna of this zone is remarkably similar in both the Atlantic and the Pacific, with the dominant fishes in trawl catches often being grenadiers (especially *Coryphaenoides*) and codlings (Moridae). The fishes of this zone and of the upper slope zone are discussed in more detail in Chapter 34.

32.10 HYPERSALINE LAGOONS AND SALT PONDS

Hypersaline lagoons are shallow embayments in arid and semiarid regions in which the combination of high evaporation rates, poor oceanic circulation, and low freshwater inflow results in salinities frequently exceeding 40 ppt. The number of fish species in a hypersaline lagoon typically is low, but if the fluctuations in temperature and salinity are not too extreme, the fish populations can be quite large. The most abundant fishes typically are juveniles of the same species that characterize local estuaries and other shallow-water environments. In the Gulf of Mexico, where such lagoons are common, characteristic species are sea catfish, sheepshead minnow, tidewater silverside, striped mullet, pinfish, southern flounder *(Paralichthys lethostigma)*, and various species of croakers (Sciaenidae). Most of the fishes are planktivores or predators on other fishes, although the sheepshead minnow, mullet, and some drum do browse on algae and detritus (Copeland and Nixon 1974).

Salt ponds, like hypersaline lagoons, are semi-isolated, shallow embayments. They are located in nonarid regions, however, so their salinities seldom are greater than—and often are less than—that of the local seawater. The fishes are the same species found in local eel grass beds such as silversides, pipefish, stickleback, killifish, and other small species, as well as the young of various larger species. Some salt ponds seem to be quite important locally as nursery grounds for inshore fishes. On the Hawaiian Islands are numerous low-salinity ponds in old lava flows. These ponds are inhabited by a wide variety of euryhaline fishes that have accidentally been washed into them by storm waves or have moved into them through outlet streams. The ponds apparently are not used as nursery areas, but they often are good producers of fish for human consumption (Brock 1977).

32.11   LESSONS FROM COASTAL FISHES

Anyone who is even vaguely familiar with the fish and habitats of the continental shelves of temperate regions knows that this chapter has not done justice to their diversity. Ideally, each major habitat type deserves a separate chapter. Describing the kelp-forest fish communities off California (as in this chapter), for example, provides only a general notion of the kelp-bed communites off the Atlantic coast of North America or off Australia. Yet no matter where fish communities on the continental shelves are studied, they prove to be diverse and widely distributed, thanks to planktonic larvae. This leads to a tremendous dynamism in response to constant changes in oceanic conditions brought about by changes in ocean currents or in climate or by local events, such as major storms. They also show strong affinities to habitat structure, with different species groups being found in association with different physical and biological substrates. We also, however, are beginning to appreciate that whereas local fish faunas can show dramatic changes in both abundance and composition, they also have a tendency to constantly reorganize themselves into the fairly predictable patterns that we recognize as fish communities. Even the recent shifts in many of these communities into new conditions caused by overfishing and pollution seem to be reversible if the insults to the natural systems are taken away, although recovery sometimes will be very slow. Naturally, human endeavor constantly pushes the coastal marine environment into more extreme conditions to which the fishes must adapt, but the limits of their ability to adapt are rapidly being discovered. Most marine fish, it turns out, are tied less to the vast, open ocean than they are to the continents, the habitat of humans.

SUPPLEMENTAL READINGS

Bigelow and Schroeder 1953; Culliney 1979; Foster and Schiel 1985; Gibson 1993; Hobson 1994; Horn et al. 1998; Scott and Scott 1988.

WEB CONNECTIONS

ALASKA MARINE COMMUNITIES
www.csc.noaa.gov/lcr/kachemak/html/intro/1hintro.htm

MARINE SANCTUARIES, USA
www.sanctuaries.nos.noaa.gov/welcome.html

C H A P T E R 33
Tropical Reefs

The diversity of fish life on tropical reefs can only be called astonishing. Somewhere between 30% and 40% of all fish species are associated with such reefs in one way or another, and anywhere from 250 to 2,200 species are likely to be found in, on, or near a major complex of reefs. The number of species in even one small area often is hard to comprehend; Smith and Tyler (1972) collected 75 species from an isolated coral dome that was only approximately 3 m in diameter and 1.6 m high. Most of the fishes found on reefs are acanthopterygians with an extraordinary array of adaptations for maintaining themselves in a crowded environment. Just as fascinating as the individual species are the complex evolutionary processes that have given rise to the complex structure of tropical reef ecosystems. Indeed, the reefs, as we see them today, are the product of millions of years of coevolution among fish, invertebrates, and algae. Arguably, fish are the movers and shakers of tropical reef ecosystems; without them, reefs as we know them would not exist.

Both the diversity of reef fishes and the complexity of reef ecosystems have been appreciated for a long time, but only recently have they become the focus of ecological studies. Sale (1991) attributes this interest to three factors:

1. *Mobility and spatial scale.* The fishes of tropical reefs are small (mostly <30 cm long) and sedentary, thereby resulting in a wealth of species concentrated in areas that can be easily studied and manipulated.

2. *Accessibility.* The development of scuba gear, of modern transportation systems, and of a tourist industry on remote islands has made reefs readily accessible to biologists. Also, reefs are located in warm, shallow water in which an observer can hover for hours, watching the activity below.

3. *Temporal scale.* Most reef fishes live from 1 to 10 years and reproduce continually, so their pattern of life fits readily within the time scales of PhD dissertation research projects and research grants.

This chapter summarizes some of the progress that has been made in the study of reef fish and their communities by describing (1) the reef habitat, (2) the types of reef fishes based on feeding habits, (3) species interactions, (4) activity patterns, (5) life cycles of reef fishes, and (6) community structure.

33.1 THE REEF ENVIRONMENT

Tropical reefs are found between latitudes 30° north and 30° south in shallow water (usually <50 m deep) that is warm enough to support the growth of corals and clear enough to allow photosynthesis at moderate depths. This means that the water in reef areas rarely drops below 18°C (it is usually around 23–25°C) and that underwater visibility usually is at least 10 to 20 m. Although corals typically are associated with tropical reefs and are famous for their reef-building habits, many so-called coral reefs have largely been built by calcareous algae. In addition, many rocky coastal areas and reefs support a complex fish and invertebrate fauna similar to those of coral and algal reefs. Most tropical reefs are surrounded by nutrient-poor oceanic waters, so their extremely high productivity is surprising. This productivity can be attributed to a combination of (1) a high degree of recycling of nutrients within the reefs; (2) the ability of reef organisms, especially fish, to concentrate nutrients that are taken in from surrounding waters; (3) the photosynthetic activity of attached and symbiotic algae under optimal light and temperature conditions; and (4) nitrogen fixation by blue-green algae on the reefs (Wiebe et al. 1975).

Four major regions of the ocean have tropical reefs: (1) the Indo-Pacific region, (2) the Eastern Pacific region, (3) the Western Atlantic (West Indian) region, and (4) the Eastern Atlantic (West African) region (see Chapter 26). These regions have had long-independent evolutionary histories, but the processes that have formed the reef communities are very similar. Thus, each reef typically has a series of habitat zones with distinctive fish and invertebrate faunas. Six major zones can be recognized, although all are not present on every reef.

The **off-reef floor** is the shallow sea bottom around a reef. It typically is sandy and often supports beds of sea grass; thus, it may be an important foraging area for reef fish. Many reefs are not surrounded by a level, sandy floor but drop off abruptly, often to great depths. The **reef drop-off**, in its upper 50 to 60 m, is favored by large numbers of fish, which can find shelter on the cliff face and abundant plankton in the water immediately off it. The **reef face** that is above either the floor or the drop-off often is the richest habitat for fish and invertebrates. Its complex growths of coral and calcareous algae provide innumerable cracks and crevices for protection, and the abundant invertebrates and epiphytic algae provide an ample source of food. The **reef surface** also is a rich habitat for life, but the organisms that live there must be able to withstand the constant surge of the waves and, in some areas, the rise and fall of tides. Behind the main reef, toward the shore, often is a sandy-bottomed **reef flat** containing scattered chunks of coral. The reef flat may be a protected area bordering a lagoon, or it may be a flat, rocky area between the reef and the shore. In the former case, the number of fish species living in the area often is the highest of any reef zone. Many coral reefs completely enclose an area, thereby creating a quiet-water **lagoon** that usually contains small patches of reef.

Although zones are useful reef descriptors, the topography of reefs is constantly changing. Each reef is made of irregular patches of algae, sessile invertebrates, and bare rock and sand. The size, shape, and relative abundance of these patches changes from year to year in response to the various factors that favor one type of patch over another. Growing coral, for example, produces constant change in the fine structure of reefs. On a larger scale, tropical storms may knock out large sections of reef and cause boulders on sandy areas to move. These changes, whether large or small, essentially are unpredictable; thus, they provide a habitat for reef fish that continually changes on a local scale, even if the overall environment seems to be quite benign and stable (Connell 1978).

33.2   TYPES OF REEF FISHES

There are many ways to classify reef fishes, but for understanding the structure of fish communities, the clearest approach is to classify them by feeding habits. Following Hobson (1974), we place reef fishes into three general feeding categories: (1) generalized carnivores, (2) specialized carnivores, and (3) herbivores.

Generalized Carnivores

These fish have the classic rover-predator body shape, with large, flexible mouths that are well suited for seizing large prey (Fig. 33.1). By and large, they prey on mobile fishes and invertebrates. The generalized carnivores are of three basic types: (1) nocturnal, (2) crepuscular, and (3) diurnal.

The *nocturnal predators* are specialized to the extent that they have large eyes and typically feed on benthic crustaceans that move about the reef at night. The *crepuscular predators* are rover-predators, representing such families as the Serranidae, Carangidae,

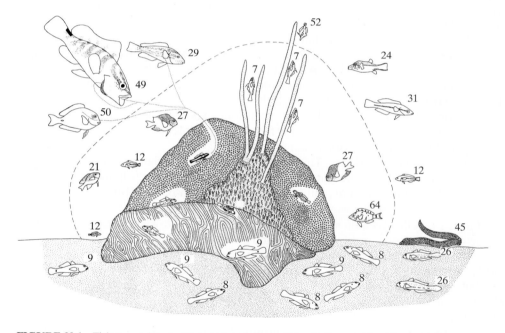

FIGURE 33.1   Fishes associated with an isolated piece of West Indian reef, in which a cleaner goby (19, *Gobiosoma*) has a station. Waiting to be cleaned are herbivorous parrotfish (29, *Scarus*), a generalized carnivore (49, a grouper, *Epinephelus*), and a herbivorous surgeonfish (50, *Acanthurus*). Associated with the reef surface are diurnal planktivores (27, a damselfish, *Pomacentrus*; 12, a wrasse, *Thalassoma*); a territorial herbivore (21, a damselfish, *Pomacentrus*); a herbivorous parrotfish (64, *Sparisoma*); small, plankton-feeding goby (7, *Coryphopterus*); and a small filefish (52, *Monacanthus*). Visiting the patch to feed on occasion are a diurnal feeder on hard-shelled invertebrates (24, a puffer, *Canthigaster*), a diurnal invertebrate-picker (31, a wrasse, *Halichoeres*), and a crevice-feeder (45, a moray eel, *Muraena*). Associated with the sand around the coral and the coral itself are a guild of gobies (8, 9, 25, 26, and 28) feeding variously on algae, detritus, and small invertebrates, often at night. From Smith and Tyler (1972).

and Lutjanidae. These fishes become active during twilight, because the low light levels offer a certain amount of concealment to the predators while still enabling them to see their prey, particularly schooling fishes. *Diurnal predators* are similar to crepuscular predators in their body shape and preferences for fish. They obtain prey by slowly cruising over the reef or by lying in concealment. In either situation, however, they must wait until a nearby fish makes a defensive mistake, such as moving too far from cover or becoming separated from a school.

Specialized Carnivores

A majority of reef fish species have highly specialized feeding habitats that reflect the complexity of the evolutionary processes characterizing reef environments. Although many of these specializations are extraordinary, it is not unusual to find several species on the same reef with approximately the same specializations. These similar species form feeding guilds whose members may interact with each other either competitively or cooperatively. On a broader basis, the specialized carnivores can be divided into eight types: (1) ambushers, (2) water-column stalkers, (3) crevice predators, (4) concealed-prey feeders, (5) diurnal predators on benthic invertebrates, (6) cleaners, (7) diurnal planktivores, and (8) nocturnal planktivores.

Ambushers   Ambush feeding is one of the principal methods of prey capture used by reef carnivores, and there are many examples of extreme use of camouflage. Among these fish are lizardfishes (Synodontidae), scorpionfishes (Scorpaenidae), and flounders (Bothidae). Such fishes match their backgrounds so well that they frequently are overlooked by prey (and divers). Scorpionfishes, including the deadly stonefish *(Synanceia)*, are particularly famous for their ability to blend into the reef environment, thereby looking more like corals than fish.

Water-column stalkers   This group of fishes also specialize in making themselves invisible to potential prey so that they can ambush them. These fish are silvery, elongate forms with long, pointed snouts full of sharp teeth. A head-on view often belies their large size. They drift through the water column toward a small fish and then seize it with a sudden lunge, propelled by a fin structure that is typical of lie-in-wait predators. Examples of this group are cornetfishes (Fistulariidae) and trumpetfishes (Aulostomidae).

Crevice predators   These fish actively seek their prey, which mostly are fish that hide in the numerous crevices and small caves that are characteristic of reefs. The main specializations of this group are elongate bodies and small heads, which allow them to penetrate into crevices to find their prey. Examples are moray eels (Muraenidae) and reef brotulas (Ophidiidae).

Concealed-prey feeders   This group of fishes also actively seek prey that are hidden on or about the reef, taking both active invertebrates and small fish. It consists mainly of the goatfishes (Mullidae), an abundant and conspicuous group of reef dwellers. Goatfish have long chin barbels, which they use to probe the reef surface or sandy flats around a reef. Once a goatfish has located a prey organism with its barbels, it sucks it up with its flexible, slightly subterminal mouth. Depending on the species, goatfishes can be either nocturnal or diurnal.

Diurnal predators on benthic invertebrates   These are among the most colorful and peculiar-looking reef fishes. Most use vision to find their prey, which typically are the more conspicuous reef invertebrates, such as sponges, corals, tunicates, sea urchins, and snails. These feeding relationships provide some of the more spectacular examples of coevolution of predators and prey. Over the course of several million years of predator pressure, the invertebrates have evolved a formidable array of defense mechanisms, mostly various combinations of spines, toxins, heavy armor, and adherence to the substrate. Simultaneously, the predatory fishes have evolved many mechanisms for overcoming these defenses. One means of overcoming heavy armor is to have strong, hard jaws, such as the heavy beaks possessed by most members of the order Tetraodontiformes (e.g., puffers, boxfishes, etc.). Another is to have strong pharyngeal teeth, such as those of many of the wrasses (Labridae). These fish (called *durophages*) crush various hard-shelled invertebrates. Other species circumvent the armed defenses of their prey by biting off exposed pieces of flesh, such as the polyps of coral and the "fans" of sessile polychaete worms. Many of the species that feed in this manner are the deep-bodied, brightly colored butterflyfishes (Chaetodontidae), which come equipped with elongate snouts that are tipped by small mouths containing tiny (but sharp) teeth, with which they can clip off pieces of invertebrates. Some butterflyfishes use the same type of feeding mechanism for seizing prey located deep in crevices, such as small crustaceans or even encrusting sponges. The depth to which a species can reach into a crevice is reflected in the length of its snout, which in some species may be more than 25% of the body length. Yet another way that some butterflyfishes feed is by scraping the surface of living coral, thereby obtaining mucus secreted by the coral along with small, attached invertebrates and algae.

Diurnal predators also have a number of behavioral specializations. Some species of wrasse with small mouths feed by carefully searching the reef surface and then picking up invertebrates that are too small for most other species to use as food. Other wrasses use their snouts to overturn chunks of rock and coral on the reef edge and so expose hidden invertebrates. Some puffers find invertebrates in sandy-bottomed areas by uncovering them with a jet of water blown from their mouth. Fish that feed in such a manner often are followed by other fish, which attempt to capture the exposed invertebrates first. *Following behavior* also is elicited by groups of herbivorous fishes, which disturb the substrate, or even by foraging octopuses and moray eels, the activities of which may send small fish fleeing from cover and into mouths of other predators (Diamant and Shpigel 1985).

Cleaners   These fishes utilize another specialized form of invertebrate feeding; picking ectoparasites and dead or diseased tissue from other fish (see Chapter 27). For some species, such as cleaner gobies of western Atlantic reefs and cleaner wrasses of Indo-Pacific reefs, cleaning is the principal feeding specialization. The distinctive behavior and bright coloration of cleaner fish attract other fish to them. Most of the attracted fish are other teleosts, but even sharks can come to be cleaned (Sazima and Moura 2000). The client fish express their receptivity to cleaning by entering into a trance-like state, with fins held rigidly, opercula flared, and mouths open. Such fish often exhibit special contrasting color patterns as well. The cleaner then swims about the fish, picking off ectoparasites from the sides and, for larger fishes, often from within the mouth cavity as well. They will also pick dead or diseased tissue from wounds, eat loose scales, consume mucus, and nibble off pieces of healthy fin. Although the removal of healthy tissue is detrimental to the fish being cleaned, the cleaning of infected tissue from wounded fish may promote healing (Foster 1985). Cleaner fish either work from stations on the reef to which other fish come to be cleaned or move freely about the reef during the day, stopping to clean territorial fish. Territorial damselfish will seek out cleaner stations and benefit from their visits by reduced parasite loads, but the frequency of visits is related to the distance they have to travel to get to the station (Cheney and Côté 2001).

   Diurnal planktivores   Rather than finding their food on the reef itself, diurnal planktivores use the reef mainly for shelter and hover above it in brilliant, shifting shoals while feeding on zooplankton. These fish require specializations for capturing prey that is either very small (copepods) or gelatinous (e.g., larvaceans, chaetognaths, fish eggs, etc.) and for escaping predators (Hobson 1974). These factors have caused fish from different families (e.g., Pomacentridae, Serranidae, Acanthuridae) to evolve quite similar morphologies: streamlined bodies, deeply forked or lunate tails, fine gill rakers, as well as small and upturned mouths. The first two features enable fish to dive quickly into the cover of the reef at the approach of a predatory fish. Their body shape, although streamlined, is more or less a compromise, however, between the rover-predator shape and the deep-bodied shape, because a deep body is useful for hovering in the water column and picking out the plankton. Some planktivores have, in fact, retained a deep body and stout spines to repel predators. Usually, the further that a fish forages from the reef, the more streamlined the body (Hobson 1991). The flexible, small mouth enables the fish to capture small prey by suction. The mouth is upturned to shorten the snout, which gives the fish binocular vision at close range for pinpointing its small prey (Hobson 1974).

   Nocturnal planktivores   These fishes have the moderately deep bodies, forked tails, and upturned mouths of the diurnal forms, but they have a larger and less flexible mouth and large eyes. Typically, their bodies are red. These species leave the reef at night to forage on large (>2 mm) zooplankton and invertebrates that also leave the reef to enter the water column at night.

Herbivores

Herbivores make up approximately 20% of all fish species that inhabit reefs (Sale 1977). What they may lack in variety, however, they make up for in numbers and conspicuousness as well as in their impact on the appearance of reefs (Horn 1989). Herbivores often are the most noticeable fish on reefs. Small, brightly colored damselfishes vigorously defend their territories, while schools of large parrotfishes (Scaridae) and surgeonfishes (Acanthuridae) browse over the reef surface. Other herbivores (or partial herbivores) are scattered through the various families of reef fishes, such as Kyphosidae, Chaetodontidae, Blenniidae, Pomacanthidae, and Siganidae. Their main sources of food are filamentous algae that coat the reef and, to a lesser extent, the sea grasses and algae that grow on reef flats. The herbivorous fishes (and invertebrates) keep the algae on the reef cropped down to a thin mat (only 1–2 mm in thickness) and can create a ring around reefs up to 10 m wide that is bare of vegetation (Randall 1965). Grazing by herbivores is important for the well-being of reef-building corals and algae, because patches of reef from which herbivores are excluded quickly become covered with thick growths of filamentous and leafy algae, which may smother the corals (Lewis 1986).

The plants are eaten in various ways by fish, such as scraping by parrotfish and plucking by damselfish. Nevertheless, the diets of the various herbivores seem to differ little from species to species, although major differences may be found in their abilities to digest different kinds of algae. Smaller herbivores, such as damselfish, may protect their share of this food supply by defending sections of reef against their own kind as well as against other species of fish. This aggressive defense results in heavier growths of algae (and invertebrates) within the territories than outside them. To exploit these growths, some herbivores will move into a territory as a group, thereby overwhelming the defender (Robertson et al. 1976). Species of sturgeonfish (Acanthuridae) that feed on detritus, however, are tolerated in the territories of herbivorous sturgeonfish species (Choat 1991).

33.3 SPECIES INTERACTIONS

Because tropical reefs are both a physically benign and ancient environment, the fish fauna is complex, with complex patterns of interactions. These interactions, especially predation, competition, symbiosis, and mimicry, are very important determinants of community structure.

Predation

The variety of ways in which reef fish act as predators has already been discussed, but reef fish also serve as prey, both for each other and for reef invertebrates. Thus, they have many means, both morphological and behavioral, of protecting themselves. Most reef fishes possess a formidable array of spines, which often are venomous or can be rigidly locked in place. Many have cryptic or disruptive color patterns as well (to confuse predators). Behaviorally, small reef fish protect themselves from predators either by shoaling or by hiding in reef crevices. Many reef fish spend most of their lives in one small area, so they have well-known hiding places instantly available. Others may roam the reef, typically in shoals, but return to specific areas to hide when they are not actively feeding. While resting, small fish are still vulnerable to crevice predators, so many fish cram themselves into the smallest possible hiding places and wedge themselves in with erect spines. Parrotfish can secrete a cocoon of mucus about themselves while resting, which seems to provide protection against predators that find hidden prey by smell or touch.

Predation probably is the most important cause of fish death on reefs, so it can have considerable effects, both direct and indirect, on the structure of fish communities (Hixon 1991). When predator densities are high and/or refuges are limited, fish abundance and diversity can be limited by predation (Hixon and Beets 1993). When predators are systematically removed from small reefs, survivorship of juvenile fish can increase dramatically (Carr and Hixon 1995). On the other hand, on some reefs predation may not be limiting, and it may even increase diversity. Predation may, for example, directly promote diversity by reducing populations of competitively superior species. Indirectly, predation may make shelters from piscivores a limiting resource and, thereby, cause interspecific competition for shelter (Shulman 1985). It also may be responsible for the general pattern of daily activity of reef fishes discussed in Section 3.4.

Competition

Competition, like predation, has undoubtedly been an important process in shaping reef fish communities. Presumably, the high degree of resource partitioning among reef fishes and the many morphological differences among related species, such as the different snout lengths of butterflyfishes, are the result of past competitive interactions that have led to co-evolved divergence. Nevertheless, the importance of competition on reefs today is difficult to demonstrate, because rarely can either shelter or food be shown to be limiting resources. For example, after a severe storm leveled much of a Hawaiian reef, little change was found in either its species composition or its population sizes (Walsh 1983). The resident fishes had moved into deeper water during the storm but then returned to occupy their greatly diminished habitat. Likewise, ecological differences could be found among co-occurring species of butterflyfish, but Findley and Findley (1985) could not find any evidence of competition among the species. Some species seem to have few ecological differences and form guilds that recognize each other's signals; for example, the damselfish, which "garden" algae, defend their territories not only against conspecifics but against other damselfish species as well (Sale 1977). Among damselfishes, however, species with larger body sizes are competitively superior compared with smaller species (Robertson 1995).

Symbiosis

Symbiotic relationships are common among reef fishes and may be more important in structuring reef habitats and fish communities than generally is thought (Box 33.1). For example, planktivores and other fishes use corals for cover from predators and deposit feces deep in their shelter areas, which are consumed by other fish and invertebrates (Robertson 1982). Meyer et al. (1983) have shown that coral heads that—during the day—shelter fishes that feed at night away from the reef grow faster than those without such fishes. Larger, more complex coral heads in turn provide refuges from predation for more fish. Some growth forms of coral, therefore, likely evolved to encourage fish to use them as shelter.

Another example of *mutualism* is shoals that are made up of two or more fish species; such shoals confer on co-operating species the advantages of large shoals. Mutualistic relationships also may form between invertebrates and fish, with the classic example being anemonefish and anemones (Fautin 1991). Anemonefish are protected from predators by the stings of anemones (from which the anemonefish are protected by a special mucus coating), whereas the anemones are protected from grazing predators by the aggressive actions of the anemonefish. The anemones also may benefit by being able to develop a more expansive shape, thereby providing more surface area for the symbiotic algae that grow in the tentacles (Fautin 1991). Presumably, the anemones also take advantage of the waste products of the fish to improve growth.

Other fish–invertebrate interactions are *commensal* in nature, such as those exhibited by fishes that hide among the spines of sea urchins and starfish or in the hollow tubes of sponges. An example of commensalism among fish species is the following behavior of fishes that seek prey stirred up by other fishes. A purely commensal relationship, however,

---

BOX 33.1

Peaceful Coexistence Among Anemonefishes

The anemonefishes (*Amphiprion* spp.) are small, colorful damselfishes that have a mutulistic relationship with anemones (see Section 3.4). The anemones in which they live can be a limited resource, which would suggest that competition among different species of anemonefishes should be intense. When Elliott and Mariscal (2001) investigated nine co-existing species in reefs off Papua, New Guinea, however, they found little evidence for strong competition among the species, despite the fact that virtually all of the suitable anemones were occupied. The anemonefishes shared the 10 species (4 common) of available anemones; different anemonefishes were primarily associated either with different species of anemone or with anemones in different reef zones. Where overlap in the use of anemones occurred, aggressive interactions took place only if the competing fish were the same size. Small species of anemonefish, however, could coexist with large species on the same anemone. When fish were removed from an anemone, more individuals of the same species quickly colonized it, thereby demonstrating both the strong selection by each species for particular anemones and the shortage of suitable anemones as hosts.

---

is that of some cornetfish (Fistularidae) and parrotfishes. The long, narrow cornetfish "rides" on top of the parrotfish, essentially using the parrotfish as a mobile cover from which it can ambush small fish.

Mimicry

As reef fish become better known, more and more cases of mimicry are being documented (Russell et al. 1976). Most of the examples currently known are either cases of ***aggressive mimicry***, in which a predator species mimics a harmless one to gain access to prey, or of ***Batesian mimicry***, in which a harmless species mimics a predatory species for its own protection. Often, several species in one area resemble each other, thereby forming a mimicry "ring" ***(Mullerian mimicry)***. Perhaps the most famous example of an aggressive mimic is the sabertooth blenny (*Aspidontus* spp.), which mimics cleaner wrasses in morphology, color, and behavior and then bites chunks out of hapless fish waiting to be cleaned. Usually, however, only young or naive fish are fooled by the blenny. Examples of Batesian mimicry are found among reef fishes that have eye spots on the caudal peduncle. Such spots may be used to temporarily startle a predator by giving the impression that the pursued fish is much larger than it really is. An extreme development of this type of mimicry occurs in the plesiopid fish, *Calloplesiops altivelus*, which mimics the moray eel, *Gymnothorax melagris* (McCosker 1977). Both species have a similar color pattern, so when the fish dashes into the reef while fleeing a predator, it may leave its rear half-exposed. When combined, the color pattern and eye spot give the impression of the head of a moray eel in an aggressive posture.

33.4 ACTIVITY PATTERNS

One way that reef species avoid interactions with other species is to have periods of activity that do not overlap with those of potential competitors or predators. Indeed, one of the striking aspects of these fish communities is the difference between night and day on a reef. During the day, the deeper cracks, crevices, and caves of a reef are filled with night-active fish, whereas during the night, they are filled with day-active fish (Fig. 33.2). The changeover from diurnal to nocturnal fish and then back again is a fascinating and complex phenomenon (Hobson 1974).

During the day, the reef's surface is a busy place because most of the brightly colored, specialized carnivores are active at this time, either feeding, defending territories, or spawning. Immediately above the reef are schools of small plankton-feeding fish, which are under

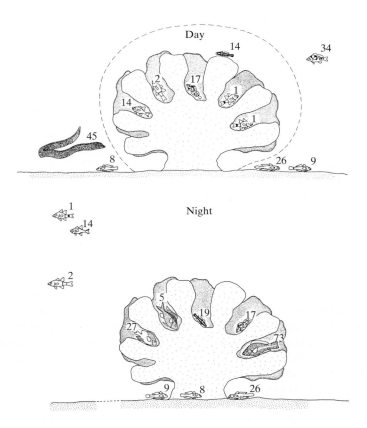

FIGURE 33.2 Day and night residents of a West Indian coral head. Nocturnal planktivores (1, 2, and 14; cardinalfishes, Apogonidae), a cleaner goby (19, *Gobisoma*), a predatory blenny (17, *Labrisomus*), a piscivore (34, *Hypoplectrus*), various gobies (8, 9, and 26; Gobiidae), a crevice feeder (45, moray eel, *Muraena*), diurnal planktivores (5 and 27; damselfishes, Pomacentridae), and a diurnal herbivore (73, a parrotfish, *Sparisoma*). From Smith and Tyler (1972).

the surveillance of various piscivorous fish. Closer to the reef are large, relatively inactive schools of nocturnal planktivores and other nocturnal fish that do not feed on the reef. Deep within the reef are other nocturnal fish, often red and having large eyes, such as squirrelfish (Holocentridae), cardinalfish (Apogonidae), and bigeyes (Priacanthidadae).

As twilight approaches, the transition between day and night fish faunas begins. The dominant behavior of the diurnal fishes is cover-seeking. As the light dims, the diurnal planktivores come closer to the reef, while the specialized carnivores become less active and seek out their nighttime hiding places. Fish that roam the reef during the day stream back to their regular resting places. Most of these fish are in place 15 to 20 minutes after sunset (Hobson 1975). For the next 20 minutes or so, the reef is remarkably quiet, because the nocturnal fish have not yet emerged. One reason for this general hiatus in fish activity is, presumably, that during this time of low light intensity, piscivorous fish are most active—and most effective—at capturing prey. The quiet period ends when the nocturnal planktivores emerge in large numbers from their hiding places and swim up into the water column.

By this time, it is nearly dark. On the surface of the reef, small, unspecialized predators begin to move about, feeding on small crustaceans that also have emerged from hiding. Nocturnal piscivores, such as conger eels (Congridae) and some goatfish, also become active at this time, seeking out hidden diurnal fish. Despite this activity, the number of active fish on the reef itself is smaller at night than during the day. As dawn approaches, the process observed in the evening reverses itself, and the diurnal fish once again resume their activities.

33.5   REPRODUCTION

For most of its life, a typical reef fish stays in a very restricted part of a reef, where it lives, spawns, and dies. For some fish, this resident period may last only a year; for others, it may last 10 or more years. Divers commonly recognize individual fish for several years in the same place, so most of the smaller reef fish likely live at least 3 to 5 years. Spawning takes place in this same general area, but embryos and larvae, with few exceptions, are planktonic.

Spawning may take place year-round in equatorial regions but may last only 4 to 6 months at higher latitudes. During the spawning period, the frequency of spawning ranges from daily to monthly, depending on the species. Most spawning takes place at dusk or at night, presumably to reduce predation on fertilized eggs by planktivores. Before spawning, many of the larger species migrate short distances to places where conditions are optimal for the newly fertilized eggs to be swept away by currents. Smaller fish cannot afford the risk of being eaten while migrating, so they tend to spawn in place (Barlow 1981). Typically, they guard developing embryos in hidden places until they hatch. Other species provide their embryos protection as bearers, such as mouth-brooding cardinalfishes (Apogonidae) and jawfishes (Opistognathidae), and a few species are livebearers. With few exceptions, the larvae of these species also are planktonic, although they do not seem to disperse as widely as the larvae of the larger, more fecund, pelagic spawners (Barlow 1981).

Regardless of the degree of parental care, spawning behavior among reef fishes frequently is complex (Thresher 1984). For example, parrotfishes (Scaridae) have elaborate mating systems that can involve the conversion of females to males, as needed (see Chapter 22). Most reef fish are mass spawners or form only temporary pairs, but a surprising number are monogamous. They include many butterflyfishes, angelfishes, and anemonefishes. The latter lay their eggs under the protective tentacles of anemones. Among the more unusual pair-spawners are the harlequin bass (*Serranus* spp.) of the Caribbean, which are hermaphroditic and take turns being male or female during the spawning sequence with the same mate.

33.6   EARLY LIFE HISTORY

One of the most remarkable aspects of tropical reef fishes is that virtually all the teleosts have pelagic larvae. Pelagic larvae are an excellent mechanism for sending out large numbers of low-cost propagules to be moved away passively from the parental area, thereby allowing new areas to be colonized quickly and promoting gene flow among populations to maintain genetic diversity. It has long been assumed that larval dispersal is the reason so many reef fishes occur throughout their biogeographic regions, even the huge Indo-Pacific region (Gill and Kemp 2002). At the same time, a problem exists with tropical reef systems being "open" systems from which most larvae leave: Most reefs are around islands and widely separated from one another. This means that the vast majority of the larvae would never find a place to settle down, even if they lived for weeks or months (as many do). There also are distinct advantages to having the larval loop being part of a "closed" population, because a species with such a strategy ultimately is likely to be better adapted to local conditions than one that depends exclusively on recruits from elsewhere. Good evidence indicates that some reef fishes do maintain closed populations (Box 33.2). Complex current systems around islands and the behavior of larvae to actively keep themselves from being swept away contribute to allowing larvae to settle in their home areas (Cowen 2002, Leis and McCormick 2002). Evidence for the universality of both open and closed populations, however, is equivocal, and most populations likely have some messy mixture of larvae that settle close to home and larvae that disperse widely (Mora and Sale 2002). In the Great Barrier Reef of Australia, evidence suggests that some island reefs have current systems around them that allow them to be self-sustaining and also to be sources of larvae for neighboring islands with less favorable conditions. These islands otherwise would be less rich in their species composition (Doherty 2002).

Regardless of the nature of the populations, the hundreds of species of reef fishes are continuously sending larvae into the waters above the reefs, and these larvae are continuously settling out onto the reef. There is a poorly understood, constantly changing assemblage of larvae at various stages of development above the reefs. Most of these larvae get eaten, swept away, or starve before they are able to settle down. The larvae that make it have to be lucky and to have the resources to grow rapidly to reach the settlement stage. Settlement is a difficult time for the fish, because they have to metamorphose from a fish that is adapted to pelagic life to one that is adapted for benthic life. Metamorphosis takes several days and involves the fish becoming more heavily pigmented to match reef backgrounds. The settling juveniles then face a "wall of mouths" of resident fishes that are eager to eat them or to chase them away from occupied habitats (Leis and McCormick 2002). Transforming fish actually may be quite fussy, therefore, about where

BOX 33.2
Stay-at-Home Cleaner Gobies

One of the key techniques used to find out if the larvae of reef fishes drift between reefs is to look at the genetics of widely separated populations. Fairly uniform populations suggest that gene flow occurs among the reefs through continuous recruitment of larvae from other reefs. Genetically divergent populations indicate isolation. Some species, such as the blueheaded wrasse (*Thalassoma bifasciatum*), seem to be fairly uniform genetically over wide areas, despite evidence that most larvae settle on the home reef (Swearer et al. 1999). In contrast, a Caribbean cleaner goby (*Elacatinus evelynae*) shows strong genetic differentiation among islands, even though the larval stage lasts 3 weeks, which is enough time for currents to carry larvae long distances (Taylor and Hellberg 2003). The gobies are unusual in that populations of different islands have different color patterns. Although the color patterns tend to be similar in populations of different geographic regions, white forms and blue forms can exist on islands separated by only 23 km. Bright color patterns have evolved in these gobies as signals to other fishes that the gobies are cleaners and, therefore, are not to be eaten. This suggests that individuals arriving from other islands with even subtle differences in color patterns from the resident gobies would be susceptible to predation (being different from the accepted cleaners). It also suggests that most of the larvae never leave their home islands. The genetic evidence shows that many of these populations have been isolated from one another for thousands of years, indicating that stay-at-home larvae and strong local selection for certain color patterns can lead to speciation (Taylor and Hellberg 2003).

they settle out, preferring areas with fewer predators and sneaking into a more favorable habitat. Most fish that settle successfully likely do so at night, when most reef predators are least active. At times, successful recruitment to a reef may result from huge numbers of larvae settling out at once, thereby overwhelming the predators (Doherty 2002).

33.7   COMMUNITY STRUCTURE

The complex interactions among tropical reef fishes make the overall structure of the fish communities very difficult to comprehend. Our perception of reef fish communities is strongly influenced by four factors: (1) most reef fish are highly specialized, having developed complex adaptations to a complex environment; (2) the number of reef fish species is very large, and their population densities are high; (3) among spatially separated reef patches is a great deal of variation in the composition of the fish fauna; and (4) reef fishes produce millions of larvae, which are continuously settling out on the reefs. In addition, repeated surveys of coral reef areas often show considerable variation from one survey to the next, although on a large scale, repeatability in species composition is found between surveys. Such surveys have been used to support a range of concepts of how tropical reef fish communities are organized, from open, nonequilibrium systems that are almost chaotic in structure to closed, highly organized equilibrium systems. Four general hypotheses have been used to explain the structure of reef fish communities: (1) the niche differentiation hypothesis, (2) the lottery hypothesis, (3) the recruitment limitation hypothesis, and (4) the predation hypothesis (Hixon 1991, 1998).

The *niche differentiation hypothesis* holds that the fish community is composed of specialized fishes that subdivide food and space as the result of past competition, which has forced species into narrow niches (see Box 33.1). Some competition still occurs within guilds of similar species, but it affects mainly fine-scale subdivision of resources. In this view, the reef communities are in a state of near-equilibrium, so they should be predictable in both composition and structure once the interactions among the species are understood.

The *lottery hypothesis* argues that a steady rain of recruits onto the reefs must occur, because reef fish spawn over long periods of time and produce large numbers of larvae. The only successful recruits, however, are those that happen to settle on a vacant place on the reef. Because the vacancies as well as the species of larvae likely to land in any given space are both unpredictable, Sale (1980) likened the process to a "lottery" for living space. Thus, the basic composition of a reef community is determined mainly by random processes, although competition and predation determine the fine structure that is observed. The randomness is driven not only by the "lottery" of reef fish recruitment but also by the constantly changing nature of the reef environment, through growth of corals, damage by storms, and other factors. This hypothesis was a major stimulus to force biologists to rethink how complex fish communities were assembled, but recent studies suggest that less randomness is involved in key processes, such as recruitment, than once thought (e.g., Sponaugle and Cowen 1997).

The *recruitment limitation hypothesis* suggests that mortality rates of the pelagic embryos and larvae of reef fish are so high that there are never enough recruits (newly settled fish) to saturate the environment. Thus, populations of each species should be determined largely by variation in recruitment success, which often would be density dependent (i.e., the more larvae available to settle, the more would likely successfully recruit). Once the fish had settled, however, mortality would no longer depend on their population density. This would result in fish assemblages that would be highly variable in both time and space. So far, the main predictions of this hypothesis have not been demonstrated experimentally (Hixon 1998).

The *predation hypothesis* argues that plenty of recruits are arriving on the reefs to fill vacant spaces but that the intensity of predation on these small fish is so high that they rarely live very long. This intense level of piscivory (even on adult fish) prevents each reef from becoming saturated with fish and limits the role of competition in structuring the communities (Hixon 1991). Perhaps the most striking evidence for the predation hypothesis is the large number of piscivores (or partial piscivores) on reefs (as much as 50% of the species in some areas) and the many morphological and behavioral mechanisms that fish possess to defend themselves against being eaten. Under this hypothesis, reef communities also tend to have a strong random component to their structure.

The four hypotheses just presented represent extreme perspectives. Presumably, reef fish communities actually are assembled by a combination of ecological processes. Sale (1991), however, provides convincing arguments that reef fish assemblages show a high degree of variability at virtually any scale (from whole reefs to small reef sections), so that persistent structure is difficult to find. The assemblage of species found at any given location results from variability in recruitment combined with intense predation and competition among the resident fish in a habitat that shows a great deal of random variation in structure (i.e., patchiness). At the same time, each species has its own preferred habitat and food and can interact flexibly with a large number of other fish species, often in ways that enhance the survival of the interacting species. These latter factors lend predictability to the busy chaos of the reef. Some of the evidence for this view includes:

- Most pelagic embryos and larvae of reef fishes remain tied to the home reef, although many can be swept away to other reefs.
- Guilds of similar fish species are common, but they reduce overlap in resource use by a variety of mechanisms, of which direct competition is only one.
- Many of the more visible reef fish have fairly long life spans and are very sedentary; thus, they give the appearance of community constancy to human observers while masking the variability in recruitment of small fish. Even so, reef fish assemblages are quite variable through time (Sale and Guy 1992).
- Artificial reefs placed near natural reefs are quickly colonized by juvenile reef fish in a highly variable fashion, although the structure of the reef can determine species composition (Caley and St. John 1996).

Ault and Johnson (1998) demonstrate that on sections of large, contiguous reefs, a certain amount of predictability is found in both space and time in the fish assemblages. They suggest that this predictability results, in part, from larger fishes of various species shifting to new locations when these spaces become vacant rather than simply from new recruits occupying these spaces, especially spaces with superior cover or food supply. Ault and Johnson argue that at least some of the variability in previous studies comes from the widespread use of small, isolated patch reefs as the unit of study, because these patches are easily colonized by postlarval fishes but not by adults. Another factor that must be considered is that not all reefs are the same. Current patterns around some reefs may sweep most of the larvae away, so the fish communities depend much more on recruits from neighboring reefs. Presumably, such reefs would have communities with a stronger random component to their structure than in those reefs where most recruits did not leave the locale.

33.8   LESSONS FROM TROPICAL REEF FISHES

Clearly, we still have much to learn about the dynamic evolutionary and ecological processes that result in the spectacular assemblages of fishes on tropical reefs. The relative importance of factors affecting distribution and abundance of reef fish likely vary from region to region, from locality to locality, and from time to time. Thus, studies and experiments on the effects of predation, competition, larval recruitment, and other factors on community structure will continue to give apparently conflicting results while also continuing to provide insights into some of the most fascinating biotic assemblages anywhere (Sale 2002). Perhaps just as important as understanding what shapes the reef assemblages on a day-to-day basis is understanding what has shaped them on a historical basis. From an evolutionary perspective, reefs have been shaped by constant interactions among the fishes and invertebrates that make up the reefs, partly as an arms race and partly as a co-operative effort.

Unfortunately, our opportunities to discover new interactions and to learn about and from tropical reefs are rapidly being diminished as humans destroy reefs through pollution, sedimentation, mining of coral, and other activities (Richmond 1993). In some reefs, such as those around Jamaica, harvesting of fish of all kinds and sizes is so extensive that the coral reefs are dying because of the loss of the intense interactions with fish that they require for survival (Hughes 1994). If such trends continue, many areas where hundreds of species once coexisted in splendid complexity will soon be only simple systems with low productivity.

SUPPLEMENTAL READINGS

Collette and Earle 1972; Hixon 1991; Sale 1991, 2002; Thresher 1984; Wilson and Wilson 1985.

WEB CONNECTIONS

General
www.coris.noaa.gov
www.reefbase.org

Great Barrier Reef
www.gbrmpa.gov.au

CHAPTER 34
Epipelagic Zone

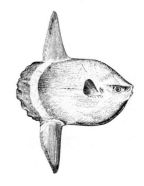

The surface waters of the oceans provide an enormous—if nearly featureless—habitat for fish. Lack of habitat diversity limits the number of fishes that occupy this region, the epipelagic zone, to less than 2% of all known fish species. The region contains vast expanses of nutrient-poor water that support few fish; as a consequence, epipelagic fishes are concentrated in areas overlying continental shelves or in oceanic areas where upwelling increases productivity of the water. They also are found almost entirely in the upper 100 m of the water column, where light can penetrate and permit phytoplankton to grow and where visual predators can see their prey. Despite the limited number of species, epipelagic fishes are, overall, the most valuable group of fish to humans, because they either occur in enormous numbers (e.g., herrings and anchovies) or are particularly favored as food fishes (e.g., tunas and salmon). As a direct consequence of their value, the biology of epipelagic fishes has been intensively studied, although the difficulty of performing such studies means that many basic questions remain unanswered. In this chapter, we summarize this information by discussing (1) adaptations for pelagic life; (2) ecological types of epipelagic fishes; (3) physical, chemical and biological factors affecting distribution and abundance; (4) migrations and movements; and (5) life history of a representative species, the Pacific sardine *(Sardinops sagax)*.

34.1 ADAPTATIONS FOR PELAGIC LIFE

Most epipelagic fish have streamlined bodies that permit continuous and often rapid swimming. Examples include (1) the galeid sharks (mackerel sharks, requiem sharks, whale sharks, etc.), (2) clupeiform teleosts (herrings and anchovies), (3) Salmonidae (salmon), (4) atheriniform teleosts (flying fishes, halfbeaks, sauries, etc.), and (5) perciform teleosts, most notably Carangidae (jacks), Coryphaenidae (dolphins), Bramidae (pomfrets), Sphyraenidae (barracudas), Scombridae (tunas), and Istiophoridae (billfish). Most of these fishes shoal, and most are visual predators on zooplankton or fish. A number of these fishes, however, are filter-feeders on plankton. The predators typically have smooth and fusiform bodies, deeply forked tails, and large mouths. Both prey and predators reduce their visibility by being silvery in color, which scatters incoming light, and by being countershaded.

Filter-feeding epipelagic fishes have the same general morphological features as the predatory fishes, and many of them are, in fact, facultative predators. They will pick individual invertebrates and small fish out of the water column or filter-feed on phytoplankton, depending on which mode is energetically the most advantageous. All fishes capable of filter-feeding have a well-developed apparatus for straining small organisms from the water, usually long, fine gill rakers. Curiously, both the largest (whale sharks and basking sharks) and the smallest (anchovies) adult epipelagic fishes are filter-feeders.

34.2 ECOLOGICAL TYPES

Epipelagic fishes can be divided, for convenience, into two basic ecological types: *oceanic forms*, or those that spend all or part of their life cycle in ocean regions that are not above the continental shelves; and *neritic forms*, or those that spend all or part of their life cycle living in the waters above the continental shelves. The two categories are not mutually exclusive, both because many epipelagic fish are both oceanic and neritic, often at different stages in their life cycle, and because the two ecological regions have no firm boundaries.

Oceanic epipelagic fish can be divided into true, partial, and accidental residents of the open ocean. The *true residents* spend their entire life cycle there and are of two basic types: those that are free-swimming, and those that are associated with drifting seaweed, jellyfish, and other objects. The number of species that are true residents is small—mainly a few species of shark, tuna, flying fishes, sauries, dolphins, swordfish, marlins, and ocean sunfish, along with commensal remoras (Echeneidae) and pilotfish (Carangidae). Most of these fishes make extensive migrations across the open ocean and, occasionally, come close to the continents. Drifting seaweed, particularly pelagic *Sargassum,* provides abundant cover and food in some areas for epipelagic fish and even supports its own unique fish fauna, including the endemic sargassum fish *(Histrio histrio)*.

The most abundant *partial residents* typically are the juveniles of neritic or benthic species that are associated with jellyfish and drifting seaweed. Such juvenile partial residents also are found free-swimming, although most partial residents are adult fish, such as salmon, dolphin, flying fish, and whale sharks. These fish spawn in the inshore areas (or in streams), and the young use similar areas as nursery grounds. Another group of partial residents are deepsea fishes, such as the lanternfishes (Myctophidae), that migrate up to the surface waters at night. Occasionally, true and partial residents of the zone are joined by adults and juveniles of species that are more properly associated with inshore and deepsea environments but have been accidentally carried into the zone by currents.

Neritic fish are among the most abundant fish in the world. They include herrings, sardines, anchovies, and menhaden as well as the predators on them, such as sharks, tunas, mackerel, jacks, billfish, and salmon. Such fish take advantage of the high productivity of inshore waters caused by upwelling and shoreline productivity, and most complete their entire life cycle in the region. Some, however, are partial residents that spawn in bays, estuaries, or streams. It also is not unusual to find among these fishes certain species that really are characteristic of other inshore habitats but have been carried into the open water by currents and storms.

34.3   PHYSICAL FACTORS

The open ocean appears to be featureless, but physical factors nevertheless are important determinants of fish distribution and abundance in the epipelagic zone. Important physical factors include (1) temperature, (2) light, (3) upwelling, (4) ocean currents, (5) islands and banks, and (6) flotsam.

Temperature

Temperature has a strong relationship to the distribution patterns of epipelagic fishes. The 8°C to 10°C isotherm, for example, loosely separates the cold ocean water, which is dominated by salmon, and the warm ocean water, which is dominated by tunas and billfish. Many epipelagic fish, but particularly the tunas, move north as the water warms and south as it cools, so the arrival of commercially important species in some waters can be predicted on the basis of sea temperatures (Blackburn 1965). Similarly, changes in average sea temperatures of only 2°C to 3°C can cause dramatic changes in fish species and abundance (Radovich 1961). In the Gulf of Alaska, for example, the catch of pink salmon by the commercial fishery is strongly correlated with sea temperature at the surface: An increase of 1°C to 2°C is associated with major increases in salmon abundance (Francis and Sibley 1991). As discussed later in this chapter, temperature is important because it interacts with other physical and chemical factors and can affect processes such as predation and competition. In discussions of global climate change, changes in

---

BOX 34.1

Ocean Conditions and the Decline of Wild Atlantic Salmon

In the North Atlantic, fish populations are affected by the North Atlantic Oscillation (NAO), a major ocean regime shift that is reflected in changes in sea temperatures at the surface. Atlantic salmon, which spend several years at sea feeding on invertebrates and fish and which migrate long distances to find their prey, have populations that are strongly influenced by the NAO. The effects seem to be strongest on survival of the salmon as they enter the ocean as smolts. When the salmon enter sea water that is 5°C to 7°C, survival rates are low, but when they enter sea water that is 8°C to 10°C survival rates tend to be high (Friedland 2002). The reasons for this are not clear but may relate to slower growth rates at lower temperatures.

Slower growth rates result in increased predation on smolts because of longer exposure to predators. Lower ocean productivity can have a similar effect by reducing the abundance of alternate prey for predators. Unfortunately, the effects of ocean conditions on Atlantic salmon survival are obscured by a general downward trend related to overfishing, large-scale raising of salmon in net pens, and heavy fishing on the small fish that are prey for wild salmon. The prey-fish fishery exists, in good part, to provide food for salmon in pens! The pen-raised fish also escape in large numbers, potentially spreading disease and parasites to wild fish or competing with them for prey (Friedland 2002).

epipelagic fisheries often are used as indicators of change, because shifts in the abundance of key species can be both sudden and dramatic. For example, during El Niño events, when the ocean warms a few degrees off central California, salmon fisheries decline, and a fishery for albacore *(Thunnus alalunga)* develops. While dramatic, however, El Niño events are relatively small changes compared to the regime shifts in the oceans that take place over many decades. In the Pacific Ocean, cooler waters off North America (California), South America (Peru), and Asia (Japan) favor anchovies (Engraulidae), whereas warmer waters favor sardines (Clupeidae). A shift from one dominant species to another and then back again takes place over an approximately 50-year cycle (Chavez et al. 2003). During the periods of shifting abundance, seabird populations and fisheries suffer (Box 34.1).

Light

Because most epipelagic fishes are visual feeders, light is an important factor. Tuna fishing often is best when water transparency, as measured by the depth at which a white disc (secchi disc) can be seen on a sunny day, is 15 to 35 m (Blackburn 1965). Water that is too clear, however, contains little food. On the other hand, some of the most productive waters are rather turbid from plankton blooms and thus favor filter-feeding planktivores as well as piscivores smaller than those found in clear water. Light also is important, because vision is a principal cue that fish use for shoaling. Thus, shoals tend to break up in the evening.

Upwelling

Although light and temperature have considerable independent influence on the ecology of epipelagic fishes, they usually act in concert with oceanographic conditions, particularly those associated with upwellings and currents. Upwelling is a phenomenon that occurs along coastlines and midocean convergences of deep, strong currents where cold, nutrient-rich deep water is brought to the surface. This water supports large blooms of phytoplankton, which in turn support zooplankton and many of the world's most important fisheries. When the upwelling fails or is reduced, the fisheries in that area fail as well. This phenomenon is well illustrated by the Peruvian anchoveta *(Engraulis ringens)* fishery, once the largest fishery in the world. When oceanic conditions change as part of the 50-year regime shift (Chavez et al. 2003) and upwelling is reduced, the anchoveta populations decline precipitously, the fishery fails, and sea birds, which also depend on the anchoveta, die by the millions. Off Japan, changes in the climate regime produce reductions in upwelling that are associated with declines in sardine *(Sardinops melanosticta)* populations. In response, fewer bluefin tuna *(Thunnus thynnus)* move to the region to feed; instead, they stay further offshore, where they presumably feed on alternate prey (Polovina 1996).

Ocean Currents

Ocean currents also play a major role in both concentrating and dispersing fish. Fish distributions may be bounded by current edges that form distinct, if fluctuating, boundaries to oceanic regions. At times, these boundaries are even visible, but more typically, they are detectable as rapid changes in temperature, salinity, and turbidity. For example, Nakamura (1969) indicates that albacore in the Asian northern Pacific have a northern limit determined by the cold North Pacific Current and a southern limit determined by the North Equatorial Current and that, within the broad area between these two current systems, their distribution is determined by the Kuroshio Current, which flows at varying strengths according to the season. Many epipelagic fish also use currents for reproduction, spawning "upstream" so that the eggs and young will drift into suitable areas for feeding and, eventually, wind up in adult feeding areas.

Islands and Banks

Islands and banks are likely to concentrate epipelagic fish because of their interactions with currents and upwelling, which often make them areas of high ocean productivity. Large eddies may form on the downwind or downcurrent side of islands, thereby concentrating plankton and, consequently, fish (Blackburn 1965). Such areas also contain concentrations of small fish and invertebrates that are associated with reefs and banks and, thereby, provide an additional source of food for pelagic predators.

BOX 34.2
Tuna FADs

Tuna and other scombrids are such highly mobile species that it often is hard for fishers to catch them without going long distances in large boats. One solution is the creation of Fish Aggregation Devices (FADs), which are large objects (e.g., buoys) anchored to float either on the surface or just below it. They are used most widely around tropical islands in the Indian and Pacific oceans. Mobile fishers also will create floating FADs out of various kinds of debris, follow them around, and capture fish that are attracted to them with purse seines. The FADs take advantage of the fascination that pelagic fish seem to have for these objects. Besides the visual stimulus hypothesis, the most likely reason for large fish to aggregate around FADs is that small prey fish use FADs for shelter. In a study using sonar of how fish associate with FADs in French Polynesia, Josse et al. (2000) found several layers of fish. Close to the devices and concentrating at 10 to 50 m deep were large shoals of juvenile (<50 cm total length) bigeye tuna (*Thunnus obesus*) and yellowfin tuna (*T. albacares*). Another, less-dense aggregation of fish was found at 50 to 150 m and apparently was made up mostly of yellowfin and albacore tuna (*T. alalunga*; 50–100 cm). Beneath this group (to 500 m) and rather dispersed were large, adult fish of several tuna species. Josse et al. (2000) noted that (1) a wide overlap occurred in distribution of the three groups, (2) the aggregations were highly variable in density and composition, and (3) the shoals tended to disperse at night. The FADs also are used by a variety of other pelagic fishes.

Flotsam

One of the more curious aspects of epipelagic fish is their tendency to associate with drifting objects (flotsam), jellyfish, floating seaweed, and other material (Box 34.2). One reason for this association appears to be that flotsam serves as "a visual stimulus in an optical void" (Hunter and Mitchell 1966, p. 27). For juvenile fish, however, flotsam may be important because it offers some protection from predators. The numbers and diversity of small fish hanging around such pelagic structures are little appreciated, and the abundance of jellyfish or drifting weeds in a region may have a significant positive impact on the year-class strength of some species (Kingsford 1993). For example, Dooley (1972) found 54 species from 23 families inhabiting floating *Sargassum* mats off Florida. Many of these fish were juveniles of neritic species that used seaweed for shelter and that fed on the abundant fish and invertebrates associated with it. Likewise, juvenile fish may use jellyfish for shelter and food, even though the young fish also may serve as prey for the jellyfish. Even marine turtles can act as mobile cover for small fish, although occasionally, the turtles become accidentally impaled by swordfish seeking to eat the fish being sheltered (Frazier et al. 1994).

34.4   CHEMICAL FACTORS

The surface waters of the oceans are well mixed and well oxygenated, so salinity and oxygen seldom have much effect on the distribution of epipelagic fishes. Pollutants are chemical factors that also may be affecting populations of epipelagic fishes in unknown ways. As Longhurst et al. (1972) point out, however, the effects of pollutants so far have been largely masked by natural fluctuations in fish populations in response to changing oceanic conditions coupled with heavy exploitation by humans.

34.5   BIOLOGICAL FACTORS

Competition, predation, and symbiosis undoubtedly are all important in regulating fish populations of the epipelagic zone. Their roles are poorly understood, however, both because of the difficulties in studying such processes in large, widespread fish populations and because the populations respond so strongly to changes in oceanographic conditions and to the intense fishing efforts directed at them.

   Competition among members of the same species appears to be one of the more important processes regulating the size of epipelagic fish populations, thereby allowing each population to adapt to changing oceanographic conditions and exploitation (see Section 34.7). Basically, intraspecific competition for limited food among adult fish leads to limited growth and smaller sizes and, hence, to decreased fecundity. Intraspecific competition for food among larval fish also leads to slower growth, which in turn results in increased mortality, particularly from predation.

   The role of competition in interspecific interactions is less certain. Many rather similar species coexist in the open oceans, which suggests that competition may have an important function at times. For example, in the tropical and subtropical Pacific Ocean are four species of tuna with wide, broadly overlapping ranges and ecological requirements. Often, two species will school together. The neritic epipelagic regions of the world usually have two or more species of clupeids in each area. Each species appears to have a competitive advantage over the other(s) under appropriate oceanic conditions, so the dominant species may change with time. For example, in the English Channel, herring *(Clupea harengus)* was replaced in the 1930s as the dominant planktivore by pilchard *(Sardinops pilchardus)*, but the situation was reversed in the 1970s (Cushing 1978). The change was accompanied by many other oceanic and biological changes, similar to the factors driving the anchovy–sardine shifts in the Pacific Ocean. The actual mechanisms that allow one species to become dominant over the other are poorly understood, however, and are complicated in this case by the large schools of other plankton-feeding fishes that may coexist with the clupeids while seeming to exploit the same food supply (Murphy 1977).

Predation has a powerful influence on epipelagic fish populations and interacts strongly with oceanic conditions and fisheries. Often, the influence is very direct. On the Atlantic coast of North America, the seasonal arrival of large numbers of voracious bluefish produces a decline in the abundance of the anchovies and sandlances on which they feed (Safina and Burger 1989). In the Bering Sea, cannibalism by adult walleye pollock *(Theragra chalcogramma)* on juveniles, combined with predation by marine birds and mammals, is a major factor regulating pollock numbers (Livingston 1993). On the other hand, because oceanic conditions exert a powerful influence on the populations of plankton-feeding fish, the populations of piscivores may be strongly affected by the environmentally induced fluctuations in their prey. Off California, changes in the abundance of brown pelicans seem to show a direct relationship to changes in the abundance of their main prey, the anchovy. On the other hand, the influences of predator and prey populations on each other are somewhat buffered, because most predators are capable of preying on a wide range of organisms and will switch prey if the population of one prey species declines (unless a fishery reduces the abundance of alternate prey).

One of the strongest indications of the importance of predation to epipelagic fish is that shoaling is the dominant mode of behavior. Large shoals of small fish offer protection from predators and facilitate feeding, whereas schools of predatory fish presumably have an advantage in finding prey (see Chapter 11).

Examples of *symbiosis* are common in the epipelagic zone and include (1) multispecies shoaling, (2) tuna and porpoise, (3) remoras and large fishes, (4) pilotfish and sharks, and (5) fish that live in association with floating invertebrates (Parin 1968). Shoals of fish that are made up of more than one species are common, especially among the tunas and clupeids. Presumably, much of the advantage of such relationships is an increase in shoal size, which results in increased protection from predators or increased probability of locating a patch of prey. These advantages would be particularly beneficial to a rare species that shoaled with an abundant one. Among the clupeids, the ability of a species to shoal with another may be one of the main methods by which a species that is at a competitive disadvantage can still maintain small populations among the much larger ones of its competitor (Radovich 1979). The tuna–porpoise relationship is a special case of two species that school in close association. It seems to be a commensal relationship, with the schools of tuna following the schools of porpoise (or, perhaps, vice versa). The relationship is not obligatory for either species. How the two species benefit is not known, but the tuna may be taking advantage of the ability of the porpoise to find shoals of small fish through the use of sonar. This relationship has been detrimental to both species, however, because tuna fishermen set nets around shoals of porpoise to capture the associated tuna and kill the porpoises in the process.

In contrast to the first two examples, in which animals of equal size (usually) are shoaling together, remoras and pilotfish are small fish that live with large pelagic fish in relationships that have elements of both commensalism and parasitism. Remoras use a special, sucker-like apparatus to attach themselves to sharks and other large fish, turtles, and whales. Remoras use their hosts for transportation and protection, leaving them only for short pursuits of small fish and invertebrates. Remoras often act as cleaners, picking parasites off the host fish, but whether this benefit compensates for the energy expended by the host in carrying the remoras around is doubtful. Pilotfish *(Naucrates)* have a relationship with their host similar to that of remoras, except that they swim in the friction layer of water around the host and, consequently, are pulled along by the host's swimming (Parin 1968).

Small epipelagic fishes also commonly are found in association with jellyfish, siphonophores, and pelagic tunicates, living either in the body cavity or among the tentacles. The man-of-war fish *(Nomeus)* is always found among the tentacles of the man-of-war siphonophore. Presumably, this fish has a mutualistic relationship with the man-of-war, much like the one that anemonefishes have with their hosts on coral reefs (see Chapter 33).

34.6   MIGRATIONS AND MOVEMENTS

Most epipelagic fish move freely from one area to another in search of food, for spawning, or in response to changing oceanographic conditions. For large predators, such as tuna and salmon, these movements can cover thousands of kilometers—from one side of an ocean to the other and back (Fig. 34.1). For example, mako sharks *(Isurus oxyrinchus)* make long, seasonal migrations, apparently following temperature gradients; tagged individuals have been recorded as moving more than 4,500 km in 1 year (Casey and Kohler 1992). The movements of bluefin tuna also are complicated and extensive, because they migrate among diverse spawning and feeding grounds (see Box 11.1).

When spawning is involved, the overall migration pattern typically follows a triangular pattern (see Chapter 11). The adults migrate in a direction opposite that of the surface currents, toward areas that are suitable for spawning and close to waters that support the high concentrations of plankton that are needed by the larval fish. The adults then return to their feeding area, while the larvae drift in the surface currents to a nursery area. As they grow, the juvenile fish actively migrate to the adult areas. This pattern reduces competition for food among the different life-history stages of a species and also reduces the incidence of cannibalism. A classic example of the triangular pattern is found in the herring of the North Sea. They migrate to spawning grounds off the coast of Great Britain, where they deposit their eggs on stones and gravel. The larvae hatch and then drift over to the coast of Germany and Denmark, where shallow waters serve as nursery areas. After 1 or 2 years, at a length of 9 to 10 cm, the herring move off the coast toward the North Sea, where they eventually join the adults (Harden-Jones 1968).

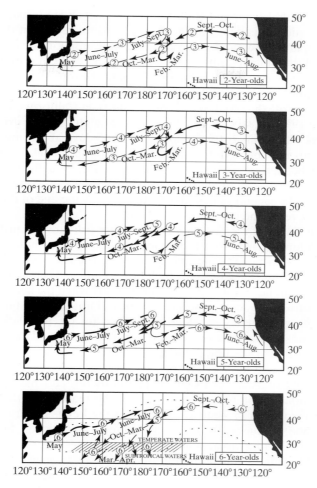

FIGURE 34.1 Presumed albacore migration patterns by age groups in the northern Pacific Ocean. From Otsu and Uchida (1963).

34.7 PACIFIC SARDINE

As commercial fishermen are well aware, the local populations of epipelagic fish undergo considerable fluctuation from year to year as well as over much longer periods of time. These fluctuations often seem to be unpredictable, but this is mainly because their causes are complex. Fish populations are simultaneously affected by oceanic conditions, interactions with other species, and fisheries. Furthermore, such factors may act in different ways and at different intensities during the various stages of a fish's life cycle. The factors also may act in concert, either in opposition to one another or synergistically. An idea regarding the complexity of the interactions can be obtained by examining the history of the Pacific sardine, one of the most studied of epipelagic fishes

(Murphy 1966, 1977; Ahlstrom and Radovich 1970; Chavez et al. 2003). This study was stimulated by the sudden collapse of the Pacific sardine fishery in the 1940s, an economic disaster apparently caused by overfishing and unfavorable oceanic conditions acting together. The ultimate result was the conversion of Monterey, California, from the fishing village made famous by John Steinbeck to the tourist mecca focusing on the Monterey Bay Aquarium.

Sardines spawn off the coast of California and Baja California at temperatures of between 13°C and 19°C (mostly 15–18°C). The larvae hatch from the drifting, translucent eggs in 2 or 3 days. They then drift for another 40 to 45 days, absorbing the yolk sac and capturing organisms they bump into, until they develop the capability to swim actively. At this stage, they begin to move inshore and grow rapidly on a diet of zooplankton and diatoms. They typically reach lengths of 11 to 12 cm by the end of the first year, of 17 to 18 cm by the end of the second year, and of 19 to 20 cm by the end of the third year. They often live for 6 to 7 years and reach lengths of 25 cm.

Before the collapse of the populations, the larger fish moved north in the summer, as far as British Columbia, and moved south in the winter, although the timing and extent of these movements depended on oceanic conditions. The main oceanic feature that affects sardines is the California Current, which flows southward along the coast, bringing with it cold water from the Gulf of Alaska. During years when the current flowed strongly, water temperatures were lower, and the sardine did not move as far north. Increased upwelling, however, presumably provided more food for the populations in the south. When the current was weak, water temperatures were warmer, and the sardine was found farther to the north. It also was noted that the "warm" years were those in which the sardine had high reproductive success, the opposite was true of the "cold" years.

During the cold years, heavy fisheries apparently turned a natural declining trend in sardine populations into a population collapse. The reasons for the collapse are not certain, but Murphy (1966) presents evidence that the key is in survival of the larvae. Each female sardine produces 100,000 to 200,000 eggs in a season, but less than 0.1% survive through the drifting larval stage. Murphy hypothesized that the main cause of death of the larval sardines was predation by invertebrates, such as arrow worms and copepods. According to this hypothesis, despite the enormous numbers of larval sardines likely to be present in a spawning area, they are a relatively small part of the total plankton and, consequently, a small part of the total diet of the invertebrate predators. Therefore, the number of sardines that are eaten depends on the density of predators and on the length of time during which the larvae are exposed to predation. In years that are unfavorable for the sardine, productivity of the inshore areas increases, and so zooplankton and invertebrate predator populations increase. At the same time, the sardine larvae take a few days longer to reach the point where they can actively avoid the invertebrate predators. These two factors result in much lower survival of the sardine larvae during the cold years. If the sardine population is large, so many eggs and larvae are produced that enough can survive to maintain the population—or at least keep the rate of population decline minimal. If the sardine population is small, however, it will stay small and become even smaller, particularly if the cold years and fishing continue (which they do periodically). Because of the high fecundity of the sardines and their high survival rates during warm years, a small adult population can lead to a large adult population if little fishing and a series of years favorable to larval survival occur (Fig. 34.2).

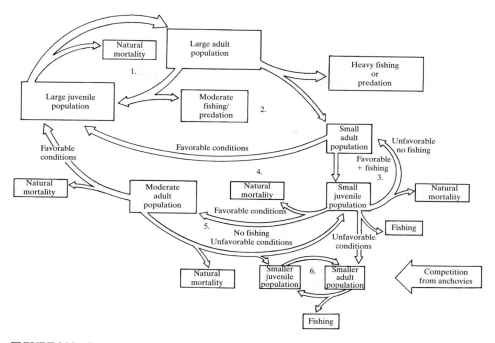

FIGURE 34.2   Some possible population cycles in the Pacific sardine. These cycles are only a few of the possible combinations, because the relationship between larval survival and adult populations is very complex and may be affected by several factors simultaneously. For example, cycle 1 can revert directly to cycle 6 even in the absence of fishing if the right combination of ocean and biological conditions exists. **1.** The no- or light-exploitation cycle, in which large populations maintain themselves under favorable ocean conditions. **2.** The heavy exploitation cycle under favorable ocean conditions. **3.** The cycle for maintaining a small but significant population. **4.** The cycle for restoration of large populations, following overfishing, during years of favorable ocean conditions. **5.** The cycle resulting from unfavorable ocean conditions following overfishing. **6.** The small population cycle resulting initially from overfishing of an already depleted population, perhaps maintained by competition from the northern anchovy population and incidental capture of sardines in other fisheries.

Murphy's predation hypothesis is an attractive one, but another hypothesis, developed by Lasker (1978) to explain population fluctuations in the ecologically similar northern anchovy, may offer an even better explanation for the fluctuations of sardine populations in response to oceanic conditions. In this hypothesis, the critical factor for survival of the larvae is the availability of food of the proper size, abundance, and nutritional value during the first few days of feeding. Sardines (and anchovies) at this stage are weak swimmers, have high metabolic rates and small mouths, and feed only during the day. This means that small organisms (probably dinoflagellates) must be available at high densities if the larvae are going to capture enough during the day not only to avoid starving to death at night but also to grow quickly to a size at which a larger size range of prey becomes available. The best conditions for early larval survival therefore seem to be warm, stable oceanic conditions that produce blooms of organisms of the proper size and nutritional value. Storms or upwelling may disperse such organisms to a point where their density is too low for the larvae to survive. Upwelling may greatly increase the total abundance of organisms, but the diatoms that result are too small and the invertebrates too large and active. Upwelling, however, may be beneficial for older larvae that are large enough to capture copepods and other invertebrates. Regardless of the mechanism, the amount of upwelling plays a crucial role in the abundance of sardines. Thus, Ware and Thomson (1991) suggest that optimal conditions for sardine larvae exist under moderate upwelling conditions produced by winds of 7 to 8 m/s for extended periods of time. Recently, there has been a modest resurgence in the abundance of sardines off the Pacific Coast, and a fishery has become reestablished. Thousands of tons are frozen and shipped to southern Australia, where they are fed to bluefin tuna raised in net pens (Wolf et al. 2001), and the tuna are sold to Japan. A distinct possibility exists that the sardines carry a viral disease that may not be present in the southern hemisphere; if the virus spreads, it could become a factor affecting nonresistant fish populations off Australia.

Whatever factors contributed to the decline of the sardine, the northern anchovy became the dominant planktivore in the California Current ecosystem. The sheer abundance of anchovies though competition and predation on larvae may have reduced the ability of the sardine to recover quickly from the presumably exceptionally small population created by the fishery. Suppression of a once-abundant species by a dominant species apparently is common in epipelagic fishes and may cause populations of the suppressed species to no longer track environmental fluctuations as strongly (Skud 1982). In the Atlantic Ocean, the effect of dominance by one species is shown in the population fluctuations of the Atlantic herring and the Atlantic mackerel *(Scomber scombrus)*. When the mackerel is dominant, its populations fluctuate in response to temperature conditions during the larval stage, but those of the herring do not. The reverse is true when the herring is dominant (once an ocean regime shift has taken place).

A new long-term test of the fisheries and regime shift interaction is now underway. Apparently, ocean conditions in the Pacific are now cycling back to the cooler conditions favoring anchovies (Chavez et al. 2003). The overall impact of the present fisheries on sardine populations probably is small, but considerable uncertainty exists regarding its potential to enhance the natural decline and to reduce the recovery potential.

34.8 LESSONS FROM EPIPELAGIC FISHES

The open ocean is full of biological surprises. Only recently, for example, have we begun to appreciate that drifting jellyfish can be a significant habitat for small fish. Considering that comparatively few species live in the open-ocean environment, interactions among the species can be quite complex and include symbiotic interactions. The open ocean also is a place of extremes. To take advantage of regional concentrations of food or desirable conditions for spawning, larger epipelagic fish make some of the longest migrations of all vertebrates, often swimming from one side of an ocean to the other. Smaller inshore species develop some of the highest "standing crops" of any vertebrate. Such large populations depend on favorable local ocean conditions, especially for their larval stages, and consequently show enormous fluctuations in numbers, with dominant species shifting back and forth in response to long-term fluctuations in the ocean conditions. Such species depend on favorable environmental conditions for their persistence in

large numbers, but it is increasingly apparent that biotic interactions, especially predation, also play a role in determining abundance. For decades, the ultimate predators,
commercial fishermen (abetted by fisheries managers), insisted that stocks of epipelagic
fishes could not be overfished, because when conditions are good, even a few fish can
produce enough young to restock the environment. We now know better—but still overfish many stocks.

SUPPLEMENTAL READINGS

Blaxter and Holliday 1963; Jennings et al. 2001; Leet et al. 2001; Longhurst 1998; Mooney 1998;
Pauly and MacLean 2003.

WEB CONNECTIONS

EL NIÑO EVENTS
www.pmel.noaa.gove/tao/elnino/

ATLANTIC SALMON
www.asf.ca/links/

HERRINGS
www.gma.org/herring/
www.pac.dfo-mpo.gc.ca/sci/herring/

TUNA
www.tunaresearch.org/

C H A P T E R 35
Deep Sea Habitats

In the surface waters of the ocean, the water is well lighted, well mixed, and capable of supporting actively photosynthesizing algae. Below this epipelagic zone, however, conditions change rapidly. Between 200 and roughly 1,000 m (*mesopelagic zone*), the light gradually fades to extinction, and temperatures fall through a more or less permanent thermocline to between 4°C and 8°C. Nutrient levels, dissolved oxygen levels, and circulation rate also fall, and pressure increases. Below 1,000 m (*bathypelagic zone*), conditions are more uniform until the bottom (*deepwater benthic zone*) is reached. The bottom is characterized by complete darkness, low temperatures, low nutrients, low dissolved oxygen levels, and great pressure. This environment is the most extensive aquatic habitat on Earth. With the mean depth of the oceans being approximately 4,000 m, roughly 98% of their water is found below 100 m, and 75% is found below 1,000 m. The vastness of this environment, coupled with its probable stability through geological time, has led to the development of a diverse—and often bizarre—fish fauna, which makes up approximately 11% of all known fish species. Probably the most numerous fish in existence are the small (<10 cm) pelagic forms, particularly bristlemouths (*Cyclothone*). These fish form much of the *deep-scattering layer* of the ocean, so called because the sonic pulses of sonar reflect off the millions of swimbladders, thereby giving the impression of a false bottom.

By the standards of the lighted world, all deep sea fishes are very peculiar in appearance, and one of the challenges of deep sea ichthyology is to determine the uses of strange structural features. Inference is the main tool in this endeavor, because trips to deep sea habitats are extraordinarily difficult. Similar problems exist in studying the distribution and ecology of deep sea fishes using trawls that are towed several kilometers beneath a ship. Nevertheless, these fishes are so fascinating that they are responsible for a large literature.

In this chapter, we summarize the diversity, adaptations, reproductive strategies, and ecology of deep sea fishes. This summary owes a debt to the writings of N. B. Marshall (1954, 1960, 1966, 1971, 1984). These works are classics of clarity that have integrated the diverse information regarding deep sea fishes with many original insights and observations.

35.1 DIVERSITY

The diversity and abundance of deep sea fishes is indicated by the results of midwater trawling. In a 12-km² area off Bermuda, 1,500 such trawl hauls produced 115,000 fish from 46 families and 220 species (Fitch and Lavenberg 1968). Similarly, Brewer (1973) reported catching 56,000 fish in 61 hauls in the eastern Pacific, representing 49 families and 113 species. In both studies, more than 90% of the fishes belonged to two families, the Gonostomatidae (bristlemouths) and the Myctophidae (lanternfishes). Most of the species were quite rare and represented in the collections by 1 to 20 individuals. Large and diverse collections of deep sea fishes are made primarily in the tropical and sub-tropical regions of the oceans, particularly near continents. The bathypelagic fish fauna especially becomes depleted in high latitudes. Deep sea fish appear to be totally absent from the Arctic region and are represented only by approximately 30 species in the Antarctic region.

One of the more curious aspects of deep sea fishes is the scarcity of acan-thopterygian species among them. This is generally interpreted to indicate that the deep sea fauna is ancient and so well adapted to its environment that invasions by more recently derived fishes have been greatly restricted (Haedrich 1996). Another taxonomic characteristic of deep sea fishes is that pelagic forms have evolutionary origins different from those of the benthic forms. The deep sea pelagic fishes belong mostly to their own superorders and orders, indicating long evolution in the deep sea environment: the Saccopharyngiformes (gulper eels), the Stenopterygii (bristle-mouths, hatchetfishes, jellynose fishes, etc), the Scopelomorpha (lanternfishes, greeneyes, telescope fishes, barracudinas, etc.), and the Lophiiformes (anglerfishes; see Chapter 20 for details). The benthic forms, in contrast, are mostly elopomorph (e.g., halosaurs, spiny eels, and cutthroat eels) and paracanthopterygian fishes (e.g., bro-tulids, cusk-eels, cods, and grenadiers) in orders that also include many related shal-low-water forms. The few acanthopterygian forms, both benthic and pelagic, are mainly in the orders Lampridiformes and Beryciformes, which are regarded as being ancient groups.

35.2 ADAPTATIONS

The fish that inhabit the mesopelagic, bathypelagic, and deepwater benthic zones differ remarkably from one another in structure. However, there may be closely related species in the different zones.

Mesopelagic fishes are adapted for an active life under low-light conditions. Most of them make extensive vertical migrations, often moving at night into the epipelagic zone, where they prey on plankton and each other, and then moving down several hun-dred meters during the day. These fishes have muscular bodies, well-ossified skeletons, scales, well-developed central nervous systems, well-developed gills (especially in forms that inhabit oceanic regions with low oxygen supplies), large hearts, large kidneys, and usually, swimbladders. The fish with swimbladders are migratory and have an extraordi-nary capacity to regulate the amount of gas in the swimbladder as they move up and down the water column (Haedrich 1996). Because they primarily are visual predators, their eyes are large, often with high concentrations of photosensitive pigments in the rods of the retina, thereby giving the fishes extreme sensitivity to light. Some of the deepest-living mesopelagic fishes have tubular eyes, typically pointing upward, with large lenses that permit binocular vision and concentration of limited light signals. Such visual capabilities enable them to pick out small planktonic organisms in dim light. Most mesopelagic fishes lack spines, so their main defense from predators is concealment. In color, they are either black (lie-in-wait predators) or silvery with countershading (migratory forms that encounter low-light levels). In addition, many have rows of photophores on their bellies; these photophores produce light that helps to break up the silhouette of the fish as seen by predators peering up into the downward-streaming light. Muntz (1976), however, suggests that the special nature of bioluminescence has permitted some mesopelagic predators to overcome this camouflage by having yellow lenses that filter out the ambient light, thereby leaving the bioluminescence visible. Because photophores are well developed in other locations on mesopelagic fishes, they also may serve other functions, particularly intraspecific signaling for shoaling and reproduction. Ecologically, mesopelagic fishes can be divided into zooplankton feeders, which have small mouths and fine gill rakers, and piscivores, which have large mouths and coarse gill rakers.

Bathypelagic fishes, in contrast to mesopelagic fishes, are largely adapted for a sedentary existence in a habitat with low levels of food and no light other than bioluminescence. These fishes generally have poorly developed and flabby muscles, weak skeletons with minimal ossification, no scales, poorly developed central nervous systems (except those parts associated with the lateral line and olfactory systems), small eyes, small gills, small kidneys, small hearts, and reduced or absent swimbladders. Marshall (1984) notes that these also are characteristics of fish larvae, indicating that these features have developed through the retention of larval characteristics in the adults. As they do in larvae, these features allow fish to remain suspended in the water column with almost no expenditure of energy, because the fish achieve nearly neutral buoyancy without maintaining a gas-filled swimbladder in a high-pressure environment. Their eyes are small and may be nonfunctional. The most important sensory system usually is the acousticolateralis system, although the olfactory system also may be well developed, particularly in males, which often find females by odors (Jumper and Baird 1991).

Bathypelagic fish are uniformly black in color and have only a few small photophores. Among anglerfish, photophores usually are confined to lures, which they use for attracting prey. Because of the scarcity of food, anglerfish and other bathypelagic predators consume whatever invertebrates or fish come close enough to be grabbed. This means that the predator must be able to capture a wide size range of prey, which is accomplished by having a large mouth with sharp teeth for the capture of large prey combined with well-developed, overlapping gill rakers to prevent the escape of small prey that have been swallowed.

Deepwater benthic fishes are similar to mesopelagic fishes in that they have muscular bodies and well-developed organ systems; however, they are more variable in many of their characteristics. Photophores may be present but usually are absent. Eyes range from well developed to absent, as do swimbladders. The fishes also are variable in size, with large species (>1 m) being fairly common. Curiously enough, many (if not most) deep sea benthic fishes are elongate, with many being eels (or at least eel-like). The most abundant forms (or at least the most conspicuous) seem to be rattails (Macrouridae) and brotulas (Ophidiidae), but a wide variety of other families, such as the Myxinidae (hagfishes), Zoarcidae (eelpouts), Chloropthalmidae (greeneyes), eels (various families), Cyclopteridae (lumpfishes), and Ogcocephalidae (batfishes), also are well represented. The commonness of eel-like forms may relate to the importance of the lateral line as a sensory system, because an elongate body results in long lateral-line canals. Haedrich (1996) notes that the lateral line may function largely to detect low-frequency sound in these fishes, and evidence suggests that some benthic fishes have special muscles that are used for drumming (the creation of sound to attract mates). The sense of smell also is important in many of these fishes, as indicated by the rapidity with which they find a trap that is baited with dead fish. Smell, combined with touch and well-developed cephalic lateral-line canals, also may be important for locating invertebrates of the deep sea benthos that, together with carrion, are the main items in their diets.

Ecologically and morphologically, deep sea benthic fish seem to be of two basic types, "true" **benthic** fish and **benthopelagic** fish. True benthic fish are in continuous contact with the bottom and are either lie-in-wait predators or species that move slowly over the bottom searching for food. The deepest-living fish known (Ophidiidae, *Abyssobrotula galatheae*) (Fig. 35.1) is in the latter category; it is a blind, elongate fish that feeds on benthic invertebrates (Nielson 1977). Benthopelagic fish live just above, but close to, the bottom; they typically have diets that are mixtures of benthic organisms and specialized benthopelagic zooplankton (Mauchline and Gordon 1986). Benthopelagic fish usually are neutrally buoyant, and Koslow (1996) notes that they seem to have two different body types, flabby and robust. Flabby benthopelagic fishes are like bathypelagic fishes in that they save energy by reducing body mass and swimming about slowly. One such predator is the cusk-eel (*Acanthonus armatus*), which has an elongate body, large otoliths, and spiny operculi; it also has a body that is 90% water and a brain, which is the smallest known in relation to body size for any vertebrate, located in a huge head (Fine et al. 1987). Robust benthopelagic swimmers are muscular, good swimmers that cruise about the bottom in search of prey or live in areas, such as seamounts, that have strong currents to swim against. The orange roughy *(Hoplostethus atlanticus)* is one such species and is abundant enough to be harvested commercially (Koslow 1996). Other fishes, however, defy such ecological classification systems. For example, members of the common and widely distributed genus *Bathypterois* (Chloropthalmidae, spiderfishes), which also are nearly blind, feed on benthopelagic zooplankton but are basically benthic fish. They have elongate rays on the caudal and pelvic fins that permit them to "stand" on the bottom, presumably facing into currents, and grab passing zooplankton (Fig. 35.2) (Sulak 1977).

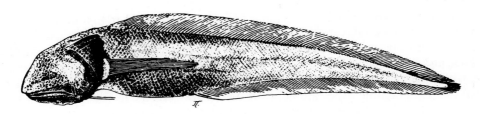

FIGURE 35.1 *Abyssobrotula galatheae*, the deepest-living fish known. From Nielson (1977), used by permission of Scandinavian Science Press Ltd.

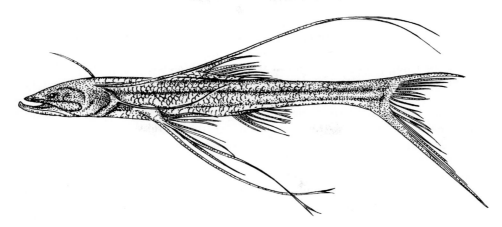

FIGURE 35.2   The spiderfish (*Bathypterois oddi*) is a predator on benthopelagic organisms. After Sulak (1977), used by permission of Scandinavian Science Press Ltd.

35.3   REPRODUCTIVE STRATEGIES

Among the more remarkable adaptations of deep sea fishes are their methods of reproduction. Because population densities of many species are extremely low, just finding a mate is a major problem. As a consequence, many unconventional reproductive strategies have evolved. Most mesopelagic and bathypelagic predators have epipelagic larvae, which they produce in large numbers. A female anglerfish, for example, may contain as many as 5 million eggs. The main problem is finding a mate to fertilize these eggs. To facilitate this process, female anglerfishes apparently release a species-specific pheromone into the water that male anglerfish detect and follow to the female. In many species, once the female is found, the male anglerfish attaches himself to her with specially equipped jaws and assumes a parasitic mode of existence. In species in which this strategy is best developed, the gonads of both sexes do not mature until the female has been parasitized by a male (Pietsch 1976). An additional advantage of this strategy is that the energy required for reproduction is reduced, but not at the expense of fecundity. Most species of mesopelagic and bathypelagic fishes have males that are smaller than females and have much larger olfactory organs, presumably also because of the energetic advantages. In some of these species (e.g., Idiacanthidae), the males do not feed, although they are free-living. Each male therefore has a limited time to find a female, which is enhanced by the male's extraordinary ability to detect female pheromones and by the behavior of the female (Jumper and Baird 1991). Another solution to the energy and mate-finding problems is to be hermaphroditic, thereby ensuring that any individual can mate with any other individual of the same species. This is characteristic of a number of deep sea predators, such as the lancetfishes (Alepisauridae) and the barracudinas (Paralepididae), although it is not known whether these fishes are capable of self-fertilization (Mead et al. 1964).

In deep sea benthic fishes, finding a mate seems to be less of a problem than assuring that the young will settle in suitable habitats. The grenadiers (Macrouridae) and brotulas (Ophidiidae) apparently produce sounds with their swimbladders and thereby attract mates over some distance. In other deep sea benthic fishes, bioluminescence also may play a role in this regard. Deep sea eels, such as longnose eels (*Synaphobranchus*), may aggregate for spawning through the use of smell or, perhaps, through homing to a hereditary spawning ground (Menzies et al. 1973). Although eels all have pelagic lepto-cephalous larvae, most deep sea benthic fishes seem to have relatively short larval stages. Many species produce large, benthic eggs that result in large, active larvae, which settle out quickly. Some of these species, especially among the eelpouts (Zoarcidae), may have parental care as well. Other species are viviparous, most prominently chi-maeras and other chondrichthyian fishes but also species of brotulas and eelpouts. Re-gardless of how the young are produced, they probably remain close to the spawning area. As a result, there is a much higher incidence of species that are confined to one ocean basin than is true for bathypelagic and mesopelagic fishes, many of which are nearly cosmopolitan in their distribution.

35.4 ECOLOGY

Physical Factors

The main physical factors affecting distribution of deepsea fishes are (1) temperature, (2) light, (3) pressure, (4) features of the ocean floor, and (5) currents.

The *temperature* of the deep ocean is rather stable, between 4°C and 8°C, so it seems unlikely that temperature has much of an impact on local distribution patterns. The great reduction in diversity of bathypelagic and mesopelagic fishes at high latitudes, however, may be partially a function of the slightly lower temperatures that exist there. Backus et al. (1977) were able to divide the mesopelagic zone of the Atlantic Ocean into zoogeographic regions, noting that faunal changes coincided with changes in physical conditions, particularly temperatures at approximately 200 m. Many mesopelagic fishes nevertheless show a great tolerance for temperature changes and may migrate through the permanent thermocline on a daily basis, thereby encountering temperature changes of 10°C to 20°C.

Light, in contrast to temperature, is extremely important in determining the local distributions of deep sea fish. Indeed, the distinction between mesopelagic and bathy-pelagic zones as well as the adaptations of the fish that live in the zones are, either di-rectly or indirectly, related in large part to light levels. In the mesopelagic zone, the movements of fish that make vertical migrations are cued to light. The fish follow the re-ceding light levels up in the evening and then move back down in the early morning as light levels in the surface waters increase.

Pressure, like light, has a major impact on the vertical distributions of deep sea fish-es, because it increases at the rate of 1 atm per 10 m. The absence or reduction of the swimbladder in so many deep sea fishes undoubtedly relates to the energy costs of filling the bladder at great depths. One indication of this is that the swimbladders of some deep sea fishes are functional when the fishes are juveniles living in the epipelagic zone, but they regress or become fat-filled when the fish change to their adult habitat (Horn 1970). Mesopelagic fishes that make extensive vertical migrations apparently expend consider-able energy deflating the swimbladder as they move up and inflating it as they move down. Even fishes that inhabit nearly anoxic waters at the bottom of the migratory cycle manage to reinflate their swimbladders with oxygen (Douglas et al. 1976). At great depths, pressure may be a limiting factor to fishes, although other factors, such as scarcity of food, also are important. Bathypelagic fishes usually are not found below 3,000 m, and the greatest depth record for benthic fishes is approximately 8,370 m (Nielson 1977).

Features of the ocean floor and *currents* often act together in determining both local and worldwide distribution patterns of deep sea fishes, particularly benthic fishes. Although much of the deep ocean floor is a featureless plain covered with sediment, many deep sea benthic fishes are associated with rock outcroppings or canyons, which have distinct invertebrate communities as well. On a broader scale, benthic fishes also are more diverse and numerous on the continental slope, presumably because of a combination of greater food supply and greater habitat diversity. On an even broader scale, undersea mountain ranges and ridges often form barriers to fish movement as well as boundaries for distinct sets of oceanic conditions. Thus, Brown (1974) found that each of three basins off southern California separated by ridges had distinctive elements in its mesopelagic fish fauna that reflected, in part, the currents that flowed into the basins. There were enough differences that 11% to 16% of the fish that Brown collected in each basin were not found in the other basins. When such patterns are examined for entire oceans, regions of endemism can be noted, particularly for benthic fishes, which do not disperse as easily as pelagic fishes over geologic barriers.

Chemical Factors

Because the ocean is so well mixed, salinity and oxygen levels are the main chemical factors that are likely to influence the distribution of deep sea fishes. Although water masses that distinct deep sea fish faunas are associated with often differ slightly in their salinities, it seems unlikely, given the small differences that exist (1–2 ppt), that salinity by itself limits the distribution of any given species. Oxygen, in contrast, probably is very important (Box 35.1). In many areas, often at depths between 100 and 1,000 m, is an *oxygen minimum layer*, where dissolved oxygen levels may be almost undetectable. These layers undoubtedly act as barriers to the movement of some fish, but they often are inhabited by large numbers of mesopelagic fish for at least part of the day. Below the oxygen minimum layer, oxygen levels typically are low enough (5–6 mg/L) so that, in combination with the low temperatures, they greatly limit fish activity. Remember that all oxygen in the deep sea was dissolved in the water when the water was in contact with the surface, so presumably, it is gradually renewed through the flow of ocean currents. At times, benthic fishes also may be limited by oxygen availability.

Biological Factors

Because the deep sea environment is so hard to study, the predatory, competitive, and symbiotic interactions among species are poorly understood. *Predation* is presumably the most common type of interaction, both because all known deep sea fishes are carnivorous and because many of the behavioral and structural adaptations of the fish can best be explained as mechanisms either to help avoid being eaten or to increase the probability of capturing suitable prey. In the bathypelagic zone, most fish do not appear to be selective in their feeding habits, so who eats whom seems to depend largely on

BOX 35.1
Lilliputian Fishes of the Southern Atlantic

The southern Atlantic Ocean along the coast of North America just north of Cape Hatteras has a number of unusual oceanographic features (Sulak and Ross 1996). In part of this area, the deep sea benthic fishes are all considerably smaller than their counterparts elsewhere, a discovery made by a combination of trawling and videotaped observations using a submersible vehicle. Thus, witch flounders (*Glyptocephalus cynoglossus*) in a reference area averaged approximately 150 mm in total length, with larger individuals being common, whereas those in the Lilliput area averaged less than 100 mm, with few individuals being larger than that. Similar patterns were noticed in two other dominant fishes, eelpout (*Lycenchelys verrillii*) and hagfish (*Myxine glutinosa*). These fishes were present at much higher densities than in other areas but were very sluggish. Active

benthopelagic species largely were absent. Sulak and Ross (1996) indicated that this odd fauna apparently developed in response to low oxygen levels related to large deposits of organic matter on the bottom combined with poor circulation. The abundant benthic invertebrates apparently deplete the oxygen near the bottom through respiration and stirring up the organic matter. To take advantage of this abundant food source, the fishes have developed the means to live in hypoxic conditions, including small size. This is a good example of how deep sea fishes adapt to local conditions. "The Lilliputian phenomenon seems to signal a unique and paradoxical set of conditions ... which simultaneously enhance population density while restricting fish size, faunal diversity, and species activity modes" (Sulak and Ross 1996, p. 109).

size. Even this rule does not always hold, however, because a number of anglerfishes, blackdragons (Idiacanthidae), and viperfishes (Chauliodontidae) are capable of swallowing fish considerably larger than themselves through the use of distensible stomachs and "hinged" heads.

Competition, like predation, has presumably been an important force in shaping deep sea fish communities, as indicated by the surprisingly high diversity of these communities and the striking morphological differences among closely related species that occupy different habitat zones. On the other hand, within zones are groups of coexisting species that all seem to do about the same thing, such as the small plankton-feeders of the mesopelagic zone, the anglerfishes of the bathypelagic zone (Box 35.2), and the various rattailed fishes of the deep water benthic zone. Tyler and Pearcy (1975) noted that the diets of three species of mesopelagic lanternfish were broadly similar but, nevertheless, that some segregation by food habits occurred, perhaps related to slightly different depth distributions. Johnson and Glodek (1975) speculated that distribution differences among similar species of mesopelagic pearleyes (Scopelarchidae) are related to different tolerances to low oxygen levels; one species has a competitive superiority only when oxygen levels are low (because of its longer gill filaments).

The deep sea floor, like the bathypelagic zone, is a stable environment with scarce food and a wide array of ecologically similar species. Most of the fishes apparently roam

BOX 35.2

How Do Anglerfishes Coexist?

In the bathypelagic zone are approximately 100 widely distributed species of anglerfishes, many of which occur together. Although these species present an extraordinary series of morphological variations on the basic anglerfish theme, they all apparently consume whatever prey is available, including (presumably) each other. How these fishes manage to share the scarce food resources of the zone is not known, but a combination of spatial (depth) preferences combined with widespread dispersal of the epipelagic larvae can be hypothesized. As Sale (1977) hypothesized for coral reef fishes, these larvae may be part of an anglerfish lottery, in which the winner is the larva that happens to settle out in an area not occupied by another anglerfish or predator (regardless of species) and so grows instead of being eaten. Given the apparent rarity of anglerfishes in general, long-term success of an individual would depend, in large part, on "winning" (i.e., consuming or avoiding) the rare encounters with other bathypelagic predators as well as encountering enough prey to avoid starvation.

the floor in search of food, mostly invertebrates (which are surprisingly diverse) or carrion that has dropped from the waters above. The latter food-falls are uncommon but important sources of energy for deep sea organisms (Stockton and DeLaca 1982). Some segregation likely occurs among the fishes based on depth and food preferences, but a considerable amount of overlap appears to occur among coexisting species. For example, photographs of areas baited with dead fish indicate that it is not unusual for five or six species of fish to be attracted to the bait in a few hours. The use of mucus secretions by hagfish to coat dead fish, however, thereby making them inedible to other species, indicates direct competition for this type of resource.

Mesopelagic Zone

The most conspicuous feature of the mesopelagic zone ecology is that many of the fishes make nightly vertical migrations into the epipelagic zone. This enables them to take advantage of the abundant zooplankton in the surface waters (which also often move upward at night) and yet avoid full-time exposure to epipelagic predators. Where an oxygen minimum layer exists, migrating mesopelagic fish may further avoid large predators by seeking refuge in it. The result of these migrations seems to be a net export of energy downward. Much of this energy may be recycled in the mesopelagic zone through predaceous fish and copepods that filter-feed, in part, on fecal material. Some is lost downward (and supports bathypelagic fish), however, and some is recycled upward (through squid and other epipelagic nocturnal predators that forage in the mesopelagic zone). The importance of energy recycling within this zone is indicated by the large populations of fish that do not make vertical migrations or that migrate on an irregular basis. Many of these species are predators that simply wait in the water column and prey on passing migratory fish and zooplankton.

The pattern of energy flow through the mesopelagic zone and its abundance and diversity of fish indicate that community structure is complex. Little is known about annual fluctuations in the abundance of mesopelagic species, so whether fish diversity is maintained through extremely complex specializations or differential responses to environmental fluctuations, which prevent any one species in a group of potential competitors from becoming dominant, remains an open question. Although the evidence is scanty, it does seem to favor the latter hypothesis. Feeding and distribution studies of similar species generally indicate some segregation by depth and food types, but a great deal of ecological overlap also is found (Dewitt and Cailliet 1972, Tyler and Pearcy 1975). The potential fluctuations in local mesopelagic fish populations are indicated by the study of Brewer (1973), who extensively sampled the mesopelagic zone of the Gulf of California but failed to capture a single specimen of bristlemouth, even though they had been abundant there only a few years earlier. Many of the smaller migratory species may live only 1 to 5 years and devote much of their energy to the continuous production of eggs and sperm (Koslow 1996). The nonmigratory species apparently have similar life spans but appear to spawn only once. Therefore, any factor that affected reproduction in these fish could have rapid consequences.

Bathypelagic Zone

If the most conspicuous feature of the mesopelagic zone is the movement of fish, the most conspicuous feature of the bathypelagic zone is the lack of movement. The dominant fish of this zone, anglerfishes and bristlemouths, seem to spend most of their time suspended in the water column, waiting for prey organisms to pass by or be lured to them by their photophores. Bathypelagic fish are extreme generalists in their feeding, which is necessary because of the scarcity of food. The limited energy available in this zone all comes from above, from fecal material, detritus, and an occasional mesopelagic fish or invertebrate. The scarcity of bathypelagic fish below 3,000 m presumably reflects, in part, the small amount of food that filters through the upper regions of the water column. Considering the small amount of energy probably flowing through the zone, it is a wonder that bathypelagic fish exist at all, much less in such diversity.

Deep Water Benthic Zone

Compared with the fishes of the bathypelagic zone, the fish of the deep water benthic zone are active and often quite abundant. Energy (in the form of organic matter) enters this zone through a number of pathways, originating either on the continents or in the water column. Organic matter from the continents can move into the zone through migrations of fish and invertebrates, through sinking of plant material (e.g., eelgrass), and through currents that flow down and along the continental shelf and slope. That these are among the most important sources of energy for deep sea benthic organisms is indicated by the gradual decrease in fish and invertebrate numbers as well as biomass as the distance from the continents increases. Benthic fishes often are abundant around seamounts as well. Organic matter from the water column enters in the form of (1) particulate matter raining down from above, particularly where the deepwater benthic zone and the mesopelagic zone meet; (2) dead fish, whales, and other large "particles" that sink to the sea floor; and (3) mesopelagic fish that approach the bottom during their vertical migrations. The organic matter from these various sources is consumed, broken down, and recycled by both invertebrates and fish. Fish are considered to be especially important in this recycling process, because they consume the dead material that falls to the bottom, crop the invertebrates, and then, through their feces, distribute the organic matter rather evenly over the ocean floor.

To play this role effectively, deep sea benthic fish should be relatively unspecialized, consuming whatever prey they encounter. This seems to be largely the case, although the fishes specialize to the extent that some species feed mostly on benthopelagic organisms, some on invertebrates on the surface of the bottom (*epifauna*), some on invertebrates that burrow into the substrate (*infauna*), and some on carrion (Sedberry and Musick 1978). On the upper continental slope, some of the more abundant fishes, such as longfinned hake and cutthroat eel, feed mostly on mesopelagic fish. What is surprising is that mesopelagic fish and invertebrates also are common in the stomachs of benthic fish that occur well below the intersection of the mesopelagic zone with the bottom. Because benthic fish rarely are captured more than 1 m or so off the bottom, they likely are taking advantage of mesopelagic forms that become resident in deep water either seasonally or at certain stages of their life cycle.

Fishes that feed on benthic invertebrates usually are the most abundant species in this zone. As indicated previously, these fish consume whatever invertebrates they can capture. Some specialization occurs based on prey size (reflected in mouth size of the fish), depth distribution, and preference for epifauna or infauna. The infauna-feeders often have large amounts of sediment in their digestive tract as well as the small infaunal invertebrates they presumably were seeking. The nutritional value of the sediments to fish is not known. The infauna also is important as a secondary food source for scavengers, such as hagfish and snubnose eels. When dead fish are placed on the bottom beneath a suspended camera, scavengers, both vertebrate and invertebrate, are seen to quickly congregate. If the dead fish are large, many of the scavengers will burrow into them and consume the dead fish from the inside out. Large numbers of other fish, such as grenadiers, often also are attracted to such bait stations, but they may be feeding primarily on scavenging amphipods and other invertebrates rather than on the bait itself.

Although deep sea benthic fishes may grow fairly large ($\approx 50-100$ cm, $1-3$ kg), they apparently are rather slow-growing and reach maturity in 15 to 20 years (Koslow 1996). This indicates that natural mortality rates of these fish are relatively low and that larger individuals may be in excess of 100 years old. Orange roughy, now being harvested in large numbers, have been aged at up to 125 years, with a mean age at first spawning of 25 years (Koslow 1996). Not surprisingly, these fish are not able to sustain high rates of harvest.

35.5 LESSONS FROM DEEP SEA FISHES

The ancient and specialized nature of deep sea fishes indicates that this environment has been one of the most stable on Earth. These fishes show some of the extremes to which the basic vertebrate plan can be pushed. Given their specialized morphology, it is not surprising that the fishes are part of complex and highly efficient systems that waste little energy. Deep sea fish also are links connecting energy pathways (food webs) of the various zones. Our understanding of how these fish live is based largely on studies of mashed individuals brought up from great depths, but the fish are so alien in appearance and ecology that they will continue to be favored for study and speculation.

SUPPLEMENTAL READINGS

Fitch and Lavenberg 1968; Grassle et al. 1975; Haedrich 1996; Herring 2002; Koslow 1996; Marshall 1954, 1971, 1984; Pietsch and Grobecker 1987.

WEB CONNECTIONS

GENERAL
www.people.whitman.edu/~yancey/fish.html

FISHERIES
www.ices.dk/marineworld/deepseafish.asp

MESOPELAGIC FISHES
www.ifm.uib.no/StaffPages/salvanes/Salvane&Krist-encycl-2001.pdf

C H A P T E R 36
Polar Regions

The aquatic environments of the Arctic and Antarctic often are compared to those of the deep sea, in part because these environments are uniformly cold and in part because polar organisms have many of the same adaptations as deepsea benthic organisms. More differences than similarities are found, however, because of the unique features of the polar environments, such as long summer days that make high levels of primary production possible and long winter nights that create and maintain thick ice cover. Temperatures also are colder than in the deep sea, typically between +1.0°C and −1.9°C. The two polar regions differ greatly from each other in their environmental characteristics, and differences in their fish faunas reflect this. The Arctic consists largely of an ice-covered sea of low productivity that experienced 85-m fluctuations in sea level during the Pleistocene (Dayton et al. 1994). It is surrounded by land that greatly limits the exchange of water with the Atlantic and Pacific oceans. The Antarctic, in contrast, is a continent surrounded by highly productive seas that have "free" interchange with the world's oceans, although unique oceanographic conditions have effectively kept the fish fauna of the continental shelf of Antarctica isolated. As a result, the Antarctic has a comparatively diverse fauna that is characterized by a high degree of endemism, whereas the Arctic shares much of its limited fauna with the northern Atlantic and Pacific oceans. Because of these differences, the Arctic is treated only briefly in this chapter, but the Antarctic is discussed in some detail.

36.1 ARCTIC FISHES

The Arctic is not rich in either species or numbers of fish. Arctic America has fewer than 110 species (McAllister 1977), and the total for the entire Arctic probably is not much greater. Presumably because of the coldness and low productivity of the environment, most of the species belong to families of predominantly sluggish, bottom-dwelling bony fishes, such as the Cyclopteridae (lumpfishes, 10 species), Cottidae (sculpins, 14 species), Zoarcidae (eelpouts, 10 species), Gadidae (cods, eight species), Stichaeidae (pricklebacks, six species), Anarhichadidae (wolffishes, four species), Agonidae (poachers, four species), Pleuronectidae (right-eye flounders, six species), and a few sharks (sleeper sharks, Squalidae, two species) (Box 36.1). Because of the grinding of arctic shores by sea ice, most of these fishes rarely venture into shallow water, and only sculpins (*Myoxocephalus*) occur on a regular basis in the intertidal or subtidal regions. Fifteen species of euryhaline or anadromous salmonids and three species of smelt (Osmeridae),

---

BOX 36.1

Contaminants in an Arctic Shark

The only sharks found in the Arctic are sleeper sharks, which include the Greenland shark (*Somniosus microcephalus*). These are large (4–5 m total length), slow-moving sharks with a poorly understood ecology and life history. The stomach contents of a few individuals suggest that they feed mainly on flounders and seals, but their position in arctic food webs was not certain until recently. Examination of the tissues from these sharks showed a high level of contamination from organochlorine compounds, which are long-lasting pollutants that concentrate as they move up food chains (Fisk et al. 2002). The high levels indicate that the sharks are feeding at least at the fourth or fifth trophic level, with seals probably being one of the most important dietary items. They also suggest the sharks are very long-lived. The effects of these contaminants on reproduction and other functions of the sharks are not known.

---

however, are known from the Arctic; these species, along with herring, can be abundant in bays during the short summer months and in the midwaters above the shelf.

Perhaps the most distinctive and abundant arctic fish are two species of endemic codfish, the polar cod *(Arctogadus glacialis)* and the arctic cod *(Boreogadus saida)*. These fish differ markedly from most other cods in that their mouths angle slightly upward (rather than being terminal or subterminal) and their barbels are much smaller in size. Such adaptations reflect the association of the two cods with pack ice and the fact that they feed largely on the peculiar amphipods and diatoms found in or on the under-surface of the ice. The only other fish regularly found in association with the ice are juvenile sandlances *(Ammodytes)*, which actually may hide in holes in the ice (McAllister 1977). The community of organisms associated with the ice (which has its parallel in the Antarctic) is known as the ***cryopelagic community*** (Andriashev 1970). The cods are abundant enough so that flocks of kittiwakes will follow icebreakers to capture fish that are thrown up in the wakes. They also are quite important in the diets of seals and whales that feed under the ice.

The presence of a cryopelagic community and the predominance of benthic fishes in the Arctic indicate the strong selective pressures that this harsh environment must exert on the fishes. McAllister (1977) has observed that arctic fishes tend to have larger eyes than equivalent species found farther south, presumably a response to reduced light under the ice, in deep water, and during winter. One of the most interesting adaptations of arctic benthic fishes, which links them with deepsea and Antarctic fishes, is their basic reproductive strategy. Most lay a small number of large eggs, which are (presumably) cared for by one or both parents until hatching or beyond. According to Marshall (1953), this strategy has the advantages of reducing competition among the larvae and of producing large, active young with a wide range of food items available to them. Such larvae are either pelagic or benthic, depending on the species.

36.2   ANTARCTIC FISHES

The southern ocean surrounding Antarctica contains at least 313 species of fish from 50 families. Depending on which taxonomic analysis is accepted and where the line separating the Antarctic from other regions is drawn, Antarctic fish are 65% to 96% endemic. DeWitt (1971), estimated that 86% of the fishes (excluding deep sea forms) south of the Antarctic Convergence were found nowhere else. The Antarctic Convergence (Antarctic Polar Front) separates the "true" Antarctic fauna (or at least the shallow-water fauna) from that of the rest of the southern ocean, because it is a remarkably stable oceanographic feature that surrounds the Antarctic continent. It is characterized by clockwise flows that tend to keep drifting organisms away from the continental area by spinning them off in a northward direction. In addition, it presents a distinct temperature boundary, marking the region where the cold Antarctic surface water sinks beneath the warmer sub-Antarctic surface water. The convergence is strictly a surface phenomenon, so the deepwater fauna shows a much lower degree of endemism than the shallow-water fauna.

The most diverse and abundant fishes in Antarctica are the 213 benthic species on the continental shelf and slope, of which 45% are from five families of the suborder Notothenioidei: (1) Nototheniidae (Antarctic cods), (2) Harpagiferidae (hook plunderfishes), (3) Artedidraconidae (dragon plunderfishes), (4) Bathydraconidae (Antarctic dragonfishes), (5) and Channichthyidae (icefishes) (Eastman and McCune 2000) (Fig. 36.1). Other families represented on the shelves of the Antarctic are Rajidae (skates, eight species), Muraenolepididae (eel cods, four species), Zoarcidae (eelpouts, 23 species), Liparidae (snail fishes, 67 species), and Achiropsettidae (southern flounders, four species). Most of these fishes show remarkable adaptations to the peculiar features of the Antarctic environment. The rest of this chapter therefore is devoted to a discussion of (1) morphological, physiological, and reproductive adaptations; (2) factors affecting distribution; and (3) fish communities.

36.3   ADAPTATIONS

Morphological Adaptations

Most Antarctic fish are bottom-oriented, visual predators, and their basic external features reflect this. The dominant notothenioids have large, flattened heads with large mouths and large eyes that often are located toward the top of the head. The bodies tend to be elongate, often tapering to a small tail, with large pectoral fins, small pelvic fins, and long dorsal and anal fins. The swimbladder is absent. With the conspicuous exception of the skates and flounders, other Antarctic fishes deviate from this theme either by having more eel-like bodies, with continuous dorsal and anal fins, or by having more fusiform bodies, such as in the pelagic Antarctic herring (Nototheniidae, *Pleuragramma antarcticum*). Despite the absence of a swimbladder, a number of Antarctic fishes, such as the "herring," are neutrally buoyant, which enables them to feed on the abundant planktonic krill. Buoyancy is achieved through fat deposits beneath the skin and in the muscles and by reduced mineralization of scales and bone (Eastman and DeVries 1982). This reinvention of neutral buoyancy is considered to result from evolutionary retention of larval characters. Curiously, retention of larval

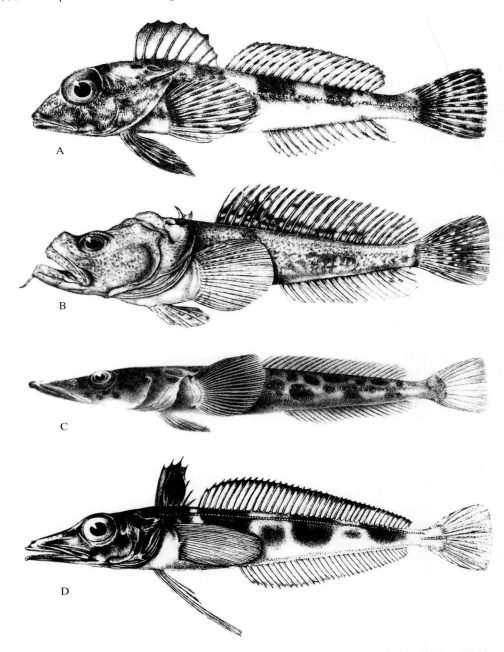

FIGURE 36.1 Representative nototheniod fishes. **A**. Thornfish, *Bovichthys angustifrons* (Bovichthidae). **B**. Plunderfish, *Pogonophyrne scotti* (Harpagiferidae). **C**. Dragonfish, *Gymnodraco acuticeps* (Bathydraconidae). **D**. Icefish, *Chaenodraco wilsoni* (Channichthyidae). **A** and **B** from Regan (1913), **C** from Boulenger (1902), and **D** from Regan (1914). Courtesy of J. T. Eastman.

characters also results in a poorly developed lateral-line system, which would seem to be disadvantageous (Montgomery and Clements 2000).

Physiological Adaptations

The remarkable nature of the physiological adaptations of Antarctic fishes is indicated by two observations. First, many of the fishes have been observed to be active at temperatures very close to the freezing point of salt water (−1.9°C); some of these species actually are associated with ice on a regular basis, either resting on anchor ice or swimming among floating ice platelets. Second, the icefishes have no hemoglobin in their blood, so their gills are cream-colored rather than pink. Some other Antarctic fishes have reduced amounts of hemoglobin.

Resistance to freezing seems to be accomplished either by living in deep water or by having "antifreeze" in the blood (Eastman 1993). Fish that live in water below 30 m actually have body fluids that will freeze at temperatures higher than the water in which they are living! These supercooled fish survive because of the absence of ice crystals in the water to "seed" body fluids and start the formation of ice internally. When such fish are caught in traps and brought up to the surface, however, they freeze as soon as they enter the ice-laden surface waters (DeVries 1970). The notothenioid fishes that live in the shallow waters where ice crystals are present have glycoproteins in the blood that give the body fluids freezing points lower than that of the surrounding water (see Chapter 6). Furthermore, some species are extremely sensitive to minor temperature changes and actively avoid "warm" temperatures (Crawshaw and Hammel 1971). Development of the distinctive antifreeze in notothenioid fishes is a key adaptation that allowed ancestral forms to invade and diversify in the Antarctic (Eastman and McCune 2000).

Reproductive Adaptations

Antarctic fishes have large, yolky eggs, so Marshall (1953) predicted that these eggs would be laid on the bottom. In a study of spiny plunderfish *(Harpagifer bispinis)*, Daniels (1978) found this was, indeed, the case and discovered further that this fish lays its eggs in a nest that is guarded for approximately 4 months. This is the longest period of nest-guarding for any known species of fish. Such behavior probably is common among Antarctic fish, because such parental care is characteristic of many Antarctic invertebrates as well. The eggs are laid during early winter, usually while there is still ice cover, and hatch when the ice breaks up and the spring plankton bloom begins. The larvae swim upward soon after hatching and join the plankton, on which they feed. The planktonic stage may last as long as 1 year.

36.4 FACTORS AFFECTING DISTRIBUTION

The known distribution limits of most Antarctic species seem to be associated with physical factors, particularly temperature, currents, depth, and ice, although some zoogeographic barriers occur as well, at least for the notothenioids (Eastman 1993). The annual range of temperatures in Antarctic waters is usually less than 7°C or 8°C, but the apparent sensitivity of the fishes to even small changes in temperature means that this is an important environmental cue. The departure of many fishes from food-rich inshore areas in autumn may be largely a response to dropping temperatures (cued by the changing light regime), and it ensures the avoidance of water containing ice crystals. On a broader scale, the restriction of most Antarctic fishes to the polar region probably reflects their stenothermy, with the Antarctic Convergence providing a particularly strong temperature barrier. The convergence also is a region of strong currents, which may limit the distribution of the fishes as well.

The depth distributions of the fish are tied to a number of factors, principally temperature and substrate. Because of the presence of anchor ice in shallow water (<33 m) and the depth of the continental shelf around Antarctica (average depth, 400–500 m), the greatest number of species occurs between 300 and 600 m. In this region, substrates are varied, benthic invertebrates are abundant, and temperatures are fairly constant.

36.5   FISH COMMUNITIES

The fish of the Antarctic fall into three broad ecological groups: (1) pelagic fish, (2) benthic fish, and (3) deepsea fish. The divisions between the groups, however, are not sharp.

Pelagic Fish

The open surface waters of the Antarctic are dominated by crustacean krill *(Euphausia superba)*, which support large populations of whales, seals, penguins, and other marine birds. Perhaps because of the presence of these large predators, few fish are found in the upper 200 m of the water column, but mesopelagic species are abundant. Eastman (1993) indicates that several distinct ecological types of pelagic fish occur among the notothenioids, including mesopelagic predators, mesopelagic zooplanktivores, cryopelagic fish, and epibenthic fish. All of these fish have more fusiform bodies than those of bottom-dwelling species. They also have reduced hard parts and large fat deposits to increase buoyancy, yet they retain large fins and other "benthic" features. The species that is most specialized for pelagic life is the zooplanktivorous Antarctic herring, which is one of the few Antarctic fishes with a forked tail and the only species to be found regularly in the epipelagic zone. Because this species is abundant, it is the principal prey of the large, piscivorous Antarctic toothfish (*Dissostichus mawsoni*) (Box 36.2),

Perhaps the most unusual species among the pelagic forms are the cryopelagic fishes, which feed on invertebrates associated with the underside of the ice sheet or the sides of icebergs. The Antarctic herring is a component of this community, which also includes two species of Antarctic cod (*Pagothenia* spp.) that are specifically adapted for cryopelagic life. These species live on the sides of icebergs by clinging to them with their pectoral fins or by resting in crevices, much as a benthic fish would use a rocky cliff. They are silvery-white in color, with inconspicuous eyes and high levels of blood antifreeze (Eastman 1993).

Epibenthic species look like the pelagic forms but are not neutrally buoyant, although they are more buoyant than benthic species. They swim slowly above the bottom, where they feed on a mixture of krill, benthic invertebrates, and mesopelagic fish (Eastman 1993).

BOX 36.2

Learning About the Behavior of Fish Under Ice

It obviously is very difficult to observe the behavior of pelagic Antarctic fishes under the ice sheets that cover such large areas, especially in winter. An ingenious solution to this problem was to trap piscivorous Weddell seals (*Leptonychotes weddelli*) coming up through holes in the ice and then mount small video cameras, along with transmitters that recorded environmental data, on their heads before release (Fuiman et al. 2002). Because the seals returned to the same holes, they could be recaptured and the cameras changed or removed. The videos revealed that loose shoals of Antarctic herring moved up and down approximately 100 m per day but rarely got too close to the under-surface of the ice. One reason is, perhaps, that large Antarctic toothfish are concentrated there, although the toothfish make forays into deeper water. The toothfish are prey for the seals, and one attempting to escape maintained a speed of 3.4 m/s for 24 s, which is remarkably fast for a fish considered to be sluggish (Fuiman et al. 2002). See *Web Connections* for links to more information and video clips.

Benthic Fish

Most Antarctic fish live on the continental shelf. Finding 10 to 20 species, many in quite large numbers, at one location is not unusual, because the shelf is a very stable environment that supports large populations of invertebrates to serve as prey. This food supply is nearly constant year-round, which stands in marked contrast to the strong seasonality exhibited by the availability of food in the pelagic regions (Fig. 36.2). As a result, the benthic communities are quite complex. The fish feed largely on invertebrates that are themselves largely carnivorous; in turn, the fish may be taken by seals, birds, and other fish. Thus, food chains with seven or eight links likely are present. The energy for this system is derived from the plankton blooms and growths of benthic algae that occur during the austral summer, coupled with (presumably) efficient recycling of nutrients (ultimately through detritus) in winter.

The efficiency of the system is indicated by the specializations of the fish. Considerable segregation by depth preferences occurs, most conspicuously the separation of fishes with high levels from those with low levels of antifreeze in the blood. Among species coexisting at the same depth seems to be considerable specialization in habitat use (Gutt and Ekau 1996), in both type and size of prey consumed, and in method of feeding. Although some species are generalists, feeding on whatever prey is available, many specialize by feeding largely on just one or two taxa of invertebrates, such as amphipods or polychaete worms. The degree of specialization is illustrated by the feeding behavior of the Magellan plunderfish (*Harpagifer bispinis*), which feeds mostly on amphipods and scaleworms. The fish captures scaleworms by ambushing a passing individual. It removes the scales from each worm by spitting it out and sucking it in repeatedly, presumably scraping off the scales on its palatine teeth (R. A. Daniels, personal communication). Likewise, an investigation of the diets of two piscivorous dragonfishes

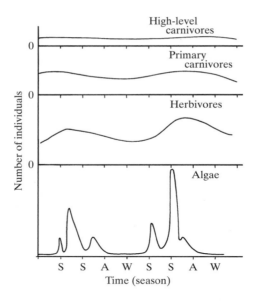

FIGURE 36.2   Seasonal changes in the flora and fauna of the Antarctic. Most fishes are high-level carnivores. Diagram by R. A. Daniels.

(Bathydraconidae) living at 200 to 400 m revealed that one fed mainly on mesopelagic myctophids (lanternfishes) and the other on benthic fishes (Pakhomov 1998a). The diets of some species nevertheless are flexible. At times, the notothenioid *Trematomus hansoni* has as a significant percentage of its diet waste from fishing vessels, such as processed fish and kitchen scraps (Pakhomov 1998b).

Deep Sea Fishes

The deep sea fishes of the Antarctic region, especially south of the convergence, are poorly known. The fish fauna of the continental slope and deeper areas, as well as of the mesopelagic fauna, is a mixture of notothenioid fishes and representatives of the typical deepsea fish families. A number of the non-notothenioid species are endemic, but many are more widely distributed. The bathypelagic and mesopelagic fish faunas are made up of approximately 80 species (Eastman 1993). Although many of these species are endemic, all of them belong to widespread families. The mesopelagic lanternfishes (Myctophidae, 35 species) are particularly numerous, taking advantage of the abundant krill as a food source.

36.6   EVOLUTION OF A SPECIES FLOCK

Species flocks are groups of closely related fishes that evolved in isolated habitats with few potential competitors at the time of invasion. Classic examples are cichlids of the Great Lakes of Africa (see Chapter 30) and sculpins of Lake Baikal, Russia (see Chapter 25). Eastman and McCune (2000) argue that the notothenioid fishes of Antarctica also represent a species flock, and one that is unusual because it is marine. They note that Antarctica has been an isolated, cold environment for 25 million years and clearly a center of evolution for both benthic fishes and invertebrates. The thornfishes (Bovichtidae) that inhabit cold coastal waters of Australia, New Zealand, and South America are apparently the sister group to the Antarctic notothenoids. They are bottom-dwelling fishes without a swimbladder. Although not very diverse (10 species), the thornfishes possess a key adaptation: glycopeptide antifreeze in their tissues. Once ancestral notothenoids made it to the Antarctic, they would have found an environment that was rich in food but poor in other fishes. This would have allowed them to radiate into the diverse niches the 95 species now occupy, especially because the dynamic nature of the Antarctic ice sheet creates plenty of opportunity for local isolation and extinction (Eastman and McCune 2000). Among of the best evidence for the species-flock nature of this radiation is the presence of several species of pelagic but swimbladder-free notothenioids. Montgomery and Clements (2000) also suggest that the icefishes (Channichthyidae) could have diversified in isolation only because the loss of hemoglobin in their blood (from a single gene mutation) seems to be disadvantageous, thereby requiring a restructuring of the circulatory system for effective delivery of oxygen to tissues. "[T]he Antarctic fauna occupies an extreme in the spectrum of habitats where fishes are found—it provides a glimpse of the scope of adaptation and evolution in one of the world's most isolated an usual habitats" (Eastman and McCune 2000, p. 97).

36.7 LESSONS FROM POLAR FISHES

The polar regions have both distinctive fish faunas adapted to low temperatures and seasonally high productivity. Most remarkable, however, is the notothenioid fish fauna of the Antarctic, which is the result of millions of years of adaptive radiation in an isolated, cold environment. This also is true of the other remarkable elements of the ecosystem, from krill to whales. The result is a system of highly structured, complex food webs with a great deal of biomass that is potentially harvestable by humans. Despite international agreements to protect biodiversity in the Antarctic, harvest of krill and fish is taking place—with unknown effects on the ecosystem. Extensive harvest of fish would almost certainly be disastrous for the fish communities. The fishes are very slow growing (e.g., a 25-cm fish likely is 5–10 years old, with the largest individuals being from 25–30 years old or older), so replacement of harvested fish would be extremely slow. Although comparatively little is known about the community structure of the continental shelf benthos, where many of the harvestable fish occur, the complexity and apparent long-term stability of the communities may mean that a new stress (i.e., fishing) could severely disrupt them. In addition, because bottom trawling is the method of fishing most likely to be used, the physical damage caused by fishing could severely disturb the reproductive success of some species by destroying nests and removing parental guardians. The Antarctic has the potential to be one place in the world where fish and other organisms are managed on a sustainable basis (if at all) through international cooperative efforts (Berkman 1992). Whether the forces of rationality can overcome the forces of greed in this instance is still problematic.

SUPPLEMENTAL READINGS

Andriashev 1970; Eastman 1993; Kock 1992.

WEB CONNECTIONS

Notothenioid Fishes
www.oucom.ohiou.edu/dbms-eastman/index.htm

Seals with Camera
www.tamug.edu/marb/davisdocs/Mar.Biol.%202002%20Fuiman%20et%20al..pdf
www.utmsi.utexas.edu/staff/fuiman/antarcticfishes/

Antarctica, General
www.astro.uchicago.edu/cara/outreach/links.html

C H A P T E R 37
Conservation

The fish faunas of the world are changing rapidly. In fresh water, extinction of endemic species is becoming increasingly common as tolerant, aggressive species invade degraded ecosystems. In marine systems, fisheries are collapsing. In all aquatic environments, decreases in the abundance and diversity of fishes are being documented. Increasingly, aquaculture is being relied on to provide fish for humanity. The direct cause of these dramatic changes is the exponential increase in human population. Not only are we exploiting fish directly at unsustainable rates, we also are competing directly with them for water, space, and food. Whether such changes are good, bad, or simply of no consequence depends both on one's value system and on one's state of hunger. Nevertheless, the changes *are* taking place, and the future of humankind *is* closely linked to that of fish (and the rest of Earth's biota). An Earth turned into a wasteland for fish will be equally a wasteland for humans. On the other hand, wise treatment of the waters and the life within them can provide benefits for humankind through the indefinite future. It therefore is important to understand the changes that are taking place, to monitor the health of our waters, and to find ways to reverse present trends (Leidy and Moyle 1997). Thus, in this chapter, we briefly discuss (1) the status of fish faunas, (2) the causes of declines in fish diversity and abundance (exploitation, alien species, habitat attention, pollution) (3) aquaculture, (4) fisheries magagement, (5) reserves, (6) economic values of fish, (7) noneconomic values of fish, and (8) fish conservation in the real world. In the final section, we present a series of examples to illustrate major problems and solutions in dealing with the conservation of fish and fisheries.

37.1   STATUS OF FISH FAUNAS

Fresh Water

The vast majority of endangered fishes live in fresh water, basically because freshwater environments are islands or ribbons of water surrounded by oceans of land filled with people. Humans confiscate much of the water flowing across the landscape, and even so, there often is not enough water for human needs—much less water to share with fish (Abramovitz 1996, Postel et al. 1996). Water that we do not directly use, we pollute. Our streams and rivers are the ultimate recipients of the by-products of human activity, from sewage to sediment. This means that although endangered fishes can be found everywhere, they are most prevalent in regions with one or more of the following characteristics: (1) highly developed economies; (2) small, isolated bodies of water; (3) high endemism; (4) arid or Mediterranean climates; (5) big rivers; and (6) big lakes.

Examples of waters with all of these characteristics can be found in the arid southwestern United States. The major watershed of this region, the Colorado River, has been dammed in many places, the water diverted, and exotic species introduced into the altered habitats. What water remains is full of salts and other pollutants (including alien fish), because the Colorado, like other big rivers, is the ultimate sump for a vast region. As a result, most of the native fishes of the Colorado River are endangered, and they are largely absent from the lower river. In the tiny springs and creeks of nearby desert regions, many endemic pupfishes and minnows are threatened by multiple factors, such as diversion of water for human use and introduction of predatory fish. Even the desert fish that live in Death Valley National Park can be regarded as endangered, because the city of Las Vegas (with one of the highest per-capita water-consumption rates of any city in the world) is pumping ancient water from aquifers that may feed the spring systems on which the fish depend. Likewise, the fish faunas of large desert lakes, such as Pyramid Lake, Nevada, are threatened because of diversion of the water that keeps the lakes fresh. The pattern in the southwestern United States is only one example of a trend that is found throughout the industrialized countries of the world. Compared to tropical countries, they have relatively few fishes, yet a high percentage of the fishes are in danger of extinction. Unfortunately, tropical regions also increasingly qualify as areas with large numbers of endangered species, as Third World countries strive to dam, divert, and pollute their rivers following the model of the Western world (Dudgeon 1992). Because tropical waters have more species to lose, the numbers of endangered species may be high even though the percentages are low.

These patterns are reflected in the numbers and percentages of species that are extinct or need special protection (i.e., those with the status of endangered, threatened, vulnerable, or special concern) in various parts of the United States and the world (Table 37.1). In North America, at least 40 fish taxa are extinct (Miller et al. 1989), and 364 need special protection (Williams et al. 1989). Together, these comprise approximately one-third of the region's entire fauna. These taxa are not evenly distributed over the continent, however. A much higher percentage of species are in trouble in states with arid or Mediterranean climates than in those of wetter regions. A similar pattern can be seen in Europe and other regions. Overall, a conservative estimate for the percentage of all freshwater fishes meriting special protection to prevent extinction in the next 25 to 50 years is 20% ($\approx$2,100 species), assuming present trends continue. As an indication of the conservative nature of this number, approximately 200 species of cichlids have been eradicated from Lake Victoria in East Africa since the 1950s, following the introduction of the predatory Nile perch and other changes to the lake ecosystem (Goldschmidt et al. 1993).

Estuaries and Inland Seas

Estuaries are naturally inhabited by ecologically resilient species of fish. Estuaries are also, however, sites of major coastal cities, are subject to having their vital freshwater inflows diverted, and are receiving waters with hundreds of exotic species that are brought

TABLE 37.1 **PERCENTAGES OF SELECTED FRESHWATER FISH FAUNAS THAT ARE EXTINCT OR MAY NEED SPECIAL PROTECTION TO AVOID EXTINCTION IN THE NEXT 25 TO 50 YEARS (CLASSIFIED AS THREATENED, ENDANGERED, VULNERABLE, OR OF SPECIAL CONCERN)**

Region	Total Taxa	% Needing Protection
North America	1174	31
California	116	69
Arkansas	147	22
Wisconsin	200	29
Virginia	190	23
Mexico	375	24
Europe		
Sweden	58	28
Great Britain	56	18
France	49	51
Spain	42	67
Czech Republic	53	47
Turkey	109	28
Sri Lanka	88	18
Australia	195	34
South Africa	98	63
Costa Rica	156	9

Information from Elvira (1995), Jenkins and Burkhead (1994), Larje (1990), Leidy and Moyle (1997), Maitland and Lyle (1990), Pethiyagoda (1994), and Pollard et al. (1990).

in annually through the ballast water of ships. Thus, most large estuaries are highly altered ecosystems. This has resulted in declines of many fish species (and their fisheries), but surprisingly few estuarine species are endangered. A major exception are the native fishes of the upper San Francisco Estuary of California, where two resident fish species already are formally recognized as threatened species and several others have been proposed for such status. In this case, the declines are caused by various interacting factors, the most prominent being freshwater diversions and a phenomenal rate of invasion by exotic species from ballast water (Bennett and Moyle 1996, Cohen and Carleton 1998).

Similar problems exist in the estuarine Baltic and Black seas of Eurasia, coupled with widespread pollution, leading to eutrophication and accumulation of toxic substances (Leppakoski and Mihnea 1996). Curiously, fish catches in the Baltic Sea have actually increased in response to eutrophication, although the species captured are of lower value than before. Many fish are regarded as unsafe for human consumption because of high levels of toxins in their tissues (Platt 1995). In the Black Sea, the fisheries have essentially collapsed. In the same region, fisheries also have collapsed in the Caspian and Aral seas, which have suffered the cumulative insults of a massive diversion of inflowing water coupled with unrestricted pollution. In the Caspian Sea, three species of sturgeon and the endemic "species flock" of shad *(Alosa)* species can be regarded as endangered

(Rozengurt and Hedgpeth 1989). In the Aral Sea, 20 of 24 native species have gone extinct (Micklin 1988).

In the face of such unrelenting bad news, it is worth noting that efforts are underway to reverse the downward trends in some major estuaries of the world, including those of the Baltic Sea, the Black Sea, San Francisco, and Chesapeake Bay. The resiliency of estuarine fishes is reflected in the ongoing recovery of striped bass populations in Chesapeake Bay following efforts to reduce fishing pressure and to clean up pollutants.

Marine Environments

Comparatively few (156) marine fishes are formally recognized as threatened (Leidy and Moyle 1998). This situation may be changing rapidly, however, as more is learned about conditions in the world's oceans and the state of its fishes. For example, only recently has it been recognized that the large barndoor skate *(Raja laevis)*, once common in the northern Atlantic Ocean, is on the verge of extinction (Casey and Myers 1998) and that pelagic sharks in the same region are declining at an alarming rate (Baum et al. 2003). The widespread decline of the skate went unrecognized because it generally was assumed that a widely distributed species that had become absent or rare in one region probably was still common in another. Indeed, marine fish are protected, to some extent, by the sheer volume of salt water and the extent of the marine environment, which helps to shield them from pollution and other human insults. Because the fish are concentrated on continental shelves and around islands, however, the negative influence of humans gradually is becoming more pervasive. Even tropical reefs have been severely degraded by sediments and pollutants washed onto them from nearby land or damaged by the mining of coral as a source of lime (Richmond 1993, Hughes 1994).

Perhaps the most obvious manifestation of the problems that marine fish face is the collapse of fisheries around the world as a result of overfishing accompanied by major changes in marine ecosystems. The world harvest of wild fish peaked at approximately 100 million metric tons (Mt) per year and currently is declining because more than 70% of the world's fish stocks are either overexploited or fully exploited (Botsford et al. 1997) (Fig. 37.1). Unfortunately, fisheries statistics typically are misreported by fishing countries, so catches may be declining much more rapidly than the official trends indicate (Watson and Paul 2001). The world catch is estimated to hover at approximately 85 to 95 million Mt/year but is increasingly made up of small, low-value species that feed at low trophic levels (Naylor et al. 2000). For example, fisheries on the Georges Bank, off the northeastern United States, have been virtually shut down in the hope of spurring recovery by the once-enormous populations of commercial "groundfish." Before the collapse of the fisheries, major species in the catch switched from valuable codfish, haddock, and flounder to the much less-valuable dogfish shark and skates (Sherman 1994). None of the overfished species in this system is in danger of extinction—as far as we know—but their enormous decline signifies major changes to the ecosystem of which they are a part, which includes many obscure species about which we know little. Fisheries can drive species to extinction, however, if they are large and valuable enough. One of the first species to be driven to near-endangered status by a fishery is the bluefin tuna *(Thunnus thynnus)* of the Atlantic Ocean. This species is so prized in Japan for its flesh that harvesting it is lucrative even when numbers are extremely low. Large pelagic sharks, such as

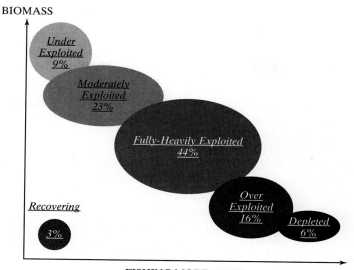

FIGURE 37.1   Diagrammatic representation of the degree to which commercial marine fish stocks are being exploited. Also shown is the general relationship between exploitation rate (fishing mortality) and size of the existing population (biomass). After Botsford et al. (1997).

the white shark, also are extremely vulnerable to extirpation from fishing, given their low reproductive and growth rates. In California, white sharks became a protected species because of lobbying by a single fisherman, who realized that his own fishing for them had seriously depleted the population in northern California. In short, we have little cause to be sanguine about the status of marine fishes. Collapsing fisheries and drastic declines of large, mobile species indicate that many poorly documented species may be in trouble (Roberts and Hawkins 1999).

37.2   CAUSES OF DECLINES IN FISH DIVERSITY AND ABUNDANCE

Aquatic environments are changing constantly, even where they are not strongly affected by humans. As the previous 10 chapters indicate, fish often are alternately abundant and rare, on an irregular basis, in response to natural environmental fluctuations. In a sense, therefore, the pressures that humans put on fish populations are nothing new. On the other hand, humans are placing stress on fish populations to an unprecedented extent. This stress, because it is in addition to the natural stresses that act on fish populations, results in greatly accelerated rates of extinction and other irreversible changes. For example, the rapid decline of coho salmon *(Oncorhynchus kisutch)* in California, Oregon, and Washington seems to be the result of accelerated human-caused stresses (logging and urbanization of watershed, damming of rivers, etc.) during a period when coho populations probably would have been naturally low because of extended droughts and less-productive conditions in the ocean (Brown et al. 1994).

The human-related stress can be divided into four components, which often act on fish populations simultaneously: (1) exploitation, (2) introduction of nonnative species, (3) habitat alterations, and (4) pollution.

Exploitation

Exploitation is a major cause of changes in fish populations and is increasingly implicated as a major factor degrading marine ecosystems. Few waters in the world today do not have some sort of fishery. The impacts of fisheries fall into three major areas: removal effects, bycatch, and habitat alteration.

Removal effects.   The most dramatic changes to marine fish populations are those brought about by removal of important components of marine ecosystems by overfishing. The direct results of such removal can be the elimination of fishable stocks, such as those of codfish in the Georges Bank and sardines off California (see Chapter 34). Such disasters are best known for the economic hardships they bring to fishing communities, but comparatively little attention is paid to their impact on the biotic communities to which overexploited species belong. The collapse of the anchoveta fishery off Peru (see Chapter 34) also was accompanied by a permanent collapse of the huge populations of sea birds that depended on the anchoveta. Overexploitation of herring in the Norwegian Berents Sea ecosystem ultimately resulted in large-scale starvation of predatory cod, sea birds, and seals that depended on the herring. There also was a great reduction in the abundance of other species of fish, because the cod devoured whatever they could before their population collapsed (Hamre 1994). The effects of industrial fisheries also have been connected with the decline of sea lion, seal, and sea otter populations in Alaska, thereby causing major changes to coastal ecosystems (Estes et al. 1998) (see Chapter 32).

Species whose populations are fished to low levels often are replaced by other species that are ecologically similar but differ in their life-history characteristics and desirability to humans. On the Georges Bank, codfish, haddock, and flounders were replaced by sharks and skates. These effects were recorded because of their spectacular nature, but they surely were accompanied by other, unrecorded effects, such as changes in the species composition of the zooplankton or benthos or changes in the abundance of unexploited species. These indirect effects are barely understood, and we have little idea if they are reversible. Overexploitation rarely, however, leads to the extinction of a species or population of fish, because fishing usually becomes uneconomic when populations get too low. Thus, many changes wrought by overexploitation should, in theory, be reversible, because the components of the original system usually are still present. The main exceptions to the "rule" that overexploitation of fish does not lead to extinction are with fishes of unusually high value that are unusually easy to catch, such as bluefin tuna, sharks, and sturgeon.

Another possible exception to the preceding rule is the fishery for tropical fishes for the aquarium trade. Because of their cash value to poverty-stricken peoples, such fishes are being intensively collected in many parts of the tropics. Especially vulnerable to collection by nets are freshwater fishes that become confined to isolated pools during the dry season; few of these fish can escape a serious collecting effort. Even more damaging is the use of poisons to collect aquarium fish; here, the basic technique is to pour some toxic substance into a stream and then collect the largest and most valuable fishes as they appear at the surface in distress. Most will survive if placed immediately in fresh water, but such collecting, of course, kills nearly everything in the stream that is not collected. Thus, it is not unusual to find streams, especially in Southeast Asia, with depleted faunas. Similar effects have been noted when coral reef fishes are collected for the salt-water aquarium trade using sodium cyanide. Likewise, collecting reef fish using dynamite kills many fish and destroys reef habitats.

Bycatch.   Fisheries rarely are entirely selective on the target species and catch many fish that are undersized or have no commercial value, called bycatch. In shrimp fisheries, 50% to 90% of the catch by weight consists of small fish that are discarded, many of them juveniles of valuable commercial fish species. None of these fish survive. Even fisheries with low bycatch rates can take huge numbers of nontarget animals. In the Bering Sea and Gulf of Alaska, where bycatch is "only" 6% by weight, approximately 1 billion individuals of 130 species are killed each year by this circumstance (National Research Council 1998). The total bycatch worldwide amounts to approximately 27 million Mt/year—approximately one-third as much as the total marine fish harvest (National Research Council 1998). Because shrimp fisheries are concentrated in some areas (e.g., Gulf of Mexico), their effects on nontarget species may be devastating. Some small benthic species may be endangered by such fisheries, but information is lacking.

Habitat alteration.   Only recently has it been widely recognized that some kinds of fishing physically alter the habitats that are being fished. The effects of this habitat alteration can be just as devastating as those of the direct harvest of the fish. In areas where bottom-trawl fisheries are active, the trawl boards plow up the bottom, destroying the burrows of invertebrates (important in food webs), killing small nontarget species, increasing suspended sediments, and generally reducing the productivity of the trawled areas (Norse 1993). Trawling, dredging, and other similar activities have the impact of destroying structure—including low reefs and sea grass beds—on the seafloor, thereby eliminating cover and food sources for many species. Waitling and Norse (1998) reviewed the effects of trawling, dredging, and other gear that is dragged over the seafloor and likened it to the clear-cutting of forest lands. In some cases, where areas are trawled repeatedly in a single year, the benthic habitats never get a chance to recover. Waitling and Norse (1998) estimate that trawling annually affects approximately half the continental shelf area of the world—or approximately 15 million km². "With the possible exception of agriculture, we doubt that any other human activity physically disturbs the biosphere to this degree" (Waitling and Norse 1998, p. 1191).

Alien Species

Fishes are frequently introduced by humans into waters to which they are not native. Deliberate introductions typically are made to "improve" local fish faunas or aquatic ecosystems or to provide economic benefits to some group. For decades, fishery agencies around the world had making such well-intentioned introductions as one of their missions. Major sport fisheries all over the world are supported by alien fishes such as brown trout and largemouth bass, as are subsistence and commercial fisheries by species such as common carp and various tilapias. Eventually, it became obvious to fishery biologists that many introduced species often had unpleasant—or completely negative—side effects, causing declines in fisheries, endangered species, and headaches for management agencies. This was termed the "Frankenstein effect" by Moyle et al. (1987). Such deliberate introductions still take place, but mainly by anglers and others who are not authorized to do so. Increasingly, new fish invasions are the result of **by-product introductions** (Moyle 1998). These are introductions that result from human activities in which organisms are released into the environment through carelessness, neglect, and lack of regard for the consequences. They include fish that "escape" from aquaculture operations, pet fish released by aquarists, fish transported through canals, and fish carried in the ballast water of ships. In some areas, such as the canal systems of central Florida and the streams of Oahu, Hawaii, a bizarre collection of exotic aquarium fishes dominate the fauna, often swimming among beds of equally exotic aquatic plants covered with exotic snails. Ecosystems as disparate as the Laurentian Great Lakes and the San Francisco Estuary have faunas that are dominated by species brought in by ballast water and canal systems. The impact of by-product introductions has been greatest in the inland waters (Box 37.1), but introductions of marine fishes are becoming increasingly common (Baltz 1991). The most pervasive marine invasions have been through the discharge of ballast water by ships, which has resulted in the establishment of hundreds of species of invertebrates and fish around the world (Carlton and Geller 1993). From a fish perspective, one of the most spectacular invasions has been that of the Mediterranean Sea, via the Suez Canal, by many species of Red Sea fishes (see Chapter 26). In the past, by-product introductions were written off as "accidental." This is not possible today, however, because managers (and even some politicians) are well aware both of the sources of new invading species and of the potential economic and ecological damage they can cause.

BOX 37.1

Biotic Homogenization: A World Trend

A major result of the combination of alien invasions and native species extinctions is **biotic homogenization**, in which freshwater fish faunas in disparate regions become increasingly similar. This is best documented for the United States (Rahel 2000, Marchetti et al. 2001), but it is a worldwide phenomenon (Rahel 2002). Thus, lakes and reservoirs in the United States, Japan, Spain, and South Africa are likely to be dominated by common carp, largemouth bass, sunfish (*Lepomis* spp.), and mosquitofish (*Gambusia* spp.), with perhaps northern pike or bullhead catfish (*Ictalurus* spp.) thrown in for good measure. Cold-water streams in these same areas are likely to contain both rainbow trout and brown trout.

Often, endemic native fishes disappear in these situations or are relegated to minor status as community members, thereby resulting in entire food webs made up of alien species. Thus, the introduction of northern pike into natural lakes in a national park in Spain resulted in the near-elimination of native fishes (nine species) and the apparent reduction in the abundance of seven other alien fishes (Elvira et al. 1996). As a result, the pike switched to feeding mainly (73% by volume) on red swamp crayfish (*Procambarus clarki*), a crustacean introduced from the United States. The number-two prey item was largemouth bass, another predator introduced from the United States, which often prefers crayfish.

Although virtually any aquatic environment that can support fish can be invaded by new species, alien species usually are most successful in environments that have been altered by humans, where native fishes are already stressed or reduced in numbers (Moyle and Light 1996). Thus, in the Central Valley of California, streams with natural hydrologic regimes are mostly dominated by native species, but streams that have been dammed tend to be dominated by introduced species. In the Laurentian Great Lakes, nonnative fish invasions were facilitated by overexploitation of native fishes and by pollution. Nevertheless, native fish often persist—and even thrive—in altered environments if alien fish are not present. This means that introduced species may be the direct cause of the disappearance of native forms through (1) predation, (2) competition, (3) disease, and (4) hybridization (Moyle 1986).

Predation can result in rapid and drastic changes in a fish fauna when an exotic piscivore is introduced into a community of fishes not adapted to its style of predation. Thus, sea lampreys were able to virtually wipe out populations of large fishes in the upper Great Lakes in less than 20 years (see Chapter 29). Peacock bass (*Cichla ocellaris*) and largemouth bass have been able to eliminate most fish species from large lakes into which they have been introduced in Central America. Nile perch have been able to devour into extinction nearly 200 species of fish from Africa's Lake Victoria. Examples of predation eliminating species from streams are less well known, but introductions of piscivorous brown trout have been implicated in the decline of stream fishes worldwide, most notably in Australia and New Zealand (Crowl et al. 1992).

Competition from alien fishes as a factor causing the decline of native fish populations is very difficult to demonstrate yet undoubtedly is important. For example, in the Great Lakes, the decline of many of plankton-feeding whitefish species (*Coregonus* spp.) was related to the establishment of alewife, which not only reduced populations of large zooplankters in the lakes but crowded into whitefish "space" during critical times of the year (see Chapter 29). Often, the competitive superiority of an introduced species over a native form is related to habitat change. Species that evolved in a stream environment typically decline in numbers if the stream is impounded and lake-adapted fishes are introduced. Some species actually may change the environment themselves. Thus, in the Mississippi drainage, native bottom-feeding species, such as buffalofish (*Ictiobus* spp.), may disappear from shallow lakes once carp become established, root up the bottom, and decrease the water clarity.

Disease is a problem when pathogens and parasites carried by introduced species infect native species. Although a poorly understood factor, it may be more important than generally is realized. An introduced species may find invading easier if potential predators and competitors have been decimated by disease. For example, the invasion of red shiner (*Cyprinella lutrensis*) into the Virgin River, Utah, may have been facilitated by the negative effects on the native woundfin (*Plagopterus argentissimus*) of the Asian tapeworm carried by the invaders. The shiners had been infected with this tapeworm, carried by grass carp brought in from Asia for weed control (in Arkansas), and apparently are fairly resistant to its effects. The woundfin is now an endangered species. One of the major concerns about rearing Atlantic salmon in pens in Europe and North America is that under crowded conditions, they are more likely to develop diseases, which can spread to wild fish.

Hybridization is a problem when an alien species interbreeds with a closely related, endemic form, thereby producing hybrids that back-cross with the parents. If the hybrid (intermediate) phenotypes then have some competitive or reproductive disadvantage compared with the phenotypes that most closely resemble the introduced form, the intermediate phenotypes gradually will be eliminated from the population, and the population increasingly will resemble the "pure" introduced species. Apparently, this mechanism has been responsible for the elimination of many interior cutthroat trout populations by rainbow trout. One of the best-documented cases, however, is the virtual elimination of the Mojave tui chub (*Siphatales mohavensis*) from the Mojave River in California through hybridization with the arroyo chub (*Gila orcutti*) (Hubbs and Miller 1942). The Mojave tui chub is now listed as an endangered species. A similar situation may exist in the Pecos River, Texas, where a single introduction of the sheepshead minnow (*Cyprinodon variegatus*) resulted in nearly complete hybridization with the endemic Pecos pupfish (*C. pecoensis*) in nearly 700 km of river in just a few years (Echelle and Connor 1989). The hybrids are expected to eventually take over the entire river system from the native form, except in isolated habitats.

Habitat Alterations

Habitat alteration by humans in the inland waters (including estuaries) is the single biggest cause of faunal change and is increasingly becoming a major factor in the marine environment as well. Every time a stream is dammed, a stream bed channelized, a watershed logged or heavily grazed, or a lakeside marsh filled in, the fish fauna changes, often very subtly. Removing large trees that shade a salmon spawning stream may increase production of juvenile salmon by increasing the amount of sunlight on the water and, hence, the stream's productivity. If such a logging operation causes the stream to become too warm, however, or is accompanied by the silting in of the spawning gravels, then minnows, suckers, and other species may increase in numbers instead.

On a more extreme scale, most of our major river systems have been so altered by dams, locks, channelization, urbanization, deforestation, and wetland drainage that their fish faunas bear little resemblance to those originally present. Once-common species become rare and uncommon, but tolerant species become abundant, especially alien species (e.g., Burr 1991, Edwards and Contreras-Balderas 1991). One of the most altered rivers in the United States is the upper Tennessee River in Virginia, North Carolina, and Tennessee (Neves and Angermeier 1990). Nearly 40% of the riverine habitat has been flooded by reservoirs or changed by discharges from 11 major dams. Reservoirs favor different species than riverine habitats, and they isolate populations of stream fishes so that recolonization following local extinction events is no longer possible. In addition, domestic, agricultural, and industrial pollution are major problems in the area. In 1967, a spill of fly ash into the Clinch River (a major tributary) killed nearly all the fish in a 144-km stretch. The upper Tennessee drainage has a rich fish fauna (224 taxa, including 25 endemic forms). Of these taxa, 58 (26%) are endangered or otherwise need special protection. The fact that many river-dwelling fishes in North America are threatened species is a tip-of-the-iceberg indication for the extent to which river fish communities have been changed. The changes in the fish communities also can have effects that ripple through the ecosystem, changing the abundances of other organisms as well (Box 37.2).

BOX 37.2

Loss of Fish Diversity Causes Loss of Mussel Diversity

Eastern North America, especially the southeastern area, not only contains a rich diversity of fishes but also a rich diversity of mussels of the family Unionidae. These mussels are closely tied to the fishes, because their main method of dispersal is through their glochidia larvae, which are parasitic on the gills of fish. The glochidia are carried by fish for days or weeks before dropping off and beginning a sedentary existence. Often, glochidia of mussel species have only a limited number of host fish species they can infect. Thus, diversity of mussels typically is highest where fish diversity also is high (Vaughn and Taylor 2000). In an Oklahoma river, the number of mussel species increased in proportion to the distance downstream from a major dam that altered flows and temperatures. A major cause of the increase in mussel species was the increase in the abundance and diversity of their host fishes (Vaughn and Taylor 1999). This example suggests that many complex interactions among species are interrupted when aquatic environments are heavily disturbed by human activity.

In the oceans, shipwrecks and other debris of civilization have created numerous new reefs where none existed before, often greatly increasing the abundance and diversity of local fish faunas. In many coastal areas, such reefs are now being created deliberately. On the other hand, the destruction of intertidal habitats by the construction of breakwaters and marinas and the filling in of coastal habitats such as salt marshes, mangrove swamps, and estuaries is causing declines not only in fishes that are unique to such areas but also in populations of offshore fishes that require such areas as nurseries for their young. The filling of these areas can be deliberate, such as for land "reclamation," or the result of upstream erosion, which fills coastal lagoons with sediment. Often just as destructive is the creation of navigation channels through marshes, mangrove forests, and shallow areas of bays. The channels and the constant dredging needed to maintain them cause major disruptions of the local ecosystems, thereby reducing biotic productivity and diversity (Norse 1993).

Pollution

Pollution really is a type of habitat alteration, but it nevertheless deserves separate consideration because of its extent and frequently subtle effects. Of course, the best-known effects of pollution are not subtle at all. Releases of toxic chemicals have caused massive kills in hundreds of miles of major river systems as well as in lakes, estuaries, and inland seas. Many streams in mining areas are devoid of fish life because of acidic water draining from the mines. Many mountain lakes are losing their fish faunas as they become acidic from material carried in the smoke from distant industries. Large amounts of sewage in lakes and streams cause depletion of the oxygen supply, killing all but the hardiest fish or forcing them to move elsewhere. Fortunately, fishes are, by and large, very adaptable. They have the reproductive potential to quickly recolonize areas from which they have been eliminated once the cause of their disappearance is resolved, provided they had a refuge. For example, Atlantic salmon returned to spawn in the Penobscot River in Maine, after a decade-long absence, following a successful effort to "clean up" polluting industries and towns. Permanent elimination of species by pollution occurs either where the populations are highly localized (endemic) or where the pollution is continual and severe, such as in the Azov and Caspian seas.

Direct kills of fish by pollution are of obvious concern, but in the long run, the changes caused by the sublethal effects of pollution are equally significant. Power plants that use water for cooling may raise the temperatures of the stream, lake, or estuarine regions from which they draw and into which they discharge the water. Such thermal pollution may kill fish directly, but more often, it causes local shifts in the composition of the fish fauna. From a fisheries point of view, these shifts may even be desirable at times, because during cool weather, game fish (and, hence, fishermen) may be attracted to the warmer water associated with power plants. Other fishes, however, may avoid the warmer water, just as they may avoid water with reduced oxygen concentrations, increased turbidity, or gooey bottoms caused by pollution. Typically, the overall effect in such situations is decreased numbers of species accompanied by increases in tolerant species, such as common carp. Similar effects may result from sublethal levels of toxic compounds such as pesticides, especially if the toxins accumulate in the flesh of the fish. In such cases, they may reduce temperature tolerances, change behavior patterns, increase vulnerability to disease, and generally reduce growth and survival rates. Because such toxins typically accumulate through food chains, predatory game fishes are especially likely to be affected. For example, mosquitofish (*Gambusia* spp.) living in waters adjacent to cotton fields may develop extraordinary tolerances to pesticides and, thus, survive. Predatory fish, however, such as largemouth bass, that prey on the mosquitofish are now largely absent from such areas, presumably poisoned by their prey.

The interaction of mosquitofish, pesticides, and bass shows why the effects of pollution often take a long time to be noticed; the disappearance of the bass was a gradual process. Likewise, the disappearance of trout and other fish from mountain lakes in Europe and eastern North America took decades, because the gradual increase in acidity caused by atmospheric pollution affected the ability of fish to reproduce long before the water actually was toxic to adult fish. Such subtle environmental changes often interact with other factors as well. The extinction of the blue pike *(Stizostedion vitreum glaucum)* from Lake Erie apparently was largely the result of massive environmental changes caused by pollution, but overfishing of the pike and other factors also were present (Miller et al. 1989).

37.3   AQUACULTURE

As wild fish populations decline, aquaculture is held up as the means to provide protein for the growing human population while also sustaining fisheries and taking exploitation pressure off of wild fish populations. Aquaculture is fish farming, the rearing of aquatic organisms under controlled conditions to maximize production. Typically, fish and other organisms used in aquaculture are either fully or partially domesticated, having traits more suited for survival in crowded ponds and raceways than in the wild. As a means of providing protein for protein-short countries, freshwater aquaculture has been remarkably successful, producing millions of tons of fish reared in ponds, canals, reservoirs, and polluted waters. Fish reared in such operations typically are herbivores, detritivores, or omnivores, such as common carp and other large cyprinids and various tilapia species. Such fish can be reared at high densities using simple techniques and food that often results from waste products of other human activities.

Unfortunately, many forms of aquaculture are not so benign, especially those operations that are designed to provide high-quality products for the Western world. The ever-expanding shrimp farms in Asia, for example, are carved out of shallow coastal habitats, especially mangrove ecosystems, which are important rearing habitats for many marine fishes (Naylor et al. 1998). As wild salmon populations have become depleted, more and more salmon sold on the market are pen-reared in bays and estuaries, an industry now valued at more than $2 billion year. These culture operations have several problems. First, they are a "sink" for ocean fishes, because approximately 70% of the food fed to the salmon is fish meal and fish oil and nearly 3 pounds of small ocean fish are caught to produce each pound of salmon (Naylor et al. 1998). Second, the net pens are set out in open waters, so the waste products from the operations are a major source of pollution, as are the pesticides and antibiotics used to control salmon diseases. Third, many salmon escape from the pens and have the potential to displace native salmon in local streams. For example, more than 1.2 million large Atlantic salmon were released from Norwegian pens in a 2-month period as the result of storm damage (Beveridge et al. 1994). These fish were from domesticated strains originating in the Baltic Sea and are quite different genetically from the strains in the local Norwegian streams.

In the United States, the principal species raised by inland aquaculture operations are channel catfish, rainbow trout, and various Pacific salmon. All of these species require food with high protein content, usually derived from fish meal, and have some of the same associated problems as those described for net-pen aquaculture, although perhaps not quite so severe. The salmon (and many of the trout) operations are a special case, because most of them are publicly funded hatcheries. A major purpose of these hatcheries is to mitigate for loss of salmon production caused by dams, diversions, and other factors. Some hatcheries, especially in Alaska, are designed to increase salmon production in local marine waters to benefit fisheries. These salmon hatcheries have been a mixed blessing. They have sustained some salmon fisheries, but they also have contributed to the decline of wild stocks, both by allowing excessive fishing on mixed hatchery and wild stocks in the ocean and by causing various kinds of behavioral, ecological, and disease interactions among hatchery and wild fish (see Chapter 19). The focus on hatcheries as the principal means of maintaining or restoring salmon populations also led to neglect in solving problems associated with maintaining wild populations with their enormous genetic and ecological diversity and high cultural values.

37.4 FISHERIES MANAGEMENT

Historically, most fish populations and species, if they have been managed at all, have been managed to support fisheries. Typically, this has meant focusing on a few species with high value to anglers and commercial fishers and neglecting other species. The complete dominance of this attitude during the first half of the twentieth century led to widespread introduction of game fishes, manipulation of habitats to favor important fishes, use of hatcheries to maintain species like salmon while natural habitats degraded, and adoption of optimistic policies toward commercial fishing that allowed many fish stocks to be depleted. Many success stories can be found, especially in inland waters, but realization is growing that traditional management strategies have failed to protect many fish populations, even valuable ones, and aquatic habitats. The destruction wrought by modern "industrial" commercial fisheries has been particularly egregious, as noted earlier. Clearly, fisheries have to be managed in a more sustainable manner than they have been in the past.

Olver et al. (1995) argue that fisheries can be managed successfully for long-term benefits to humans if the ecosystems of which the harvested fish are a part are not severely disrupted. They also argue that for practical and evolutionary reasons, large fisheries, such as those for Atlantic cod, are best managed at the stock level, where a stock is a reproductively isolated subpopulation with the potential to become genetically distinct (if it is not already). Olver et al. state their fundamental principle of fisheries management as "aquatic ecosystems should be managed to ensure long-term sustainability of native fish stocks." These authors then go on to list and justify a series of supporting principles that can be used to guide management actions in fisheries:

- "The sustainability of a fish stock requires protection for the specific physical and chemical habitats utilized by the individual members of that stock" (p. 1589).
- "The sustainability of a fish stock requires maintenance of its supporting native community" (p. 1589).
- "Vulnerable, threatened, and endangered species must be rigidly protected from all anthropogenic stresses" (p. 1590).
- "Exploitation of populations or stocks undergoing rehabilitation will delay, and may preclude, full rehabilitation" (p. 1590).
- "Harvest must not exceed the regeneration rate of a population or its individual stocks" (p. 1590).
- "Direct exploitation of spawning aggregations increases the risk to sustainability of fish stocks" (p. 1590).

These principles recognize that exploited fish populations are part of larger ecosystems that must be considered in managing the fishery. Implementation of such principles requires use of the *precautionary approach* to fisheries management: Set harvest rates or other actions to levels that can be proven will do no harm (Dayton 1998). This puts the burden of proof on the exploiters, not the agencies and the public, and focuses on the intergenerational equity: The resource should be just as available to future generations as it is to the present one (Hilborn et al. 2001). Thus, a harvest management plan for a commercial fishery would have to demonstrate that it would not significantly further deplete the harvested species or have a negative effect on other species through bycatch or habitat alteration. Thus, management requires an assessment of risk to the fish population and to the humans who depend on it. If the precautionary principle was used to manage fisheries worldwide, it is likely that after damaged fish populations were rehabilitated, the total true sustainable yield of the world's fisheries would be as high—or higher—than it has ever been. The problem, of course, is that such holistic management requires adoption, by the fishers and the fishing industry, of ecologically based ethics that may reduce catches or economic gains in the short term. The National Research Council (1998) provides general recommendations for implementing changes in commercial fishing to make the industry sustainable.

37.5   RESERVES

Because of the failure of many traditional management practices for marine fisheries, there is a new push for a fairly old concept: reserves. Reserves are areas of coastline or ocean from which fishing and other human activities are either banned or greatly restricted. The fundamental idea of reserves is that the complexities and uncertainties in traditional fisheries management almost inevitably lead to declines in fisheries, so reserves provide a hedge against management failures by following the precautionary principle (Lauck et al. 1998).

In a reserve, fish populations serve a number of purposes, including providing a source of fish for fisheries outside the reserve (as adults, juveniles, or larvae) and as a comparison for the health of fished populations and ecosystems. To be successful, reserves have to be (1) large in size, (2) able to serve as a source of fish for outside areas (e.g., by protecting spawning areas), and (3) vigorously protected from poaching and other negative influences (Lauck et al. 1998). Naturally, considerable controversy exists about how many reserves are needed, how big they should be, and where they should be located. Reserves are most popular for tropical or temperate reef habitats, where they can be tourist attractions for divers and a relatively small protected area most likely will be successful. Generally, fishers do not want them to be in the places where they normally fish, nor do they want them to be too large. A problem with marine reserves is that they take a long time to really work, especially if the habitat as well as the fish populations need to recover, and their success often is hard to evaluate (Russ 2002). Allison et al. (1998) point out that whereas marine reserves obviously are an important management technique, they cannot be relied on as the principal way to protect fish populations, both because they are subject to natural and human-caused environmental changes (e.g., regime shifts and oil spills) that may keep large populations of fish from developing and because not all species will thrive in reserves. Some fishes, for example, have larvae that naturally disperse long distances; others have larvae that stay fairly close to where they were spawned. The former species may require fairly dispersed populations or multiple reserves to keep going, and the latter species may not disperse well into surrounding areas to improve fisheries. Regardless of these potential problems, marine reserves are being established all over the world.

In fresh water, reserves are equally hard to establish—and equally likely to be controversial. This is because to work, they must encompass entire watersheds (Moyle and Sato 1991). There often is the perception that large, protected terrestrial areas (e.g., national parks) completely protect the fishes that live in them. This is true only if the park contains the entire watershed. More typically, they contain only the uppermost headwaters (mountains), with their limited faunas, or the middle reaches of streams, with the headwaters and the lower reaches unprotected. Thus, sediments or pollutants can wash down through the protected area from upstream, and dams can prevent fish from entering from downstream. Because stream and river systems are so fragmented worldwide (Dynesius and Nilsson 1994), other forms of management typically must be used in conjunction with reserves to protect fish diversity (e.g., Collares-Pereria et al. 2002). For both freshwater and marine fishes, part of the conservation strategies has to lie in promoting the economic and noneconomic values of fish.

37.6 ECONOMIC VALUES OF FISH

In a way, it is astonishing that fish conservation is so difficult to achieve, because the overall value of fish is obvious, as indicated by their enormous importance for food and recreation (e.g., aquarium fish, sport fisheries, diving). The basic problem is the economic system under which we operate, in which the value of short-term economic gains predominates over the value of more sustainable long-term gains. "Wealth or the prospect of wealth generates political and social power that is used to promote unlimited exploitation of resources" (Ludwig et al. 1993, p. 17). Four alternative economic values that can be offered for fish are (1) market values, (2) ecosystem values, (3) existence values, and (4) intergenerational values (Moyle and Moyle 1995).

Market Values

Most fish species have at least some direct monetary value, are part of food webs that support economically important species, or live in environments whose good health is essential for the support of fisheries. Furthermore, commercial and subsistence fisheries provide a good share of the protein needed by humans. Yet despite this high value, fish populations are in decline everywhere. Major fisheries around the world, such as salmon fisheries of the Pacific Northwest and Atlantic cod fisheries, have collapsed. Other declining economic values of wild-caught fish include sport fisheries, both of which depend on large populations of healthy fish in clean water. In most cases, the loss of fisheries and fish populations is predictable, and often, it is blatantly obvious when the decline is taking place. Unfortunately, corrective action rarely is taken in time, despite the economic, social, and ecological losses that can be seen looming on the horizon. Clearly, therefore, traditional economic values do not protect fish or the ecosystems of which they are a part.

Ecosystem Values

Functioning aquatic ecosystems provide many benefits to human society (even beyond edible fish) called *ecosystem services* (Daily 1997). For example, aquatic systems have an enormous capacity to absorb and detoxify pollutants, a "free" service that has largely been taken for granted until recently. As aquatic ecosystems become degraded, however, they lose this capacity. The costs of purifying water for drinking or other uses then become much higher, the quality of water is reduced, and the likelihood of toxic episodes through equipment failure increases. The value of healthy ecosystems often can be measured in terms of human health and of the long-term economic health of the human systems to which they are connected. Fish provide an important service here, because they often are the most sensitive and conspicuous indicators of ecosystem change and, thus, can give early warning if conditions detrimental to human health are emerging (Holm-lund and Hammer 1999). Other examples of ecosystem services are in Figure 37.2.

The problem with ecosystem values is that they are very difficult to quantify, except when disaster strikes as a result of their having been overlooked. For example, high levels of logging in the Pacific Northwest have resulted in widespread loss of the ability

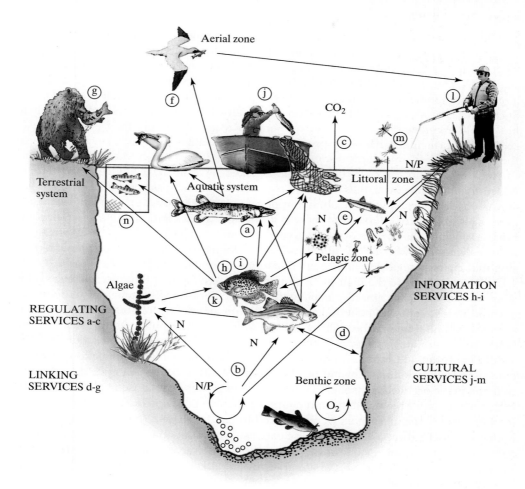

FIGURE 37.2 Ecosystem services produced by fishes. ***Regulating services*** include (*a*) top-down ecosystem regulation by predators, (*b*) disturbance of sediment to keep nutrients in circulation, and (*c*) carbon exchange with the air. ***Linking services*** include transport of nutrients, carbon, and energy among (*d*) pelagic and benthic and (*e*) pelagic and littoral parts of the ecosystem as well as transport of nutrients (via predation) to (*f*) birds and (*g*) mammals. ***Informational services*** include fish as (*i*) indicators of ecosystem health and (*j*) monitors of ecosystem conditions. ***Cultural services*** include (*j*) income/food, (*k*) improved water quality through top-down predation, (*l*) recreation, (*m*) control of vectors for disease, and (*n*) aquaculture. After Holmlund and Hammer (1999).

of streams to sustain large salmon populations. If the many warnings about the effects on salmon of clear-cutting and other logging practices had been heeded, not only would there be harvestable, naturally produced populations of salmon available for the indefinite future, the forest ecosystems themselves would have been able to produce a more constant income for local communities.

Existence Values

The development of this concept is an effort by economists to put a monetary value on what Norton (1987) calls *amenity values*: the recreational and aesthetic values derived from natural systems, the joy of learning from studying the behavior of a fish, the appreciation of salmon and cod for their cultural and historic symbolism, and the near-religious feelings that anglers can experience when connecting with wild fish in a wild place. Such experiences can be highly valued even by individuals who never actively seek them but just like knowing that such an experience is available should they ever choose to seek it. The principal method for providing existence values is through surveys in which the public is asked, in regard to some natural feature, "How much would you be willing to pay to maintain this?" If the question is asked in terms of a few cents per month added to water or power bills (e.g., to protect anadromous fish), the answer often is a strong yes. The probability of obtaining negative replies, however, increases with the price. This implies, of course, that there is a price on species and habitats and that if the price is too high, the species or habitats can be eliminated. Despite this serious weakness in the calculation of existence values, such calculations are increasingly common in decision-making with regard to management actions for endangered species.

One way around the high-price dilemma is to assume that the true existence value of species or habitats cannot be calculated and that, because species belong to people as a whole, individuals who benefit economically from degrading a habitat or endangering a species should pay for the right to do so. In this case, the public-value surveys, combined with estimates of direct economic losses, can indicate how much the payment should be. If the determined value of a species or habitat is high, then actions severely damaging to it presumably would be altered or not allowed. For example, the introduction of aquatic organisms through the ballast water of ships has caused billions of dollars in damage to city water systems in the Great Lakes and disrupted estuarine ecosystems in many places. If ship owners were required to pay for even a fraction of the damage, both direct and indirect, caused by these expensive introductions, they would quickly become willing to adopt techniques to reduce them. Politically, of course, implementing such a value system for species in a market-oriented economy is nearly impossible, despite its obvious benefits.

Intergenerational Values

Existence values have received a great deal of attention from economists, because in theory, they can be fit into present economic systems. A basic problem is that most economists today treat resources, such as fish, as if they belonged only to the present generation (i.e., the present "owners"). If there is no sense of these resources as also belonging to future generations, then they can be spent, gambled, or traded to produce immediate wealth—until they are gone. The increasingly frequent collapse of major fisheries certainly demonstrates the lack of intergenerational values in the present economic system.

An opposite approach is to place a high value on **sustainability**, the equitable distribution of resources between generations, to make sure that future generations are no worse off than the present one (Daily and Ehrlich 1992, Norgaard 1994). Unfortunately, it is difficult to place an economic value on sustainability, in part because it is impossible to know just what future generations will value (Box 37.3). Desert pupfish largely are regarded as a scientific curiosity today, but they might be regarded as a valuable and highly nutritional delicacy in the future (as they once were by the Panamint Indians). Thus, protective actions designed to make sure that a species or habitat does not disappear may take many years to provide a direct economic contribution—assuming that it ever does. In contrast, development projects that might eliminate a species or a run of salmon may have an immediate economic and political payoff. This results in decisions such as the introduction of Nile perch into Lake Victoria, which is now the subject of a profitable export fishery. The cost of this introduction was the elimination of the small cichlids that once were mainstays of a subsistence fishery as well as the deforestation of the region around the lake to provide the wood necessary for kiln-drying of large perch. The native cichlids, in contrast, had been sun-dried. The lake has become more turbid from erosion and eutrophication, reducing further the cichlid populations. Clearly, future generations of people around the lake will be worse off because of the introduction, which has resulted in a depletion of local noncash resources.

The best way to demonstrate the need for intergenerational values is to look at how decisions made by past generations for short-term gain are now reaping enormous costs for present generations. The most obvious examples are the many dams installed on large rivers to divert water and generate power. The effects of these dams on downstream ecosystems usually were not considered to be a cost when they were built, yet

---

BOX 37.3

Fish in the Bank

Although the need for intergenerational values in our economic system is clear, it is not clear how this change can be accomplished. One approach is to use a **safe minimum standard**, which is based on the idea that if information on a species is limited and its potential for extinction is high, then we should assume that it has—or will have—high economic value. Another approach is to consider features of the environment, such as salmon runs, to be **natural capital**, the value of which should be placed on the same basis as monetary capital in investment funds. The idea is that capital should be maintained to provide income for present and future generations. Endangered species or runs of salmon, under this scenario, are regarded as depleted natural capital that needs to be restored as an investment in the future (Moyle and Moyle 1995). A further extension of this idea is to protect natural diversity as **genetic capital**, on the assumption that genetic diversity is needed to adapt valuable species to unpredictable climate change or for use in selective breeding programs for aquaculture.

these costs are very real today, as demonstrated by the disappearance of sturgeon fisheries in Russia, salmon fisheries in the Pacific Northwest, and sardine fisheries in the Mediterranean (from the construction of the Aswan Dam on the Nile River).

The substitutes for the loss of wild fish have never really worked as promised (if they were promised at all), despite efforts to promote hatchery programs and reservoir fisheries. When dams were built on the Colorado River during the 1950s and early 1960s, who could have foreseen the passage of the federal Endangered Species Act, which would ultimately mandate that large quantities of water should be devoted to saving obscure native fishes? These same fish were the target of a major poisoning program in 1962, for fear that they would compete with alien game fish introduced into the newly built Flaming Gorge Reservoir.

37.7 NON-ECONOMIC VALUES OF FISH

The nontraditional economic values just discussed often are recognized as valid, but usually not by traditional market-oriented economists. These values therefore have little impact on the decisions made by businesses and governments today. An alternative approach waives economic arguments and presents various presumptive noneconomic values as reasons for protecting fish and aquatic ecosystems. The kinds of values often put forth in this area include aesthetic values, cultural values, scientific research values, and teaching values.

The *aesthetic values* of small, obscure species have long been recognized by aquarists, and this very recognition has turned a number of small or unusually attractive fishes into resources with definite economic value. Although the demand by aquarists for exotic fishes seems to be insatiable, often to the detriment of wild populations, aquarists are rather selective in their tastes. Sterba (1959) lists more than 1,600 species that have been kept in aquaria in Europe at one time or another, and a generous estimate would put the worldwide total at perhaps double that number. Many of these species, of course, have been kept by a few individuals or public aquaria only for their curiosity value and, thus, would be unlikely to have the steady market demand necessary for making them a true resource. Unfortunately, many fishes are not likely to have even fleeting value as curiosities (except to scientists). Examples might be the many rather similar species of minnows and darters in North American streams, the small benthic fishes of the Antarctic, or the numerous similar species of sculpins, blennies, and other forms in tide pools.

In recent years, the popularity of snorkeling and scuba diving, particularly on tropical reefs and in rocky coastal areas of temperate regions, has greatly increased public awareness and appreciation of fish (at least in Western countries). Yet even in popular diving areas, the majority of species rarely are seen, because they are either nocturnal, cryptically colored, or very small. In any case, often the most popular fishes for viewing are the large predatory fishes that also have commercial value. Nevertheless, many fishes do have great aesthetic value. Efforts therefore are made to attach dollar values to the aesthetic values in terms of money spent on recreation, travel, and equipment. Such values, however, generally are much less than other economic values (e.g., the value of a reef as the base for an oil refinery).

Because of their aesthetic and culinary properties, many fish have high ***cultural value*** as icons or symbols of tradition. In many cultures, they are a frequent subject of art, often with considerable symbolic significance (Moyle and Moyle 1991). In the United States and Canada, the importance of some fish to Indian tribes has been a major force in conservation. In the Pacific Northwest, runs of salmon have been conserved or restored to provide traditional fishing opportunities. Flows in the lower Truckee River, Nevada, have been partially restored to save the cui-ui *(Chasmistes cujus)* from extinction; one of the most effective arguments in favor of doing this has been the importance of the fish in the culture of the Paiute people, who still live in the area.

Currently, the cultural values of fish include ***scientific research*** and ***educational values***. The scientific research values of fish are obvious to ichthyologists, who find all fish intrinsically interesting. Fortunately for ichthyologists, who are relatively few in number (at least in their pure form), many other scientists find fish useful as well. Sticklebacks and poeciliids have proven invaluable for studies of evolution, behavior, and genetics that are of general interest. The study of reef fish ecology has provided considerable insight regarding the structure and function of complex ecosystems. Lampreys have proven invaluable for studies of neurophysiology. As important as fish already are in basic research, however, their potential for experimental and in situ studies is just beginning to be realized.

The educational values of fish also are obvious. They are the one group of vertebrates in which it really is possible to maintain large populations in the laboratory for observation and experimentation. Many varieties are readily available for dissection, thereby giving an easy overview of the diverse adaptations that are possible in vertebrates. Most important, they typically can be found in large numbers and diversity in waters fairly close to most educational institutions, so many types of ecological demonstrations can be made. Fish have the right combination of availability, size, and ease of handling to make them the most desirable group of vertebrates for comparative classroom studies of ecology, behavior, evolution, and anatomy; fish are very valuable in other areas, such as genetics and physiology, as well.

The problem with these arguments is that, directly or indirectly, they are an effort to turn noneconomic values into economic values (Ehrenfeld 1976). Unfortunately, these values are so indirect or tenuous that they usually will seem small when compared with the very real economic values that can be gained through the short-term exploitation of an environment or a species. A well-known confrontation between these values came with the efforts to stop the construction of Tellico Dam on the Little Tennessee River to protect the endangered snail darter *(Percina tanasi)*. The Endangered Species Act of 1973 (ESA) was successfully invoked, passing all the way through to the U.S. Supreme Court to protect the species. Congress subsequently changed the law to make exemptions possible through the creation of a "God Committee." Even this committee, however, was unwilling to condemn a species to apparent extinction, so a special exemption from the rules of the ESA had to be obtained through congressional subterfuge. The dam was finally built. Subsequently, two rather ironic facts emerged: First, the snail darter existed in other streams, and so was not condemned to extinction after all. Second, the dam has fallen far short of its projected economic value. Thus, if the darter had prevailed in this argument, a great deal of real money would have been saved.

Unfortunately, the ESA is under continuous siege to weaken its protections. The original ESA was passed after a majority of Congress was persuaded to save the more spectacular symbols of the disappearing American wilderness, such as bald eagles, wolves, and whooping cranes. Once the law went into effect, however, it quickly became evident that it had opened a Pandora's box of problems for many elements of the ever-expanding U.S. economy. The list of endangered or threatened species is rapidly growing longer and longer as not only small, obscure fish but also invertebrates and plants are added. This may be one of the main reasons why other countries have not followed the lead of the United States in passing such a stringent law. Thus, it increasingly appears that whereas economic and some noneconomic arguments can be used to save species and habitats, they can easily be overcome in most cases. This means that the preservation of biotic diversity ultimately will depend on developing arguments that appeal to a sense of morality—or even to the feeling of religious awe that natural systems can invoke. One way that such feelings can be expressed is in the basic idea that all species and, therefore, the habitats in which they live have a right to continued existence; that as humans, we do not have the right to unthinkingly terminate evolutionary lines and wondrously complex natural systems.

Norton (1987) argues that this line of reasoning can be regarded as too extreme or dogmatic by many people and suggests that it is better to base the preservation of natural variety on the ***transformative value*** of natural systems. This argument maintains that interactions with natural systems have the potential to profoundly and positively transform our attitudes toward existence, so that we can live on this planet in a less materialistic and more spiritual way, and that such a transformation provides the moral foundation for living on the Earth as if it belonged to future generations.

37.8   FISH CONSERVATION IN THE REAL WORLD

Given the enormous short-term economic pressures that drive world politics and, therefore, conservation decisions, how well are we doing in resolving the conflicts between economic and noneconomic values, between the sustainable and nonsustainable use of natural resources? Are any of these values being applied to fish conservation? These questions are best answered by examining examples of contemporary conflicts in fish conservation. Five case histories, covering a wide range of problems, are examined here; (1) Atlantic cod, (2) Pacific salmon, (3) desert fishes, (4) Sri Lankan stream fishes, and (5) the Kissimmee River, Florida. These examples were chosen because they indicate some of the ways in which the problems are—or can be—solved.

Atlantic Cod

The collapse during the early 1990s of the fishery for Atlantic cod (*Gadus morhua*) and other "groundfish" off Newfoundland, Labrador, and New England is a spectacular example of the loss of a major fishery because of overexploitation (Hutchings and Meyers 1994, 1995). The fishery for cod began during the late fifteenth century and quickly became important in the economies of Canada and a number of European countries. Because of their size, abundance, and palatability, cod became important in the cuisines of many nations, from the fish and chips of England to the glutinous lutefisk of Sweden and

Norway. During much of its nearly 500-year history, the fishery appeared to be sustainable, and cod was regarded as an example of a species that was nearly impossible to overfish because of its high fecundity and rapid growth rates. Nevertheless, annual catches varied widely. Starting in the nineteenth century, evidence appeared that indicated local stock depletion, causing fishermen to move to new areas to fish. In 1954, immense stern trawlers that could both catch and process fish entered the fishery. Soon, the factory trawlers of 20 nations were fishing the North Atlantic, resulting in a maximum catch of cod (in 1968) of 810,000 Mt. In 1977, Canada extended its jurisdictional limit out to 200 miles and attempted to regulate the declining fishery. Although scientific evidence did exist to the contrary, in 1986 the Canadian government accepted optimistic estimates of cod stock abundances that indicated heavy exploitation was still possible. The fishing continued, and the cod population abruptly collapsed to commercial extinction. Attempts were made to blame poor ocean conditions for growth and survival of young fish as the ultimate cause of the collapse, but this proved to be wishful thinking (Hutchings and Meyers 1994). In 1994, the U.S. National Marine Fisheries Service closed approximately 6,000 square miles of the Georges Bank—the most productive region—to fishing, despite the protests from fishermen.

While the Atlantic cod populations off Canada collapsed because of the direct effects of exploitation, those off Norway collapsed because of the overexploitation of a principal prey species, the Atlantic herring *(Clupea harengus)* (Hamre 1994). When stocks of an alternate prey species, the capelin *(Malotus villosus)* also were reduced because of natural factors, the then-abundant cod ate every available fish—including their own young. This was followed by starvation of large cod as well as of sea birds and seals. The fisheries for cod, herring, and capelin all collapsed. This example illustrates another little-appreciated fact of modern ocean fisheries: They alter not just the populations of the species that are the objects of fisheries but also the ecosystems of which the fish are a part. Fisheries alter ecosystems not only through the capture of fish but through physical alteration of the benthic habitat. In particular, the trawl doors of large trawls literally dig into the soft bottom like plows, thereby altering an area's ability to support benthic invertebrates. In some shallow ocean regions, the bottom is so churned up that it looks like a field that has been cultivated by a mad plowman (Norse 1993). The result can be a devastating loss in the abundance and diversity of benthic organisms and fish (Watling and Norse 1998). Because these organisms are important in the diet of cod, their decline may have increased the susceptibility of the cod populations to overfishing by making their recovery more difficult.

The best thing that can be said about these (and other) collapses of fisheries, and the resulting changes in oceanic ecosystems, is that we can learn from them. Many other fisheries in the world, if examined closely, show signs of imminent collapse because of overexploitation (Botsford et al. 1997) (Fig. 37.1). Such collapses can be prevented if the fishery is reduced or halted for a short period. Of course, this requires the nations and fishing industries responsible to admit that their activities have caused the declines. Given the economic reality that industrial fisheries are unlikely to halt until the fish are gone, there is some hope in the analysis of Myers et al. (1995) that most exploited fish stocks can recover from low population levels if left alone. Presumably, the supporting ecosystems can recover as well, although perhaps in different configurations. The collapse of fisheries and subsequent moratoriums on fishing provide an opportunity for restructuring fisheries to work in a more sustainable fashion, through tools such as limited entry and by returning to fishing methods that are less capital intensive. Recovery of the damaged ecosystems on which Atlantic cod depends will require more than just a short moratorium and new fishing regulations. Brailovskaya (1998) proposes that large areas of the continental shelves and banks off New England be set aside as marine wilderness reserves, where no fishing is allowed. Such reserves not only would protect regional biodiversity but would also provide a major source of recruits to commercial fisheries taking place outside the protected areas, because the reserves presumably would support substantial populations of the large female fish that contribute so much to sustaining fish populations.

Unfortunately, despite a moratorium on fishing since 1993, the cod populations have not recovered, thereby suggesting that reserves would have had little impact on recovery. The reasons for the failure to recover are hypothesized to be high predation rates on the remaining adults by seals and poor oceanographic conditions, which limit larval and juvenile survival (Fu et al. 2001).

Pacific Salmon

The seven species of Pacific salmon (genus *Oncorhynchus*) have tremendous value as commercial and sport fishes along the Pacific coast, from Alaska to southern California. They are major players in the oceanic and freshwater ecosystems they move between to complete their life cycles. They also have high aesthetic and cultural values, especially to the native peoples who once depended on salmon for their very existence. The annual runs of salmon are one of the principal features defining the Pacific Northwest (National Research Council 1996). In recent years, salmon catches in Alaska have been at record highs in some years, thanks to a combination of favorable environmental conditions in the Gulf of Alaska and good harvest management. In Washington, Oregon, and California, however, salmon populations are in rapid decline. Some populations have been listed as endangered, and many others have been proposed for endangered status. In California, for example, coho salmon (*O. kisutch*) are now gone from half their native spawning streams and are at less than 10% of their historic numbers (Brown et al. 1994). In the Central Valley of California, chinook salmon (*O. tshawytscha*) are extinct in the San Joaquin River and are denied access to approximately 80% of their historic spawning grounds by dams in the Sacramento River (Yoshiyama et al. 1996, 1998). These declines have resulted from a combination of loss of freshwater habitat from dams and watershed abuse as well as inadequate harvest management (including hatchery operations).

To make matters worse, the decline in salmon has been exacerbated by poor conditions in the ocean for salmon survival. When ocean conditions are favorable for Alaska salmon, they tend to be unfavorable for salmon on the U.S. West Coast, following a 20- to 30-year climatic cycle (Hare et al. 1999). This has led to the interesting contradiction of, at times, exceptionally low market prices for Alaskan salmon (exacerbated by the influx of cheap salmon reared in aquaculture operations) and very expensive measures being taken to protect and rehabilitate the dwindling salmon runs in the southern part of their range (National Research Council 1996). Mundy (1997) points out that this contradiction results, in good part, from the ability of humans, mainly fishermen, to alter their behavior in response to natural fluctuations in salmon numbers in the northern part of the salmon's range, and their inability to do so in the southern part. Thus, in Alaska, fishing may be banned completely for some runs if ocean conditions result in poor survival of the salmon and relatively small numbers return to the streams. As Mundy states:

> By contrast, in the Columbia River basin, when El Niño Southern Oscillation reduces ocean productivity and drought reduces freshwater survival, human behavior remains unaltered. The hydroelectric system, commercial barge transportation system, irrigation systems, and timber industry continue to consume or perturb salmon as they operate. Without direct consequences to all those sectors of the economy that take salmon as these animals dwindle, it remains unlikely that the wide-scale geographic effort necessary to prevent salmon from being extirpated can be successfully mounted. (Mundy 1997, p. 327)

Despite Mundy's pessimistic conclusion, major efforts are being made to rehabilitate wild salmon populations in the Pacific Northwest, especially through operating water projects in a more fish-friendly manner. Efforts at improved watershed management and rehabilitation are underway in a few areas. In some streams, dams that block salmon passage have been removed, and others have been proposed for removal. A number of regional scientific assessments, sponsored by the U.S. federal government, have been made during the past decade and have strongly recommended developing a means of reversing the downward trends in salmon populations. The recommendations of these assessments have largely been not been implemented so far, however, because of strong opposition from entrenched economic interests. Even so, considerable popular sentiment for doing so exists, especially because inhabitants of the region realize that their long-term well-being is connected to the health of the salmon populations and their watersheds.

The most successful arguments for rehabilitating salmon runs are essentially noneconomic, such as the need to protect Native American tribal rights to salmon in rivers, to maintain the lifestyles of commercial fishers, to provide for salmon-dependent ecosystems, or to prevent unique races of fish from going extinct. These arguments have been successful mainly because they also are grounded in laws, such as Indian treaties and the ESA, that reflect a general desire of the American people to protect their natural heritage. Whether the protective laws will change as the perceived economic consequences of protection become more severe is problematic. In the meantime, economic justifications that reflect intergenerational obligations also are being developed, based on the idea of natural capital and other concepts central to ecological economics (Constanza 1991, Akire 1993, Moyle and Moyle 1995). The success of such efforts to protect salmon in the Pacific Northwest will tell us much about how large-scale efforts to protect natural resources and biodiversity can work. Ultimately, protection and rehabilitation of salmon will involve thousands of streams and rivers, millions of people, huge sums of money, and immense political and cultural will.

Desert Fishes

The specialized fishes that inhabit the arid regions of the world present a special problem in conservation (Rinne et al. 1996). Most are small species endemic to isolated springs, lakes, and small streams, and most have no (or very limited) economic value. In contrast, the water in which they live is enormously valuable to humans in these water-short areas. The growing conflicts over water in desert areas (and extinction of fishes) is a harbinger of conflicts over fresh water that are developing in more water-rich parts of the continents (Postel et al. 1996). As a result, most fishes endemic to desert regions are threatened with extinction (Contreras-Balderas and Lozano-Vilano 1994, Elvira 1995, Rinne et al. 1996). These fishes, and the aquatic ecosystems they represent, are likely to persist only if noneconomic arguments for their preservation prevail.

In the United States, an organization made up largely of biologists (the Desert Fishes Council) invoked nearly every noneconomic argument discussed in this chapter to successfully protect scattered populations of desert fishes, using the ESA and other legal tools that are available to conservationists (Pister 1991). Popular support for their protection was engendered, in part, by publicizing the plight of the pupfishes (Cyprinodontidae), whose name implies they are both cute and helpless. Nevertheless, the long-term future of these fishes and their habitats is far from secure because of increased demand for the groundwater that feeds the springs and streams in which they live. The pumping of ancient aquifers by growing cities such as Las Vegas, Nevada, ultimately may dry up aquatic habitats hundreds of miles away. This is already happening in Mexico, where at least 11 species of fish have gone extinct, largely from drying of their habitats in combination with pollution and introduction of predatory fishes (Contreras-Balderas and Lozano-Vilano 1994). Given Mexico's rapidly growing human population, this trend is unlikely to change, despite the presence of protective laws.

In Spain, most of the endemic fishes in arid areas are endangered, yet so far, no systematic efforts to protect these fishes have been undertaken (Elvira 1995). In 1990, however, a law providing some protection for the native flora and fauna was passed by the Spanish government, which at least recognizes that the problem exists (Rinne et al. 1996). Likewise, protective actions for the desert fishes of Australia are in their infancy, in part because the threats to many desert fishes are perceived as still being low even though 34% of Australia's 192 freshwater fish species are at risk (Rinne et al. 1996).

The protection of desert fishes, more than most fishes, depends on noneconomic arguments for preservation being translated into policy and laws by governments. Because the modern conservation movement largely developed in the United States, Canada, and northern Europe, it is not surprising that the strongest laws for the protection of creatures like desert fishes—and the strongest will to enforce the laws—exist in these areas. That this conservation ethic is spreading to other countries is indicated by the passage elsewhere of protective laws, although as yet, they often are unenforced. A movement also exists to develop more international and regional agreements to protect biodiversity, including obscure organisms like desert fishes, such as those efforts in Europe discussed by Biber-Klemm (1995).

Sri Lankan Stream Fishes

Sri Lanka is a tropical, densely populated country, with a human population that may have been almost as high a thousand years ago as it is today. It also has a rich endemic fish fauna in its fresh waters (Pethiyagoda 1991, 1994). Most of these fishes are small and have economic value mainly in the aquarium trade, although some are harvested for food as well. They also are of considerable value in controlling malaria-carrying mosquitoes. Despite dense human populations and the civil wars that have ravaged the country, the endemic biota still seems largely intact, if declining. The reasons for this are complex, but a major contributing factor has been the dominance of Theraveda Buddhism as the religious/philosophical system among the Sinhalese people and of Hinduism among the Tamil people. Both religions have strong respect for life of all types as a fundamental tenant, including moral restrictions on the killing of other creatures. Buddhism in particular has had a major role in the protection of biodiversity since at least 2,200 years ago, when Buddhist kings began establishing parks and other protected areas, usually associated with cities or temples (de Silva 1980, Moyle and Senanayake 1982). The combination of protected areas and respect for other creatures has produced a culture in which humans and wild creatures live in surprisingly close contact, even in big cities like Colombo.

The imposition of British rule in the nineteenth century did not change the religious beliefs of the people, but it did result in the clearing of enormous tracts of formerly wild land for tea plantations, adoption of Western technology, and increased growth of the human populations. The British also began the establishment of a system of parks and preserves that give Sri Lanka, proportionately, one of the largest systems of protected areas of any country in the world. Unfortunately, 95% of the protected areas are in the dry zone of the island, which is home to leopards and elephants but not to many endemic species, especially fishes (Pethiyagoda 1994).

Despite this history of moral and political protection, many endemic fishes of Sri Lanka are now in danger of extinction as the result of large water projects, deforestation, and other symptoms of a rapidly expanding population and westernization of the economy (Senanayake and Moyle 1982, Pethiyagoda 1994). Some of the species may have been saved through translocation (Wikramanayake 1990), but their long-term survival depends on adequate protection of watersheds, especially in the rain forests of the wet zone of the island. For example, a major problem for the fishes was created with the introduction of mechanized logging of the rain forests, thereby resulting in tractor and truck roads on steep slopes. These roads are a now a major source of sediment, so the once-clear streams flow muddy much of the time, with adverse consequences to the native fishes. If these same slopes are logged with traditional methods that use elephants instead of tractors, the streams remain clear, even after the heaviest rains. Elephant logging also has the advantage of providing economic incentives to protect the wild elephant populations from which the worker elephants are recruited.

Sri Lanka, more than most countries, has a philosophical and historical basis for protecting watersheds and aquatic biodiversity. So far, however, it has made little formal effort to do so. The persistence of its imperiled native fish fauna will be a major test of whether Sri Lanka's ancient conservation ethic will survive its clash with western economic imperatives. The success of conservation in modern Sri Lanka has implications far beyond Sri Lanka itself. The continued success of an ethical system that historically has supported conservation is likely to make it increasingly attractive to Westerners seeking a philosophical basis for conservation.

Kissimmee River, Florida

Increasingly, aquatic biologists and managers recognize that the key to restoring native stream fishes is restoring riverine ecosystems through recreating natural flow regimes

(Poff et al. 1997). One of the most ambitious projects to restore a natural flow regime is that of the Kissimmee River in central Florida (Toth et al. 1997). Historically, the Kissimmee was a rather vague, low-gradient river that flowed through a 1,633-square-mile watershed, much of it lakes, sloughs, and wetlands. The river was approximately 103 miles long and emptied into Lake Okeechobee, the second-largest lake in the contiguous United States. Overflow from Lake Okeechobee was the main source of water for the Everglades, which covered much of south Florida. This vast system, which flooded on a massive scale at times, was deemed a nuisance to the farmers and urbanites of Florida. As a result, the lakes and streams of the Kissimmee watershed were converted to a system of canals and reservoirs during the period from 1962 to 1971, and the Kissimmee River itself became canal C-38, which was 35 miles shorter than the original river. Water from annual rains was held in headwater reservoirs, which gradually were emptied during the dry season. This control of the flow allowed most of the wetlands on the flood plain to be drained and large amounts of organic matter to accumulate in the remaining wet areas. Not surprisingly, the riverine and wetland ecosystems were severely degraded, as signified by major declines in aquatic bird and fish populations.

Even while the project was being completed, many people became disturbed over the aesthetic and environmental damage, and restoration became first a cause, then a real possibility (Toth et al. 1997). By 1988, the goal became to reestablish "a river-floodplain ecosystem that is capable of supporting and maintaining a balanced, integrated, adaptive community of organisms having a species composition, diversity, and functional organization comparable to that of the natural habitat of the region" (Toth et al. 1997, p. 432). A key to achieving this goal is to restore the natural flow regime, which was justified, in part, because many of the native fishes (including important sport fishes) required the regime for successful reproduction. The ecological integrity of the Kissimmee River system is slowly being restored by allowing seasonal flooding of restored wetlands, filling in canals, and reconnecting meanders of the original river channel. When finished, the project will have restored 40 square miles of river and floodplain ecosystem and reestablished 43 miles of the original river channel. Obviously, this project is not a complete restoration of the Kissimmee River ecosystem. When finished, however, it will be one of the largest such projects ever attempted and likely will be held up as an example of a large-scale project with large-scale benefits, not only to fish and birds but to people.

37.9   CONCLUSIONS

The diverse examples presented here have one common theme: They all feature fish populations and aquatic ecosystems that have been degraded by human activities. They also indicate that many—and perhaps most—aquatic systems can be at least partly rehabilitated, with long-term benefits to humans. Two major (and strongly interactive) factors working against ecosystem rehabilitation are unchecked human population growth and the overwhelming importance of western economic values in decision making at all levels, from the individual to the global. In many ways, the present world economy depends on continuous economic growth, fueled by resource extraction (including fishing) and human population growth (Daily and Ehrlich 1992). The economic, scientific, legal, and moral tools are widely available to reverse this trend, but so far, the political will largely is not. Rarely is action taken until a disaster occurs, as in the Atlantic cod and Pacific salmon fisheries. On the other hand, in a growing number of examples, such as the Kissimmee River, groups have found ways to resist what seem to be implacable forces of degradation. An important tool for reversing trends is the increased information now available about the benefits derived from properly functioning ecological systems and how unregulated human activity reduces those benefits. With this knowledge, decision makers and politicians can be held responsible for ecological damage and its social and economic costs.

Increasingly, people interested in protecting fish and aquatic ecosystems have realized that major shifts in thinking are needed if we are to develop large-scale solutions to these problems. Some examples include:

- The Ecological Society of America has developed a set of principles for ecosystem-based management that provide guidelines for a more sustainable approach to conservation (Christensen et al. 1996). Ecosystem-based management is increasingly recognized as being necessary for marine fisheries in particular (Botsford et al. 1997, National Research Council 1998).
- Recognition is growing that protection of aquatic diversity in fresh water requires appropriate management of watersheds, a natural unit on the landscape (Williams et al. 1997). At least in the United States, a growing movement of citizens is demanding watershed-based management.
- Large-scale marine reserves, often involving hundreds or thousands of square kilometers, are being proposed as an effective solution to the problem of sustainable management of marine resources (Lauck et al. 1998).
- Invasions of alien species, as by-products of human activity (especially international trade), are being recognized as a global environmental problem that needs to be solved, especially in aquatic systems (Moyle 1998, Mooney and Hobbs 2000).
- Recognition is growing that many government policies result in subsidies that encourage pollution, overfishing, by-product introductions, and other practices with high long-term costs (Roodman 1996). For example, the U.S. Forest Service recently has been increasingly pressured to eliminate or reduce logging, mining, off-road vehicle use, and other land-damaging activities on public land because of the harm that such practices often do to aquatic ecosystems and fisheries.

37.10   THE FINAL LESSON

Ultimately, the only way that most of the world's fishes and aquatic ecosystems will be maintained in the future is through major changes in the world's ethical climate. Individuals, especially those in social classes and developed countries with inordinately high rates of resource consumption, must become morally responsible for the ecological effects of their actions. In our view, convincing people of their ethical responsibilities to the environment and, therefore, to the future should become a major focus of religious and philosophical institutions. We recognize that teaching such religious scruples is enormously difficult, given the rampant materialism of today. Nevertheless, the world's religions all have a strong conservation ethic entwined in their doctrines (Callicott 1994), and the Sri Lankan example indicates that such ethics can help to protect biodiversity for thousands of years. Even Christianity, which some have blamed for the ecological crisis (White 1967), can be seen as having fundamental teachings that support an ecological worldview (DeWitt 1995). One group in particular that needs to be made better aware of the ecological consequences of their actions are scientists, engineers, and technicians. As Orr (1996) states, "Knowledge has social, political, and ecological consequences as surely as any act of Congress ... " The generation and rapid spread of new knowledge untempered by wisdom has spurred irresponsible technological and economic growth. The scientific community now has the responsibility, working with our religious and ethical leaders, to develop the wisdom to help humanity reverse its present destructive course. This process, alas, will be slow and painful—and too slow to save many species and ecosystems. Therefore, political action also is necessary to protect enough pieces of our natural systems to ensure that widespread restoration will be possible when humanity finally realizes that its own welfare ultimately depends on environmental integrity. "Nature never did betray/The heart that loved her" (Wordsworth 1798).

SUPPLEMENTAL READINGS

Collares-Pereira et al. 2002; Ehrenfeld 1976; Meffe and Carroll 1997; Mooney and Hobbs 2000; Orr 1996; Rolston 1985; Snyder 1990.

WEB CONNECTIONS

CONSERVATION ORGANIZATIONS
www.conservefish.org

MARINE RESERVES
www.marine-reserves.org.nz
www.fknms.nos.noaa.gov/tortugas/benefits/biblio

Bibliography

Abel, D. C., Graham, J. B., Lowell, W. R., and Shabetai, R. 1986. Elasmobranch pericardial function. 1. Pericardial pressures are not always negative. *Fish Physiol. Biochem.* 1:75–83.

Abell, R. A., and 10 others. 2000. *Freshwater ecoregions of North America.* Covelo, CA: Island Press. 319 pp.

Abramovitz, J. N. 1996. Sustaining freshwater ecosystems. Pages 60–77 *in* L. Starke, ed. *State of the world 1996.* New York: W. W. Norton.

Adams, S. M. 1976a. Ecology of eelgrass, *Zostera marina* (L.) fish communities. I. Structural analysis. *J. Exp. Mar. Biol. Ecol.* 22:269–291.

Adams, S. M. 1976b. Feeding ecology of eelgrass fish communities. *Trans. Amer. Fish. Soc.* 105(4):514–519.

Adams, S. R., and Parsons, G. R. 1998. Laboratory-based measurements of swimming performance and related metabolic rates of field-sampled smallmouth buffalo (*Ictiobus bubalus*): A study of seasonal changes. *Physiol. Zool.* 71:350–358.

Ahlberg, P. E. 2001. Major events in vertebrate evolution: paleontology, phylogeny, genetics, and development. London: Taylor and Francis.

Ahlstrom, E. H., and Radovich, J. 1970. Management of the Pacific sardine. Pages 183–193 *in* N. C. Benson, ed. *A century of fisheries in North America.* Amer. Fish. Soc. Spec. Publ. 7.

Akahito, and 12 others. 2000. Evolutionary aspects of gobioid fishes based upon a phylogenetic analysis of mitochondrial cytochrome *b* genes. *Gene* 259:5–15.

Akire, C. 1993. Wild salmon as natural capital: Accounting for sustainable use. Washington, D.C.: The Wilderness Society. 39 pp.

Alevizon, W. S. 1975. Spatial overlap and competition in congeneric surfperches (Embiotocidae) off Santa Barbara, California. *Copeia* 1975:352–355.

Alexander, R. M. 1966a. Structure and function in catfishes. *J. Zool.* (London) 148:88–152.

Alexander, R. M. 1966b. Physical aspects of swimbladder function. *Biol. Rev.* 41:141–176.

Alexander, R. M. 1967. *Functional design in fishes.* London: Hutchinson Lib. 160 pp.

Alexander, R. M. 1993. Buoyancy. Pages 77–97 *in* D. H. Evans, ed., *The physiology of fishes.* Boca Raton, Fla.: CRC Press.

Allan, J. D. 1995. *Stream ecology: structure and function of running waters.* London: Chapman and Hall. 388 pp.

Allen, A. P., and 10 others. 1999. Concordance of taxonomic composition patterns across multiple lake assemblages: effects of scale, body size, and land use. *Can. J. Fish. Aquat. Sci.* 56:2029–2040.

Allen, G. R. 1989. *Freshwater fishes of Australia.* Neptune City, N.J.: TFH Publ. 240 pp.

Allen, G. R., and Coates, D. 1990. An ichthyological survey of the Sepik River, New Guinea. *Rec. West. Aust. Mus. Suppl.* 34:31–116.

Allen, M. J., Pecorelli, H., and Word, J. 1976. Marine organisms around outfall pipes in Santa Monica Bay. *J. Water Poll. Cont. Fed.* 48(8):1881–1893.

Allison, G. W., Lubchenco, J., and Carr, M. H. 1998. Marine reserves are necessary but not sufficient for marine conservation. *Ecol. App.* 8(Suppl.):79–92.

Alm, G. 1949. Influence of heredity and environment on various forms of trout. *Ann. Rpt. Inst. Freshw. Res.* (Drottningholm) 29:29–34.

Altinok, I., Galli, S. M., and Chapman, F. A. 1998. Ionic and osmotic regulation capabilities of juvenile Gulf of Mexico sturgeon, *Acipenser oxyrinchus desotoi. Comp. Biochem. Physiol.* 120A:609–619.

Altringham, J. D., and Block, B. A. 1997. Why do tuna maintain elevated slow muscle temperatures? Power output of muscle isolated from endothermic and ectothermic fish. *J. Exp. Biol.* 200:2617–2627.

Altringham, J. D., and Johnston, I. A. 1986. Energy cost of contraction in fast and slow muscle fibers isolated from an elasmobranch and an antarctic teleost fish. *J. Exp. Biol.* 121:239–250.

Altringham, J. D., Wardle, C. S., and Smith, C. I. 1993. Myotomal muscle function at different locations in the body of a swimming fish. *J. Exp. Biol.* 182:191–206.

Alves, M. J., Collares-Pereira, M. J., Dowling, T. E., and Coelho, M. M. 2002. The genetics of maintenance of an all-male lineage in the *Squalius alburnoides* complex. *J. Fish Biol.* 60:649–662.

Alves-Gomes, J. A. 2001. The evolution of electroreception and bioelectrogenesis in teleost fish: A phylogenetic perspective. *J. Fish Biol.* 58:1489–1511.

Amundsrud, J. R., Faber, D. J., and Keast, A. 1974. Seasonal succession of free-swimming perciform larvae in Lake Opinicon, Ontario, *J. Fish. Res. Bd. Canada* 31(10):1661–1665.

Andrews, J. W., and Stickney, R. R. 1972. Interactions of feeding rates and environmental temperature on growth, food conversion, and body composition of channel catfish. *Trans. Am. Fish. Soc.* 101:94–99.

Andriashev, A. P. 1970. Cryopelagic fishes of the Arctic and Antarctic and their significance in polar ecosystems. Pages 297–304 *in* M. W. Holdgate, ed., *Antarctic ecology.* New York: Academic Press.

Angermeier, P. L., and Schlosser, I. J. 1989. Species–area relationships for stream fishes. *Ecology* 70:1450–1462.

Aoyama, J., Mochioka, N., Otake, T., Ishikawa, S., Kawakami, Y., Castle, P., Nishida, M., and Tsukamoto, K. 1999. Distribution and dispersal of anguillid leptocephali in the western Pacific Ocean revealed by molecular analysis. *Mar. Ecol. Prog. Ser.* 188:193–200.

Aoyama, J., Nishida, M., and Tsukamoto, K. 2001. Molecular phylogeny and evolution of the freshwater eel, genus *Anguilla. Mol. Phylo. Evol.* 20:450–459.

Aparicio, S., and 40 others. 2002. Whole-genome shotgun assembly and analysis of the genome of *Fugu rubripes. Science* 297:1301–1310.

Applegate, V. C. 1950. *Natural history of the sea lamprey (Petromyzon marinus) in Michigan.* U.S. Fish. Wildl. Serv. Spec. Sci. Rept. 555:1–237.

Armstrong, J. D., Priede, I. G., and Lucas, M. C. 1992. The link between respiratory capacity and changing metabolic demands during growth of northern pike, *Esox lucius* L. *J. Fish Biol.* 41(Suppl. B):65–75.

Astrup, J., and Mohl, B. 1998. Discrimination between high and low repetition rates of ultrasonic pulses by the cod. *J. Fish Biol.* 52:205–208.

Atema, J. 1971. Structures and functions of the sense of taste in the catfish *(Ictalurus natalis). Brain, Behavior, and Evolution* 25(4):273–294.

Attrill, M. J., and Power, M. 2002. Climatic influence on a marine fish assemblage. *Nature* 417:275–278.

Ault, T. R., and Johnson, C. R. 1998. Spatially and temporally predictable fish communities on coral reefs. *Ecol. Monogr.* 68:25–50.

Avise, J. C., Helfman, G. S., Saunders, N. C., and Hales, L. S. 1986. Mitochondrial DNA differentiation in North Atlantic eels: Population genetic consequences of an unusual life history pattern. *Proc. Nat. Acad. Sci. U. S. A.* 83:4350–4354.

Avise, J. C., Smith J. J., and Ayala, F. J. 1975. Adaptive differentiation with little genic change in two native California minnows. *Evolution* 29:411–426.

Backus, R. H., Craddock, J. E., Haedrich, R. L., and Robison, B. H. 1977. Atlantic mesopelagic zoogeography. Pages 266–287 *in* B. F. Nafpaktitus et al., eds. *Fishes of the western North Atlantic, pt 7.* Mem. 1, Sears Found. Mar. Res.

Bagenal, T. B. 1978. Aspects of fish fecundity. Pages 75–101 *in* S. D. Gerking, ed., *Ecology of freshwater fish production.* New York: Wiley.

Bailey, H. C., and Doroshov, S. I. 1995. The duration of the interval associated with successful inflation of the swimbladder in larval striped bass. *Aquaculture* 131:135–143.

Bailey, R. G. 1998. *Ecosystems: The ecosystem geography of the oceans and continents.* New York: Springer-Verlag.

Bainbridge, R. 1958. The speed of swimming of fish as related to size and to the frequency and amplitude of the tail beat. *J. Exp. Biol.* 35:109–133.

Baines, G. W. 1975. Blood pH effects in eight fishes from the teleostean family Scorpaenidae. *Comp. Biochem. Physiol.* 51A:833–843.

Baird, I., Hogan, Z., Phylaivanh, B., and Moyle, P. 2001. A communal fishery for migratory catfish *Pangasius macronema* in the Mekong River. *Asian Fish. Sci.* 14:25–41.

Bakker, T. C. M., and Milinski, M. 1993. The advantages of being red: sexual selection in the stickleback. *Mar. Behav. Physiol.* 23:287–300.

Balik, S. 1995. Freshwater fish in Anatolia, Turkey. *Biol. Cons.* 72:213–223.

Balon, E. K. 1975a. Reproductive guilds in fishes: A proposal and definition. *J. Fish. Res. Bd. Canada* 32(6):821–864.

Balon, E. K. 1975b. Terminology of intervals in fish development. *J. Fish. Res. Bd. Canada* 32(9):1663–1670.

Balon, E. K. 1977. Fish gluttons: The natural ability of some fishes to become obese when food is in extreme abundance. *Hydrobiologia* 52:239–241.

Balon, E. K. 1978. Kariba: The dubious benefits of large dams. *Ambio* 7(2):40–48.

Balon, E. K. 1979. The theory of saltation and its application in the ontology of fishes: Steps and thresholds. *Env. Biol. Fish* 4:97–101.

Balon, E. K. 1981a. Additions and amendments to the classification of reproductive styles in fishes. *Env. Biol. Fish* 6:377–389.

Balon, E. K. 1981b. Saltatory processes and altricial to precocial forms in the ontogeny of fishes. *Amer. Zool.* 21:573–596.

Balon, E. K. 1984. Patterns in the evolution of reproductive styles in fishes. Pages 35–53 *in* G. W. Potts and R. J. Wootton, eds., *Fish reproduction: Strategies and tactics.* London: Academic Press.

Balon, E. K., and Coche, A. G. 1974. *Lake Kariba: A man-made tropical ecosystem in Central Africa.* The Hague: W. Junk. 767 pp.

Balsano, J. S., Rasch, E. M., and Monaco, P. J. 1989. The evolutionary ecology of *Poecilia formosa* and its triploid associate. Pages 277–297 *in* G. K. Meffe and F. F. Snelson, Jr., eds., *Ecology and evolution of livebearing fishes (Poeciliidae).* Englewood Cliffs, N.J.: Prentice Hall.

Baltz, D. M. 1984. Life history variation in female surfperches (Perciformes, Embiotocidae). *Env. Biol. Fish* 10:159–171.

Baltz, D. M. 1991. Introduced fishes in marine systems and seas. *Biol. Cons.* 56:151–178.

Baltz, D. M., Feelger, J. W., Rakocinski, C. F., and McCall, J. N. 1998. Food, density, and microhabitat: factors affecting growth and recruitment potential of juvenile saltmarsh fishes. *Env. Biol. Fish.* 53:89–103.

Baltz, D. M., and Moyle, P. B. 1982. Life history characteristics of tule perch (*Hysterocarpus traski*) populations in contrasting environments. *Env. Biol. Fish.* 7:229–242.

Baltz, D. M., Rakocinski, C., and Fleeger, J. W. 1993. Microhabitat use by marsh-edge fishes in a Louisiana estuary. *Env. Biol. Fish.* 36:109–126.

Barber, W. E., and Minckley, W. L. 1966. Fishes of Aravaipa Creek, Graham and Pinal Counties, Arizona. *Southw. Nat.* 11(3):313–324.

Bardach, J. E., and Todd, J. H. 1970. Chemical communication in fish. Pages 205–240 *in* J. W. Johnson et al., eds. *Advances in chemoreception*. Vol. 1. New York: Appleton-Century-Crofts.

Barlow, G. W. 1961. Causes and significance of morphological variation in fishes. *Syst. Zool.* 10(1):105–117.

Barlow, G. W. 1972. The attitude of fish eye-lines in relation to body shape and to stripes and bars. *Copeia* 1972(1):5–12.

Barlow, G. W. 1973. Competition between color morphs of the polychromatic Midas cichlid *Cichlasoma citrinellum. Science* 179:806–807.

Barlow, G. W. 1976. The Midas cichlid in Nicaragua. Pages 333–358 *in* T. B. Thorson, ed., *Investigations of the ichthyofauna of Nicaraguan Lakes.* Lincoln: School of Life Sci., Univ. Nebraska.

Barlow, G. W. 1981. Patterns of parental investment, dispersal, and size in coral reef fishes. *Env. Biol. Fish.* 6:65–85.

Barlow, G. W. 2000. *The cichlid fishes: Nature's grand experiment in evolution.* Cambridge, Mass.: Perseus Publishing.

Barrett, B. A., and McKeown, B. A. 1988a. Sustained exercise increases plasma growth hormone concentration in two anadromous salmonids. *Can. J. Fish. Aquat. Sci.* 45:747–749.

Barrett, B. A., and McKeown, B. A. 1988b. Growth hormone response to sustained swimming in exercise-acclimated steelhead trout, *Salmo gairdneri. J. Fish. Biol.* 32:799–800.

Barrett, B. A., and McKeown, B. A. 1988c. Sustained exercise augments long-term starvation increases in plasma growth hormone in the steelhead trout, *Salmo gairdneri. Can. J. Zool.* 66:853–855.

Barrett, I., and Williams, A. A. 1965. Hemoglobin content of the blood of fifteen species of marine fishes. *Calif. Fish and Game* 51:2116–2181.

Barthem, R., and Goulding, M. 1997. *The catfish connection.* New York: Columbia Univ. Press.

Barton, B. A. 1988. Endocrine and metabolic responses of fish to stress. *Proc. Int. Assoc. Aquat. An. Med.* 19:41–55.

Barton, B. A. 2002. Stress in fishes: A diversity of responses with particular reference to changes in circulating corticosteroids. *Integ. and Comp. Biol.* 42:517–525.

Barton, B. A., and Iwama, G. K. 1991. Physiological changes in fish from stress in aquaculture with emphasis on the response and effects of corticosteroids. *Ann. Rev. Fish Diseases* 1:3–26.

Baskin, J. N., Zaret, T. M., and Mago-Leccia, F. 1980. Feeding of reportedly parasitic catfishes (Trichomycteridae and Cetopsidae) in the Rio Portuguesa basin, Venezuela. *Biotropica* 12:182–186.

Basu, N., Nakano, T., Grau, E. G., and Iwama, G. K. 2001. The effects of cortisol on heat shock protein 70 levels in two fish species. *Gen. and Comp. Endocrin.* 124:97–105.

Batty, R. S., and Wardle, C. S. 1979. Restoration of glycogen from lactic acid in the anaerobic muscle of plaice, *Pleuronectes platessa. L. J. Fish Biol.* 15:509–519.

Baum, J. K., Myers, R. A., Kehler, D. G., Worm, B., Harley, S. J., and Doherty, P. A. 2003. Collapse and conservation of shark populations in Northwest Atlantic. *Science* 299:390–392.

Bayley, P. B. 1995. Understanding larger river-flood plain ecosystems. *Bioscience* 45:153–158.

Bayley, P. B., and Li, H. W. 1993. Riverine fishes. Pages 251–281 *in* P. Calow and G. E. Petts, eds. *The rivers handbook.* Vol. 1. Oxford: Blackwell Scientific Pubs.

Beamish, F. W. H. 1964. Respiration of fishes with special emphasis on standard oxygen consumption. II. Influence of weight and temperature in respiration of several species. *Can. J. Zool.* 42:177–188.

Beamish, F. W. H. 1970. Oxygen consumption of largemouth bass, *Micropterus salmoides,* in relation to swimming speed and temperature. *Can. J. Zool.* 48:1221–1228.

Beamish, F. W. H. 1993. Environmental sex determination in southern brook lamprey, *Ichthyomyzon gagei. J. Fish. Aquat. Sci.* 50:1299–1307.

Beamish, F. W. H., and Tandler, A. 1990. Ambient ammonia, diet and growth in lake trout. *Aquat. Toxicol.* 17:155–166.

Beaudet, L., Brownman, H. I., and Hawryshyn, C. W. 1993. Optic nerve response and retinal structure in rainbow trout of different sizes. *Vision Res.* 33:1739–1746.

Beaudet, L., Novales Flamarique, I., and Hawryshyn, C. W. 1997. Cone photoreceptor topography in the retina of sexually mature Pacific salmonid fishes. *J. Comp. Neurol.* 383:49–59.

Becker, C. D., and Fugihara, F. G. 1978. The bacterial pathogen, *Flexibacter columnaris,* and its epizoology among Columbia River fish: A review and synthesis. *Amer. Fish. Soc. Monogr.* 2.

Beddow, T. A., and McKinley, R. S. 1999. Importance of electrode positioning in biotelemetry studies estimating muscle activity in fish. *J. Fish Biol.* 54:819–831.

Behnke, R. J. 1992. Native trout of western North America. *Amer. Fish. Soc. Monogr.* 6:275 pp.

Bell, A. M. 2001. Effects of an endocrine disruptor on courtship and aggressive behavior of male three-spined stickleback, *Gasterosteus aculeatus. Animal Behavior* 62:775–780.

Bell, M. A. 1976a. Evolution of phenotypic diversity in *Gasterosteus aculeatus* superspecies on the Pacific coast of North America. *Syst. Zool.* 25(3):211–227.

Bell, M. A. 1976b. Reproductive character displacement in threespine sticklebacks. *Evolution* 30(4):847–850.

Bellamy, D., and Chester-Jones, I. 1961. Studies on *Myxine glutinosa.* I. The chemical composition of the tissues. *Comp. Biochem. Physiol.* 3:175–183.

Bemis, W. E., Burggren, W. W., and Kemp, N. E., eds. 1986. *The biology and evolution of lung-fishes.* New York: A. R. Liss. (Also *J. Morph. Suppl. 1*).

Bemis, W. E., Findeis, E. K., and Grande, L. 1997. An overview of Acipenseriformes. *Envir. Biol. Fish* 48:25–71.

Bemis, W. E., and Kynard, B. 1997. Sturgeon rivers: An introduction to sturgeon biogeography and life history. *Env. Biol. Fish.* 48:167–183.

Ben-Tuvia, A. 1978. Immigration of fishes through the Suez Canal. *NOAA Fish. Bull.* 76(1):249–255.

Bennett, B. A., and Attwood, C. G. 1991. Evidence for recovery of a surf-zone fish assemblage following the establishment of a marine reserve on the southern coast of South Africa. *Mar. Ecol. Prog. Ser.* 75:173–181.

Bennett, R. O., Kraeuter, J. N., Woods, III, L. C., Lipsky, M. M., and May, E. B. 1987. Histological evaluation of swim bladder non-inflation in striped bass larvae *Morone saxatilis. Dis. Aquat. Org.* 3:91–95.

Bennett, W. A., and Moyle, P. B. 1996. Where have all the fishes gone? Interactive factors producing fish declines in the Sacramento–San Joaquin estuary. Pages 519–541 *in* J. T. Hollibaugh, ed., *San Francisco Bay: The Ecosystem.* Pacific Division, AAAS.

Bennion, G. R. 1968. *The control of the function of the heart in teleost fish.* M.S. thesis, Univ. of British Columbia, Vancouver. 53 pp.

Berendzen, P. B., and Dimmick, W. W. 2002. Phylogenetic relationships of Pleuronectiformes based on molecular evidence. *Copeia* 2002:642–652.

Berg, L. S. 1940. *Classification of fishes both recent and fossil.* Ann Arbor, Mich.: J. W. Edwards. 517 pp.

Berg, L. S. 1949. *Freshwater fishes of the U.S.S.R. and adjacent countries.* Jerusalem: Israel Prog. Sci. Transl. 510 pp.

Berg, T., and Steen, J. B. 1965. Physiological mechanisms for aerial respiration in the eel. *Comp. Biochem. Physiol.* 15:469–484.

Berglund, A. 1995. Many mates make male pipefish choosy. *Behaviour* 32:213–218.

Berglund, A. 1997. Mating systems and sex allocation. Pages 237–265 *in* J. J. Godin, ed. *Behavioural ecology of teleost fishes*. Oxford: Oxford Univ. Press.

Bergman, E. N. 1990. Energy contributions of volatile fatty acids from the gastrointestinal tract in various species. *Physiol. Rev.* 70:567–590.

Berkman, P. A. 1992. The antarctic marine ecosystem and humankind. *Rev. Aquat. Sci.* 6:295–333.

Bernal, D., Dickson, K. A., Shadwick, R. E., Graham, J. B. 2001. Review: Analysis of the evolutionary convergence for high performance swimming in lamnid sharks and tunas. *Comp. Biochem. Physiol. Part A* 129:695–726.

Bernstein, B. B., and Jung, N. 1979. Selective pressures and coevolution in a kelp-canopy community in Southern California. *Ecol. Mono.* 49:335–355.

Berra, T. M. 2001. *Freshwater fish distribution*. San Diego: Academic Press.

Berra, T. M., and Allen, G. R. 1989. Burrowing, emergence, behavior, and functional morphology of the Australian salamanderfish, *Lepidogalaxias salamandroides*. *Fisheries* (Bethesda) 14(5):2–10.

Bevelhimer, M. S., and Adams, S. M. 1993. A bioenergetics analysis of diel vertical migration by kokanee salmon, *Oncorhynchus nerka. Can. J. Fish. Aquat. Sci.* 50:2336–2349.

Beveridge, M. C. M., Ross, L. G., and Kelly, L. A. 1994. Aquaculture and biodiversity. *Ambio* 23:497–502.

Biber-Klemm, S. 1995. Legal aspects of the conservation of freshwater fish in the northern Mediterranean region. *Biol. Cons.* 72:321–334.

Bigelow, H. B., and Schroeder, W. C. 1953. Fishes of the Gulf of Maine. *U.S. Fish. Wildl. Ser. Fish. Bull.* 74:1–577.

Billard, R., and G. Lecointre. 2001. Biology and conservation of sturgeon and paddlefish. *Rev. Fish Biol. Fisheries* 10:355–392.

Binotti, I., Giovenco, S., Giardina, B., Antonini, E., Brunori, M., and Wyman, J. 1971. Studies on the functional properties of fish hemoglobins. II. The oxygen equilibrium of the isolated hemoglobin components from trout blood. *Arch. Biochem. Biophys.* 142:274–280.

Birstein, V., Waldman, J. R., and Bemis, W. E., eds. 1997. Sturgeon biodiversity and conservation. *Envir. Biol. Fish.* 48:440 pp.

Bjerring, H. C. 1985. Facts and thoughts on piscine phylogeny. Pages 179–206 *in* P. H. Greenwood, R. S. Miles, and C. Patterson, eds. *Interrelationships of fishes*. London: Academic Press.

Björnsson, B. Th. 1997. The biology of salmon growth hormone: From daylight to dominance. *Fish Physiol. Biochem.* 17:9–24.

Blaber, S. J. M. 2000. *Tropical estuarine fishes: Ecology, exploitation, and conservation*. Oxford: Blackwell Science.

Blaber, S. J. M. 2002. Fish in hot water: the challenges facing fish and fisheries research in tropical estuaries. *J. Fish Biol.* 61(Suppl. A):1–20

Blackburn, M. 1965. Oceanography and the ecology of tunas. *Oceanogr. Mar. Biol. Ann. Rev.* 3:299–322.

Blake, B. F. 1977. The effect of the impoundment of Lake Kainji, Nigeria, on the indigenous species of mormyrid fishes. *Freshw. Biol.* 7:37–42.

Blake, R. W. 1979. The energetics of hovering in the mandarin fish *(Synchropus punctatus). J. Exp. Biol.* 82:25–33.

Blake, R. W. 1983a. *Fish locomotion*. Cambridge: Cambridge Univ. Press.

Blake, R. W. 1983b. Hovering performance of a negatively buoyant fish. *Can. J. Zool.* 61:2629–2630.

Blaxhall, P. C., and Daisley, K. W. 1973. Routine haematological methods for use with fish blood. *J. Fish Biol.* 5:771–782.

Blaxter, J. H. S. 1974. *The early life history of fish*. New York: Springer-Verlag. 765 pp.

Blaxter, J. H. S. 1985. The herring: A successful species? *Can. J. Fish. Aquatic Sci.* 42:21–30.

Blaxter, J. H. S., and Fuiman, L. A. 1990. The role of sensory systems of herring larvae in evading predatory fishes. *J. Mar. Biol. Assoc. U.K.* 70:413–427.

Blaxter, J. H. S., and Holliday, F. G. T. 1963. The behavior and physiology of herring and other clupeids. *Adv. Mar. Biol.* 2:261–393.

Block, B. A. 1991a. Endothermy in fish: thermogenesis, ecology and evolution. Pages 269–311 *in* P. W. Hochchka and T. P. Mommsen, eds., *Biochemistry and Molecular Biology of Fishes.* Vol. 1. Amsterdam: Elsevier Science Publishers B. V. (Academic Publishing Division).

Block, B. A. 1991b. Evolutionary novelties: How fish have built a heater out of muscle. *Amer. Zool.* 31:726–742.

Block, B. A. 1994. Thermogenesis in muscle. *Annu. Rev. Physiol.* 56:535–577.

Block, B. A., and 10 others. 2001. Migratory movements, depth preferences, and thermal biology of Atlantic bluefin tuna. *Science* 293:1310–1314.

Block, B. A., and Finnerty, J. R. 1994. Endothermy in fishes: A phylogenetic analysis of constraints, predispositions, and selection pressures. *Environ. Biol. Fish.* 40:283–302.

Block, B. A., Finnerty, J. R., Stewart, A. F. R., and Kidd, J. 1993. Evolution of endothermy in fish: Mapping physiological traits on a molecular phylogeny. *Science* 260:210–214.

Blumer, L. S. 1979. Male parental care in the bony fishes. *Quart. Rev. Biol.* 54:149–161.

Bodznik, D. 1978. Calcium ion: An odorant for natural water discriminations and the migratory behavior of sockeye salmon. *J. Comp. Physiol.* 127:157–166.

Boehlert, G. W. 1996. Biodiversity and the sustainability of marine fisheries. *Oceanography* 9:28–35.

Bone, Q. 1966. On the function of the two types of myotomal muscle fibre in elasmobranch fish. *Fish. J. Mar. Biol. Assoc. U.K.* 46:321–349.

Booth, J. H. 1979. The effects of oxygen supply, epinephrine, and acetylcholine on the distribution of blood flow in trout gills. *J. Exp. Biol.* 83:31–39.

Borkholder, B. D., and Edwards, A. J. 2001. Comparing the use of dorsal fin spines with scales to back-calculate length-at-age estimates in walleyes. *North Amer. J. Fish. Manag.* 21:935–942.

Bortone, S. A., and Davis, W. P. 1994. Fish intersexuality as indicator of environmental stress. *Bioscience* 44:165–172.

Botsford, L. W., Castilla, J. C., and Peterson, C. H. 1997. The management of fisheries and marine ecosystems. *Science* 277:509–515.

Boudouresque, C. F. 1999. The Red Sea–Mediterraean link: Unwanted effects of canals. Pages 213–228 *in* T. Sandlund, P. J. Schei, and A. Viken, eds. *Invasive species and biodiversity management.* Dordrect: Kluwer.

Boulenger, G. A. 1907. *Zoology of Egypt: Fishes of the Nile.* London: Hugh Rees. 578 pp.

Boulenger, G. A. 1910. Ichthyology. I. History of literature down to 1880. Pages 243–250 *in Encyclopedia Brittanica.* 11th ed. Vol. 14. New York: Encyclopedia Brittanica Co.

Bouton, N., Seehausen, O., and van Alphen, J. M. 1997. Resource partitioning among rock-dwelling haplochromines (Pisces: Cichlidae) from Lake Victoria. *Ecol. Freshw. Fish* 6:225–240.

Bowen, B. W., Bass, A. L., Rocha, L. A., Grant, W. S., and Robertson, D. R. 2001. Phylogeography of the trumpetfishes (*Aulostomus*): Ring species complex on a global scale. *Evolution* 55: 1029–1039.

Bowering, W. R. 1976. Distribution, age and growth, and sexual maturity of witch flounder (*Glyptocephalus cynoglossus*). *J. Fish. Res. Bd. Canada* 33(7):1574–1584.

Boyce, S. J., and Clarke, A. 1997. Effect of body size and ration on specific dynamic action in the Antarctic plunderfish, *Harpagifer antarcticus,* Nybelin 1947. *Physiol. Zool.* 70:679–690.

Brailovskaya, T. 1998. Obstacles to protecting marine biodiversity through marine wilderness preservation: Examples from the New England region. *Cons. Biol.* 12:1236–1240.

Braum, E. 1978. Ecological aspects of fish eggs, embryos, and larvae. Pages 102–136 *in* S. Gerking, ed., *Ecology of freshwater fish production.* New York: Wiley.

Bray, R. N., and Ebeling, A. W. 1975. Food, activity, and habitat of three "picker-type" microcarnivorous fishes in the kelp forests off Santa Barbara, California. *NOAA Fish. Bull.* 73(4):815–829.

Bray, R. N., and Hixon, M. A. 1978. Nightshocker: Predatory behavior of the Pacific electric ray *(Torpedo californica)*. *Science* 200:333–334.

Bray, R. N., Miller, A. C., and Geesey, D. G. 1981. The fish connection: A trophic link between planktonic and rock reef communities. *Science* 214:204–205.

Brecka, B. J., Wahl, D. H., and Hooe, M. L. 1996. Growth, survival, and body composition of largemouth bass fed various commercial diets and protein concentrations. *Prog. Fish-Cult.* 58:104–110.

Breder, C. M. 1934. Ecology of an oceanic freshwater lake, Andros Island, Bahamas, with special reference to its fishes. *Zoologica* 18:57–88.

Breder, C. M., and Rosen, D. E. 1966. *Modes of reproduction in fishes*. Garden City, N.Y.: Nat. Hist. Press. 941 pp.

Brett, J. R. 1964. The respiratory metabolism and swimming performance of young sockeye salmon. *J. Fish. Res. Bd. Canada* 21:1183–1226.

Brett, J. R. 1971. Energetic responses of salmon to temperature. A study of some thermal relations in the physiology and freshwater ecology of sockeye salmon *(Oncorhynchus nerka)*. *Am. Zool.* 11:99–113.

Brett, J. R. 1979. Environmental factors and growth. Pages 599–675 *in* W. S. Hoar, D. J. Randall, and J. R. Brett, eds., *Fish physiology*. Vol. 9. New York: Academic Press.

Brett, J. R., and Groves, T. D. 1979. Physiological energetics. Pages 279–352 *in* W. S. Hoar, D. J. Randall, and J. R. Brett, eds., *Fish physiology*. Vol. 9. New York: Academic Press.

Brett, J. R., Shelbourn, J. E., and Shoop, C. T. 1969. Growth rate and body composition of fingerling sockeye salmon, *Oncorhynchus nerka*, in relation to temperature and ration size. *J. Fish. Res. Bd. Canada* 26:2363–2394.

Brewer, G. D. 1973. Midwater fishes from the Gulf of California and the adjacent eastern tropical Pacific. *Los Angeles Co. Mus. Nat. Hist. Contrs. in Sci.* 242:1–47.

Brick, M. E., and J. J. Cech, Jr. 2002. Metabolic responses of juvenile striped bass to exercise and handling stress with various recovery environments. *Trans. Amer. Fish. Soc.* 131:855–864.

Bridges, W. W., Cech, J. J., Jr., and Pedro, D. N. 1976. Seasonal hematological changes in winter flounder, *Pseudopleuronectes americanus*. *Trans. Am. Fish. Soc.* 105:596–600.

Briggs, C. T., and Post, J. R. 1997. Field metabolic rates of rainbow trout estimated using electromyogram telemetry. *J. Fish Biol.* 51:807–823.

Briggs, D. E. G. 1992. Conodonts: A major extinct group added to the vertebrates. *Science* 256:1285–1286.

Briggs, J. C. 1974. *Marine zoogeography*. New York: McGraw-Hill. 475 pp.

Briggs, J. C. 1979. Ostariophysan zoogeography: An alternative hypothesis. *Copeia* 4:111–118.

Briggs, J. C. 2000. Centrifugal speciation and centres of origin. *J. Biogeog.* 27:1183–1188.

Briggs, P. T., and O'Connor, J. S. 1971. Comparison of shore-zone fishes over naturally vegetated and sand-filled bottoms in Great South Bay. *N.Y. Fish. Game J.* 18(1):15–41.

Brill, R. W., and Bushnell, P. G. 1991. Metabolic and cardiac scope of high energy demand teleosts—The tunas. *Can. J. Zool.* 69:2002–2009.

Brill, R. W., Dewar, H., and Graham, J. B. 1994. Basic concepts relevant to heat transfer in fishes, and their use in measuring physiological thermoregulatory abilities of tunas. *Env. Biol. Fishes* 40:109–124.

Brock, R. E. 1977. Occurrence and variety of fishes in mixohaline ponds of the Kona, Hawaii, coast. *Copeia* 1977 (1):134–139.

Brocksen, R. W., and Cole, R. E. 1972. Physiological responses of three species of fishes to various salinities. *J. Fish. Res. Bd. Canada* 29:399–405.

Brodal, A., and Fänge, R., eds. 1963. *The biology of the myxine*. Oslo: Scand. Univ. Books. 588 pp.

Bromley, P. J. 1994. The role of gastric evacuation experiments in quantifying the feeding rates of predatory fish. *Rev. in Fish Biol. and Fisheries* 4:36–66.

Bronseth, T., and Folstad, I. 1997. The effects of parasites on courtship dance in threespine stickle-backs: More than meets the eye? *Can. J. Zool.* 75:589–594.

Brooks, J. L. 1968. The effects of prey size selection by lake planktivores. *Syst. Zool.* 17(3):272–291.

Brooks, J. L., and Dodson, S. 1965. Predation, body size, and composition of plankton. *Science* 150:28–35.

Brooks, R. 2000. Negative genetic correlation between male sexual attractiveness and survival. *Nature* 406:67–70.

Brooks, R. 2002. Variation in female mate choice within guppy populations: population divergence, multiple ornaments, and the maintenance of polymorphism. *Genetica* 116:343–358.

Brothers, E. B., Mathews, C. P., and Lasker, R. 1976. Daily growth increments in otoliths from larval and adult fishes. *NOAA Fish. Bull.* 74:1–8.

Brown, C. L., Doroshov, S. I., Nunez, J. M., Hadley, C., Vaneenennaam, J., Nishioka, R. S., and Bern, H. A. 1988. Maternal triiodothyronine injections cause increases in swimbladder inflation and survival rates in larval striped bass, *Morone saxatilis. J. Exp. Zool.* 248:168–176.

Brown, C. R., and Cameron, J. N. 1991. The induction of specific dynamic action in channel catfish by infusion of essential amino acids. *Physiol. Zool.* 64(1):276–297.

Brown, D. W. 1974. Hydrography and midwater fishes of three contiguous oceanic areas off Santa Barbara, California. *Los Angeles Co. Mus. Nat. Hist. Conts. Sci.* 261:1–30.

Brown, G. E., Chivers, D. P., and Smith, R. J. F. 1995. Localized defection by pike: A response to labelling by cyprinid alarm pheromone? *Behav. Ecol. Sociobiol.* 36:105–110.

Brown, L. R., Moyle, P. B., and Yoshiyama, R. M. 1994. Historical decline and current status of coho salmon in California. *North Amer. J. Fish. Mgmt.* 14:237–261.

Brown-Peterson, N. J. 1993. Fish assemblages in natural versus well-established recolonized seagrass meadows. *Estuaries* 16:177–189.

Brownman, H. I., Novales Flamarique I., and Hawryshyn, C. W. 1994. Ultraviolet photoreception contributes to the foraging performance of two species of zooplanktivorous fishes. *J. Exp. Biol.* 186:187–198.

Bruton, M. N. 1996. Alternative life-history strategies of catfishes. *Aquat. Living Resour.* 9 (Hors serie):35–41.

Buckley, J. A. 1977. Heinz body hemolytic anemia in coho salmon *(Oncorhynchus kisutch)* exposed to chlorinated wastewater. *J. Fish Res. Bd. Canada* 34:215–224.

Buckup, P. A. 1998. Relationships of the Characidinae and the phylogeny of characiform fishes (Teleostei: Ostariophysi). Pages 123–144 *in* L. R. Malabarba et al., eds. *Phylogeny and classification of neotropical fishes.* EDIPUCRS, Porto Alegre, Brazil.

Buddington, R. K., and Doroshov, S. I. 1986. Development of digestive secretions in white sturgeon juveniles *(Acipenser transmontanus). Comp. Biochem. Physiol.* 83A:233–238.

Budelli, R., and Caputi, A. 2000. The electric image in weakly electric fish: Perception of objects of complex impedance. *J. Exp. Biol.* 203:481–492.

Budker, P. 1971. *The life of sharks.* New York: Columbia Univ. Press, 222 pp.

Bullock, T. H. 1973. Seeing the world through a new sense; electroreception in fish. *Amer. Sci.* 61(3):316–325.

Bullock, T. H., Bodznick, D. A., and Northcutt, R. G. 1983. The phylogenetic distribution of electroreception: Evidence for convergent evolution of a primitive vertebrate sense modality. *Brain Research Reviews* 6:25–46.

Bulow, F. J. 1987. RNA/DNA ratios as indicators of growth in fish: A review. Pages 45–64 *in* R. C. Summerfelt and G. E. Hall, eds., *Age and growth of fish.* Ames: Iowa State Univ. Press.

Bulow, F. J. 1970. RNA/DNA ratios as indicators of recent growth rates of a fish. *J. Fish. Res. Bd. Canada* 27:2343–2349.

Burger, J. W. 1962. Further studies on the function of the rectal gland in the spiny dogfish. *Physiol. Zool.* 35:205–217.

Burger, J. W., and Hess, W. 1960. Function of the rectal gland in the spiny dogfish. *Science* 131:670–671.

Burgess, T. J. 1978. The comparative ecology of two sympatric polychromatic populations of *Xererpes fucorum* Jordan and Gilbert (Pisces, Pholididae) from the rocky intertidal zone of central California. *J. Exp. Mar. Biol. Ecol.* 35:43–58.

Burggren, W. W. 1978. Gill ventilation in the sturgeon, *Acipenser transmontanus*: unusual adaptations for bottom dwelling. *Resp. Physiol.* 34:153–170.

Burkholder, J. M., Glasgow, H. B., Jr., and Hobbs, C. W. 1995. Fish kills linked to a toxic ambush predator dinoflagellate: Distribution and environmental conditions. *Mar. Ecol. Prog. Ser.* 124:43–61.

Burr, B. M. 1991. The fishes of Illinois: An overview of a dynamic fauna. *Bull. Ill. Nat. Hist. Surv.* 34:417–427.

Burr, B. M., and Mayden, R. L. 1992. Phylogenetics and North American freshwater fishes. Pages 18–75 *in* R. L. Mayden, ed., *Systematics, historical ecology, and North American freshwater fishes.* Stanford, Calif.: Stanford Univ. Press.

Busacker, G. P., Adelman, I. A., and Goolish, E. M. Growth 1990. Pages 363–387 *in* C. B. Schreck and P. B. Moyle, eds. *Methods for fish biology.* Bethesda, Md.: American Fisheries Society.

Bushnell, P. G. 1988. Cardiovascular and respiratory responses to hypoxia in three species of obligate ram ventilating fishes, skipjack tuna, *Katsuwonis pelamis*, yellowfin tuna, *Thunnus albacares*, and bigeye tuna, *Thunnus obesus*. Ph.D. thesis. University of Hawaii, Honolulu.

Bussing, W. A. 1985. Patterns of distribution of the Central American ichthyofauna. Pages 453–473 *in* F. G. Stehli and S. D. Webb, eds., *The great American biotic interchange.* New York: Plenum Pub.

Bussing, W. A. 1993. Fish communities and environmental characteristics of a tropical rain forest river in Costa Rica. *Rev. Trop. Biol.* 41:791–809.

Bussing, W. A. 1998. *Freshwater fishes of Costa Rica.* 2nd ed. San Jose, Calif.: Univ. Costa Rica. 468 pp.

Butler, P. J., Taylor, E. W., Capra, M. F., and Davison, W. 1978. The effect of hypoxia on the levels of circulating catecholamines in the dogfish *Scyliorhinus canicula. J. Comp. Physiol.* 127:325–330.

Cailliet, G. M. 1992. Demography of the central California population of the leopard shark (*Triakis semifasciata*). *Aust. J. Mar. Freshwater Res.* 43:183–193.

Cailliet, G. M., Love, M. S., and Ebeling, A. W. 1986. *Fishes: A field and laboratory manual on their structure, identification, and natural history.* Belmont, Calif.: Wadsworth Publ.

Caley, M. J., and St. John, J. 1996. Refuge availability: structures assemblages of tropical reef fishes. *J. Anim. Ecol.* 65:414–428.

Callicott, J. B. 1994. Conservation values and ethics. Pages 24–49 *in* G. K. Meffe and C. R. Carroll, eds., *Principles of conservation biology.* Sunderland, Mass: Sinauer Assoc.

Cameron, J. N. 1970a. Blood characteristics of some marine fishes of the Texas gulf coast. *Tex. J. Sci.* 21:275–283.

Cameron, J. N. 1970b. The influence of environmental variables on the hematology of pinfish (*Lagodon rhomboides*) and striped mullet (*Mugil cephalus*). *Comp. Biochem. Physiol.* 32:175–192.

Cameron, J. N. 1971a. Oxygen dissociation characteristics of the blood of rainbow trout, *Salmo gairdneri. Comp. Biochem. Physiol.* 38A:600–704.

Cameron, J. N. 1971b. Methemoglobin in erythrocytes of rainbow trout blood. *Comp. Biochem. Physiol.* 40A:743–749.

Cameron, J. N. 1975. Morphometric and flow indicator studies of the teleost heart. *Can. J. Zool.* 53:691–698.

Cameron, J. N. 1976. Branchial ion uptake in Arctic grayling: Resting values and effects of acid base disturbances. *J. Exp. Biol.* 64:711–725.

Cameron, J. N. 1978. Chloride shift in fish blood. *J. Exp. Zool.* 206:289–295.

Cameron, J. N., and Davis, J. C. 1970. Gas exchange in rainbow trout *(Salmo gairdneri)* with varying blood oxygen capacity. *J. Fish Res. Bd. Canada* 27:1069–1085.

Cameron, J. N., and Iwama, G. K. 1987. Compensation of progressive hypercapnia in channel catfish and blue crabs. *J. Exp. Biol.* 133:187–197.

Cameron, J. N., and Iwama, G. K. 1989. Compromises between ionic regulation and acid base regulation in aquatic animals. *Can. J. Zool.* 67:3078–3084.

Cameron, J. N., Randall, D. J., and Davis, J. C. 1971. Regulation of the ventilation-perfusion ratio in the gills of *Dasyatis sabina* and *Squalus suckleyi. Comp. Biochem. Physiol.* 39A:505–519.

Carey, F. G. 1982a. Warm fish. Pages 216–233 *in* C. R. Taylor, K. Johansen, and L. Bolis, eds., *A companion to animal physiology.* Cambridge: Cambridge University Press.

Carey, F. G. 1982b. A brain heater in the swordfish. *Science* 216:1327–1329.

Carey, F. G., and Teal, J. M. 1966. Heat conservation in tuna fish muscle. *Proc. Nat. Acad. Sci. U. S. A.* 56:1461–1469.

Carey, F. G., Teal, J. M., Kanwisher, J. W., Lawson, K. D., and Beckett, J. S. 1971. Warm-bodied fish. *Am. Zool.* 11:137–145.

Carline, R. F., Gagen. C. J., and Sharpe, W. E. 1994. Brook trout *(Salvelinus fontinalis)* population dynamics and mottled sculpin *(Cottus bairdi)* occurrence in relation to acidic episodes in streams. *Ecol. Freshw. Fish* 3:107–115.

Carlton, J. T., and Geller, J. 1993. Ecological roulette: The global transport and invasion of nonindigenous marine organisms. *Science* 261:78–82.

Carmichael, G. J., Tomasso, J. R., Sinnco, B. A., and Davis, K. B. 1984. Characterization and alleviation of stress associated with hauling largemouth bass. *Trans. Amer. Fish. Soc.* 113:778–785.

Carpenter, S. R., and Kitchill, J. F. 1993. *The trophic cascade in lakes.* New York: Cambridge University Press. 385 pp.

Carr, M. H., and Hixon, M. A. 1995. Predation effects on early post-settlement survivorship of coral-reef fishes. *Mar. Ecol. Prog. Ser.* 124:31–42.

Carrier, J. C., and Evans, D. H. 1976. The role of environmental calcium in freshwater survival of the marine teleost *Lagodon rhomboides. J. Exp. Biol.* 65:529–538.

Carvalho, G. R. 1993. Evolutionary aspects of fish distribution: Genetic variability and adaptation. *J. Fish. Biol.* 43(Suppl. A):53–73.

Carvalho, M. R. 1996. Higher level elasmobranch phylogeny, basal squaleans, and paraphyly. Pages 35–62 *in* M. L. J. Stiassny, L. R. Parenti, and G. D. Johnson, eds. *Interrelationships of fishes.* San Diego, Calif.: Academic Press.

Casey, J. G., and Kohler, N. E. 1992. Tagging studies on the shortfin mako shark *(Isurus oxyrinchus)* in the western Atlantic ocean. *Aust. J. Mar. Freshwater Res.* 43:45–60.

Casey, J. M., and Myers, R. A. 1998. Near extinction of a larger, widely distributed fish. *Science* 281:690–692.

Castro, J. I. 1983. *The sharks of North American waters.* College Station: Texas A&M Press. 180 pp.

Catlett, R. H., and Millich, D. R. 1976. Intracellular and extracellular osmoregulation of temperature-acclimated goldfish: *Carassius auratus* L. *Comp. Biochem. Physiol.* 55A:261–269.

Cech, J. J., Jr. 1990. Respirometry. Pages 335–362 *in* C. B. Schreck and P. B. Moyle, eds., *Methods for fish biology.* Bethesda, Md.: American Fisheries Society.

Cech, J. J., Jr., Bartholow, S. D., Young, P. S., and Hopkins, T. E. 1996. Striped bass exercise and handling stress in freshwater: Physiological responses to recovery environment. *Trans. Amer. Fish. Soc.* 125:308–320.

Cech, J. J., Jr., Bridges, D. W., Rowell, D. M., and Balzer, P. J. 1976. Cardiovascular responses of winter flounder, *Pseudopleuronectes americanus* (Walbaum), to acute temperature increase. *Can. J. Zool.* 54:1383–1388.

Cech, J. J., Jr., Campagna, C. G., and Mitchell, S. J. 1979. Respiratory responses of largemouth bass *(Micropterus salmoides)* to environmental changes in temperature and dissolved oxygen. *Trans. Amer. Fish. Soc.* 108:166–171.

Cech, J. J., Jr., Laurs, R. M., and Graham, J. B. 1984. Temperature-induced changes in blood gas equilibria in the albacore, *Thunnus alalunga,* a warm-bodied tuna. *J. Exp. Biol.* 109:21–34.

Cech, J. J., Jr., and Massingill, M. J. 1995. Tradeoffs between respiration and feeding in Sacramento blackfish, *Orthodon microlepidotus. Env. Biol. Fish.* 44:157–163.

Cech, J. J., Jr., Mitchell, S. J., and Massingill, M. J. 1979. Respiratory adaptations of Sacramento blackfish, *Orthodon microlepidotus* (Ayres), for hypoxia. *Comp. Biochem. Physiol.* 63A:411–415.

Cech, J. J., Jr., and Wohlschlag, D. E. 1973. Respiratory responses of the striped mullet, *Mugil cephalus* L. *Contr. Mar. Sci.* 19:91–100.

Cech, J. J., Jr., and Wohlschlag, D. E. 1975. Summer growth depression in the striped mullet, *Mugil cephalus* L. *Contr. Mar. Sci.* 19:91–100.

Cech, J. J., Jr. and Wohlschlag, D. E. 1982. Seasonal patterns of respiration, gill ventilation, and hematological characteristics in the striped mullet, *Mugil cephalus* L. *Bull. Mar. Sci.* 32:130–138.

Chan, S. T. H., and Yeung, W. S. B. 1983. Sex control and sex reversal in fish under natural conditions. Pages 171–222 *in* W. S. Hoar, D. J. Randall, and E. M. Donaldson, eds., *Fish physiology 9B: Reproduction, behavior, and fertility control.* London: Academic Press.

Chao, L. N., and Musick, J. A. 1977. Life history, feeding habits, and functional morphology of juvenile sciaenid fishes in the York River estuary, Virginia. *NOAA Fish. Bull.* 75(4):657–702.

Chavez, F. P., Ryan, J., Lluch-Cota, S. E., Ñiquen, M. 2003. From anchovies to sardines and back: multidecadal changes in the Pacific Ocean. *Science* 299: 217–221.

Chen, L., and Martinich, R. L. 1975. Pheromonal stimulation and metabolite inhibition of ovulation in zebrafish, *Brachydanio rerio. NOAA Fish. Bull.* 73(4):889–894.

Cheney, K. L., and I. M. Côté 2001. Are Caribbean cleaning symbioses mutualistic? Costs and benefits of visiting cleaning stations to longfin damselfish. *Anim. Behav.* 62:927–933.

Cherry, D. S., and Guthrie, R. K. 1975. Significance of detritus or detritus-associated invertebrates to fish production in a new impoundment. *J. Fish. Res. Bd. Canada* 32(10):1799–1804.

Chesley, L. C. 1934. The concentrations of proteases, amylase, and lipase in certain marine fishes. *Biol. Bull.* 66:133–144.

Choat, J. H. 1991. The biology of herbivorous fishes on coral reefs. Pages 120–155 *in* P. F. Sale, ed. *The ecology of fishes on coral reefs.* San Diego: Academic Press.

Choat, J. H., Axe, L. M., and Lou, D. C. 1996. Growth and longevity in fishes of the family Scaridae. *Mar. Ecol. Prog. Ser.* 145:33–41.

Christensen, N. L., and 12 others. 1996. The report of the Ecological Society of America Committee on the Scientific Basis for Ecosystem Management. *Ecol. Appl.* 6:665–691.

Christiansen, J. S., Ringo, E., and Jobling. M. 1989. Effects of sustained exercise on growth and body composition of first-feeding fry of Arctic charr, *Salvelinus alpinus* L. *Aquaculture* 79:329–335.

Claiborne, J. B. 1998. Acid base regulation. Pages 177–198 *in* D. H. Evans, ed., *The physiology of fishes.* 2nd ed. Boca Raton, Fla.: CRC Press.

Claiborne, J. B., and Heisler, N. 1986. Acid-base regulation and ion transfers in the carp *(Cyprinus carpio)*: pH compensation during graded long- and short-term environmental hypercapnia, and the effect of bicarbonate infusion. *J. Exp. Biol.* 126:41–61.

Clouthier, R., and Ahlberg, P. E. 1996. Morphology, characters, and the interrelationships of the basal Sarcopterygians. Pages 445–480 *in* M.L. J. Stiassny, L. R. Parenti, and G. D. Johnson, *Interrelationships of fishes.* San Diego, Calif.: Academic Press.

Coates, M. I. 2000. Fins to limbs: What the fossils say. *Am. Zool.* 40:976–977.

Cohen, A. N., and Carlton, J. T. 1998. Accelerating invasion rate in a highly invaded estuary. *Science* 279:555–558.

Cohen, D. M. 1970. How many recent fishes are there? *Proc. Calif. Acad. Sci.* 37(17):341–346.

Colclough, S. R., Gray, G., Bark, A., and Knights, B. 2002. Fish and fisheries of the tidal Thames: Management of the modern resource, research aims, and future pressures. *J. Fish Biol.* 61(Suppl. A): 64–73.

Collares-Pereira, M. J. 1989. Hybridization in European cyprinids: evolutionary potential of unisexual populations. Pages 281–288 *in* R.M. Dawley & J.P. Bogart, eds., *Evolution and ecology of unisexual vertebrates*. Albany: New York State Museum.

Collares-Pereira, M. J., Coelho, M. M., and Cowx, I. G. 2002. *Conservation of freshwater fishes: options for the future*. London: Fishing News Books. 462 pp.

Collette, B. B., and Earle, S. A., eds. 1972. Results of the Tektite Program: Ecology of coral reef fishes. *Bull. Los Angeles Co. Nat. Hist. Mus. 14.* 180 pp.

Colletti, A. E., and Olson, K. R. 1988. Catecholamine metabolism by the perfused rainbow trout gill. *J. Exp. Zool.* 248:177–184.

Collie, N. L., and Ferraris, R. P. 1995. Nutrient fluxes and regulation in fish intestine. Pages 221–239 *in* P.W. Hochachka and T. P. Mommsen, eds., *Biochemisty and molecular biology of fishes.* Vol. 4—Metabolic biochemistry. Amsterdam: Elsevier Science B.V.

Collins, T. M., Trexler, J. G., Nico, L. G., and Rawlings, T. A. 2002. Genetic diversity in a morphologically conservative invasive taxon: Multiple introductions of swamp eels into the southeastern United States. *Cons. Biol.* 16:1024–1035.

Colman, J. G. 1997. A review of the biology and ecology of the whale shark. *J. Fish Biol.* 51:1219–1234.

Colt, J., and Tchobanoglous, G. 1978. Chronic exposure of channel catfish, *Ictalurus punctatus,* to ammonia: Effects on growth and survival. *Aquaculture* 15:353–372.

Compagno, L. J. V. 1973. Interrelationships of living elasmobranchs. Pages 15–62 *In* P. H. Greenwood, R. S. Miles, and C. Patterson, eds., *Interrelationships of fishes.* New York: Academic Press.

Compagno, L. J. V. 1977. Phyletic relationships of living sharks and rays. *Amer. Zool.* 17(2):303–322.

Compagno, L. J. V. 1990. Alternative life-history styles of cartilaginous fishes in time and space. *Env. Biol. Fish* 28:33–76.

Compagno, L. J. V. 1999. Systematics and body form. Pages 1–42 *in* W. C. Hamlett, ed. *The biology of elasmobranch fishes.* Baltimore: Johns Hopkins University Press.

Connell, J. R. 1978. Diversity in tropical rain forests and coral reefs. *Science* 199:1302–1310.

Conover, D. O., and Kynard, B. E. 1981. Environmental sex determination: Interaction of temperature and genotype in a fish. *Science* 213:577–579.

Constanza, R. 1991, ed. *Ecological economics: The science and management of sustainability*. New York: Columbia University Press.

Contreras-Balderas, S., and Lozano-Vilano, M. 1994. Water, endangered fishes, and development perspectives in arid lands of Mexico. *Conservation Biology* 8:379–387.

Coombs, S., and Popper, A. N. 1979 Hearing differences among Hawaiian squirrelfish (family Holocentridae) related to differences in the peripheral auditory system. *J. Comp. Physiol. A* 132:203–213.

Copeland, B. J., and Bechtel, T. J. 1974. Some environmental limits of six Gulf coast estuarine organisms. *Cont. Mar. Sci.* 18:169–204.

Copeland, B. J., and Nixon, S. W. 1974. Hypersaline lagoons. Pages 312–330 *in* H. T. Odum, et al., eds., *Coastal ecological systems of the United States.* Washington, D.C.: The Conservation Foundation.

Cortes, E. 1995. Demographic analysis of the Atlantic sharpnose shark, *Rhizoprionon terraenovae,* in the Gulf of Mexico. *NOAA Fish. Bull.* 93:57–66.

Cossins, A. R., and Kilby, R. V. 1989. The seasonal modulation of Na$^+$ exchanger activity in trout erythrocytes. *J. Exp. Biol.* 144:463–478.

Courtney, W. R., Sahlman, H. F., Miley, W. W., and Herrma, D. J. 1974. Exotic fishes in fresh and brackish waters of Florida. *Biol. Cons.* 6 (4):292–302.

Cousins, K. L., and Farrell, A. P. 1996. Stretch induced release of atrial natriuretic factor from the heart of rainbow trout (*Oncorhynchus mykiss*). *Can. J. Zool.* 74:380–387.

Cowen, R. K. 2002. Larval dispersal and retention and consequences for population connectivity. Pages 149–170 *in* P. F. Sale, ed. *Coral reef fishes: dynamics and diversity in a complex ecosystem.* Amsterdam: Academic Press.

Cowey, C. B., and Walton, M. J. 1989. Intermediary metabolism. Pages 259–329 *in* J. Halver, ed. *Fish Nutrition.* New York: Academic Press.

Cracraft, J. 1974. Continental drift and vertebrate distribution. *Ann. Rev. Ecol. Syst.* 5:215–261.

Cracraft, J. 1994. Species diversity, biogeography, and the evolution of biotas. *Amer. Zool.* 34:33–47.

Craig, J. F. 1992. Human-induced changes in the composition of fish communities in the African Great Lakes. *Rev. Fish Biol. Fisheries* 2:93–124.

Craig, J. F., ed. 1996. *Pike: Biology and exploitation.* London: Chapman and Hall.

Crawshaw, L. I., and Hammel, H. T. 1971. Behavioral thermoregulation in two species of antarctic fish. *Life Sci.* 10(17):1009–1020.

Crawshaw, L. I., Wollmuth, L. P., and O'Connor, C. S. 1989. Intracranial ethanol and ambient anoxia elicit selection of cooler water by goldfish. *Amer. J. Physiol.* 256:R133–R137.

Crespi, B. J., and Teo, R. 2002. Comparative phylogenetic analysis of the evolution of semelparity and life history of salmonid fishes. *Evolution* 56:1008–1020.

Crivelli, A., and Maitland, P. S. 1995. Introduction: Endemic freshwater fishes of the northern Mediterranean region. *Biol. Cons.* 72:121–122.

Crocker, C. E., and Cech, J. J., Jr. 1996. The effects of hypercapnia on the growth of juvenile white sturgeon, *Acipenser transmontanus. Aquaculture* 147:293–299.

Crocker, C. E., and Cech, J. J., Jr. 1998. Effects of hypercapnia on blood-gas and acid-base status in the white sturgeon, *Acipenser transmontanus. J. Comp. Physiol. B* 168:50–60.

Crossman, E. J., and McAllister, D. E. 1986. Zoogeography of freshwater fishes of the Hudson Bay drainage, Ungava Bay, and the Arctic archipelago. Pages 53–105 *in* C. H. Hocutt and E. O. Wiley, eds., *The zoogeography of North American freshwater fishes.* New York: Wiley.

Crowder, L. B. 1984. Character displacement and habitat shift in a native cisco in southeastern Lake Michigan: Evidence for competition? *Copeia* 1984:878–883.

Crowl, T. A., Townsend, C. R., and McIntosh, A. R. 1992. The impact of introduced brown and rainbow trout on native fish: The case of Australasia. *Rev. Fish Biol. Fisheries* 2:217–241.

Culliney, J. L. 1979. *The forests of the sea.* Garden City, N.Y.: Doubleday.

Cushing, D. H. 1968. *Fisheries biology: A study in population dynamics.* Madison: Univ. of Wisconsin Press. 200 pp.

Cushing, D. H. 1978. Biology of fishes of the pelagic community. Pages 317–340 *in* D. H. Cushing and J. J. Walsh, eds., *The ecology of the seas.* Philadelphia: Saunders.

Cuthbert, A. W., and Maetz, J. 1972. The effects of calcium and magnesium on sodium fluxes through gills of *Carassius auratus. J. Physiol.* 221:633–643.

Currie, S., Moyes, C. D., and Tufts, B. L. 2000. The effects of heat shock and acclimation temperature on hsp70 and hsp30 mRNA expression in rainbow trout: *in vivo* and *in vitro* comparisons. *J. Fish Biol.* 56:398–408.

D'Amico Martell, A. L., and Cech, J. J., Jr. 1978. Peripheral vascular resistance in the gills of the winter flounder, *Pseudopleuronectes americanus. Comp. Biochem. Physiol.* 59A:419–423.

Dahlberg, M. D., and Odum, E. P. 1970. Annual cycles of species occurrence, abundance, and diversity in Georgia estuarine fish populations. *Amer. Midl. Nat.* 83(2):382–392.

Daily, G. C., ed. 1997. *Nature's services: Societal dependence of natural ecosystems.* Washington, D.C.: Island Press.

Daily, G. C., and Ehrlich, P. R. 1992. Population, sustainability, and Earth's carrying capacity. *Bioscience* 42:761–771.

Dana, T. F. 1975. Development of contemporary eastern Pacific coral reefs. *Mar. Biol.* 33:355–374.

Daniels, R. A. 1978. Nesting behavior of *Harpagifer bispinis* in Arthur Harbour, Antarctic Peninsula. *J. Fish Biol.* 12:465–474.

Daniels, R. A. 2001. Untested assumptions: The role of canals in the dispersal of sea lamprey, alewife, and other fishes of the Eastern United States. *Env. Biol. Fish.* 60:309–329.

Danzmann, R. G., Ferguson, M. M., Skulason, S., Snorrason, S. S., and Noakes, D. L. G. 1991. Mitochondrial DNA diversity among four sympatric morphs of Arctic charr, *Salvelinus alpinus* L., from Thingvallavatn, Iceland. *J. Fish Biol.* 39:649–659.

Darken, R. S., Martin, K. L. M., and Fisher, M. C. 1998. Metabolism during delayed hatching in terrestrial eggs of a marine fish, the grunion *Leuresthes tenuis. Physiol. Zool.* 71:400–406.

Darlington, P. J. 1957. *Zoogeography: The geographical distribution of animals.* New York: Wiley. 673 pp.

Davenport, J. 1994. How and why do flying fish fly? *Rev. Fish Biol. Fisheries* 4:184–214.

Davenport, J., and Sayer, M. D. J. 1993. Physiological determinants of distribution in fish. *J. Fish Biol.* 43(Suppl. A):121–145.

Davie, P. S., Wells, R. M. G., and Tetens, V. 1986. Effects of sustained swimming on rainbow trout muscle structure, blood oxygen transport, and lactate dehydrogenase isozymes: Evidence for increased aerobic capacity of white muscle. *J. Exp. Zool.* 237:159–171.

Davies, P. L., Ewart, K. V., and Fletcher, G. L. The diversity and distribution of fish antifreeze proteins: New insights into their origins. Pages 279–305 *in* P. W. Hochachka and T. P. Mommsen, eds. *Biochemisty and molecular biology of fishes.* Vol. 2: Molecular biology frontiers. Amsterdam: Elsevier.

Davies, P. R., Hanyu, I., Furukawa, K., and Nomura, M. 1986. Effect of temperature and photoperiod on sexual maturation and spawning of the common carp. III. Induction of spawning by manipulating photoperiod and temperature. *Aquaculture* 52:137–144.

Davison, W., and Goldspink, G. 1984. The cost of swimming for two teleost fish. *N. Z. J. Zool.* 11:225–232.

Daxboeck, C., and Holeton, G. F. 1978. Oxygen receptors in the rainbow trout, *Salmo gairdneri. Can. J. Zool.* 56:1254–1259.

Day, F. 1878. *The fishes of India.* Vol. 2. London: Bernard Quaritch. 195 pp.

Day, D. S., and Pearcy, W. G. 1968. Species associations of benthic fishes on the continental shelf and slope off Oregon. *J. Fish. Res. Bd. Canada* 25(12):2665–2675.

Dayton, P. K. 1998. Reversal of the burden of proof in fisheries management. *Science* 279:821–822.

Dayton, P. K., Mordida, B. J., and Bacon, F. 1994. Polar marine communities. *Amer. Zool.* 34:90–99.

de Pinna, M. C. C. 1996. Teleostean monophyly. Pages 147–162 *in* M.L. J. Stiassny, L. R. Parenti, and G. D. Johnson, eds. *Interrelationships of fishes.* San Diego: Academic Press.

de Silva, A. 1980. The impact of Buddhism on the conservation of the flora and fauna in ancient Sri Lanka. *FAO Tigerpaper* 7(4):21–25.

De Carvalho, M. R. 1996. Higher level elasmobranch phylogeny, basal squaleans, and paraphyly. Pages 35–62 *in* M. L. J. Stiassny, L. R. Parenti, and G. D. Johnson, eds. *Interrelationships of fishes.* San Diego, Califo.: Academic Press.

De Geoij, P., Luttikhuizen, P., van der Meer, J., and Piersma, T. 2001. Facilitation on an intertidal mudflat: The effect of siphon nipping by flatfish on burying depth of the bivalve *Macoma balthica. Oecologia* 126:500–506.

Deacon, J. E., and Minckley, W. L. 1974. Desert fishes. Pages 385–488 *in* R. W. Brown, ed., *Desert biology.* Vol. 2. New York: Academic Press.

Dean, B. 1895. *Fishes, living and fossil*. New York: Macmillan, 300 pp.

DeMartini, E. E. 1969. A correlative study of the ecology and comparative feeding mechanism morphology of the Embiotocidae (surfperches) as evidence of the family's adaptive radiation into available ecological niches. *Wassman J. Biol.* 27(2):177–247.

Demski, L. S., and Northcutt, R. G. 1983. The terminal nerve: A new chemosensory system in vertebrates? *Science* 220:435–437.

Denton, E. J., and Rowe, D. M. 1994. Reflective communication between fish, with special reference to the greater sand eel, *Hyperoplus lanceolatus*. *Phil. Trans. R. Soc. London Ser. B* 334:221–237.

Denton, E. J., and Shaw, T. J. 1963. The visual pigments of some deep-sea elasmobranchs. *J. Mar. Biol. Assoc. U.K.* 43:65–70.

Denton, E. J., and Warren, F. J. 1956. Visual pigments of deep-sea fish. *Nature* 178:1059.

DeVlaming, V. L. 1972a. The effects of temperature and photoperiod on reproductive cycling in the estuarine gobiid fish *Gillichthys mirabilis*. *NOAA Fish. Bull.* 70:1137–1152.

DeVlaming, V. L. 1972b. Reproductive cycling in the estuarine gobiid fish *Gillichthys mirabilis*. *Copeia* 1972(2):278–291.

DeVlaming, V. L. 1972c. The role of the endocrine system in temperature-controlled reproductive cycling in the estuarine gobiid fish *Gillichthys mirabilis*. *Comp. Biochem. Physiol.* 41A:697–713.

Devlin, R. H., Yesaki, T. Y. Biagi, C. A., Donaldson, E. M., Swanson, P., and Chan, W. K. 1994. Extraordinary growth in salmon. *Nature* 371:1738–1748.

DeVries, A. L. 1970. Freezing resistance in Antarctic fishes. Pages 320–328 *in* M. W. Holdgate, ed., *Antarctic ecology*, I. New York: Academic Press.

DeVries, A. L. 1984. Role of glycopeptides and peptides in inhibition of crystallization of water in polar fishes. *Phil. Trans. R. Soc. London B.* 304:575–588.

DeVries, A. L., and Wohlschlag, D. E. 1969. Freezing resistance in some Antarctic fishes. *Science* 163:1073–1075.

DeWitt, C. B. 1995. Ecology and ethics: The relation of religious belief to ecological practice in the Biblical tradition. *Biodiv. Cons.* 4:838–848.

DeWitt, F. A., and Cailliet, G. M. 1972. Feeding habits of two bristlemouth fishes, *Cyclothone acclinidens* and *C. signata* (Gonostomatidae). *Copeia* 1972(4):868–871.

DeWitt, H. H. 1971. *Coastal and deep-water benthic fishes of the Antarctic*. Amer. Geogr. Soc. Antarctic Map Folio Series 15. 10 pp.

DeWoody, J. A., and Avise, J. C. 2000. Microsatellite variation in marine, freshwater, and anadromous fishes compared to other animals. *J. Fish Biol.* 56:61–473.

Diamant, A., and Shpigel, M. 1985. Interspecific feeding associations of groupers (Teleostei, Serranidae) with octopuses and moray eels in the Gulf of Eilat (Aqaba). *Env. Biol. Fish.* 13:153–159.

Diana, J. S. 1995. *Biology and ecology of fishes*. Carmel, Ind.: Cooper Publishing Group. 441 pp.

Dickson, K. A. 1994. Tunas as small as 207 mm fork length can elevate muscle temperatures significantly above ambient water temperature. *J. Exp. Biol.* 190:79–93.

Dietz, T. J., and Somero, G. N. 1992. The threshold induction temperature of the 90-kDa heat shock protein is subject to acclimatization in eurythermal goby fishes (genus *Gillichthys*). *Proc. Natl. Acad. Sci. U. S. A.* 89:3389–3393.

Dietz, T. J., and Somero, G. N. 1993. Species- and tissue-specific synthesis patterns for heat-shock proteins HSP 70 and HSP 90 in several marine teleost fishes. *Physiol. Zool.* 66:863–880.

Dijkgraaf, S. 1962. The functioning and significance of the lateral-line organs. *Biol. Rev.* 38:51–105.

Dijkgraaf, S., and Kalmijn, A. J. 1963. Untersuchungen über die Funktion der Lorenzinischen Ampullen an Haifischen. *Z. Vgl. Physiol.* 47:438–456.

Dingle, H. 1996. *Migration: the biology of life on the move*. Oxford: Oxford University Press.

Dittman, A. H., Quinn, T. P., Dickhoff, W. W., and Larsen, D. A. 1994. Interactions between novel water, thyroxine and olfactory imprinting in underyearling coho salmon, *Oncorhynchus kisutch* (Walbaum). *Aqua. Fish. Manage.* 25(Suppl. 2):157–169.

Dittman, A. H., Quinn, T. P., and Nevitt, G. A. Timing of imprinting to natural and artificial odors by coho salmon (*Oncorhynchus kisutch*). *Can. J. Fish. Aquat. Sci.* 53:434–442.

Dobbs, G. H., Lin, Y., and DeVries, A. L. 1974. Aglomerularism in Antarctic fish. *Science* 185:793–794.

Doherty, P. J. 2002. Variable replenishment and the dynamics of reef fish populations. Pages 327–358 *in* P. F. Sale, ed. *Coral reef fishes: dynamics and diversity in a complex ecosystem.* Amsterdam: Academic Press.

Dominey, W. J., and Blumer, L. S. 1984. Cannibalism of early life stages of fishes. Pages 43–64 *in* G. Hausfater and S. Blaffer-Hrdy, eds. *Infanticide: comparative and evolutionary perspectives.* New York: Aldine.

Donald, J. A. 1998. Autonomic nervous system. Pages 407–439 *in* D. H. Evans, ed. *The physiology of fishes.* 2nd ed. Boca Raton, Fla.: CRC Press.

Donald, J., and Campbell, G. 1982. A comparative study of the adrenergic innervation of the teleost heart. *J. Comp. Physiol.* 147:85–91.

Donaldson, E. M., Fagerlund, U. H. M., Higgs, D. A., and McBride, J. R. 1979. Hormonal enhancement of growth. Pages 455–597 *in* W. S. Hoar, D. J. Randall, and J. R. Brett, eds. *Fish Physiology.* Vol. 9. New York: Academic Press.

Donoghue, P. C. J., and Aldridge, R. J. 2001. Origin of the mineralized skeleton. Pages 85–105 *in* P. E. Ahlberg, ed. *Major events in vertebrate evolution: paleontology, phylogeny, genetics, and development.* London: Taylor and Francis.

Donoghue, P. C. J., Forey, P. L., and Aldridge, R. J. 2000. Conodont affinity and chordate phylogeny. *Biol. Rev.* 75:191–251.

Dooley, J. K. 1972. Fishes associated with the pelagic *Sargassum* complex, with a discussion of the *Sargassum* community. *Contr. Mar. Sci.* 16:1–32.

Doroshev, S. I., and Cornacchia, J. W. 1979. Initial swim bladder inflation in the larvae of *Tilapia mossambica* (Peters) and *Morone saxatilis* (Walbaum). *Aquaculture* 16:57–66.

Douglas, E. L., Friedl, W. A., and Pickwell, G. V. 1976. Fishes in oxygen-minimum zones: Blood oxygenation characteristics. *Science* 191:957–959.

Douglas, M. E., and Matthews, W. J. 1992. Does morphology predict ecology? Hypothesis testing within a freshwater fish assemblage. *Oikos* 65:213–224.

Douglas, R. H. 1983. Spectral sensitivity of rainbow trout (*Salmo gairdneri*). *Rev. Can. Biol. Exptl.* 42:117–122.

Downing, J. A., and Plante, C. 1993. Production of fish populations in lakes. *Can. J. Fish. Aquat. Sci.* 50:110–120.

Driedzic W. R., West, J. L., Sephton, D. H., and Raymond, J. A. 1998. Enzyme activity levels associated with the production of glycerol as an antifreeze in liver of rainbow smelt (*Osmerus mordax*). *Fish Physiol. Biochem.* 18:125–134.

Drucker, E. G., and Jensen, J. S. 1996. Pectoral fin locomotion in the striped surfperch. I. Kinematic effects of swimming speed and body size. *J. Exp. Biol.* 199:2235–2242.

Drummond, R. A., Spoor, W. A., and Olson, G. F. 1973. Some short-term indicators of sublethal effects of copper on brook trout, *Salvelinus fontinalis. J. Fish. Res. Bd. Canada* 30:698–701.

Duan, C. 1997. The insulin-like growth factor system and its biological actions in fish. *Amer. Zool.* 37:491–503.

Dudgeon, D. 1992. Endangered ecosystems: A review of the conservation status of tropical Asian rivers. *Hydrobiologia* 248:167–191.

Duman, J. G., and DeVries, A. L. 1974a. Freezing resistance in winter flounder, *Pseudopleuronectes americanus. Nature* 247:237–238.

Duman, J. G., and DeVries, A. L. 1974b. The effects of temperature and photoperiod on antifreeze production in cold-water fishes. *J. Exp. Zool.* 190:89–97.

Duncan, A., and Kubecka, J. 1995. Land/water ecotone effects in reservoirs on fish fauna. *Hydrobiologia* 303:11–30.

Dyer, S. D., Dickson, K. L., and E. G. Zimmerman. 1991. Tissue-specific patterns of synthesis of heat-shock proteins and thermal tolerance of the fathead minnow (*Pimephales promelas*) *Can. J. Zool.* 69:2021–2027.

Dynesius, M., and Nilsson, C. 1994. Fragmentation and flow regulation of river systems in the northern third of the world. *Science* 266:753–762.

East, P., and Magnan, P. 1987. The effect of locomotor activity on the growth of brook charr, *Salvelinus fontinalis* Mitchell. *Can. J. Zool.* 65:843–846.

Eastman, J. T. 1993. *Antarctic fish biology: Evolution in a unique environment.* San Diego, Calif.: Academic Press. 322 pp.

Eastman, J. T., and DeVries, A. L. 1982. Buoyancy studies of notothenioid fishes in McMurdo Sound, Antarctica. *Copeia* 1982:385–393.

Eastman, J. T., and McCune, A. R. 2000. Fishes on the Antarctic continental shelf: Evolution of a marine species flock? *J. Fish Biol.* 57(Suppl. A):84–102.

Ebeling, A. W., Larsen, R. J., and Alevizon, W. S. 1980. Habitat groups and island–mainland distribution of kelp-bed fishes off Santa Barbara, California. Pages 403–431 *in* D. M. Power, ed., *Multidisciplinary symposium on the California islands.* Santa Barbara Museum Nat. Hist.

Echelle, A. A., and Connor, P. J. 1989. Rapid, geographically extensive introgression after secondary contact between two pupfish species (*Cyprinodon,* Cyprinodontidae). *Evolution* 43:716–727.

Echelle, A. A., and Kornfield, I., eds. 1984. *Evolution of fish species flocks.* Orono: University of Maine Press.

Echelle, A. A., Echelle, A. F., and Crozier, C. D. 1983. Evolution of an all-female fish, *Menidia clarkhubbsi* (Atherinidae). *Evolution* 37:772–784.

Eddy, F. B. 1971. Blood gas relationships in the rainbow trout, *Salmo gairdneri. J. Fish. Res. Bd. Canada* 24:2267–2281.

Eddy, F. B. 1973. Oxygen dissociation curves of the blood of the tench, *Tinca tinca. J. Exp. Biol.* 58:281–283.

Eddy, F. B. 1982. Osmotic and ionic regulation in captive fish with particular reference to salmonidae. *Comp. Biochem. Physiol.* 73B:125–141.

Eddy, S., Moyle, J. B., and Underhill, J. C. 1963. The fish fauna of the Mississippi River above St. Anthony Falls as related to the effectiveness of this falls as a migration barrier. *J. Minn. Acad. Sci.* 30(2):111–115.

Edwards, D. G., and Cech, J. J., Jr. 1990. Aquatic and aerial metabolism of juvenile monkeyface prickleback, *Cebidichthys violaceus,* an intertidal fish of California. *Comp. Biochem. Physiol.* 96A(1):61–65.

Edwards, R. J., and Contreras-Balderas, S. 1991. Historical changes in the ichthyofauna of the lower Rio Grande (Rio Bravo del Norte) Texas and Mexico. *Southwest. Nat.* 36:201–212.

Egginton, S., and Sidell, B. D. 1989. Thermal acclimation induces adaptive changes in subcellular structure of fish skeletal muscle. *Am. J. Physiol.* 256 *(Regul., Integr., Comp. Physiol.* 25):R1–R9.

Ehrenfeld, D. W. 1976. The conservation of nonresources. *Amer. Sci.* 64:648–656.

Ehrlich, P. R., Talbot, F. H., Russell, B. C., and Anderson, G. R. V. 1977. The behavior of chaetodontid fishes with special reference to Lorenz's "poster colouration" hypothesis. *J. Zool. Soc.* (London) 183:213–228.

Eisler, R. 1965. Erythrocyte counts and hemoglobin content in nine species of marine teleosts. *Chesapeake Sci.* 6:119–120.

Ellis, A. E. 1977. The leucocytes of fish: A review. *J. Fish Biol.* 11:453–491.

Elliott, J. K., and Mariscal, R. N. 2001. Coexistence of nine anemonefish species: Differential host and habitat utilization, size, and recruitment. *Mar. Biol.* 138:23–36.

Elliott, M., and Hemingway, K. L. 2002. *Fishes in estuaries.* Oxford: Blackwell Science.

Elvira, B. 1995. Conservation status of endemic freshwater fish in Spain. *Biol. Cons.* 72:129–136.

Elvira, B., Nicola, G. G., and Almodovar, A. 1996. Pike and red swamp crayfish: A new case on predatory–prey relationship between aliens in central Spain. *J. Fish Biol.* 48:437–446.

Emery, A. R. 1973. Preliminary comparisons of day and night habits of freshwater fish in Ontario lakes. *J. Fish. Res. Bd. Canada* 30:761–774.

Emery, S. H. 1986. Hematological comparisons of endothermic vs. ectothermic elasmobranch fishes. *Copeia* 1986:700–705.

Endler, J. A. 1978. A predator's view of animal color patterns. Pages 319–364 *in* M. K. Hecht, W. C. Steere, and B. Wallace, eds., *Evolutionary biology.* Vol. 11. New York: Plenum Press.

Erdmann, M. V. 2000. New home for "old fourlegs": how the coelacanth was discovered on the other side of the Indian Ocean. *California Wild.* Spring:8–13.

Erdmann, M. V., Caldwell, R. L., and Moosa, M. K. 1998. Indonesian "King of the Sea" discovered. *Nature* 395:335.

Eschmeyer, W. N., ed. 1998. *Catalog of Fishes.* 3 Vols. San Francisco: California Academy of Sciences.

Estes, J. A., Tinker, M. T., Williams, T. M., and Doak, D. F. 1998. Killer whale predation on sea otters linking oceanic and nearshore systems. *Science* 282:473–476.

Etnier, D. A., and Starnes, W. C. 1993. *The fishes of Tennessee.* Knoxville: Univ. of Tennessee Press. 681 pp.

Evans, D. H. 1967a. Sodium, chloride, and water balance of the intertidal teleost *Xiphister atropurpureus.* 1. Regulation of plasma concentration and body water content. *J. Exp. Biol.* 47:513–518.

Evans, D. H. 1967b. Sodium, chloride and water balance in the intertidal teleosts, *Xiphister atropurpureus.* II. The role of the kidney and the gut. *J. Exp. Biol.* 47:519–524.

Evans, D. H. 1975. Ionic exchange mechanisms in fish gills. *Comp. Biochem. Physiol.* 51A:491–495.

Evans, D. H. 1977. Further evidence for Na/NH_4 exchange in marine teleost fish. *J. Exp. Biol.* 70:213–220.

Evans, D. H. 1979. Ionic and osmotic regulation in fish. Pages 305–390 *in* G. M. D. Maloiy, ed., *Osmotic and ionic regulation in animals.* New York: Academic Press.

Evans, D. H. 1982. Mechanisms of acid extrusion by two marine fishes: The teleost, *Opsanus beta,* and the elasmobranch, *Squalus acanthias. J. Exp. Biol.* 97:289–299.

Evans, D. H. 1984. The roles of gill permeability and transport merchanisms in euryhalinity. Pages 239–283 *in* W. S. Hoar and D. J. Randall, eds., *Fish physiology.* Vol. 10, Pt. B. New York: Academic Press.

Evans, D. H. 1990. An emerging role for a cardiac peptide hormone in fish osmoregulation. *Annu. Rev. Physiol.* 52:43–60.

Evans, D. H. 1993. Osmotic and ionic regulation. Pages 315–341 *in* D. H. Evans, ed., *The Physiology of Fishes.* Boca Raton, Fla.: CRC Press.

Evans, D. H. 1998. ed. *The physiology of fishes.* 2nd ed. Boca Raton, Fla.: CRC Press, 519 pp.

Evans, D. H. 2001. Vasoactive receptors in abdominal blood vessels of the dogfish shark, *Squalus acanthias. Physiol. Biochem. Zool.* 74(1):120–126.

Evans, D. H. 2002. Cell signaling and ion transport across the fish gill epithelium. *J. Exp. Zool.* 293:336–347.

Evans, D. H., and Gunderson, M. P. 1999. Characterization of an endothelin ET_B receptor in the gill of the dogfish shark *Squalus acanthias. J. Exp. Biol.* 202:3605–3610.

Evans, D. H., and Takei, Y. 1992. A putative role for natriuretic peptides in fish osmoregulation. *News Physiol. Sci.* 7:15–18.

Evans, D. H., Toop, T., Donald, J., and Forrest, J. N., Jr. 1993. C-type natiruretic peptides are potent dilators of shark vascular smooth muscle. *J. Exp. Zool.* 265:84–87.

Evans, D. L., and Jaso-Friedmann, L. 1982. Nonspecific cytotoxic cells as effectors of immunity in fish. *Annu. Rev. Fish Dis.* 2:109–121.

Everhart, W. H., and Youngs, W. D. 1981. *Principles of fishery science.* 2nd ed. Ithaca, N. Y.: Cornell University Press, 349 pp.

Fänge, R. 1968. The formation of eosinophilic granulocytes in the esophageal lymphomyeloid tissue in the elasmobranchs. *Acta Zool. Stockh.* 49:155–161.

Fänge, R. 1976. Gas exchange in the swimbladder. Pages 189–211 *in* G. M. Hughes, ed., *Respiration of amphibious vertebrates.* London: Academic Press.

Fänge, R. 1982. Exogenous otoliths of elasmobranchs. *J. Mar. Biol. Ass. U.K.* 62:225.

Fänge, R. 1983. Gas exchange in fish swim bladder. *Rev. Physiol. Biochem. Pharmac.* 97:111–158.

Fänge, R. 1984. Lymphomyeloid tissues in fishes. *Vidensk. Meddr. dansk naturh. Foren.* 145:143–152.

Fänge, R., and Mattisson, A. 1981. The lymphomyeloid (hemopoietic) system of the Atlantic nurse shark, *Ginglymostoma cirratum. Biol. Bull.* 160:240–249.

Fänge, R., and Nilsson, S. 1985. The fish spleen: Structure and function. *Experientia* 41:152–158.

Farlinger, S., and Beamish, F. W. H. 1978. Changes in blood chemistry and critical swimming speed of largemouth bass, *Micropterus salmoides,* with physical conditioning. *Trans. Am. Fish. Soc.* 107:523–527.

Farmer, G. J., and Beamish, F. W. H. 1969. Oxygen consumption of *Tilapia nilotica* in relation to swimming speed and salinity. *J. Fish. Res. Bd. Canada* 26:2807–2821.

Farrell, A. P. 1984. A review of cardiac performance in the teleost heart: Intrinsic and humoral regulation. *Can. J. Zool.* 62:523–536.

Farrell, A. P. 1991. From hagfish to tuna: A perspective on cardiac function in fish. *Physiol. Zool.* 64:1137–1164.

Farrell, A. P. 1993. Cardiovascular system. Pages 219–250 *in* D. H. Evans, ed., *The physiology of fishes.* Boca Raton, Fla.: CRC Press.

Farrell, A. P., and Jones, D. R. 1992. The heart. Pages 1–88 *In* W. S. Hoar, D. J. Randall, and A. P. Farrell, eds., *Fish physiology.* Vol. 12, Pt. A. New York: Academic Press.

Fausch, K. D. 1984. Profitable stream positions for salmonids: Relating specific growth rate to net energy gain. *Can. J. Zool.* 62:441–451.

Fausch, K. D. 1988. Competition between native and introduced salmonids in streams: What have we learned? *Can. J. Fish. Aquat. Sci.* 45:2238–2246.

Fausch, K. D., and Northcote, T. G. 1992. Large woody debris and salmonid habitat in a small coastal British Columbia stream. *Can. J. Fish. Aquat. Sci.* 49:682–693.

Fausch, K. D., Power, M. E., and Murakami, M. 2002. Linkages between stream and forest food webs: Shigeru Nakano's legacy for ecology in Japan. *TREE* 17:429–434.

Fautin, D. G. 1991. The anemonefish symbiosis: What is known and what is not. *Symbiosis* 10:23–46.

Fay, R. R., Kendall, J. I., Popper, A. N., and Tester, A. L. 1974. Vibration detection by the macula neglecta of sharks. *Comp. Biochem. Physiol.* 47A:1235–1240.

Febry, R., and Lutz, P. 1987. Energy partitioning in fish: The activity related cost of osmoregulation in a euryhaline cichlid. *J. Exp. Biol.* 128:63–85.

Feldmeth, C. R., and Jenkins, T. M., Jr. 1973. An estimate of the energy expenditure by rainbow trout *(Salmo gairdneri)* in a small mountain stream. *J. Fish. Res. Bd. Canada* 30:1755–1759.

Fernald, R.D. 1993. Vision Pages 161–189 *in* Evans, D. H. ed., *The physiology of fishes.* Boca Raton, Fla.: CRC Press.

Ferry-Graham, L. A. 1999. Mechanics of ventilation in swellsharks, *Cephaloscyllium ventriosum* (Scyliorhinidae). *J. Exp. Biol.* 202:1501–1510.

Findeis, E. K. 1997. Osteology and phylogenetic interrelationships of sturgeons (Acipenseridae). *Envir. Biol. Fish* 48:73–126.

Findley, J. S., and Findley, M. T. 1985. A search for pattern in butterfly fish communities. *Amer. Nat.* 126:800–816.

Fine, M. L., Horn, M. H., and Cox, B. 1987. *Acanthonus armatus*, a deep-sea teleost with a minute brain and large ears. *Proc. R. Soc. Lond. B* 230:257–265.

Fine, M. L., Winn, H. E., and Olla, B. L. 1977. Communication in fishes. Pages 472–518 *in* T. A. Sebeok, ed., *How animals communicate*. Terre Haute: Indiana Univ. Press.

Fines, G. A., Ballantyne, J. S., and Wright, P. A. 2001. Active urea transport and an unusual basolateral membrane composition in the gills of a marine elasmobranch. *Am. J. Physiol. Regulatory Integrative Comp. Physiol.* 280:R16–R24.

Fink, S. V., and Fink, W. L. 1981. Interrelationships of the ostariophysan fishes (Teleostei). *Zool. J. Linn. Soc.* 72:297–353.

Finn, J. P., and Nielson, N. O. 1971. Inflammatory response in rainbow trout. *J. Fish Biol.* 3:463–478.

Fischer, E. A., and Peterson, C. W. 1987. The evolution of sexual patterns in the seabasses. *Bioscience* 37:482–489.

Fisk, A. T., Tittlemier, S. A., Pranschke, J. L., and Norstrom, R. J. 2002. Using anthropogenic contaminants and stable isotopes to assess the feeding ecology of Greenland sharks. *Ecology* 83:2162–2172.

Fitch, J. E., and Lavenberg, R. J. 1968. *Deep-water fishes of California*. Berkeley: Univ. of Cal. Press. 155 pp.

Fitzsimmons, J. M. 1972. A revision of two genera of goodeid fishes (Cyprinodontiformes, Osteichthyes) from the Mexican Plateau. *Copeia* 1972(4):728–756.

Flecker, A. S. 1992. Fish trophic guilds and the structure of a tropical stream: weak direct vs. strong indirect effects. *Ecology* 73:927–940.

Fletcher, G. L., Idler, D. R., Vaisius, A., and Hew, C. L. 1989. Hormonal regulation of antifreeze protein gene expression in winter flounder. *Fish Physiol. Biochem.* 7:387–393.

Flik, G., Fenwick, J. C., and Wendelaar Bonga, S. E. 1989. Calcitropic actions of prolactin in freshwater North American eel (*Anguilla rostrata* LeSueur). *Am. J. Physiol.* 257:74–79.

Foreman, R. E., Gorbman, A., Dodd, J. M., and Olsson, R., eds. 1985. *Evolutionary biology of primitive fishes*. NATO ASI Series, Series A, Life Sciences 103. 463 pp.

Forey, P., and Janvier, P. 1993. Agnathans and the origin of jawed vertebrates. *Nature* 361:129–134.

Forey, P., and Janvier, P. 1994. Evolution of the early vertebrates. *Amer. Scientist* 82:554–565.

Forney, J. L. 1974. Interactions between yellow perch abundance, walleye predation, and survival of alternate prey in Oneida Lake, New York. *Trans. Amer. Fish. Soc.* 103(1):15–24.

Forrest, J. N., Silva, P., Epstein, A., and Epstein, F. H. 1973a. Effect of rectal gland extirpation on plasma sodium in the spiny dogfish. *Bull. Mt. Des. Biol. Lab.* 13:41–42.

Forrest, J. N., Jr., Mackay, W. C., Gallagher, B., and Epstein, F. H. 1973b. Plasma cortisol response to saltwater adaptation in the American eel *Anguilla rostrata*. *Am. J. Physiol.* 224:714–717.

Forsell, J., P. Ekström, I. Novales Flamarique, and B. Holmqvist. 2001. Expression of pineal ultraviolet- and green-like opsins in the pineal organ and retina of teleosts. *J. Exp. Biol.* 204:2517–2525.

Forster, M. E., Axelsson, M., Farrell, A. P., and Nilsson, S. 1991. Cardiac function and circulation in hagfishes. *Can. J. Zool.* 69:1985–1992.

Forster, R. P., and Berglund, F. 1956. Osmotic diuresis and its effect on total electrolyte distribution in plasma and urine of the aglomerular teleost *Lophius americanus*. *J. Gen. Physiol.* 39:349–359.

Foskett, J. K., Logsdon, C. D., Turner, T., Machen, T. E., and Bern, H. A. 1981. Differentiation of the chloride extrusion mechanisms during seawater adaptation of a teleost fish, the cichlid *Sarotherodon mossambicus*. *J. Exp. Biol.* 93:209–224.

Fossi, M. C., Casini, S., Marsili, L., Neri, G., Mori, G., Ancora, S., Moscatelli, A., Ausili, A., and No-tarbertolo-di-Sciara, G. 2002. Biomarkers for endocrine disruptors in three species of Mediterranean larger pelagic fish. *Marz. Environ. Res.* 54:667–671.

Foster, S. A. 1985. Wound healing: A possible role of cleaning stations. *Copeia* 1985:875–880.

Foster, M. S., and Schiel, D. R. 1985. The ecology of giant kelp forests in California: A community profile. *U.S. Fish Wildl. Serv. Biol. Rept.* 85(7.2).

Fourney, R. M., Fletcher, G. L., and Hew, G. L. 1984. Accumulation of winter flounder antifreeze messenger RNA after hypophysectomy. *Gen. Comp. Endocrinol.* 54:392–401.

Fox, D. L. 1978. *Animal biochromes and structural colors.* Berkeley: Univ. of Calif. Press.

Francis, R. C., and Sibley, T. H. 1991. Climate change and fisheries: What are the real issues? *Northw. Env. J.* 7:295–307.

Frantzen, M., Johnsen, H. K., and Mayer, I. 1997. Gonadal development and sex steroids in a female Arctic charr broodstock. *J. Fish Biol.* 51:697–709.

Fratzer, C., Dörr S., and Neumeyer, C. 1994. Wavelength discrimination of the goldfish in the ultraviolet spectral range. *Vision Res.* 34:1515–1520.

Frazier J. G., Fierstine, H.L., and Beavers, S. C. 1994. Impalement of marine turtles (Reptilia, Chelonia, Chelonildae and Dermochekyidae) by billfishes (Osteichthyes, Perciformes, Istiophoridae and Xiphiiidae). *Env. Biol. Fish.* 39:85–96.

Freadman, M. A. 1979. Role partitioning of swimming musculature of striped bass, *Morone saxatilis* Walbaum, and bluefish, *Pomatomus saltatrix* L. *J. Fish Biol.* 15:417–423.

Freadman, M. A. 1981. Swimming energetics of striped bass *(Morone saxatilis)* and bluefish *(Pomatomus saltatrix)*: Hydrodynamic correlates of locomotion and gill ventilation. *J. Exp. Biol.* 90:253–265.

Fricke, H. W. 1970. Ecological and ethological field observations on colonies of the garden eels *Gorgasia sillneri* and *Taenioconger hassi* (in German). *Zeit. Tierpysch.* 27(9):1076–1099.

Fricke, H. 2001. Coelacanths: A human responsibility. *J. Fish Biol.* 59(Suppl. A): 332–338.

Fricke, H. W., and Fricke, S. 1977. Monogamy and sex change by aggressive dominance in coral reef fish. *Nature* 266:830–832.

Fricke, H., and Hissmann, K. 1994. Home range and migrations of the living coelcanth. *Mar. Biol.* 120:171–180.

Fricke, H. W., Hissmann, K., Schauer, J., Reinicke, O., Kasang, L., and Armstrong, M. J. 1991. Habitat and population size of the coelacanth *Latimeria chalumnae* at Grand Comoro. *Env. Biol. Fish* 32:287–300.

Friedland, K. D. 2002. Emerging science and management issues with Atlantic salmon. Pages 117–136 *in* K. D. Lynch, M. L. Jones, and W. W. Taylor, eds. *Sustaining North American salmon: Perspectives across regions and disciplines.* Bethesda, Md.: Amer. Fish. Soc.

Friel, J.P., and Lundberg, J. G. 1996. *Micromyzon akamai*, Gen. et. Sp. Nov., a small and eyeless banjo catfish (Siluriformes: Aspredinidae) from the river channels of the lower Amazon Basin. *Copeia* 1996:61–648.

Frisch, A. J., and Anderson, T. A. 2000. The response of coral trout (*Plectropomus leopardus*) to capture, handling and transport and shallow water stress. *Fish Physiol. Biochem.* 23:23–34.

Froute, E., Magyary, I., Lehoczky, I., and Weiss, S. 2002. mtDNA sequence data supports Asian ancestry and single introduction of common carp into the Danube Basin. *J. Fish Biol.* 61:301–304.

Fryer, G., and Iles, T. D. 1972. *The cichlid fishes of the Great Lakes of Africa.* Edinburgh: Oliver and Boyd. 641 pp.

Fu, C., Mohn, R., and Fanning, L. P. 2001. Why the Atlantic cod (*Gadus morhua*) stock off eastern Nova Scotia has not recovered. *Can. J. Fish. Aquat. Sci.* 58:1613–1623.

Fuentes, J., and Eddy, F. B. 1996. Drinking in freshwater-adapted rainbow trout fry, *Oncorhynchus mykiss* (Walbaum), in response to angiotensin I, angiotensin II, angiotensin-converting enzyme inhibition, and receptor blockade. *Physiol. Zool.* 69:1555–1569.

Fuentes, J., McGeer, J. C., and Eddy, F. B. 1996. Drinking rate in juvenile Atlantic salmon, *Salmo salar* L. fry in response to a nitric oxide donor, sodium nitroprusside and an inhibitor of angiotensin converting enzyme, enalapril. *Fish Physiol. Biochem.* 15:65–69.

Fuiman, L. A., Davies, R. W., and Williams, T. M. 2002. Behavior of midwater fishes under the Antarctic ice; Observations by a predator. *Mari. Biol.* 140:815–822.

Fyhn, U. E. H., Clarke, W. C., and Withler, R. E. 1991. Hemoglobins in smoltifying chinook salmon, *Oncorhynchus tshawytscha*, subjected to photoperiod control. *Aquaculture* 95:359–372.

Galis, F., and Metz, J. A. J. 1998. Why are there so many cichlid species? *TREE* 13:1–2.

Gallagher, M. L., Luczkovich, J. J., and Stellwag, E. J. 2001. Characterization of the ultrastructure of the gastrointestinal tract mucosa, stomach contents and liver enzyme activity of the pinfish during development. *J. Fish Biol.* 58:1704–1713.

Gamperl, A. K., Axelsson, M., and Farrell, A. P. 1995. Effects of swimming and environmental hypoxia on coronary blood flow in rainbow trout. *Am. J. Physiol.* 269:1258–1266.

Gannon, B. J., and Burnstock, G. 1969. Excitatory adrenergic innervation of the fish heart. *Comp. Biochem. Physiol.* 29:765–773.

Gardiner, B. C., Maisey, J. G., and Littlewood, D. T. J. 1996. Interrelationships of basal neopterygians. Pages 117–146 *in* M. L. J. Stiassny, L. R. Parenti, and G. D. Johnson, eds. *Interrelationships of fishes*. San Diego, Calif.: Academic Press.

Garey, W. F. 1962. Cardiac response of fishes in asphyxic environments. *Biol. Bull.* 122:362–368.

Garrison, L. P., and Link, J. S. 2000. Dietary guild structure of the fish community in the Northeast United States continental shelf ecosystem. *Mar. Ecol. Prog. Ser.* 202:231–240.

Gascon, D., and Leggett, W. C. 1977. Distribution, abundance, and resource utilization of littoral fishes in response to a nutrient/production gradient in Lake Memphremagog. *J. Fish. Res. Bd. Canada* 34(8):1105–1117.

Gatz, A. J., Jr. 1979. Community organization in fishes as indicated by morphological features. *Ecology* 60:711–718.

Gee, J. H. 1980. Respiratory pattern and antipredator responses in the central mudminnow, *Umbra limi*, a continuous, facultative, air-breathing fish. *Can. J. Zool.* 58:819–827.

Gee, J. H. 1983. Ecologic implications of buoyancy control in fish. Pages 140–176 *in* P. W. Webb and D. Weihs, eds., *Fish biomechanics*. New York: Praeger.

Gee, J. H., and Northcote, T. G. 1963. Comparative ecology of two sympatric species of dace *(Rhinichthys)* in the Fraser River System, British Columbia. *J. Fish. Res. Bd. Canada* 20(1):105–118.

Gelwick, F. P., and Matthews, W. J. 1990. Temporal and spatial patterns in littoral-zone fish assemblages of a reservoir (Lake Texoma, Oklahoma–Texas, U.S.A.). *Env. Biol. Fish* 27:107–120.

Gerald, J. W. 1971. Sound production during courtship in six species of sunfish (Centrarchidae). *Evolution* 25:75–87.

Gerking, S. D. 1966. Annual growth cycle, growth potential, and growth compensation in the bluegill sunfish in northern Indiana lakes. *J. Fish. Res. Bd. Canada* 23:1923–1956.

Gery, J. 1969. The fresh-water fishes of South America. Pages 828–848 *in* E. Fittkau et al., eds. *Biogeography and Ecology in South America*. Monogr. Biol. 19(2). The Hague: W. Junk.

Gesser, H., Anderson, P., Brams, P., and Sund-Laursen, N. 1982. Isotopic effects of adrenaline on the anoxic or hypercapnic myocardium of rainbow trout and eel. *J. Comp. Physiol.* 147:123–128.

Gibbs, A., and Somero, G. N. 1990. Na^+-K^+-adenosine triphosphatase activities in gills of marine teleost fishes: Changes with depth, size and locomotory activity level. *Mar. Biol.* 106:315–321.

Gibson, R. N. 1993. Intertidal teleosts: Life in a fluctuating environment. Pages 513–536 *in* T. J. Pitcher, ed., *Behaviour of teleost fishes*. 2nd ed. New York: Chapman and Hall.

Gibson, R. N., and Yoshiyama, R. M. 1999. Intertidal fish communities. Pages 264–296 *in* M. H. Horn, K. L. M. Martin, and M. A. Chotkowski, eds., *Intertidal fishes: Life in two worlds*. San Diego, Calif.: Academic Press.

Giles, M. A., and Vanstone, W. E. 1976. Ontogenetic variation in the multiple hemoglobin of coho salmon *(Oncorhynchus kisutch)* and the effect of environmental factors on their expression. *J. Fish. Res. Bd. Canada* 33:1144–1149.

Gill, A. C., and Kemp, J. M. 2002. Widespread Indo-Pacific shore-fish species: A challenge for taxonomists, biogeographers, ecologists, and fishery and conservation managers. *Env. Biol. Fish* 65:165–174.

Gill, A. C., and R. D. Mooi. 2002. Phylogeny and systematics of fishes. Pages 15–41 *in* P. J. B. Hart and J. D. Reynolds, eds. *Handbook of fish biology and fisheries*. Vol. 1—Fish biology. London: Backwell.

Gillen, R. G., and Riggs, A. 1971. The hemoglobins of a fresh-water teleost, *Cichlasoma cyanoguttatum* (Baird and Girard). I. The effects of phosphorylated organic compounds upon the oxygen equilibria. *Comp. Biochem. Physiol.* 38B:585–595.

Gilmore, R. G. 1993. Reproductive biology of lamnoid sharks. *Env. Biol. Fish* 38:95–116.

Glebe, B. D., and Leggett, W. C. 1981. Latitudinal differences in energy allocation and use during the freshwater migrations of American shad *(Alosa sapidissima)* and their life history consequences. *Can. J. Fish. Aquat. Sci.* 38:806–820.

Goddard, S. V., Kao, M. H., and Fletcher, G. L. 1999. Population differences in antifreeze production cycles of juvenile Atlantic cod (*Gadus morhua*) reflect adaptations to overwintering environment. *Can. J. Fish. Aquat. Sci.* 56:1991–1999.

Godin, J. J., ed. 1997. *Behavioural ecology of teleost fishes.* Oxford: Oxford Univ. Press, 384 pp.

Goldschmidt, T., Witte, F., and Wanink, J. 1993. Cascading effects of the introduced Nile perch on the detritivorous/phytoplanktivorous species in the sublittoral areas of Lake Victoria. *Cons. Biol.* 7:686–700.

Goldstein, L., Janssen, P. A., and Forster, R. P. 1967. Lungfish *Neoceratodus forsteri:* Activities of ornithine-urea cycle and enzymes. *Science* 157:316–317.

Gong, Z., Fletcher, G. L., and Hew, C. L. 1992. Tissue distribution of fish antifreeze protein messenger RNAs. *Can. J. Zool.* 70:810–814.

Gong, Z., King, M. J., Fletcher, G. L., and Hew, C. L. 1995. The antifreeze protein genes of winter flounder are differentially regulated in liver and non-liver tissues. *Biochem. Biophys. Res. Comm.* 206:387–392.

Gonzalez, R. J., and McDonald, D. G. 1994. The relationship between oxygen uptake and ion loss in fish from diverse habitats. *J. Exp. Biol.* 190:95–108.

Gonzalez, R. J., Wilson, R. W., Wood, C. M., Patrick, M. L., and Val, A. L. 2002. Diverse strategies for ion regulation in fish collected from the ion-poor, acidic Rio Negro. *Physiol. Biochem. Zool.* 75(1):37–47.

Gonzalez, R. J., Wood, C. M., Wilson, R. W., Patrick, M. L., Bergman, H. L., Narahara, A., and Val, A. L. 1998. Effect of water pH and calcium concentration on ion balance in fish of the Rio Negro, Amazon. *Physiol. Zool.* 71:15–22.

Goode, G. B., and Bean, T. M. 1895. *Oceanic ichthyology. Smithson. Inst. Spec. Bull.* 2. 540 pp.

Goolish, E. M., and Okutake, K. 1999. Lack of gas bladder inflation by the larvae of zebrafish in the absence of an air–water interface. *J. Fish Biol.* 55:1054–1063.

Gordon, B. L. 1977. *The secret lives of fishes.* New York: Grosset & Dunlap. 305 pp.

Gorlick, D. L., Atkins, P. O., and Losey, G. S., Jr. 1978. Cleaning stations as water holes, garbage dumps, and sites for the evolution of reciprocal altruism. *Amer. Nat.* 112:341–353.

Gorr, T., and Kleinschmidt, T. 1993. Evolutionary relationships of the coelacanth. *Amer. Sci.* 81:72–82.

Goujet, D. 2001. Placoderms and basal gnathostome apomorphies. *In* P. E. Ahlberg, ed. *Major events in vertebrate evolution: paleontology, phylogeny, genetics, and development*. London: Taylor and Francis.

Goulding, M. 1980. *The fishes and the forest.* Berkeley: Univ. of Calif. Press.

Graham, J. B. 1983. Heat transfer. Pages 248–279 *in* P. W. Webb and D. Weihs, eds., *Fish biomechanics.* New York: Praeger Publ.

Graham, J. B. 1997. *Air breathing fishes: Evolution, diversity, and adaptation.* San Diego, Calif.: Academic Press. 299 pp.

Grande, L., and Bemis, W. E. 1996. Interrelationships of Acipenseriformes, with comments on "Chondrostei." Pages 85–116 *in* M. L. J. Stiassny, L. R. Parenti, and G. D. Johnson, eds., *Interrelationships of fishes.* San Diego, Calif.: Academic Press.

Grant, W. S., and Leslie, R. W. 2001. Inter-ocean dispersal is an important mechanism in the zoogeography of hakes (Pisces: *Merluccius*). *J. Biogeog.* 28:699–721.

Grassle, J. F., Sanders, H. L., Hessler, R. R., Rowe, G. T., and McLellan, T. 1975. Pattern and zonation; A study of the bathylmegafauna using the research submersible *Alvin. Deep-sea Res.* 22:457–481.

Grau, E. G., Dickhoff, W. W., Nishioka, R. S., Bern, H. A., and Folmar, L. C. 1981. Lunar phasing of the thyroxine surge preparatory to seaward migration of salmonoid fish. *Science* 211:607–609.

Grayton, B. D., and Beamish, F. W. H. 1977. Effects of feeding frequency on food intake, growth and body composition of rainbow trout *(Salmo gairdneri). Aquaculture* 11:159–172.

Greaney, G. S., and Powers, D. A. 1978. Allosteric modifiers of fish hemoglobin: In vitro and in vivo studies of the effect of ambient oxygen and pH on erythrocyte ATP concentrations. *J. Exp. Zool.* 203:339–349.

Greenwood, P. H. 1976. Fish fauna of the Nile. Pages 127–141 *in* J. Rzoska, ed., *The Nile: Biology of an ancient river.* The Hague: W. Junk.

Greenwood, P. H., Miles, R. S., and Patterson, C., eds. 1973. *Interrelationships of fishes.* New York: Academic Press. 536 pp.

Greenwood, P. H., Rosen, D. E., Weitzman, S. H., and Myers, G. S. 1966. Phyletic studies on teleostean fishes, with a provisional classification of living forms. *Bull. Amer. Mus. Nat. Hist.* 131(4):341–455.

Grigg, G. C. 1969. Temperature-induced changes in the oxygen equilibrium curve of the blood of the brown bullhead, *Ictalurus nebulosus. Comp. Biochem. Physiol.* 28:1203–1223.

Grigg, G. C. 1974. Respiratory function of blood of fishes. Pages 331–368 *in* N. Florkin and B. T. Scheer, eds., *Chemical zoology.* Vol. 8. New York: Academic Press.

Grizzle, J. M., Mauldin II, A. C., Young, D., and Henderson, E. 1985. Survival of juvenile striped bass (*Morone saxatilis*) and *Morone* hybrid bass (*Morone chrysops* × *Morone saxatilis*) increased by addition of calcium to soft water. *Aquaculture* 46:167–171.

Groot, C., and Margolis, L. eds., 1991. *Pacific salmon life histories.* Vancouver: Univ. of B.C. Press. 564 pp.

Groot, C., Quinn, T. P., and Hara, T. J. 1986. Responses of migrating adult sockeye salmon (*Onchorhynchus nerka*) to population-specific odours. *Can. J. Zool.* 64:926–932.

Gross, M. R. 1982. Sneakers, satellites, and parentals: Polymorphic mating strategies in North American sunfishes. *Z. Tierpsychol.* 60:1–26.

Gross, M. R. 1984. Sunfish, salmon, and the evolution of alternative reproductive strategies and tactics in fishes. Pages 55–75 *in* G. W. Potts and R. J. Wootton, eds. *Fish reproduction: Strategies and tactics.* London: Academic Press.

Gross, M. R., Coleman, R. M., and McDowall, R. M. 1988. Aquatic productivity and the evolution of diadromous fish migration. *Science* 239:1291–1293.

Gross, M. R., and Sargent, R. C. 1985. The evolution of male and female parental care in fishes. *Amer. Zool.* 25:807–822.

Grossman, G. D. 1982. Dynamics and organization of a rocky intertidal fish assemblage: The persistence and resilience of taxocene structure. *Amer. Nat.* 119:611–637.

Grossman, G., Coffin, R., and Moyle, P. 1980. Feeding ecology of the bay goby (Pisces, Gobiidae). Effect of behavioral, ontogenetic, and temporal variation on diet. *J. Exp. Mar. Biol. Ecol.* 44:47–59.

Grossman, G. D., Moyle, P. B., and Whitaker, J. O., Jr. 1982. Stochasticity in structural and functional characteristics of an Indiana stream fish assemblage: A test of community theory. *Amer. Nat.* 120:423–454.

Grossman, G. D., Ratajczak, R. E., Jr., Crawford, M., and Freeman, M. C. 1998. Assemblage organization in stream fishes: Effects of environmental variation and interspecific interactions. *Ecol. Monogr.* 68:395–420.

Grothe, D. R., and Eaton, J. W. 1975. Chlorine-induced mortality in fish. *Trans. Am. Fish. Soc.* 104:800–802.

Grove, J. S., and Lavenberg, R. J. 1997. *The fishes of the Galapagos Islands.* Stanford, Calif.: Stanford Univ. Press. 863 pp.

Gruber, S. H. 1977. The visual system of sharks: Adaptations and capability. *Amer. Zool.* 17:453–470.

Gruber, S. H., and Cohen, J. L. 1978. Visual system of the elasmobranchs: State of the art 1960–1975. Pages 391–417 *in* E. S. Hodgson and R. F. Mathewson, eds., *Sensory biology of sharks, skates, and rays.* Washington, D.C.: Off. Naval Res. Publ.

Gruber, S. H., and Compagno, L. J. V. 1981. Taxonomic status and biology of the bigeye thresher, *Alopias superciliosus. NOAA Fish. Bull.* 79:617–640.

Gruber, S. H., Gulley, R. L., and Brandon, J. 1975a. Duplex retina in seven elasmobranch species. *Bull. Mar. Sci.* 25:353–358.

Gruber, S. H., Hamasaki, D. I., and Davis, B. L. 1975b. Window to the epiphysis in sharks. *Copeia* 1975(2):378–380.

Gruber, S. J., and Dickson, K. A. 1997. Effects of endurance training in the leopard shark, *Triakis semifasciata. Physiol. Zool.* 70:481–492.

Guegan, J.-F., Labert, A., Leveque, C., Combes, C., and Euzet, L. 1992. Can host body size explain the parasite species richness in tropical freshwater fishes? *Oecologia* 90:197–204.

Guppy, M., and Hochachka, P. W. 1978. Skipjack tuna white muscle: A blueprint for the integration of aerobic and anaerobic carbohydrate metabolism. Pages 175–181 *in* G. D. Sharp and A. E. Dizon, eds., *The physiological ecology of tunas.* New York: Academic Press.

Gutt, J., and Ekau, W. 1996. Habitat partitioning of dominant high Antarctic demersal fish in the Wedell Sea and Lazarev Sea. *J. Exp. Mar. Biol. Ecol.* 206:25–37.

Haas, R. 1976a. Sexual selection in *Nothobranchius guentheri* (Pisces, Cyprinodontidae). *Evolution* 20:614–622.

Haas, R. 1976b. Behavioral biology of the annual killifish, *Nothobranchius guentheri. Copeia* 1976(1):80–91.

Haedrich, R. L. 1983. Estuarine fishes. Pages 183–207 *in* B. H. Ketchum, ed., *Estuaries and enclosed seas.* New York: Elsevier.

Haedrich, R. L. 1996. Deep water fishes: Evolution and adaptation in the earth's largest living spaces. *J. Fish Biol.* 49(Suppl. A):4–53.

Haedrich, R. L., Rowe, G. T., and Polloni, P. T. 1975. Zonation and faunal composition of epibenthic populations on the continental slope south of New England. *J. Mar. Res.* 33(3):191–212.

Haglund, T. R., Buth, D. G., and Lawson, R. 1992. Allozyme variation and phylogenetic relationships of Asian, North American, and European populations of the ninespine stickleback, *Pungitius pungitius.* Pages 438–452 *in* R. L. Mayden, ed., *Systematics, historical ecology, and North American freshwater fishes.* Stanford, Calif.: Stanford Univ. Press.

Hahn, G. 1960. Ferntastsinn and Stomungssinn beim augenlosen Hohlenfisch *Anoptichthys jordani* Hubbs and Innes im Vergleich zu einigen Teleosteern. *Naturwissenschaften.* 47:611.

Haines, T. A. 1980. Seasonal patterns of muscle RNA/DNA ratio and growth in black crappie, *Pomoxis nigromaculatus. Env. Biol. Fish* 5:67–70.

Hall, B. K. 2001. John Samuel Budgett (1872–1904): In pursuit of *Polypterus. Bioscience* 51:399–407.

Hall, F. G., and McCutcheon, F. H. 1938. The affinity of hemoglobin for oxygen in marine fishes. *J. Cell. Comp. Physiol.* 11:205–212.

Halstead, B. W. 1967. *Poisonous and venomous marine animals of the world: vertebrates.* Washington, D.C: U.S. Gov. Printing Office. 1500 pp.

Halstead, L. B. 1968. *Patterns of vertebrate evolution.* San Francisco: W. H. Freeman. 209 pp.

Halver, J. E. 1957. Nutrition of salmonoid fishes. IV. An amino acid test diet for chinook salmon. *J. Nutr.* 62:245–254.

Halver, J. E. 1972. The vitamins. Pages 29–103 *in* J. E. Halver, ed., *Fish nutrition.* New York: Academic Press.

Halver, J. E. 1976. Formulating practical diets for fish. *J. Fish. Res. Bd. Canada* 33:1032–1039.

Halver, J. E., and Shanks, W. E. 1960. Nutrition and salmonoid fishes. VIII. Indispensable amino acids for sockeye salmon. *J. Nutr.* 72:340–346.

Hamasaki, D. I., and Streck, P. 1971. Properties of the epiphysis cerebri of the small spotted dogfish shark, *Scyliorhinus caniculus* L. *Vision Res.* 11:189–198.

Hamlett, W. C. 1993. Ontogeny of umbilical cord and placenta in the Atlantic sharpnose shark, *Rhizoprion terraenovae. Env. Biol. Fish.* 38:253–267.

Hamlett, W. C., ed. 1999. *The biology of elasmobranch fishes.* Baltimore, Md.: Johns Hopkins University Press.

Hamre, J. 1994. Biodiversity and exploitation of the main fish stocks in the Norwegian Barents ecosystem. *Biodiv. Cons.* 3:473–492.

Handy, R. D., and Poxton, M. G. 1993. Nitrogen pollution in mariculture: Toxicity and excretion of nitrogenous compounds by marine fish. *Rev. Fish Biol. Fisheries* 3:205–241.

Hankin, D. G. 1978. New fluorescent fish scale marker. *Prog. Fish. Cult.* 40:163–164.

Hansen, K., and Herring, P. J. 1977. Dual bioluminescent systems in the anglerfish genus *Linophryne* (Pisces, Ceratoidea). *J. Zool.* (London) 182:103–124.

Hara, T. J. 1986. Role of olfaction in fish behavior. Pages 152–176 *in* T. J. Pritcher, ed., *The behavior of teleost fishes.* Baltimore, Md.: Johns Hopkins Univ. Press.

Hara, T. J. 1993. Chemoreception. Pages 191–218 *in* D. H. Evans, ed., *The physiology of fishes.* Boca Raton, Fla.: CRC Press.

Hara, T. J., Macdonald, S., Evans, R. E., Marui, T., Arai. S. 1984. Morpholine, bile acids, and skin mucus as possible chemical cues in salmonid homing: Electrophysiological re-evaluation. Pages 363–378 *in* J. D. McCleave, G. P. Arnold, J. J. Dodson, and W. H. Neill, eds., *Mechanisms of migration in fishes.* New York: Plenum Press.

Harbison, G. R. 1987. Encounters with a swordfish (*Xiphias gladius*) and sharptail mola (*Masturus lanceolatus*) at depths greater than 600 meters. *Copeia* 1987:511–513.

Harden-Jones, F. R. 1968. *Fish migration.* London: E. Arnold Pub. 325 pp.

Hardisty, M. W., and Potter, I. C., eds. 1971. *The biology of lampreys.* Vols. 1 and 2. New York: Academic Press.

Hare, S. R., Mantua, N. J., and Francis, R. C. 1999. Inverse production regimes: Alaska and West Coast Pacific salmon. *Fisheries* (Bethesda) 24(1):6–14.

Harrington, R. W. 1961. Oviparous hemaphroditic fish with internal self-fertilization. *Science* 134:1749–1750.

Harris, A. J. 1965. Eye movements of the dogfish, *Squalus acanthias* 1. *J. Exp. Biol.* 43:107–130.

Harris, L. G., Ebeling, A. W., Laur, D. R., and Rowley, R. J. 1984. Community recovery after storm damage: A case of facilitation in primary succession. *Science* 224:1336–1338.

Hart, J. L. 1973. *Pacific fishes of Canada.* Fish. Res. Bd. Canada Bull. 180. 740 pp.

Hartman, W. L., and Burgner, R. L. 1972. Limnology and fish ecology of sockeye salmon nursery lakes of the world. *J. Fish. Res. Bd. Canada* 29(6):699–715.

Harvey, B. C. 1991. Interactions among stream fishes: Predator induced habitat shifts and larval survival. *Oecologia* 87:29–36.

Harvey, B. C., and Stewart, A. J. 1991. Fish size and habitat depth relationships in headwater streams. *Oecologia* 87:336–342.

Hasler, A. D. 1966. *Underwater guideposts.* Madison: Univ of Wisc. Press. 155 pp.

Hasler, A. D., Horrall, R. N., Wisby, W. J., and Braemer, W. 1958. Sun orientation and homing in fishes. *Limnol. Oceanogr.* 3(4):353–361.

Hasler, A. D., and Scholtz, A. T. 1978. Olfactory imprinting in coho salmon (*Oncorhynchus kisutch*). Pages 356–369 *in* K. Schmid-Koenig and W. T. Keeton, eds., *Animal migration, navigation, and homing.* Berlin: Springer-Verlag.

Hasler, A. D., and Scholtz, A. T. 1983. Olfactory imprinting and homing in salmon: Investigations into the mechanism of the imprinting process. *Zoophysiology.* Vol. 14. Berlin: Springer-Verlag. 134 pp.

Hatfield, T. 1997. Genetic divergence in adaptive characters between sympatric species of stickleback. *Amer. Nat.* 149:1009–1029.

Hawkins, A. D. 1993. Underwater sound and fish behavior. Pages 129–170 *in* T. J. Pitcher, ed., *The behaviour of teleost fishes.* 2nd ed. New York: Chapman and Hall.

Hawryshyn, C. W. 1998. Vision. Pages 345–374 *in* D. H. Evans, ed., *The physiology of fishes.* 2nd ed. Boca Raton, Fla.: CRC Press.

Hawryshyn, C. W., and McFarland, W. N. 1987. Cone photoreceptor mechanisms and the detection of polarized light in fish. *J. Comp. Physiol. A* 160:459–465.

Hayden, J. B., Cech, J. J., Jr., and Bridges, D. W. 1975. Blood oxygen dissociation characteristics of the winter flounder, *Pseudopleuronectes americanus* (Walbaum). *J. Fish. Res. Bd. Canada* 32:1539–1544.

Hayward, R. S., Noltie, D. B., and Wang, N. 1997. Use of compensatory growth to double hybrid sunfish growth rates. *Trans. Am. Fish. Soc.* 126:316–322.

Hazel, J. R. 1993. Thermal biology. Pages 427–467 *in* D. H. Evans, ed., *The physiology of fishes.* Boca Raton, Fla.: CRC Press.

Hecht, E., and Zajac, A. 1974. *Optics.* Reading, Mass.: Addison-Wesley. 565 pp.

Heckmann, R. A., Deacon, J. E., and Greger, P. D. 1986. Parasites of the woundfin minnow, *Plagopterus argentissimus,* and other endemic fishes from the Virgin River, Utah. *Great Basin Nat.* 46:662–676.

Hedrick, R. P. 1998. Relationships of the host, pathogen, and environment: Implications for diseases of cultured and wild fish populations. *J. Aquat. Anim. Health* 10:107–111.

Heiligenberg, W. 1993. Electrosensation. Pages 137–160 *in* D. H. Evans, ed., *The physiology of fishes.* Boca Raton, Fla.: CRC Press.

Heisler, N. 1988. Acid–base regulation. Pages 215–252 *in* T. J. Shuttleworth, ed., *Physiology of elasmobranch fishes.* Berlin: Springer-Verlag.

Heisler, N. 1993. Acid–base regulation. Pages 343–378 *in* D. H. Evans, ed., *The physiology of fishes.* Boca Raton, Fla.: CRC Press.

Heist, E. J. 1999. A review of population genetics in sharks. *Amer. Fish. Soc. Symposium.* 23:161–168.

Helfield, J. M., and Naiman, R. J. 1992. Effects of salmon-derived nitrogen on riparian forest growth and implications of stream productivity. *Ecology* 82:2403–2409.

Helfman, G. S. 1981a. The advantage to fishes of hovering in shade. *Copeia* 1981:392–399.

Helfman, G. S. 1981b. Twilight activities and temporal structure in a freshwater fish community. *Can. J. Fish. Aq. Sci.* 38:1405–1420.

Helfman, G. S., and Clark, J. B. 1986. Rotational feeding: Overcoming gape-limited foraging in anguillid eels. *Copeia* 1986:679–685.

Henderson, P. A. 1990. Fish of the Amazonian Igaop: Stability and conservation in a high diversity–low biomass system. *J. Fish Biol.* 37(Suppl. A):61–66.

Herald, E. S. 1959. From pipefish to seahorse—A study of phylogenetic relationships. *Proc. Calif. Acad. Sci.* 26:465–473.

Herald, E. S. 1961. *Living fishes of the world.* New York: Doubleday. 304 pp.

Herbold, B. 1984. Structure of an Indiana stream fish assemblage: choosing an appropriate model. *Amer. Nat.* 124:561–572.

Herring, P. J. 2002. *The biology of the deep ocean.* Oxford: Oxford Univ. Press. 314 pp.

Hershey, J. 1987. *Blues.* N.Y.: Alfred Knopf. 205 pp.

Hey, J. 2001. The mind of the species problem. *TREE* 16:326–329.

Higgs, D. A., Fagerlund, U.H.M., McBride, J. R., and Eales, J. G. 1979. Influence of orally administered L-thyroxine or 3,5,3′-triiodo-L-thyronine on growth, food consumption, and food conversion of underyearling coho salmon (*Oncorhynchus kisutch*). *Can. J. Zool.* 57:1974–1979.

Hilborn, R., Maguire, J., Parma, A. M., and Rosenberg, A. A. 2001. The precautionary approach and risk management: Can they increase the probability of successes in fisheries management? *Can. J. Fish. Aquat. Sci.* 58:99–107.

Hildebrand, M. 1974. *Analysis of vertebrate structure.* New York: John Wiley.

Hill, J., and Grossman, G. D. 1993. An energetic model of microhabitat use for rainbow trout and rosyside dace. *Ecology* 74:685–698.

Hirano, T. 1986. The spectrum of prolactin action in teleosts. Pages 53–74 *in* C. L. Ralph, ed., *Comparative endocrinology: Developments and directions.* New York: Alan R. Liss, Inc.

Hixon, M. A. 1980. Competitive interactions between California reef fishes of the genus *Embiotoca. Ecology* 61:918–931.

Hixon, M. A. 1991. Predation as a process structuring coral reef fish communities. Pages 475–508 *in* P. F. Sale, ed., *The ecology of fishes on coral reefs.* San Diego, Calif.: Academic Press.

Hixon, M. A. 1998. Population dynamics of coral-reef fishes: Controversial concepts and hypotheses. *Aust. J. Ecol.* 23:192–2001.

Hixon, M. A., and Beets, J. P. 1993. Predation, prey refuges, and the structure of coral-reef fish assemblages. *Ecol. Monogr.* 63:77–101.

Hoagland, T. M., Weaver, L. W., Jr., Conlon, J. M., Wang, Y., and Olson, K. R. 2000. Effects of endothelin-1 and homologous trout endothelin on cardiovascular function in rainbow trout. *Am. J. Physiol. Regulatory Integrative Comp. Physiol.* 278:R460–R468.

Hoar, W. S. 1976. Smolt transformation: Evolution, behavior, and physiology. *J. Fish. Res. Bd. Canada* 33(6):1234–1252.

Hoar, W. S., and Randall, D. J., eds. 1984. *Fish physiology.* Vol. 10, Pt. A. New York: Academic Press. 456 pp.

Hoar, W. S., Randall, D. J., and Farrell, A. P. 1992a. *Fish physiology.* Vol. XII, Part A. The Cardiovascular System. 340 pp. San Diego, CA

Hoar, W. S., Randall, D. J., and Farrell, A. P. 1992b. *Fish physiology.* Vol. XII, Part B. The Cardiovascular System. 474 pp. San Diego, CA

Hoare, D. J., Paxton, C. D., and Godin, G. J. D. 2000. Social organization of free-ranging fish schools. *Oikos* 89:546–554.

Hobson, E. S. 1974. Feeding relationships of teleostean fishes on coral reefs in Kona, Hawaii. *NOAA Fish. Bull.* 72(4):915–1031.

Hobson, E. S. 1975. Feeding patterns among tropical reef fishes. *Amer. Sci.* 63:382–392.

Hobson, E. S. 1994. Ecological relations in the evolution of acanthopterygian fishes in warm-temperate communities of the northeastern Pacific. *Env. Biol. Fish.* 40:49–90.

Hochachka, P. W., and Somero, G. N. 1984. *Biochemical adaptation.* Princeton, N.J.: Princeton Univ. Press. 537 pp.

Hocutt, C. H., Jenkins, R. E., and Stauffer, J. R., Jr. 1986. Zoogeography of the fishes of the central Appalachians and central Atlantic coastal plain. Pages 161–211 *in* C. H. Hocutt and E. D. Wiley, eds., *The zoogeography of North American freshwater fishes.* New York: Wiley.

Hocutt, C. H., and Stauffer, J. R. 1975. Influence of gradient on the distribution of fishes in Conowingo Creek, Maryland and Pennsylvania. *Ches. Sci.* 16(1):143–147.

Hocutt, C. H., and Wiley, E. O., eds. 1986. *The zoogeography of North American freshwater fishes.* New York: Wiley.

Hoese. H. D., and Moore, R. H. 1998. *Fishes of the Gulf of Mexico.* 2nd Ed. College Station, TX: Texas A & M Press. 440 pp.

Hogan, Z. 2003. *Ecology, genetics, and conservation of migratory catfish (Pangastidae) in the Mekong River.* Ph.D. dissertation. Univ. of Calif. at Davis.

Hogman, W. J. 1968. Annulus formation of scales of four species of coregonids reared under artificial conditions. *J. Fish. Res. Bd. Canada* 25:2111–2112.

Holeton, G. F. 1970. Oxygen uptake and circulation by hemoglobinless Antarctic fish (*Chaenocephalus aceratus* Lonnberg) compared with three red-blooded Antarctic fish. *Comp. Biochem. Physiol.* 34:457–471.

Holland, K. N., and Sibert, J. R. 1994. Physiological thermoregulation in bigeye tuna, *Thunnus obesus. Env. Biol. Fishes* 40:319–327.

Holmgren, S. 1977. Regulation of the heart of a teleost, *Gadus morhua,* by autonomic nerves and circulating catecholamines. *Acta Physiol. Scand.* 99:62–74.

Holmgren, S., Grove, D. J., and Fletcher, D. J. 1983. Digestion and the control of gastrointestinal motility. Pages 23–40 *in* J. C. Rankin, T. J. Pitcher, and R. Duggan, eds., *Control processes in fish physiology.* New York: Wiley-Interscience.

Holmlund, C. M., and Hammer, M. 1999. Ecosystem services generated by fish populations. *Ecol. Econ.* 29:253–268.

Honma, Y. 1998. Asian hagfishes and their fisheries biology. Pages 45–56 *in* J. M. Jorgenson, J. P. Lomholt, R. E. Weber, and H. Malte, eds., *The biology of hagfishes.* London: Chapman and Hall.

Hopkins, C. D. 1974. Electric communication in fish. *Amer. Sci.* 62(4):426–437.

Hopkins, T. E., and Cech, J. J., Jr. 1994. Effect of temperature on oxygen consumption of the bat ray, *Myliobatis californica* (Chondrichthyes, Myliobatidae). *Copeia* 1994:529–532.

Hopkins, T. E., Eldridge, M. B., and Cech, J. J., Jr. 1995. Metabolic costs of viviparity in yellowtail rockfish, *Sebastes flavidus. Env. Biol. Fish.* 43:77–84.

Hopkins, T. E., and Larson, R. J. 1990. Gastric evacuation of three food types in the black and yellow rockfish *Sebastes chrysomelas* (Jordan and Gilbert). *J. Fish Biol.* 36:673–681.

Horn, M. H. 1970. The swimbladder as a juvenile organ in stomateiod fishes. *Breviora* 359:1–9.

Horn, M. H. 1972. The amount of space available for marine and freshwater fishes. *NOAA Fish. Bull.* 70(4):1295–1297.

Horn, M. H. 1984. Stromateoidei: Development and relationships. Pages 620–628 *in* G. Moser et al., eds., *Ontogeny and systematics of fishes.* ASIH Spec. Publ. 1.

Horn, M. H. 1989. Biology of marine herbivorous fishes. *Oceanogr. Mar. Biol. Ann. Rev.* 27:167–272.

Horn, M. H., Grimes, P. W., Pfleger, C. F., and McClanahan, L. L. 1978. Buoyancy function of the enlarged fluid-filled cranium in the deep-sea ophidiid fish *Acanthonus armatus. Mar. Biol.* 46:335–339.

Horn, M. H., Martin, K. L. M., and Chotkowski, M. A., eds., 1999. *Intertidal fishes: Life in two worlds.* San Diego, Calif.: Academic Press. 399 pp.

Horn, M. H., and Riegle, K. C. 1981. Evaporative water loss and intertidal vertical distribution in relation to body size and morphology of stichaeoid fishes from California. *J. Exp. Mar. Biol. Ecol.* 50:273–288.

Horn, M. H., and Riggs, C. D. 1973. Effects of temperature and light on the rate of air breathing of the bowfin, *Amia calva. Copeia* 1973(4):653–657.

Houde, E. D., and Rutherford, E. S. 1993. Recent trends in estuarine fisheries: Predictions of fish production and yield. *Estuaries* 16:161–176.

Houlihan, D. F., and Laurent, P. 1987. Effects of exercise training on the performance, growth, and protein turnover of rainbow trout (*Salmo gairdneri*). *Can. J. Fish. Aquat. Sci.* 44:1614–1621.

Houssay, S. F. 1912. *Forme, puissance, et stabilité des poissons.* Paris: Herman. 372 pp.

Houston, A. H., and Cyr, D. 1974. Thermoacclimatory variation in the hemoglobin systems of goldfish (*Carassius auratus*) and rainbow trout (*Salmo gairdneri*). *J. Exp. Biol.* 61:455–461.

Houston, A. H., and DeWilde, M. A. 1968. Thermacclimatory variations in the hematology of the common carp, *Cyprinus carpio. J. Exp. Biol.* 49:71–81.

Houston, A. H., and Gingras-Bedard, J. H. 1994. Variable versus constant temperature acclimation regimes: Effects on hemoglobin isomorph profile in goldfish, *Carassius auratus. Fish Phys. Biochem.* 13:445–450.

Houston, A. H., and Rupert, R. 1976. Immediate response of the hemoglobin system of the goldfish, *Carassius auratus,* to temperature change. *Can. J. Zool.* 54:1737–1741.

Howland, K. L., Tonn, W. M., Babaluk, J. A., and Tallman, R. F. 2001. Identification of freshwater and anadromous inconnu in the Mackenzie River system by analysis of otolith strontium. *Trans. Amer. Fish. Soc.* 130:725–741.

Hubbs, C. L. 1955. Hybridization between fish species in nature. *Syst. Zool.* 4:1–20.

Hubbs, C. L. 1964. History of ichthyology in the United States after 1850. *Copeia* 1964(1):42–60.

Hubbs, C. L. 1974. Review of: *Marine Zoogeography* by J. C. Briggs. *Copeia* 1974:1002–1005.

Hubbs, C. L., and Hubbs, L. C. 1932. Apparent parthenogenesis in nature, in a form of fish of hybrid origin. *Science* 76:628–630.

Hubbs, C. L., and Lagler, K. F. 1964. *Fishes of the Great Lakes region.* Ann Arbor: Univ. of Mich. Press. 213 pp.

Hubbs, C. L., and Miller, R. R. 1942. Mass hybridization between two genera of cyprinid fishes in the Mohave Desert, California. *Papers Mich. Acad. Sci. Arts, Letters* 28:343–378.

Hubbs, C. L., and Potter, I. C. 1971. Distribution, phylogeny, and taxonomy. Pages 1–65 *in* M. W. Hardisty and I. C. Potter, eds., *The biology of lampreys.* Vol. 1. New York: Academic Press.

Huckabee, J. W., Goodyear, C. P., and Jones, R. D. 1975. Acid rock in the Great Smokies: Unanticipated impact on aquatic biota of road construction in regions of sulfide mineralization. *Trans. Amer. Fish. Soc.* 104(4):677–684.

Huet, M. 1959. Profiles and biology of Western European streams as related to fish management. *Trans. Amer. Fish. Soc.* 88:155–163.

Hughes, G. M. 1963. *Comparative physiology of vertebrate respiration.* Cambridge, Mass.: Harvard Univ. Press. 146 pp.

Hughes, G. M., and Grimstone, A. V. 1965. The fine structure of the secondary lamellae of the gills of *Gadus pollachius. Quart. J. Micro. Sci.* 106:343–353.

Hughes, T. P. 1994. Catastrophes, phase shifts, and large-scale degradation of a Caribbean coral reef. *Science* 265:1547–1551.

Humphries, P., Potter, I. C., and Loneragan, N. R. 1992. The fish community in the shallows of a temperate Australian estuary: Relationships with the aquatic macrophyte *Ruppia megacarpa* and environmental variables. *Est. Coast. Shelf Sci.* 34:325–346.

Hunter, J. R., and Mitchell, C. T. 1966. Association of fishes with flotsam in the offshore waters of Central America. *NOAA Fish. Bull.* 66(1):13–30.

Huntingford, F. A., and Torricelli, P., eds. 1993. *Behavioural ecology of fishes*. Switzerland: Harwood Acad. Publ.

Huse, G. 1998. Sex-specific life history strategies in capelin (*Mallotus villosus*)? *Can. J. Fish. Aquat. Sci.* 55:631–638.

Hutchings, J. A., and Myers, R. A. 1994. What can be learned from the collapse of a renewable resource: Atlantic cod, *Gadus morhua*, of Newfoundland and Labrador. *Can. J. Fish. Aquat. Sci.* 51:2126–2146.

Hutchings, J. A., and Myers, R. A. 1995. The biological collapse of Atlantic cod off New Foundland and Labrador: An exploration of historical changes, in exploitation, harvesting technology, and management. Pages 39–92 *in* R. Arnason and L. Felt, eds., *The North Atlantic fisheries: Successes failures, and challenges*. Charlottetown, PEI: The Institute of Island Studies.

Hutchinson, G. E. 1957. *A treatise on limnology*. Vol. 1. Geography, physics, and chemistry. New York: Wiley, 1015 pp.

Hyatt, K. D., and Stockner, J. G. 1985. Responses of sockeye salmon (*Oncorhynchus nerka*) to fertilization of British Columbia coastal lakes. *Can. J. Fish. Aquat. Sci.* 42:320–331.

Hyde, D. A., and Perry, S. F. 1989. Differential approaches to blood acid–base regulation during exposure to prolonged hypercapnia in two freshwater teleosts: The rainbow trout (*Salmo gairdneri*) and the American eel (*Angilla rostrata*). *Physiol Zool.* 62:1164–1186.

Hyvärinen H., Holopainen, I. J., and Piironen. J. 1985. Anaerobic wintering of crucian carp (*Carassius carassius* L.)—Annual dynamics of glycogen reserves in nature. *Comp. Biochem. Physiol.* 82A:797–803.

Idler, D. R., and Bitners, I. 1959. Biochemical studies on sockeye salmon during spawning migration. V. Cholesterol, fat, protein and water in the body of the standard fish. *J. Fish. Res. Bd. Canada* 16:235–241.

Imsland, A. K., Brix, O., Nævdal, G., and Samuelsen, E. N. 1997. Hemoglobin genotypes in turbot (*Scophthalmus maximus* Rafinesque), their oxygen affinity properties, and relations with growth. *Comp. Biochem. Physiol.* 116A:157–165.

Ishimatsu, A, Hishida, Y., Takita, T., Kanda, T., and others. 1998. Mudskippers store air in their burrows. *Nature* 391:237–238.

Ivlev, V. S. 1961. *Experimental ecology of the feeding of fishes*. New Haven, Conn.: Yale Univ. Press.

Iwama, G., and Nakanishi, T., eds. 1996. *The fish immune system: Organism, pathogen, and environment*. San Diego, Calif.: Academic Press. 380 pp.

Iwama, G. K., and Heisler, N. 1991. Effect of environmental water salinity on acid–base regulation during environmental hypercapnia in the rainbow trout (*Oncorhynchus mykiss*). *J. Exp. Biol.* 158:1–18.

Iwama, G. K., Thomas, P. T., Forsyth, R. B., and Vijayan, M. M. 1998. Heat shock protein in fish. *Rev. Fish Biol. Fish.* 8:35–56.

Iwata, M. 1995. Downstream migratory behavior of salmonids and its relationship with control and thyroid hormones: A review. *Aquaculture* 135:131–139.

Jampol, I. M., and Epstein, F. H. 1970. Sodium-potassium–activated adenosine triphosphatase and osmotic regulation by fishes. *Am. J. Physiol.* 218:607–611.

Jansen, P. A., Slettvold, H., and Finstad, A.G. 2001. Niche segregation between arctic charr. (*Salvelinus alpinus*) and brown trout (*Salmo trutta*): an experimental study. *Can. J. Fish. Aquat. Sci.* 59:6–11.

Janssen, R. G., and Randall, D. J. 1975. The effects of changes in pH and PCO_2 in blood and water on breathing in rainbow trout, *Salmo gairdneri*. *Resp. Physiol.* 25:235–245.

Janssens, P. A., and Cohen, P. P. 1966. Ornithine–urea cycle enzymes in the African lungfish, *Protopterus aethiopicus*. *Science* 152:358–359.

Janvier, P. 1981. The phylogeny of the Craniata, with particular reference to the significance of fossil Agnathans. *J. Vert. Paleont.* 1(2):121–159.

Jara, Z. 1957. On the morphology and function of the so-called palatal organ in the carp (*Cyprinus carpio* L.). *Preeglad Zoologiczny* 1:110–112.

Jarvik, E. 1977. The systematic position of acanthodian fishes. Pages 199–225 *in* S. M. Andrews, R. S. Miles, and A. D. Walker, eds., *Problems in vertebrate evolution*. Linn. Soc. Lond. Symp. 4.

Jayaram, K. C. 1974. Ecology and distribution of fresh-water fishes, amphibia, and reptiles. Pages 517–584 *in* M. S. Mani, ed., *Ecology and biogeography of India*. The Hague: W. Junk.

Jenkins, R. E., and Burkhead, M. M. 1994. *Freshwater fishes of Virginia*. Bethesda, Md.: Amer. Fish. Soc. 1079 pp.

Jenkins, R. E., Lachner, E. A., and Schwartz, F. J. 1972. Fishes of central Appalachian drainages: their distribution and dispersal. Pages 43–117. *in* P. C. Holt, ed. *The distributional history of the biota of the southern Appalachians*. Blacksburg, VA: Virg. Poly. Inst. Res. Div, Monogr. 4.

Jenkins, T. M. 1969. Social structure, position choice, and distribution of two trout species (*Salmo trutta* and *Salmo gairdneri*) resident in mountain streams. *Anim. Behav. Monogr.* 2(2):57–123.

Jensen, A. C. 1966. Life history of the spiny dogfish. *NOAA Fish. Bull.* 65(3):527–554.

Jensen, F. B. 1991. Multiple strategies in oxygen and carbon dioxide transport by haemoglobin. Pages 55–78 *in* A. J. Woakes, M. K. Grieshaber, and C. R. Bridges, eds., *Physiological strategies for gas exchange and metabolism*. Cambridge: Cambridge Univ. Press.

Jensen, F. B., Nikinmaa, M., and Weber, R. E. 1993. Environmental perturbations of oxygen transport in teleost fishes: Causes, consequences and compensations. Pages 161–179 *in* J. C. Rankin and F. B. Jensen, eds., *Fish ecophysiology*. London: Chapman and Hall.

Jia, Z., DeLuca, C. I., Chao, H., and Davies, P. L. 1996. Structural basis for the binding of a globular antifreeze protein to ice. *Nature* 384:285–288.

Jobling, M. 1981. Temperature tolerance and the final preferendum: Rapid methods for the assessment of optimum growth temperatures. *J. Fish Biol.* 19:439–455.

Johannes, R. E., and Larkin, P. A. 1961. Competition for food between redside shiners (*Richardsonius balteatus*) and rainbow trout (*Salmo gairdneri*) in two British Columbia lakes. *J. Fish. Res. Bd. Canada* 18(2):203–220.

Johansen, K., and Hanson, D. 1968. Functional anatomy of the hearts of lungfishes and amphibians. *Am. Zool.* 8:191–210.

Johansen, K., Lenfant, C., Schmidt-Nielsen, K., and Peterson, J. A. 1968. Gas exchange and control of breathing in the electric eel, *Electrophorus electricus* Z. *Vergl. Physiol.* 61:137–163.

Johanson-Sjöbeck, M., and Stevens, J. D. 1976. Hematological studies on the blue shark, *Prionace glauca* L. *J. Mar. Biol. Assn. U.K.* 55:237–240.

John, K. R. 1964. Survival of fish in intermittent streams of the Chiricahua mountains, Arizona. *Ecology* 45(1):112–119.

Johnson, D. L., and Stein, R. A. 1979. *Response of fish to habitat structure in standing water*. N. Cent. Div. Amer. Fish. Soc. Spec. Publ. 6.

Johnson, G. D. 1993. Percomorph phylogeny: Progress and problems. *Bull. Mar. Sci.* 52:3–28.

Johnson, G. D., and Patterson, C. 1993. Percomorph phylogeny: A survey of acanthomorphs and a new proposal. *Bull. Mar. Sci.* 52:554–626.

Johnson, G. D., and Patterson, C. 1996. Relationships of lower teleostean fishes. Pages 251–232 *in* M. L. J. Stiassny, L. R. Parenti, and G. D. Johnson, eds. *Interrelationships of fishes*. San Diego, Calif.: Academic Press.

Johnson, L. 1976. Ecology of arctic populations of lake trout (*Salvelinus namaycush*), lake whitefish (*Coregonus clupeaformis*), arctic char (*S. alpinus*), and associated species in unexploited lakes of the Canadian northwest territories. *J. Fish. Res. Bd. Canada* 33(11):2459–2488.

Johnson, R. H., and Nelson, D. R. 1973. Agonistic display of the gray reef shark, *Carcharhinus menisorrah*, and its relationship to attacks on man. *Copeia* 1973(1):76–83.

Johnson, R. K., and Barnett, M.A. 1975. An inverse correlation between meristic characters and food supply in mid-water fishes: Evidence and possible explanations. *NOAA Fish. Bull.* 73(2):284–298.

Johnson, R. K., and Glodek, G. S. 1975. Two new species of *Evermanella* from the Pacific Ocean, with notes on other midwater species endemic to the Pacific Central or the Pacific Equatorial water masses. *Copeia* 1975(4):715–730.

Johnson, T. P., Cullum, A. J., and Bennett, A. F. 1998. Partitioning the effects of temperature and kinematic viscosity on the C-start performance of adult fishes. *J. Exp. Biol.* 201:2045–2051.

Johnston, I. A. 1982. Capillarization, oxygen diffusion distances and mitochondrial content of carp muscles following acclimation to summer and winter temperatures. *Cell Tissue Res.* 222:325–337.

Johnston, J. A., Davison, W., and Goldpink, G. 1977. Energy metabolism of carp swimming muscles. *J. Comp. Physiol.* 114:203–216.

Jones, A. G., and Avise, J. C. 1997. Polygynandry in the dusky pipefish *Syngnathus floridae* revealed by microsatellite DNA markers. *Evolution* 51:1611–1622.

Jones, E. C. 1971. *Isistius brasiliensis,* a squaloid shark, the probable cause of crater wounds on fishes and cetaceans. *NOAA Fish. Bull.* 69(4):791–798.

Jonsson, B., and Jonsson, N. 2001. Polymorphism and speciation in Arctic charr. *J. Fish Biol.* 58:605–638.

Jordan, D. S. 1895. The fishes of Sinaloa. *Proc. Calif. Acad. Sci.* 5:378–513.

Jordan, D. S. 1922. *The days of a man.* Vols. I, II. New York: World Book Co. 1616 pp.

Jordan, D. S., and Evermann, B. W. 1900. *The fishes of North and Middle America. Bull. U. S. Nat. Mus.* 47, pts. I–IV. 3313 pp. +392 plates.

Jordan, D. S., and Evermann, B. W. 1903. The aquatic resources of the Hawaiian Islands. *Bull. U. S. Fish. Comm.* 23:1–765.

Jordan, D. S., and Starks, E. C. 1895. The fishes of Puget Sound. *Proc. Calif. Acad. Sci.* 5:785–855.

Jordan, H. E., and Speidel, C. C. 1930. Blood formation in cyclostomes. *Am. J. Anat.* 46:355–392.

Jordan, J. 1976. The influence of body weight on gas exchange in the airbreathing fish *Clarius batrachus. Comp. Biochem. Physiol.* 53A:305–310.

Jorgensen, J. M., Lomholt, J. P., Weber, R. E., and Malte, H. 1998. *The biology of hagfishes.* London: Chapman and Hall.

Josse, E., Dagorn, L., and Bertrand, A. 2000. Typology and behaviour of tuna aggregations around fish aggregating devices from acoustic surveys in French Polynesia. *Aquat. Living Resour.* 13:183–192.

Joyeux, J.-C., Floeter, S. R., Ferreira, C. E. L., and Gasparini, J. L. 2001. Biogeography of tropical reef fishes: the South Atlantic puzzle. *J. Biogeog.* 28:831–841.

Jumper, G. Y., Jr., and Baird, R. C. 1991. Location by olfaction: A model and application to the mating problem in the deep-sea hatchetfish *Argyropelecus hemigymnus. Amer. Nat.* 138:1431–1458.

Jürss, K., and Bastrop, R. 1995. Amino acid metabolism in fish. Pages 159–189 *in* P. W. Hochachka, and T. P. Mommsen, eds., *Biochemistry and molecular biology of fishes.* Vol. 4. Metabolic biochemistry. Amsterdam: Elsevier Science B.V.

Kah, O., Pontet, A., Danger, J. M., Dubourg, P., Pelletier, G., Vaudry, H., and Calas, A. 1989. Characterization, cerebral distribution and gonadotropin release activity of neuropeptide Y (NPY) in the goldfish. *Fish Physiol. Biochem.* 7:69–79.

Kajiura, S. M., and Holland, K. N. 2002. Electroreception in juvenile scalloped hammerhead and sandbar sharks. *J. Exp. Biol.* 205:3609–3621.

Kalmijn, A. J. 1971. The electric sense of sharks and rays. *J. Exp. Biol.* 55:371–383.

Kalmijn, A. J. 1974. The detection of electric fields from inanimate and animate sources other than electric organs. Pages 147–200 *in* A. Fessard, ed., *Handbook of sensory physiology.* Vol. 3 (3). Berlin: Springer-Verlag.

Kalmijn, A. J. 1978a. Electric and magnetic sensory world of sharks, skates, and rays. Pages 507–528 *in* E. S. Hodgson and R. F. Mathewson, eds., *Sensory biology of sharks, skates, and rays.* Washington, D.C.: Off. Naval Res. Publ.

Kalmijn, A. J. 1978b. Experimental evidence of geomagnetic orientation in elasmobranch fishes. Pages 347–353 *in* K. Schmidt-Koenig and W. T. Keeton, eds., *Animal migration, navigation, and homing.* Berlin: Springer-Verlag.

Kamohara, R. H. 1964. Revised catalog of fishes of Kochi Prefecture, Japan. *U. S. A. Mar. Biol. Stat.* 11(1):1–99.

Kandel, J. S., Horn, M. H., and Van Antwerp, W. 1994. Volatile fatty acids in the hindguts of herbivorous fishes from temperate and tropical marine waters. *J. Fish Biol.* 45:527–529.

Kapoor, B. G., Evans, H. E., and Pevzner, R. A. 1975a. The gustatory system in fish. *Adv. Mar. Biol.* 13:53–108.

Kapoor, B. G., Smit, H., and Verighina, I. A. 1975b. The alimentary canal and digestion in teleosts. *Adv. Mar. Biol.* 13:109–239.

Karnaky, K. J., Jr. 1986. Structure and function of the chloride cell of *Fundulus heteroclitus* and other teleosts. *Amer. Zool.* 26:209–224.

Karnaky, K. J., Jr. 1998. Osmotic and ionic regulation. Pages 159–178 *in* D. H. Evans, ed., *The physiology of fishes.* 2nd ed. Boca Raton, Fla.: CRC Press.

Katzman, S., and J. J. Cech, Jr. 2001. Juvenile coho salmon locomotion and mosaic muscle are modified by 3′, 3′, 5′-tri-iodo-L-thyronine (T_3). *J. Exp. Biol.* 204:1711–1717.

Kavaliers, M. 1979. Pineal involvement in the control of circadian rhythmicity in the lake chub, *Couesius plumbeus. J. Exp. Zool.* 209:33–40.

Kavanau, J. L. 1998. Vertebrates that never sleep: Implications for sleep's basic function. *Brain Res. Bull.* 46:269–279.

Kay, I. 1998. *Introduction to animal physiology.* New York: BIOS Scientific Publishers. 214 pp.

Keast, A. 1965. Resource subdivision amongst cohabiting fish species in a bay, Lake Opinicon, Ontario. *Univ. of Mich. Great Lakes Res. Div. Publ.* 13:106–132.

Khanna, S. S., and Singh, H. R. 1966. Morphology of the teleostean brain in relation to feeding habits. *Proc. Nat. Acad. Sci. India* 336:306–316.

Kieffer, J. D., Alsop, D., and Wood, C. M. 1998. A respiratory analysis of fuel use during aerobic swimming at different temperatures in rainbow trout. *J. Exp. Biol.* 201:3123–3133.

Kingsford, M. J. 1993. Biotic and abiotic structure in the pelagic environment: Importance to small fishes. *Bull. Mari. Sci.* 53:393–415.

Kinne, O. 1960. Growth, food intake, and food conversion in a euryplastic fish exposed to different temperatures and salinities. *Physiol. Zool.* 33:288–317.

Kisch, B. 1948. Electrocardiographic investigation of the heart of fish. *Expl. Med. Surg.* 6:31–62.

Klimley, A. P. 1985. Schooling in *Sphyrna lewini,* a species with low risk of predation: A non-egalitarian state. *Z. Tierpsychol.* 70:297–319.

Klimley, A. P. 1994. The predatory behavior of the white shark. *Amer. Sci.* 82:122–133.

Klimley, A. P. 1995. Hammerhead city. *Natural History* 104(10):33–39.

Klimley, A. P. 2000. Sharks beware. *Amer. Scientist* 87:488–491.

Klimley, A. P. 2003. *The secret life of sharks.* New York: Simon and Shuster. 290 pp.

Klimley, A. P., Beavers, S. C., Curtis, T. H., and Jorgensen, S. J. 2002. Movements and swimming behavior of three species of sharks in La Jolla Canyon, California. *Env. Biol. Fish* 63:117–135.

Klimley, A. P., and Chinley, D. G., eds. 1996. *Great white sharks: The biology of Carcharodon carcharias.* San Diego, Calif.: Academic Press. 928 pp.

Klimley, A. P., and Holloway, C. F. 1999. School fidelity and homing synchronicity of yellowfin tuna, *Thunnus albacares. Marine Biol.* 133(2):307–317.

Kline, K. 1978. *Aspects of digestion in stomachless fishes.* Ph.D. dissertation. Univ. of Calif., Davis. 78 pp.

Klontz, G. W. 1972. Haematological techniques and immune response in rainbow trout. Pages 89–99 *in* L. E. Mawdesley-Thomas, ed., *Diseases of fish.* Symp. *Zool. Soc. Lond.* No. 30. London:Academic Press.

Koban, M., Yup, A. A., Agellon, L. B., and Powers, D. A. 1991. Molecular adaptation to environmental temperature: Heat-shock response of the eurythermal teleost *Fundulus heteroclitus. Mol. Mar. Biol. Biotech.* 1:1–17.

Kobayashi, H. 1990. CO_2 back-diffusion in the rete aids O_2 secretion in the swimbladder of the eel. *Resp. Physiol.* 79:231–242.

Kobayashi, S., Yamada, J., Maekawa, K., and Ouchi, K. 1972. Calcification and nucleation in fish scales. Pages 84–90 *in* H. K. Erban, ed., *Biomineralization research reports.* Stuttgart: Schattauer-Verlag.

Kock, K.-H. 1992. *Antarctic fish and fisheries.* Cambridge: Cambridge Univ. Press.

Kohda, M., Tanimura, M., Kikue-Nakamura, M., and Yamagishi, S. 1995. Sperm drinking by female catfishes: A novel mode of insemination. *Env. Biol. Fish.* 42:1–6.

Koslow, J. A. 1996. Energetic and life history patterns of deep-sea benthic, benthopelagic, and seamount associated fish. *J. Fish Biol.* 49(Suppl. A): 54–74.

Kottelat, M., and Chu, X. 1988. Revision of *Yunnanilus* with descriptions of a minature species flock and six new species from China (Cypriniformes, Homalopteridae). *Env. Biol. Fish.* 23:65–93.

Kozhov, M. 1963. *Lake Baikal and its life.* Monogr. Biologicae 11. The Hague: W. Junk. 344 pp.

Kramer, D. L., and Bryant, M. J. 1995. Intestine length in the fishes of a tropical stream: 1. Ontogenetic allometry. *Env. Biol. Fish* 42:115–127.

Kramer, D. L., and Graham, J. B. 1976. Synchronous air breathing, a social component of respiration in fishes. *Copeia* 1976:689–697. Vol. 1976. no. 4

Kramer, D. L., and McClure M. 1982. Aquatic surface respiration, a widespread adaptation to hypoxia in tropical freshwater fishes. *Env. Biol. Fish* 7:47–55.

Kramer, D. L., and R. L. McLaughlin. 2001. The behavioral ecology of intermittent locomotion. *Amer. Zool.* 41:137–153.

Kramer, D. L., Manley, D., and Bourgeois, R. 1983. The effect of respiratory mode and oxygen concentration on the risk of aerial predation in fishes. *Can. J. Zool.* 61:653–665.

Krause, J., and Godin, J.-G. J. 1996. Influence of parasitism on shoal choice in the banded killifish (*Fundulus diaphanus,* Teleostei, Cyprinodontidae). *Ethology* 102:40–49.

Krause, J., Godin, J.-G. J., and Brown, D. 1996. Size-assortiveness in multi-species fish shoals. *J. Fish Biol.* 49:221–225.

Krogh, A. 1939. *Osmotic regulation in aquatic animals.* London: Cambridge Univ. Press.

Krumschnabel, G., Schwarzbaum, P. J., Lisch, J., Biasi, C., and Weiser, W. 2000. Oxygen-dependent energetics of anoxia-tolerant and anoxia-intolerant hepatocytes. *J. Exp. Biol.* 203:951–959.

Kubota, T., Shiobara, Y., and Kubodera, T. 1991. Food habits of the frilled shark, *Chlamydoselachus anguineus,* collected from Suruga Bay, Central Japan. *Nippon Suisan Gakkaishi* 57:15–20.

Kuchnow, K. P. 1971. The elasmobranch pupillary response. *Vision Res.* 11:1395–1406.

Kuehne, R. A. 1962. A classification of streams illustrated by fish distribution in an eastern Kentucky creek. *Ecology* 43(4):608–614.

Kuhn, W., Ramel, A., Kuhn, H. J., and Marti, E. 1963. The filling mechanism of the swimbladder, generation of high gas pressures through hairpin countercurrent multiplication. *Experientia* 19:497–511.

Kupschus, S., and Termain, D. 2001. Associations between fish assemblages and environmental factors in nearshore habitats of a subtropical estuary. *J. Fish Biol.* 58:1383–1403.

Kurlansky, M. 1997. *Cod.* New York: Penguin Books. 294 pp.

Kuwamura, T., and Nakashima, Y. 1998. New aspects of sex change among reef fishes: Recent studies in Japan. *Env. Biol. Fish* 52:125–135.

LaBounty, J. F., and Deacon, J. E. 1972. *Cyprinodon milleri,* a new species of pupfish (family Cyprinodontidae) from Death Valley, California. *Copeia* 1972(4):769–780.

Laegdsgaard, P., and Johnson, C. 2001. Why do juvenile fish utilise mangrove habitats? *J. Exp. Mar. Biol. Ecol.* 257:229–253.

Lafferty, K. D., and Kuris, A. M. 2002. Trophic strategies, animal diversity, and body size. *TREE* 17:507–513.

Lagler, K. F., Bardach, J. E., Miller, R. R., and Passino, D. R. M. 1977. *Ichthyology.* 2d ed. New York: Wiley. 506 pp.

Lai, N. C., Graham, J. B., Dalton, N., Shabetai, R., and Bhargava, V. 1998. Echocardiographic and hemodynamic determinations of the ventricular filling pattern in some teleost fishes. *Physiol. Zool.* 71:157–167.

Lamouroux, N., Poff, N. L., and Angermeier, P. L. 2002. Intercontinental convergence of stream fish community traits along geomorphic and hydraulic gradients. *Ecology* 53:1792–1807.

Landa, J. T. 1998. Bioeconomics of schooling fishes: Selfish fish, quasi-free riders, and other fishy tales. *Env. Biol. Fish.* 53:353–364.

Lanzing, W. J. R., and Bower, C. C. 1974. Development of colour patterns in relation to behaviour in *Tilapia mossambica* (Peters). *J. Fish Biol.* 6:29–41.

Lapennas, G. N., and Schmidt-Nielsen, K. 1977. Swimbladder permeability to oxygen. *J. Exp. Biol.* 67:175–196.

Larimore, R. W., Childers, W. F., and Hekcrotte, C. 1959. Destruction and reestablishment of stream fish and invertebrates affected by drought. *Trans. Amer. Fish. Soc.* 88:261–285.

Larje, R. 1990. Rare fish in Sweden: *Nemacheilus* survey and public reactions. *J. Fish Biol.* 37(Suppl. A):219–221.

Larkin, P. A. 1956. Interspecific competition and population control in freshwater fish. *J. Fish. Res. Bd. Canada* 13(2):327–342.

Larsen, R. J. 1980. Competition, habitat selection, and the bathymetric distribution of two rockfish (*Sebastes*) species. *Ecol. Monogr.* 50:221–239.

Lasker, R. 1978. The relation between oceanographic conditions and larval anchovy food in the California current: Identification of factors contributing to recruitment failure. *Rapp. P.-v. Reun. Const. Int. Explor. Mer.* 173:212–230.

Lauck, T., Clark, C. W., Mangel, M., and Munro, G. R. 1998. Implementing the precautionary principal in fisheries management through marine reserves. *Ecol. Appls.* 8(Suppl. 1):72–78.

Lauder, G. V. 1983. Prey capture hydrodynamics in fishes: Experimental tests of two models. *J. Exp. Biol.* 104:1–13.

Lauder, G. V. 1985. Aquatic feeding in lower vertebrates. Pages 210–229 *in* M. Hildebrand, D. Bramble, K. Liem, and D. Wake, eds., *Functional vertebrate morphology.* Cambridge: Harvard Univ. Press.

Lauder, G. V. 2000. Function of the caudal fin during locomotion in fishes: Kinematics, flow visualization, and evolutionary patterns. *Amer. Zool.* 40:101–122.

Lauder, G. V., and Liem, K. F. 1983. The evolution and interrelationships of the actinopterygian fishes. *Bull. Mus. Comp. Zool. Harvard* 150:95–197.

Laurent, P., Hobe, H., and Dunel-Erb, S. 1985. The role of environmental sodium chloride relative to calcium in gill morphology of freshwater salmonid fish. *Cell Tissue Res.* 240:675–692.

Laurent, P., Holmgren, S., and Nilsson, S. 1983. Nervous and humoral control of the fish heart: Structure and function. *Comp. Biochem. Physiol.* 76A:525–542.

Le Gac, F., Thomas, J. L., Mourot, B., and Loir, M. 2001. In vivo and in vitro effects of prochloraz and nonylphenol ethoxylated on trout spermatogenesis. *Aquat. Toxicol.* 53:187–200.

Leary, R. F., and Booke, H. E. 1990. Starch gel electrophoresis and species distinctions. Pages 141–170 *in* C. B. Schreck and P. B. Moyle, eds., *Methods for fish biology.* Bethesda, Md.: Amer. Fish. Soc.

Lecklin, T., and M. Nikinmaa. 1999. Seasonal and temperature effects on the adrenergic responses of arctic charr (*Salvelinus alpinus*) erythrocytes. *J. Exp. Biol.* 202:2233–2238.

LeCren, E. D., Kipling, C., and McCormack, J. C. 1972. Windemere: Effects of exploitation and eutrophication of the salmonid community. *J. Fish. Res. Bd. Canada* 29(6):819–832.

Lecointre, G., and Nelson, G. 1996. Clupeomorpha, sister-group of Ostariophysi. Pages 193–208 *in* M.L. J. Stiassny, L. R. Parenti, and G. D. Johnson, eds., *Interrelationships of fishes*. San Diego, Calif.: Academic Press.

Lee, D. S., Gilbert, C. R., Hocutt, C. H., Jenkins, R. E., McCallister, D. E., and Stauffer, J. R. 1980. *Atlas of North American freshwater fishes*. Raleigh, N.C.: State Mus. Nat. Hist.

Lee, R. F., Phleger, C. F., and Horn, M. H. 1975. Composition of oil in fish bones: Possible function in neutral buoyancy. *Comp. Biochem. Physiol.* 50B:13–16.

Lee, R. M. 1920. A review of the methods of age and growth determination in fishes by means of scales. *Fish. Invest.* Ser. II. 4(2):1–32.

Leet, W. S., Dewees, C. M., Kleingbeil, R., and Larson, E. J., eds. 2001. *California's living marine resources: A status report*. Sacramento: Calif. Dept. Fish and Game. 592 pp.

Leggett, W. C. 1977. The ecology of fish migrations. *Ann. Rev. Ecol. Syst.* 8:285–308.

Leggett, W. C., and Carscadden, J. E. 1978. Latitudinal variation in reproductive characteristics of American shad *(Alosa sapidissima)*: Evidence for population specific life history strategies in fish. *J. Fish. Res. Bd. Canada* 35:1469–1478.

Leidy, R. A., and Moyle, P. B. 1998. Conservation status of the world's fish fauna: An overview. Pages 187–227 *in* P.L. Fiedler and P. M. Kareiva, eds., *Conservation biology for the coming decade*. New York: Chapman and Hall.

Leigh, E. G., Jr., Paine, R. T., Quinn, J. F., and Suchanek, T. H. 1987. Wave energy and intertidal productivity. *Proc. Nat. Acad. Sci. U. S. A.* 84:1314–1318.

Leis, J. M., and McCormick, M. I., 2002. The biology, behavior, and ecology of the pelagic, larval stage of coral reef fishes. Pages 171–200 *in* P. F. Sale, ed. *Coral reef fishes: Dynamics and diversity in a complex ecosystem*. Amsterdam: Academic Press

Leivestad, H., and Muniz, I. P. 1976. Fish kill at low pH in a Norwegian river. *Nature* 259:391–392.

Lelek, A. 1973. Sequence of changes in fish populations of the new tropical man-made lake, Kainji, Nigeria, West Africa. *Arch. Hydrobiol.* 71:381–420.

Lemly, A. D. 1985. Suppression of native fish populations by green sunfish in first-order streams of piedmont North Carolina. *Trans. Amer. Fish. Soc.* 114:705–712.

Leppäkoski, E., and Mihnea, P. E. 1996. Enclosed seas under man induced change: A comparison between the Baltic and Black seas. *Ambio* 25:380–389.

Lévêque, C. 1997. *Biodiversity dynamics and conservation: the freshwater fishes of Africa*. Cambridge: Cambridge Univ. Press.

Lever, C. 1996. *Naturalized fishes of the world*. San Diego, Calif.: Academic Press. 408 pp.

Lewis, D. S. 1974. The effects of the formation of Lake Kainji, Nigeria, upon the indigenous fish population. *Hydrobiologica* 45(1–3):281–301.

Lewis, S. M. 1986. The role of herbivorous fishes in the organization of a Caribbean reef community. *Ecol. Monogr.* 56:183–200.

Lewis, W. M., Jr., Hamilton, S. K., Rodriguez, M. A., Saunders, J. F., and Lasi, M. A., 2001. Foodweb analysis of the Orinoco floodplain based on production estimates and stable isotope data. *J. N. Am. Benthol. Soc.* 20:241–254.

Li, W., Sorensen, P. W., and Gallaher, D. D. 1995. The olfactory system of migratory adult sea lamprey *(Petromyzon marinus)* is specifically and acutely sensitive to unique bile acids released by conspecific larvae. *J. Gen. Physiol.* 105:569–587.

Liem, K. F. 1963. *The comparative osteology and phylogeny of the Anabantoidei (Teleostei, Pisces)*. Ill. Biol. Monogr. 30. 149 pp.

Liem, K. F. 1974. Evolutionary strategies and morphological innovations: Cichlid pharyngeal jaws. *Syst. Zool.* 22:425–441.

Liem, K. F. 1980. Adaptive significance of intra- and interspecific differences in the feeding repertoires of cichlid fishes. *Amer. Zool.* 20:295–314.

Liem, K. F. 1981. Larvae of air-breathing fishes as countercurrent flow devices in hypoxic environments. *Science* 211:1177–1179.

Lighthill, M. J. 1969. Hydromechanics of aquatic animal propulsion. *Ann. Rev. Fluid Mech.* 1:413–446.

Lin, H., and Randall, D. J. 1991. Evidence for the presence of an electrogenic proton pump on the trout gill epithelium. *J. Exp. Biol.* 161:119–134.

Lindsey, C. C., and McPhail, J. D. 1986. Zoogeography of fishes of the Yukon and Mackenzie basins. Pages 639–674 *in* C. H. Hocutt and E. O. Wiley, eds. *The zoogeography of North American freshwater fishes.* New York: Wiley.

Lineaweaver, T. H., and Backus, R. H. 1969. *The natural history of sharks.* New York: Lippincott. 256 pp.

Lissman, H. W. 1963. Electric location in fishes. *Sci. Amer.* 152:1–12.

Livingston, P. A. 1993. Importance of predation by groundfish, marine mammals and birds on walleye pollock *Theragra chalcogramma* and Pacific herring *Clupea pallasi* in the eastern Bering Sea. *Mar. Ecol. Prog. Ser.* 102:205–215.

Livingston, R. J. 1976. Diurnal and seasonal fluctuations or organisms in a north Florida esturary. *Est. Coast. Mar. Sci.* 4(3):373–400.

Livingston, R. J., Kobylinski, G. J., Lewis, F. G., and Sheridan, P. F. 1975. Long-term fluctuations of epibenthic fish and invertebrate populations in Apalachicola Bay, Florida. *NOAA Fish. Bull.* 74(2):311–321.

LoBue, C. P., and Bell, M. A. 1993. Phenotypic manipulation by the cestode parasite, *Schistocephalus solidus* of its intermediate host, *Gasaterosteus aculeatus*, the threespine stickleback. *Amer. Nat.* 142:725–735.

Loew, E. R., and Lythgoe, J. N. 1978. The ecology of cone pigments in teleost fishes. *Vision Res.* 18:715–722.

Loew, E. R., McFarland, W. N., Mills E. L., and Hunter, D., 1993. A chromatic action spectrum for planktonic predation by juvenile yellow perch, *Perca flavescens. Can. J. Zool.* 71:384–386.

Loftus, K. H., and Regier, H. A., eds. 1972. Symposium of salmonid communities in oligotrophic lakes. *J. Fish. Res. Bd. Canada* 29(6):613–986.

Lone, K. P., and Matty, A. J. 1982. The effect of feeding 11-ketotestosterone on the food conversion efficiency and tissue protein and nucleic acid contents of juvenile carp, *Cyprinus carpio. J. Fish Biol.* 20:93–104.

Long, J. A. 1995. *The rise of fishes: 500 million years of evolution.* Baltimore, Md.: The Johns Hopkins Univ. Press, 233 pp.

Longhurst, A. R. 1998. *Ecological geography of the sea.* San Diego, Calif.: Academic Press. 398 pp.

Longurst, A., Coldbrook, M., LeBasseur, R., Lorenzen, C., and Smith, P. 1972. The instability of ocean populations. *New Sci.* 1972:1–5.

Loos, J. L., and Woolcott, W. S. 1969. Hybridization and behavior in two species of *Percina* (Percidae). *Copeia* 1969(2):374–385.

Loretz, C. A. 1995. Atrial natriuretic peptide regulation of vertebrate intestinal ion transport. *Amer. Zool.* 35:490–502.

Loretz, C. A., Collie, N. L., Richman, N. H., III, and Bern, H. A. 1982. Osmoregulatory changes accompanying smoltification in coho salmon. *Aquaculture* 28:67–74.

Losey, G. S., Jr. 1978. The symbiotic behavior of fishes. Pages 1–31 *in* D. I. Mostofsky, ed., *The behavior of fish and other aquatic animals.* New York: Academic Press.

Losey, G. S., Nelson, P. A., and Zamzow, J. 2000. Ontogeny of spectral transmission in the eye of the tropical damselfish, *Dascyllus albisella* (Pomacentridae), and possible effects on UV vision. *Environ. Biol. Fish.* 59:21–28.

Lotrich, V. A. 1973. Growth, production, and community composition of fishes inhabiting a first-, second-, and third-order stream of eastern Kentucky. *Ecol. Monogr.* 43(3):377–397.

Love, R. M. 1970. *The chemical biology of fishes.* London: Academic Press. 547 pp.

Love, M. S., and Caillet, G. M. 1979. *Readings in ichthyology.* Santa Monica, Calif.: Goodyear, 525 pp.

Love, M. S., Yoklavich, M., and Thorsteinson, L. 2002. *The rockfishes of the northeast Pacific.* Berkeley: Univ. of Calif. Press. 404 pp.

Lowe-McConnell, R. H. 1975. *Fish communities in tropical freshwaters.* London: Longman. 337 pp.

Lowe-McConnell, R. H. 1987. *Ecological studies in tropical fish communities.* Cambridge: Cambridge Univ. Press.

Lowe McConnell, R. 1994. The roles of ecological and behaviour studies of cichlids in understanding fish diversity and speciation in African Great Lakes: A review. Pages 335–345 *in* K. Martens, B. Goddeeris, and G. Coulter, eds., *Speciation in ancient lakes.* Arch. Hydrobiol. Beih. Ergeben. Limnol. 44.

Lucas, M. C., and Baras, E. 2001. *Migration of freshwater fishes.* Oxford: Blackwell Science. 420 pp.

Ludwig, D, Hillborn, R., and Walters, C. 1993. Uncertainty, resource exploitation, and conservation: Lessons from history. *Science* 260:17–36.

Lugo, A. E., and Snedaker, S. C. 1974. The ecology of mangroves. *Ann. Rev. Ecol. Syst.* 5:39–64.

Lund, R., and Grogan, E. D. 1997. Relationships of the Chimaeriformes and the basal radiation of the Chondrichthyes. *Rev. Fish Biol. Fisheries* 7:65–123.

Lundberg, J. G. 2001. Freshwater riches of the Amazon. *Nat. Hist.* 110(7):36–43.

Lundberg, J. G., Kottelat, M., Smith, G. R., Stiassny, M. L. J., and Gill, A. C. 2000. So many fishes, so little time: An overview of recent ichthyological discovery in continental waters. *Ann. Missouri Bot. Gard.* 87:26–62.

Lundberg, J. G., and McDade, L. A. 1990. Systematics. Pages 65–108 *in* C. B. Schreck and P. B. Moyle, eds., *Methods for fish biology.* Bethesda, Md.: American Fisheries Soc.

Lurie, E. 1960. *Louis Agassiz: A life in science.* Chicago: Univ. of Chicago Press. 390 pp.

Lushchak, V. I., Lushchak, L. P., Mota, A. A., and Hermes-Lima, M. 2001. Oxidative stress and antioxidant defenses in goldfish *Carassius auratus* during anoxia and reoxygenation. *Am. J. Physiol. (Regulatory Integrative Comp. Physiol.)* 280:R100–R107.

Lythgoe, J., and Lythgoe, G. 1971. *Fishes of the sea.* New York: Doubleday. 320 pp.

Lythgoe, J., and Lythgoe, G. 1991. *Fishes of the sea: The North Atlantic and Mediterranean.* London: Blandford. 256 pp.

MacGintie, G. E. 1939. The natural history of the blind goby *(Typhlogobius californiensis*, Steindachner). *Amer. Midl. Nat.* 21:489–508.

Macpherson, E., and Duarte, C. M. 1994. Patterns in species richness, size, and latitudinal range of East Atlantic fishes. *Ecography* 17:242–248.

Maddock, L., Bone, Q., and Rayner, J. M. V., eds. 1994. *Mechanics and physiology of animal swimming.* Cambridge: Press Syn. Univ. of Cambridge. 250 pp.

Madsen, S. S. 1990. The role of cortisol and growth hormone in seawater adaptation and development of hypoosmoregulatory mechanisms in sea trout parr *(Salmo trutta trutta).* Gen. Comp. Endocr. 79:1–11.

Maetz, J., and Garcia Romeu, F. 1964. The mechanism of sodium and chloride uptake by the gills of a freshwater fish, *Carassius aurâtus.* II. Evidence of NH_4^+ Na^+ and HCO_3^- CL^- exchanges. *J. Gen. Physiol.* 50:391–422.

Maetz, J., Sawyer, W. H., Dickford, G. E., and Mayer, N. 1967. Évolution de la balance minérale du sodium chez *Fundulus heteroclitus* au cours du transfert d'eau de mer en eau douce: Effets l'hypophysectomie et de la prolactine. *Gen. Comp. Endocrinol.* 8:163–176.

Magnien, R. E. 2001. The dynamics of science, perception, and policy during the outbreak of *Pfiesteria* in the Chesapeake Bay. *Bioscience* 51:843–852.

Maisey, J. G. 1996. *Discovering fossil fishes.* New York: Henry Holt and Co. 223 pp.

Maitland, P. S., and Lyle, A. A. 1990. Practical conservation of British fishes: Current action on six declining species. *J. Fish Biol.* 40(Suppl. A):255–256.

Major, P. F. 1978. Predator–prey interaction in two schooling fishes, *Caranx ignobilis* and *Stolephorus purpureus. Anim. Behav.* 26:760–777.

Mallatt, J. 1982. Pumping rates and particle retention efficiencies of the larval lamprey, an unusual suspension feeder. *Biol. Bull.* 163:197–210.

Mallatt, J. 1984. Feeding ecology of the earliest vertebrates. *Zool. J. Linn. Soc.* 82:261–272.

Mallatt, J., Sullivan, J., and Winchell, C. J., 2001. The relationship of lampreys to hagfishes: A spectral analysis of ribosomal DNA sequences. Pages 106–118 *In* P.K. Ahlberg, ed. *Major events in early vertebrate evolution.* London: Taylor and Francis.

Marchetti, M. P., Light, T., Feliciano, J., Armstrong, T., Hogan, Z., Viers, J., and Moyle, P. B. 2001. Homogenization of California's fish fauna through abiotic change. Pages 259–278 *in* J. Lockwood and M. McKinney, eds., *Biotic homogenization.* Dordrecht: Kluwer

Markert, J. R., Higgs, D. A., Dye, H. M., and MacQuarrie, D. W. 1977. Influence of bovine growth hormone on growth rate, appetite, and food conversion of yearling coho salmon *(Oncorhynchus kisutch)* fed two diets of different composition. *Can. J. Zool.* 55:74–83.

Markle, D. F. 1989. Aspects of the character homology and phylogeny of the Gadiformes. *Los Angeles Co. Mus. Nat. Hist. Sci. Ser.* 32:59–88.

Markle, D. F., and Musick, J. A. 1974. Benthic-slope fishes found at 900 m depth along a transect in the western North Atlantic ocean. *Mar. Biol.* 16:225–233.

Marsh-Matthews, E., Skierkowski, P., and DeMarais, A. 2001. Direct evidence for mother-to-embryo transfer of nutrients in the livebearing fish, *Gambusia geiseri. Copeia* 2001(1):1–6.

Marshall, N. B. 1953. Egg size in arctic, antarctic, and deep-sea fishes. *Evolution* 7:328–341.

Marshall, N. B. 1954. *Aspects of deepsea biology.* London: Hutchinson's. 380 pp.

Marshall, N. B. 1960. Swimbladder structure of deep-sea fishes in relation to their systematics and biology. *Discovery Rept.* 31:1–122.

Marshall, N. B. 1962. Some convergences between the benthic fishes of polar seas. Pages 273–278 *in* R. Carrick, M. W. Holdgate, and J. Prost, eds., *Symposium on Antarctic biology.* Paris: Hermann.

Marshall, N. B. 1966. *The life of fishes.* New York: Universe Books. 402 pp.

Marshall, N. B. 1971. *Explorations in the life of fishes.* Cambridge: Harvard Univ. Press. 204 pp.

Marshall, N. B. 1979. *Developments in deep-sea biology.* Dorset: Blandford Press.

Marshall, N. B. 1984. Progenetic tendencies in deep-sea fishes. Pages 91–101, *in* G. W. Potts and R. J. Wooton, eds., *Fish reproduction: Strategies and tactics.* New York: Academic Press.

Marshall, W. S. 2002. Na^+, Cl^-, Ca^+, and Zn^{2+} transport by fish gills: Retrospective review and prospective synthesis. *J. Exp. Zool.* 293:264–283.

Martin, L. K., and Cailliet, G. M. 1983. Age and growth determination of the bat ray, *Myliobatis californica* Gill, in central California. *Copeia* 1983(3):762–773.

Martini, F. H. 1998a. The ecology of hagfishes. Pages 57–77 *in* J. M. Jorgenson, J. P. Lomholt, R. E. Weber, and H. Malte, eds., *The biology of hagfishes.* London: Chapman and Hall.

Martini, F. H. 1998b. Secrets of the slime hag. *Sci. Amer.* 279(4):70–75.

Marty, G. D., Cech, J. J., Jr., Hinton, D. E. 1990. Effect of incubation temperature on oxygen consumption and ammonia production by Japanese medaka, *Oryzias latipes*, eggs and newly hatched larvae. *Env. Tox. Chem.* 9:1397–1403.

Matern, S. A., Moyle, P. B., and Pierce, L. C. 2002. Native and alien fishes in a California estuarine marsh: Twenty-one years of changing assemblages. *Trans. Amer. Fish. Soc.* 131:797–816.

Matthews, W. J. 1998. *Patterns in freshwater fish ecology.* London: Chapman and Hall. 756 pp.

Matthews, W. J., and Heins, D. C. 1987. *Community and evolutionary ecology of North American stream fishes.* Norman: Univ. of Okla. Press.

Matthews, W. J., Hough, D. J., and Robison, H. W. 1992. Similarities in fish distribution and water quality patterns in streams of Arkansas: Congruence of multivariate analysis. *Copeia* 1992:296–305.

Mattisson, A., and Fänge, R. 1982. The cellular structure of the Leidig organ in the shark *Etmopterus spinax* (L.). *Biol. Bull.* 62:182–194.

Mattsoff, L., and Nikinmaa, M. 1988. Effects of external acidification on the blood acid–base status and ion concentrations of lamprey. *J. Exp. Biol.* 136:351–361.

Matty, A. J. 1985. *Fish endocrinology.* Portland, Ore.: Timber Press. 267 pp.

Matty, A. J. 1986. Nutrition, hormones, and growth. *Fish Physiol. Biochem.* 2:141–150.

Matty, A. J., and Cheema, I. R. 1978. The effect of some steroid hormones on the growth and protein metabolism of rainbow trout. *Aquaculture* 14:163–178.

Mauchline, J., and Gordon, J. D. M. 1986. Foraging strategies of deep-sea fish. *Mar. Ecol. Prog. Ser.* 27:227–238.

Mayden, R. L. 1991. Cyprinids of the New World. Pages 240–263 *in* I. J. Winfield and J. S. Nelson, eds., *Cyprinid fishes: Systematics, biology and exploitation.* New York: Chapman and Hall.

Mayden, R. L., ed. 1992a. *Systematics, historical ecology, and North American freshwater fishes.* Stanford, Calif.: Stanford Univ. Press. 969 pp.

Mayden, R. L. 1992b. Explorations of the past and the dawn of systematics and historical ecology. Pages 3–17 *in* R. L. Mayden, ed., *Systematics, historical ecology, and North American freshwater fishes.* Stanford, Calif.: Stanford Univ. Press.

Mayden, R. L., and Wiley, E.O. 1992. The fundamentals of phylogenetic systematics. Pages 114–185 *in* R. L. Mayden, ed. *Systematics, historical ecology, and North American freshwater fishes*. Stanford, Calif.: Stanford Univ. Press.

Mazik, P. M., Simco, B. A., and Parker, N. C. 1991. Influence of water hardness and salts on survival and physiological characteristics of striped bass during and after transport. *Trans. Amer. Fish. Soc.* 120:121–126.

Mazumder, A., Taylor, W. D., McQueen, D. J., and Lean, D. R. S. 1990. Effects of fish and plankton on lake temperature and mixing depth. *Science* 247:312–315.

McAllister, D. E. 1963. *A revision of the smelt family, Osmeridae.* Bull. Nat. Mus. Canada 91. 53 pp.

McAllister, D. E. 1977. Ecology of the marine fishes of arctic Canada. Pages 49–65 *in* Section II. *Marine ecology.* Ottawa: Proc. Circumpolar Conf. on Northern Ecology.

McCarthy, J. E., and Conte, F. P. 1996. Determination of the volume of vascular and extravascular fluids in Pacific hagfish (*Polistotrema stouti*). *Amer. Zool.* 6:605 [abstract].

McCleave, J.D. 1993. Physical and behavioural controls on the oceanic distribution and migration of leptocephali. *J. Fish Biol.* 43(supplement A):243–273.

McCormick, S. D. 1990. Cortisol directly stimulates differentiation of chloride cells in tilapia opercular membrane. *Am. J. Physiol.* 259:R857–R863.

McCormick, S. D. 2001. Endocrine control of osmoregulation in teleost fish. *Amer. Zool.* 41:781–794.

McCormick, S. D., Cunjak, R. A., Dempson, B., O'Dea, M. F., and Carey, J. B. 1999. Temperature-related loss of smolt charateristics in Atlantic salmon (*Salmo salar*) in the wild. *Can. J. Fish. Aquat. Sci.* 56:1649–1658.

McCormick, S. D., Sakamoto, T., Hasegawa, S., and Hirano, T. 1991. Osmoregulatory actions of insulin-like growth factor-I in rainbow trout (*Oncorhynchus mykiss*). *J. Endocrinol.* 130:87–92.

McCosker, J. 1977. Fright posture of the plesiopid fish *Calloplesiops altivelis:* An example of Batesian mimicry. *Science* 197:400–401.

McCosker, J. E., and Lagios, M. D., eds. 1979. The biology and physiology of the living coelacanth. *Occ. Pap. Calif. Acad. Sci.* 134. 175 pp.

McDowall, R. M. 1978. *New Zealand freshwater fishes.* Auckland: Heinemann.

McDowall, R. M. 1988. *Diadromy in fishes.* London: Croom Helm. 308 pp.

McDowall, R. M. 1990. *New Zealand fishes: A natural history and guide.* Auckland, N.Z.: Heinemann Reed, 553 pp.

McDowall, R. M. 1997. The evolution of diadromy in fishes (revisited) and its place in phylogenetic analysis. *Rev. Fish Biol. Fisheries* 7:443–462.

McDowall, R. M. 1998. Fighting the flow: Downstream–upstream linkages in the ecology of diadromous fish fauna in West Coast New Zealand rivers. *Freshw. Biol.* 40:111–122.

McDowall, R. M. 2002. Accumulating evidence for a dispersal biogeography of southern cool temperate freshwater fishes. *J. Biogeog.* 29:207–219.

McDowall, R. M., and Robertson, D. A. 1975. Occurrence of galaxiid larvae and juveniles in the sea. *N. Z. J. Mar. Freshw. Res.* 9(1):1–9.

McEachran, J. D., Dunn, K. A., and Miyake, T. 1996. Interrelationships of the batoid fishes (Chondrichthyes: Batoidea). Pages 63–84 *in* M. L. J. Stiassny, L. R. Parenti, and G. D. Johnson, eds. *Interrelationships of fishes.* San Diego, Calif.: Academic Press.

McElman, J. F., and Balon, E. K. 1979. Early ontogeny of walleye, *Stizostedion vitreum,* with steps of saltatory development. *Env. Biol. Fish* 4:309–348.

McEnroe, M., and Cech, J. J., Jr. 1985. Osmoregulation in juvenile and adult white sturgeon, *Acipenser transmontanus. Env. Biol. Fish.* 14:23–40.

McFarland, W. N., and Munz, F. W. 1965. Regulation of body weight and serum composition by hagfish in various media. *Comp. Biochem. Physiol.* 14:383–398.

McHugh, J. L. 1967. Estuarine nekton. Pages 581–620 *in* G. H. Lauff, ed., *Estuaries.* AAAS Publ. 83.

McKaye, K. R. 1977. Defense of a predator's young by herbivorous fish: An unusual strategy. *Amer. Nat.* 11(978):310–315.

McKaye, K. R. 1983. Ecology and breeding behavior of a cichlid fish, *Cyrtocara eucinostomus,* on a large lek in Lake Malawi, Africa. *Env. Biol. Fish.* 8:81–96.

McKaye, K. R. 1985. Cichlid–catfish mutualistic defense of young in Lake Malawi. *Oecologia* 66:358–363.

McKaye, K. R., and Barlow, G. W. 1976. Chemical recognition of young by the Midas cichlid, *Cichlasoma citrinellum. Copeia* 1976:276–282.

McKaye, K. R., and Kocher, T. 1983. Head-ramming behavior by three paedophageous cichlids in Lake Malawi, Africa. *Anim. Behav.* 31:206–210.

McKeown, B. A. 1984. *Fish migration.* London: Croom Helm. 224 pp.

McKnight, T. L. 1998. *Physical geography.* 6th ed. Upper Saddle River, N.J.: Prentice Hall.

McLaughlin, R. H., and O'Gower, A. K. 1971. Life history and underwater studies of a heterodont shark. *Ecol. Monogr.* 41:271–289.

McLeave, J. D. 1993. Physical and behavioral controls on the oceanic distribution and migration of leptocephali. *J. Fish. Biol.* 43(Suppl. A):243–273.

McNeely, J. A. 1996. Economic incentives for conserving biodiversity. Pages 647–655 *in* R. C. Szaro and D. W. Johnston, eds., *Biodiversity in managed landscapes: Theory and practice.* New York: Oxford Univ. Press.

McPhail, J. D. 1977a. A possible function of the caudal spot in characid fishes. *Can. J. Zool.* 55(7):1063–1066.

McPhail, J. D. 1977b. Sons and lovers: The functional significance of sexual dichromatism in a fish, *Neoheterandria tridentiger* (Garman). *Behaviour* 64:329–339.

McPhail, J. D., and Lindsey, C. C. 1970. *Freshwater fishes of northwestern Canada and Alaska. Fish. Res. Bd. Canada Bull.* 173. 379 pp.

McQueen, D. J. 1990. Manipulating lake community structure: Where do we go from here? *Freshw. Biol.* 23:613–620.

Mead, G. W., Bertelsen, E., and Cohen, D. M. 1964. Reproduction among deep-sea fishes. *Deepsea Res.* 11(4):569–596.

Meade, I. L., and Perrone, S. J. 1980. Selective haematological parameters in steelhead trout, *Salmo gairdneri* Richardson. *J. Fish Biol.* 17:9–12.

Mearns, A. J. 1973. Southern California's inshore demersal fishes: Diversity, distribution, and disease as responses to environmental quality. *CalCOFI Rpt.* 17:141–148.

Mearns, A. J., and Smith, L. 1975. Benthic oceanography and the distribution of bottom fish off Los Angeles. *CalCOFI Rpt.* 18:118–124.

Mecklenburg, C. W., Mecklenburg, T. A., and Thorsteinson, L. K. 2002. *Fishes of Alaska.* Bethesda, Md.: Amer. Fish. Soc. 1037 pp.

Meffe, G. K., and Carroll, C. R. 1997. *Principles of conservation biology.* 2nd ed. Sunderland, Mass.: Sinauer Associates. 729 pp.

Meffe, G. K., and Snelson, F. F., Jr., eds. 1989. *Ecology and evolution of livebearing fishes (Poeciliidae).* Englewood Cliffs, N.J.: Prentice Hall. 453 pp.

Mendelson, J. 1975. Feeding relationships among species of *Notropis* (Pisces: Cyprinidae) in a Wisconsin stream. *Ecol. Monogr.* 45(3):199–230.

Menon, A. G. K. 1973. Origin of the freshwater fish fauna of India. *Curr. Sci.* 42(16):553–556.

Menzies, R. J., George, R. Y., and Rowe, G. T. 1973. *Abyssal environment and ecology of the world oceans.* New York: Wiley. 488 pp.

Methvenm D. A., Haedrich, R. L., and Rose, G. A. 2001. The fish assemblage of a Newfoundland estuary: diel, monthly, and annual variation. *Est. Coast, Mar. Sci.* 52:669–687.

Metscher, B. D., and Ahlberg, P. E. 2001. Origin of the teleost tail: phylogenetic framework for genetic studies, *in* P. E. Ahlberg, ed., *Major events in vertebrate evolution: paleontology, phylogeny, genetics, and development.* London: Taylor and Francis.

Mettee, M. F., O'Neil, P. E., and Pierson, J. M. 1996. *Fishes of Alabama and the Mobile Basin.* Birmingham, Ala.: Oxmoor House. 820 pp.

Meyer, A. 1990. Ecological and evolutionary consequences of the trophic polymorphism in *Cichlasoma citrinellum* (Pisces: Cichlidae). *Biol. J. Linn. Soc.* 39:279–299.

Meyer, A., Kocher, T. D., Basasibwaki, P., and Willson, A. C. 1990. Monophyletic origin of Lake Victoria cichlid fishes suggested by mitochondrial DNA sequences. *Nature* 347:550–553.

Meyer, J. L., Schultz, E. T., and Helfman, G. S. 1983. Fish schools: An asset to corals. *Science* 220:1047–1049.

Meyer, A., and Wilson, A. C. 1990. Origin of tetrapods inferred from their mitochondria DNA affiliation to lungfish. *J. Mol. Evol.* 31:359–364.

Micklin, P. O. 1988. Dessication of the Aral Sea: A water management disaster in the Soviet Union. *Science* 241:1170–1176.

Milinski, M. 1985. Risks of predation of parasitized sticklebacks *(Gasterosteus aculeatus)* under competition for food. *Behaviour* 93:203–216.

Miller, D. J., and Lea, R. N. 1972. *Guide to the coastal marine fishes of California.* Calif. Dept. Fish, *Game Fish Bull.* 157. 249 pp.

Miller, D. S., Masereeuw, R., Henson, J., and Karnaky, K. J., Jr. 1998. Excretory transport of xenobiotics by dogfish shark rectal gland tubules. *Am. J. Physiol. (Reg. Integr. Comp. Physiol.* 44) 275:R697–R705.

Miller, R. G. 1951. *The natural history of Tahoe fishes.* Ph.D. diss. Stanford University. 160 pp.

Miller, R. J., and Evans, H. E. 1965. External morphology of the brain and lips in catostomid fishes. *Copeia* 1965(4):467–487.

Miller, R. R. 1958. Origin and affinities of the freshwater fish fauna of western North America. Pages 187–222 *in* C. L. Hubbs, ed., *Zoogeography.* Washington, D.C.: AAAS.

Miller, R. R. 1966. Geographical distribution of Central America freshwater fishes. *Copeia* 1966(4):773–802.

Miller, R. R., and Smith, M. L. 1986. Origin and geography of the fishes of central Mexico. Pages 487–518 *in* C. H. Hocutt and E. O. Wiley, eds., *The zoogeography of North American fishes.* New York: Wiley.

Miller, R. R., Williams, J. D., and Williams, J. E. 1989. Extinctions of North American fishes during the past century. *Fisheries* 14:22–38.

Millikin, M. R. 1982. Qualitative and quantitative nutrient requirements of fishes: A review. *Fish. Bull.* 80:655–686.

Minckley, W. K., and Deacon, J. 1968. Southwestern fishes and the enigma of endangered species. *Science* 159:1424–1431.

Minckley, W. L., and Deacon, J. E. eds., 1991. *Battle against extinction: Native fish management in the American West.* Tucson: Univ. Ariz. Press. 517 pp.

Minckley, W. L., Henrickson, D. A., and Bond, C. E. 1986. Geography of western North American freshwater fishes: Description and relationships to intercontinental tectonism. Pages 519–614 *in* C. H. Hocutt and E. O. Wiley, eds., *The zoogeography of North American freshwater fishes.* New York: Wiley.

Mittlebach, G. G. 1984. Predation and resource partitioning in two sunfishes (Centrarchidae). *Ecology* 65:499–513.

Mochioka, N., and Iwamizu, M. 1996. Diet of anguilloid larvae: Leptocephali feed selectively on larvacean houses and fecal pellets. *Mar. Biol.* 125:447–452.

Möller, H. 1984. Reduction of a larval herring population by jellyfish predator. *Science* 224:621–622.

Möller, P. 1995. *Electric fishes: History and behavior.* New York: Chapman and Hall.

Möller, P. R., Nelson, J. G., and Fossen, I. 2003. Fish migration: A Patagonian toothfish found off Greenland. *Nature* 421:599.

Mommsen, T. P., and Walsh, P. J. 1991. Urea synthesis in fishes: Evolutionary and biochemical perspectives. Pages 137–163 *in* P. W. Hochachka and T. P. Mommsen, eds., *Biochemistry and molecular biology of fishes.* Vol 1. New York: Elsevier.

Montgomery, J. C., Baker, C. F., and Carton, A. G. 1997. The lateral line can mediate rheotaxis in fish. *Nature* 389:960–963.

Montgomery, J., and Clements, K. 2000. Disaptation and recovery in the evolution of Antarctic fishes. *TREE* 15:267–271.

Montgomery, J. C., and MacDonald, J. A. 1987. Sensory tuning of lateral line receptors in antarctic fish to the movements of planktonic prey. *Science* 235:195–196.

Moodie, G. E. E., and Reimchen, T. E. 1976. Phenetic variation and habitat differences in *Gasterosteus* populations of the Queen Charlotte Islands. *Syst. Zool.* 25:49–61.

Mooi, R. D., and Gill, A. C. 2002. Historical biogeography of fishes. Pages 43–68 *in* P. J. B. Hart and J. D. Reynolds, eds. *Handbook of fish biology and fisheries,* Vol. 1. Fish biology. London: Backwell.

Mooney, H. A., ed. 1998. Ecosystem management for sustainable marine fisheries. *Ecol. Apps.* 8 (Suppl 1):1–174.

Mooney, H. A., and Hobbs, R. J. 2000. *Invasive species in a changing world.* Covelo, Calif.: Island Press. 457 pp.

Moore, A., Freake, S. M., and Thomas, I. M. 1990. Magnetic particles in the lateral line of the Atlantic salmon (*Salmo salar* L.). *Phil. Trans. R Soc. Lond.* 329:11–15.

Mora, C., and Sale, P. F. 2002. Are populations of coral reef fish open or closed? *TREE* 17:422–428.

Morgan, J. D., and Iwama, G. K. 1991. Effects of salinity growth, metabolism, and ion regulation in juvenile rainbow and steelhead trout *(Oncorhynchus mykiss)* and fall chinook salmon *(Oncorhynchus tshawytscha). Can. J. Fish. Aquat. Sci.* 48:2083–2094.

Morin, J. A., Harrington, A., Nealson, K., Krieger, N., Baldwin, T., and Hastings, J. 1975. Light for all reasons: Versatility in the behavioral repertoire of the flashlight fish. *Science* 190:74–76.

Morri, H. 1979. Changes with time of ammonia and urea concentrations in the blood and tissue of mudskipper fish, *Periophthalmodon cantonensis* and *Boleophthalmus pectinirostris* kept in water and on land. *Comp. Biochem. Physiol. A Physiol.* 64A:235–243.

Morse, D. H. 1977. Feeding behavior and predator avoidance in heterospecific groups. *Bioscience* 27(5):332–339.

Moseley, F. N., and Copeland, B. J. 1969. A portable dropnet for representative sampling of nekton. *Contrib. Mar. Sci. Univ. Texas* 14:37–45.

Moss, S. A. 1977. Feeding mechanisms in sharks. *Amer. Zool.* 17:355–364.

Moss, S. A. 1984. Sharks: An introduction for the amateur naturalist. Englewood Cliffs, N.J.: Prentice Hall. 246 pp.

Motta, P. J. 1985. Functional morphology of the head of Hawaiian and mid-Pacific butterflyfishes (Perciformes, Chaetodontidae). *Env. Biol. Fish.* 13:253–276.

Moy-Thomas, J., and Miles, W. S. 1971. *Paleozoic fishes.* New York: J. Saunders. 259 pp.

Moyle, J. B. 1956. Relationships between the chemistry of Minnesota surface waters and wildlife management. *J. Wildl. Mgmt.* 20(3):303–320.

Moyle, J. B., and Clothier, W. D. 1959. Effects of management and winter oxygen levels on the fish population of a prairie lake. *Trans. Amer. Fish Soc.* 88:178–185.

Moyle, P. B. 1969. *Ecology of the fishes of a Minnesota lake, with special reference to the Cyprinidae.* Ph.D. diss. Univ. of Minn., Minneapolis. 169 pp.

Moyle, P. B. 1973. Ecological segregation among three species of minnows (Cyprinidae) in a Minnesota lake. *Trans. Amer. Fish Soc.* 102(4):794–805.

Moyle, P. B. 1976a. *Inland fishes of California.* Berkeley: Univ. of Calif. Press. 405 pp.

Moyle, P. B. 1976b. Some effects of channelization on the fishes and invertebrates of Rush Creek, Modoc County, California. *Calif. Fish Game* 62(3):179–186.

Moyle, P. B. 1977. In defense of sculpins. *Fisheries* 2(1):20–23.

Moyle, P. B. 1986. Fish introductions into North America: Patterns and ecological impact. Pages 27–43 *in* H. A. Mooney and J. A. Drake, eds., *Ecology of biological invasions of North America and Hawaii.* New York: Springer-Verlag.

Moyle, P. B. 1993. *Fish: An enthusiasts guide.* Berkeley: Univ. of Calif. Press. 272 pp.

Moyle, P. B. 1998. Effects of invading species on freshwater and estuarine ecosystems. Pages 174–194, *in* O. T. Sandlund, P. J. Schei, and Å Viken, eds., *Invasive species and biodiversity management.* London: Chapman and Hall.

Moyle, P. B. 2002. *Inland fishes of California.* Berkeley: Univ. of Calif. Press. 502 pp.

Moyle, P. B., Daniels, R. A., Herbold, B., and Baltz, D. M. 1985. Patterns in distribution and abundance of a noncoevolved assemblage of estuarine fishes in California. *Fish. Bull.* 84:105–117.

Moyle, P. B., and Herbold, B. 1987. Life history patterns and community structure in stream fishes of western North America: Comparisons with eastern North America and Europe. Pages 25–32 *in* W. J. Matthews and D. C. Heins, eds., *Community and evolutionary ecology of North American stream fishes.* Norman: Univ. of Okla. Press.

Moyle, P. B., and Leidy, R. L. 1992. Loss of biodiversity in aquatic systems: Evidence from fish faunas. Pages 128–169 *in* P. L. Fiedler and S. K. Jain, eds., *Conservation biology: The theory and practice of nature conservation, preservation, and management.* New York: Chapman and Hall.

Moyle, P. B., and Li, H.W. 1979. Community ecology and predator–prey relationships in warmwater streams. Pages 171–180 *in* H. Clepper, ed., *Predator–prey systems in fisheries management.* Washington, D.C.: Sport Fish. Inst.

Moyle, P. B., Li, H. W., and Barton, B. 1987. The Frankenstein effect: Impact of introduced fishes on native fishes of North America. Pages 415–426 *in* R. H. Stroud, ed., *The role of fish culture in fisheries management.* Bethesda, Md.: Amer. Fish. Soc.

Moyle, P. B., and Moyle, M. A. 1991. Introduction to fish imagery in art. *Env. Biol. Fish* 31:5–23.

Moyle, P. B., and Moyle, P. R. 1995. Endangered fishes and economics: Intergenerational obligations. *Env. Biol. Fish* 42, pages 29–37.

Moyle, P. B., and Sato, G. M. 1991. On the design of preserves to protect native fishes. Pages 155–169. *In* W. L. Minckley and J. E. Deacon, eds., *Battle against extinction: Native fish management in the American West.* Tucson: Univ. of Arizona Press.

Moyle, P. B., and Senanayake, F. R. 1982. Wildlife conservation in Sri Lanka: A Buddhist dilemma. *FAO Tigerpaper* 9(4):1–4.

Moyle, P. B., and Senanayake, F. R. 1984. Resource partitioning among the fishes of rainforest streams in Sri Lanka. *J. Zool. Lond.* 202:195–223.

Moyle, P. B., and Vondracek, B. 1985. Persistence and structure of the fish assemblage in a small California stream. *Ecology* 66:1–13.

Muir, B. S., and Kendall, J. I. 1968. Structural modifications in the gills of tunas and other oceanic fishes. *Copeia* 1968(2):388–398.

Mundy, P. R. 1997. The role of harvest management in the future of Pacific salmon populations: Shaping human behavior to enable the persistence of salmon. Pages 315–329 *in* D. Stouder, P. A. Bisson, and R. J. Naiman eds., *Pacific salmon and their ecosystems: Status and future options.* New York: Chapman and Hall.

Muntz, W. R.A. 1976. On yellow lenses in mesopelagic animals. *J. Mar. Biol. Ass. U. K.* 56:963–976.

Munz, F. W. 1971. Vision: Visual pigments. Pages 1–32 *in* W. S. Hoar and D. J. Randall, eds. *Fish physiology.* Vol. 5. New York: Academic Press.

Munz, F. W., and McFarland, W. 1964. Regulatory function of a primitive vertebrate kidney. *Comp. Biochem. Physiol.* 13:381–400.

Murdaugh, H. V., and Robin, E. D. 1967. Acid–base metabolism in the dogfish sharks. Pages 249–264 *in* P. W. Gilbert, R. F. Mathewson, and D. P. Rall, eds., *Sharks, skates, and rays.* Baltimore, Md.: Johns Hopkins Press.

Murphy, G. I. 1966. Population biology of the Pacific sardine *(Sardinops caerulea). Proc. Calif. Acad. Sci.* 34(1):1–84.

Murphy, G. I. 1968. Pattern in life history and the environment. *Amer. Nat.* 102:390–404.

Murphy, G. I. 1977. Clupeioids. Pages 283–308 *in* J. A. Gulland, ed., *Fish population dynamics.* New York: Wiley.

Murray, A. M. 2001. The fossil record and biogeography of the Cichlidae (Actinopterygii: Labroidei). *Biol. J. Linn. Soc.* 74:517–532.

Musick, J. A., Bruton, M. N., and Balon, E. K., eds. 1991. The biology of *Latimeria chalumnae* and evolution of coelacanths. *Env. Biol. Fish* 32. 436 pp.

Muss, A., Robertson, D. R., Stephen, C. A., Wirtz, P., and Bowen, B. W. 2001. Phylogeography of *Ophioblennius*: The role of ocean currents and geography in reef fish evolution. *Evolution* 55:561–572.

Muus, B. J. 1967. *Freshwater fish of Britain and Europe.* London: Collins. 222 pp.

Myers, G. S. 1951. Fresh-water fishes and East Indian zoogeography. *Stanf. Ichthy. Bull.* 4(1):11–21.

Myers, G. S. 1964. A brief sketch of the history of ichthyology in America to the year 1850. *Copeia* 1964(1):33–41.

Myers, R. A., Barrowman, N. J., Hutchings, J. A., and Rosenberg, A. A. 1995. Population dynamics of exploited fish stocks at low population levels. *Science* 269:1106–1108.

Myrberg, A. A., Jr. 1978. Underwater sound: Its effect on the behavior of sharks. Pages 391–417 *in* E. S. Hodgson and R. F. Mathewson, eds., *Sensory biology of sharks, skates, and rays.* Washington, D.C.: Off. Naval Res. Publ.

Nafpaktitis, B. G. 1978. Systematics and distribution of laternfishes of the genera *Lobianchia* and *Diaphus* (Myctophida) in the Indian Ocean. *Los Ang. Co. Nat. Hist. Mus. Sci. Bull.* 30:1–92.

Naiman, R. J. 1976. Productivity of a herbivorous pupfish population *(Cyprinodon nevadensis)* in a warm desert stream. *J. Fish Biol.* 9:125–137.

Naiman, R. J., and Soltz, D. L. 1981. *Fishes in North American deserts.* New York: Wiley.

Nakamura, H. 1969. *Tuna distribution and migration.* London: Fishing News Books. 76 pp.

Nakamura, R. 1976. Experimental assessment of factors influencing microhabitat selection by the two tidepool fishes *Oligocottus masculosus* and *O. snyderi. Mar. Biol.* 37:97–104.

Nakano, S., and Murakami, M. 2001. Reciprocal subsidies: dynamic interdependence between terrestrial and aquatic food webs. *PNAS* 98:166–170.

Nakano, T., and Tomlinson, N. 1967. Catecholamine and carbohydrate concentrations in rainbow trout *(Salmo gairdneri)* in relation to physical disturbance. *J. Fish. Res. Bd. Canada* 24:1701–1715.

Nakatsuru, K., and Kramer, D. L. 1982. Is sperm cheap? Limited male fertility and female choice in the lemon tetra (Pisces, Characidae). *Science* 216:753–755.

Nakaya, K. 1995. Hydrodynamic function of the head of the hammerhead shark (Elasmobranchi. Sphyrnidae). *Copeia* 1995:330–336.

Narnaware, Y. K., Peyon, P. P., Lin, X., and Peter, R. E. 2000. Regulation of food intake by neuropeptide Y in goldfish. *Am. J. Physiol. Regulatory Integrative Comp. Physiol.* 279:R1025–R1034.

National Research Council. 1996. *Upstream: Salmon and society in the Pacific Northwest.* Washington, D.C.: National Academy Press. 452 pp.

National Research Council 1998. *Sustaining marine fisheries.* Washington, D.C.: National Academy Press.

Nauen, J. C., and Lauder, G. U. 2002. Hydrodynamics of caudal fin locomotion by chub mackerel, *Scomber japonicus* (Scombridae). *J. Exp. Biol.* 205:1709–1724.

Naylor, R. L., and 9 others. 1998. Nature's subsidies to shrimp and salmon farming. *Science* 282:883–884.

Naylor, R. L., and 9 others. 2003. Effects of aquaculture on world fish supplies. *Nature* 405:1017–1024.

Neill, W. H., Chang, R. K. C., and Dizon, A. E. 1976. Magnitude and ecological implications of thermal inertia in skipjack tuna, *Katsuwonus pelamis* (L.). *Env. Biol. Fish* 1:61–80.

Neill, W. H., and Magnuson, J. J. 1974. Distributional ecology and behavioral thermoregulation of fishes in relation to heated effluent from a power plant at Lake Monona, Wisconsin. *Trans. Am. Fish. Soc.* 103:663–710.

Neill, W. H., Magnuson, J. J., and Chipman, G. G. 1972. Behavioral thermoregulation by fishes: A new experimental approach. *Science* 176:1443–1455.

Nekvasil, N. P., and Olson, K. R. 1986a. Plasma clearance, metabolism, and tissue accumulation of ^{3}H-labelled catecholamines in trout. *Am. J. Physiol.* 259:R519–R525.

Nekvasil, N. P., and Olson, K. R. 1986b. Extraction and metabolism of circulating catecholamines in the trout gill. *Am. J. Physiol.* 259:R526–R531.

Nelson, D. R., Johnson, R. R., McKibben, J. N., and Pittenger, G. G. 1986. Agonistic attacks on divers and submersibles by gray reef sharks, *Carcharhinus amblyrhynchos:* Antipredatory or competitive? *Bull. Mar. Sci.* 38:68–88.

Nelson, J. S. 1968. Hybridization and isolating mechanisms between *Catostomus commersoni* and *C. macrocheilus* (Pisces, Catostomidae). *J. Fish. Res. Bd. Canada* 25:101–150.

Nelson, J. S. 1984. *Fishes of the world.* 2nd ed. New York: Wiley-Interscience. 521 pp.

Nelson, J. S. 1994. *Fishes of the world.* 3rd ed. New York: Wiley. 600 pp.

Nemeth, D. H. 1997a. Modulation of buccal pressure during prey capture in *Hexagrammos decagrammus* (Teleostei: Hexagrammidae). *J. Exp. Biol.* 200:2145–2154.

Nemeth, D. H. 1997b. Modulation of attack behavior and its effect on feeding performance in a trophic generalist fish, *Hexagrammos decagrammus. J. Exp. Biol.* 200:2155–2164.

Neumeyer, C. 1985. An ultraviolet receptor as a fourth receptor type in goldfish colour vision. *Naturwissenchaften* 72:162–163.

Neverman, D., and Wurtsbaugh, W. A. 1994. The thermoregulatory function of diel vertical migration for a juvenile fish, *Cottus extensus. Oecologia* 98:247–256.

Neves, R. J., and Angermeier, P. L. 1990. Habitat alteration and its effects on native fishes in the upper Tennessee River system, east-central USA. *J. Fish Biol.* 37 (suppl. A):45–52.

Neville, C. M. 1979. Sublethal effects of environmental acidification on rainbow trout *(Salmo gairdneri). J. Fish. Res. Bd. Canada* 36:84–87.

Nevitt, G. A., Dittman, A. H., Quinn, T. P., and Moody, W. J., Jr. 1994. Evidence for a peripheral olfactory memory in imprinted salmon. *Proc. Natl. Acad. Sci. U.S.A.* 91:4288–4292.

Nichols, J. T. 1943. *The fresh-water fishes of China.* New York: Amer. Mus. Nat. Hist. 322 pp.

Nicol, J. A. C., and Zyznar, E. S. 1973. The tapetum lucidum in the eye of the big-eye, *Priacanthus arenatus* Cuvier. *J. Fish Biol.* 5:519–522.

Nielson, J. G. 1977. The deepest-living fish, *Abyssobrotula galatheae,* a new genus and species of oviparous ophidioids (Pisces, Brotulidae). *Galathea Rpt.* 14:41–48.

Nikinmaa, M. 1986. Control of red cell pH in teleost fishes. *Ann. Zool. Fennica* 23:223–235.

Nikinmaa, M. 1990. *Vertebrate red blood cells.* Berlin: Springer-Verlag.

Nikinmaa, M., Cech, J. J., Jr., and McEnroe, M. 1984. Blood oxygen transport in stressed striped bass *(Morone saxatilis)*: Role of beta-adrenergic response. *J. Comp. Physiol. B* 154:365–369.

Nikolsky, G. V. 1954. *Special ichthyology.* Jerusalem (1961): Israel Prog. Sci. Trans. 538 pp.

Nikolsky, G. V. 1963. *The ecology of fishes.* New York: Academic Press. 352 pp.

Nilsson, G. 1996. Brain and body oxygen requirements of *Gnathonemus petersii,* a fish with an exceptionally large brain. *J. Exp. Biol.* 199:603–607.

Nilsson, G. E., Lofman, C. O., and Block, M. 1995. Extensive deformation in fish gills observed by *in vivo* microscopy: Apparent adaptations for enhancing oxygen uptake. *Exp. Bio.* 198:1151–1156.

Nilsson, N. A. 1963. Interaction between trout and char in Scandinavia. *Trans. Amer. Fish. Soc.* 92(3):276–285.

Nilsson, N. A. 1967. Interactive segregation between fish species. Pages 295–313 *in* S. D. Gerking, ed., *The biological basis of freshwater fish production.* New York: Wiley.

Nilsson, S., and Grove, D. J. 1974. Adrenergic and cholinergic innervation of the spleen of the cod *Gadus morhua, Eur. J. Pharmac.* 28:135–143.

Nilsson, S., and Holmgren, S. 1993. Autonomic nerve functions. Pages 279–313 *in* D. H. Evans, ed., *The physiology of fishes.* Boca Raton, Fla.: CRC Press.

Nishida, M. 1991. Lake Tanganyika as an evolutionary reservoir of old lineages of East African cichlid fishes: Inferences from allozyme data. *Experientia* 47:974–979.

Nixon, S. W., and Oviatt, C. A. 1973. Ecology of a New England salt marsh. *Ecol. Monogr.* 43(4):463–498.

Noakes, D. L. G. 1992. Behaviour and rhythms in fishes. Pages 39–50 *in* M.A.Ali. ed., *Rhythms in fishes.* New York.: Plenum Press.

Nonnotte, G. 1981. Cutaneous respiration in six freshwater teleosts. *Comp. Biochem. Physiol.* 70A:541–543.

Norgaard, R. B. 1994. Ecology, politics, and economics: Finding the common ground for decision making in economics. Pages 439–465 *in* G. K. Meffe and C. R. Carroll, eds., *Principles of conservation biology.* Sunderland, Mass.: Sinauer Associates.

Norman, J. R., and Greenwood, P. H. 1975. *A history of fishes.* 3rd ed. New York: Halstead Press, 467 pp.

Norse, E. A. 1993. *Global marine biodiversity: A strategy for building conservation into decision making.* Washington, D.C.: Inland Press, 383 pp.

Northcote, T. G., ed. 1969. *Symposium on salmon and trout in streams.* H. R. MacMillan lectures in fisheries. Vancouver: Univ. of B. C.

Northcote, T. G., Arcifa, M. S., and Froehlich, O. 1987. Fin-feeding by the piranha (*Serrasalmus spilopluera* Kner): The cropping of a novel renewable resource. *Proc. V. Congr. Europe Ichthyol. Stockholm* 1985:133–143.

Northcutt, R. G. 1978. Brain organization in the cartilaginous fishes. Pages 117–193 *in* E. S. Hodgson and R. F. Mathewson, eds., *Sensory biology of sharks, skates, and rays.* Washington, D.C.: Off. Naval Res. Publ.

Norton, B. G. 1987. *Why preserve natural variety?* Princeton N.J.: Princeton Univ. Press. 281 pp.

Novacek, M. J., and Marshall, L. G. 1976. Early biogeographic history of the ostariophysan fishes. *Copeia* 1976(1):1–12.

Novales Flamarique, I., and Browman, H. I. 2001. Foraging and prey-search behavior of small juvenile rainbow trout (*Oncorhynchus mykiss*) under polarized light. *J. Exp. Biol.* 204:2415–2422.

Novales Flamarique, I., and Hawryshyn, C. W. 1997. Is the use of underwater polarized light by fish restricted to crepuscular time periods? *Vision Res.* 37:975–989.

Novales Flamarique, I., and Hawryshyn, C. W. 1998. The common white sucker (*Catastomus commersoni*): A fish with ultraviolet sensitivity that lacks polarization sensitivity. *J. Comp. Physiol. A* 182:331–341.

Nshombo, M. 1994. Foraging behavior of the scale-eater *Plecodus straeleni* (Cichlidae, Teleostei) in Lake Tanganyika, Africa. *Env. Biol. Fish* 39:59–67.

Nursall, J. R. 1981. Behavior and habitat affecting the distribution of five species of sympatric mudskippers in Queensland. *Bull. Mar. Sci.* 31:730–735.

Oberdorff, T., Hugueny, B., Compin, A., and Belkessam, D. 1998. Noninteractive fish communities in the coastal streams of Northwestern France. *J. Anim. Ecol.* 67:472–484.

Ochi, H., Onchi, T., and Yanagisawa, Y. 2001. Alloparental care between catfishes in Lake Tanganyika. *J. Fish Biol.* 59:1279–1286.

O'Day, W. T. 1974. Bacterial luminescence in the deepsea fish *Oneirodes acanthias* Gilbert (1915). *Los Ang. Nat. Hist. Mus. Contr. Sci.* 255:1–12.

Ogura, M., and Ishida, Y. 1995. Homing behavior and vertical movements of four species of Pacific salmon (*Oncorhynchus* spp.) in the central Bering Sea. *Can. J. Fish. Aquat. Sci.* 52:532–540.

Oguri, M. 1964. Rectal glands of marine and fresh-water sharks: Comparative histology. *Science* 144:1151–1152.

Ohno, S. 1974. Protochordata, Cyclostomata, and Pisces. Pages 1–91 *in* B. John, ed., *Animal cytogenetics*. Vol. 4. Berlin: Bortraeger.

Oliveira, K., McCleave, J. D., and Wippelhauser, G. S. 2001. Regional variation and effect of lake: River area on sex distribution of American eels. *J. Fish Biol.* 58:943–952.

Olsen, P. E., and McCune, A. R. 1991. Morphology of the *Semionotus elegans* species group from the early Jurassic part of the Newark supergroup of eastern North America with comments on the family Semionotidae (Neopterygii). *J. Vert. Paleo.* 11:269–292.

Olson, K. R. 1992. Blood and extracellular fluid volume regulation: Role of the reninangiotensin system, kallikrein-kinin system, and atrial natriuretic peptides. Pages 135–154 *in* W. S. Hoar, D. J. Randall, and A. P. Farrell, eds., *Fish physiology*. Vol. XII, Part B. The cardiovascular system. San Diego, Calif.: Academic Press.

Olson, K. R. 1996. Secondary circulation in fish: Anatomical organization and physiological significance. *J. Exp. Zool.* 275:172–185.

Olson, K. R. 1998. The cardiovascular system. Pages 129–154 *in* D. H. Evans, ed., *The physiology of fishes*. 2nd ed. Boca Raton, Fla.: CRC Press.

Olson, K. R. 2002a. Gill circulation: Regulation of perfusion distribution and metabolism of regulatory molecules. *J. Exp. Zool.* 293:320–335.

Olson, K. R. 2002b. Vascular anatomy of the fish gill. *J. Exp. Zool.* 293:214–231.

Olson, K. R., and Duff, D. W. 1992. Cardiovascular and renal effects of eel and rat atrial natriuretic peptide in rainbow trout, *Salmo gairdneri. J. Comp. Physiol.* 162:408–415.

Olver, C. H., Shuter, B. J., and Minns, C. K. 1995. Toward a definition of conservation principles for fisheries management. *Can. J. Fish. Aquat. Sci.* 52:1584–1594.

Orr, D. W. 1996. Slow knowledge. *Cons. Biol.* 10:699–702.

Orti, G., and Meyer, A. 1997. The radiation of characiform fishes and the limits of resolution of mitochondrial ribosomal DNA sequences. *Syst. Biol.* 46:75–100.

Orvig, T., ed. 1968. *Current problems of lower vertebrate phylogeny*. New York: Wiley-Interscience. 539 pp.

Ostrand, K. G., Thies, M. L., Hall, D. D., and Carpenter, M. 1996. Gar ichthyotoxin: Its effects on natural predators and the toxin's evolutionary function. *Southw. Nat.* 41:375–377.

Otsu, T., and Uchida, R. N. 1963. Model of the migration of albacore in the North Pacific Ocean. *NOAA Fish. Bull.* 63:33–44.

Ottaway, E. M., and Simkiss, K. 1977. Instantaneous growth rates of fish scales and their use in studies of fish populations. *J. Zool. (Lond.)* 181:407–419.

Oviatt, C. A., Gall, A. L., and Nixon, S. W. 1972. Environmental effects of Atlantic menhaden on surrounding waters. *Chesapeake Sci.* 13:321–323.

Oviatt, C. A., and Nixon, S. W. 1973. The demersal fish of Narraganset Bay: An analysis of community structure, distribution, and abundance. *Est. Coast. Mar. Sci.* 1:361–378.

Owen, R. B., Crossley, R., Johnson, T. C., Tweddle, D., Kornfield, I., Davison, S., Eccles, D. H., and Engstom, D. E. 1990. Major low levels of Lake Malawi and their implications for speciation rates in cichlid fishes. *Proc. R. Soc. Lond. B* 240:519–553.

Paine, R. T., and Palmer, A. R. 1978. *Sicyases sanguineus:* A unique trophic generalist from the Chilean intertidal zone. *Copeia* 1978(1):75–81.

Pakhomov, E. A. 1998a. Diet of two Antarctic dragonfishes (Pisces: Bathydraconidae) from the Indian sector of the Southern Ocean. *Antar. Sci.* 10:55–61.

Pakhomov, E. A. 1998b. Feeding plasticity of the Antarctic fish *Trematomus hansoni* Boulenger, 1902 (Pisces: Nototheniidae): The influence of fishery waste on the diet. *Polar Biol.* 19:289–292.

Paller, M. H., Gladden, J. B., and Heuer, J. H. 1992. Development of the fish community in a new South Carolina reservoir. *Am. Midl. Nat.* 128:95–114.

Panella, G. 1971. Fish otoliths: Daily growth layers and periodical patterns. *Science* 173:1124–1127.

Pankhurst, N. W., and Van Der Kraak, G. 1997. Effects of stress on reproduction and growth of fish. Pages 73–93 *in* G. K. Iwama, A. D. Pickering, J. P. Sumpter, and C. B. Schreck eds., *Fish Stress and health in aquaculture.* Cambridge: Cambridge University Press.

Parenti, L. R. 1981. A phylogenetic and biogeographic analysis of the cyprinodontiform fishes (Teleostei, Atherinomorpha). *Bull. Amer. Mus. Nat. Hist.* 168:335–557.

Parenti, L. R. 1984. A taxonomic revision of the Andean killifish genus *Orestias* (Cyprinodontiformes, Cyprinodontidae). *Bull. Amer. Mus. Nat. Hist.* 178:107–214.

Parenti, L. R. 1989. A revision of the phallostethid fishes (Atherinomorpha, Phallostethidae). *Proc. Calif. Acad. Sci.* 46:243–277.

Parenti, L. R. 1993. Relationships of atherinomorph fishes (Teleostei). *Bull. Mar. Sci.* 52:170–196.

Parin, N. V. 1968. *Ichthyofauna of the epipelagic zone.* Jerusalem: Israel Prog. Sci. Trans. 206 pp.

Parkyn, D. C., and Hawryshyn, C. W. 1993. Polarized-light sensitivity in rainbow trout (*Oncorhynchus mykiss*). *J. Comp. Physiol. A* 172:493–500.

Parrish, J. K. 1993. Comparison of hunting behavior of four piscine predators attacking schooling prey. *Ethology* 95:233–246.

Parrish, J. K., and Edelstein-Keshet, L. 1999. Complexity, pattern, and evolutionary trade-offs in animal aggregation. *Science* 284:99–104.

Patrick, M. L., Pärt, P., Marshall, W. S., and Wood, C. M. 1997. Characterization of ion and acid–base transport in the fresh water adapted mummichog (*Fundulus heteroclitus*). *J. Exp. Zool.* 279:208–219.

Patrick, M. L., Wood, C. M., and Marshall, W. S. 1997. Calcium regulation in the freshwater adapted mummichog. *J. Fish Biol.* 51:135–145.

Patten, B. C. 1964. The rational decision process in salmon migration. *J. Cons. Perma. Int. Explor. Mer.* 8:410–417.

Patterson, C. 1977. The contribution of paleontology to teleostean phylogeny. Pages 579–643 *in* M. K. Hecht, P. G. Goody, and B. M. Hecht, eds., *Major patterns in vertebrate evolution.* New York: Plenum Press.

Patterson, C., and Rosen, D. E. 1977. Review of the ichthyodectiform and other mesozoic teleost fishes and the theory and practice of classifying fossils. *Bull. Amer. Mus. Nat. Hist.* 158:81–172.

Paul, A. J., Paul, J. M., and Smith, R. L. 1988. Respiratory energy requirements of the cod *Gadus macrocephalus* Tilesius relative to body size, food intake, and temperature. *J. Exp. Mar. Biol. Ecol.* 122:83–89.

Pauly, D., and MacLean, J. 2003. *In a perfect ocean: The state of fisheries and ecosystems in the North Atlantic Ocean.* Covelo, Calif.: Island Press.

Payne, A. I. 1986. *The ecology of tropical lakes and rivers.* New York: Wiley.

Pelis, R. M., Zydlewski, J., and McCormick, S. D. 2001. Gill Na^+-K^+-$2Cl^-$ cotransporter abundance and location in Atlantic salmon: Effects of seawater and smolting. *Am. J. Physiol. (Regulatory Integrative Comp. Physiol.)* 280:R1844–R1852.

Pelster, B. 1998. Buoyancy. Pages 25–42 *in* D. H. Evans, ed., *The physiology of fishes.* 2nd ed. Boca Raton, Fla.: CRC Press.

Pelster, B., and Niederstätter, H. 1997. pH dependent proton secretion in cultured swim bladder gas gland cells. *Am. J. Physiol.* 273:1719–1725.

Pelster, B., and Scheid, P. 1992. Countercurrent concentration and gas secretion in the fish swim bladder. *Physiol. Zool.* 65(1):1–16.

Peng, K. W., Chew, S. F., Lim, C. B., Kok, T. W. K., Kuah, S. S. L., and Ip, Y. K. 1998. The mudskipper *Periophthalmodon schloseri* and *Boleophthalmus boddaerti* can tolerate environmental ammonia of 446 μM and 36 μM, respectively. *Fish Physiol. Biochem.* 19:59–69.

Percy, L. R., and Potter, I. C. 1976. Blood cell formation in the river lamprey, *Lampetra fluviatilis. J. Zool. (Lond.)* 178:319–340.

Pérez-España, H., Gálvan-Magaña, F., and Abitia-Cárdenas, L. A. 1998. Growth, consumption, and productivity of the California killifish in Ojo de Liebre Lagoon, Mexico. *J. Fish Biol.* 52:1068–1077.

Perrot, M. N., Grierson, C. E., Hazon, N., and Balment, R. J. 1992. Drinking behavior in sea water and fresh water teleosts, the role of the renin-angiotensin system. *Fish Physiol. Biochem.* 10:161–168.

Perret, W. S., and Caillouet, C. W. 1974. Abundance and size of fishes taken by trawling Vermillion Bay, Louisiana. *Bull. Mar. Sci.* 24:52–75.

Perry, S. F., Malone, S., and Ewing, D. 1987a. Hypercapnic acidosis in the rainbow trout (*Salmo gairdneri*). II. Renal ionic fluxes. *Can. J. Zool.* 65:896–902.

Perry, S. F., Malone, S., and Ewing, D. 1987b. Hypercapnic acidosis in rainbow trout (*Salmo gairdneri*). I. Branchial ionic fluxes and blood acid–base status. *Can. J. Zool.* 65:888–895.

Perry, S. F., and Tuffs, B., eds. 1998. *Fish Respiration.* San Diego, Calif.: Academic Press, 365 pp.

Persson, L., Bystrom, P., and Wahlstrom, E. 2000. Cannibalism and competition in Eurasian perch: population dynamics of an ontogenetic omnivore. *Ecology* 81:1058–1071.

Peter, R. E., and Chang, J. P. 1999. Brain regulation of growth hormone secretion and food intake in fish. Pages 55–67 *in* D. S. Prasada Rao and R. E. Peter eds. *Neural regulation in the vertebrate endocrine system: neuroendocrine regulation.* New York: Kluver Academic/Plenum.

Peter, R. E., and Yu, K. L. 1997. Neuroendocrine regulation of ovulation in fishes: Basic and applied aspects. *Rev. Fish Biol. Fisheries* 7:173–197.

Peterson, M. S., and Ross, S. T. 1991. Dynamics of littoral fishes and decapods along a coastal river-estuarine gradient. *Est. Coast Shelf Sci.* 33:467–483.

Peterson, R. H., Coombs, K., Power, J., and Paim, U. 1989. Responses of several fish species to pH gradients. *Can. J. Zool.* 67:1566–1572.

Pethiyagoda, R. 1991. *Freshwater fishes of Sri Lanka.* Colombo, Sri Lanka: Wildlife Heritage Trust. 362 pp.

Pethiyagoda, R. 1994. Threats to the indigenous freshwater fishes of Sri Lanka and remarks on their conservation. *Hydrobiol.* 285:189–201.

Petrovic, M., Solé, M., López De Alda, M. J., and Barceló, D. 2002. Endocrine disruptors in sewage treatments plants, recieving river waters, and sediments: integration of chemical analysis and biological effects on feral carp. *Environ. Toxicol. Chem.* 21(10):2146–2156.

Pfeiler, E. 1986. Towards an explanation of the developmental strategy in leptocephalus larvae of marine teleost fishes. *Env. Biol. Fish* 15:3–13.

Pflieger, W. L. 1975. *The fishes of Missouri.* Missouri Dept. of Fish and Game. 343 pp.

Pic, P., Mayer-Gostan, N., and Maetz, J. 1974. Branchial effects of epinephrine in the seawater-adapted mullet. I. Water permeability. *Am. J. Physiol.* 226:698–702.

Piermarini, P. M., and Evans, D. H. 2000. Effects of environmental salinity on Na$^+$/K$^+$-ATPase in the gills and rectal gland of a euryhaline elasmobranch (*Dasyatis sabina*). *J. Exp. Biol.* 203:2957–2966.

Pietsch, T. W. 1974. Osteology and relationships of ceratioid anglerfishes of the family Oneirodidae, with a review of the genus *Oneirodes* Lutken. *L. A. Co. Nat. Hist. Mus. Sci. Bull.* 18:1–113.

Pietsch, T. W. 1976. Dimorphism, parasitism, and sex: Reproductive strategies among deepsea ceratoid anglerfishes. *Copeia* 1976(4):781–793.

Pietsch, T. W. 1978a. Evolutionary relationships of the seamoths (Teleostei, Pagasidae) with a classification of gasterosteiform families. *Copeia* 1978(3):517–529.

Pietsch, T. W. 1978b. The feeding mechanism of *Stylephorus chordatus* (Teleostei, Lampridiiformes): Functional and ecological implications. *Copeia* 1978(2):255–262.

Pietsch, T. W., ed. 1995. *Georges Cuvier (1828): Historical portrait of the progress of ichthyology from its origins to our own time.* Trans. from French by A. J. Simpson. Baltimore: Johns Hopkins Univ. Press.

Pietsch, T. W., and Grobecker, D. B. 1978. The compleat angler: Aggressive mimicry in an antennariid anglerfish. *Science* 201:369–370.

Pietsch, T. W., and Grobecker, D. B. 1987. *Frogfishes of the world: Systematics, zoogeography, and behavioral ecology.* Stanford, Calif.: Stanford Univ. Press. 420 pp.

Pilley, C. M., and Wright, P. A. 2000. The mechanisms of urea transport by early life stages of rainbow trout (*Oncorhynchus mykiss*). *J. Exp. Biol.* 203:3199–3207.

Pisam, M., Caroff, A., and Rambourg, A. 1987. Two types of chloride cells in the epithelium of a freshwater-adapted fish: *Lebistes reticulatus*; their modifications during adaptations to salt water. *Amer. J. Anat.* 179:40–50.

Pister, E. P. 1991. The Desert Fishes Council: Catalyst for change. Pages 55–68 *in* W. L. Minckley and J. E. Deacon, eds., *Battle against extinction: Native fish management in the American West.* Tucson: Univ. of Ariz. Press.

Pitcher, T. J., ed. 1993. *The behaviour of teleost fishes.* 2nd ed. London: Chapman and Hall. 715 pp.

Pitcher, T. J., and Parrish, J. K. 1993. Functions of shoaling behavior in fishes. Pages 363–440 *in* T. J. Pitcher, ed., *The behaviour of teleost fishes.* 2nd ed. London: Chapman and Hall.

Pitcher, T. J., Partridge, B. L., and Wardle, C. S. 1976. A blind fish can school. *Science* 194:963–65.

Platt, A. E. 1995. Dying seas. *World Watch* 8:10–19.

Poff, N. L., and 7 others. 1997. The natural flow regime: A paradigm for river conservation and restoration. *Bioscience* 47:769–784.

Pohlmann, K., Graasso, F. W., and Breithaupt, T. 2001. Tracking wakes: The nocturnal predatory strategy of piscivorous catfish. *PNAS* 98:7371–7374.

Polis, G. A., and Holt, R. D. 1992. Intraguild predation: The dynamics of complex trophic interactions. *TREE* 7:151–154.

Pollard, D. A., Ingram, B. A., Harris, J. H., and Reynolds, L. F. 1990. Threatened fishes in Australia: An overview. *J. Fish Biol.* 57(Suppl. A):67–78.

Polovina, J. J. 1996. Decadal variation in the trans-Pacific migration of northern bluefin tuna (*Thunnus thynnus*) coherent with climate induced-change in prey abundance. *Fish. Oceanogr.* 5:114–119.

Poluhowich, J. J. 1972. Adaptive significance of eel multiple hemoglobins. *Physiol. Zool.* 45:215–222.

Pombo, L., Elliott, M., and Rebelo, J. E. 2002. Changes in the fish fauna of the Ria de Aveiro estuarine lagoon (Portugal) during the twentieth century. *J. Fish Biol.* 61(Suppl. A):167–181.

Popper, A. N., and Coombs, S. 1980. Acoustic detection by fishes. Pages 403–430 *in* M. A. Ali, ed. *Environmental physiology of fishes.* NATO Advanced Study Institutes Series A: Life Sciences 35. New York: Plenum Press.

Popper, A. N., and Fay, R. R. 1973. Sound detection and processing by teleost fishes: A critical review. *J. Acoust. Soc. Amer.* 53:1515–1529.

Popper, A. N., and Fay, R. R. 1977. Structure and function of the elasmobranch auditory system. *Amer. Zool.* 17:443–452.

Popper, A. N., and Platt, C. 1993. Inner ear and lateral line. Pages 99–136 *in* D. H. Evans, ed. *The physiology of fishes,* 1st ed. Boca Raton, Fla.: CRC Press.

Pörtner, H. O. 2002. Physiological basis of temperature-dependent biogeography: Trade-offs in muscle design and performance in polar ectotherms. *J. Exp. Biol.* 205:2217–2230.

Postel, S. L., Dailey, G. C., and Ehrlich, Paul R. 1996. Human appropriation of renewable fresh water. *Science* 271:785–788.

Potter, I. C., Beckley, L. E., Whitfield, A. K., and Lenaton, R. C. J. 1990. Comparisons between the roles played by estuaries in the life cycles of fishes in temperate western Australia and Southern Africa. *Env. Biol. Fish* 28:143–187.

Potts, G. W., and Wootton, R. J., eds. 1984. *Fish reproduction: Strategies and tactics.* London: Academic Press.

Potts, W. T. W., and Fleming, W. R. 1971. The effect of environmental calcium and ovine prolactin on sodium balance in *Fundulus kansae. J. Exp. Biol.* 54:63–75.

Poulson, T. L., and White, W. B. 1969. The cave environment. *Science* 165:971–981.

Power, M. 1987. Predator avoidance by grazing fishes in temperate and tropical streams: Importance of stream depth and prey size. Pages 333–353 *in* W. C. Kerfoot and A. Sih, eds., *Predation: Direct and indirect effects on aquatic communities.* Hanover, N.H.: Univ. of New England Press.

Power, M., and Matthews, W. J. 1983. Algae-grazing minnows (*Campostoma anomalum*), piscivorous bass (*Micropterus* spp.) and the distribution of attached algae in a small prairie margin stream. *Oecologia* 60:328–332.

Power, M. E. 1983. Grazing responses of tropical freshwater fishes to different scales of variation in their food. *Env. Biol. Fish.* 9:103–115.

Powers, D. A. 1972. Hemoglobin adaptation for fast and slow water habitats in sympatric catostomid fishes. *Science* 177:360–362.

Prejs, A., and Blaszczyk, M. 1977. Relationships between food and cellulase activity in freshwater fishes. *J. Fish Biol.* 11:447–452.

Prem, C., and Pelster, B. 2001. Swimbladder gas gland cells cultured on permeable supports regain their characteristic polarity. *J. Exp. Biol.* 204:4023–4029.

Prem, C., Salvenmoser, W., Würtz, J., and Pelster, B. 2000. Swimbladder gas gland cells produce surfactant: In vivo and in culture. *Am. J. Physiol.* (*Regulatory Integrative Comp. Physiol.*) 279:R2336–R2343.

Preston, J. L. 1978. Communication systems and social interactions in a goby-shrimp symbiosis. *Anim. Behav.* 26:791–802.

Priede, I. G. 1976. Functional morphology of the bulbus arteriosus of rainbow trout (*Salmo gairdneri* Richardson). *J. Fish Biol.* 9:209–216.

Purnell, M. A. 2001. Scenarios, selection, and the ecology of early vertebrates, *in* P. E. Ahlberg, ed. *Major events in vertebrate evolution: paleontology, phylogeny, genetics, and development.* London: Taylor and Francis.

Qian, X., Cui, Y., Xiong, B., and Yang, Y. 2000. Compensatory growth, feed utilization and activity in gibel carp, following feed deprivation. *J. Fish Biol.* 56:228–232.

Rabalais, N. N., Turner, R. E., and Scavia, D. 2002. Beyond science into policy: Gulf of Mexico hypoxia and the Mississippi River. *Bioscience* 52:129–142.

Radakov, D. V. 1972. *Schooling in the ecology of fish.* New York: Wiley. 173 pp.

Radovich, J. 1961. Relationships of some marine organisms of the northeast Pacific to water temperatures, particularly during 1957–1959. *Calif. Dept. Fish Game Fish Bull.* 112:1–62.

Radovich, J. 1979. Managing pelagic schooling prey species. Pages 365–376 *in* H. Clepper, ed., *Predator–prey systems in fisheries management.* Washington, D.C.: Sport Fishing Inst.

Rahel, F. J. 2000. Homogenization of freshwater faunas across the United States. *Science* 288:854–856.

Rahel, F. J. 2002. Homogenization of freshwater faunas. *Ann. Rev. Ecol. Syst.* 33:291–315.

Rahel, F. J., Lyons, J. D., and Cochran, P. A. 1984. Stochastic or deterministic regulation of assemblage structure: It may depend on how the assemblage is defined. *Amer. Nat.* 124:583–589.

Randall, D. J. 1968. Functional morphology of the heart in fishes. *Am. Zool.* 8:179–189.

Randall, D. J. 1970. The circulatory system. Pages 133–172 *in* W. S. Hoar and D. J. Randall, eds., *Fish physiology.* Vol. 4. New York: Academic Press.

Randall, D. J. 1982. The control of respiration and circulation in fish during exercise and hypoxia. *J. Exp. Biol.* 100:275–288.

Randall, D. J., and Cameron, J. N. 1973. Respiratory control of arterial pH as temperature changes in rainbow trout, *Salmo gairdneri. Amer. J. Physiol.* 225(4):997–1002.

Randall, D. J., and Perry, S. F. 1992. Catecholamines. Pages 255–300 *in* W. S. Hoar, D. J. Randall, and A. P. Farrell, eds. *Fish Physiology.* Vol. XIIB. San Diego, Calif.: Academic Press.

Randall, D. J., Wilson, J. M., Peng, K. W., Kok, T. W. K., Kuah, S. S. L., Chew, S. F., Lam, T. J., and Ip, Y. K. 1999. The mudskipper, *Periophthalmodon schlosseri*, actively transports NH_4^+ against a concentration gradient. *Am. J. Physiol. (Regulatory Integrative Comp. Physiol.)* 46:R1562–R1567.

Randall, D. J., Wood, C. M., Perry, S. F., Bergman, H., Maloiy, G. M. O., Mommsen, T. P., and Wright, P. A. 1989. Urea excretion as a strategy for survival in a fish living in a very alkaline environment. *Nature* 337(6203):165–166.

Randall, J. E. 1965. Grazing effect on sea grasses by herbivorous reef fishes in the West Indies. *Ecology* 46(2):255–260.

Raymond, J. A. 1994. Seasonal variation of trimethylamine oxide and urea in the blood of a coldadapted marine teleost, the rainbow smelt. *Fish Phys. Biochem.* 13:13–22.

Regan, C. T. 1913. The classification of the percoid fishes. *Proc. Zool. Soc. London* 12:111–145.

Regan, C. T. 1929. Fishes. *Encyclopedia Brittanica.* 14:305–329.

Regan, C. T., and Trewavas, E. 1932. *Deep-sea angler fishes (Ceratoidea).* Dana Rpt. 2. 123 pp.

Reid, S. D., Moon, T. W., and Perry, S. F. 1992. Rainbow trout hepatocyte beta-adrenoreceptors, catecholamine responsiveness, and effects of cortisol. *J. Exp. Biol.* 158:217–240.

Reid, S. D., and Perry, S. F. 1991. The effects and physiological consequences of raised levels of cortisol on rainbow trout *(Oncorhynchus mykiss)* erythrocyte beta-adrenoreceptors. *J. Exp. Biol.* 158:217–240.

Reidy, S. P., Kerr, S. R., and Nelson, J. A. 2000. Aerobic and anaerobic swimming performance of individual Atlantic cod. *J. Exp. Biol.* 203:347–357.

Reimchen, T. E., and Nosil, P. 2001. Ecological causes of sex-biased parasitism in threespine stickleback. *Biol. J. Linn. Soc.* 73:51–63.

Reisenbichler, R. R. 1988. Relation between distance transferred from natal stream and recovery rate for hatchery coho salmon. *N. Amer. J. Fish. Mgmt.* 8:172–184.

Renaud, C. B. 1997. Conservation status of northern hemisphere lampreys (Petromyzontidae). *J. Appl. Ichthyol.* 13:143–148.

Renfro, J. L., and Hill, L. G. 1970. Factors influencing the aerial breathing and metabolism of gars *(Lepisosteus). Southw. Nat.* 15(1):45–54.

Reynolds, W. W., and Casterlin, M. E. 1979. Behavioral thermoregulation and the "final referendum" paradigm. *Amer. Zool.* 19:211–224.

Ribbink, A. J. 1990. Alternative life styles of some African cichlid fishes. *Env. Biol. Fish.* 37:87–100.

Ribbink, A. J., and Lewis, D. S. C. 1982. *Melanochromis crabro* sp. nov.: A cichlid fish from Lake Malawi which feeds on ectoparasites and catfish eggs. *Neth. J. Zool.* 32:72–87.

Richards, J. G., Heigenhauser, G. J. F., and Wood, C. M. 2002a. Lipid oxidation fuels recovery from exhaustive exercise in white muscle of rainbow trout. *Am. J. Physiol.* (*Regulatory Integrative Comp. Physiol.*) 282:R89–R99.

Richards, J. G., Mercado, A. J., Clayton, C. A., Heigenhauser, G. J. F., and Wood, C. 2002b. Substrate utilization during graded aerobic exercise in rainbow trout. *J. Exp. Biol.* 205:2067–2077.

Richmond, R. H. 1993. Coral reefs: Present problems and future concerns resulting from anthropogenic disturbance. *Amer. Zool.* 33:524–536.

Ricker, W. E. 1979. Growth rates and models. Pages 677–743 *in* W. S. Hoar, D. J. Randall, and J. R. Brett, eds., *Fish physiology.* Vol. 8. New York: Academic Press.

Riggs, A. 1970. Properties of fish hemoglobins. Pages 209–252 *in* W. S. Hoar and D. J. Randall, eds., *Fish physiology.* Vol. 4. New York: Academic Press.

Rimmer, D. W., and Wiebe, W. J. 1987. Fermentative microbial digestion in herbivorous fishes. *J. Fish Biol.* 31:229–236.

Ringler, N. H. 1979. Selective predation by drift-feeding brown trout (*Salmo trutta*). *J. Fish. Res. Bd. Canada* 46:392–403.

Rinne, J. N., Ivanstoff, W., Crowley, L. E. M., and Lobon-Cervia, J. 1996. Conservation of desert fishes: Spain, Australia, and the United States. Pages 377–400 *in* R. C. Szaro and D. W. Johnston, eds., *Biodiversity in managed landscapes: Theory and practice.* New York: Oxford Univ. Press.

Roberts, B. L. 1978. Mechanoreceptors and the behaviour of elasmobranch fishes with special reference to the acoustico-lateralis system. Pages 331–390 *in* E. S. Hodgson and R. F. Mathewson, eds. *Sensory biology of sharks, skates, and rays.* Office of Naval Research, Arlington, Virginia: U. S. Navy.

Roberts, C. M., and Hawkins, J. P. 1999. Extinction risk in the sea. *TREE* 14:241–246.

Roberts, J. L. 1975. Active branchial and ram gill ventilation in fishes. *Biol. Bull.* 148:85–105.

Roberts, J. L., and Graham, J. B. 1979. Effect of swimming speed on the excess temperatures and activities of heart and red and white muscles in the mackerel *Scomber japonicus. NOAA Fish. Bull.* 76:861–867.

Roberts, T. R. 1969. Osteology and relationships of characoid fishes, particularly in the genera *Lepsetus, Salminus, Hoplias, Ctenolucius,* and *Acestrophynchus. Proc. Calif. Acad. Sci. Ser. 4* 36(15):391–500.

Roberts, T. R. 1971. Osteology of the Malaysian phallosteid fish, *Ceratostethus bicornis,* with a discussion of the evolution of remarkable structural novelties in its jaws and external genitalia. *Bull. Mus. Comp. Zool. Harvard* 142:393–418.

Roberts, T. R. 1972. Ecology of fishes in the Amazon and Congo basins. *Bull. Mus. Comp. Zool. Harvard* 143(2):117–147.

Roberts, T. R. 1976. Geographic distribution of African freshwater fishes. *Zool. J. Linn. Soc.* 57:249–319.

Roberts, T. R. 1982. Unculi (horny projections rising from single cells), an adaptive feature of the epidermis of ostariophysan fishes. *Zool. Scripta* 11:55–76.

Roberts, T. R. 1984. Skeletal anatomy and classification of the neotenic Asian salmoniform superfamily Salangoidea (icefishes or noodlefishes). *Proc. Calif. Acad. Sci.* 43(13):179–220.

Roberts, T. R. 1986. *Danionella translucida,* a new genus and species cyprinid fish from Burma, one of the smallest living vertebrates. *Env. Biol. Fish.* 16:231–241.

Roberts, T. R. 1989. The freshwater fishes of western Borneo (Kalimantan Barat, Indonesia). *Mem. Calif. Acad. Sci.* 14. 210 pp.

Roberts, T. R. 1992. Systematic revision of the Old World freshwater fish family Notopteridae. *Ichthyol. Explor. Freshwaters* 2:361–383.

Roberts, T. R. 1993. Artisanal fisheries and fish ecology below the great waterfalls of the Mekong River in southern Laos. *Nat. Hist. Bull. Siam.* 41:31–62.

Roberts, T. R., and Vidhayanon, C. 1991. Systematic revision of the Asian catfish family Pangasi-idae, with biological observations and descriptions of three new species. *Proc. Acad. Nat. Sci. Phila.* 143:97–144.

Robertson, D. R. 1982. Fish feces and fish food on a Pacific coral reef. *Mar. Ecol. Prog. Ser.* 7:253–265.

Robertson, D. R. 1995. Competitive ability and the potential for lotteries among territorial reef fishes. *Oecologia* 103:180–190.

Robertson, D. R., Sweatman, H. P. A., Fletcher, E. A., and Cleland, M. G. 1976. Schooling as a mechanism for circumventing the territoriality of competitors. *Ecology* 57:1208–1220.

Robertson, D. R., and Warner, R. R. 1978. Sexual patterns in the labroid fishes of the Western Caribbean. II: The parrotfishes (Scaridae). *Smithson. Contr. Zool.* 255:1–26.

Robertson, J. C., and Hazel, J. R. 1999. Influence of temperature and membrane lipid composition on the osmotic water permeability of teleost gills. *Physiol. Bichem. Zool.* 72(5):623–632.

Robertson, J. D. 1954. The chemical composition of the blood of some aquatic chordates including members of the Tunicata, Cyclostomata, and Osteichthyes. *J. Exp. Biol.* 31:424–442.

Robins, C. R., Bailey, R. M., Bond, C. E., Brooker, J. R., Lachner, E. A., Lea, R. N., and Scott, W. B. 1991. *Common and scientific names of fishes from the United States and Canada. Amer. Fish. Soc. Spec. Publ.* 20. 182 pp.

Robinson, B. W., and Wilson, D. S. 1994. Character release and displacement in fishes: A neglected literature. *Amer. Nat.* 144:596–627.

Robison, H. W., and Buchanan, T. M. 1988. *Fishes of Arkansas.* Fayetteville: Univ. of Ark. Press. 536 pp.

Rodriguez, M. A., and Lewis, W. M., Jr. 1994. Regulation and stability is fish assemblages of neotropical floodplain lakes. *Oecologia* 99:166–180.

Rolston, H. 1985. Duties to endangered species. *Bioscience* 35:718–726.

Rolston, H., III. 1994. *Conserving natural value.* New York: Columbia Univ. Press.

Rombough, P. J., and Moroz, B. M. 1990. The scaling and potential importance of cutaneous and branchial surfaces in respiratory gas exchange in young chinook salmon *(Oncorhynchus tshawytscha). J. Exp. Biol.* 154:1–12.

Rombough, P. J., and Moroz, B. M. 1997. The scaling and potential importance of cutaneous and branchial surfaces in respiratory gas exchange in larval and juvenile walleye, *Stizostedion vitreum. J. Exp. Biol.* 200:2459–2468.

Romer, A. S. 1966. *Vertebrate paleontology.* Chicago: Univ. of Chicago Press. 468 pp.

Ronan, M., and Bodznick, D. 1991. Behavioral and neurophysiological demonstration of a lateralis skin photosensitivity in larval sea lampreys. *J. Exp. Biol.* 161:97–117.

Rønnestad, I., Rojas-Garcia, C. R., and Skadal, J. 2000. Retrograde peristalsis; a possible mechanism for filling the pyloric caeca? *J. Fish Biol.* 56:216–218.

Roodman, D. M. 1996. Paying the piper: Subsidies, politics, and the environment. *World Watch Paper* 133. 80 pp.

Root, R. W. 1931. The respiratory function of the blood of marine fishes. *Biol. Bull.* 61:427–456.

Rosen, D. E. 1964. The relationships and taxonomic position of the halfbeaks, killifishes, silversides, and their relatives. *Bull. Amer. Mus. Nat. Hist.* 127(5):217–268.

Rosen, D. E. 1975. A vicariance model of Caribbean biogeography. *Syst. Zool.* 24(4):437–454.

Rosen, D. E., Forey, P. L., Gardiner, B. G., and Patterson, C. 1981. Lungfishes, tetrapods, paleontology, and plesiomorphy. *Bull. Amer. Mus. Nat. Hist.* 167:159–276.

Rosen, D. E., and Greenwood, P. H. 1970. Origin of the Weberian apparatus and the relationships of ostariophysan and gonorynchiform fishes. *Amer. Mus. Novit.* 2428:1–25.

Rosen, M. W., and Cornford, N. E. 1971. Fluid friction of fish slimes. *Nature* 234:49–51.

Rosenberger, L. J., and M. W. Westneat. 1999. Functional morphology of undulatory pectoral fin locomotion in the stingray *Taeniura lymma* (Chondrichthyes: Dasyatidae). *J. Exp. Biol.* 202:3523–3539.

Ross, S. T. 1986. Resource partitioning in fish assemblages: A review of field studies. *Copeia* 1986:353–388.

Ross, S. T. 2001. *Inland fishes of Mississippi*. Jackson: Univ. of Mississippi Press.

Ross, S. T., McMichael, R. H., and Ruple, D. L. 1987. Seasonal and diel variation in the standing crop of fishes and macroinvertebrates from a Gulf of Mexico surf zone. *Est. Coast. Shelf Sci.* 25:391–412.

Rossi-Fanelli, A., and Antonini, E. 1960. Oxygen equilibria of hemoglobin from *Thunnus thynnus. Nature* 186:895–896.

Rothschild, B. J. 1965. Hypotheses on the origin of exploited skipjack tuna *(Katsuwonus pelamis)* in the eastern and central Pacific Ocean. *U. S. Fish Wildl. Serv. Spec. Sci. Rept. Fish.* 512:1–20.

Rounsefell, G. A. 1957. Fecundity of the North American Salmonidae. *U. S. Fish Wildl. Serv. Fish. Bull.* 57:451–468.

Rovainen, C. M. 1979. Neurobiology of lampreys. *Physiol. Rev.* 59(4):1007–1077.

Royce, W. F. 1972. *Introduction to fishery science*. New York: Academic Press. 351 pp.

Rozengurt, M.A., and Hedgpeth, J. W. 1989. The impact of altered river flow on the ecosystem of the Caspian Sea. *Rev. Aquat. Sci.* 1:337–364.

Rulifson, R.A. 1985. Distribution and abundance of fishes in tributaries of South Creek Estuary, North Carolina. *J. Elisha Mitchill Soc.* 101:160–176.

Russ, G. R. 2002. Yet another review of marine reserves as reef fishery management tools. Pages 421–443 *in* P. F. Sale, eds., *Coral reef fishes: dynamics and diversity of a complex system*. Amsterdam: Academic Press.

Russell, F. S. 1976. *The eggs and planktonic stages of British marine fishes*. London: Academic Press. 524 pp.

Russell, B. C., Allen, G. R., and Lubbock, H. R. 1976. New cases of mimicry in marine fishes. *J. Zool. (Lond.)* 180:407–425.

Russell, D. F., Wilkens, L. A., and Moss, F. 1999. Use of behavioural stochastic resonance by paddle fish for feeding. *Nature* 402:291–294.

Ryder, R. A. 1977. Effects of ambient light variations on behavior of yearling, subadult, and adult walleyes *(Stizostedion vitreum vitreum)*. *J. Fish. Res. Bd. Canada* 34(10):1481–1491.

Safina, C., and Burger, J. 1989. Population interactions among free-living bluefish and prey fish in an ocean environment. *Oecologia* 79:91–95.

Saila, S. B. 1961. A study of winter flounder movements. *Limnol. Oceanogr.* 6:292–298.

Sale, P. F. 1971. Extremely limited home range in a coral reef fish, *Dascyllus aruanus* (Pisces: Pomacentridae). *Copeia* 1971:324–327.

Sale, P. F. 1977. Maintenance of high diversity in coral reef fish communities. *Amer. Nat.* 111(978):337–359.

Sale, P. F. 1980. The ecology of fishes on coral reefs. *Oceanogr. Mar. Biol. Ann. Rev.* 18:367–421.

Sale, P. F., ed. 1991. *The ecology of fishes on coral reefs*. San Diego, Calif.: Academic Press. 754 pp.

Sale, P. F., ed. 2002. *Coral reef fishes: dynamics and diversity in a complex ecosystem*. Amsterdam: Academic Press. 549 pp.

Sale, P. F., and Guy, J. A. 1992. Persistence of community structure: What happens when you change taxonomic scale? *Coral Reefs* 11:147–154.

Sanderson, S. L., and Cech, J. J., Jr. 1992. Energetic cost of suspension feeding versus particulate feeding by juvenile Sacramento blackfish. *Trans. Am. Fish. Soc.* 121:149–157.

Sanderson, S. L., Cech, J. J., Jr., and Patterson, M. R. 1991. Fluid dynamics in suspension-feeding blackfish. *Science* 251:1346–1348.

Sanderson, S. L., Chestnutt, C. R., and Lobel, P. S. 1996. Evidence for ram suspension feeding by the piscivore, *Seriola dumerili. Env. Biol. Fish* 46:365–373.

Sanford, C. P. J. 2001. Kinematic analysis of a novel feeding mechanism in the brook trout *Salvelinus fontinalis* (Teleostei: Salmonidae): behavioral modulation of a functional novelty. *J. Exp. Biol.* 204:3905–3916.

Sansom, I. J., Smith, M. P., Armstrong, H. A., and Smith, M. M. 1992. Presence of the earliest verte-brate hard tissues in conodonts. *Science* 256:1308–1311.

Santer, R. M., and Greer Walker, M. 1980. Morphological studies on the ventricle of teleost and elasmobranch hearts. *J. Zool. (Lond.)* 190:259–272.

Santini, F., and Winterbottom, R. 2002. Historical biogeography of Indo-western Pacific coral reef biota: Is the Indonesian region a centre of origin? *J. Biogeog.* 29:189–205.

Sarawatari, T., and Okiyama, M. 1992. Life history of *Salangichthys microdon,* Salangidae, in a brackish lake, Lake Hinuma, Japan. *Nippon Suisan Gakkaishi* 58:235–248.

Sargent, R. C., and Gross, M. R. 1986. Williams principle: An explanation of parental care in teleost fishes. Pages 275–293 *in* T. J. Pitcher, ed., *The behaviour of teleost fishes*. London: Croom Helm.

Satchell, G. H. 1965. Blood flow through the caudal vein of elasmobranch fish. *Aust. J. Sci.* 27:241–242.

Satchell, G. H. 1971. *Circulation in fishes*. London: Cambridge Univ. Press. 131 pp.

Satchell, G. H. 1976. The circulatory system of air-breathing fish. Pages 105–123 *in* G. M. Hughes, ed., *Respiration of amphibious vertebrates*. London: Academic Press.

Satchell, G. H. 1991. *Physiology and form of fish circulation*. Cambridge: Cambridge Univ. Press. 235 pp.

Sato, T. 1986. A brood parasitic catfish of mouthbrooding cichlid fishes in Lake Tanganyika. *Nature* 323:58–59.

Saul, W. G. 1975. An ecological study of fishes at a site in upper Amazonian Ecuador. *Proc. Phila. Acad. Sci.* 127(12):93–134.

Saunders, M. W., and McFarlane, G. A. 1993. Age and length of the female spiny dogfish, *Squalus acanthias,* in the Strait of Georgia, British Columbia, Canada. *Env. Biol. Fish* 38:49–58.

Savitz, J. 1971. Effects of starvation on body protein utilization of bluegill sunfish (*Lepomis macrochirus* Rafinesque), with a calculation of caloric requirements. *Trans. Amer. Fish. Soc.* 100:18–21.

Sawada, Y. 1982. Phylogeny and zoogeography of the superfamily Cobitoidea (Cyprinoidea, Cypriniformes). *Mem. Faculty of Fisheries Hokkaido U.* 28:65–223.

Sawyer, W. H., Blair-West, J. R., Simpson, P. A., and Sawyer, M. K. 1976. Renal responses of Aus-tralian lungfish to vasotocin, angiotensin II, and NaCl infusion. *Amer. J. Physiol.* 231:593–602.

Sazima, I., and Moura, R. L. 2000. Shark (*Carcharinus perezi*) cleaned by the goby (*Elacatinus randalli*) at Fernando de Noronha Archipelago, western South Atlantic. *Copiea* 2000:297–299.

Scavia, D., et al. 1986. Influence of salmonid predation and weather on long-term water quality trends in Lake Michigan. *Can. J. Fish. Aquat. Sci.* 43:435–443.

Schaefer, S. A., and Lauder, G. V. 1986. Historical transformation of functional design: Evolution-ary morphology of feeding mechanisms in loricarioid catfishes. *Syst. Zool.* 35:489–508.

Schaeffer, B. 1973. Interrelationships of chondrosteans. Pages 207–227 *in* P. G. Greenwood, R. S. Miles, and C. Patterson, eds., *Interrelationships of fishes*. London: Academic Press.

Schaeffer, B., and Rosen, D. 1961. Major adaptive levels in the evolution of the actinopterygian feeding mechanism. *Amer. Zool.* 1:187–204.

Schalles, J. F., and Wissing, T. E. 1976. Effects of dry pellet diets on the metabolic rates of bluegill (*Lepomis macrochirus*). *J. Fish. Res. Bd. Canada.* 33:2243–2249.

Schellart, N. A. M., and Wubbels, R. J. 1998. The auditory and mechanosensory lateral line system. Pages 285–312 *in* D. H. Evans, ed., *The physiology of fishes,* 2nd ed. Boca Raton, Flao: CRC Press.

Schindler, D. W., et al. 1990. Effects of climatic warming on lakes of the central boreal forest. *Science* 250:967–970.

Schliewen, U. K., Tautz, D., and Paabo, S., 1994. Sympatric speciation suggested by monophyly of crater lake cichlids. *Nature* 368:629–632.

Schlosser, I. J. 1982. Fish community structure and function along two habitat gradients in the headwater stream. *Ecol. Monogr.* 52:395–414.

Schluter, D. 1996. Ecological speciation in postglacial fishes. *Phil. Trans. R. Soc. Lond. B* 351:807–814.

Schluter, D. 2001. Ecology and the origin of species. *TREE* 16:372–380.

Schmidt, J. 1922. The breeding places of the eel. *Phil. Trans. Roy. Soc. Lond. B.* 211:179–208.

Schmidt, P. Y. 1950. *Fishes of the Sea of Okhotsk*. Jerusalem: Prog. Sci. Trans. 392 pp.

Schmidt-Nielsen, K. 1975. *Animal physiology, adaptation, and environment*. London: Cambridge Univ. Press. 699 pp.

Schönweger, G., Schwerte, T., and Pelster, B. 2000. Temperature-dependent development of cardiac activity in unrestrained larvae of the minniw *Phoxinus phoxinus. Am. J. Physiol. (Regulatory Integrative Comp. Physiol.)* 279:R1634–R1640.

Schreer, J. F., Cooke, S. J., and McKinley, R. S. 2001. Cardiac response to variable forced exercise at different temperatures: An angling simulation for smallmouth bass. *Trans. Amer. Fish. Soc.* 130:783–795.

Schultz, L. P. 1960. *Fishes of the Marshall and Marianas Islands*. Vol 2. Smithsonian Inst., U. S. Nat. Mus. Bull. 202. 438 pp.

Schultz, R. J. 1989. Origins and relationships of unisexual poeciliids. Pages 69–90 *in* G. K. Meffe and F. F. Snelson, Jr., eds., *Ecology and evolution of livebearing fishes (Poeciliidae)*. Englewood Cliffs, N. J.: Prentice Hall.

Schwartz, F. J. 1972. World literature on fish hybrids with an analysis by family, species, and hybrid. *Gulf Coast Mar. Lab. Mus.* 33:1–328.

Scott, T. D. 1962. *The marine and freshwater fishes of South Australia*. Adelaide, Australia: Govt. Printer. 338 pp.

Scott, W. B., and Crossman, E. J. 1973. *Freshwater fishes of Canada. Fish. Res. Bd. Canada Bull.* 184:1–966.

Scott, W. B., and Scott, M. G. 1988. *Atlantic fishes of Canada. Can. Bull. Fish. Aquat. Sci.* 219. 731 pp.

Seaburg, K. G., and Moyle, J. B. 1964. Feeding habits, digestive rates, and growth of some Minnesota warmwater fishes. *Trans. Amer. Fish. Soc.* 93:269–282.

Secombes, C. J. 1996. The nonspecific immune system: Cellular defenses. Pages 63–105 *in* G. Iwama, and T. Nakanishi, eds., *The fish immune system: Organism, pathogen, and environment*. San Diego, Calif.: Academic Press.

Sedberry, G. R., and Musick, J. R. 1978. Feeding strategies of some demersal fishes of the continental slope and rise off the mid-Atlantic coast of the USA. *Marine Biol.* 44:337–375.

Sedberry, G. R., and Van Dolah, R. F. 1984. Demersal fish assemblages associated with hard bottom habitat in the South Atlantic bight of the U.S.A. *Env. Biol. Fish* 251–258.

Seehausen, O., van Alpern, J. J. M., and Witte, F. 1997. Cichlid fish diversity threatened by eutrophication that curbs sexual selection. *Science* 277:1808–1811.

Senanayake, F. R., and Moyle, P. B. 1981. Conservation of the freshwater fishes of Sri Lanka. *Biol. Cons.* 22:181–195.

Serafy, J. E., Hopkins, T. E., and Walsh, P. J. 1997. Field studies on the ureogenic gulf toadfish in a subtropical bay. 1. Patterns of abundance, size composition and growth. *J. Fish Biol.* 50:1258–1270.

Shand, J. 1997. Ontogenetic changes in retinal structure and visual acuity: A comparative study of coral reef teleosts with differing post-settlement lifestyles. *Env. Biol. Fish* 49:302–322.

Shapiro, D. Y. 1984. Sex reversal and sociodemographic processes in coral reef fishes. Pages 103–116 *in* G. W. Potts and R. J. Wootton, eds., *Fish reproduction: Strategies and tactics*. London: Academic Press.

Sharp, G. D., and Dizon, A. E. 1978. *The physiological ecology of tunas*. New York: Academic Press. 485 pp.

Shaw, E. 1978. Schooling fishes. *Amer. Sci.* 66:166–175.

Shell, E.W. 1959. *Chemical composition of the blood of smallmouth bass.* Ph.D. diss. Cornell Univ.

Sherman, K. 1994. Sustainability, biomass yields, and health of coastal ecosystems: An ecological perspective. *Mar. Ecol. Prog. Ser.* 112:277–301.

Sherwood, N. M., and Hew, C. L., eds. 1994. *Fish physiology.* Vol. XIII, Molecular endocriniology of fish. San Diego, Calif.: Academic Press. 518 pp.

Shim, K. F., Landesman, L., and Lam, T. J. 1989. Effect of dietary protein on growth, ovarian development and fecundity in the dwarf gourami, *Colisa lalia* (Hamilton). *J. Aqua. Trop.* 5:111–123.

Shirai, S. 1996. Phylogenetic interrelationships of Neoselachians (Chondrichthyes: Euselachii). Pages 9–34 *in* M. L. J. Stiassny, L. R. Parenti, and G. D. Johnson, eds., *Interrelationships of fishes.* San Diego, Calif.: Academic Press.

Shoubridge, E. A., and Hochachka, P. W. 1980. Ethanol: Novel endproduct of vertebrate anaerobic metabolism. *Science* 209:308–309.

Shu, D., Chen, L., Zhang, X., Han, J., and Li, Y. 2001. Chengjiang Lagerstatte and earliest known chordates. *Zool. Sci.* 18:447–448.

Shulman, M. J. 1985. Coral reef fish assemblages: Intra- and interspecific competition for shelter sites. *Env. Biol. Fish* 13:81–92.

Shulman, M. J., and Bermingham, E. 1995. Early life histories, ocean currents, and the population genetics of Caribbean reef fishes. *Evolution* 49:897–910.

Shurcliff, W. A. 1962. *Polarized light, production and use.* Cambridge, Mass.: Harvard, University Press. 31 pp.

Shuttleworth, T. J. 1988. Salt and water balance: Extrarenal mechanisms. Pages 171–199 *in* T. J. Shuttleworth, ed., *Physiology of elasmobranch fishes.* Berlin: Springer-Verlag.

Sicheri, F., and Yang, D. S. C. 1995. Ice-binding structure and mechanisms of an antifreeze protein from winter flounder. *Nature* 375:427–431.

Sidell, B. D., and Driedzic, W. R. 1985. Relationship between cardiac energy metabolism and cardiac work demand in fishes. Pages 386–401 *in* R. Gilles, ed., *Circulation, respiration, and metabolism.* Berlin: Springer-Verlag.

Siebert, G., Schmitt, A., and Bottke, I. 1964. Enzymes of the amino acid metabolism in cod musculature. *Arch. Fisch. Wiss.* 15:233–244.

Sih, A. 1994. Predation risk and the evolutionary ecology of reproductive behavior. *J. Fish Biol* 45(Suppl. A):111–130.

Silva, P., Solomon, R., Spokes, K., and Epstein, F. H. 1977a. Mechanism of active chloride secretion by shark rectal gland: Role of Na-K-ATPase in chloride transport. *J. Physiol.* 233:298–306.

Silva, P., Solomon, R., Spokes, K., and Epstein, F.H. 1977b. Ouabain inhibition of gill Na^+-K^+-ATPase: Relationship to active chloride transport. *J. Exp. Zool.* 199:419–426.

Silva, P., Solomon, R. J., and Epstein, F. H. 1999. Mode of activation of salt secretion by C-type natriuretic peptide in the shark rectal gland. *Am. J. Physiol.* (*Regulatory Integrative Comp. Physiol.*) 277:R1725–R1732.

Simkova, A., Morand, S., Matejusova, I., Jurajda, P., and Gelnar, M. 2001. Local and regional influences on patterns of parasite species richness of central European fishes. *Biodiv. Conserv.* 10:510–525.

Singh, B. N. 1976. Balance between aquatic and aerial respiration. Pages 125–164 *in* G. M. Hughes, ed., *Respiration of amphibious vertebrates.* London: Academic Press.

Sinha, V. R. P., and Jones, J. W. 1975. *The European freshwater eel.* Liverpool: Liverpool Univ. Press. 146 pp.

Sioli, H. 1975. Tropical river: The Amazon. Pages 461–488 *in* B. A. Whitton, ed., *River ecology.* Berkeley: Univ. of Calif. Press.

Sire, J. 1990. From ganoid to elasmoid scales in the actinopterygian fishes. *Netherlands J. Zool.* 40:75–92.

Sisneros, J. A., Tricas, T. C., and Luer, C. A. 1998. Response properties and biological function of the skate electrosensory system during ontogeny. *J. Comp. Physiol. A* 183:87–99.

Skud, B. E. 1982. Dominance in fishes: The relation between environment and abundance. *Science* 216:144–149.

Sloman, K. A., Desforges, P. R., and Gilmour, K. M. 2001. Evidence for a mineralocorticoid-like receptor linked to branchial chloride cell proliferation in freshwater rainbow trout. *J. Exp. Biol.* 204:3953–3961.

Smart, G. R., Knox, D., Harrison, J. G., Ralph, J. A., Richards, R. H., and Cowey, C. B. 1979. Nephrocalcinosis in rainbow trout *Salmo gairdneri* Richardson; The effect of exposure to elevated CO_2 concentrations. *J. Fish. Dis.* 2:279–289.

Smith, C. L., and Powell, C. R. 1971. The summer fish communities of Brier Creek, Marshall County, Oklahoma. *Amer. Mus. Nov.* 2458:1–30.

Smith, C. L., Smith, C. S., Schaeffer, B., and Atz, J. W. 1975. *Latimeria,* the living coelacanth, is ovoviviparous. *Science* 190:1105–1106.

Smith, C. L., and Tyler, J. C. 1972. Space sharing in a coral reef fish community. *Bull. L. A. Co. Mus. Nat. Hist.* 14:125–170.

Smith, C. L., Tyler, J. C., and Feinberg, M. N. 1981. Population ecology and biology of the pearlfish *(Carapus bermudensis)* in the lagoon at Bimini, Bahamas. *Bull. Mar. Sci.* 31:876–902.

Smith, D. G. 1970. *Notacanthiform leptocephali* in the western North Atlantic. *Copeia* 1970(1):1–9.

Smith, F. M., and Jones, D. R. 1978. Localization of receptors causing hypoxic bradycardia in trout *Salmo gairdneri. Can. J. Zool.* 56:1260–1265.

Smith, G. R., and Todd, T. N. 1984. Evolution of species flocks in north temperate lakes. Pages 45–68 in A. A. Echelle and I. Kornfield, eds., *Evolution of fish species flocks.* Orono, Me.: Univ. Maine Press.

Smith, G. R., and Stearley, R. F. 1989. The classification and scientific names of rainbow and cutthroat trouts. *Fisheries (Bethesda)* 14(1):4–10.

Smith, H. M. 1945. The freshwater fishes of Siam, or Thailand. *Bull. U. S. Nat. Mus.* 188:1–622.

Smith, H. W. 1929. The excretion of ammonia and urea by the gills of fish. *J. Biol. Chem.* 81:727–742.

Smith, H.W. 1961. *From fish to philosopher.* New York: Doubleday. 293 pp.

Smith, J. L. B. 1949. *The sea fishes of Southern Africa.* Capetown, South Africa: Central News Agency. 580 pp.

Smith, J. L. B. 1956. *The search beneath the sea.* New York: H. Holt. 260 pp.

Smith, L. S., and Bell, G. R. 1975. *A practical guide to the anatomy and physiology of Pacific salmon.* Ottowa, Canada: Fish. Mar. Serv. Misc. Spec. Publ. 27. 14 pp.

Smith, M. L., and Miller, R. R. 1986. The evolution of the Rio Grande Basin as inferred from its fish fauna. Pages 457–486 in C. H. Hocutt and E. O. Wiley, eds., *The zoogeography of North American freshwater fishes.* New York: Wiley.

Smith, M. M., and Heemstra, P. C., eds. 1986. *Smiths' sea fishes.* Berlin: Springer

Smith, M. P., I. J. Sansom, and Cochrane, K. D. 2001. The Cambrian origin of vertebrates. Pages 72–84 in P. E. Ahlberg, ed. *Major events in vertebrate evolution: paleontology, phylogeny, genetics, and development.* London: Taylor and Francis.

Smith, P. W. 1971. Illinois streams: Classification based on their fishes and an analysis of factors responsible for the disappearance of native species. *Ill. Nat. Hist. Surv. Biol. Note* 76:1–14.

Smith, R. J. F. 1982. The adaptive significance of the alarm substance-fright reaction system. Pages 327–342 in T. J. Hara, ed., *Chemoreception in fishes.* Amsterdam: Elsevier.

Smith, S. H. 1968. Species succession and fishery exploitation of the Great Lakes. *J. Fish. Res. Bd. Canada* 25(4):667–693.

Smith, T. B., Wahl, D. H., and Mackie, R. I. 1996. Volatile fatty acids and anaerobic fermentation in temperature piscivorous and ominvorous freshwater fish. *J. Fish Biol.* 48:429–841.

Snedecor, G. W., and Cochran, W. G. 1967. *Statistical models.* Ames: Iowa St. Univ. Press. 593 pp.

Snyder, G. 1990. *The practice of the wild.* San Francisco, Calif.: North Point Press.

Snyder, R. J. 1991. Migration and life histories of the threespine stickleback: Evidence for adaptive variation in growth rate between populations. *Env. Biol. Fishes* 31:381–388.

Somero, G. N. 1986. Protons, osmolytes, and fitness of internal milieu for protein function. *Am. J. Physiol.* 251(*Regulatory Integrative Comp. Physiol.*) 20:197–213.

Sorensen, P. W., and Scott, A. P. 1994. The evolution of hormonal sex pheromones in teleost fish: Poor correlation between the pattern of steroid release by goldfish and olfactory sensitivity suggests that these cues evolved as a result of chemical spying rather than signal specialization. *Acta Physiol. Scand.* 152:191–205.

Sorensen, P. W., and Caprio, J. 1998. Chemoreception. Pages 375–405 *in* D. H. Evans ed., *The physiology of fishes.* 2nd ed. Boca Raton, Fla.: CRC Press.

Sötz, E., Niederstätter, H., and Pelster, B. 2002. Determinants of intracellular pH in gas gland cells of the swimbladder of the European eel *Anguilla anguilla. J. Exp. Biol.* 205:1069–1075.

Southern California Coastal Water Research Project. 1973. *The ecology of the Southern California bight: Implications for water quality management.* El Segundo, Calif.: SCCWRP TR 104. 531 pp.

Specker, J. L., and Schreck, C. B. 1982. Changes in plasma corticosteroids during smoltification of coho salmon, *Oncorhynchus kisutch. Gen. Comp. Endocrinol.* 46:53–58.

Spencer, C. N., McClelland, B. R., and Stanford, J. 2001. Shrimp stocking, salmon collapse, and eagle displacement: cascading interactions in the food web of a large aquatic ecosystem. *Bioscience* 41:14–21.

Sponaugle, S., and Cowen, R. K. 1997. Early life history traits and recruitment patterns of Caribbean wrasses (Labridae). *Ecol. Monogr.* 67:177–202.

Springer, S. 1979. A revision of the catsharks, family Scyliorhinidae. *NOAA Tech Rpt. NMFS Cir.* 422:1–152.

Springer, V. G. 1982. Pacific plate biogeography, with special reference to fishes. *Smithson. Contrib. Zool.* 367:1–182.

Springer, V. G., and Gold, J. P. 1989. *Sharks in question.* Washington, D.C.: Smithsonian Inst. Press. 187 pp.

Stacey, N. E. 1984. Control and timing of ovulation by exogenous and endogenous factors. Pages 207–221 *in* G. W. Potts and R. J. Wootton, eds., *Fish reproduction: Strategies and tactics.* London: Academic Press.

Stager, J. C., Reinthal, P. N., and Livingston, D. A. 1986. A 25,000 year history for Lake Victoria, East Africa, and some comments on its significance for the evolution of cichlid fishes. *Freshw. Biol.* 16:15–19.

Starnes, W. C., and Etnier, D. A. 1986. Drainage evolution and fish biogeography of the Tennessee and Cumberland drainage realm. Pages 325–362 *in* C. H. Hocutt and E. O. Wiley, eds. *The zoogeography of North American freshwater fishes.* New York: John Wiley.

Starrett, W. 1951. Some factors affecting the abundance of minnows in the Des Moines River, Iowa. *Ecology* 32(1):13–27.

Stauffer, J. R., Jr., Arnegard, M. E., Cetron, M. E., Sullivan, J. J., Chistsulo, L. A., Turner, G. F., Chiotha, S., and McKaye, K. R. 1997. Controlling vectors and hosts of parasitic diseases using fishes: A case history of schistosomiasis in Lake Malawi. *Bioscience* 47:41–49.

Stearley, R. F. 1992. Historical ecology of the Salmoninae, with special reference to *Oncorhynchus.* Pages 622–658 *in* R. L. Mayden, ed., *Systematics, historical ecology, and North American freshwater fishes.* Stanford, Calif.: Stanford Univ. Press.

Stearns, S. C. 1976. Life history tactics: A review of the ideas. *Quart. Rev. Biol.* 51(1):3–47.

Steen, J. B. 1970. The swim bladder as a hydrostatic organ. Pages 413–443 *in* W. S. Hoar and D. J. Randall, eds., *Fish physiology.* Vol. 4. New York: Academic Press.

Steen, J. B. 1971. *Comparative physiology of respiratory mechanisms.* New York: Academic Press. 182 pp.

Steffensen, J. F., and Lomholt, J. P. 1985. Cutaneous oxygen uptake and its relation to skin blood perfusion and ambient salinity in the plaice, *Pleuronectes platessa. Comp. Biochem. Physiol.* 81A:373–375.

Steffensen, J. F., and Lomholt, J. P. 1992. The secondary vascular system. Pages 185–213 *in* W. S. Hoar, D. J. Randall, and A. P Farrell, eds., *Fish Physiology.* Vol. XII, Part A, The cardiovascular system. San Diego, Calif.: Academic Press.

Stein, R. A., DeVries, D. R., and Dettmers, J. M. 1995. Food-web regulation by a planktivore: Exploring the generality of the trophic cascade hypothesis. *Can. J. Fish. Aquat. Sci.* 52:2518–2526.

Stein, R. A., and Kitchell, J. F. 1975. Selective predation by carp *Cyprinus carpio* (L.) on benthic mollusks in Skadar Lake, Yugoslavia. *J. Fish Biol.* 7(2):391–399.

Stenslokken, K. O., Sundin, L., and Nilsson, G. E. 1999. Cardiovascular effects of endothelin-1 in Atlantic cod: Evidence for pillar cell contraction. *J. Exp. Biol.* 202:1151–1157.

Stephens, J. S., Johnson, R. K., Key, G. S., and McCosker, J. E. 1970. The comparative ecology of three sympatric species of California blennies of the genus *(Hypsoblennius)* Gill (Teleostomi, Blenniidae). *Ecol. Monogr.* 40:213–233.

Sterba, G. 1959. *Freshwater fishes of the world.* London: Vista Books. 878 pp.

Stevens, E. D. 1968. The effect of exercise on the distribution of blood to various organs in rainbow trout. *Comp. Biochem. Physiol.* 25:615–625.

Stevens, E. D. 1979. The effect of temperature on tail beat frequency of fish swimming at constant velocity. *Can. J. Zool.* 57:1628–1635.

Stevens, E. D., Bennion, G. R., Randall, D. J., and Shelton, G. 1972. Factors affecting arterial pressures and blood flow from the heart in intact, unrestrained lingcod (*Ophiodon elongatus*). *Comp. Biochem. Physiol.* 43A:681–695.

Stevens, E. D., and Carey, F. G. 1981. One why of the warmth of warm-bodied fish. *Amer. J. Physiol. (Regulatory Integrative Comp. Physiol.*) 240:151–155.

Stevens, E. D., and Devlin, R. H. 2000. Intestinal morphology in growth hormone transgenic coho salmon. *J. Fish Biol.* 56:191–195.

Stevens, E. D., and Randall, J. R. 1967. Changes in blood pressure, heart rate, and breathing rate during moderate swimming activity in rainbow trout. *J. Exp. Biol.* 46:307–315.

Stevens, E. D., and Sutterlin, A. M. 1976. Heat transfer between fish and ambient water. *J. Exp. Biol.* 65:131–145.

Stevens, G. C. 1996. Extending Rapoport's rule to Pacific marine fishes. *J. Biogeography* 23:149–154.

Stevenson, M. M., Schnell, G. D., and Black, R. 1974. Factor analysis of fish distribution patterns in western and central Oklahoma. *Syst. Zool.* 23(2):202–218.

Stewart, N. E., Shumway, D. L., and Doudoroff, P. 1967. Influence of oxygen concentration on the growth of juvenile largemouth bass. *J. Fish. Res. Bd. Canada* 24:475–494.

Stiassny, M. L. 1996. An overview of freshwater biodiversity: with some lessons from African fishes. *Fisheries* 21(9):7–13.

Stiassny, M. L. J., Parenti, L. R., and Johnson, G. D. 1996. *Interrelationships of fishes.* San Diego, Calif.: Academic Press.

Stickney, R. R., and Shumway, S. E. 1974. Occurrence of cellulase activity in the stomachs of fishes. *J. Fish Biol.* 6:779–790.

Stockley, P., Gage, M. J. G., Parker, G. A., and Möller, A. P. 1997. Sperm competition in fishes: The evolution of testis size and ejaculate characteristics. *Amer. Nat.* 149:933–954.

Stockton, W. L., and DeLaca, T. E. 1982. Food falls in the deep sea: Occurrence, quality, and significance. *Deep-sea Research* 29:157–169.

Stoiber, W., Haslett, J. R., and Sänger, A. M. 1999. Myogenic patterns in teleosts: What does the present evidence really suggest? *J. Fish Biol.* 55(Suppl. A):84–99.

Strange, E. M., Moyle, P. B., and Foin, T. C. 1992. Interactions between stochastic and deterministic processes in stream fish community assembly. *Env. Biol. Fish.* 36:1–15.

Strauss, R. E., and Bond, C. E. 1990. Taxonomic methods: Morphology. Pages 109–140 *in* C. B. Schreck and P. B. Moyle, eds., *Methods for fish biology.* Bethesda, Md.: Amer. Fish. Soc.

Strong, W. R., Snelson, F. F., and Gruber, S. H. 1990. Hammerhead shark predation on stingrays: An observation of prey handling by *Sphyrna mokarran. Copeia* 1990:836–840.

Subrahmanyam, C. B., and Drake, S. H. 1975. Studies of animal communities of two north Florida salt marshes. Part I. Fish communities. *Bull. Mar. Sci.* 25(4):445–465.

Sulak, K. J. 1977. The systematics and biology of *Bathypterois* (Pisces, Chloropthalmidae) with a revised classification of benthic myctophiform fishes. *Galathea Rpt.* 14:49–108.

Sulak, K. J., and Ross, S. W. 1996. Lilliputian bottom fish fauna of the Hatteras upper middle continental slope. *J. Fish Biol.* 49(Suppl. A):91–113.

Summerfelt, R. C., and Hall, G. E. 1987. *Age and growth in fish.* Ames: Iowa State Univ. Press. 544 pp.

Summers, A. P., and Ferry-Graham, L. A. 2001. Ventilatory modes and mechanics of the hedgehog skate (*Leucoraja erinacea*): testing the continuous flow model. *J. Exp. Biol.* 204:1577–1587.

Sundin, L., and Nilsson, G. E. 1998. Endothelin redistributes blood flow through the lamellae of rainbow trout gills: Evidence for pillar cell contraction. *J. Comp. Physiol. B* 168:619–623.

Sundin, L., and Nilsson, G. E. 2002. Branchial innervation. *J. Exp. Zool.* 293:232–248.

Sutterlin, A. M. 1969. Effects of exercise on the cardiac and ventilation frequency in three species of freshwater teleosts. *Physiol. Zool.* 42:36–52.

Sutterlin, A. M. 1975. Chemical attraction of some marine fish in their natural habitat. *J. Fish. Res. Bd. Canada* 32:729–738.

Svardson, G. 1949. Natural selection and egg number in fish. *Ann. Rpt. Inst. Freshw. Res. Drottningholm* 29:115–122.

Swanson, C. 1998. Interactive effects of salinity on metabolic rate, activity, growth and osmoregulation in the euryhaline milkfish (*Chanos chanos*). *J. Exp. Biol.* 201:3355–3366.

Swanson, C., Cech, J. J., Jr., and Piedrahita, R. H. 1996. *Mosquitofish: Biology, culture, and use in mosquito control.* Elk Grove, Calif.: Mosquito and Vector Cont. Assn. Calif. 87 pp.

Swanson, C., Mager, R. C., Doroshov, S. I., and Cech, J. J., Jr. 1996. Use of salts, anesthetics, and polymers to minimize handling and transport mortality in delta smelt. *Trans. Amer. Fish. Soc.* 125:326–329.

Swanson, C., Young, P. S., and Cech, J. J., Jr. 1998. Swimming performance of delta smelt: Maximum performance, and behavioral and kinematic limitations on swimming at submaximal velocities. *J. Exp. Biol.* 201:333–345.

Swearer, S. E., Caselle, J. E., Lea, D. W., and Warner, R. R. 1999. Larval retention on an island of population of coral reef fish. *Nature* 402:799–802.

Swingle, H. S., and Smith, E. V. 1940. Experiments on the stocking of fish ponds. *Trans. N. Amer. Wild. Conf.* 15:267–276.

Symons, P. E. K. 1976. Behavior and growth of juvenile Atlantic salmon *(Salmo salar)* and three competitors at two stream velocities. *J. Fish. Res. Bd. Canada* 33(12):1766–2773.

Taborsky, M. 1998. Sperm competition in fish: "Bourgeois" males and parasitic spawning. *TREE* 13:222–224.

Takeda, T. 1996. Effects of experimental ventilation and ambient PO_2 on O_2 uptake of isolated cutaneous tissue in the carp, *Cyprinus carpio. Comp. Biochem. Physiol.* 113A:107–111.

Tanaka, J. N., and Grizzle, J. M. 2002. Effects of nonylphenol on the gonadal differentiation of the hermaphroditic fish, *Rivulus marmoratus. Aquat. Toxicol.* 57:117–125.

Tavolga, W. A. 1956. Visual, chemical, and sound stimuli as cues in the sex discriminatory behavior of the gobiid fish, *Bathygobius soporator. Physiol. Zool.* 31:259–271.

Taylor, L. R., Compagno, L. J. V., and Struhsaker, P. J. 1983. Megamouth: A new shark species, genus, and family of lamnoid shark (*Megachasma pelagios,* Family Megachasmidae) from the Hawaiian Islands. *Proc. Calif. Acad. Sci.* 43:87–110.

Taylor, M. S., and Hellberg, M. E. 2003. Genetic evidence for local retention of pelagic larvae in a Caribbean reef fish. *Science* 299:107–109.

Teichmann, H. 1962. Was leistet der Geruchssinn bei Fischen? *Umschau Wiss. Tech.* 62:588–591.

Templeman, W. 1976. Transatlantic migrations in spiny dogfish (*Squalus acanthias*). *J. Fish. Res. Bd. Canada* 33(11):2605–2609.

Tesch, F. W. 1977. *The eel: Biology and management of anguillid eels.* London: Chapman and Hall. 434 pp.

Teugels, G. G. 1996. Taxonomy, phylogeny, and biogeography of catfishes (Ostariophysi, Siluroidei): An overview. *Aquat. Living Resour.* 9(Hors serie):9–34.

Teylaud, A. R. 1971. Food habits of the goby *Ginsburgellus novemlinaetus* and the clingfish, *Acros rubiginosis*, associated with echinoids in the Virgin Islands. *Caribb. J. Sci.* 11:41–45.

Thiel, R., Sepulveda, A., Kafemann, R., and Nellen, W. 1995. Environmental factors as forces structuring the fish community of the Elbe Estuary. *J. Fish Biol.* 46:47–69.

Thompson, E. D., Olsson, P., Mayer, G. D., Haux, C., Walsh, P. J., Burge, E., and Hogstrand, C. 2001. Effects of 17β-estradiol on levels and distribution of metallothionein and zinc in squirrelfish. *Am. J. Physiol.* (*Regulatory Integrative Comp. Physiol.*) 280:R527–R535.

Thompson, W. F., and Van Cleve, R. 1936. Life history of the Pacific halibut. *Rpt. Int. Fish. Comm.* 9:1–184.

Thomson, D. A., Findley, L. T., and Kerstitch, A. N. 2000. *Reef fishes of the Sea of Cortez: the rocky shore fishes of the Gulf of California.* Austin: Univ. Texas Press. 353 pp.

Thomson, D. A., and Lehner, C. E. 1976. Resilience of a rocky intertidal fish community in a physically unstable environment. *J. Exp. Mar. Biol. Ecol.* 22(1):1–29.

Thomson, K. S. 1972. The adaptation and evolution of early fishes. *Quart. Rev. Biol.* 46(2):139–166.

Thomson, K. S. 1973. New observations on the coelacanth fish, *Latimeria chalumnae. Copeia* 1973(4):813–814.

Thomson, K. S. 1977. On the individual history of cosmine and a possible electroreceptive function of the pore-canal system in fossil fishes. Pages 247–269 *in* R. M. Andrews, R. S. Miles, and A. D. Walker, eds., *Problems in vertebrate evolution. Linn. Soc. Lond. Symp.* Series 4.

Thomson et al. 2000. Austin: Univ. Texas Press. 353 pp.

Thorarensen, H., Gallaugher, P., and Farrell, A. P. 1996. Cardiac output in swimming rainbow trout, *Oncorhynchus mykiss*, acclimated to seawater. *Physiol. Zool.* 69(1):139–153.

Thorrold, S. R., Latkoczy, C., Swart, P. K., and Jones, C. M. 2001. Natal homing in a marine fish metapopulation. *Science* 291:297–299.

Thorson, F. B., Cowan, C. M., and Watson, D. E. 1973. Body fluid solutes of juveniles and adults of the euryhaline bull shark (*Carcharinus leucas*) from freshwater and saline environments. *Physiol. Zool.* 46:29–42.

Thorson, T. 1958. Measurement of the fluid compartments of four species of marine Chondrichthyes. *Physiol. Zool.* 31:16–23.

Thorson, T. B., ed. 1976. *Investigations of the ichthyofauna of Nicaraguan lakes.* Lincoln: School of Life Sci., Univ. of Nebraska. 663 pp.

Thorstad, E. B., Økland, F., Koed, A., and McKinley, R. S. 2000. Radio-transmitted electromyogram signals as indicators of swimming speed in lake trout and brown trout. *J. Fish Biol.* 57:547–561.

Thresher, R. E. 1977. Eye ornamentation of Caribbean reef fishes. *Z. Tierpsychol.* 43:152–158.

Thresher, R. E. 1984. *Reproduction in reef fishes.* Neptune City, N.J.: TFH Pub.

Tibbets, C. A., and Dowling, T. E. 1996. Effects of intrinsic and extrinsic factors on population fragmentation in three species of North American minnows (Teleostei: Cyprinidae). *Evolution* 50:1280–1292.

Tippetts, W., and Moyle, P. 1978. Epibenthic feeding by rainbow trout *(Salmo gairdneri)* in the McCloud River, California. *J. Anim. Ecol.* 47:549–559.

Tipsmark, C. K., and Madsen, S. S. 2001. Rapid modulation of Na^+/K^+-ATPase activity in osmoregulatory tissues of a salmonid fish. *J. Exp. Biol.* 204:701–709.

Todd, E. S. 1968. Terrestrial sojourns of the longjaw mudsucker, *Gillichthys mirabilis. Copeia* 1968(1):192–194.

Tohyama, Y., Ichimiya, T., Kasama-Yoshida, H., Cao, Y., Hasegawa, M., Kojima, H., Tamai, Y., and Kurihara, T. 2000. Phylogenetic relation of lungfish indicated by the amino acid sequence of myelin DM20. *Molec. Brain Res.* 80:256–259.

Tonn, W. M., Magnuson, J. J. Rask, M., and Toivonen, J. 1990. Intercontinental comparison of small-lake fish assemblages: The balance between local and regional processes. *Amer. Nat.* 136:345–375.

Tort, L., Gonzalez-Arch, F., Hidalgo, J. 1991. On the blood volume of the Mediterranean dogfish, *Scyliorhinus canicula. Fish. Physiol. Biochem.* 9:173–177.

Toth, L. A., Arrington, D. A., and Begue, G. 1997. Headwater restoration and reestablishment of natural flow regimes: Kissimmee River of Florida. Pages 425–444 *in* J. E. Williams, C. A. Wood, and M. P. Dombeck, eds., *Watershed restoration: Principles and practices.* Bethseda, Md.: Amer. Fish. Soc.

Townsend, C. R. 1996. Invasion bioloy and ecological impacts of brown trout, *Salmo trutta,* in New Zealand. *Biol. Cons.* 78:13–22.

Tricas, T. C. 2001. The neuroecology of the elasmobranch electrosensory world: Why peripheral morphology shapes behavior. *Env. Biol. Fish.* 60:77–92.

Trippel, E. A., and Beamish, F. W. H. 1993. Multiple trophic level structuring in *Salvelinus-Coregonus* assemblages in boreal forest lakes. *Can. J. Fish. Aquat. Sci.* 50:1442–1455.

Trott, L. B. 1981. A general review of the pearlfishes (Pisces, Carapidae). *Bull. Mar. Sci.* 31:623–629.

Tsai, C. 1973. Water quality and fish life below sewage outfalls. *Trans. Amer. Fish Soc.* 102(4):281–292.

Tsai, C., and Zeisel, R. B. 1969. Natural hybridization of cyprinid fishes in Little Patuxent River, Maryland. *Ches. Sci.* 102(2):69–74.

Tsukamoto, K., Nakai, I., and Tesch, W. 1998. Do all freshwater eels migrate? *Nature* 396:635.

Tullis, A., Block, B. A., and Sidell, B. D. 1991. Activities of key metabolic enzymes in the heater organs of scombroid fishes. *J. Exp. Biol.* 161:383–403.

Turingan, R. G, Wainwright, P. C., and Hensley, D. A. 1995. Interpopulation variation in prey use and feeding biomechanics in Caribbean triggerfishes. *Oecologia* 102:296–304.

Tyler, A. V. 1971. Periodic and resident components in communities of Atlantic fishes. *J. Fish. Res. Bd. Canada* 28(7):935–946.

Tyler, A. V. 1972. Food resource division among northern marine demersal fishes. *J. Fish. Res. Bd. Canada* 29(7):997–1003.

Tyler, H. R., and Pearcy, W. G. 1975. The feeding habits of three species of lantern-fishes (Family Myctophidae) off Oregon, USA. *Mari. Biol.* 32:7–11.

Tzeng, W. N., and Tsai, Y. C. 1994. Changes in otolith microchemistry of the Japanese eel, *Anguilla japonica,* during its migration from the ocean to the rivers of Taiwan. *J. Fish. Biol.* 45:671–683.

Underhill, J. 1986. The fish fauna of the Laurentian Great Lakes, the St. Lawrence Highlands, Newfoundland, and Labrador. Pages 105–136 *in* C. H. Hocutt and E. O. Wiley, eds., *The zoo-geography of North American freshwater fishes.* New York: Wiley.

Unmack, P. J. 2001. Biogeography of Australian freshwater fishes. *J. Biogeogr.* 28:1053–1089.

Utida, S., and Hirano, T. 1973. Effects of changes in environmental salinity on salt and water movement in the intestine and gills of the eel *Anguilla japonica*. Pages 240–278 *in* W. Chavin, ed., *Responses of fish to environmental changes*. Springfield, Ill.: Chas. C. Thomas.

Uyeno, T., and Smith, G. R. 1972. Tetraploid origin of the karyotype of catostomid fishes. *Science* 175:644–646.

Vaisius, A., Martin-Kearley, J., and Fletcher, G. L. 1989. Antifreeze protein gene transcription in winter flounder is not responsive to temperature. *Cell. Mol. C. Biol.* 35:547–554.

Val, A. L., de Menezes, G. C., and Wood, C. M. 1998. Red blood cell adrenergic responses in Amazonian teleots. *J. Fish Biol.* 52:83–93.

Valerio, P. F., Kao, M. H., and Fletcher, G. L. 1992. Fish skin: An effective barrier to ice crystal propagation. *J. Exp. Biol.* 164:135–151.

van Dam, L. 1938. *On the utilization of oxygen and regulation of breathing in some aquatic animals.* Ph.D. diss. Univ. of Groningen, The Netherlands.

Vander Zanden, M. J., and Vadeboncouer, Y. 2002. Fishes as integrators of benthic and pelagic food webs in lakes. *Ecology* 83:2152–2161.

Vanicek, C. D., and Kramer, R. H. 1969. Life history of the Colorado squawfish, *Ptychocheilus lucius,* and the Colorado chub, *Gila robusta,* in the Green River in Dinosaur National Monument, 1964–1966. *Trans. Amer. Fish. Soc.* 98(2):193–208.

Vanni, M. J., Luecke, C., Kitchell, J. F., and Magnuson, J. J. 1990. Effects of planktivorous fish mass mortality on the plankton community of Lake Mendota, Wisconsin: Implications for biomanipulation. *Hydrobiologia* 200/201:329–336.

Vaughn, C. C., and Taylor, C. M. 1999. Impoundments and the decline of freshwater mussels: A case study of an extinction gradient. *Cons. Biol.* 13:912–920

Vaughn, C. C., and Taylor, C. M. 2000. Macroecology of a host–parasite relationship. *Ecography* 23:11–20.

Verbost, P. M., Bryson, S. E., Wendelaar Bonga, S. E., and Marshall, W. S. 1997. Na^+-dependent Ca^{2+} uptake in isolated epithelium of *Fundulus heteroclitus. J. Comp. Physiol. B* 167:205–212.

Via, S. 2001. Sympatric speciation in animals: the ugly duckling grows up. *TREE* 16:381–390.

Vidar Helvik, J., Drivenes, Ø., Harboe, T., and Seo, H. 2001. Topography of different photoreceptor cell types in the larval retina of Atlantic halibut (*Hippoglossus hippoglossus*). *J. Exp. Biol.* 204:2553–2559.

Videler, J. J. 1993. *Fish swimming.* London: Chapman and Hall. 260 pp.

Vijayan, M. M., Pereira, C., Kruzynski, G., and Iwama, G. K. 1998. Sublethal concentrations of contaminant induce the expression of hepatic heat shock protein 70 in two salmonids. *Aquat. Toxicol.* 40:101–108.

Vince, S., Valiela, I., Backus, N., and Teal, J. M. 1976. Predation by the salt-marsh killifish *Fundulus heteroclitus* (L.) in relation to prey size and habitat structure: Consequences for prey distribution and abundance. *J. Exp. Mar. Biol. Ecol.* 23:255–266.

Vincent, R. E., and Miller, W. H. 1969. Altitudinal distribution of brown trout and other fishes in a headwater tributary of the South Platte River. *Ecology* 50(3):464–466.

Virnstein, R. W. 1977. The importance of predation by crabs and fishes on benthic infauna in Chesapeake Bay. *Ecology* 58(6):1199–1217.

Vladykov, V. D., and Kott, E. 1977. Satellite species among the holarctic lampreys (Petromyzonidae). *Can. J. Zool.* 57(4):860–867.

Vogel, S. J. 1983. *Life in moving fluids.* Princeton, N.J.: Princeton Univ. Press. 352 pp.

Volya, G. 1966. Some data on digestive enzymes in some Black Sea fishes and a micromodification of a method for the identification of a trypsin, amylase, and lypase. In *"Fisiologia morskikhyzhitvotnykh" Nauka.* (In Russian.)

von der Emde, G. 1998. Electroreception. Pages 313–343 *in* D. H. Evans, ed., *The physiology of fishes*. 2nd ed. Boca Raton, Fla.: CRC Press.

von Arx, W. S. 1962. *An introduction to physical oceanography*. Reading, Mass.: Addison-Wesley. 422 pp.

von Bertalanffy, L. 1938. A quantitative theory of organic growth. *Hum. Biol.* 10:181–213.

von Geldern, C., and Mitchell, D. F. 1975. Largemouth bass and threadfin shad in California. Pages 436–449 *in* H. Clepper, ed., *Black bass biology and management*. Washington, D.C.: Sport Fish. Inst.

Wainwright, S. A., Vosburgh, F., and Hebrank, J. H. 1978. Shark skin: Function in locomotion. *Science* 202:747–749.

Walker, M. W., Kirschvinke, J. L., Chang, S. R., and Dizon, A. E. 1984. A candidate magnetic sense organ in the yellowfin tuna, *Thunnus albacares*. *Science* 224:751–753.

Walker, R. M., and Johansen, P. H. 1977. Anaerobic metabolism in goldfish (*Carassius auratus*). *Can. J. Zool.* 55:1304–1311.

Walls, G. L. 1942. *The vertebrate eye and its adaptive radiation*. Bloomfield Hills, Mich.: Cranbrook Inst. Sci.

Walsh, P. J. 1998. Nitrogen excretion and metabolism. Pages 201–214 *in* D. H. Evans, ed., *The physiology of fishes*, 2nd ed. Boca Raton, Fla.: CRC Press.

Walsh, P. J., Tucker, B. C., and Hopkins, T. E. 1994. Effects of confinement/crowding on ureogenesis in the gulf toadfish, *Opsanus beta*. *J. Exp. Biol.* 191:195–206.

Walsh, W. J. 1983. Stability of a coral reef fish community following a catastrophic storm. *Coral Reefs* 2:49–63.

Waples, R. S. 1991. Genetic interactions between hatchery and wild salmonids: Lessons from the Pacific Northwest. *Can. J. Fish. Aquat. Sci.* 48(Suppl. 1):124–133.

Wardle, C. S. 1975. Limits of fish swimming. *Nature* 255:725–727.

Wardle, C. S., and Reid, A. 1977. The application of large-amplitude elongated body theory to measure swimming power in fish. Pages 171–191 *in* J. H. Steel, ed., *Fisheries mathematics*. London: Academic Press.

Ware, D. M., and Thomson, R. E. 1991. Link between long-term variability in upwelling and fish production in the northeast Pacific Ocean. *Can. J. Fish. Aquat. Sci.* 48:2296–2306.

Warner, R. R. 1997. Sperm allocation in coral reef fishes. *Bioscience* 47:561–564.

Warner, R. R. 1982. Mating systems, sex change, and sexual demography in the rainbow wrasse, *Thalassoma lucasanum*. *Copeia* 1982:653–661.

Warner, R. R., and Robertson, D. R. 1978. Sexual patterns in the labroid fishes of the western Caribbean. I. The wrasses (Labridae). *Smithson. Contrib. Zool.* 254:1–27.

Watanabe, T., Kiron, V., and Satoh, S. 1997. Trace minerals in fish nutrition. *Aquaculture* 151:185–207.

Waters, J. M., Craw, D., Youngson, J. H., and Wallis, G. P. 2001. Genes meet geology: Fish phylogeographic pattern reflects ancient rather than modern drainage connections. *Evolution* 55:1844–1851.

Waters, J. M., Lopez, J. A., and Wallis, G. P. 2000. Molecular phylogenetics and biogeography of galaxiid fishes (Osteichthyes: Galaxiidae): Dispersal, vicariance, and the position of *Lepidogalaxius salamandroides*. *Syst. Zool.* 49:777–795.

Waters, J. M., and Wallis, G. P. 2001. Cladogenesis and loss of the marine life-history phase in freshwater galaxiid fishes (Osmeriformes: Galaxiidae). *Evolution* 55:587–597.

Watling, L., and Norse, E. A. 1998. Disturbance of the seabed by mobile fishing gear: A comparison to forest clearcutting. *Cons. Biol.* 12:1180–1197.

Watson, R., and Pauly, D. 2001. Systematic distortion in world fisheries catch trends. *Nature* 414:534–536.

Weatherley, A. H. 1972. *Growth and ecology of fish populations*. New York: Academic Press. 293 pp.

Webb, P. W. 1971. The swimming energetics of trout. I. Thrust and power at cruising speeds. *J. Exp. Biol.* 55:489–500.

Webb, P. W. 1975a. Hydrodynamics and energetics of fish propulsion. *Bull. Fish. Res. Bd. Canada* 190:1–159.

Webb, P. W. 1975b. Acceleration performance of rainbow trout (*Salmo gairdneri*) and green sunfish *(Lepomis cyanellus). J. Exp. Biol.* 63:451–465.

Webb, P. W. 1978a. Fast-start performance and body form in seven species of teleost fish. *J. Exp. Biol.* 74:211–226.

Webb, P. W. 1978b. Temperature effects on acceleration of rainbow trout, *Salmo gairdneri. J. Fish. Res. Bd. Can.* 35:1417–1422.

Webb, P. W., and Brett, J. R. 1972. The effects of sublethal concentration of whole bleached kraft-mill effluent on the growth and food conversion efficiency of underyearling sockeye salmon *(Oncorhynchus nerka). J. Fish. Res. Bd. Canada* 29:1555–1563.

Webb, P. W., and Weihs, D. 1983. *Fish biomechanics.* New York: Praeger Publ. 398 pp.

Webber, R. E., Wells, R. M. G., and Rossetti, J. E. 1983. Allosteric interactions governing oxygen equilibria in the haemoglobin system of the spring dogfish, *Squalus acanthias. J. Exp. Biol.* 103:109–120.

Weber, D. D., and DeWilde, J. A. M. 1975. Oxygenation properties of haemoglobins from the flatfish plaice (*Pleuronectes platessa*) and flounder *(Platichthys flesus). J. Comp. Physiol.* 128:127–137.

Weber, D. D., and Ridgway, G. J. 1962. The deposition of tetracycline drugs on bones and scales of fish and its possible use for marking. *Prog. Fish. Cult.* 24:150–155.

Weber, J., and Shanghavi, D. S. 2000. Regulation of glucose production in rainbow trout: Role of epinephrine in vivo and in isolated hepatocytes. *Am. J. Physiol.* (*Regulatory Integrative Comp. Physiol.*) 278:R956–R963.

Weber, R. E. 1979. Erythrocytic nucleoside triphosphates inhibit oxygen-linked binding of carbon dioxide by carp hemoglobin. *Acta Physiol. Scand.* 105:16A–17A.

Weber, R. E. 1990. Functional significance and structural basis of multiple hemoglobins with special reference to exothermic vertebrates. *Comp. Physiol.* 6:58–75.

Weber, R. E., and Jensen, F. B. 1988. Functional adaptations in hemoglobins from ectothermic vertebrates. *A. Rev. Physiol.* 50:161–179.

Weber, R. E., and Lykkeboe, G. 1978. Respiratory adaptations in carp blood: Influences of hypoxia, red blood cell organic phosphates, divalent cations, and CO_2 on hemoglobin–oxygen affinity. *J. Comp. Physiol.* 128:127–137.

Wedemeyer, G. A. 1996. *Physiology of fish in intensive culture systems.* New York: Chapman and Hall. 232 p.

Weihs, D. 1974. Energetic advantages of burst swimming of fish. *J. Theoret. Biol.* 48:215–229.

Weiner, G. S., Shreck, C. R., and Li, H. W. 1986. Effects of low pH on reproduction in rainbow trout. *Trans. Amer. Fish. Soc.* 115:75–82.

Weinreb, E. L. 1958. Studies on the histology and histopathology of the rainbow trout, *Salmo gairdneri iridius.* I. Haematology under normal and experimental conditions of inflammation. *Zoologica* 43:145–154.

Weinreb, E. L., and Weinreb, S. 1969. A study of experimentally induced endocytosis in a teleost. I. Light microscopy of peripheral blood cell response. *Zoologica* 54:25–34.

Welcomme, R. L. 1967. The relationship between fecundity and fertility in the mouthbrooding cichlid, *Tilapia leucosticta. J. Zool. (Lond.)* 151:453–468.

Welcomme, R. L. 1985. *River fisheries.* FAO Fish. Tech. Pap. 262. 330 pp.

Wells, L. 1968. Seasonal depth distribution of fish in southeastern Lake Michigan. *NOAA Fish. Bull.* 67(1):1–15.

Wells, L. 1970. Effects of alewife predation on zooplankton populations in Lake Michigan. *Limnol. Oceanogr.* 15(4):556–565.

Wendelaar Bonga, S. E., Löwik, C. J. M., and Van Der Meij, J. C. A. 1983. Effects of external Mg^{2+} and Ca^{2+} on branchial osmotic water permeability and prolactin secretion in the teleost fish *Sarotherodon mossambicus. Gen. Comp. Endocrinol.* 52:222–231.

Wendelaar Bonga, S. E., Flik, G. Löwik, C. W. G. M., and Van Eyes, G. J. J. M. 1985. Environmental control of prolactin synthesis in the teleost fish *Oreochromis* (formerly *Sarotherodon*) *mossambicus. Gen. Comp. Endocrinol.* 57:352–359.

Wendt, C. A. G., and Saunders, R. L. 1973. Changes in carbohydrate metabolism in young Atlantic salmon in response to various forms of stress. Pages 55–82 *in* M. W. Smith and W. M. Carter, eds., *Proceedings of the International Symposium on the Atlantic Salmon: Management, biology and survival of the species.* Fredericton, N.B.: Unipress.

Werner, E. E., and Hall, D. J. 1974. Optimal foraging and the size selection of prey by the bluegill sunfish *(Lepomis macrochirus). Ecology* 55:1042–1052.

Werner, E. E., and Hall, D. J. 1976. Niche shifts in sunfishes: Experimental evidence and significance. *Science* 191:404–406.

Werner, E. E., Hall, D. J., Laughlin, D. R., Wagner, D. J., Wilsmann, L. A., and Funk, F. C. 1977. Habitat partitioning in a freshwater fish community. *J. Fish. Res. Bd. Canada* 34(3):360–370.

Werns, S., and Howland, H. C. 1976. Size and allometry of the saccular air bladder of *Gnathonemus petersi* (Pisces, Mormyridae): Implications for hearing. *Copeia.* 1976(1):200–202.

Westby, G. W. M. 1988. The ecology, discharge activity, and predatory behavior of gymnotiform electric fish in the coastal streams of French Guiana. *Behav. Ecol. Sociobiol.* 22:341–354.

Westoll, T. S. 1949. On the evolution of the Dipnoi. Pages 121–184 *in* G. L. Jepsen et al., eds., *Genetics, paleontology, and evolution.* Princeton, N.J.: Princeton Univ. Press.

White, B. N. 1986. The isthmian line, antitropicality and American biogeography: Distributional history of the Atherinopsinae (Pisces: Atherinidae). *Syst. Zool.* 35:176–194.

White, L. W. 1967. The historical roots of our ecologic crisis. *Science* 155:1203–1207.

Whiteley, N. M., and Egginton, S. 1999. Antarctic fishes have a limited capacity for catecholamine synthesis. *J. Exp. Biol.* 202:3623–3629.

Whitfield, A. K. 1994. Fish species diversity in southern African estuaries: An evolutionary perspective. *Environ. Biol. Fish.* 40:37–90.

Wickler, W. 1968. *Mimicry in plants and animals.* New York: McGraw-Hill. 255 pp.

Widder, E. A. 1998. A predatory use of counterillumination by the squaloid shark, *Isistius brasiliensis. Environ. Biol. Fish.* 53:267–273.

Wiebe, W. J., Johannes, R. E., and Webb, K. L. 1975. Nitrogen fixation in a coral reef community. *Science* 188:257–259.

Wiesel, G. F. 1973. Anatomy and histology of the digestive system of the paddlefish, *Polyodon spathula. J. Morph.* 145:143–150.

Wiest, F. C. 1995. The specialized locomotory apparatus of the freshwater hatchetfish family, Gasteropelecidae. *J. Zool. (Lond.)* 236:571–592.

Wikramanayake, E. D. 1990. Conservation of endemic rain forest fishes of Sri Lanka: Results of a translocation experiment. *Cons. Biol.* 4:32–37.

Wikramanayake, E. D., and Moyle, P. B. 1989. Ecological structure of tropical fish assemblages in wet-zone streams of Sri Lanka. *J. Zool. (Lond.)* 218:503–526.

Wilcox, W. A. 1887. A man killed by a swordfish. Bull. *U. S. Fish Comm.* 6:417.

Wiley, B., and Collette, B. 1970. Breeding tubercles and contact organs in fishes: Their occurrence, structure, and significance. *Bull. Amer. Mus. Nat. Hist.* 143(3):145–216.

Wilga, C. D., and Lauder, G. V. 2999. Locomotion in sturgeon: Function of the pectoral fins. *J. Exp. Biol.* 202:2413–2432.

Williams, G. C. 1966. *Adaptation and natural selection.* Princeton, N.J.: Princeton Univ. Press. 307 pp.

Williams, G. C. 1975. *Sex and evolution. Monogr. Pop. Biol.* 8. Princeton Univ. 200 pp.

Williams, J. E., and 7 others. 1989. Fishes of North America: Endangered, threatened, or of special concern, 1989. *Fisheries* 14:2–20.

Williams, J. E., Wood, C. A., and Dombeck, M. P., eds. 1997. *Watershed restoration: Principles and practices.* Bethesda, Md.: Amer. Fish. Soc. 561 pp.

Willis, T. V., and Magnuson, J. J. 2001. Patterns in fish species composition across the interface between streams and lakes. *Can. J. Fish. Aquat. Sci.* 57:1042–1052.

Willson, M. F., Gende, S. M., and Marston, B. H. 1998. Fishes and the forest. *Bioscience*: 48:455–462.

Wilson, D. S., Muzzall, P. M., and Ehlinger, T. J. 1996. Parasites, morphology, and habitat use in a bluegill sunfish (*Lepomis macrochirus*) population. *Copeia* 1996:348–354.

Wilson, E. O. 1975. *Sociobiology.* Cambridge: Harvard Univ. Press. 1005 pp.

Wilson, J. M., Randall, D. J., Vogl, A. W., and Iwama, G. K. 1997. Immunolocation of proton-ATPase in the gills of a marine elasmobranch, *Squalus acanthias. J. Exp. Zool.* 278:78–86.

Wilson, J. P., and Sheaves, M. 2001. Short-term temporal variations in taxonomic composition and trophic structure of a tropical estuarine fish assemblage. *Mar. Biol.* 19:787–796.

Wilson, M. H. V., and Caldwell, M. D., 1993. New Silurian and Devonian fork-tailed "athelodonts" are jawless vertebrates with stomachs and deep bodies. *Nature* 361:442–444.

Wilson, R., and Wilson, J. Q. 1985. *Watching fishes: Life and behavior on coral reefs.* New York: Harper and Row.

Winemiller, K. O. 1990. Spatial and temporal variation in tropical fish trophic networks. *Ecol. Monogr.* 60:331–367.

Winemiller, K. O., and Jepsen, D. B. 1998. Effects of seasonality and fish movement on tropical river food webs. *J. Fish Biol.* 53(Suppl. A):267–296.

Winfield, I. J., and Nelson, J. S. 1991. *Cyprinid fishes: Systematics, biology, and exploitation.* New York: Chapman and Hall. 667 pp.

Winterbottom, R., and Emery, A. R. 1981. A new genus and two new species of gobiid fishes (Perciformes) from the Chagos Archipelago, central Indian Ocean. *Environ. Biol. Fish.* 6:139–149.

Wintrobe, M. M. 1934. Variations in the size and hemoglobin content of erythrocytes in the blood of various vertebrates. *Folia Hemat. Lpz.* 51:32–49.

Wirgin, I. I., and Waldman, J. R., 1994. What DNA can do for you. *Fisheries (Bethesda)* 19(7):16–27.

Wittenberg, J. B., and Haedrich, R. L. 1974. The choroid rete mirabile of the fish eye. II. Distribution and relation to the pseudobranch and to the swimbladder rete mirabile. *Biol. Bull.* 145:137–156.

Wittenberg, J. B., and Wittenberg, V. A. 1962. Active secretion of oxygen into the eye of the fish. *Nature* 194:106–107.

Wohlschlag, D. E., Cameron, J. N., and Cech, J. J., Jr. 1968. Seasonal changes in the respiratory metabolism of the pinfish (*Lagodon rhomboides*). *Contr. Mar. Sci.* 13:89–104.

Wohlschlag, D. E., and Wakeman, J. M. 1978. Salinity stresses, metabolic responses, and distribution of the coastal spotted seatrout, *Cynoscion nebulosus. Contr. Mar. Sci.* 21:173–185.

Wolf, P., Smith, P. E., and Bergen, D. R. 2001. Pacific sardine. Pages 299–302 *in* W. S. Leet, C. M. Dewees, R. Kleingbeil, and E. J. Larson, eds. *California's living marine resources: a status report.* Calif. Dept. Fish and Game, Sacramento.

Wood, C. M. 1993. Ammonia and urea metabolism and excretion. Pages 379–425 *in* D.H. Evans, ed., *The physiology of fishes.* Boca Raton, Fla.: CRC Press.

Wood, C. M., Gilmour, K. M., Perry, S. F., Part, P., Laurent, P., and Walsh, P. J. 1998. Pulsatile urea excretion in gulf toadfish (*Opsanus beta*): Evidence for activation of a specific facilitated diffusion transport system. *J. Exp. Biol.* 201:805–817.

Wood, C. M., and Marshall, W. S. 1994. Ion balance, acid–base regulation, and chloride cell function in the common killifish, *Fundulus heteroclitus*: A euryhaline estuarine teleost. *Estuaries* 17(1A):34–52.

Wood, C. M., Perry, S. F., Wright, P. A., Bergman, H. L., and Randall, D. J. 1989. Ammonia and urea dynamics in the Lake Magadi tilapia, a ureotelic teleost fish adapted to an extremely alkaline environment. *Resp. Physiol.* 77:1–20.

Wood, C. M., and Shelton, G. 1975. Physical and adrenergic factors affecting systemic vascular resistance in the rainbow trout: A comparison with branchial vascular resistance. *J. Exp. Biol.* 63:505–523.

Wood, C. M., and Shuttleworth, T. J., eds. 1995. *Cellular and molecular approaches to fish ionic regulation.* Vol. XIV, Fish Physiology. San Diego, Calif.: Academic Press. 352 pp.

Wood, S. C., Johansen, K., and Weber, R. E. 1972. Haemoglobin of the coelacanth. *Nature* 239:283–285.

Woodward, J. J., and Smith, I. S. 1985. Exercise training and the stress response in rainbow trout, *Salmo gairdneri* Richardson. *J. Fish Biol.* 26:435–447.

Wooton, R. J. 1973. The effect of size of food ration on egg production in the female threespined stickleback. *J. Fish Biol.* 5:89–96.

Wooton, R. J. 1977. *The biology of sticklebacks.* New York: Academic Press. 388 pp.

Wooton, R. J. 1984. *A functional biology of sticklebacks.* London: Croom Helm. 400 pp.

Wooton, R. J. 1990. *Ecology of teleost fishes.* London: Chapman and Hall. 404 pp.

Wootton, J. T., and Oemke, M. P. 1992. Latitudinal differences in fish community trophic structure and the role of herbivory in a Costa Rican stream. *Environ. Biol. Fish.* 35:311–319.

Wourms, J. P. 1977. Reproduction and development in chondrichthyan fishes. *Amer. Zool.* 17(2):379–410.

Wourms, J. P. 1981. Viviparity: the maternal-fetal relationship in fishes. *Amer. Zool.* 21:473–515.

Wourms, J. P. 1993. Maximization of evolutionary trends for placental viviparity in the Atlantic sharpnose shark, *Scolidon laticaudus. Environ. Biol. Fish.* 38:269–294.

Wourms, J. P., Atz, J. W., and Stribling, M. D. 1991. Viviparity and the maternal-embryonic relationship in the coelacanth, *Latimeria chalumnae. Environ. Biol. Fish.* 32:225–248.

Wourms, J. P., and Demski, L. S., eds. 1993. The reproduction and development of sharks, skates, rays, and ratfishes. *Environ. Biol. Fish.* 38. 294 pp.

Wright, P. A., and Land, M. D. 1998. Urea production and transport in teleost fishes. *Comp. Biochem. Physiol.* 119A:47–54.

Wright, P. A., Part P., and Wood, C. M. 1995. Ammonia and urea excretion in the tidepool sculpin (*Oligocottus maculosus*)—Sites of excretion, effects of reduced salinity and mechanisms of urea transport. *Fish Physiol. Biochem.* 14:111–123.

Wu, T. Y., and Yates, G. T. 1978. A comparative mechanophysiological study of fish locomotion with implications for tuna-like swimming mode. Pages 313–337 *in* G. Sharp and A. E. Dizon, eds., *The physiological ecology of tunas.* New York: Academic Press.

Wurtsbaugh, W. A., and Cech, J. J., Jr. 1983. Growth and activity of juvenile mosquitofish: Temperature and ration effects. *Trans. Am. Fish. Soc.* 112:653–660.

Wurtsbaugh, W. A., and Neverman, D. 1988. Post-feeding thermotaxis and daily vertical migration in a larval fish. *Nature* 333:846–848.

Wurtsbaugh, W. A., and Tapia, R. A. 1988. Mass mortality of fishes in Lake Titicaca (Peru–Bolivia) associated with the protozoan parasite *Ichthyophthirius multifilis. Trans. Amer. Fish. Soc.* 117:213–217.

Yamamoto, K. 1988. Contraction of spleen in exercised freshwater teleosts. *Comp. Biochem. Physiol.* 89A:65–66.

Yancey, P. H., and Somero, G. N. 1980. Methylamine osmoregulatory solutes of elasmobranch fishes counteract urea inhibition of enzymes. *J. Exp. Zool.* 212:205–213.

Yant, P. R., Karr, J. R., and Angermeier, P. L. 1984. Stochasticity in stream fish communities: An alternative interpretation. *Amer. Nat.* 124:573–582.

Yap, S. Y. 2002. On the distributional patterns of Southeast Asian freshwater fish and their history. *J. Biogeogr.* 29:1187–1199.

Yoklavich, M. M., Greene, H.G., and Cailliet, G. M. 2000. Habitat associations of deep-water rockfishes in a submarine canyon: an example of a natural refuge. *NOAA Fish. Bull.* 98:625–641.

Yoshiyama, R. M. 1981. Distribution and abundance patterns of rocky intertidal fishes in central California. *Environ. Biol. Fish.* 6:315–332.

Yoshiyama, R. M., and Cech, J. J, Jr. 1994. Aerial respiration by rocky intertidal fishes of California and Oregon. *Copeia* 1994(1):153–158.

Yoshiyama, R. M., and 8 others. 1992. Homing behavior and site fidelity in intertidal sculpins (Pisces: Cottidae). *J. Exp. Mar. Biol. Ecol.* 160:115–130.

Yoshiyama, R. M., Gerstung, E. R., Fisher, F. W., and Moyle, P. B. 1996. Historical and present distribution of chinook salmon in the Central Valley drainage of California. Pages 309–362 *in* Sierra Nevada Ecosystem Project, final report to Congress, Vol. III. Davis: University of California Centers for Water and Wildland Resources.

Yoshiyama, R. M., Fisher, F. W., and Moyle, P. B. 1998. Historical abundance and decline of chinook salmon in the Central Valley region of California. *N. Amer. J. Fish. Mgmt.* 18:487–521.

Young, G. C. 1986. The relationships of placoderm fishes. *Zool. J. Linn. Soc.* 88:1–57.

Young, P. S., and Cech, J. J., Jr. 1993a. Effects of exercise conditioning on stress responses and recovery in cultured and wild young-of-the-year striped bass, *Morone saxatilis. Can. J. Fish. Aquat. Sci.* 50:2094–2099.

Young, P. S., and Cech, J. J., Jr. 1993b. Improved growth, swimming performance, and muscular development in exercise-conditioned young-of-the-year striped bass (*Morone saxatilis*). *Can. J. Fish. Aquat. Sci.* 50:703–707.

Young, P. S., and Cech, J. J., Jr. 1994. Optimum exercise conditioning velocity for growth, muscular development, and swimming performance in young-of-year striped bass (*Morone saxatilis*). *Can. J. Fish. Aquat. Sci.* 51:1519–1527.

Zakaria-Ismail, M. 1994. Zoogeography and biodiversity of the freshwater fishes of Southeast Asia. *Hydrobiologia* 285:41–48.

Zardoya, R., and Meyer, A. 1997. The complete DNA sequence of the mitochondrial genome of a "living fossil," the coelacanth (*Latimeria chalumnae*). *Genetics* 146:995–1010.

Zaret, T. M. 1972. Predators, invisible prey, and the nature of polymorphism in the Cladocera (Class Crustacea). *Limnol. Oceanogr.* 17(2):171–184.

Zaret, T. M., and Paine, R. T. 1973. Species introduction in a tropical lake. *Science* 182:421–437.

Zhadin, V. I., and Gerd, S. V. 1961. *Fauna and flora of the lakes and reservoirs of the U.S.S.R.* Jerusalem: Israel Prog. Sci. Trans. 626 pp.

Zimmer, C. 1999. The mystery of the mermaid's purse. *Natural History* 108(4):24–25.

Zydlewski, J., and McCormick, S. D. 1997. The ontogeny of salinity tolerance in the American shad, *Alosa sapidissima. Can. J. Fish. Aquat. Sci.* 54:182–189.

Zyznar, E. S., and Nicol, J. A. C. 1973. Reflecting materials in the eyes of three teleosts, *Orthopristes chrysopterus, Dorosoma cepedianum,* and *Anchoa mitchilli. Proc. R. Soc. Lond. B* 184:15–27.

Index